T0134977

# Advances in Intelligent Systems and Computing

Volume 867

**Series editor**

Janusz Kacprzyk, Systems Research Institute, Polish Academy of Sciences,
Warsaw, Poland
e-mail: kacprzyk@ibspan.waw.pl

The series "Advances in Intelligent Systems and Computing" contains publications on theory, applications, and design methods of Intelligent Systems and Intelligent Computing. Virtually all disciplines such as engineering, natural sciences, computer and information science, ICT, economics, business, e-commerce, environment, healthcare, life science are covered. The list of topics spans all the areas of modern intelligent systems and computing such as: computational intelligence, soft computing including neural networks, fuzzy systems, evolutionary computing and the fusion of these paradigms, social intelligence, ambient intelligence, computational neuroscience, artificial life, virtual worlds and society, cognitive science and systems, Perception and Vision, DNA and immune based systems, self-organizing and adaptive systems, e-Learning and teaching, human-centered and human-centric computing, recommender systems, intelligent control, robotics and mechatronics including human-machine teaming, knowledge-based paradigms, learning paradigms, machine ethics, intelligent data analysis, knowledge management, intelligent agents, intelligent decision making and support, intelligent network security, trust management, interactive entertainment, Web intelligence and multimedia.

The publications within "Advances in Intelligent Systems and Computing" are primarily proceedings of important conferences, symposia and congresses. They cover significant recent developments in the field, both of a foundational and applicable character. An important characteristic feature of the series is the short publication time and world-wide distribution. This permits a rapid and broad dissemination of research results.

More information about this series at http://www.springer.com/series/11156

Marcus Strand · Rüdiger Dillmann
Emanuele Menegatti · Stefano Ghidoni
Editors

# Intelligent Autonomous Systems 15

Proceedings of the 15th International
Conference IAS-15

 Springer

*Editors*
Marcus Strand
Baden-Wuerttemberg Cooperative State
University
Karlsruhe, Germany

Rüdiger Dillmann
Humanoids and Intelligence Systems Lab
KIT - Karlsruher Institut für Technologie
Karlsruhe, Germany

Emanuele Menegatti
University of Padua
Padua, Italy

Stefano Ghidoni
University of Padua
Padua, Italy

ISSN 2194-5357        ISSN 2194-5365   (electronic)
Advances in Intelligent Systems and Computing
ISBN 978-3-030-01369-1        ISBN 978-3-030-01370-7   (eBook)
https://doi.org/10.1007/978-3-030-01370-7

Library of Congress Control Number: 2018955480

This Springer imprint is published by the registered company Springer Nature Switzerland AG
The registered company address is: Gewerbestrasse 11, 6330 Cham, Switzerland

# Preface

This volume contains the papers presented at IAS-15, the 15th International Conference on Intelligent Autonomous Systems, one of the premier forums for the exchange of latest research results in the field of autonomous systems, its applications, and its impact on our society held on June 11–15, 2018, in Baden-Baden, Germany.

IAS-15 belongs to the most traditional robotics events, and we are proud to host it in Baden-Baden again. The goal of the IAS conference is to bring together leading researchers interested in all aspects of autonomy and adaptivity of artificial systems. One of the driving forces of this conference is the observation that intelligence and autonomy are best studied and demonstrated using various mobile robots acting autonomously in large rough terrain, underwater, air, or space under challenging conditions.

For the second time, the International Conference on Intelligent Autonomous Systems (IAS) is held in Baden-Baden and for the first time is organized by the three institutions DHBW, KIT, and FZI in Karlsruhe and strongly supported by the University of Padova.

The total amount of 97 papers have been submitted to IAS-15 with latest research accomplishments, innovations, and visions in the field of robotics, artificial intelligence, machine learning, and its applications. Each paper was evaluated by two to four reviewers, and finally, 78 papers have been accepted for presentation at the conference after several rebuttal iterations. Additionally, two invited talks are included in the conference proceedings.

The papers contained in the proceedings cover a wide spectrum of research in intelligent autonomous systems including multi-agent systems, motion planning, robot control, handling uncertainty with probabilistic logic, navigation, human–machine interaction, computer vision and perception as well as unmanned aerial systems and unmanned driving.

The proceedings include all accepted papers and reflect a variety of topics concerning intelligent autonomous systems and their impact to industry and society. The organizers would like to express their gratitude to all contributors in the preparation phase and during the evaluation phase and the conference. Thanks to this additional assistance, IAS-15 could be such a success. We would especially like to thank the program committee members for their valuable support and for the preparation of the reviews, which allowed to make a proper selection of high-quality papers. The staff of the three organizing institutions took a great part in planning and organizing the conference. Thanks to VDI/VDE for supporting the organization of the conference and to the sponsoring companies Sick AG, Bosch AG, and Intel for their financial support making this event possible.

We hope that this proceedings, and especially the participation in the IAS-15 conference and its related events, inspires new ideas, fosters new research, and creates new friendships which grow into fruitful collaborations.

Finally, we thank the city of Baden-Baden for hosting IAS for the second time. Organizing IAS-15 was a wonderful experience, and we hope that the conference gives its participants as much pleasure as we had when preparing it. Enjoy the proceedings of IAS-15.

Marcus Strand
Rüdiger Dillmann
Emanuele Menegatti
Stefano Ghidoni

# Organization

## Program Committee

| | |
|---|---|
| Ilya Afanasyev | Innopolis University |
| H. Levent Akin | Bogazici University |
| Luis Almeida | University of Porto |
| Nancy Amato | Texas A&M University |
| Francesco Amigoni | Politecnico di Milano |
| Marco Antonelli | IT+Robotics |
| Salvatore Anzalone | Université Paris 8 |
| Shinya Aoi | Kyoto University |
| Tamim Asfour | Karlsruhe Institute of Technology |
| Sven Behnke | University of Bonn |
| Davide Brugali | Università degli Studi di Bergamo |
| Simone Ceriani | Joint Research Centre |
| Antonio Chella | Università degli Studi di Palermo |
| Li Chen Fu | National Taiwan University, Taiwan |
| Andrea Cherubini | LIRMM, Universite de Montpellier |
| Mohamed Chetouani | Sorbonne |
| Ryosuke Chiba | Asahikawa Medical University |
| Grazia Cicirelli | Institute of Intelligent Systems for Automation |
| Rita Cucchiara | Dipartimento di Ingegneria dell'Informazione, University of Modena and Reggio Emilia |
| Davide Cucci | Politecnico di Milano |
| Antonio D'Angelo | Department of Mathematics and Computer Science, University of Udine |
| Tiziana D'Orazio | Institute of Intelligent Systems for Automation, National Research Council |
| Rüdiger Dillmann | Karlsruhe Institute of Technology, Germany |
| Gregory Dudek | McGill University, Canada |
| Esra Erdem | Sabanci University |

Keiji Suzuki                    Hokkaido University, Japan
Pierluigi Taddei                Joint Research Centre
Masaki Takahashi                Department of System Design Engineering,
                                    Keio University
Wataru Takano                   The University of Tokyo
Mohan Trivedi                   UCSD, US
Takashi Tsubouchi               University of Tsukuba
Ryuichi Ueda                    Chiba Institute of Technology
Tijn van der Zant               University of Groningen, Netherlands
Markus Vincze                   Vienna University of Technology
Oskar von Stryk                 TU Darmstadt
Hesheng Wang                    Shanghai Jiao Tong University
Jing-Chuan Wang                 Shanghai Jiao Tong University
Qining Wang                     Peking University
Yangsheng Xu                    CMU , US
Hiroaki Yamaguchi               The University of Tokyo, Japan
Katsu Yamane                    CMU
Atsushi Yamashita               The University of Tokyo
Hiroshi Yokoi                   The University of Electro-Communications
Primo Zingaretti                Università Politecnica delle Marche, DII
Henning Zoz                     Centro de Investigación e Innovación Tecnológica,
                                    Mexico

# Contents

Human Detection and Interaction

# Path Planning, Localization
# and Navigation

# A Viterbi-like Approach for Trajectory Planning with Different Maneuvers

Jörg Roth$^{(\boxtimes)}$

Department of Computer Science, Nuremberg Institute of Technology,
Kesslerplatz 12, 90489 Nuremberg, Germany
joerg.roth@th-nuernberg.de

**Abstract.** The task of a trajectory planning tries to find a sequence of driving commands that connects two configurations, whereas we have to consider nonholonomic constraints, obstacles and driving costs. In this paper, we present a new approach that supports arbitrary primitive trajectories, cost functions and constraints. The vehicle's driving capabilities are modeled by a list of supported maneuvers. For maneuvers there exist equations that map configurations to driving commands. From all possible maneuver sequences that connect start and target, we compute the optimum with a Viterbi-like approach.

## 1 Introduction

Trajectory planning is a fundamental function of a mobile robot. When executing complex tasks such as transporting goods, the robot has to drive trajectories that meet certain measures of optimality. For this, a cost function may consider driving time, energy consumption, mechanical wear or buffer distance to obstacles.

Whereas a geometric *route planning* tries to find a line string with minimal costs that does not cut an obstacle (with respect to the robot's driving width), the *trajectory planning* also considers nonholonomic constraints such as curve angles or orientations. Our approach is based on three key ideas:

- A first route planning step reduces the overall complexity of the problem, as it only considers workspace dimensions.
- We model the measure of optimality as arbitrary cost function that considers obstacles. It may use different mechanisms to check for obstacles, i.e. grids or spatial indexing to speed up execution.
- The robot's driving capabilities are modeled by a list of *primitive trajectories* (e.g. straight forward, arc, clothoid) and *maneuvers* (small sequences of primitive trajectories). These lists are also considered as black box for our algorithm. The benefit: our trajectory planning is able to support different types of vehicles without changing the overall algorithm.

Our approach is able to compute suitable trajectories in short runtime that *linearly* depends on the number of intermediate points. Moreover, the first driving command converges very fast. In practice it is available in *constant time*. This means, a mobile

© Springer Nature Switzerland AG 2019
M. Strand et al. (Eds.): IAS 2018, AISC 867, pp. 3–14, 2019.
https://doi.org/10.1007/978-3-030-01370-7_1

robot is able to start driving very quickly with the first driving command (e.g. in emergency situations) whereas the remaining sequence is computed while driving.

The approach is successfully realized and evaluated on the *Carbot* platform.


## 2  Related Work

Early work investigates shortest paths for vehicles that can drive straight forward and circular curves [3, 5, 17]. Without obstacles, we can connect two configurations with only three primitive trajectories. Further work addresses the problem of discontinuities in the curvature and tries to integrate the clothoid as further primitive trajectory [2, 19]. Instead of clothoids, also cubic spirals or splines were considered [4, 12].

More related to our approach is work that investigates longer paths that go through an environment of obstacles. As the space of possible trajectory sequences gets very large, probabilistic approaches are a suitable method to find at least a suboptimal solution [7, 9, 10]. They randomly connect configurations by primitive trajectories and are able to search on the respective graph to plan an actual path. Further work uses potential fields [1] or visibility graphs [14]. With the help of geometric route planners, the overall problem of trajectory planning can be reduced. In [8], the route planning step and a local trajectory planning step are recursively applied.

Random sampling can also be used to improve generated trajectories. E.g. CHOMP [21] uses functional gradient techniques based on Hamiltonian Monte Carlo to iteratively improve the quality of an initial trajectory. The approach in [13] represented the continuous-time trajectory as a sample from a Gaussian process generated by a linear time-varying stochastic differential equation. Then gradient-based optimization technique optimizes trajectories with respect to a cost function.

Our approach was inspired by the *state lattice* idea introduced in [16]. State lattices are discrete graphs embedded into the continuous state space. Vertices represent states that reside on a hyperdimensional grid, whereas edges join states by trajectories that satisfy the robot's motion constraints. The original approach is based on equivalence classes for all trajectories that connect two states and perform inverse trajectory generation to compute the result trajectory. [6] introduced a two-step approach, with coarse planning of states based on Dynamic Programming, and a fine trajectory planning that connects the formerly generated states.

### Discussion

Many prior approaches impose limitations on the set of primitive trajectories. Often, this is a result of the optimization approach or of its stochastic nature. We be-lieve the set of primitive trajectories should be considered as black box.

Approaches that directly plan on the configuration space have to deal with many dimensions and thus large spaces for solutions. We may use probabilistic methods or discretization of configuration space parameters. However, a large environment (e.g. many rooms in a building) may already lead to an undesirable large set of possibilities, even if we ignore further configuration dimensions. Several weaknesses of random sampling were shown in the context of probabilistic roadmaps in [11]. Due to its stochastic nature runtime can hardly determined, but this is required in critical situations. We thus selected a method that gives complete control of the scale of search space and allows to pre-determine runtime.

State lattice approaches are widely used in on-road planning. States are usually embedded into a grid of workspace dimensions. This however limits planning steps to next neighbors. The original approach allows to 'skip' multiple neighbors. However, we believe, a wider segmentation that is a result of a prior route planning is more suitable. This, e.g., would simplify the planning of long straight trajectories.

We introduce a new planning approach that is able to integrate arbitrary sets of primitive trajectories, arbitrary cost functions and is able respond in linear runtime.

## 3 The Trajectory Planning

### 3.1 Basic Considerations

We now assume the robot drives in the plane in a workspace $\mathcal{W}$ with positions $(x, y)$. The configuration space $\mathcal{C}$ covers an additional dimension for the orientation angle, i.e. a certain configuration is defined by $(x, y, \theta)$. The goal is to find a collision-free sequence of trajectories that connects two configurations, meanwhile minimizes a cost function. To merge collisions and costs, we require the cost function to produce infinite costs for trajectories that cut obstacles.

This problem has many degrees of freedom. Whereas even small distances can be connected by an infinite number of trajectories, the problem gets worse for larger environments with many obstacles. We thus introduce the following concepts:

- A route planning that solely operates on workspace $\mathcal{W}$ computes a sequence of collision-free lines of sight (with respect to the robot's width) that minimize the costs.
- As the route planning only computes route points in $\mathcal{W}$, we have to specify additional variables in $\mathcal{C}$ (here orientation $\theta$). From the infinite assignments, we only consider a small finite set.
- From the infinite set of trajectories between two route points, we only consider a finite set of *maneuvers*. Maneuvers are trajectories, for which we know formulas that derive the respective parameters (e.g. curve radii) from start and target configurations.
- Even though these concepts reduce the problem space to a finite set of variations, this set would by far be too large for complete checks. We thus apply a Viterbi-like approach that significantly reduces the number of checked variations to find an optimum.

We carefully separated the cost function from all planning components. We assume there is a mapping from a route or trajectory sequence to a cost value according to two rules: first, we have to assign a single, scalar value to a trajectory sequence that indicates its costs. If costs cover multiple attributes (e.g. driving time *and* battery consumption), the cost function has to weight these attributes and create a single cost value. Second, a collision with obstacles has to result in infinite costs.

Cost values may take into account, e.g., the path length, the amount of turn-in-place operations, the amount of backward trajectories, changes between forward and backward driving, the amount of changes in curvature or the expected energy consumption. Also the distance to obstacles could be considered, if, e.g., we want the robot to keep a safe distance where possible.

## 3.2   Primitive Trajectories and Maneuvers

The basic capabilities of movement are defined by a set of *primitive trajectories*. The respective set can vary between different robots. E.g. the *Carbot* [18] is able to execute the following primitive trajectories:

- $L(\ell)$: linear (straight) driving over a distance of $\ell$ (may be negative for back-ward);
- $T(\Delta\theta)$: turn in place over $\Delta\theta$;
- $A(\ell, r)$: drive a circular arc with radius $r$ (sign distinguishes left/right) over a distance of $\ell$ (that may be negative for backward);
- $C(\ell, \kappa s, \kappa t)$: clothoid over a distance of $\ell$ with given start and target curvatures.

We are able to map primitive trajectories directly to driving commands that are natively executed by the robot's motion subsystem. Implicitly, they specify functions $c_s \rightarrow c_t$ that map two configurations. For a certain configuration in $\mathcal{C}$ that defines a pose $(x, y, \theta)$, the linear trajectory, e.g., specifies a function

$$L(\ell): (x, y, \theta) \rightarrow (x+\ell \cdot \cos(\theta), y+\ell \cdot \sin(\theta), \theta) \qquad (1)$$

Due to non-holonomic constraints, it usually is not possible to reverse this mapping. i.e., for given $c_s, c_t \in \mathcal{C}$ there is in general no primitive trajectory that maps $c_s$ to $c_t$.

At this point, we introduce *maneuvers*. Maneuvers are small sequences of primitive trajectories (usually 2–5 elements) that *are* able to map given $c_s, c_t \in \mathcal{C}$. More specifically:

- A maneuver is defined by a sequence of primitive trajectories (e.g. denoted *ALA* or *AA*) and further constraints. Constraints may relate or restrict the respective primitive trajectory parameters.
- For given $c_s, c_t \in \mathcal{C}$ there exist formulas that specify the parameters of the involved primitive trajectories, e.g. $\ell$ for $L$, $A$ and $C$, $r$ for $A$, $\kappa s$, $\kappa t$ for $C$.
- Sometimes, the respective equations are underdetermined. As a result, multiple maneuvers of a certain type (sometimes an infinite number) map $c_s$ to $c_t$. Thus, we need further parameters, we call *free parameters* to get a unique maneuver.

We explain the concept with the help of two maneuvers we call *S-Arcs* and *Wing-Arc* (Fig. 1). S-Arcs (Fig. 1 left) is a maneuver of type *AA*. The idea of S-Arcs is to drive two arcs with equal arc radii but with opposite left/right direction.

Due to geometric relations, we get equations for $r$, $\ell_1$ and $\ell_2$. The calculating method: To simplify the computation, we first roto-translate start and target to move the start to $(0, 0, 0)$ and target to $(x'_t, y'_t, \theta'_t)$. This leads to the formulas.

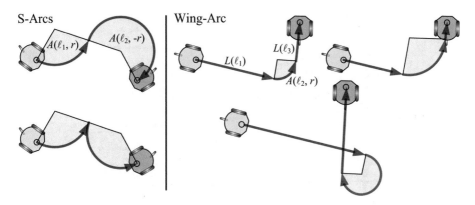

**Fig. 1.** Example maneuvers

$$r = \frac{y'_t(1+ct) - x'_t st - \sqrt{x'^2_t(st^2 - 2ct + 2) + y'^2_t(3 + ct^2) - 2x'_t y'_t st(1+ct)}}{2(ct - 1)} \qquad (2)$$

$$\ell_1 = r \cdot \arccos\left(\frac{(1+ct)}{2} - \frac{y'_t}{2r}\right), \ell_2 = \ell_1 - r \cdot \theta'_t \qquad (3)$$

where $st = \sin(\theta'_t)$, $ct = \cos(\theta'_t)$. Actually, we get two solutions for $\ell_1, \ell_2$ parameters as arcs can always be driven in two directions (clockwise or counter-clockwise). For S-Arcs we get a total of 4 combinations. Thus, the free parameters for S-Arcs define the arc senses of rotation. It depends on the application to evaluate the respective combinations, e.g. if we want to avoid to change forward to backward or vice versa while driving.

Wing-Arc (Fig. 1 right) is a maneuver of type *LAL*. As we have a total of four parameters ($r$, $\ell_1, \ell_2$ and $\ell_3$) we have an underdetermined set of equations, thus we have a single free parameter $r$ (besides the arc sense of rotation).

For the Carbot project, we identified 8 maneuvers (Table 1). We assigned names that illustrate the maneuver's shape, e.g. the J-Arc drives a path that looks like the letter 'J'. The Dubins-Arcs correspond to the combination with three arcs of Dubins original approach [5]. From the maneuvers above, $\int$ - Arcs can be considered as a 'Swiss knife': it allows reaching any target configuration without a turn in place whereas the middle linear trajectory spans a reasonable distance to the target.

We are able to replace a single arc by two clothoids that are connected with the same curvature and end with curvature zero [15]. With these maneuvers, trajectories can be planned that did not have to deal with discontinuous curvatures.

Let $\Pi$ denote the set of maneuver types relevant for a certain scenario or application. Note that $\Pi$ can easily be extended or reduced. E.g. a certain application could avoid all maneuvers that contain a *T*.

We also may invent new maneuvers to increase the overall driving capabilities. For a new maneuver, we only have to set up equations that assign the respective trajectory parameters from start and target configuration.

**Table 1.** Available maneuvers in the Carbot project

| Maneuver | Pattern | Free parameters |
|----------|---------|-----------------|
| 1-Turn | *LTL* | *No* |
| 2-Turns | *TLT* | *No* |
| J-Bow | *LA, LCC* | Single arc sense |
| J-Bow2 | *AL, CCL* | Single arc sense |
| ∫ - Arcs | *ALA, CCLCC* | Two arc radii and senses |
| S-Arcs | *AA, CCCC* | Two arc senses |
| Wing-Arc | *LAL, LCCL* | Single arc radius |
| Dubins-Arcs | *AAA, CCCCCC* | Arc radius for all three arcs |

## 3.3   Finding an Optimal Variation

Let $p_1 = (x_1, y_1)...p_n = (x_n, y_n)$ denote a route found by the route planning component for a start $(p_1, \theta_1)$ and target $(p_n, \theta_n)$. Our problem is to find a sequence of maneuvers (and thus primitive trajectories) that connects start, target and all route points in-between. We have to face two problems:

- Apart from $\theta_1, \theta_n$ the orientation angles are not specified. If our route has more than two route points, we thus have an infinite set of possible intermediate orientations.
- Some maneuver types have free parameters with an infinite set of possible assignments, e.g. $r$ for Wing-Arc.

Of the infinite number of intermediate orientations and maneuver parameters we define a finite set of promising candidates. This obviously leads to sub-optimal results. However, in reality, it does not significantly affect the overall trajectory costs. Let $O_i$ denote all orientation candidates for a route point $i \geq 2$. In our experiments $|O_i| = 5$; it contains variations of angles from the previous and to the next route point.

For free parameters we distinguish the small set of variations for arc senses (e.g. for J-Bow) and the infinite set for arc radii (e.g. for Wing-Arc). For the first type we are able to iterate through all variations. For the second type, we select a small set of candidates, similar to orientation angles. This set could be $\{r_{min}, 3 \cdot r_{min}, 5 \cdot r_{min}\}$, where $r_{min}$ is the minimal curve radius. In our algorithm, $params(M)$ denotes the set of parameters for a maneuver type $M \in \Pi$. For maneuver types with no free parameters (e.g. 1-Turn), we define $params(M) = \{\varnothing\}$, whereas $\varnothing$ is an empty parameter setting.

Even though we now 'only' have a finite set of variations $\Pi \times O_i \times params(M)$ for a single route step, the number of variations still is by far too large for a complete check. To give an impression: for 5 route points we get a total number of 20 million, for 20 route points $2 \cdot 10^{37}$ permutations (for an average of 20 maneuvers and 5 intermediate angles). Obviously, we need an approach that computes an appropriate result without iterating through all permutations.

Our approach is inspired by the Viterbi algorithm [20] that tries to find the most likely path through hidden states. To make use of this approach, we replace 'most likely' by 'least costs', and 'hidden states' by 'unknown parameters'. We thus look for a sequence of maneuvers/orientations/free parameters that connect them with minimal costs.

This approach is suitable, because optimal paths have a characteristic: the interference between two primitive trajectories in that path depends on their distance. If they are close, a change of one usually also causes a change of the other, in particular, if they are connected. If they are far, we may change one trajectory of the sequence, without affecting the other. Viterbi reflects this characteristic, as it checks all combinations of neighboring (i.e. close) maneuvers to get the optimum.

We now are able to formulate our trajectory planning algorithm. We have

- the input: route points $p_1, \ldots p_n$, two orientation angles $\theta_1$ and $\theta_n$,
- a cost function to evaluate trajectories,
- maneuver types $\Pi$, candidates $O_i$ and $params(M)$,
- the output: a sequence of maneuvers with specified free parameters and the orientation angles $\theta_2, \ldots \theta_{n-1}$.

Figure 2 sketches the algorithm. Starting with $(p_1, \theta_1)$ it iteratively finds optimal maneuvers to $(p_i, \theta_{ij})$. As we have multiple intermediate angles $O_i = \{\theta_{i1}, \theta_{i2}, \ldots\}$, we keep optimal trajectory sequences to each of these angles in $S$. Because the number of trajectories in S only depends on the last route point (and in particular not on the route before), the runtime and memory usage is of $O(n)$.

| Algorithm getOptimalTrajectorySequence | if such $m$ exists { |
|---|---|
| **Algorithm** | if such *m* exists { |
| **getOptimalTrajectorySequence** |    $s' \leftarrow s$ extended by $m$; |
| $S \leftarrow \{(p_1, \theta_1)\}$; |    compute costs $c$ of $s'$ |
| for $(i \leftarrow 2$ to $n)$ { |      by the *Evaluator* |
|   $S' \leftarrow \{\}$; |    if $c < minC$ { |
|   for each $\theta_{ij} \in O_i$ { |      $minC \leftarrow c$; $minS \leftarrow s'$; } |
|    $minC \leftarrow \infty$; $minS \leftarrow undef$; |   } // end if exists |
|    for each optimal sequence $s \in S$ { |    } // end for each *param* |
|     for each maneuver type $M \in \Pi$ { |   } // end for each *M* |
|      for each $param \in params(M)$ { |  } // end for each *s* |
|       compute $m$ that |  if $minS \neq undef$ {$S' \leftarrow S' \cup \{minS\}$; }; |
|       - is of type $M$ |  } // end for each $\theta_{ij}$ |
|       - has *param* as free |  $S \leftarrow S'$; |
|        parameters | } // end for $i$ |
|       - extends $s$ to $(p_i, \theta_{ij})$ | if $S = \{\}$ return *no trajectory found* |
| | else  return first element of $S$ |

**Fig. 2.** The trajectory planning algorithm

For a new route step $p_{i+1}$, we again have to check multiple orientation angles of $O_{i+1}$. For this, we try to extend all trajectory sequences in $S$ by maneuvers. I.e. we compute all possible maneuver types in $\Pi$ with possible *params* to get to $(p_{i+1}, \theta_{i+1,j})$. We store the optimal trajectory sequences to each of the orientation angles in $S'$ that forms the $S$ for the next iteration.

In the last step, $O_n = \{\theta_n\}$, i.e. we only have to check the single target angle. Thus, if there is at least one trajectory sequence found, we have $|S| = 1$ and the optimal sequence is the single element of $S$.

## 3.4     Examples and Evaluation

We implemented the approach on our Carbot robot ([18], Fig. 3). It has a size of 35 cm × 40 cm × 27 cm and a weight of 4.9 kg. It is able to run with a speed of 31 cm/s. The wheel configuration allows to independently steer two wheels (Fig. 3 right). It is possible to drive curves like a car, but also to turn in place.

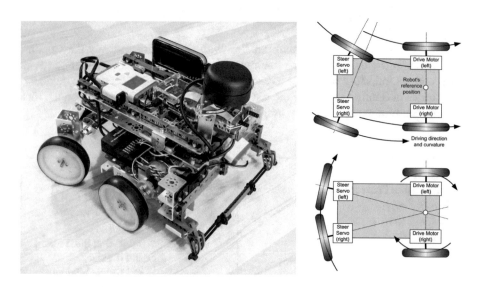

**Fig. 3.** The Carbot

For curves, both the different numbers of revolutions of the powered front wheels as well as the steering angles of the steered rear wheels are adapted to follow the respective curve geometry.

To recognize the environment, Carbot's camera is able to capture images in driving direction – the respective processing is executed onboard. A 360° Lidar sensor on top is able to recognize obstacles with cm resolution. In addition, Carbot has tactile and ultrasonic sensors. The entire world modeling, including SLAM-correction of internal poses is performed on the robot. We use a Raspberry Pi (3 Model B) to execute these tasks.

The Carbot platform also comes with a simulation environment that is able to simulate the Carbot on sensor- and motor-level, including appropriate error models and driving effects such as slippery floors. Software binaries can be tested in the simulator without the requirement to recompile them for the real robot. One important benefit: we can quickly construct very complex environments that would be costly to create in reality.

We created a number of environments to test our trajectory planning. One is a complex environment with walls, pillars and triangular obstacles; further environments are rectangular labyrinths. Prior to the trajectory planning, we performed a route planning that creates the route points. Currently, Carbot is able to execute route planning tasks with three approaches:

- a grid-based approach called *GAA* (*Grid-based A\* Advanced*),
- based on *Extended Voronoi Diagrams*,
- based on *Visibility Graphs*.

The examples (Fig. 4) show trajectories planned by our approach whereas the route planning used GAA. Note that, even the Carbot is able to turn in place, we assigned higher costs for them, thus the planning algorithm tries to find arc maneuvers instead. The top and middle images are created by our simulation tool. The bottom images show real runs using the Lidar device – we used our debugging environment to paint the results.

To measure the performance, we randomly distributed 23 start/target positions and processed the $23 \cdot 22 = 506$ routes between these positions and plan trajectories by our approach. We used two execution environments to measure the runtime:

- a desktop PC with i7-4790 CPU, 3.6 GHz that represents a high computational power (e.g. of future robot platforms),
- a Raspberry Pi 3 Model B, 1.2 GHz 64-bit ARMv8 CPU that represents the current reduced computational power on the real Carbot.

Table 2 presents the major results. The numbers indicate the average time to compute the trajectories for a single route step.

An amount of approx. 30–40% is spent for evaluating a trajectory by the cost function. Here, a time consuming part is to check collisions of trajectories with obstacles.

We divided the statistics in two packages. 'All' computed the set of candidates as introduced before, i.e. 8 maneuvers, 3 candidates for arc radii and the 5 candidates for orientations. Here, the algorithm checked 137.7 variations on average per route step.

A second package 'Reduced' works with a reduced set of candidates: only the maneuvers 2-Turns and $\int$ - Arcs, only a single arc radius, only three orientations and only a single arc sense of rotation were considered. As a result, the average number of variations is significantly reduced to 11.56. As only few patterns were checked, the resulting trajectories usually were much simpler. This illustrates: we may adapt the approach to reduced computations platform in a fine-grained manner, if required.

There exists another approach to deal with lower computational power. The algorithm keeps $|O_i|$ optimal variations for an iteration, but the leading maneuvers in S converge very fast to a single one. In our experiments, in 44.7% of all cases, the first maneuver for all considered sequences in $S$ was fixed, if the algorithm checked route point 3. For route point 4, the first maneuver was fixed in 88.5% of all cases, for route point 5, 95.7%. In our experiments, not more than 6 route points were required to fix the first maneuver. On average 3.7 route points were required.

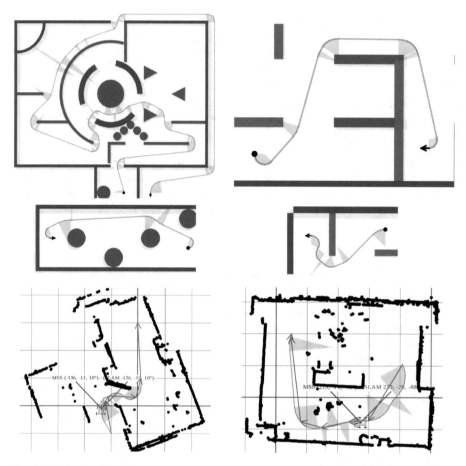

**Fig. 4.** Examples for planned trajectory sequences (top, middle: simulated environments, bottom: real environments)

**Table 2.** Execution statistics

| Amount per route step | All | | Reduced | |
|---|---|---|---|---|
| | Desktop | Carbot | Desktop | Carbot |
| Exec. time without cost function (ms) | 0.542 | 8.86 | 0.0297 | 0.274 |
| Cost function computation (ms) | 0.336 | 5.53 | 0.0121 | 0.125 |
| Total exec. time (ms) | 0.878 | 14.39 | 0.0418 | 0.399 |
| Avg. checked variations | 137.7 | | 11.56 | |

We make use of this observation: if we had to plan long routes, we iterate through the route points until the first maneuver was fixed, usually in constant time – according to our experience this takes 3–6 routing points. We then could start driving this maneuver; meanwhile, we compute the next one.

This approach significantly reduces the computation time to find the first driving command. Usually, during driving, the robot has to react to position correction by a SLAM component or detects new obstacles that may affect further maneuvers anyway. Thus, it is not reasonable to plan all trajectories to the target in a single step.

## 4    Conclusions and Future Work

Our approach computes a sequence of trajectories that minimize costs. To reduce the complexity of the problem, we first compute optimal routes that only take into account the space dimensions, ignoring non-holonomic constraints. The key idea is then to produce a sufficiently large set of suitable trajectory candidates that undergo an evaluation by the cost function. The candidate set is generated by a list of maneuvers – for each we know a closed solution to map start and target configuration to a sequence of driving commands. To reduce the overall number of checked permutations, we apply a Viterbi-like mechanism. The approach is efficient. The computation time only linearly increases with the route length, whereas the first trajectory usually is available in constant time. There exist sets of parameters that allow execution even on small computing platforms.

For future work, we want to address a problem: As the Viterbi-like approach does not check all combinations, we have to accept a suboptimal trajectory sequence. This usually is tolerable. However, in rare cases a solution requires a non-optimal decision *in the middle* of the route. As Viterbi tries to optimize step by step, such paths will not be found. In the future, we want to investigate the class of scenarios that cause unacceptable solutions by our approach. A solution could artificially integrate additional route points into the optimal route or partly reverse early decisions, if such scenarios were detected.

## References

1. Barraquand, L., Langlois, B., Latombe, J.-C.: Numerical potential field techniques for robot path planning. IEEE Trans. Syst. Man. Cybern. **22**(2), 224–241 (1992)
2. Boissonnat, J.D., Cerezo, A., Leblond, J.: A note on shortest paths in the plane subject to a constraint on the derivative of the curvature, Research Report 2160 Inst. Nat. de Recherche en Informatique et an Automatique (1994)
3. Bui, X.N., Boissonnat, J.D., Soueres, P., Laumond, J.P.: Shortest path synthesis for dubins nonholonomic robot. In: IEEE Conference on Robotics and Automation, San Diego, CA, pp. 2–7 (1994)
4. Delingett, H., Hebert, M., Ikeuchi, K.: Trajectory generation with curvature constraint based on energy minimization. In: Proceedings of the International Workshop on Intelligent Robots and Systems IROS91, Osaka, Japan (1991)
5. Dubins, L.E.: On curves of minimal length with a constraint on average curvature and with prescribed initial and terminal positions and tangents. Am. J. Math. **79**(3), 497–516 (1957)
6. Gu, T., Dolan, J.M.: On-road motion planning for autonomous vehicles. In: ICIRA 2012, pp. 588–597. Springer, Heidelberg (2012)

7. Karaman, S., Walter, M.R., Perez, A., Frazzoli, E., Teller, S.: Anytime motion planning using the RRT*. In: IEEE International Conference on Robotics and Automation (ICRA), May 9–13, 2011, Shanghai, China, pp. 4307–4313 (2011)
8. Laumond, J.-P., Jacobs, P.E., Taïx, M., Murray, R.M.: A motion planner for nonholonomic mobile robots. IEEE Trans. Robot. Autom. **10**(5), 577–593 (1994)
9. LaValle, S.M.: Rapidly-exploring random trees: a new tool for path planning, TR 98-11, Computer Science Dept., Iowa State University (1998)
10. LaValle, S.M., Kuffner, J.J.: Randomized kinodynamic planning. Int. J. Robot. Res. **20**(5), 378–400 (2001)
11. LaValle, S.M., Branicky, M.S., Lindemann, S.R.: Classical grid search and probabilistic roadmaps. Int. J. Robot. Res. **23**(7–8), 673–692 (2004)
12. Liang, T.-C., Liu, J.-S., Hung, G.-T., Chang, Y.-Z.: Practical and flexible path planning for car-like mobile robot using maximal-curvature cubic spiral. Robot. Auton. Syst. **52**, 312–335 (2005)
13. Mukadam, M., Yan, X., Boots, B.: Gaussian process motion planning. In: IEEE International Conference on Robotics and Automation (ICRA), Stockholm, Sweden, May 16–21, 2016, pp. 9–15 (2016)
14. Muñoz, V.F., Ollero, A.: Smooth trajectory planning method for mobile robots. In: Proceedings of the Congress on Computer Engineering in System Applications, Lille, France, pp. 700–705 (1995)
15. Nelson, W.L.: Continuous curvature paths for autonomous vehicles. In: Proceedings of the IEEE International Conference on Robotics and Automation, Scottsdale, AZ, pp. 1260–1264 (19890
16. Pitvoraiko, M., Kelly, A.: Efficient constrained path planning via search in state lattices. In: International Symposium on Artificial Intelligence, Robotics, and Automation in Space (2005)
17. Reeds, J.A., Shepp, L.A.: Optimal paths for a car that goes both forwards and back-wards. Pacific J. Math. **145**, 367–393 (1990)
18. Roth, J.: A novel development paradigm for event-based applications. In: International Conference on Innovations for Community Services (I4CS), Nuremberg, Germany, July 8–10, 2015, IEEE xplore, pp. 69–75 (2015)
19. Scheuer, A., Fraichard, T.: Continuous-curvature path planning for car-like vehicles. In: IEEE RSJ International Conference on Intelligent Robots and Systems, 7–11 September, Grenoble, France (1997)
20. Viterbi, A.: Error bounds for convolutional codes and an asymptotically optimum de-coding algorithm. IEEE Trans. Inf. Theory **13**(2), 260–269 (1967)
21. Zucker, M., Ratliff, N., Dragan, A., Pivtoraiko, M., Klingensmith, M., Dellin, C., Bagnell, J. A., Srinivasa, S.: CHOMP: Covariant Hamiltonian Optimization for Motion Planning. Int. J. Robot. Res. **32**, 1164–1193 (2013)

# Improving Relaxation-Based Constrained Path Planning via Quadratic Programming

Franco Fusco[1], Olivier Kermorgant[1(✉)], and Philippe Martinet[1,2]

[1] Centrale Nantes, Laboratoire des Sciences du Numérique de Nantes LS2N,
Nantes, France
`olivier.kermorgant@ec-nantes.fr`
[2] Inria Sophia Antipolis, Valbonne, France

**Abstract.** Many robotics tasks involve a set of constraints that limit the valid configurations the system can assume. Some of these constraints, such as loop-closure or orientation constraints to name some, can be described by a set of implicit functions which cause the valid Configuration Space of the robot to collapse to a lower-dimensional manifold. Sampling-based planners, which have been extensively studied in the last two decades, need some adaptations to work in this context.

A proposed approach, known as *relaxation*, introduces constraint violation tolerances, thus approximating the manifold with a non-zero measure set. The problem can then be solved using classical approaches from the randomized planning literature. The relaxation needs however to be sufficiently high to allow planners to work in a reasonable amount of time, and violations are counterbalanced by controllers during actual motion. We present in this paper a new component for relaxation-based path planning under differentiable constraints. It exploits Quadratic Optimization to simultaneously move towards new samples and keep close to the constraint manifold. By properly guiding the exploration, both running time and constraint violation are substantially reduced.

## 1 Introduction

Sampling-based planning techniques have been successfully exploited to solve a number of problems involving a wide variety of systems, such as mobile robots and manipulators. They rely on the construction of a graph, either in the form of networks [7] or trees [10], trying to approximate the valid Configuration Space (CS) of a system. Nodes, corresponding to configurations, are generated randomly and validated in a further step by checking for collision. In many situations the sampling process and the construction of edges – which represent motions between pairs of configurations – require simple operations. Random samples can be obtained by drawing each component independently from a given random distribution, while local motions are often created using linear interpolation.

© Springer Nature Switzerland AG 2019
M. Strand et al. (Eds.): IAS 2018, AISC 867, pp. 15–26, 2019.
https://doi.org/10.1007/978-3-030-01370-7_2

However, many robotics tasks impose a number of constraints on the system. As an example, a domestic robot carrying a tray loaded with objects should ensure that during its motion the platter remains horizontal. These constraints cause the valid configuration space to reduce to a lower-dimensional manifold implicitly defined by constraint equations. This introduces many challenges in the planning problem, mainly due to the fact that classical samplers and local routers cannot be exploited any more.

In order to deal with this added complexity, new tools have been developed to generate and connect samples inside the constraint manifold. Early studies focused on closed kinematic chains [17] and used the Gradient Descent algorithm to enforce loop-closure on random invalid samples. Cortes *et al.* introduced the *Random-Loop Generator* [4], a sampling technique designed to produce valid samples for closed-loop mechanisms.

Planning under task-space constraints was investigated in [13], which introduces *Tangent Space Sampling* and *First-Order Retraction* in order to sample feasible joint configurations. The *CBiRRT* (Constrained Bi-directional Rapidly-exploring Random Tree) planner [1] uses the Jacobian pseudo-inverse in a similar way to the Tangent Space Sampling in order to project an infeasible sample on the valid manifold defined by end-effector pose constraints.

Further methods have been designed to explore an approximation of the constraint manifold, based on high-dimensional continuation. *AtlasRRT* [5] focuses on the joint construction of Atlases and of a bi-directional RRT to approximate and explore the constraint manifold. A similar strategy is exploited in [8], defining a set of tangent spaces that locally approximate the manifold. A generalized framework based on Atlases is proposed in [16]. It extends several existing randomized planners to the exploration of constrained manifolds.

Other recent works [2,3] use the concept of *relaxation*, consisting in allowing a small constraint violation during planning. This technique has been mainly exploited to plan motions for compliant systems, using a control action during trajectory execution to steer robot's state close to the constraint manifold. The planning phase is therefore solved via standard techniques from the sampling based domain, since the valid CS is no longer a zero-measure set.

Such approach requires a trade-off between the quality of a planned path and planning time. In fact, if the allowed constraint violation is too small, the topology of the valid Configuration Space changes to a set of extremely narrow passages and the planning time increases significantly. Furthermore, the technique is highly dependent on system's compliance and on controller's ability to reconfigure a robot in feasible states during the motion. Thus, many practical scenarios involving rigid robots cannot exploit relaxation, since they would require a higher quality path directly from the planning step.

In this paper we propose a new approach inspired by relaxation techniques that allows to generate in a short amount of time paths with lower constraint violation. The algorithm uses Quadratic Programming (QP) to locally perform motions in the relaxed CS, with the objective of keeping configurations close to the original manifold while extending to a random sample. We exploit the

Jacobian matrix of the constraints to locally linearize the manifold and guide the exploration towards valid states. Thanks to smaller violation tolerances, the necessity of a control action during execution can be reduced, allowing a broader range of robots to exploit these techniques.

The remainder of this paper is organized as follows: differentiable constraints and the concept of relaxed Configuration Space are introduced in Sect. 2.1, while in Sect. 2.2 the technique used to perform local motions inside the relaxed CS is detailed. The router is integrated in a randomized planner, which is presented in Sect. 2.3. Experiments have been conducted in simulation considering different setups, to demonstrate the generality of our approach. We show in Sect. 3 that the technique allows to rapidly find paths featuring small constraint violation.

## 2   Planning Algorithm

In this section we present the proposed planner based on QP to enforce a set of constraints while planning. In order to connect two configurations, the local router is asked to accomplish two tasks concurrently: drive the system toward a given sample and keep the error associated with constraints as small as possible.

We formulate such problem as a Sequential Quadratic Optimization, wherein each step aims at changing the current configuration to a sample that is nearer to the target one, while enforcing the constraints. In addition, the displacement of each coordinate of the configuration vector is bounded. This allows to obtain a discrete set of intermediate configurations, and to consistently check for collisions along the path.

We introduce the constraints in Sect. 2.1, while the local path planner is detailed in Sect. 2.2. The connection routine is integrated in a complete Sampling-Based Planner based on the RRT-Connect algorithm [9], which is presented in Sect. 2.3. It attempts to generate new samples and to connect them to the existing graph using the local planner. Post-processing operations can finally be performed to enhance the quality of the resulting path, if any is found.

### 2.1   Differentiable Constraints

We consider the case of an $n$-dimensional Configuration Space $\mathcal{C} \subset \mathbb{R}^n$, the configuration vector being denoted as $\mathbf{q} = \begin{bmatrix} q_1 \cdots q_n \end{bmatrix}^T$. Each coordinate is assumed to be bounded in a range $[q_{i,\min}, q_{i,\max}]$. A set of $n_c$ constraints is also considered, each one being described by a differentiable function $C_i : \mathcal{C} \to \mathbb{R}$ ($i = 1, \cdots, n_c$). Altogether, they define the constrained Configuration Space as:

$$\mathcal{C}_C = \{\mathbf{q} \in \mathcal{C} : C_i(\mathbf{q}) = 0 \quad \forall i\} \tag{1}$$

To approximate the valid set defined by all constraints we introduce $n_c$ constants $\varepsilon_i > 0$, which quantify the maximum allowed violation of each constraint at a given configuration $\mathbf{q}$. These constants need to be set manually by the user depending on the required quality of the planned path. We then define the relaxed Configuration Space $\mathcal{C}_R$:

$$\mathcal{C}_R = \{\mathbf{q} \in \mathcal{C} : -\varepsilon_i \leq C_i(\mathbf{q}) \leq \varepsilon_i \quad \forall i\} \tag{2}$$

In order to shorten the notation in following sections, we stack all constraints in the $n_c$-dimensional vector $\mathbf{e} = \left[C_1(\mathbf{q}) \cdots C_{n_c}(\mathbf{q})\right]^T$. We finally recall that the constraints are assumed to be differentiable. Under this assumption, we introduce the Jacobian matrix $\mathbf{J}_e = \frac{\partial \mathbf{e}}{\partial \mathbf{q}}$, whose $i$-th row is given by the gradient of the constraint $C_i$.

The original planning problem is then formulated as finding a discrete sequence of points in $\mathcal{C}_R$ which approximate a valid path in $\mathcal{C}_C$. The exploration, as detailed in the following section, is guided by the use of $\mathbf{J}_e$ to locally approximate the manifold implicitly defined by the constraints.

## 2.2   Local Motions Using Quadratic Programming

In order to evaluate a local path between two configurations, we solve sequentially a number of QP problems. Our iterative algorithm exploits a single step to perform a short motion towards the goal while keeping $\mathbf{e}$ as close to zero as possible. The problem is formalized as follows: two samples $\mathbf{q}_a \in \mathcal{C}_R$ (initial configuration) and $\mathbf{q}_b \in \mathcal{C}$ (final state) are considered. During the process, we denote with $\mathbf{q}^{(j)}$ the configuration obtained after the $j$-th iteration (such that $\mathbf{q}^{(0)} = \mathbf{q}_a$). We then select a proper value of $\mathbf{q}^{(j+1)}$, i.e., the next way-point of the local path from $\mathbf{q}_a$ to $\mathbf{q}_b$, by solving the quadratic optimization problem

$$\mathbf{q}^{(j+1)} = \arg\min_{\mathbf{x} \in \mathcal{C}} \left\| \mathbf{Q}^{(j)} \mathbf{x} - \mathbf{v}^{(j)} \right\|^2 \quad \left( \mathbf{Q}^{(j)} \in \mathbb{R}^{m \times n}, \ \mathbf{v}^{(j)} \in \mathbb{R}^m \right) \tag{3}$$

subject to a set of $p$ linear inequalities in the form:

$$\mathbf{A}^{(j)} \mathbf{x} \leq \mathbf{b}^{(j)} \quad \left( \mathbf{A}^{(j)} \in \mathbb{R}^{p \times n}, \ \mathbf{b}^{(j)} \in \mathbb{R}^p \right) \tag{4}$$

We use the superscript $(j)$ to underline that $\mathbf{Q}$, $\mathbf{v}$, $\mathbf{A}$ and $\mathbf{b}$ are evaluated using only values coming from the previous iteration, thus being constant quantities during the $(j+1)$-th step.

In the sequel, we firstly show how to derive the expression of matrices contained in (3) and (4), and afterwards the iterative scheme is detailed.

**Objective and Linear Inequalities.** Since the final goal of the iterative scheme is to reach $\mathbf{q}_b$ – or at least to move as close as possible to it – the considered objective function should contain a term that reaches its minimum in correspondence of the given configuration. In addition, a further contribution should be considered to enforce the constraints. A candidate function satisfying both requirements could be:

$$f\left(\mathbf{q}^{(j+1)}\right) = \left\| \mathbf{q}^{(j+1)} - \mathbf{q}_b \right\|^2 + \left\| \alpha \, \mathbf{e}^{(j+1)} \right\|^2 \tag{5}$$

wherein the constant parameter $\alpha \in \mathbb{R}^{n_c \times n_c}$ is a positive definite diagonal matrix used as a weighting factor to modulate the "priority" associated to the second task. The term $\mathbf{e}^{(j+1)}$, corresponding to constraints violation evaluated at $\mathbf{q}^{(j+1)}$, is however generally non-linear in the configuration vector, and does not fit the quadratic formulation of (3). However, the function can be linearized around $\mathbf{q}^{(i)}$, using the Jacobian matrix $\mathbf{J}_e$ introduced in Sect. 2.1:

$$\mathbf{e}^{(j+1)} - \mathbf{e}^{(j)} \simeq \mathbf{J}_e^{(j)} \left( \mathbf{q}^{(j+1)} - \mathbf{q}^{(j)} \right) \tag{6}$$

After injecting the linearized error in (5), the objective can be re-written in a matrix form compatible with (3), thus obtaining:

$$\mathbf{Q}^{(j)} = \begin{bmatrix} \mathbf{I}_n \\ \alpha \, \mathbf{J}_e^{(j)} \end{bmatrix} \qquad \mathbf{v}^{(j)} = \begin{bmatrix} \mathbf{q}_b \\ \alpha \left( \mathbf{J}_e^{(j)} \, \mathbf{q}^{(j)} - \mathbf{e}^{(j)} \right) \end{bmatrix} \tag{7}$$

Regarding the set of linear inequalities, we choose to reduce the search interval according to two factors. The first one imposes upper and lower bounds to each component of the configuration vector:

$$\mathbf{q}_{min} \leq \mathbf{q}^{(j+1)} \leq \mathbf{q}_{max} \tag{8}$$

The second set of inequalities that is considered limits the local motion of each coordinate to a symmetric interval centered in $\mathbf{q}^{(j)}$:

$$\mathbf{q}^{(j)} - \beta^k \Delta \mathbf{q} \leq \mathbf{q}^{(j+1)} \leq \mathbf{q}^{(j)} + \beta^k \Delta \mathbf{q} \quad (\beta \in (0,1) \text{ and } k \in \mathbb{N}) \tag{9}$$

In this relation, the entries of $\Delta \mathbf{q} \in \mathbb{R}^n$, all being positive, correspond to the allowed step of each coordinate, while $\beta^k$ is used to tune the size of the allowed interval. This constraint is justified by two reasons. On one hand the sequential optimization relies on the linearized error dynamics. The approximation must be kept consistent, and therefore the configurations $\mathbf{q}^{(i)}$ and $\mathbf{q}^{(i+1)}$ should not differ too much. If the new configuration is too far away from the initial one, the constraint error could exceed the tolerance, even if the linearized one is null, as depicted in Fig. 1. The use of the coefficient $\beta^k$ allows to resize the step size during the optimization, and its use is detailed in next section. On the other hand, in many practical planning problems some non-differentiable constraints could be considered as well. They should be verified at each iteration, and a large step size could bear the system to "jump" over small invalid regions. As an example, in our approach we perform discrete collision checking, by verifying the validity of a configuration at each iteration. With the set of inequalities (9) we try to avoid situations wherein collisions with small obstacles are not detected.

**Sequential Optimization.** The overall local motion routine, called *QPMove*, is reported in Algorithm 1. It performs a Sequential Optimization by solving at each step a QP instance as formulated in the previous section.

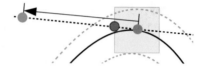

**Fig. 1.** From an initial configuration (orange) lying on the constraint – the black continuous line – the optimization would move the system to an invalid configuration (in blue). If the motion is limited to the small yellow rectangle, there are higher chances to still fall inside the relaxed region (surrounded by the dashed gray lines).

An iteration starts with the evaluation of the matrices defining the objective and the inequalities. Then, the inner cycle (between Algorithms 1 and 1) is executed to solve the current QP instance. The obtained solution is tested by checking constraints violation at the new configuration. This is a fundamental step: since constraints are linearized during the procedure, a resulting sample will only ensure an approximation of the error to be optimized. On the other hand, a component of the actual error could fall outside its valid range $[-\varepsilon_i, \varepsilon_i]$. To better enforce bounding constraints when such situation occurs, $k$ is incremented by one unit and the optimization step is repeated. This shrinks the valid range of motion, possibly leading $\mathbf{q}^{(j+1)}$ to lie inside $\mathcal{C}_\mathrm{R}$. Nonetheless, the procedure

---

**Algorithm 1.** QP-based Motion Validator (QPMove)

---

1: QPMove($\mathbf{q}_a$, $\mathbf{q}_b$) :
2:   $\mathbf{q}^{(0)} \leftarrow \mathbf{q}_a$
3:   $f^{(0)} \leftarrow +\infty$
4:   $f^{(1)} \leftarrow +\infty$
5:   **for** $j$=0 to $j_\mathrm{max}$ **do**
6:       $k \leftarrow 0$
7:       $\mathbf{Q}^{(j)}$, $\mathbf{v}^{(j)}$, $\mathbf{A}^{(j)} \leftarrow$ INIT_QP_ITERATION($\mathbf{q}^{(j)}$, $\mathbf{q}_b$)
8:       **do**
9:           **if** $k > k_\mathrm{max}$ **then**
10:               **return** "failure", $\mathbf{q}(j)$
11:           **end if**
12:           $\mathbf{b}^{(j)} \leftarrow$ GET_QP_B_VECTOR($\mathbf{q}^{(j)}$, $k$)
13:           $\mathbf{q}^{(j+1)}$, $f^{(j+1)} \leftarrow$ SOLVE_QP_INSTANCE($\mathbf{Q}^{(j)}$, $\mathbf{v}^{(j)}$, $\mathbf{A}^{(j)}$, $\mathbf{b}^{(j)}$)
14:           $\mathbf{e}^{(j+1)} \leftarrow$ EVALUATE_ERROR($\mathbf{q}^{(j+1)}$)
15:           $k \leftarrow k + 1$
16:       **while not**($-\varepsilon \leq \mathbf{e}^{(j+1)} \leq \varepsilon$)
17:       **if** IN_COLLISION($\mathbf{q}^{(j+1)}$) **then**
18:           **return** "failure", $\mathbf{q}(j)$
19:       **end if**
20:       **if** $f^{(j+1)} \leq f_{min}$ **then**
21:           **return** "success", $\mathbf{q}_b$
22:       **end if**
23:       **if** $f^{(j+1)} \leq f^{(j)}$ and $f^{(j+1)} \geq f^{(j)} - \Delta f^{(-)}$ **then**
24:           **return** "failure", $\mathbf{q}(j + 1)$
25:       **end if**
26:       $\mathbf{q}^{(j)} \leftarrow \mathbf{q}^{(j+1)}$
27:       $f^{(j)} \leftarrow f^{(j+1)}$
28: **end for**
29: **return** "failure", $\mathbf{q}(j_\mathrm{max})$

---

is limited up to a maximum value of $k$, in order to prevent the algorithm from spending too much time on some critical samples.

The remaining part of the main loop is instead used to check the progresses done between two iterations. As briefly mentioned in the previous section, we propose to handle non-differentiable constraints – in particular, collision detection – by checking them at each new sample. Thus, after having solved the current QP instance, the validity of the sample is tested. If any violation is detected, the algorithm stops and returns the last validated sample.

A second criterion verifies instead if the objective value $f^{(j+1)}$ has become small enough, and in case the algorithm is stopped since the goal configuration has been reached with the error being sufficiently small.

A known problem of quadratic minimization is the existence of local minima. To detect stationary points, the algorithm computes the difference between the objective after subsequent iterations: if the improvement is below a given threshold $\Delta f^{(-)}$, the algorithm returns with the status "failure". It must be noted that due to the linearization the objective could get worse between successive iterations, and thus local minima are checked only if the objective is improving.

Since a proper tuning of the objective thresholds could be hard in practice, a maximum number of iterations is exploited as the last strategy to prevent the routine to waste too much computation time.

The proposed motion component requires more computation efforts and longer run time compared to the simpler technique of linear interpolation, usually exploited in the relaxation context. Nonetheless, its ability of following constraint manifold's curvature proved to effectively counter-balance the drawbacks. Figure 2 shows some comparative examples in a 2-Dimensional Configuration Space. In the first case (Fig. 2(a)) the valid set defined by the constraint is a circle. Even with a high relaxation factor, linear interpolation would fail to connect the two shown samples, since it would try to follow a straight line path that is incompatible with the curved constraint. In practical scenarios, the relaxation would be way smaller, making it even harder to connect configurations. The example depicted in Fig. 2(b) shows another feature of QPMove: even when attempting to reach an invalid sample, a valid motion can still be performed inside $\mathcal{C}_R$. We also show in Fig. 2(c) the motion that would be produced by a projection approach like the one exploited in the CBiRRT planner. To reach the goal more samples are necessary, with a more irregular spacing between adjacent samples.

## 2.3   QPlan

The component QPMove described in the previous section was used as part of a complete path-planner named *QPlan*. The algorithm exploits a bidirectional RRT [9] to explore the constraint manifold, and runs post-processing techniques to enhance the quality of a path. We detail in the following how the typical components of a randomized planner were integrated.

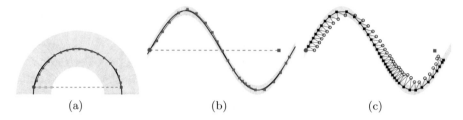

**Fig. 2.** Motions generated in $\mathcal{C}_R$ by QPMove (red), linear interpolation (green) and a projection approach (black). Start configurations are represented as blue circles, while the goals as blue squares. In (a), the Constrained Configuration Space is represented by a circular arc, while in (b) and (c) by a sinusoidal wave. Linear interpolation can only produce a short path in the first case thanks to a large relaxation, while no new samples can be generated in the sinusoidal region. QPMove (b) produces less samples, which are better distributed than in the case of a projection approach (c).

**Sampling.** A State Sampler produces new configurations which could potentially become leaves of the trees being expanded by the planner. When moving on a manifold, samples should satisfy the constraints imposed on the system.

However, QPMove does not necessarily need valid samples to extend one tree. In fact, an infeasible goal could be passed as $\mathbf{q}_b$. The algorithm would not converge to the given configuration, but it could still produce valid motions toward new points in the Constraint Manifold. The advantage of this choice is a faster sampler, since the simple uniform sampling technique can be used.

**Trees Extension.** The extension step of a bidirectional RRT works by selecting a random configuration and its nearest neighbor from the current tree. A motion is then attempted from the latter to the random sample, but limiting its length to a maximum value.

We rather adopt the greedy version of this algorithm, sometimes referred to as *RRT-ConnConn* [11]. This variant tries to extend a branch until either an invalid state is reached or the connection is successfully performed.

Since QPlan is able to find a discrete set of way-points, it could be useful to insert all intermediate configurations in the tree. However, this could bear to an over-populated tree. Therefore, only one generated configuration out of $N$ is inserted in the tree, and only if its distance from the previously added sample is higher than a given threshold.

**Post-processing.** As we handle constraints using a relaxation approach, their violation will not be completely nullified along local paths. Post-processing is used to refine an obtained solution by enforcing the constraints at each intermediate sample. If a path is found by the planner, the sequence of joint way-points is re-built, and a local QP optimization is run on each sample. The procedure is similar to the one exploited in QPMove, with two differences: rather than setting inequalities that enforce space bounds, a much smaller range is considered.

In practice, inequalities (8) are replaced by:

$$\mathbf{q}_{raw} - \mathbf{dq} \le \mathbf{q}^{(j+1)} \le \mathbf{q}_{raw} + \mathbf{dq} \tag{10}$$

where $\mathbf{q}_{raw}$ is the path configuration before optimization, and $\mathbf{dq}$ is a vector containing small values allowing the error to be minimized without moving too far away from $\mathbf{q}_{raw}$. In addition, we set as goal configuration the raw initial sample itself. It further constrain the system to only locally change its state.

In addition, a certain number of short-cutting attempts is performed, in order to reduce path's length and simplify the solution.

## 3   Simulation Results

We present in this section results obtained from simulations on several different setups with the implemented algorithm. The planner has been implemented using the components provided by OMPL, the *Open Motion Planning Library* [15], and integrated in the ROS [12] framework in the form of a MoveIt! [14] planning plugin.

During all experiments, the selection of a proper value of the parameter $\beta$ was done by trial and error. A value between 0.7 and 0.85 gave the best results in practice. Relaxation constants $\varepsilon_i$ were set to ensure a reasonably small violation of constraints. Their numerical value is reported later for each experiment.

A first set of tests was performed considering a 3-Dimensional point $(x, y, z)$ constrained to the surface of a sphere (Fig. 3(a)) and of a torus (Fig. 3(b)),

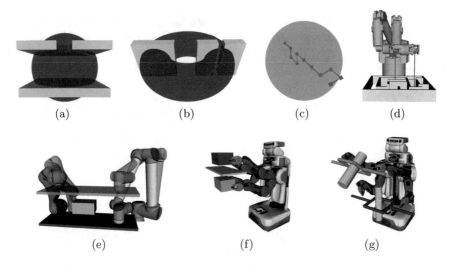

(a)            (b)            (c)            (d)

(e)            (f)            (g)

**Fig. 3.** Different tests with QPlan: a 3D point moving respectively (a) on a sphere and (b) on a torus, (c) a five-links chain whose tip is constrained to a sphere, (d) a Barrett arm solving a maze, (e) a UR10 and a Kuka LWR4 moving a plate, (f) PR2 moving a box, (g) PR2 displacing an object with a hole.

in presence of obstacles forming narrow passages. The constraints are written in the two setups as $C(x, y, z) = x^2 + y^2 + z^2 - R^2$ and $C(x, y, z) = \left(\sqrt{x^2 + y^2} - r_1\right)^2 + z^2 - r_2^2$, $R$ being the radius of the sphere and $r_1$, $r_2$ the radii defining the torus. We compared the performances of QPlan with a classical relaxation technique, considering different tolerances. Although the algorithm presented in [2] uses RTT* [6], we implemented it using the RRT-Connect algorithm for a meaningful comparison of run times. We exploited a sampler that uniformly generates samples over the manifold in order to reproduce the proposed setup. As reported in Table 1, our approach is considerably faster than a simple relaxation technique when the allowed tolerance becomes small[1].

Some tests were run on a more complex test-case, which had been proposed in [16]. A five-links kinematic chain rooted at the origin is considered (see Fig. 3(c)). Each link is parametrized by a 3D point, for a total number of 15 degrees of freedom. Five constraints are set to fix the distance between pairs of adjacent joints, while a further constraint requires chain's tip to move on a spherical surface. An additional constraint can be set to fix the vertical coordinate of the first point. While QPlan can effectively find paths for this system, standard relaxation techniques require either a higher tolerance or longer execution time. The results shown in Table 2 were obtained considering all links having a length of 0.2 m, and the constraint sphere a radius of 0.6 m. The tolerances have been fixed to 0.005 m$^2$ for the five distance constraints, to 0.025 m$^2$ for the tip and 1 mm for the vertical constraint. It is also worth to note that the run-time is at the same order of magnitude of the Atlas-based planners, according to the results given in [16].

**Table 1.** Average planning time in seconds over 100 runs of sphere and torus setups. We used $R = 1$ m, $r_1 = 1$ m, $r_2 = 0.5$ m. The relaxation factor $\varepsilon$ is given in m$^2$.

| Algorithm | $\varepsilon$ | Sphere | Torus |
|---|---|---|---|
| Relaxation | $10^{-2}$ | 0.214 | 0.477 |
| | $10^{-3}$ | 8.785 | 14.632 |
| QPlan | $10^{-2}$ | 0.104 | 0.027 |
| | $10^{-3}$ | 0.120 | 0.218 |

**Table 2.** Results with the five-links chain, under 6 or 7 constraints. The actual runtime of the first algorithm is unknown, since no plans were found before a time limit of 10 s.

| Algorithm | $n_c = 6$ | $n_c = 7$ |
|---|---|---|
| Relaxation | >10 | >10 |
| QPlan | 0.080 | 0.087 |

Other tested scenarios involve a Barrett arm solving a maze (Fig. 3(d)) and some dual arm systems cooperatively displacing an object (Figs. 3(e), (f) and (g)). We do not report running times obtained with standard relaxation techniques, since they require a much higher planning time for equal relaxation factors.

---

[1] A relaxation factor $\varepsilon = 10^{-3}$ m$^2$ on a sphere with unitary radius corresponds to constrain the points to lay at most 0.5 mm from the surface. The same factor gives a maximum distance of 1 mm from a torus having $r_1 = 1$ m, $r_2 = 0.5$ m.

During the maze-solving test three constraints are applied to the end-effector so as to keep the stick grasped by the arm vertical and in contact with labyrinth's floor. This scenario is particularly challenging for the planner, since many obstacles are encountered during planning. As explained in Sect. 2.3, we do not use a specific Sampler to generate new configurations. This might lead the robot to attempt many motions toward configurations that are completely unreachable either due to constraints or to obstacles. Using a more involved sampling technique may instead improve the performances. Another factor that greatly influences the planning time is the choice of the relaxation constants. In a first instance, we considered very strict tolerances on both stick position and orientation: 1 cm of error for the altitude and 0.05 rad for roll and pitch constraints. The chances to find a solution in a short amount of time are quite low, as shown in Table 3. However, further tests were conducted by allowing orientation constraints to be violated with at most 0.2 rad, and optimizing them in the post-processing phase. With this more tolerant setup, the planning time is slightly reduced.

In the dual-arm setups, relative translation and rotation of the end-effectors are forbidden in order to maintain a fixed relative transformation between tip frames. The challenge here comes from the high number of both Degrees of Freedom and constraints.

**Table 3.** Planning time of the tests involving different manipulators.

| Scene | Avg. | Min. | Max. |
|---|---|---|---|
| Barrett maze (0.05 rad) | 4.034 | 0.669 | 13.828 |
| Barrett maze (0.2 rad) | 3.446 | 0.426 | 13.162 |
| UR10 + LWR4 | 0.819 | 0.084 | 4.085 |
| PR2 box | 4.338 | 1.036 | 10.395 |
| PR2 cylinders | 0.866 | 0.202 | 2.322 |

## 4   Conclusions

We have presented a new approach based on Quadratic Programming that can effectively generate paths while dealing with a set of constraints. Our approach takes advantage of relaxation to enlarge the range of valid configurations while planning and of analytic description of constraints to guide local motions. As a drawback, the proposed algorithm features many parameters which need to be tuned in order to guarantee good performances. The relaxation factor plays a relevant role: if constraint violations are not too strict, the search can proceed faster. Nonetheless, with an increased tolerance robots would need higher control action to re-project the samples back to the manifold.

Finally, in our contribution we focused only on the planning step, verifying that paths generated by the algorithm can be found quickly and with better constraint enforcement. As a future line of work, experiments with a real robot should be performed to confirm the validity of the proposed approach.

# References

1. Berenson, D., Srinivasa, S.S., Ferguson, D., Kuffner, J.J.: Manipulation planning on constraint manifolds. In: IEEE International Conference on Robotics and Automation. IEEE (2009)
2. Bonilla, M., Farnioli, E., Pallottino, L., Bicchi, A.: Sample-based motion planning for soft robot manipulators under task constraints. In: IEEE International Conference on Robotics and Automation (2015)
3. Bonilla, M., Pallottino, L., Bicchi, A.: Noninteracting constrained motion planning and control for robot manipulators. In: IEEE International Conference on Robotics and Automation. IEEE (2017)
4. Cortes, J., Simeon, T.: Sampling-based motion planning under kinematic loop-closure constraints. In: Algorithmic Foundations of Robotics VI. Springer, Heidelberg (2004)
5. Jaillet, L., Porta, J.M.: Path planning under kinematic constraints by rapidly exploring manifolds. IEEE Trans. Robot. (2013)
6. Karaman, S., Frazzoli, E.: Sampling-based algorithms for optimal motion planning. Int. J. Robot. Res. (2011)
7. Kavraki, L.E., Svestka, P., Latombe, J.-C., Overmars, M.H.: Probabilistic roadmaps for path planning in high-dimensional configuration spaces. IEEE Trans. Robot. Autom. (1996)
8. Kim, B., Um, T.T., Suh, C., Park, F.C.: Tangent bundle RRT: a randomized algorithm for constrained motion planning. Robotica (2016)
9. Kuffner, J.J., LaValle, S.M.: RRT-connect: an efficient approach to single-query path planning. In: IEEE International Conference on Robotics and Automation. IEEE (2000)
10. LaValle, S.M.: Rapidly-exploring random trees: A new tool for path planning. Technical report, Department of Computer Science, Iowa State University (1998)
11. LaValle, S.M., Kuffner Jr., J.J.: Rapidly-exploring random trees. Progress and Prospects (2000)
12. Quigley, M., Conley, K., Gerkey, B., Faust, J., Foote, T., Leibs, J., Wheeler, R., Ng, A.Y.: ROS: an open-source robot operating system. In: ICRA Workshop on Open Source Software (2009)
13. Stilman, M.: Task constrained motion planning in robot joint space. In: IEEE/RSJ International Conference on Intelligent Robots and Systems. IEEE (2007)
14. Sucan, I.A., Chitta, S.: Moveit! http://moveit.ros.org. Accessed 30 Jun 2017
15. Sucan, I.A., Moll, M., Kavraki, L.E.: The Open Motion Planning Library. IEEE Robotics and Automation Magazine (2012). http://ompl.kavrakilab.org. Accessed 30 Jun 2017
16. Voss, C., Moll, M., Kavraki, L.E.: Atlas+ x: Sampling-based planners on constraint manifolds. Technical report. Rice University (2017)
17. Yakey, J.H., LaValle, S.M., Kavraki, L.E.: Randomized path planning for linkages with closed kinematic chains. IEEE Trans. Robot. Autom. (2001)

# Robust Path Planning Against Pose Errors for Mobile Robots in Rough Terrain

Yuki Doi[1(✉)], Yonghoon Ji[1], Yusuke Tamura[1], Yuki Ikeda[2],
Atsushi Umemura[2], Yoshiharu Kaneshima[2], Hiroki Murakami[2],
Atsushi Yamashita[1], and Hajime Asama[1]

[1] The University of Tokyo, Tokyo, Japan
doi@den.t.u-tokyo.ac.jp
[2] IHI Corporation, Tokyo, Japan

**Abstract.** We propose a novel path planning method considering pose errors for off-road mobile robots based on 3D terrain map information. Mobile robots navigating on rough terrain cannot follow a planned path perfectly because of uncertainties such as pose errors. In this work, we represent such pose errors as error ellipsoids to use on collision check with obstacles in a map. The error ellipsoids are estimated based on extended Kalman filter (EKF) that integrates motion errors and global positioning systems (GPS) observation errors. Simulation and experiment results show that the proposed method enables mobile robots to generate a robust path against pose errors in a large-scale rough terrain map.

**Keywords:** Path planning · Rough terrain · Random sampling
Extended Kalman Filter · Error ellipsoid

## 1 Introduction

Disaster response activities are important to save human lives and resources. In recent years, mobile robots equipped with sensors such as a camera are used when disasters occur. Mobile robots are usually operated by a remote control when human cannot enter environments that are damaged by disasters. However, a remote control has limitations because an operator is hard to recognize a surrounding environment with information through a camera. Therefore, autonomous mobile robots that are able to plan a safe path avoiding dangerous areas are needed. In 2004 and 2005, DARPA (Defense Advanced Research Projects Agency) held DARPA ground challenge [1]. In this challenge, a lot of mobile robots tried to run a long way of 150 miles in rough terrain and five robots finished. These robots were provided the navigation path in advance. In case of a disaster situation, feasible path for mobile robots cannot be provided before a disaster occurs and given that we can only get very limited environment information. Hence, the path planning method which generates safe paths for mobile robots in rough terrains is very important.

© Springer Nature Switzerland AG 2019
M. Strand et al. (Eds.): IAS 2018, AISC 867, pp. 27–39, 2019.
https://doi.org/10.1007/978-3-030-01370-7_3

Kuwata et al. proposed a real time motion planning method for autonomous cars [2]. They classified environment maps according to a risk of collision. Moreover, they reduced calculation time by preserving a former result and using it during a next motion planning. Richter et al. proposed a path planning method for autonomous cars with a self-position estimation [3]. They set high risk allowances in areas where the cars did not construct environment maps in order not to plan paths in uncertain areas. Nevertheless, these methods mentioned above assumed 2D environments that cannot deal with rough terrain. In this respect, Ji et al. proposed the broad path planning method on a 3D environment map for mobile robots traveling rough terrain [4]. In this study, a motion model of a mobile robot was considered; hence, it was possible to generate feasible paths for navigation. Moreover, they avoided the problem of generating a path that could cause the robots to fall down by restricting robot angles of a inclination and radii of a gyration during the path planning. However, robot pose errors that should be managed for actual operation of the robot were not considered.

Generally, mobile robots are affected by pose errors in many respects while navigating in the real environment. Thus, the robot cannot follow the generated path perfectly due to the pose errors. If the generated path is close to obstacles, the robot may collide with the obstacles. To take the pose errors into account during path planning, van den Berg et al. proposed path a planning method that considers motion uncertainty and imperfect sensors [5]. They calculated probability of collision with generated paths and map to judge validations of paths. However, they also assumed 2D environments. Blackmore et al. and Lee et al. proposed optimal robust path planning methods that use the random sampling integrated with chance constraints [6,7]. Chance constraints enable to restrict probabilities of collision with obstacles by satisfying constraint equations. However, chance constraints can be calculated only when obstacles are convex polyhedrons and they assumed only 2D environments. There are many obstacles of various shapes in rough terrain; thus, this method cannot be applied when disaster occurs.

In this study, we propose a safe path planning method for mobile robots in a rough terrain when pose errors affect robots. The remainder of this paper is organized as follows. Section 2 introduces the problem definition and approach of the proposed robust path planning. Then, the method of an error estimation is presented in Sect. 3. Section 4 introduces the sampling method considering the robot acceleration. The validity of the proposed path planning method is evaluated with the simulation and real experimental results in Sect. 5. Finally, Sect. 6 gives conclusions of this paper.

## 2    Approach

In this study, we propose a novel path planning scheme that can manage pose errors under the assumption that a four-wheeled vehicle as a mobile robot has a map of an entire environment consisting of 3D point cloud data in advance. An environment map is measured by unmanned aerial vehicles (UAV) in advance.

Figure 1(a) shows an example of a 3D point cloud of a rough terrain. This robot is expressed in six dimensional configurations, position information $(x, y, z)$ and orientation angles $(\phi, \theta, \psi)$ around axes, as shown in Fig. 1(b). Moreover, accelerations are restricted so as not to make movements that robots cannot handle. Maximum inclination angles of slopes are also restricted according to robot velocities; hence paths must select routes with lower angle than maximums. We also assume that the robot is equipped with two global positioning systems (GPS); hence, it can obtain its position and orientation information. In our previous work [4], we assumed that GPS sensors obtain true values of robot poses and a mobile robot can follow generated path, perfectly. However, as mentioned in Sect. 1, the robot is generally affected by observation errors of GPS and motion errors in real environments. In this study, we regard these two errors as pose errors. By taking the pose errors into consideration, the generated path produced by our proposed scheme ensures safety even if the mobile deviates from the planed path.

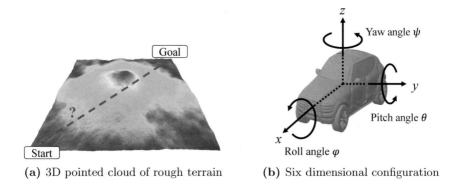

(a) 3D pointed cloud of rough terrain        (b) Six dimensional configuration

**Fig. 1.** Assumptions of proposed path planning method.

We use the random sampling algorithm [8] in the path planning method. The random sampling-based method cannot find optimal path; however, it can explore a map of large environment, quickly. The random sampling algorithm performs a path planning by generating robot configurations as new nodes and connecting them to existing ones. In the proposed method, robot velocities and angular velocities are sampled and nodes corresponding to robot poses are generated using a robot motion model. We defined paths avoiding collisions with obstacles even pose errors affect as robust paths. To consider effects of pose errors during a path planning, we adopted to estimate pose errors after each node are generated and to converted pose errors to error ellipsoids. We assumed average of GPS data equal generated node because mobile robots correct its position by the local control. Figure 2 shows the flowchart of our random sampling algorithm. In a conventional random sampling method, generated nodes are judged without the error estimation. On the other hand, in our random sampling algorithm, we estimate motion errors and observation errors by converting them into one error ellipsoid after generating each node. Error ellipsoids are used to judge validation of nodes by checking collisions with a 3D environment map. We also propose a new sampling method of robot velocities. This method enables to plan safe paths that mobile robots can follow.

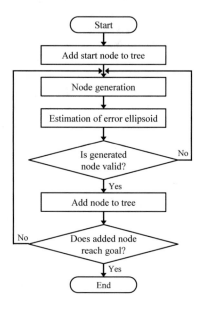

**Fig. 2.** Flowchart of proposed random sampling algorithm.

## 3    Error Estimation

### 3.1    Pose Error Estimation of Each Node

The mobile robot generally obtains pose information from GPS and controlled by control input data to moves to the target position. In our random sampling algorithm, nodes corresponding to robot poses are generated assuming control inputs data and connected to existing ones. Within components of nodes, the robot position $(x_t, y_t)$ and the yaw angle $\psi_t$ are sampled by using the motion model. The position $z_t$, the roll angle $\phi_t$ and the pitch angle $\theta_t$ are determined from map information. When control input data $\mathbf{u}_t = (v_t, \omega_t)$ is given, the robot pose at time $t$ $\mathbf{x}_t = (x_t, y_t, \psi_t)$ is given as the following motion model.

$$\mathbf{x}_t = \mathbf{f}(\mathbf{x}_{t-1}, \mathbf{u}_t) = \begin{pmatrix} x_{t-1} + v_t \Delta t \cos\left(\psi_{t-1} + \frac{\omega_t \Delta t}{2}\right) \\ y_{t-1} + v_t \Delta t \sin\left(\psi_{t-1} + \frac{\omega_t \Delta t}{2}\right) \\ \psi_{t-1} + \omega_t \Delta t \end{pmatrix}. \tag{1}$$

However, the mobile robot cannot perform the expected operation because of the pose errors consisting of observation errors and motion errors. Therefore, estimating pose errors is important to plan safe path for mobile robots. The observation errors occur when the mobile robot obtains pose information from the GPS and the motion errors occur when the mobile robot moves based on the control input data. We integrate these motion errors and observation errors into pose errors represented by an error ellipse. We estimate pose errors by using extended Kalman filter (EKF) [9] during path planning because it can integrate errors with different characteristics. EKF process is divided into two steps and the motion error and observation error are calculated in each step.

**EKF Prediction Step.** In prediction step, a robot motion error is estimated. The motion error is determined by using Jacobian $\mathbf{J}$ derived from the motion model of the mobile robot $\mathbf{f}$. The Jacobians $\mathbf{J_x}$ corresponding to the robot pose and $\mathbf{J_u}$ corresponding to the control input are given as follows:

$$\mathbf{J_x} = \frac{\partial \mathbf{f}}{\partial \mathbf{x}}, \tag{2}$$

$$\mathbf{J_u} = \frac{\partial \mathbf{f}}{\partial \mathbf{u}}. \tag{3}$$

The motion error at time $t$ is expressed as the covariance matrix as follows:

$$\bar{\mathbf{\Sigma}}_{\mathbf{x}t} = \mathbf{J}_{\mathbf{x}t}\mathbf{\Sigma}_{\mathbf{x}t-1}\mathbf{J}_{\mathbf{x}t}{}^T + \mathbf{J}_{\mathbf{u}t}\mathbf{\Sigma}_{\mathbf{u}t}\mathbf{J}_{\mathbf{u}t}{}^T, \tag{4}$$

where $\bar{\mathbf{\Sigma}}_{\mathbf{x}t}$ is robot covariance matrix after applying motion error and $\mathbf{\Sigma}_{\mathbf{u}t}$ is input covariance. Each $\mathbf{\Sigma}_{\mathbf{u}t}$ is determined based on a robot position and a satellite arrangement. $\mathbf{\Sigma}_{\mathbf{x}t-1}$ denotes covariance matrix corresponding to the pose error at previous time step $t$-1.

**EKF Update Step.** In this step, the covariance matrix $\bar{\mathbf{\Sigma}}_{\mathbf{x}}$ calculated in prediction step is updated by using GPS observation errors. The updated covariance matrix $\mathbf{\Sigma}_{\mathbf{x}}$ is given as follows:

$$\mathbf{S}_t = \mathbf{H}_t\bar{\mathbf{\Sigma}}_{\mathbf{x}t}(\mathbf{H}_t)^T + \mathbf{Q}_t, \tag{5}$$

$$\mathbf{K}_t = \bar{\mathbf{\Sigma}}_{\mathbf{x}t}(\mathbf{H}_t)^T(\mathbf{S}_t)^{-1}, \tag{6}$$

$$\mathbf{\Sigma}_{\mathbf{x}t} = (\mathbf{I} - \mathbf{K}_t\mathbf{H}_t)\bar{\mathbf{\Sigma}}_{\mathbf{x}t}. \tag{7}$$

Here, $\mathbf{Q}$ is as covariance matrix of the observation error and $\mathbf{H}$ is Jacobian of GPS observation. $\mathbf{H}$ is matrix which defines relationship between the observed value and the robot pose. In this case, observed values represent robot pose directly; thus, $\mathbf{H}$ is same to the identity matrix. $\mathbf{S}$ is observation uncertainty and $\mathbf{K}$ is Kalman gain. During the path planning, we use $\mathbf{\Sigma}_{\mathbf{x}t}$ in collision check and it is described in detail in next sebsection.

## 3.2   Collision Check Using Error Ellipsoid

The random sampling method judge validations of generated nodes by checking collisions with a map and angle values of a robot inclination. Our previous study conducted collision checks between a map and a robot model with the fixed size, as shown in Fig. 3(a) [4]. Therefore, this method often generated nodes that were close to obstacles. In order to solve this problem, we propose the method that converts estimated pose errors to error ellipsoids and uses them in collision check. Figure 3(b) shows our collision check of the generated node during the random sampling. The proposed method checks collision with error ellipsoids and 3D object, directly. Error ellipsoids changes their sizes feasibly based on pose errors of a robot. Therefore, the proposed method can delete dangerous nodes that may collide because of pose errors. In the collision check, we use a flexible collision library (FCL) [10], the C++ library of collision detection.

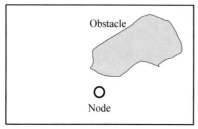

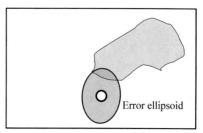

(a) Previous method that uses fixed robot size [4]

(b) Proposed method that uses error ellipsoid

**Fig. 3.** Collision check of generated node.

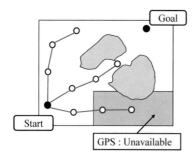

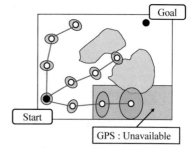

(a) Conventional random sampling method

(b) Proposed random sampling method

**Fig. 4.** Random sampling using error ellipsoids.

Figure 4 shows the concept of our proposed random sampling scheme. In gray areas, GPS signals are not available and robots cannot identify their own poses. Figure 4(a) shows a conventional random sampling method checking collisions with maps and robot models fixed their sizes. Thus, this method often plans paths that are close to obstacles. On the other hand, we can plan a path considering the observation error and the motion error of mobile robots all at once. Hence, for example, safer path planning can be performed depending on different motion errors and a GPS situation, as shown in Fig. 4(b).

## 4    Sampling Method of Velocity

In this section, we describe the novel sampling method of robot velocities (i.e., control input data at node sampling). Our method samples the robot velocities and generate nodes that correspond to robot poses using the motion model. In the previous method [4], velocities are determined based on fixed probability distribution. They did not consider accelerations; hence, it often samples velocities exceeding the limit. Figure 5 shows the proposed sampling method of robot velocities. We determine a position of probability distribution of velocity based on the former one velocity value and determine a range based on the robot acceleration as shown in Fig. 5(a). Probability distribution of each sampling is shown in Fig. 5(b). It is mixed with normal distribution and uniform distribution. This method enables to restrict accelerations and plan paths that robots can follow.

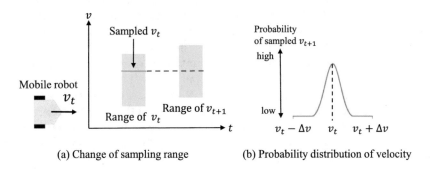

(a) Change of sampling range          (b) Probability distribution of velocity

**Fig. 5.** Proposed sampling method.

## 5    Simulation

### 5.1    Simulation Setting

In order to verify the effectiveness of the proposed path planner, we conducted simulations. A desktop computer with Intel core i7-6700 (3.40 GHz) CPU and 16.0 GB RAM memory was used to execute the proposed path planner. We used two maps in our simulation experiments The size of the map 1 that has rough terrain was $200\,\mathrm{m}^2$. There were a $5\,\mathrm{m}$ narrow route to the left and a $12\,\mathrm{m}$ broad route to the right. The size of the map 2 was $100\,\mathrm{m}^2$. There was a tunnel with a height of $7\,\mathrm{m}$ and a length of $30\,\mathrm{m}$ to the left. We assumed that the mobile robot was not able to get GPS information in the tunnel. We used path-directed subdivision tree (PDST) as a random sampling planner that determines how nodes are generated [11]. The robot was not able to follow if its acceleration was greater than $2.78\,\mathrm{m/s}^2$. Moreover, we assumed that the standard deviation of the GPS observation was $1.0\,\mathrm{m}$ in position and $1.0°$ in orientation. Velocity errors of 10% occurred. In addition, we also conducted the simulation with small velocity errors of 1% in map 2. The sizes of the error ellipsoids were determined so that the mobile robot existed in ellipsoids with a probability of 95%.

## 5.2   Simulation Result

Figure 6(a) shows the generated path using the proposed method on map 1. The generated path is expressed in yellow line and error ellipsoids are expressed in blue lines. The proposed method was able to plan the safe path that error ellipsoid did not collide with the obstacles in the rough terrain. Further, the acceleration of the mobile robot kept within the limit as shown in Fig. 6(b). Figure 7 shows generated paths through five repeated simulations on map 1. The proposed methods generated each paths in less than five seconds. The previous method planed two paths that passed through a dangerous narrow route. On the other hand, all five paths generated by the proposed method avoided the dangerous. Moreover, the proposed method did not generate detour paths.

Figure 8 shows the difference of generated paths depending on motion error values on map 2. When the set motion error was small, it planed the path that passed through a tunnel because pose errors did not accumulate so much in the tunnel. On the other hand, when the set motion error was large, it planed the path that avoid a tunnel because the path planning method was not able to find a path avoiding collision in the tunnel. Figure 9 shows the difference in area where nodes are generated on map 2. Here, nodes generated during path planning are expressed in red points. Nodes generated during path planning are expressed in red points. When the mobile robot cannot get GPS information, EKF cannot conduct the update step. For this reason, the motion error accumulated in the tunnel. Error ellipsoids will be larger than the tunnel when the motion error is large; hence, the proposed method did not generated nodes in the tunnel. However, error ellipsoids will not become very large when the motion error is small; hence, it generated nodes there. Therefore, the proposed method planed feasible paths that suitable for mobile robots errors.

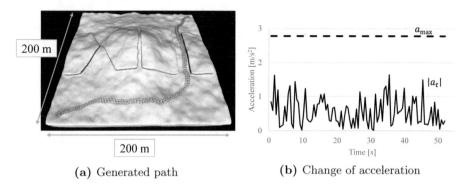

(a) Generated path                (b) Change of acceleration

**Fig. 6.** Generated path by using proposed method.

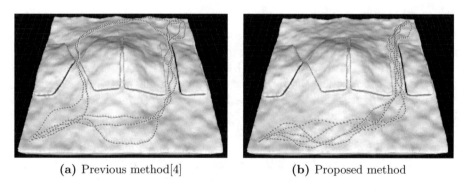

(a) Previous method[4]              (b) Proposed method

**Fig. 7.** Generated paths through five repeated simulations on map 1.

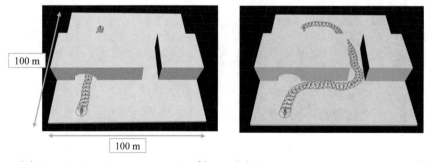

(a) Result for velocity errors of 1 %     (b) Result for velocity errors of 10 %

**Fig. 8.** Difference of generated paths depending on motion error values on map 2.

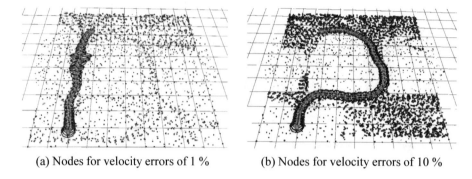

(a) Nodes for velocity errors of 1 %        (b) Nodes for velocity errors of 10 %

**Fig. 9.** Difference of generated nodes depending on motion error values on map 2.

## 6 Experiment

### 6.1 Experiment Setting

Figure 10 shows path planning result for the field experiment. The size of the map was $50\,\mathrm{m}^2$. There were a $2\,\mathrm{m}$ narrow route to the left and a $12\,\mathrm{m}$ broad route to the right. As shown in Fig. 10(b), the proposed method generated a safe path that passes through the broad route. The experimental environment and the mobile robot used for the field experiment are shown in Fig. 11. In order to install obstacles on the map as shown in Fig. 10, we used cones as obstacles in the field. The mobile robot was equipped with two GPS sensors; thus, it can obtain pose information. We assumed that observation errors of GPS followed a normal distribution and its standard deviations were $1.0\,\mathrm{m}$ in position and $1.0°$ in angle Moreover, we assumed that standard deviations of motion errors set at 10 % of robot input velocity.

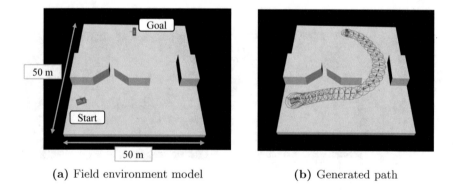

(a) Field environment model        (b) Generated path

**Fig. 10.** Path planning for field experiment.

**Fig. 11.** Overview of field experiment.

## 6.2   Experiment Result

Figure 12 shows the result of field experiment. The mobile robot traveled in order of (a) to (d). The robot was able to travel safely along the path generated in Fig. 10(b). Thus, we confirmed that the proposed path planning method is able to plan safe paths in real environments.

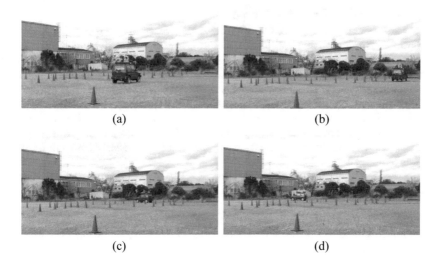

**Fig. 12.** Result of field experiment.

# 7  Conclusion

We proposed the robust path planning method against pose errors for mobile robots. In the proposed method, we used EKF to estimate pose errors and checked collisions with error ellipsoid. This enables to find feasible paths taking the pose errors into consideration. Moreover, the proposed velocity sampling method can generate the paths based on the acceleration that the actual mobile robots can deal with.

The future work related to this study is to conduct online path planning. In this study, we assumed that the map information were given in advance. However, real terrain information may change while mobile robots are navigating. Therefore, in case of real applications, it is necessary to propose an effective method which processes surrounding environment maps and generate feasible path simultaneously.

**Acknowledgement.** This work was in part funded by ImPACT Program of Council for Science, Technology and Innovation (Cabinet Office, Government of Japan).

# References

1. DARPA (Defense Advanced Research Projects Agency): Grand Challenge 2005 Report to Congress (2006). http://archive.darpa.mil/grandchallenge/docs/grand_challenge_2005_report_to_congress.pdf
2. Kuwata, Y., Teo, J., Fiore, G., Karaman, S., Frazzoli, E., How, J.P.: Real-time motion planning with applications to autonomous urban driving. IEEE Trans. Control. Syst. Technol. **15**, 1105–1118 (2009)
3. Richter, C., Ware, J., Roy, N.: High-speed autonomous navigation of unknown environments using learned probabilities of collision. In: Proceedings of the 2014 IEEE International Conference on Robotics and Automation, pp. 6114–6121 (2014)
4. Ji, Y., Tanaka, Y., Tamura, Y., Kimura, M., Umemura, A., Kanashima, Y., Murakami, H., Yamashita, A., Asama, H.: Adaptive motion planning based on vehicle characteristics and regulations for off-road UGVs. IEEE Trans. Ind. Inform. (under review)
5. van den Berg, J., Abbeel, P., Goldberg, K.: LQG-MP: optimized path planning for robots with motion uncertainty and imperfect state information. Int. J. Robot. Res. **30**(7), 895–913 (2011)
6. Blackmore, L., Li, H., Williams, B.: A Probabilistic approach to optimal robust path planning with obstacles. In: Proceedings of the American Control Conference, pp. 7–13 (2011)
7. Lee, S.U., Iagnemma, K.: Robust motion planning methodology for autonomous tracked vehicles in rough environment using online slip estimation. In: Proceedings of the 2016 IEEE/RSJ International Conference on Intelligent Robots and Systems, pp. 3589–3594 (2016)
8. LaValle, S.M.: Rapidly-exploring random trees: a new tool for path planning. Computer Science Department, Iowa State University, Technical Report TR 98–11, pp. 1–4 (1998)
9. Greg, W., Gary, B.: An introduction to the kalman filter. In: Proceedings of ACM SIGGRAPH, Course 8 (2001)

10. Pan, J., Chitta, S., Manocha, D.: FCL: a general purpose library for collision and proximity queries. In: Proceedings of the 2012 IEEE International Conference on Robotics and Automation, pp. 3859–3866 (2012)
11. Ladd, A.M., Kavraki, L.E.: Fast tree-based exploration of state space for robots with dynamics. In: Algorithmic Foundations of Robotics VI, pp. 297–312 (2004)

# Stereo Vision-Based Optimal Path Planning with Stochastic Maps for Mobile Robot Navigation

Taimoor Shakeel Sheikh and Ilya M. Afanasyev[✉]

Institute of Robotics, Innopolis University,
Universitetskaya Str. 1, 420500 Innopolis, Russia
{t.sheikh,i.afanasyev}@innopolis.ru

**Abstract.** This paper addresses the problem of stereo vision-based environment mapping and optimal path planning for an autonomous mobile robot by using the methodology of getting 3D point cloud from disparity images, its transformation to 2D stochastic navigation map with occupancy grid cell values assigned from the set {obstacle, unoccupied, occupied}. We re-examined and extended this methodology with a combination of A-Star with binary heap algorithms for obstacle avoidance and indoor navigation of the Innopolis autonomous mobile robot with ZED stereo camera.

**Keywords:** Stereo vision · Optimal path planning · Stochastic map
Occupancy grid · Disparity image · Mobile robot navigation

## 1 Introduction

The most essential tasks for autonomous mobile robot navigation are localization and environment mapping with efficient exploration of 3D scene and recognition of surrounding objects, obstacles and free space. Besides applying 2D laser rangefinder and depth sensors, one of the most popular and cost-effective SLAM solution is to use a stereo vision camera, which can provide a near distance information with the comparable accuracy relatively to other sensors [8]. The demonstration of stereo vision-based SLAM has been addressed to the papers [5,16]. Naturally, stereo vision-related navigation is based on disparity images and occupancy grids [2]. Let's consider the difference between stereo vision-based method and traditional occupancy grid mapping. When moving a robot in a real world, traditional occupancy grids are used as common frameworks for creating an environment map. But in real world the problem is not only with the ability to detect occupied areas, but also with the capability to understand a dynamic environment. While simple mapping approaches use grid arrays where only occupancy information is counted for each measurement, advanced methods include free space information with noise reduction and a separation between unknown and free map regions [2]. Traditional occupancy grid mapping algorithms assume

© Springer Nature Switzerland AG 2019
M. Strand et al. (Eds.): IAS 2018, AISC 867, pp. 40–55, 2019.
https://doi.org/10.1007/978-3-030-01370-7_4

that map voxels are independent of each other and utilize a map representation, wherein each voxel stores a single number representing an occupancy probability. It can lead to conflicts in the map, and inconsistency between a map error and the reported confidence values. Such inconsistencies are a challenge for planners that rely on the generated map for collision avoidance [1].

Occupancy grid mapping is widely used technique in probabilistic robotics for mobile robots which address the problem of generating a consistent metric map from noisy and incomplete sensor data, with the assumption that the robot pose is known [19]. For example, 3D occupancy grids from stereo-based aerial images for UAV navigations for obstacle avoidance are presented in [2,7]. The main challenge is to process 3D data acquired from stereo camera in real time. For this purpose, there exist two algorithms based on disparity and 3D space, which directly compute a disparity map and a depth map correspondingly. The disparity-based algorithm is more efficient and accurate. A grid-based mapping method designed for stereo camera-based object tracking in static environment presented in [11]. The paper [3] considers an approach with full-3D voxel-based dynamic obstacle detection from a 3D point cloud for urban scenario using stereo vision, which can differ stationary and moving obstacles, taking into account vehicle's egomotion computed through a visual odometry approach [18,20]. The obstacle avoidance for outdoor microaerial vehicles using forward-looking stereo cameras and 3D disparity space is described in [14]. A stereo odometry for indoor/outdoor navigation of a flying quadrotor robot in unknown environment based on on-board 2D path planning and collision avoidance with on-board 3D mapping using stereo depth images is considered in [15]. A stereo vision-based framework for creation of a dynamic occupancy grid map from a planning perspective in the presence of obstacles for a mobile robot was used in [12], where occupancy grids store richer data at each voxel, including an accurate estimate of an occupancy variance. Besides representing the surroundings as occupancy grids, dynamic occupancy grid mapping can provide the motion information in the grids. This framework consists of two variants: (1) Vehicle's motion estimation independently on moving objects. (2) Dynamic occupancy grid mapping based on the estimated motion information and the dense disparity map. The main benefit of this framework is the ability of mapping occupied areas and moving objects simultaneously, which is more accurate than traditional methods due to higher level of consistency between its error and the confidence. Another approach with an attractive obstacle avoidance algorithm for autonomous mobile robots based on binocular stereo vision for stochastic representation of a navigation map with reducing 3D point cloud from disparity images was proposed in [10], and inspired the authors of this paper to contribute and update this methodology for optimal path planning with A-Star based navigation.

The methodology examined in this research addresses the problem of stereo vision-based environment mapping and optimal path planning for an autonomous mobile robot. The set of stereo images are inputs for environment reconstruction into 3D point cloud model, where various objects that are a major problem for disparity maps, will be removed during processing stage. Further, a 2D stochastic navigation map is generated with occupancy grid cell values assigned from the set {obstacle, unoccupied, occupied}. Finally, the optimal path planning is created based on these occupancy values and an integration of A-Star and binary heap algorithms. To test and verify the methodology with real data acquired by a wheeled mobile robot in indoor environment, we used the following procedure.

1. Capture images (left and right) from ZED stereo camera.
2. Calibrate camera and extract epipolar geometry to get the accurate coordinates of each image and camera geometry properties.
3. Convert the pair of rectified images (left and right) to edge images (left and right).
4. Establishes a stereo correspondence between two views (left and right) and calculates disparity as the relative displacement feature which is used to extract 3D information about the environment.
5. Generate the 3D Point cloud model from disparity map.
6. To run algorithm effectively and smoothly,3D data from point cloud model have to be trimmed.
7. Build the 2D navigation map from 3D Point cloud.
8. Scoring of 2D navigation map cells on the basis of these parameters (average point density value, weight and cell status {obstacle, unoccupied, occupied}).
9. To find the optimal path for the autonomous mobile robot, pass the navigation map information to the combination of A-Star with binary heap algorithms used for obstacle avoidance and indoor navigation.

The rest of the paper is organized as follows. Section 2 explains the methodology for optimal path planning with stochastic maps for a mobile robot navigation, using A-Star algorithm and binary heap. Section 3 describes the experimental setup and demonstrates the methodology workability for obstacle avoidance and indoor navigation of the autonomous mobile robot with ZED stereo camera. Section 4 concludes the paper. Finally acknowledgement shows who supported this research work.

## 2  Methodology

A block scheme of the proposed methodology is shown in Fig. 1.

### 2.1  3D Point Cloud Model from Stereo Vision

**Basics of Depth Extraction from Stereo Camera.** The theory of stereo vision has been well understood for years, providing depth information extraction from digital images obtained by two cameras at different instances and points of view by comparing the relative positions of objects in two pictures of a scene [20]. Stereo image processing establishes a correspondence between two views (left and right) and calculates disparity as the relative displacement feature. Accurate coordinates of each image and camera geometry properties are used to

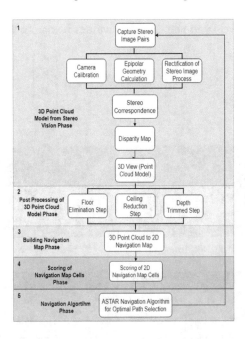

**Fig. 1.** The block scheme of stereo vision-based algorithm for mobile robot navigation

extract 3D information about the environment as shown in Fig. 2. The disparity is the distance between points in image plane, which is defined for Point 1 by Eq. (1), where $I$ is the feature point coordinate in left (L) and right (R) images. If $\mathbf{D}(P2) > \mathbf{D}(P3)$, it means that Point 2 is closer to Point 1.

$$\mathbf{D}(P1) = (\mathbf{I}(P1_L) - \mathbf{I}(P1_R)) \tag{1}$$

The pair of edge images obtained from the pair of rectified left and right images are shown in Fig. 2(e, f). The disparity of a Point with coordinates $XYZ$ is computed by Eq. (2), which will be used to generate the disparity image Fig. 2(g).

$$\mathbf{D}(P_{XYZ}) = \frac{B_{line}}{2p} * \frac{focLen}{Pixel_i} \tag{2}$$

where $\mathbf{D}(P_{XYZ})$ is the disparity value for a Point with coordinates $XYZ$, $B_{line}$ is a Baseline (i.e. a distance between centers of two cameras), $focLen$ is a camera focal length, and $Pixel_i$ is a horizontal pixel size of a camera chip.

**Basics of 3D Point Cloud Model.** The 3D Point cloud model generated from the disparity map as shown in Fig. 2(g) is according to the following Fig. 3, where Y represents the height, Z represents the depth of the object from the camera.

(X, Y, Z) is calculated from the disparity values using the following equation

$$\mathbf{x_i} = \begin{bmatrix} u \\ v \\ 1 \end{bmatrix} = \mathbf{C_{3x3}}[\mathbf{R}|\mathbf{T}]_{3x4} \begin{bmatrix} X \\ Y \\ Z \\ 1 \end{bmatrix} \tag{3}$$

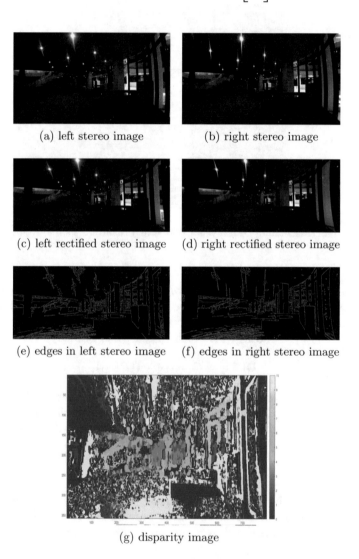

(a) left stereo image        (b) right stereo image

(c) left rectified stereo image    (d) right rectified stereo image

(e) edges in left stereo image    (f) edges in right stereo image

(g) disparity image

**Fig. 2.** 3D data extraction from Stereolabs ZED camera. (a, b) Stereo images (left and right shots of monocular cameras). (c, d) Rectified stereo images (left and right). (e, f) Edges in stereo Images (left and right) detected by disparities computation for feature points (using Eq. 1). (g) Disparity image generated after combining disparity values in Edge-related stereo images.

where $\boldsymbol{x_i}$ are image points with coordinates $(\boldsymbol{u}, \boldsymbol{v})$, $\boldsymbol{C}_{3x3}$ is the 3x3 intrinsic camera calibration matrix, and $[\boldsymbol{R}|\boldsymbol{T}]_{3x4}$ is the 3x4 projection matrix from Euclidean 3D-space with coordinates $(X, Y, Z)$ to an image [10, 18]

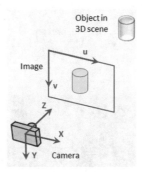

**Fig. 3.** 3D Point Cloud Model(XYZ) generated from disparity map, where Y and Z represent object's height and depth from a camera according to Eq. 3.

**Data Structure for Building Navigation Map.** Navigation map is a 2D matrix, where each cell has three parameters [10]: (a) total number of points in a cell (i.e point density for a cell); (b) the mean weights of all points in a particular cell, where a number $N$ is the point density. Each point has a weight $W_N$ that means a cell weight in the Eq. (4):

$$\frac{(W_1 + W_2 + W_3 + \ldots\ldots\ldots + W_N)}{N} \tag{4}$$

(c) the occupancy of a cell, i.e whether it is occupied (0), or it is unoccupied (1) or unknown (2). Initially all cells have unknown values. However, taking into account the first two parameters, the third one is evaluated by the algorithm Fig. 1.

## 2.2 Post Processing of 3D Point Cloud Model Phase

In our environment the reconstruction of the stereo images contains the large amount of points to obtain the 3D point cloud model. So in order to build the real time 2D navigation map with the 3D point cloud model requires a large computation power and it would not be optimal case. So before the algorithm process the 3D point cloud model to build the real time navigation map, data from point cloud model have to be trimmed to make this algorithm run effectively and smoothly. The point cloud data information have been pre-defined and will be used further in the navigation map build function. To trim the data from the point cloud following steps are considered.

**Floor Data Trimmed Step.** The prediction model of the disparity for the ground is well explained in [4] and as shown in Fig. 4, our autonomous mobile robot generates the disparity map while navigating in indoor environment using ZED camera. Thus, the obstacles or objects which are placed on the floors can be neglected since this information creates a problem in generating efficient disparity map.

$$Z_p = (\frac{H. \cos \gamma}{\cos \beta}) \tag{5}$$

**Fig. 4.** Geometry to compute an expected disparity in image using the pixel depth $Z_p$, robot's height $H$, the angle $\gamma$ between pixel depth and central point $P$, and the angle $\beta$ between camera's view direction at point $P$ and a normal to the ground

$Z_p$ is the depth of the pixel in Eq. (5) which can be calculated using $H$ gives the height of the robot on which stereo vision camera is mounted, $\gamma$ is the angle between depth of the pixel and center point P, $\beta$ is the angle viewing direction of stereo vision camera's at point P and it's angle normal to ground.

$$\gamma = (\arctan \frac{(v_p.p_y)}{f}) \tag{6}$$

Where in Eq. (6) $\gamma$ can be calculated using $v_p$ which is the vertical pixel coordinate in the image relative to optics center and $p_y$ is the vertical pixel size of the camera.

$$\beta = (\Theta + \gamma) \tag{7}$$

$\Theta$ in Eq. (7) is the angle viewing direction of stereo vision camera's and it's angle normal to ground. In this algorithm $\Theta$ is $90°$, So $Z_p$ the depth of the pixel result is shown in Eq. (8)

$$Z_p = (-H. \cos \gamma) \tag{8}$$

By using the Eq. (5) for different $\gamma$ values the pixels which are near to the floors can be neglected and the obstacles or objects which are placed on the floors wouldn't be considered as obstacle otherwise such scenario was a problem in disparity map based algorithm. The results of floor data trimmed step are shown in Fig. 5.

**Ceiling Data Trimming Step.** For this trimming step we need to avoid the obstacles or objects which are above the robot height because robot will never hit them, and it is the useless information for processing. The explained technique in the previous step would be applied along the Y axis of images. The reduce of the height information will significantly reduce the computation required to generate 3D point Cloud Model. The objects of the upper part of the room (i.e. fans, lights, etc.) can be eliminated from the environment data as shown in Figs. 6 and 7. By using the Eq. (5) for different $\gamma$ values, pixels for Y images axis which are near to the ceiling can be ignored and objects which are above certain height will not be considered as obstacles otherwise it creates a problem for disparity map-based algorithm. The results of ceiling data trimming step, which improve the computation efficiency of algorithm in generating 3D point cloud model, are shown in Fig. 7.

**Depth Data Trimming Step.** Stereo vision camera can look up to infinity, but the depth of view gets corrupted with noise after a certain depth limit. This data trimmed step is the most important because it significantly improves the computational efficiency in comparison with other steps of data trimming. Depth data trimming sets limits for depth of view since the resolution increase leads to the depth decrease and vice versa. Usually for indoor environment the robot speed is much lower than for outdoor one. Therefore, the algorithm computation efficiency can be improved by tuning the depth of view up to the limit where generating 3D point cloud model can be easily computed. The results of depth data trimming step are shown in Fig. 8.

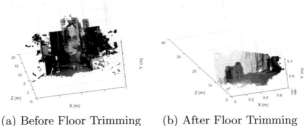

(a) Before Floor Trimming     (b) After Floor Trimming

**Fig. 5.** 3D Point cloud with floor trimmed data for a specific $\gamma$ value along the X image's axis using Eq. (5) to avoid the floor pixels.

**Fig. 6.** The visualization of ceiling trimming above certain height to avoid the obstacles/objects above the robot height since robot never hits them.

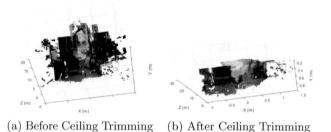

(a) Before Ceiling Trimming    (b) After Ceiling Trimming

**Fig. 7.** 3D Point cloud with ceiling trimmed data for a specific $\gamma$ value along the Y image's axis using Eq. (5) to avoid the obstacles which are above the robot height.

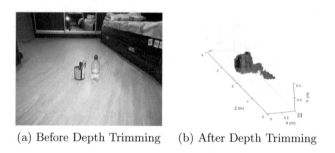

(a) Before Depth Trimming    (b) After Depth Trimming

**Fig. 8.** 3D Point cloud with a specific depth view limit factor, which helps to improve the algorithm computation efficiency for generating 3D point cloud model.

## 2.3    Building Navigation Map Phase

In order to run and build 2D navigation map computationally smooth and efficient we require the 3D point cloud model data after the post processing trimming steps. The 2D navigation map in X-Z plane is visualized from the 3D point cloud model in Fig. 1. The selected size of grid for 2D map is 50*50 since the field set in front of robot in 2D map is 10 m*10 m and each cell size of the 2D map is 5 cm*5 cm (although it can be varied and set according to the user requirement).

Each cell of 2D map contains the mean weight value of all 3D points in the volume (i.e. point density) above the cell given by Eq. (4). The next step of the algorithm is the scoring of 2D navigation map cells on the basis of average point density value. The depth of view is increased with the average point density growth and vice versa.

## 2.4  Scoring Of 2D Navigation Map Cells Phase

This is the algorithm phase, where each value of 2D navigation map cells are updated according to an average point density. Initially unknown values are assigned to each cell in 2D navigation map and then each cell is considered for scoring if a cell count is greater than average point density then

1. If mean weight of a cell is greater than a minimum threshold weight then the cell value is scored as 'Obstacle' or (0).
2. If mean weight for a cell is less than a minimum threshold weight then the cell value is scored as 'unoccupied' or (1).
3. If mean weight for a cell is equal to a minimum threshold weight then the cell value is scored as 'Occupied' or (2).

Thus, according to this approach for each cell in 2D navigation map will be assigned a value from the set {0, 1 and 2}.

## 2.5  The Navigation Algorithm

For path planning we need to give the navigation map information to A-Star algorithm for optimal path finding. As far as algorithm finds the lowest optimal cost, a path from the initial node to the goal node will pass through one or multiple possible node. A heuristic cost function based on distance is represented by F(k) in Eq. (9) that determines the order in which it searches the nodes in complete tree. It checks the vertex k that has the lowest F(k) each time in the main loop.

$$F(k) = G(K) * R(K) \tag{9}$$

In Eq. (9) the path cost from a starting point to any node k is represented by R(k) and means the heuristically estimated cost from the node k to the goal denoted by R(k). When passing through the navigation map, A-Star algorithm follows the specific path which has the lowest cost while storing the alternative nodes in a sorted priority queue. The passing algorithm discards a node with higher cost and chooses a lower cost node instead. This process will be repeated until the desired goal is reached. To enhance the computational power of A-Star algorithm the binary heap step is used. The binary heap is a tree with collection of items where either the highest or the lowest value item is at the top of the heap [13]. For the implementation of A-Star algorithm we require three input parameters: (1) The Navigation Map, (2) Initial (Start) Node, and (3) Final (Goal) node. The initial navigation map is defined with predefined parameters. The map is subdivided into a user defined numbers of grid cells where each cell represents either a free node or an obstacle. The user sets the initial node and final node for the A-Star algorithm for calculation if the virtual path according to the binary heap information. Figure 9 illustrates the block scheme of implemented A-Star algorithm with binary heap step.

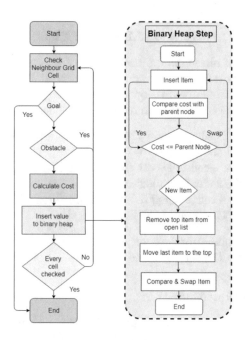

**Fig. 9.** Block scheme of A-Star algorithm with binary heap step

## 3    Experimental Setup and Comparative Analysis of Test Results

Our Innopolis autonomous mobile robot is a small size wheeled robot based on Traxxas 7407 Radio-Controlled (RC) Car Model [17] shown in Figs. 10(a) and (b), which we used to verify and test the robot navigation algorithm through different indoor environment. The basic Traxxas 7407 RC car model (Fig. 10(b)) was upgraded to an autonomous mobile robot with adding onboard controllers, ZED stereo camera and sensors. To have high computational power we added NVidia Jetson TX1 embedded controller, which supports CUDA technology for on-board parallel calculations. Then an optical incremental Lynx-motion quadrature motor encoder was mounted on the robot to control electric motors, and ZED camera was installed to capture stereo images. According to the methodology (described in Sect. 2) stereo images are rectified to create a disparity map with following building of 3D point cloud data and construction of a navigation map with cell values corresponding to obstacles and mean heights to compute an optimal path for the robot as shown in Fig. 10(c). The navigation map grid size can be set by user requirement (in our case, it is 16*16 cells as shown in Fig. 10(c). The cells marked in green, red and black colors are the starting point, the goal point and obstacles respectively, whereas the purple color represents the optimal path computed by A-star navigation algorithm integrated with binary heap. Our experiments with obstacles of different shapes like boxes and cylinders, which we set in real time, are shown in Fig. 11.

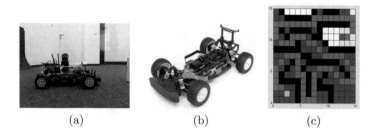

(a)                          (b)                          (c)

**Fig. 10.** (a) Innopolis Autonomous Mobile Robot, (b) Traxxas 7407 Radio-Controlled (RC) Car Model, (c) The example of a navigation map based on A-Star algorithm with binary heap obtained from 3D Point cloud for XZ cells of occupancy grid. The cells marked in green, red and black colors are the starting point, the goal point and obstacles respectively, whereas the purple color represents the optimal path computed by our algorithm.

We compared two methods of disparity map generation based on (1) block matching [9], and (2) Semi-global block matching [6]. The Table 1 presents the results of evaluation for images shown in Fig. 2(a, b) on the basis of two important parameters: (a) block size, and (b) pixel matches (found or not found) in frames per second. We used the parameters with image size of 1280*720 pixels and a range to generate disparity (varied from 32–80). The number of pixel matches (found/not found) relates to the depth information which is useful for the 3D information extraction for any type of environment.

The performance of the block matching method is better than for the semi global block matching if we use the block size larger than 11. However, a significant increase in block size will blur object boundaries in image. Another important factor is the algorithm speed. Thus, the block matching is fast as the semi-global one, which is suitable for most real time applications. Nevertheless, semi-global block matching can generate the better disparity map using a smaller block size that is important for information extraction from environment. The optimal parameters selected for the disparity map generation, using both block and semi-global block matching, are shown in Table 2. The disparity level is used to identify the closer/further objects using colors as shown in Fig. 2(g).

**Table 1.** Comparison between the block and semi global block matching methods to generate the disparity maps on basis of parameters (Block size and Frame Per Second)

| Block Size | Block Matching Method | | Semi Global Block Matching Method | |
|---|---|---|---|---|
| | Unknown Match Frame Per Second | Match Found Frame Per Second | Unknown Match Frame Per Second | Match Found Frame Per Second |
| 7 | 28 | 32 | 14 | 22 |
| 9 | 26 | 38 | 12 | 22 |
| 11 | 26 | 36 | 12 | 24 |
| 13 | 24 | 34 | 10 | 26 |
| 17 | 23 | 38 | 8 | 24 |
| 19 | 18 | 40 | 8 | 28 |

The windows size is the block size as evaluated in Table 1. The average time is better for block matching because of speed and pixels matching that allows to have more robust and improved performance in each iteration. The speed (frame per second) is dependent of the pixels match found in each block. Table 3 shows the comparison of optimal parameters obtained by A-Star navigation algorithm using both block and semi global block matching methods. The average time of A-Star navigation algorithm with block matching method is better than for semi-global block matching due to the speed, pixels matches and performance that allows better avoids obstacles in real time conditions. Our autonomous mobile

**Table 2.** Comparison of optimal parameters selected to generate the disparity maps using block and semi global block matching methods.

| Methods | Optimal Parameters to Generate Disparity Map | | | | | |
|---|---|---|---|---|---|---|
| | Disparity Level | Window Size | Average Time(sec) | Image Size | Speed(FPS) | Performance |
| Block Matching | 32 | 13*13 | 0.028 | 1280*720 | 34 | High robustness & improved performance |
| Semi Global Block Matching | 48 | 17*17 | 0.065 | 1280*720 | 24 | Improved efficiency, accuracy & sharped object boundaries |

robot navigation software was written in MATLAB and C++. The navigation algorithm was tested in indoor environment, processing ZED stereo camera video frames and demonstrating smooth work and efficiency in real time as shown in Fig. 12. We have obtained some real time results with generating 3D point cloud data, a navigation map and optimal path planning for obstacle avoidance and navigation of autonomous mobile robot through different indoor environment.

**Table 3.** Comparison of optimal parameters obtained by A-Star navigation algorithm using block and semi global block matching methods.

| Methods | Optimal Parameters for A-Star Navigation Algorithm | | | |
|---|---|---|---|---|
| | Grid Size | Cell Size | Obstacles Avoided | Average Time(sec) |
| Block Matching | 20*20 | 2*2 | Circle, Square & Triangle | 340 |
| Semi Global Block Matching | 16*16 | 1*1 | Circle Square | 410 |

 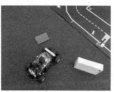 

(a) Red & white box obstacles in Frame 1    (b) White box obstacle in Frame 2    (c) White box obstacle in Frame 3

**Fig. 11.** Indoor Navigation of the Innopolis Mobile Robot

(a) White Box Obstacle in Frame 1    (b) Yellow Cylinder Obstacle in Frame 2    (c) White Box Obstacle from different view in Frame 3    (d) White Box Obstacle from different view in Frame 4

**Fig. 12.** Indoor Navigation of the Innopolis Mobile Robot

## 4  Conclusion

This paper concerns the problem of stereo vision-based environment mapping and optimal path planning for an autonomous mobile robot through different indoor environment. Stereo vision is performed with ZED stereo vision camera mounted on Innopolis mobile robot based on Traxxas 7407 Radio-Controlled Car Model. Sparse stereo images are utilized because they have lower computational complexity that is more suitable for real time application. We accumulated depth data from sparse stereo for map building, investigating indoor environment since it is more confined. To navigate an autonomous mobile robot, a 2D stochastic map environment with occupancy grid cell values assigned from the set {obstacle, unoccupied, occupied} is required. Depth map is converted to real distances with the triangulation technique. Detected indoor environment points are clustered and added to the list of obstacles. Occupancy grids are updated accordingly to obstacle presence in the immediate and adjacent cells. Based on occupancy grid information, the path planner selects the best path for an autonomous mobile robot. Since the autonomous mobile robot has non-holonomic constraints, the hierarchical collision detection technique was utilized to improve computational efficiency.

Thus, we re-examined the methodology of getting disparity images, 3D point cloud and 2D stochastic navigation map with occupancy grid cells, which represent the information about obstacles. Our contribution is in extension of this methodology with a combination of A-Star algorithm with binary heap for obstacle avoidance and optimal path planning of an autonomous mobile robot. The methodology was tested with real data acquired by Innopolis mobile robot with ZED stereo camera in indoor environment, demonstrating encouraging results.

**Acknowledgement.** This work has been supported by the Russian Ministry of education and science with the project "Development of anthropomorphic robotic complexes with variable stiffness actuators for movement on the flat and the rugged terrains" (agreement: No14.606.21.0007, ID: RFMEFI60617X0007).

## References

1. Agha-Mohammadi, A.A.: Smap: Simultaneous mapping and planning on occupancy grids. arXiv preprint arXiv:1608.04712 (2016)
2. Andert, F.: Drawing stereo disparity images into occupancy grids: measurement model and fast implementation. In: IEEE/RSJ International Conference on Intelligent Robots and Systems (IROS), pp. 5191–5197. IEEE (2009)
3. Broggi, A., Cattani, S., Patander, M., Sabbatelli, M., Zani, P.: A full-3d voxel-based dynamic obstacle detection for urban scenario using stereo vision. In: IEEE Intelligent Transportation Systems Conference (ITSC), pp. 71–76. IEEE (2013)
4. Burschkal, D., Lee, S., Hager, G.: Stereo-based obstacle avoidance in indoor environments with active sensor re-calibration. In: IEEE International Conference on Robotics and Automation (ICRA), vol. 2, pp. 2066–2072. IEEE (2002)

5. Elinas, P., Sim, R., Little, J.J.: /spl sigma/slam: stereo vision slam using the rao-blackwellised particle filter and a novel mixture proposal distribution. In: IEEE International Conference on Robotics and Automation (ICRA), pp. 1564–1570. IEEE (2006)

6. Hirschmuller, H.: Accurate and efficient stereo processing by semi-global matching and mutual information. In: IEEE Computer Society Conference on Computer Vision and Pattern Recognition, CVPR 2005, vol. 2, pp. 807–814. IEEE (2005)

7. Hrabar, S.: 3d path planning and stereo-based obstacle avoidance for rotorcraft uavs. In: International Conference on Intelligent Robots and Systems (IROS), pp. 807–814. IEEE (2008)

8. Ibragimov, I.Z., Afanasyev, I.M.: Comparison of ros-based visual slam methods in homogeneous indoor environment. In: 14th Workshop on Positioning, Navigation and Communications (WPNC), pp. 1–6. IEEE (2017)

9. Konolige, K.: Small vision systems: Hardware and implementation. In: Robotics Research, pp. 203–212. Springer (1998)

10. Kumar, S.: Binocular stereo vision based obstacle avoidance algorithm for autonomous mobile robots. In: IEEE International Advance Computing Conference (IACC), pp. 254–259. IEEE (2009)

11. Lategahn, H., Graf, T., Hasberg, C., Kitt, B., Effertz, J.: Mapping in dynamic environments using stereo vision. In: IEEE Intelligent Vehicles Symposium (IV), pp. 150–156. IEEE (2011)

12. Li, Y., Ruichek, Y.: Occupancy grid mapping in urban environments from a moving on-board stereo-vision system. Sensors 14(6), 10454–10478 (2014)

13. Ma, H.: Research on interactive segmentation algorithm based on search path optimization. In: International Conference on Intelligent Human Machine Systems and Cybernetics (IHMSC), vol. 2, pp. 286–289. IEEE (2009)

14. Matthies, L., Brockers, R., Kuwata, Y., Weiss, S.: Stereo vision-based obstacle avoidance for micro air vehicles using disparity space. In: IEEE International Conference on Robotics and Automation (ICRA), pp. 3242–3249. IEEE (2014)

15. Schmid, K., Tomic, T., Ruess, F., Hirschmüller, H., Suppa, M.: Stereo vision based indoor/outdoor navigation for flying robots. In: IEEE/RSJ International Conference on Intelligent Robots and Systems (IROS), pp. 3955–3962. IEEE (2013)

16. Se, S., Lowe, D., Little, J.: Mobile robot localization and mapping with uncertainty using scale-invariant visual landmarks. Int. J. Robot. Res. 21(8), 735–758 (2002)

17. Shimchik, I., Sagitov, A., Afanasyev, I., Matsuno, F., Magid, E.: Golf cart prototype development and navigation simulation using ros and gazebo. In: MATEC Web of Conferences, vol. 75, p. 09005. EDP Sciences (2016)

18. Siegwart, R., Nourbakhsh, I.R., Scaramuzza, D.: Introduction to Autonomous Mobile Robots. MIT Press, Cambridge (2011)

19. Thrun, S., et al.: Robotic mapping: a survey. Explor. Artif. Intell. New Millenn. 1, 1–35 (2002)

20. Yousif, K., Bab-Hadiashar, A., Hoseinnezhad, R.: An overview to visual odometry and visual slam: applications to mobile robotics. Intell. Ind. Syst. 1(4), 289–311 (2015)

# Variance Based Trajectory Segmentation in Object-Centric Coordinates

Iori Yanokura$^{(\boxtimes)}$, Masaki Murooka, Shunichi Nozawa, Kei Okada,
and Masayuki Inaba

Graduate School of Information Science and Technology, The University of Tokyo,
7-3-1 Hongo, Bunkyo-city, Tokyo 113-8656, Japan
{yanokura,murooka,s-nozawa,k-okada,inaba}@jsk.imi.i.u-tokyo.ac.jp
http://www.jsk.t.u-tokyo.ac.jp/

**Abstract.** Human imitation is suggested as a useful method for humanoid robots achieving daily life tasks. When a person does demonstration such as housework many times, there is a property that the variance of the hand trajectory in object-centric coordinates becomes small at the stage of acting on the object. In this paper, we focused on human demonstration in object-centric coordinates, and proposed a task segmentation method and motion generation of a robot. In fact, we conducted a experiment (by taking a kettle and moving it to a cup) with the life-sized humanoid robot HRP-2.

**Keywords:** Humanoid robot · Imitation learning
Task segmentation · Motion generation · Change point detection

## 1 Introduction

Researchers has been conducting research to execute various tasks by describing real world recognition and motion so far [1,2]. However, these methods require expert knowledge of programming and robots. To reduce those, researchers have been working on learning from demonstration [3,4]. In this paper, we focused on human demonstration in object-centric coordinates, and propose a task segmentation method, and implemented it in a real robot system.

## 2 Trajectory's Variance of Human Demonstration

### 2.1 Preliminaries

Tasks that people perform can be distinguished into two, one related to trajectory and one indirectly related. The former includes the action of opening the door, the action of grasping the kettle and pouring water into the cup, and the like. In the action of grasping the kettle and pouring water into the cup, the trajectory of hand in the kettle centric coordinate passes near the origin

© Springer Nature Switzerland AG 2019
M. Strand et al. (Eds.): IAS 2018, AISC 867, pp. 56–66, 2019.
https://doi.org/10.1007/978-3-030-01370-7_5

(See also Fig. 2). After grasping the kettle the change in trajectories in the kettle centric coordinates continues to be small. Variances of the trajectories in the cup centric coordinates tends to be large at the stage of reaching to the kettle. This is because the trajectory of the hand is not related to the cup. After grasping the kettle, the trajectory of hand in the cup centric coordinate passes near the origin (See also Fig. 2).

An example in which the trajectory is indirectly related is a remote controller of a television. This is not dealt with in this paper.

When a person does demonstration such as housework many times, there is a property that the variance of the hand trajectory becomes small at the stage of acting on the object. We focused on the variance of the trajectory in the object-centric coordinates, and we thought that we can segment task and generate robot's motion from human demonstration (Fig. 1).

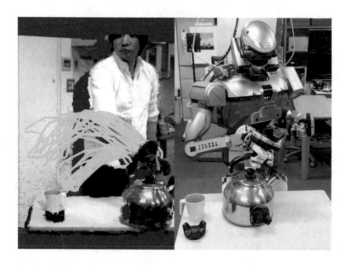

**Fig. 1.** HRP-2 imitating trajectories from human demonstration. The light blue line represents the trajectory of the hand.

## 2.2   Related Works

There is what is called Behavior Cloning as a way to learn behavior from expert's demonstration. Ross et al. [5] uses human operations as a training data, a method called DAGGER as a search for a supervised reinforcement learning strategy. Using that method, they operated the car on the simulator. Those methods fit the operation to the expert's pattern.

Yan et al. [6] created a framework called one-shot imitation learning using Neural Network. Those methods can tracking objects by performing one demonstration process by humans. Understanding the task by remembering the arrangement of markers from person's demonstration, but they have done the same tasks several thousand epoch times on simulator beforehand.

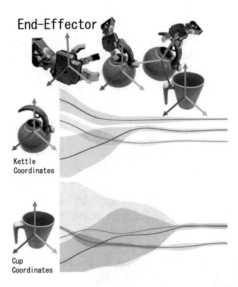

**Fig. 2.** End Effector's trajectory in object-centric coordinates. The solid line represents the center of the trajectory and the shadow line represents the variance of the trajectory.

Maeda et al. [7] generated trajectory of robot from human demonstration using Probabilistic Movement Primitives as action generation by Primitive and performed cooperative operation. In this research it is done only with tasks without specific stages. We can not directly solve tasks that have to go through the stages. In this research, by transforming the coordinates of the trajectory obtained from human demonstration and changing the point of view, even for complex tasks including procedures.

Schneider et al. [8] taught coffee makers' task movements with a robot with a one-person perspective. They gather trajectory data of the object-centric coordinate and use the Gaussian Process for their position to bind trajectory constraints derived condition range.

## 2.3   Paper Contribution

In this paper, we propose a trajectory segmentation method and generate robot's motion. In fact, we conducted a experiment (by taking a kettle and moving it to a cup) with the life-sized humanoid robot HRP-2.

# 3   Task Segmentation and Motion Generation

## 3.1   Align Time Series Data

The trajectories obtained from human demonstration should be aligned so as to maintain consistency, because those varies in time. To align trajectories, we used Dynamic Time Warping algorithm [9] which is widely used in general. This algorithm is a method that calculates an optimal match between two given sequences. By searching minimum cost paths with dynamic programming and considering the expansion and contraction in the time direction, the correspondence of time series data with the most appropriate length is obtained. The computational complexity of DTW is $O(N^2)$, but the length of time series included in one task is about $O(10^3)$, so this case is sufficient.

## 3.2   Task Segmentation

Figure 3 shows a case in which an object is grasped, taken to the target point, released, and returned to the starting point. Figure 3a shows trajectories in object-centric coordinates. There is a section where the trajectory passes near the origin in the middle and the variance becomes small. This is the section holding a object. Segmentation of the trajectory is performed by detecting the point where this variance decreases. After that, we generates a motion trajectory. The trajectory variance collected from human demonstration is shown in the Fig. 4. The point where the trajectory converges is considered as a change point, and Bayesian Changepoint Detection [10] is used for detecting those points. This is a method of detecting the location of the change point with respect to unknown time series data. Figure 4 is a figure showing the variance value of the orbit of Fig. 3 and its change point. However, the detected change point is just before the gradient changes abruptly. For this reason, a point after the change point is detected and after the gradient falls below a certain threshold value is defined as a new change point (indicated by a red line in Fig. 4). The reason why the change point is simply not lower than the threshold is not used as a true change point is because when the end effector is in contact with the object, the variance is maintained in a small state, and erroneous detection occurs is there. See also Algorithm 1.

## 3.3   Motion Generation

We indicates the flow of trajectory generation in Algorithm 2. We consider a linear regression model with weight $\mathbf{w}$ and basis function $\phi(\mathbf{x})$ as in Eq. 1.

$$y(\mathbf{x}, \mathbf{w}) = \mathbf{w}^T \phi(\mathbf{x}) \tag{1}$$

When we consider the trajectory obtained from a person's presentation as $\mathbf{t_i}$ and solve the maximization problem of the likelihood function under Gaussian noise,

$$\mathbf{W}_{ML} = (\Phi^T \Phi)^{-1} \Phi \mathbf{t} \tag{2}$$

---

**Algorithm 1.** Detect Change Points

---
1: INPUT: T as List of Time Series
2: Initialize C as Empty List of Change Point
3: **for** Trajectory = 1, T **do in parallel**
4:     Calculates Change Point Using Bayesian Changepoint Detection. Assume that Change Point is c
5:     After c, the place where the value falls below the threshold is set as a change point.
6:     Push c into C
7: **end for**
8: return C

---

---

**Algorithm 2.** Generate Trajectory by Variance based Segmentation.

---
1: Sample Trajectory from Human Demonstration T
2: Time Alignment by Dyanamic Time Warping
3: Initialize Change Point memory D as Empty Array
4: **for** coordinate = 1, Coordinates **do in parallel**
5:     Calculates Variance of Trajectory forcus on the coordinate
6:     Detect Change Points d (Algorithm 1)
7:     Push d into D
8: **end for**
9: Split T based on D
10: Calculates trajectory parameter(See also 3.3)
11: **while** true **do**
12:     Generate Trajectory until next change point
13:     Action
14:     **if** Task End **then** break
15:     **end if**
16: **end while**

---

We obtain the mean $\mu_{\mathbf{W}}$ of $\mathbf{W}_{ML}$ and the weight $\Sigma_W$ of covariance obtained by this maximum likelihood estimation. From those parameters, we generate trajectories according to $N(\mu_W, \Sigma_W)$. The following Gaussian basis is used as a basis function.

$$b_i^G(z) = \exp(-\frac{(z_t - c_i)^2}{2h}) \tag{3}$$

$$\phi_i(z_t) = \frac{b_i(z)}{\sum_j b_j(z)} \tag{4}$$

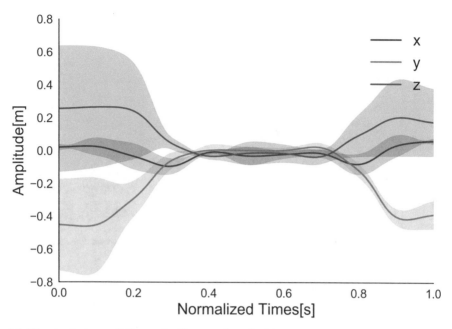

(a) The trajectory of the end effector when looking at a target object as central coordinates

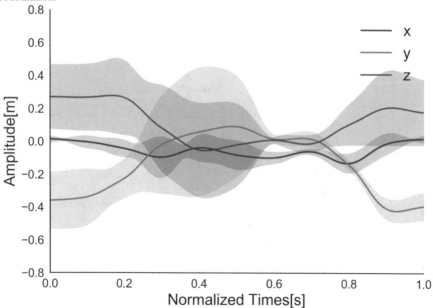

(b) The trajectory of the end effector when looking at a goal point as central coordinates

**Fig. 3.** Trajectory. The blue, red and green shaded area represents the learned trajectory distribution. The bold solid lines represent the mean trajectory.

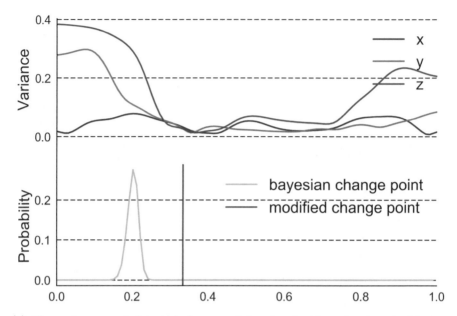

(a) The variances and detected change points when looking at a target object as central coordinates.

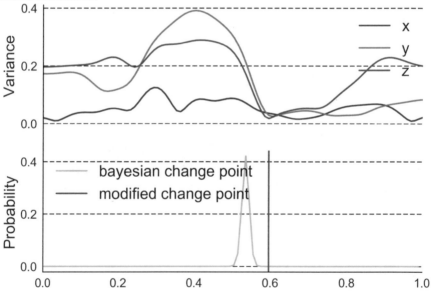

(b) The variances and detected change points when looking at a goal poind as central coordinates.

**Fig. 4.** Top: The variances of the xyz-axis trajectories. Down: Detected change points using bayesiean change point detection [10] and Modified change point (See Algorithm 1).

## 4    Experiments with REAL ROBOT

In this chapter, we conducted experiments using a life-sized humanoid robot, HRP2-JSK [11]. The trajectory of human demonstration was collected by attaching a tracker to the hand using HTC Vive. Tracking accuracy of Vive is approximately 0.3 [mm] and 6DOF data is obtained in 90 Hz.

The experiment consists of the following steps:

(1) A person performs a demonstration of taking a kettle, taking action to pour water into the cup, and returning it to a fixed position ten times.

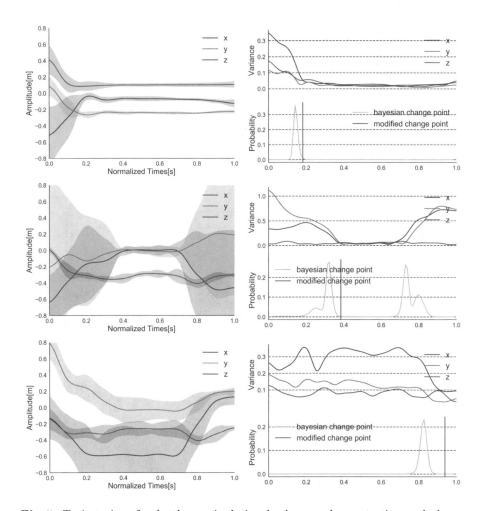

**Fig. 5.** Trajectories of a kettle manipulation by human demonstration and change points. Left: Trajectories of a kettle centric. Between 0.2-1.0[s], a person is grasping the kettle, so the variances are small. Center: The trajectory of a cup centric. Between 0.4-0.7[s], he does pouring. Right: The trajectory of a hand (a goal point as central coordinates). In final time, he placed the kettle at a goal point.

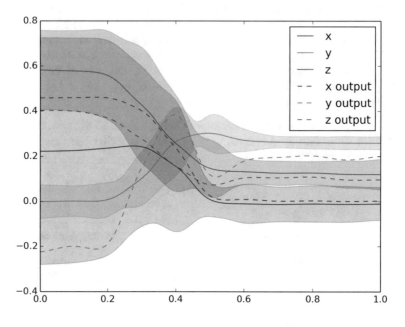

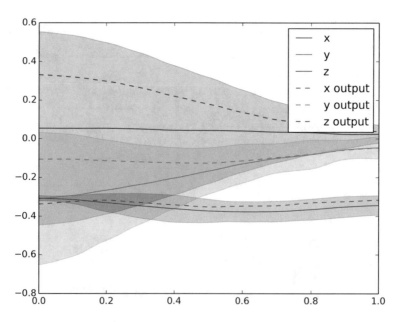

**Fig. 6.** Segmented Trajectories by proposed method. Top: Trajectories of reaching to a kettle. Bottom: Trajectories of forwarding to a cup. An each dot line indicates the automatically generated trajectory and the robot execute it in with respect to the each object's coordinates.

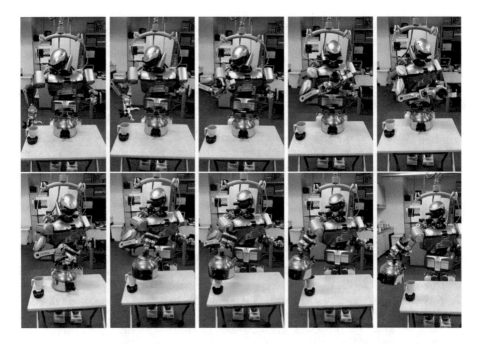

**Fig. 7.** Snapshot of HRP2-JSKNTS operating a kettle by proposed method.

(2) We acquire the trajectory sequence of the hand of the person seen from the position of the kettle, the position of the cup and the position of the goal, and learn by the proposed method.
(3) Finally, the trajectory is generated and executed by the robot.

The trajectories, variances and detected change points by human demonstration are indicated in Fig. 5. The scenes the imitation experiment by HRP-2 is shown in Fig. 7. The system successfully recognized change points for each target objects. It is important to note that the system has no built-in knowledge of points of switching tasks. In this task, there are three subtasks, reaching to a kettle, pour it into a cup, place the kettle to a fixed point. The robot is actually recognizing the subtasks.

Therefore, the robot can correctly segment the trajectory with respect to each object-centric coordinate. The real robot trajectories were generated, see also Fig. 6. Dot lines indicate the generated trajectory and the robot execute it in with respect to the each object-centric coordinate.

## 5   Conclusion

In this paper, we focused on human demonstration in object-centric coordinates, proposed a task segmentation method, and implemented it in a real robot system. This is worked well on humanoid robot HRP-2. Our robot first learned the change points of trajectories from human demonstration without knowledge about features and number of tasks. This means that he can understand the subtasks. After that, he can learn the trajectory of each task and execute it in a real world.

In future work, we should consider the whole body manipulation focusing on contact states. There are many actions to use the whole body in daily life. For example, you have a big basket, or you put a hand on a bookshelf and reach to book. These tasks must contact somewhere in the bodies. We think that it will be able to acquire the task with hints of the variance of trajectories for such whole body movement.

# References

1. Furuta, Y., Wada, K., Murooka, M., Nozawa, S., Kakiuchi, Y., Okada, K., Inaba, M.: Transformable semantic map based navigation using autonomous deep learning object segmentation. In: IEEE-RAS International Conference on Humanoid Robots, pp. 614–620 (2016)
2. Nozawa, S., Kumagai, I., Kakiuchi, Y., Okada, K., Inaba, M.: Humanoid full-body controller adapting constraints in structured objects through updating task-level reference force. In: IEEE/RSJ International Conference on Intelligent Robots and Systems, pp. 3417–3424 (2012)
3. Argall, B.D., Chernova, S., Veloso, M., Browning, B.: A survey of robot learning from demonstration. Robot. Auton. Syst. **57**(5), 469–483 (2009)
4. Kuniyoshi, Y., Yorozu, Y., Inaba, M., Inoue, H.: From visuo-motor self learning to early imitation – a neural architecture for humanoid learning. In: IEEE International Conference on Robotics and Automation, pp. 3132–3139 (2003)
5. Ross, S., Gordon, G.J., Bagnell, J.A.: No-regret reductions for imitation learning and structured prediction. In: International Conference on Artificial Intelligence and Statistics (2011)
6. Duan, Y., Andrychowicz, M., Stadie, B.C., Ho, J., Schneider, J., Sutskever, I., Abbeel, P., Zaremba, W.: One-shot imitation learning. arXiv, Vol. abs/1703.07326 (2017)
7. Maeda, G., Ewerton, M., Lioutikov, R., Amor, H.B., Peters, J., Neumann, G.: Learning interaction for collaborative tasks with probabilistic movement primitives. In: IEEE-RAS International Conference on Humanoid Robots, pp. 527–534 (2014)
8. Schneider, M., Ertel, W.: Robot learning by demonstration with local Gaussian process regression. In: IEEE/RSJ International Conference on Intelligent Robots and Systems, pp. 255–260 (2010)
9. Sakoe, H., Chiba, S.: Dynamic programming algorithm optimization for spoken word recognition. IEEE Trans. Acoust. Speech Signal Process. **26**(1), 43–49 (1978)
10. Fearnhead, P.: Exact and efficient bayesian inference for multiple changepoint problems. Stat. Comput. **16**(2), 203–213 (2006)
11. Okada, K., Ogura, T., Haneda, A., Fujimoto, J., Gravot, F., Inaba, M.: Humanoid motion generation system on hrp2-jsk for daily life environment. In: IEEE International Conference on Mechatronics and Automation, pp. 1772–1777 (2005)

# MILP-Based Dual-Arm Motion Planning Considering Shared Transfer Path for Pick-Up and Place

Jun Kurosu[1], Ayanori Yorozu[1], and Masaki Takahashi[2(✉)]

[1] Graduate School of Science and Technology, Keio University,
3-14-1 Hiyoshi, Kohoku-ku, Yokohama 223-8522, Japan
cross125k@gmail.com, ayanoriyorozu@keio.jp
[2] Department of System Design Engineering, Keio University,
3-14-1 Hiyoshi, Kohoku-ku, Yokohama 223-8522, Japan
takahashi@sd.keio.ac.jp

**Abstract.** One of the most basic tasks that a dual-arm robot does is pick-up and place work. Pick-up and place work consists of tasks in which the robot carries objects from a start position (initial position) to a goal position. The following three important points should also be considered when the dual-arm robot does this work efficiently: (1) collision avoidance of the arms, (2) which arm should move an object, and (3) the order in which the objects should be picked up and placed. In addition, dual-arm robot has operation range constraints. Depending on the position relationship between a start position and goal position, unless both arms are used, the object may not be transferred to a goal position. In this paper, we define the transfer path which must use both arms as "shared transfer path". First, we use mixed integer linear programming (MILP) based planning for the pick-up and place work to determine which arm should move an object and in which order these objects should be moved while considering the dual-arm robot's operation range. Second, we plan the path using the rapidly exploring random tree (RRT) so that the arms do not collide, enabling the robot to perform efficient pick-up and place work based on the MILP planning solution. The effectiveness of proposed method is confirmed by simulations and experiments using the HIRO dual-arm robot.

**Keywords:** Advanced motion planning · Manipulator
Mixed integer linear programming

## 1 Introduction

Dual-arm robots have attracted much attention in recent years [1]. One of the most basic tasks that an industrial dual-arm robot does is pick-up and place work. Pick-up and place work consists of tasks in which the robot carries objects from a start position (initial position) to a goal position. The dual-arm robot is expected to do this work in a dynamic environment, such as a house or a shop. It is necessary to do the pick-up and place work efficiently in a dynamic environment. The following three important points

© Springer Nature Switzerland AG 2019
M. Strand et al. (Eds.): IAS 2018, AISC 867, pp. 67–77, 2019.
https://doi.org/10.1007/978-3-030-01370-7_6

should also be considered when the dual-arm robot does this work: (1) collision avoidance of the arms, (2) which arm should move an object, and (3) the order in which the objects should be picked up and placed.

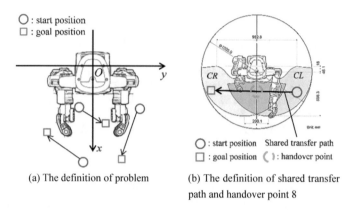

(a) The definition of problem     (b) The definition of shared transfer path and handover point 8

**Fig. 1.** Pick-up and place problem

In related studies, Srivastava et al. proposed a motion planning method for grasping a target object for a dual-arm robot. Specifically, they successfully moved two robot arms simultaneously and efficiently grasped a target object from among multiple objects [2]. Moreover, Saut et al., using a dual-arm robot, proposed a method of transporting an object from one end of the workspace to the other by the handover of an object [3]. In an environment where there are multiple objects, they performed motion and path planning so that neither arm collides with any objects during handover. Both of the above studies proposed a motion or path planning method that can work efficiently in environments where there are multiple objects. However, the operation time of a robot was not taken into account explicitly. Huang et al. proposed a motion planning method that considers operation time. This method enables a single-arm robot to minimize the time taken to finish the work when a multiple determined-point manipulator is moved [4]. Considering the above-mentioned three points described earlier, it takes a long time to search for the optimal combination without collision if we use a typical path-planning algorithm that uses an ordinary method like the rapidly exploring random tree (RRT) [5] and probabilistic roadmaps [6]. This is because we can only decide which arms can move an object and in which order the objects can be picked up and placed. There are also combinations when objects exist in the workspace.

In our previous study, we formulated a pick-up and placement task for a dual-arm robot, in which the robot carries an object in each arm to its respective goal position [7]. The circles in Fig. 1(a) indicate the start position of each object, and the squares represent each goal position. An object must be moved to its goal position from its start position by either of the two arms. However, considering the operating range of the dual-arm robot, shown in Fig. 1(b), there are object placement patterns difficult to formulate as MILP. For example, in Fig. 1(b), we can see that a start position exists within the range where only the left arm can reach (CL) and that a goal position exists within the range where only the right arm can reach (CR). When transporting an object from CL to CR, it is necessary to pick up the object from the start position using the left arm, place it at a point set within the range where both arms can reach, and then pick up the object again and bring it to the goal position with the right arm.

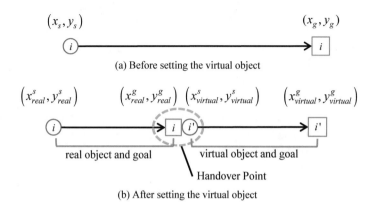

(a) Before setting the virtual object

real object and goal          virtual object and goal

Handover Point

(b) After setting the virtual object

**Fig. 2.** The definition of real and virtual object

In this study, we define the path used to transfer an object using both the left and right arms as a "shared transfer path", as shown in Fig. 1(b). Because a shared transfer path was not taken into consideration in the previous study, the present study proposes a motion planning method that explicitly takes into account a shared transfer path and operation time of pick-up and placement task.

Accordingly, we employ an MILP-based motion planning method for the pick-up and placement task to determine which arm should move which object and in which order the objects should be moved. We plan a path to avoid collisions using the RRT, enabling the robot to conduct an efficient pick-up and placement task based on the task planning solution determined from the MILP-based motion planning method. Finally, experiments using a real dual-arm robot were conducted to verify the effectiveness of the proposed method.

## 2 MILP-Based Dual-Arm Motion Planning Submission

### 2.1 Virtual Objects and Goals

The proposed method realizes a handover by setting a "virtual object" and a "virtual goal" at the handover point. Figure 2 shows a method used to make a new task by setting a virtual object and a virtual goal for a task requiring a handover.

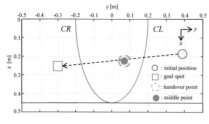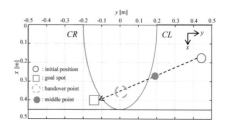

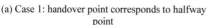

(a) Case 1: handover point corresponds to halfway point

(b) Case 2:handover point does NOT corresponds to halfway point

**Fig. 3.** The way to set the handover point

A new task $i'$ becomes $i' = n+3$ when there are $n+2$ tasks, that is, $n$ objects exist. Therefore, when there are $m$ objects requiring a handover, $i' = (n+2)+1$ to $i' = (n+2)+m$ are allocated to each object.

$$(x^s_{real}, y^s_{real}) = (x_s, y_s) \tag{1}$$

$$(x^g_{real}, y^g_{real}) = (\frac{x_s+x_g}{2}, \frac{y_s+y_g}{2}) \tag{2}$$

$$(x^s_{virtual}, y^s_{virtual}) = (\frac{x_s+x_g}{2}, \frac{y_s+y_g}{2}) \tag{3}$$

$$(x^g_{virtual}, y^g_{virtual}) = (x_g, y_g) \tag{4}$$

Where, $(x_s, y_s)$ is the start position, and $(x_g, y_g)$ is the goal position, before setting the virtual object and virtual goal. In addition, $(x^s_{real}, y^s_{real})$ is the set start position of a real object, and $(x^g_{real}, y^g_{real})$ is the set goal position of the real object. Finally, $(x^s_{virtual}, y^s_{virtual})$ is the set start position of the virtual object, and $(x^g_{virtual}, y^g_{virtual})$ is the set goal position. The coordinates shown in Eqs. (2) and (3) coincide with the handover point. As shown in Fig. 3(a) and through Eqs. (2) and (3), a handover point is defined as the middle point between the start position and goal position before the virtual object and virtual goal are set. However, when considering the operating range of the dual arm robot, it is difficult to set a handover point within the middle of an object placement pattern. It is necessary to set a handover point within a range where both arms can reach. When the middle point is out of the range where both arms can reach, as shown in Fig. 3(b), a point where both arms can reach that exists on a straight line connecting the start and goal positions is set as a handover point. In addition, when setting a handover point, it is necessary to set it where it does not overlap other objects or goals within the workspace. In this study, we assume that the object is within a 10 cm square and do not set the handover point within a 10 cm radius of the start or goal position of another object.

## 2.2 Formulation of Pick-Up and Place Work

We formulated the pick-up and place work for a dual-arm robot that carries each object to the goal positions. This problem converts into the problem of sharing multiple transfer tasks, which is regarded as the task of moving an object to a goal position from a start position. This problem is also similar to the travelling salesman problem. Constraints, such as the operating range of the dual-arm robot, mean that we can formulate the pick-up and place work of the dual-arm robot as MILP. Additionally, to achieve a handover, it is necessary to formulate the time window constraints through which one arm reaches the handover point before the other arm. However, it is difficult to formulate the time window constraints in a linear because it is difficult to formulate them through an MILP. We update the solution under the constraint conditions and derive the optimal solution that guarantees the minimum amount of operation time. For this reason, although the calculation cost increases with the proposed method, we obtain a solution that guarantees the minimized operation time by updating the solution under a certain constraint.

$$\min_{x_{ijk} \in \{0,1\}} \max_{k \in D} \left( \sum_{i \in V} \sum_{j \in V} c_{ijk} x_{ijk} \right) \tag{5}$$

$$\text{s.t.} \quad \sum_{i \in V} x_{iik} = 0, \ k \in D \tag{6}$$

$$x_{121} + x_{211} = 0, \ x_{122} + x_{212} = 0 \tag{7}$$

$$\sum_{j \in V'} x_{kjk} = 1, \ k \in D \tag{8}$$

$$\sum_{i \in V'} x_{ikk} = 1, \ k \in D \tag{9}$$

$$\sum_{i \in V'} x_{i21} = 0, \ \sum_{j \in V'} x_{2j1} = 0, \ \sum_{i \in V'} x_{i12} = 0, \ \sum_{j \in V'} x_{1j2} = 0 \tag{10}$$

$$\sum_{k \in D} \sum_{i \in V} x_{ijk} = 1, j \in V', \quad \sum_{k \in D} \sum_{j \in V} x_{ijk} = 1, i \in V' \tag{11}$$

$$x_{kjk} + \sum_{i \in V'} x_{ijk} - x_{jkk} - \sum_{i \in V'} x_{jik} = 0, \quad k \in D, j \in V' \tag{12}$$

$$u_i + (L-2) \sum_{k \in D} x_{kik} - \sum_{k \in D} x_{ikk} \le L - 1, \ i \in V' \tag{13}$$

$$u_i + \sum_{k \in D} x_{kik} + (2 - K) \sum_{k \in D} x_{ikk} \le 2, \quad i \in V' \tag{14}$$

$$\sum_{k \in D} x_{kj1} + \sum_{i \in V'} x_{ij1} = 1, \quad j \in CL, \quad \sum_{k \in D} x_{ik1} + \sum_{j \in V'} x_{ij1} = 1, \quad i \in CL \qquad (15)$$

$$\sum_{k \in D} x_{kj2} + \sum_{i \in V'} x_{ij2} = 1, \quad j \in CR, \quad \sum_{k \in D} x_{ik2} + \sum_{j \in V'} x_{ij2} = 1, \quad i \in CR \qquad (16)$$

$$x_{ijk} \in \{0, 1\}, \quad \forall i, j \in V, \ k \in D \qquad (17)$$

$$\sum_{H_{Lk} \in i,j} c_{ij1} x_{ij1} - \sum_{H_{Rk} \in i,j} c_{ij2} x_{ij2} \geq 0, \quad H_{Lk}, H_{Rk} \in H, k \in H' \qquad (18)$$

$$\sum_{H_{Rk} \in i,j} c_{ij2} x_{ij2} - \sum_{H_{Lk} \in i,j} c_{ij1} x_{ij} \geq 0, \quad H_{Lk}, H_{Rk} \in H, k \in H' \qquad (19)$$

$$\sum_{H \in i,j,k} x_{ijk} \leq l - 1 \qquad (20)$$

where $V = \{1, 2, \ldots, n+2\}$ is a set of manipulator tasks, $D = \{1, 2\}$ is the set of manipulators, $D = 1$ is the left arm, $D = 2$ is the right arm, $V' = \{3, 4, \ldots, n+2\}$ is a set of tasks, $c_{ijk}$ is the cost from task $i$ to task $j$ for manipulator $k$, and $x_{ijk}$ is the binary variable indicating whether or not the manipulator moved from task $i$ to task $j$. The decision variable $x_{ijk} = 1$ if manipulator $k$ is assigned to move from task $i$ to task $j$, and 0 otherwise. Moreover, $u_i$ is the number of tasks executed by the manipulator, while $L$ is the maximum number of tasks a manipulator may execute. Let $K$ be the minimum number of tasks a manipulator may execute.

Equation (5) is the objective function. The minimax type of objective function to minimize the maximum path length of each manipulator is set. In other words, the time to finish the pick-up and place work is minimized because it depends on the arm, which has to move along a longer path if the manipulator moves at a constant velocity.

Equations (6) to (20) are the linear constraints of this formulation. Equation (6) forbids the manipulator $k$ from moving between the same task (from task $i$ to task $i$). Equation (7) also forbid movement from task 1 to task 2 or from task 2 to task 1 because task 1 indicates the left arm, and task 2 indicates the right arm. Equation (8) ensures that manipulator $k$ moves from the start position. Equation (9) ensures that manipulator $k$ finally moves back to the start position. Equation (10) forbids the paths that manipulator $k$ cannot select because tasks 1 and 2 denote the left and right arms, respectively. Equation (11) means that each task must be executed by either arm. Equation (12) is a path continuity constraint, which means that manipulator $k$ must execute task $j$ and any other task after executing one task. Equations (13) and (14) are the subtour elimination constraints of the formulation proposed by Kara et al. and Bektas et al. [9, 10]. A subtour is a path that does not include tasks 1 and 2. The constraints herein allow only two subtours because the dual-arm robot has two manipulators. Equation (15) ensures that a task belonging to $CL$ must be executed by the left arm. Set $CR$ is also a set of tasks that should be executed by the right arm. Equation (16) ensures that a task belonging to $CR$ must be executed by the right arm. Figure 1(b) shows the operation range of HIRO, a real dual-arm robot produced by

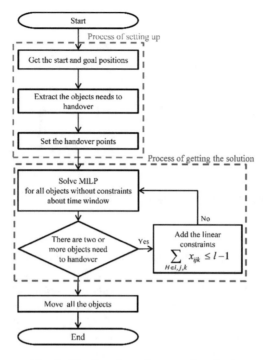

**Fig. 4.** The flowchart of proposed method.

KAWADA Robotics Corporation [8], as well as the definition of $CL$ and $CR$. Equation (17) defines the binary variable and the range of integers $i$, $j$, and $k$. $i, j, k$ and $c_{ijk}$ means cost that manipulator $k$ moves from tasks $i$ to $j$.

Equations (18) or (19) are time window constraints. As described before, to realize a handover operation $l$, it is necessary to conduct a motion planning that considers the time window constraint in which a real object is carried before a virtual object. Because time window constraints are not taken into consideration, the solutions using this process may not achieve a handover. Therefore, when a solution that does not satisfy the time window is acquired, we search for a new solution, excluding the present one. In Eqs. (18) and (19), $H$ is a combined set of integers for the acquired solution, $H'$ is a set of integers expressed through $H' = \{1, 2, \ldots, m\}$, and $m$ is the number of handover points. In addition, $H_{Lk}$ is a set of integers that indicates the solution of the left arm to handover point $k$. Specifically, it shows the order of transfer of an object in the left arm from its initial position to the handover point. Similarly, $H_{Rk}$ is a set of integers that indicate the solution for the left arm to reach handover point $k$. Therefore, the term $\sum c_{ij1}x_{ij1}$ in Eq. (18) indicates the path length of the left arm until it reaches the handover point from its initial position. Therefore, the term $\sum c_{ij2}x_{ij2}$ in Eq. (18) indicates the path length of the right arm until it reaches the handover point from its initial position. Thus, Eqs. (18) and (19) express the difference in path length between the left and right arms to the handover point.

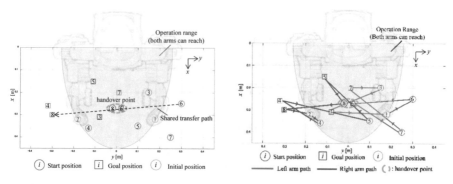

**Fig. 5.** Initial configuration after setting virtual object and goal.

**Fig. 6.** Solution of the proposed method.

When the solution acquired does not satisfy the time window constraints expressed through Eqs. (18) and (19), we add the linear constraints expressed in Eq. (20). We search for solutions other than the acquired solution that satisfy the objective function and its constraints. In Eq. (20), $l$ is the number of elements included in set $H$.

The solution is updated by sequentially adding the linear constraint expressed in Eq. (20). The solution acquired through this process satisfies the linear constraint added in Eq. (20), in addition to the objective function and constraints. Therefore, the solution satisfying Eq. (19) or Eq. (20) is the optimal solution that satisfies the time window constraints and minimizes the working time of the dual-arm robot.

The flowchart of the proposed method is shown in Fig. 4.

### 2.3    Path Planning with RRT

The solution is derived from the MILP. The path to avoid a collision is planned with the RRT using a solution similar one that decides which arm moves an object and in which order these objects are to be picked up and placed. The arm has a longer path length defined as a dynamic obstacle to prevent the finish time from increasing. An arm defined as a dynamic obstacle will linearly move to its destination. A collision-free path is planned to exist outside of areas that can possibly collide with each arm. This area is defined as the collision area (CA) on the $xy$ plane. The arm has a shorter path length that avoids the CA. The proposed method can then avoid a collision with each arm.

## 3    Result Simulation

Six objects were used for the dual-arm robot's workspace in the simulations which the coordinates of the virtual object and handover point are set to achieve a handover. Figure 5 shows initial configuration after setting virtual object and goal. Object 6 requires a handover from the left to right arm. In this simulation, object 8 is added as a virtual object to conduct a handover for object 6.

We apply the proposed method for the workspace, which includes both objects 6 and 8. As shown in Fig. 6, we confirmed that object 6 is carried before object 8 (virtual object). This result indicates that Eq. (5) through Eq. (17) of the MILP formulation, and Eq. (20) used for updating the solution with the proposed method, are effective for the pick-up and placement to minimize the operation time.

## 4    Experiment

Experiments using the dual-arm robot HIRO produced by Kawada Robotics Co. were conducted to verify the effectiveness of the proposed method. Sponge blocks were used for verification. We used six colored blocks: red, green, yellow, grey, brown, and black. The first experiment confirmed that the dual-arm robot is able to utilize the proposed method when a shared transfer path is applied. The second experiment tested the collision avoidance for each arm and confirmed that the path planning with RRT is effective when possible collisions may occur with each moving arm.

### 4.1    Experiment 1

We conducted experiments using the proposed method. The motion of the dual-arm robot for these experiments is shown in Fig. 7. The squares in the images indicate the goal position, which corresponds to the respective color of the sponge block used. The relationships between color and object ID are shown in Table 1. We confirmed that both arms operated at the same time from the beginning of the pick-up and placement task. During this process, we also confirmed that the grey block was transferred to the handover point by the left arm, and finally transferred by the right arm.

**Table 1.**  Relationship between Object ID and Color

| Object ID | 3 | 4 | 5 | 6 | 7 |
|-----------|--------|-----|-------|------|-------|
| Color | Yellow | Red | Green | Grey | Brown |

### 4.2    Collision Avoidance Using Each Arm

In the experiments shown in Fig. 7, an operation including collision avoidance obtained using the RRT method described in Sect. 2.2 was demonstrated during a pick-up and placement task. As shown in Fig. 7, the dual-arm robot carried a yellow block using its right arm, and a green block using its left arm. As shown in Fig. 8, we confirmed that the right arm passed the path indicated by the red arrow, avoided the CA, and carried the object. The results of this experiment confirm that, through the proposed method, the dual-arm robot is able to pick up and place an object without a collision with either arm.

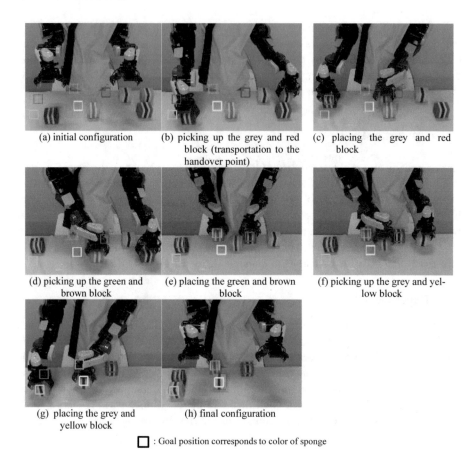

(a) initial configuration

(b) picking up the grey and red block (transportation to the handover point)

(c) placing the grey and red block

(d) picking up the green and brown block

(e) placing the green and brown block

(f) picking up the grey and yellow block

(g) placing the grey and yellow block

(h) final configuration

☐ : Goal position corresponds to color of sponge

**Fig. 7.** Verification experiment.

**Fig. 8.** Collision avoidance.

## 5 Conclusion

In this study, we focused on a pick-up and placement task of a dual-arm robot, and proposed a motion planning method for performing such a task when a shared transfer path is applied. With the proposed method, the order of transferring an object is determined using an algorithm that explicitly considers the shared transfer path. In addition, we are able to plan a path using an RRT through which the arms do not collide with each other, enabling the robot to efficiently conduct a pick-up and placement task based on the MILP planning solution. In experiments conducted to verify our proposal, we confirmed that using a path planning with RRT, objects are able to be transferred based on the solution without a collision with each arm.

In the future, we plan to extend this work to a three-dimensional space, that is, the xyz coordinates along a plane.

**Acknowledgment.** This study was supported by "A Framework PRINTEPS to Develop Practical Artificial Intelligence" of the Core Research for Evolutional Science and Technology (CREST) of the Japan Science and Technology Agency (JST) under Grant Number JPMJCR14E3.

## References

1. Smith, C., Karayiannidis, Y., Nalpantidis, L., Gratal, X., Qi, P., Dimarogonas, D.V., Kragic, D.: Dual arm manipulation – a survey. Robot. Auton. Syst. **60**(1), 1340–1353 (2012)
2. Srivastava, S., Fang, E., Riano, L., Chitnis, R., Russell, S., Abbeel, P.: Combined task and motion planning through an extensible planner-independent interface layer. In: IEEE International Conference on Robotics and Automation (ICRA), pp. 639–646 (2014)
3. Saut, J., Gharbi, M., Corte, J., Sidobre, D., Simeon, T.: Planning pick and place tasks with two-hand regrasping. In: IEEE/RSJ International Conference on Intelligent Robots and Systems (IROS), pp. 4528–4533 (2010)
4. Huang, Y., Gueta, B., Chiba, R., Arai, T., Ueyama, T., Ota, J.: Selection of manipulator system for multiple-goal task by evaluating task completion time and cost with computational time constraints. Adv. Robot. **27**(4), 233–245 (2013)
5. La Valle, S.M., Kuffer, J.: Randomized kinodynamic planning. Int. J. Robot. Res. **20**(5), 378–400 (2001)
6. Kavraki, L.E., Svestka, P., Latombe, J.-C., Overmars, M.H.: Probabilistic roadmaps for path planning in high-dimensional configuration space. IEEE Trans. Robot. Autom. **12**(4), 566–580 (1996)
7. Kurosu, J., Yorozu, A., Takahashi, M.: Simultaneous dual-arm motion planning for pick-up and place. In: International Conference on Control, Automation and Systems (ICCAS), pp. 80–85 (2016)
8. Kawada Robotics Corporation: "Nextage". http://robot-support.kawada.jp/support/hiro/. Accessed June 2017
9. Kara, I., Bektas, T.: Integer linear programming formulations of multiple salesman problem and its variations. Eur. J. Oper. Res. **174**(3), 1449–1458 (2006)
10. Bektas, T.: The multiple traveling salesman problem: an overview of formulations and solution procedures. Omega **34**(3), 209–219 (2006)

# Multiple Path Planner Integration for Obstacle Avoidance: MoveIt! and Potential Field Planner Synergy

Emanuele Sansebastiano and Angel P. del Pobil[✉]

Robotic Intelligence Lab, Jaume I University, Castellón de la Plana, Spain
emanuele.sansebastiano@outlook.com, pobil@uji.es

**Abstract.** Nowadays, robots are more and more autonomous, they are able to investigate autonomously the surrounding environment, take decisions, and, of course, accomplish elaborated tasks receiving simple inputs by users. Generally, every robot has to move to interact with the environment and human users replicating human-like motions and decisions. The literature already gives a lot of documentation about mobile/flying robots spanning paths in 2D and 3D environments, while the motion of articulated arms in 3D environments still requires investigation and experiments. This paper investigates MoveIt!, one of the most famous motion planner software, understanding its limits and trying to extend its capabilities. Integrating another planner with MoveIt!'s planning routine to split longer actions into smaller ones increases planning robustness. Due to the participation to Amazon Robotics Challenge 2017, every experiment has been carried on in automatic packing line scenarios, in which Baxter, a humanoid robot sporting two 7 DOF arms, had to perform actions avoiding known obstacles.

## 1 Introduction

RoboCup[1] is a challenge started in 1997 with the aim of producing a humanoid football team able to defeat the human world champion football team in 2050. year by year, other disciplines were introduced: Junior, Rescue, @Home, and Industrial. Amazon Robotics[2] decided to celebrate the success of previous editions of Amazon Picking Challenge renovating the competition name in Amazon Robotics Challenge (ARC) [1] and introducing it in the $20^{th}$ edition of RoboCup. The ARC aims to fully automate the task of customer order placement and delivery of products. The last challenge Amazon's automated warehouses are facing is picking and manipulating objects which are placed randomly in storage boxes. Every team participating in the competition has been called to entirely develop a system to accomplish such a tough task. The system developed by Robotic Intelligence Lab at Jaume I University [2] used Baxter as manipulator

---

[1] http://www.robocup.org/.

[2] https://www.amazonrobotics.com.

© Springer Nature Switzerland AG 2019
M. Strand et al. (Eds.): IAS 2018, AISC 867, pp. 78–85, 2019.
https://doi.org/10.1007/978-3-030-01370-7_7

and MoveIt! has been proposed among the available open-source motion planner software to accomplish the path planning task [3]. MoveIt! integrates directly with Open Motion Planning Library (OMPL) [4] and uses the motion planners from that library as its set of planners. In order to evaluate which algorithm suits better our necessities, many experiments have been performed, but none of those algorithms appeared to be enough to fully accomplish the planning task. Sometimes, MoveIt! could not find feasible solutions even if it was evident there were many possible solutions. Moreover, this phenomena seemed to be random because the same initial conditions and settings gave various results. Finding different paths derives from the nature of the planning algorithms: every tested algorithm is based on random trees investigation. In our experiments, none of those algorithms had an accomplishing rate higher than 90% for solvable problems. In addition, the found path was not always suitable for the competition aim because, instead of guiding the end-effector straight forward to the goal location, it sometimes spans a very irregular trajectory changing direction several times [5]. This behaviour increases the risk of involuntary object releasing, the possibility of not reaching the effective end-effector goal location due to error accumulation, and the execution time: Baxter's actions last 4 times more than the normal execution time during those irregular motion plans. The accomplishing rate decreases and the presence of irregular trajectories increases according to the number of obstacles in the scene and how far the goal pose is from the initial end-effector pose. In the name of reducing those disturbs, we have decided to implement a side path planner to split every action in many sub-actions. Then, MoveIt! had to deal just with short paths increasing its robustness. Due to its large usage and consistency properties, the potential field planner has been used as "pre-planner", while MoveIt! has been still kept in charge of the majority of the work. This strategy helps significantly the accomplishing rate, which becomes higher than 96% in average. Even if this procedure extended the planning time, the results are really promising in terms of planning robustness. Future investigations oriented to this direction should improve motion planning's state-of-the-art. Further details about this research are available in Sansebastiano's master thesis [6].

## 2   MoveIt! and OMPL

MoveIt!'s first commit was October 2011. MoveIt![3] is state of the art software for mobile manipulation, incorporating the latest advances in motion planning, manipulation, 3D perception, kinematics, control and navigation. MoveIt! comes with a plugin for the RViz. The plugin allows the user to set up scenes in which the robot will work, generate plans, visualize the output and interact directly with a visualized robot. RViz is a 3D visualizer for displaying sensor data and state information from ROS. MoveIt! planning is based on the set of motion planning algorithms implemented in the OMPL[4]. Right now, MoveIt! offers the

---

[3] MoveIt! official website: http://moveit.ros.org/.

[4] OMPL official website: http://ompl.kavrakilab.org/.

possibility to use eight different algorithms belonging to RRT, KPIECE, and EST families: RRT [7], RRT Connect [8], RRT Star [9], KPIECE [10], BKPIECE, LBKPIECE [11], EST [12], and SBL [13].

MoveIt! has many interesting functions: IK service, no-stuck behaviour[5], robust obstacle avoidance, and quick path planning computation under some hypothesis.

At the very beginning, the team planned to evaluate which algorithm was the fastest to perform safe obstacle avoidance among the ones supported by MoveIt!. Eventually, none of them appeared as reliable as required. Every algorithm is based on randomly exploration which does not propose always the same solution. In principles, this behaviour should not affect the final result, but, sometimes MoveIt! makes Baxter span a very dangerous path: the end-effector changes its direction several times, increasing end-effector shakes and wasting a lot of time. Generally, every tested motion required between 1 to 4 seconds to be performed properly, while, those dangerous motions spent up to 25 seconds. Our system had to use less than 7 seconds to be competitive in ARC and those sudden changes of end-effector direction increased chances to involuntary object realising. In addition, MoveIt! sometimes quitted the computation reporting that no feasible solution has been found even if there were many possible paths to span. This lack of solutions appeared randomly: MoveIt! cannot ensure finding solutions for a problem even though he previously found a solution for the exact same problem. In the end, MoveIt! appeared unusable for our purpose, but a interesting trend has been noticed. The lack of consistent solutions uses to increase proportionally to the number of obstacles in the scene and the distance between the initial end-effector pose and the end-effector goal pose. So, short motions have higher chances to be performed by spanning safe and fast trajectories. According to those facts, a side planner has been implemented to split every action in a set of way-points to convert complex problems in the sum of many simple problems. The potential field planner has been chosen to be the "pre-planner" due to its well known theory and implementation. The aim of this paper is investigating if there are evidences to go further in integrating multiple planners to improve planning results.

## 3   Potential Field Planner

The potential field planner is member of the geometric planner family. The map defined by this planner accounts for the geometric constraints of the environment and the system. The map is divided in cells, which represents a portion of space and the potential value of that space portion. The potential values contained in the cells are the result of the sum of two potential functions: an attractive function, which "attracts" the robot to the goal position, and a repulsive function, which creates trajectories to avoid obstacles:

$$U_{att}(\mathbf{q}) + U_{rep}(\mathbf{q}) = U(\mathbf{q}) \tag{1}$$

---

[5] If the robot cannot reach one of the way-points in the list, it will skip that action going for the next one. This special function avoids that systems get stuck.

where:

- $\mathbf{q}$ is the vector of coordinates defining the position of cell;
- $U_{att}(\mathbf{q})$ is the function describing the attractive potential field;
- $U_{rep}(\mathbf{q})$ is the function describing the repulsive potential field;
- $U(\mathbf{q})$ is the function describing the potential field.

As Konen and Borenstein described in their paper [14], robot trapping situations due to local minima occurs pretty often decreasing the reliability of such algorithm. In particular, this problem depends on the strategy adopted to generate the attractive field. So, a variant of the classical approach has been used: the wave-front approach [15]. This approach is strongly reliable, but larger in terms of computation. According to the classical approach, the attractive field depends on the absolute distance of every cell from the goal location. Evidently, the computation is quick and linearly dependent on the number of cells contained in the map. The wave-front approach, instead, is non-linearly dependent on the number, shape, and location of the obstacles. Like a wave produced by a stone falling into a lake, the front-wave approach starts from the goal cell and propagates the attractive field cell by cell for the whole map. The wave-front approach never get stuck in local minima, but it is definitely heavier in terms of computation.

It is understandable that this strategy will lead to long computation solutions, but the main aim of this paper is proving there is reason to investigate on synergy of multiple path planners in order to improve planning reliability.

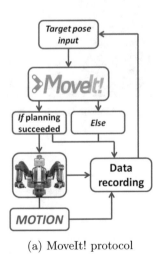

(a) MoveIt! protocol

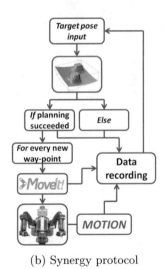

(b) Synergy protocol

**Fig. 1.** Protocols

## 4    Synergy Protocol

Since none of MoveIt!'s path planning algorithms ensured 100% success to find
a feasible path even though there were clearly many of them, a new strategy has
been proposed: MoveIt! receives a "pre-defined" path to span made of way-points
instead of just an end-effector goal position. A side planner is supposed to divide
the whole action in the sum of shorter ones, considering the end-effector as an
isolated object. Afterwords, MoveIt! will be in charge elaborate a feasible path
for every short action. MoveIt! has to perform the inverse kinematics, generating
the obstacle avoidance trajectory, and, if there is no feasible solution for one of
the way-points, skip it and proceed for the next one. Short actions means high
chances to find a suitable path, but many way-points means long computation.
A previous analysis on the mesh size to use for the map is mandatory not to
instantiate too many cells and consequently way-points to span. Figure 1 represents the tested protocols by means of block graphs: the MoveIt! protocol is
slimmer than the synergy one.

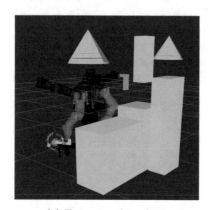

(a) Empty packing lane

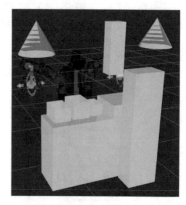

(b) Classic packing lane

**Fig. 2.** Scenes

## 5    Experiments

In order to compare the MoveIt! protocol and the synergy protocol many scenarios have been designed with varying difficulty. Every scene comes up with the
system developed by Robotic Intelligence Lab at Jaume I University to compete
in Amazon Robotics Challenge 2017 [2]. The basic packing lane is composed
by a shelf, two tables having different height, two lights and a camera located
over the tables to detect what is places on those tables. Some of the scenarios
are equipped with just basic furnitures (Fig. 2a), while others are equipped with
boxes containing objects. The system is supposed to pack various items in boxes
ready to be delivered (Fig. 2b). Data deriving from every experiment have been

post-processed to make them comparable regardless of the scene characteristics. Both protocols have been tested on 8 scenarios in which Baxter had to perform 7 consecutive actions in average. Every experiment has been repeated 5 times to ensure result consistency. Every action has been planned to have many safe and feasible paths, so a robust planner should be able to find at least one of them.

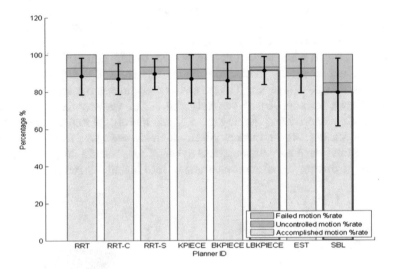

**Fig. 3.** MoveIt! algorithms - accomplishing rate comparison

Before proceeding with the protocol comparison, a previous set of experiments has been carried on the MoveIt! protocol in order to find which OMPL planner had the highest accomplishing rate. An action is considered accomplished if:

1. A feasible path have been found by MoveIt!.
2. The end-effector pose is close enough to the desired end-effector pose at the end of the motion.
3. The motion is fluent and do not change direction suddenly.

LBKPIECE algorithm resulted to be the best in terms of accomplishing rate (Fig. 3), while there was no evidence to prefer one planner among the others in terms of planning time and execution time. So, LBKPIECE algorithm has been used to compare the new protocol against the MoveIt! protocol to guarantee result consistency.

The main set of experiments has been performed on 8 difficult varying scenarios and it was repeated 5 times to check repeatability and increase result consistency. One of the most disturbing issues we met during MoveIt! tests and evaluation process has been its lack of repeatability: even though a feasible path has been previously found for a action, MoveIt! does not ensure it will repeat finding a feasible path for the exact same action.

# 6　Results and Conclusions

MoveIt! provides an easy-to-use platform for developing advanced robotics applications, evaluating new robot designs and building integrated robotics products for industrial applications. However, it does not appears very reliable if the scene is not trivial. In the name of those facts, another strategy has been adopted to keep all the functions which MoveIt! provides and incrementing its planning robustness. As it was expected, using a side planner to splitting tasks in many shorter subtasks increases a lot the planning robustness: the accomplishing rate almost reaches 96%, while it was never even close to 90% (Fig. 4a). Unfortunately, this protocol exponentially increased the planning time (Fig. 4b), decreasing the effectiveness of this approach. This drawback is mostly related with the side planning algorithm adopted to generate the set of way-points out of the initial action. This enormous planning time gap is mainly related to the technique used for generating the attractive potential field; the wave-front technique is computationally extremely expensive. Anyway, those results are very promising and encourage further investigations oriented into multiple path planner synergy strategies.

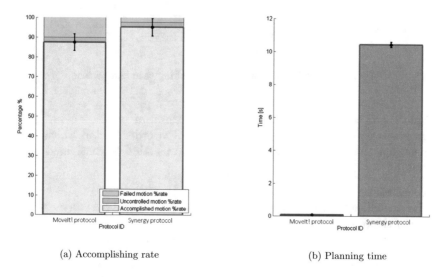

(a) Accomplishing rate　　　　　　　　　　(b) Planning time

**Fig. 4.** Accomplishing rate and planning time comparison

# 7　Code Repository

https://github.com/emanuelesansebastiano/moveit_user_support
https://github.com/emanuelesansebastiano/potential_field_planner
https://github.com/emanuelesansebastiano/ARC_thesis

# References

1. https://www.amazonrobotics.com/#/roboticschallenge
2. del Pobil, A.P., Kassawat, M., Duran, A. J., Arias, M.A., Nechyporenko, N., Mallick, A., Cervera, E., Subedi, D., Vasilev, I., Cardin, D., Sansebastiano, E., Martinez-Martin, E., Morales, A., Casañ, G.A., Arenal, A., Goriatcheff, B., Rubert, C., Recatala, G.: UJI RobInLab's Approach to the Amazon Robotics Challenge 2017. In: IEEE International Conference on Multisensor Fusion and Integration for Intelligent Systems (MFI 2017), Daegu, Korea, pp. 318–323 (2017)
3. Coleman, D., Sucan, I.A., Chitta, S., Correll, N.: Reducing the barrier to entry of complex robotic software: a MoveIt! Case study. Robotics 5(1), 3–16 (2014)
4. Sucan, I.A., Moll, M., Kavraki, L.E.: The open motion planning library. IEEE Robot. Autom. Mag. 19(4), 72–82 (2012). http://ompl.kavrakilab.org
5. https://www.youtube.com/watch?v=SZu30J6Eb_w
6. Sansebastiano, E.: Evaluation of various path planning algorithms performing obstacle avoidance for manipulators, M.Sc. thesis, Jaume I University (2017)
7. LaValle, S. M., Steven, M.: Planning Algorithms (2006)
8. Kuffner, J.J., LaValle, S.M.: RRT-connect: an efficient approach to singlequery path planning. In: IEEE International Conference on Robotics & Automation, vol. 2, pp. 995–1001 (2000)
9. Karaman, S., Frazzoli, E.: Sampling-based Algorithms for Optimal Motion Planning. arXiv:1105.1186 [cs.RO] (2011)
10. Sucan, I.A., Kavraki, L.E.: Kinodynamic motion planning by interior-exterior cell exploration. In: Workshop on the Algorithmic Foundations of Robotics (2008)
11. Bohlin, R., Kavraki, L.E.: Path planning using lazy PRM. In: IEEE International Conference on Robotics & Automation, pp. 521–528 (2000)
12. Hsu, D., Latombe, J.-C., Motwani, R.: Path planning in expansive configuration spaces. Int. J. Comput. Geom. Appl. 9(4–5), 495–512 (1999)
13. Sánchez, G., Latombe, J.-C.: A single-query bi-directional probabilistic roadmap planner with lazy collision checking. Int. J. Robot. Res. 6(1), 403–417 (2003)
14. Koren, Y., Borenstein, J.: Potential field methods and their inherent limitations for mobile robot navigation. In: IEEE Conference on Robotics and Automation, pp. 1398–1404 (1991)
15. Zelinsky, A., Jarvis, R.A., Byrne, J.C., Yuta, S.: Planning paths of complete coverage of an unstructured environment by a mobile robot. In: Proceedings of International Conference on Advanced Robotics, pp. 533–538 (1993)

# An Integrated Planning of Exploration, Coverage, and Object Localization for an Efficient Indoor Semantic Mapping

Diar Fahruddin Sasongko and Jun Miura$^{(\boxtimes)}$

Department of Computer Science and Engineering,
Toyohashi University of Technology, Toyohashi, Japan
`jun.miura@tut.jp`

**Abstract.** This paper describes an integrated viewpoint planner for indoor semantic mapping. Mapping of an unknown environment can be viewed as an integration of various activities: exploration, (2D or 3D) geometrical mapping, and object detection and localization. An efficient mapping entails selecting good viewpoints. Since a good viewpoint for one activity and that for another could be shared or conflicting, it is desirable to deal with all such activities at once, in an integrated manner. We use a frontier-based exploration, an area coverage approach for geometrical mapping, and object recognition model-based verification for generative respective viewpoints, and get the best next viewpoint by solving a travelling salesman problem. We carry out experiments using a realistic 3D robotic simulator to show the effectiveness of the proposed integrated viewpoint planning method.

**Keywords:** Viewpoint planning · Semantic mapping · Mobile robot

## 1 Introduction

Mapping is one of the fundamental functions of mobile robots. A map is used for various purposes such as localizing a robot and finding a specific object. The process of mapping in an unknown environment can be divided into two aspects. One is *where to get* new information and the other is *how to integrate* all pieces of information into a consistent representation. The second one is related to so-called SLAM (simultaneous localization and mapping) methods and numerous approaches have been proposed [1] and matured in a certain range of mapping problems. The first one is related to *viewpoint planning* and is still a hot research area [2–4]. This paper focuses the first aspect in semantic map generation.

Viewpoint planning methods differ in objectives, that is, what information will be obtained from the observations at selected viewpoints. For example, pure exploration tries to increase the area of known space [2], while in object search, a robot selects a viewpoint to increase the probability of a detected object being a target object [5]. Multiple, sometimes conflicting, objectives can also be considered in an integrated manner [3].

© Springer Nature Switzerland AG 2019
M. Strand et al. (Eds.): IAS 2018, AISC 867, pp. 86–98, 2019.
https://doi.org/10.1007/978-3-030-01370-7_8

Our previous viewpoint planner for geometric mapping [6] considers exploration and geometric data acquisition in an integrated manner. This paper extends the planner for semantic mapping, by additionally considering viewpoints for object verification. We show that an integrated viewpoint planning is better than non-integrated sequential planning by experiments in a realistic simulator.

The rest of the paper is organized as follows. Section 2 describes related work. Section 3 describes the target task and the overview of the proposed method. Section 4 briefly explains our previous planner. Section 5 describes object recognition and viewpoint generation for object verification. Section 6 shows experimental results to compare the sequential and the integrated approaches. Section 7 summarizes the paper and discusses future work.

## 2   Related Work

### 2.1   Exploration

Exploration of an unknown environment has been an important ability of mobile robots. In usual mapping algorithms, a whole environment is divided into three categories: free, occupied, and unknown. Yamauchi [2] proposed to use *frontiers*, which are boundaries between free and unknown spaces, for viewpoint generation in exploration. Several criteria are possible in choosing one among frontiers such as closest to the robot [2] and maximizing the information gain [3]. The idea of frontier is very useful in viewpoint planning for exploration and has been expanded to, for example, multi-robot exploration [7] and/or multi-criteria exploration [8].

### 2.2   Coverage

When making an entire map of an environment, a robot needs to plan a set of viewpoints so that the whole environment is collectively observed. This is sometimes called as *coverage planning* and methods for Art Gallery Problem (AGP) [9] can be adopted. In robotics context, sensor coverage problems have been discussed (e.g., [10–12]). Ardiyanto and Miura [13] proposed a generalized coverage solver that can take into account of the cost of each viewpoint imposed by the environment and/or problem settings.

### 2.3   Object Search

Ye and Tsotsos [5] proposed a framework for solving a visual object search problem. Using the probabilistic distribution of the target position and the probabilistic detection functions, the object search problem is formulated as a statistical optimization problem. Saidi et al. [14] proposed a similar approach to a 3D object search using a humanoid robot. Aydemir et al. [15] used domain knowledge on spatial relations between objects to guide a object search behavior. Masuzawa and Miura [4] treat an object verification problem as a viewpoint planning, which optimizes viewpoint sequences using a distance- and orientation-dependent object recognition model.

## 2.4   Integrated Viewpoint Planning

Multiple objectives are sometimes considered in exploration and/or mapping. Makarenko et al. [3] consider three utilities, information gain, navigability, and localizability, and choose a viewpoint which maximizes the total utility. Masuzawa and Miura [4] proposed a two-level hierarchical planner, in which the high-level part deals with determining the order of visits to viewpoint candidates, while the low-level part determines an actual viewpoint sequence to verify object candidates. They also introduced a loss function representing a deadline and showed that the robot's behavior changes for different loss functions. Diar and Miura [6] proposed a viewpoint planner which considers viewpoints for exploration and those of coverage in an integrated manner. However, they did not consider viewpoints for object verification.

## 3   Target Task and Overview of Integrated Viewpoint Planning

The task of the robot is to make a semantic map. A map describes not only 3D geometric information but also object types and locations. We use a realistic robot simulator V-REP [16] for experiments. Figure 1 shows a simulated world and a mapping result. A simulated robot is equipped with an omnidirectional laser range finder (LRF) for free space recognition and an RGB-D camera for 3D measurement, object candidate detection, and object verification.

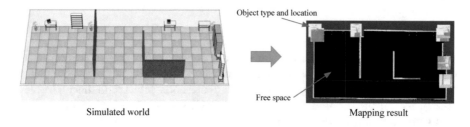

Simulated world                    Free space                    Mapping result

Object type and location

**Fig. 1.** Semantic mapping task.

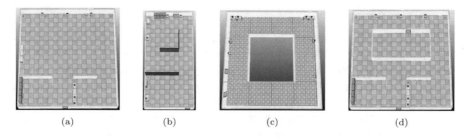

(a)                    (b)                    (c)                    (d)

**Fig. 2.** Four environments used for experiments. From left to right, Environment A, B, C, and D. The size of Environment A, B, and D is 20 m × 20 m while that of Environment B is 20 m × 20 m.

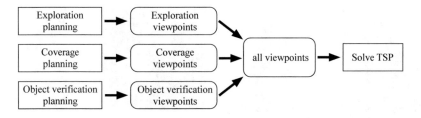

**Fig. 3.** Process of the integrated planning.

We consider the following three aspects in viewpoint planning for semantic mapping of unknown environments:

- *exploration planning* chooses viewpoints, measurements by the LRF from which can provide a large expansion of free spaces. Such free space information is necessary for the following two aspects of planning.
- *coverage planning* chooses viewpoints, observations by the camera from which can provide 3D space shape of the entire environment and object candidate information.
- *object verification planning* chooses viewpoints, observations by the camera from which can provide a more reliable information on the identity of object candidates.

Since the coverage planning is performed in already-explored regions and the object verification planning is performed for object candidates found in the coverage planning, one natural way to solve these planning problems is to perform them sequentially. That is, the robot first explores the whole environment and makes a 2D free space map. Then it observes the environment at planned viewpoints and makes a 3D description of the environment and enumerates detected object candidates. Finally it moves near to each object candidate and verify it. This sequential manner is obviously inefficient and an integrated approach is necessary.

Therefore we take an approach that the robot first generates viewpoints for these three planning aspects and chooses the best next viewpoint from a set of all generated viewpoints. The second step is done by solving a travelling salesman problem (TSP) and choosing the very first viewpoint. Figure 3 illustrates the process of planning.

## 4  Integration of Exploration and Mapping Viewpoint Planning

This section explains an integrated exploration and coverage planning by Sasongko and Miura [6]. The planning is composed of viewpoint generation for exploration, that for coverage, and integrated viewpoint selection.

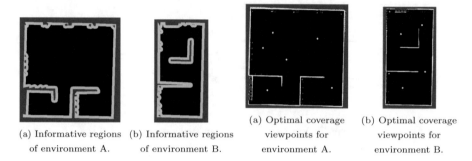

(a) Informative regions   (b) Informative regions          (a) Optimal coverage      (b) Optimal coverage
of environment A.          of environment B.                viewpoints for            viewpoints for
                                                            environment A.            environment B.

**Fig. 4.** Example informative regions shown **Fig. 5.** Optimal set of coverage viewpoints. in green color.

## 4.1  Viewpoint Generation for Exploration

We use a 2D occupancy grid map [17] to represent the free space of the environment. As a new LRF scan is input, a SLAM module updates the free space map. We adopt the frontier-based method [2]. A frontier point is a point which is in the free space region and adjacent to an unknown region. All of the frontier points are partitioned into clusters with a distance threshold $th_d$. We then calculate the centroid of each cluster of frontier points whose number of points is larger than a threshold $th_n$ and determine the frontier point closest to the centroid as an exploration viewpoint. Finally we determine the exploration viewpoint $V_e$ using the closest-frontier strategy [2]. We set the resolution of the grid map to 0.05 m and use the thresholds $th_d = 0.25$ m and $th_n = 13$.

## 4.2  Viewpoint Generation for Coverage

**Informative Region.** In usual environments, objects are, for example, on tables and shelves and not on the floor. We would therefore like to limit the target regions for coverage to such *informative regions*. We currently use a simple heuristic by Okada and Miura [18]; that is, assuming that outline of 2D free spaces corresponds to such tables and shelves, the informative regions are defined as the ones which are within a certain distance to the free space boundaries. We currently set the distance to 1 m. Figure 4 shows 2D maps and the corresponding informative regions for environments A and B shown in Fig. 2.

**Coverage Viewpoint Generation.** We need to generate a set of viewpoints which covers all currently-known informative regions. The measurable range of the RGB-D camera and the incident angle limitation[1] are considered in coverage calculation. We adopt a visibility-based viewpoint planning [19]. This method exploits a topological property of the free space, where coverage viewpoints can

---

[1] The line-of-sight of the camera and the surface normal of an observed area should be within a certain angle. Currently, we use 80° as the threshold.

be put. A skeletonization technique is applied to the free space followed by a detection of junctions, which are then used as a set of viewpoint candidates $V'_{c\_can}$. Since the skeletonization method is not always complete in a complex environment, we add a certain number of auxiliary viewpoint candidates $V_{c\_aux}$ to the set to get an updated set $V_{c\_can}$, and then obtain an optimized set $V_c$ by solving the Set Coverage Problem [13]. Figure 5 shows examples of an optimized set of coverage viewpoints.

### 4.3   Integrated Planning

The free space region expands as the robot moves and gets more data from the LRF. The robot calculates viewpoints and selects one of them at a certain timing, since it is not computationally efficient to update the set of viewpoints continuously and a frequent change of target viewpoint may make the robot move on a longer path. Therefore the robot generates a new set of the coverage viewpoints ($V_c$) for the currently-known informative regions, if the increment of the free space region is larger than a threshold $th_f$ or if the entire environment has been already explored. Note that the exploration viewpoint ($V_e$) is updated every time as the robot gets data from the LRF. The value of $th_f$ is determined experimentally as explained in Sect. 6.1.

We consider the union of $V_e$ and $V_c$:

$$V = V_e \cup V_c$$

and choose the one from the set $V$. For this purpose, we solve the Travelling Salesman Problem (TSP) using 2-opt method [20], and the first viewpoint in the tour is used as the target viewpoint. The robot moves towards that viewpoint, and when the robot reaches there or the coverage viewpoint set is updated as mentioned above, a new TSP is solved.

## 5   Object Verification and Viewpoint Generation

### 5.1   Object Candidate Detection and Verification

The robot searches the informative regions for objects. We use YOLOv2 [21] enhanced by COCO dataset [22] for object detection and verification. There are 80 classes for object candidates. YOLOv2 provides the confidence and a bounding frame of each detected object. We use the centroid of the frame as the detected position of the object. Figure 6 shows an example object detection result.

We divide the detection results to high-confidence and low-confidence, using a threshold $th_o = 0.5$. Low-confidence objects must be verified by observing them again. In the case of Fig. 6, for example, the laptop and the vase are verified.

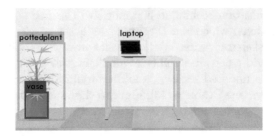

**Fig. 6.** YOLOv2 detection result. The confidence values of potted plant, vase, and laptop are 0.68, 0.35, and 0.27, respectively.

**Table 1.** Maximum verification distances for four objects.

| No | Object name | Object picture | Max. verification dist. |
|----|-------------|----------------|-------------------------|
| 1 | Laptop | | 4 m |
| 2 | Chair | | 4 m |
| 3 | Potted plant | | 4 m |
| 4 | Vase | | 3 m |

## 5.2   Verification Viewpoint Generation

In general, a closer observation increases the probability of correctly identifying an object, and such an idea should be considered in viewpoint generation for verification. Masuzawa and Miura [23] used a probabilistic observation model, which estimates the probability of successful verification from a set of observation parameters, for generating an observation plan. The model is for SIFT-based specific object recognition and was made for each object by actually observing the object from various viewpoints.

We take a similar empirical approach but use a much simpler model. That is, we assume that verification results depend only on the distance to each target object, and determine the maximum distance for which verification always succeeds. Table 1 summarizes the maximum verification distances for four objects used in this paper. The verification region of an object is thus defined as a circle centered at the object and with the radius of the maximum verification distance.

The object verification viewpoint ($V_v$) for an object is the point in the corresponding verification region and closest to the robot. If there are more than one object verification regions and they are overlapping, the object verification viewpoint is generated in the intersection region. Figure 7 shows the verification viewpoint for an example scene shown in Fig. 6. Two among three objects are low-confidence ones. The intersection of the laptop and the vase verification region is examined to choose the nearest position in it to the robot as the verification viewpoint (shown in yellow).

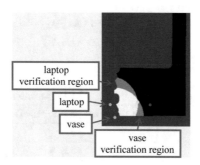

**Fig. 7.** Verification viewpoint for two objects. The white region is the intersection of two verification regions; light blue points are low-confidence objects and red point is the robot position. Yellow point is the chosen verification viewpoint.

### 5.3    Integrated Planning

We consider the object verification viewpoint $(V_v)$ in addition to the exploration and the coverage viewpoints ($V_e$ and $V_c$, respectively). The method of viewpoint planning is similar to the one described before (see Sect. 4.3), that is, we consider the union of three types of viewpoints, $V_e$, $V_c$, and $V_v$:

$$V = V_e \cup V_c \cup V_v$$

and choose the one from the set $V$ by solving the TSP (see Fig. 3).

The timing of updating viewpoints is slightly changed. In addition to the above-mentioned two conditions, we also consider if there is at least one newly detected low-confidence object.

## 6    Experimental Results

### 6.1    Determining the Threshold for the Increment of Free Space Area

The proposed integrated planning method generates a new set of the coverage viewpoints $(V_c)$ for the currently-known informative regions if the increment of the free space area is larger than a threshold $th_f$ or the entire environment has been explored. If the threshold is too low, the robot updates $V_c$ too often, thereby making the robot change the target subgoal frequently. If the threshold is too high, on the other hand, the robot rarely updates $V_c$, thereby making the robot ignore useful information of newly explored regions and newly detected objects. In either case, the robot behavior could be inefficient.

Using the integrated exploration and coverage planning explained in Sect. 4.3, we compared three values for $th_f$: $50\,\mathrm{m}^2$, $100\,\mathrm{m}^2$, and $150\,\mathrm{m}^2$ in terms of the total time and the travelled distance for environments A and B. Table 2 summarizes the results, showing a tendency that too low and too high $th_f$ values are not desirable. From the results, we decided to use $th_f=100\,\mathrm{m}^2$.

**Table 2.** Effect of $th_f$ values on the total performance.

| $th_f$ | Environment A | | Environment B | |
|---|---|---|---|---|
| | Time (min) | Distance (m) | Time (min) | Distance (m) |
| $50\,\mathrm{m}^2$ | 11 | 72 | 9 | 79 |
| $100\,\mathrm{m}^2$ | 7 | 56 | 3.5 | 25 |
| $150\,\mathrm{m}^2$ | 9 | 64 | 5 | 34 |

## 6.2   Mapping Results and Comparison Among Strategies

We show the results of the following two strategies:

– Sequential strategy ($ST_{seq}$) which performs a geometric mapping (i.e., exploration and coverage) and a semantic one (i.e., object verification and localization) sequentially.
– Integrated strategy ($ST_{int}$) which perform both the geometric and the semantic mapping in an integrated manner. This is the proposed strategy.

Figures 8 and 9 show the mapping process of the sequential and the integrated strategy for Environment A and D shown in Fig. 2. In the sequential strategy, the robot first explores the environment to make a geometric map, followed by another round of navigation for verifying low-confidence objects found in the way of geometric mapping. As a result, there are redundant movements in many places in the environments.

Table 3 summarizes the quantitative data for the strategies. The table also includes the results of geometric mapping (not including object detection and verification) by two strategies for comparison; one strategy is sequential exploration and coverage ($ST_{seq}^g$) and the other is the integrated exploration and coverage ($ST_{int}^g$). Comparison results for strategies $ST_{seq}$ and $ST_{int}$ show the effectiveness of integrated viewpoint planning. Note that strategy $ST_{int}$ performs extra object recognition tasks compared to those for geometric mapping, but requires a shorter traveling distance than their sequential version ($ST_{seq}^g$).

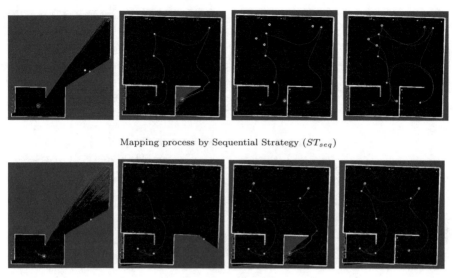

Mapping process by Sequential Strategy ($ST_{seq}$)

Mapping process by Integrated Strategy ($ST_{int}$)

**Fig. 8.** Mapping process of the sequential and the integrated strategy for Environment D. Red lines indicate the robot path; blue points indicate visited viewpoints; yellow points indicate viewpoint candidates and those with green circles indicate the current target viewpoint.

**Table 3.** Quantitative comparison of four strategies.

| Strategy | Environment A | | Environment B | | Environment C | | Environment D | |
|---|---|---|---|---|---|---|---|---|
| | Time (min) | Distance (m) | Time (min) | Distance (m) | Time (min) | Distance (m) | Time (min) | Distance (m) |
| $ST_{seq}$ | 32 | 113.8 | 22 | 70.7 | 16 | 80.9 | 26 | 146.6 |
| $ST_{int}$ | 13 | 64.2 | 10 | 45.1 | 9 | 50.9 | 20 | 105.5 |
| $ST_{seq}^{g}$ | 14 | 77.4 | 8 | 47.3 | 13 | 88.4 | 15 | 121.8 |
| $ST_{int}^{g}$ | 7 | 56.0 | 3.5 | 25.0 | 7 | 50.5 | 9.5 | 98.4 |

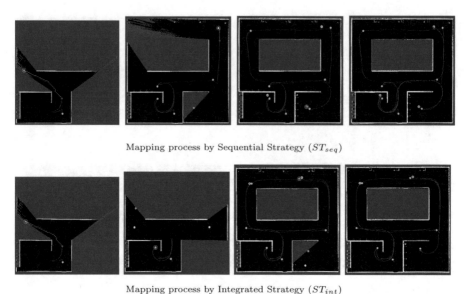

Mapping process by Sequential Strategy ($ST_{seq}$)

Mapping process by Integrated Strategy ($ST_{int}$)

**Fig. 9.** Mapping process of the sequential and the integrated strategy for Environment A. Red lines indicate the robot path; blue points indicate visited viewpoints; yellow points indicate viewpoint candidates and those with green circles indicate the current target viewpoint.

## 7   Conclusions and Future Work

This paper has presented an integrated viewpoint planning for semantic mapping of unknown environments, which includes exploration, coverage for 3D mapping, and object detection and localization. Since generating the optimal sequence of viewpoints is hard without any prior information of the environment, we repeatedly generate a locally-optimal viewpoints candidates based on the information of newly explored regions and newly detected object candidates. We experimentally determine a good interval for viewpoint updates. We tested the proposed algorithm in a realistic robotic simulation environment to show the efficiency of our integrated planning strategy.

Implementing the method on a real robot for evaluating in real situations is future work. To do this, the object recognition part needs to be enhanced largely, especially in the observation models. The current model considers only the distance to an object but a more variety of factors such as orientation and lighting conditions should also be taken into account.

**Acknowledgment.** This work is in part supported by JSPS KAKENHI Grant Number 17H01799 and the Hibi Science Foundation.

# References

1. Thrun, S., Burgard, W., Fox, D.: Probabilistic Robotics. The MIT Press, Cambridge (2005)
2. Yamauchi, B.: A frontier-based approach for autonomous navigation. In: Proceedings of the 1997 IEEE International Conference on Computational Intelligence in Robotics and Automation, pp. 146–151 (1997)
3. Makarenko, A.A., Williams, S.B., Bourgault, F., Durrant-Whyte, H.F.: An experiment in integrated exploration. In: Proceedings of 2002 IEEE/RSJ International Conference on Intelligent Robots and Systems, pp. 534–539 (2002)
4. Masuzawa, H., Miura, J.: Observation planning for environment information summarization with deadlines. In: Proceedings of 2010 IEEE/RSJ International Conference on Intelligent Robots and Systems, pp. 30–36 (2010)
5. Ye, Y., Tsotsos, J.K.: Sensor planning for 3D object search. Comput. Vis. Image Underst. **73**(2), 145–168 (1999)
6. Sasongko, D.F., Miura, J.: An integrated exploration and observation planning for an efficient indoor 3D mapping. In: Proceedings of 2017 International Conference on Mechatronics and Automation, pp. 1924–1929 (2017)
7. Faigl, J., Kulich, M.: On benchmarking of frontier-based multi-robot exploration strategies. In: Proceedings of 2015 European Conference on Mobile Robots, pp. 1–8 (2015)
8. Basilico, N., Amigoni, F.: Exploration strategies based on multi-criteria decision making for search and rescue autonomous robots. Auton. Robots **31**(4), 401–417 (2011)
9. O'Rourke, J.: Art Gallery Theorems and Algorithms. Oxford University Press, New York (1987)
10. González-Baños, H., Latombe, J.-C.: A randomized art-gallery algorithm for sensor placement. In: Proceedings of 7th Annual Symposium on Computational Geometry, pp. 232–240 (2001)
11. Agarwal, P.K., Ezra, E., Ganjugunte, S.K.: Efficient sensor placement for surveillance problems. In: Proceedings of the 5th IEEE International Conference on Distributed Computing in Sensor Systems, pp. 301–314 (2009)
12. Ramaswamy, V., Marden, J.R.: A sensor coverage game with improved efficiency guarantees. In: Proceedings of 2016 American Control Conference, pp. 6399–6404 (2016)
13. Ardiyanto, I., Miura, J.: Generalized coverage solver using hybrid evolutionary optimization. Int. J. Innovative Comput. Inf. Control **13**(3), 921–940 (2017)
14. Saidi, F., Stasse, O., Yokoi, K., Kanehiro, F.: Online object search with a humanoid robot. In: Proceedings of 2007 IEEE/RSJ International Conference on Intelligent Robots and Systems, pp. 1677–1682 (2007)
15. Aydemir, A., Sjöö, K., Folkesson, J., Pronobis, A., Jensfelt, P.: Search in the real world: active visual object search based on spatial relations. In: Proceedings of 2011 IEEE International Conference on Robotics and Automation, pp. 2818–2824 (2011)
16. Rohemr, E., Singh, S.P., Freese, M.: V-REP: a versatile and scalable robot simulation framework. In: Proceedings of 2013 IEEE/RSJ International Conference on Intelligent Robots and Systems, pp. 1321–1326 (2013)
17. Elfes, A.: Sonar-based real-world mapping and navigation. Int. J. Robotics Automat. **3**(3), 249–265 (1987)

18. Okada, Y., Miura, J.: Exploration and observation planning for 3D indoor mapping. In: Proceedings of 2015 IEEE/SICE International Symposium on System Integration, pp. 599–604 (2015)
19. Ardiyanto, I., Miura, J.: Visibility-based viewpoint planning for guard robot using skeletonization and geodesic motion model. In: Proceedings of the 2013 IEEE International Conference on Robotics and Automation, pp. 652–658 (2013)
20. Mersmann, O., Bischl, B., Bossek, J., Trautmann, H., Wagner, M., Neumann, F.: Local search and the traveling salesman problem: a feature-based characterization of problem hardness. In: Learning and Intelligent Optimization, vol. LNCS 7219, pp. 115–129 (2012)
21. Redmon, J., Farhadi, A.: YOLO9000: Better, Faster, Stronger arXiv:1612.08242 (2016)
22. Belongie, S., Bourdev, L., Girshick, R., Hays, J., Perona, P., Ramanan, D., Zitnick, C.L., Lin, T.-Y., Maire, M., Dollar, P.: Microsoft COCO: Common Objects in Context arXiv:1405.0312 (2014)
23. Masuzawa, H., Miura, J.: Observation planning for efficient environment information summarization. In: Proceedings of 2009 IEEE/RSJ International Conference on Intelligent Robots and Systems, pp. 5794–5800 (2009)

# Unsupervised Hump Detection for Mobile Robots Based On Kinematic Measurements and Deep-Learning Based Autoencoder

Oliver Rettig[1]([✉]), Silvan Müller[1], Marcus Strand[1], and Darko Katic[2]

[1] Department for Computer Science,
Baden-Wuerttemberg Cooperative State University, 76133 Karlsruhe, Germany
rettig@dhbw-karlsruhe.de
[2] ArtiMinds Robotics GmbH, 76139 Karlsruhe, Germany

**Abstract.** Small humps on the floor go beyond the detectable scope of laser scanners and are therefore not integrated into SLAM based maps of mobile robots. However, even such small irregularities can have a tremendous effect on the robot's stability and the path quality. As a basis to develop anomaly detection algorithms, example kinematics data is collected for an overrun of a cable channel and a bulb plate. A recurrent neuronal network (RNN), based on the autoencoder principle, could be trained successfully with this data. The described RNN architecture looks promising to be used for realtime anomaly detection and also to quantify path quality.

**Keywords:** Neural networks · Anomaly detection · Path planning
Kinematic measurement · Mobile robotics · Deep learning

## 1 Introduction

The navigation of mobile robots typically relies on maps of the robot's environment, created from laser scanner data by simultaneous localization and mapping (SLAM) techniques. Small humps on the floor, e.g. cable channels go beyond the detectable scope of laser scanners and are therefore not integrated into the maps. However, even such small irregularities can have a tremendous effect on the robot's stability and the path quality. In this paper we seek to integrate the detection of small anomalies into dynamic adaptation during the execution of a path and into path planning itself.

Commercial mobile platforms like the Mir-100 allow the definition of driving routes by defining manually a few target points in the map. Then, subsequent path planning is done automatically considering several boundary conditions, e.g. distances to walls. Such a map based path planning can be extended by dynamic path planning in order to adjust to temporary changes in the environment [1]. By driving around or stopping in front of unpredicted and potentially

© Springer Nature Switzerland AG 2019
M. Strand et al. (Eds.): IAS 2018, AISC 867, pp. 99–110, 2019.
https://doi.org/10.1007/978-3-030-01370-7_9

dynamic obstacles collisions can be avoided. If multiple paths connect the start with the target goal position, path selection is typically based on the path length.

Small humps like cable channels, doorsills, floor unevenness or other environmental anomalies usually cannot be seen by laser-scanners (because their heights are beneath the scanners scope). Therefore, they are typically neglected by path planning algorithms. Even though such anomalies might not impede the robot in moving forward, they reduce path quality. Induced vibrations can impact cargo or can reduce the storage life of the robot or its mechanical components. One solution to reduce such vibrations is to decelerate in front of obstacles. However, deceleration elongates the time the robot needs to pass through the route. Hence, it could be beneficial in terms of time efficiency to avoid paths with environmental anomalies.

In big factories several possible paths equal in length might only differ with regard to environmental anomalies. Thus especially in long-term applications, it can be crucial to take such differences into account for path selection.

## 2   Concept

The concept of this project is to focus on environmental anomalies, which cannot be detected by laser scanners and therefore are undocumented so far. Based on measurements of the robots kinematics, e.g. by inertial sensing, such anomalies can be detected and used with a twofold purpose: First, we want to apply anomaly detection for dynamic adaptions in order to reduce disturbances of the mobile platform e.g. vibrations. Second, based on anomaly detection we want to develop a measure assessing the quality of complete paths. This measure could be used to select a path when multiple paths are possible.

A big part of this project is the acquisition of kinematics data, exemplarily for the over runnable of a cable channel and a bulb plate. To have a good data basis for training and testing anomaly detection algorithms, the measurements are done with big effort in a motion lab with high accuracy.

Thereafter an appropriate anomaly detection algorithm is implemented, trained and tested.

## 3   Anomaly Detection and Deep Learning

Anomalies in a mathematical meaning are patterns in data that do not conform to a well defined notion of normal behaviour [2]. Already in the 19th century their detection has been studied in the statistics community [3]. "Over time, a variety of anomaly detection techniques have been developed in several research communities. Many of these techniques have been specifically developed for certain application domains, while others are more generic" [4]. In the context of this work we are only interested in anomaly detection in timeseries. Specifically, we focus on timeseries of kinematic data with the following requirements:

- Unsupervised (this excludes architectures with predefined classifiers) or at least Semi-supervised (training data without or with just a few anomalies).

- Usage of scoring techniques: Assignment of an anomaly score to each time point, depending on the degree to which this time point is considered as an anomaly. Then, integration of the score over a path can be used as a measure for the quality of the path.
- Working with one- and also with multidimensional data. A framework, which allows easy and flexible extension of the count of sensory input data channels.
- No or as little as possible expert knowledge about the data should be necessary. Thereby adjustment effort is reduced when the technique is applied to new data.
- Finding anomalous subsequences (not only single outliers).
- Working with streaming data instead of already collected data to be able to adapt dynamically for big anomalies, e.g. to slow down in front of an obstacle.

In robotics typcially high-dimensional sensory data with application specific configurations are in use. To make an anomaly detection component reusable without expensive adaptions from specialists, it is desirable to base on a flexible architecture and not to use much domain knowledge about the data. This and the need to work with streaming data to find anomalous subsequences instead only single outliers, quantifiable by a score, exclude many anomaly detection methods available in the literature.

On the other side, artificial neuronal networks in general have been used to solve a large range of problems in the field of robotics processing [5]. Particularly, deep-learning networks are identified as the leading breakthrough technique in the field of mobile robots [6]. They might be used to overcome important challenges in perception and control of mobile robots. For example in [7,8] a novelty detection in visual data to analyze the robot's environment is described.

There are several approaches to use deep-learning technologies for anomaly detection in timeseries. One way is to implement a classifier, based on slacking recurrent hidden layers building a recurrent neural network (RNN), with only two classes: normal and anomalous. This works fine, if sufficient labelled instances of both normal and anomalous instances are available. It is a supervised method, so it is inappropriate for our purpose. In practice it is impossible to collect a complete set of labelled trajectories for all possible anomalies.

Another way is to use the RNN to predict the next few time frames in the timeseries, based on the current and past values. The prediction error gives an indication of anomaly [9]. This technique is unsupervised. However, it only works properly if the normal timeseries are to some extent predictable.

A further class of unsupervised methods combines RNN with an encoder/decoder used as a reconstruction model, where some form of reconstruction error is used, as a score measure of anomaly. The so called autoencoders are trained to reconstruct the normal time-series and it is assumed that such a model would do badly to reconstruct anomalies, having not seen as often during training [9]. To decide, if a score value indicates an anomaly, a threshold have to be defined. The limiting factor of this approach is that this threshold value depends on the input data. It must be re-adjusted for any change in the structure of the input data.

A variant of the autoencoder architecture is the variational autoencoder (VAE) introduced in [10,11] and amongst others used for anomaly detection [12]. It determines a reconstruction probability instead a reconstruction error as a score, which is a more objective anomaly measure. Different input data should not result in the need of a re-adjustment of the threshold [13]. To take into account the temporal structure of timeseries in such an architecture, an additional LSTM [15] layer can be precided. Instead, the so called variational recurrent autoencoder (VRAE) architecture [14] takes into account the dynamic temporal behavior from the scratch.

## 4     Data Acquisition

### 4.1     Experimental Measurement Settings and Preparation

A commercially available mobile platform Mir-100 with differential drive in the middle and four carrying wheels around is used. A map of the room (Fig. 1) is created with the build-in algorithm (laser scanner based SLAM). The data needed for the creation of the map is acquired by driving the platform a few minutes through the room. The platform is thereby controlled by hand using the manufactures web application.

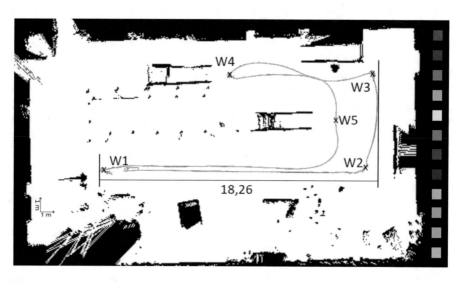

**Fig. 1.** The map of the room and the obstacle course defined by the points W1–W5. Shown are 12 repetitions of the course in different colors over each other.

An obstacle course is defined with five points (W1–W5) in the map. Two humps are placed between the points W1 and W2. Both humps cannot be detected by the build-in laser scanners and ultrasound sensors but can be overrun (Fig. 2). The first hump is a cable channel and the second a bulb plate (Fig. 2).

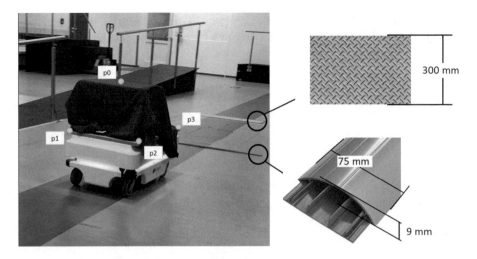

**Fig. 2.** Left: The Mir-100 driving between the points W1 and W2, one second before overrunning the cable channel. The spherical markers $\vec{p_0} - \vec{p_3}$ are tracked by an optical system. The red cover blinds installation from wooden material. Top right: A bulb plate. On the bottom right: A cable channel.

## 4.2 Measurement of Kinematics

The measurement of the robots kinematics are done in the Heidelberg Gait- and Motionlab, where a fix mounted 12-camera Vicon system (Vicon Motion Systems, Oxford, UK) for optical 3D motion capture is available. Three-dimensional positions of markers in sight of more than one camera simultaneously are calculated via triangulation. Based on four spherical markers ($d = 9$ mm diameter, $\vec{p_0} - \vec{p_3}$, Fig. 2), mounted on the mobile platform, which passively reflect infrared light emitted by strobes around the cameras, position and orientation of the mobile platform is tracked with a sample rate of 120 Hz. The accuracy of this system is around 1 mm. The installation on the mobile platform is covered to reduce the strong reflections in infrared light of untreated wooden materials (Fig. 2).

## 4.3 Preprocessing and Visualization of Kinematic Data

Preprocessing is done with the build-in filters of Vicon Nexus (Vicon Motion Systems, Oxford, UK). Markers are automatically labeled, small gaps are filled and the position timeseries of the markers are saved in c3d-files. The software LynxAnalyser (ORAT Software Entwicklung) is used for kinematic modelling and data visualization. The lab's (right-handed) coordinate system has its z-axis upward, the x-axis is aligned with the moving direction of the robot and the y-axis is orthogonal the moving direction (Fig. 2). The robots coordinate system is defined from the tracked marker positions by cross products as follows:

$$\vec{o}_{Mir}(t) = \begin{pmatrix} p_0^1 \\ p_1^1 \\ \frac{1}{3}(p_1^3 + p_2^3 + p_3^3) - \frac{d}{2} \end{pmatrix} \tag{1}$$

$$\vec{x} = \vec{p}_3 \tag{2}$$

$$\vec{z} = (\vec{p}_1 - \vec{p}_2) \times \vec{x} \tag{3}$$

$$\vec{y} = \vec{z} \times \vec{x} \tag{4}$$

$$\Sigma_{Mir}(t) = \{\vec{o}_{Mir}(t), \vec{x}(t), \vec{y}(t), \vec{z}(t)\} \tag{5}$$

Cardan angles (roll, pitch, yaw) are calculated with rotation order xyz to describe the orientation of the robot in the lab's coordinate system. Additionally a nick angle to describe the tilting of the robots z axis against the lab's z-axis is calculated by arcus cosine between the robots and the lab's z-axis. Second derivative of motion in z direction is determined based on quadratic polynomials and a moving window with a width of five.

## 5   Experiments

### 5.1   RNN Architecture

An autoencoder based anomaly detector was implemented based on Keras, a high-level neural network framework written in Python. Its architecture consists of a sequence of four network layers. The first three of type LSTM [15] with 64, 256 and 100 nodes and hyperbolic tangent as activation function, followed by a dense layer with 100 nodes and linear activation. For fitting the weights, mean squared error is chosen as loss function and RMSPROP, which keeps a moving average of the squared gradient for each weight, as optimizer. To detect anomalies the predicted data is compared with the input. The reconstruction error, determined as mean squared difference, is used as a measure of anomaly.

A nearly identical architecture is implemented for test purposes, based on DL4J, an industry-focused, commercially supported distributed deep-learning framework, which supports multiple CPUs and GPUs. In contrast to Keras, DL4J uses the build-in GravesLSTM [16] implementation. A special feature of DL4J is the support of streaming data. This allows the realization of a real-time anomaly detection architecture with a step-by-step calculation (with keeping track of the internal state of the RNN layers) needed for dynamic adaptation to anomalies.

## 5.2   Training and Anomaly Detection

The RNN was trained unsupervised, based on a training set of 25 example trials with about 15000 time frames each. Parts are cut, including one or none overrun of the cable channel or the bulb plate each. Over each trial a time window of width 100 frames is moved step by step and the resulting 100 * trial length sequences are mixed up to build up the training set. Data is preprocessed by subtracting the mean and dividing by the standard deviation. The hyper parameters of the RNN are adjusted, following roughly the recommendations of [17] and further hints from the documentation of the DL4J framework. Anomaly detection then derives a scalar score from the output of the model. A threshold value is defined to discriminate normal from anomalous data.

# 6   Results

## 6.1   Kinematic Measurements

While the robot moved through the course high repeatability ($<2$ cm) was found. Differences were visible only around the predefined points W1–W5 (Fig. 1), where the mobile platform had to realign. The results of the marker-based measurements of the humps produced by overrun of the cable channel are shown in Fig. 3. Only small deviations were visible in the transversal plane (Fig. 3: First row, first column). An unevenness of the ground was visible in a range of about 6 mm (Fig. 3: first diagram, second row) in the same order as the peaks during the overrun of the cable channel, visible between 70%–90% of the shown section. The first two smaller peaks are induced by the overrun of the front supporting wheels (1) and the driven wheel in the middle of the platform (2). The third highest peak (3) results from the rear wheels, which even take off the floor (visible in the captured videos). The spatial distances between the peaks correspond perfectly to 300 mm and 320 mm measured with a folding ruler manually.

The effect of bump is best be seen at the acceleration along the vertical axis (Fig. 3: first diagram in the last row). With the weight of the platform of 65 kg and maximum accelerations about 2 m/s$^2$, vertical forces can be estimated to about 130 N in maximum. The nick angle (Fig. 3: Second column, first row) and from the cardan sequence the roll- and pitch-angles show similar distinct the time position of the overrun of the cable channel.

## 6.2   Anomaly Detection

Training of the neuronal networks converges with and without anomalies included in the training data, with a mini batch size of 50 and a learning rate of 0.2%. Tests with mirroring the training- or the test-data show no differences. Based on the vertical position component only as input Fig. 4 shows input data (1), the data predicted by the RNN (2) and the anomaly score (3) for a test trial including an overrun of the cable channel. The horizontal line shows the

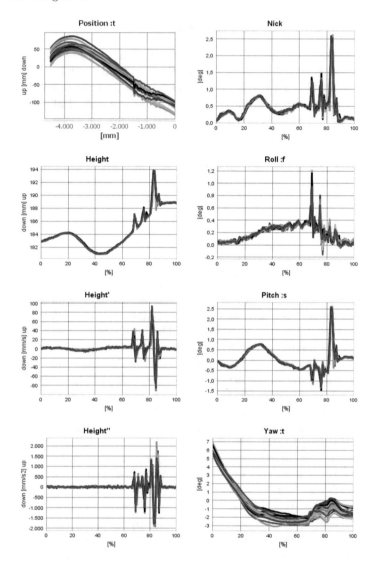

**Fig. 3.** Kinematics: 25 repetitions of the course; A section is shown, starting from the point W1 to a fix point after the first obstacle and before the second (Fig. 1). Data based on the markers ($\vec{p_0} - \vec{p_3}$); Position x,y in the lab system; Nick is the angle against upright z-axis; Roll, Pitch and Yaw are cardan angles; additional height and two derivatives are shown.

threshold defined a way that the overrun of any wheel is detected as an anomaly. The RNN was trained with one epoch only and the loss reached a value of 0.1311. With five epochs a loss of 0.0286 could be reached.

If the second derivative of the position is used as input, after one epoch of training a loss of only 0.8572 was reached. The results based on training with five epochs, reaching a loss of 0.3801 are shown in Fig. 5. The threshold value, shown as horizontal gray line is quasi unchanged.

For overrun of a bulb plate, the corresponding graphs based on vertical position component only are shown in Fig. 6. To detect the bulb plate only as an anomaly the threshold value (shown as a horizontal gray line) must be set to a value about 0.5, which is much higher than for detecting a cable channel.

Instead of one-dimensional input only, in last experiment all nine timeseries of the direction cosine matrix, describing the orientation of the mobile platform, are used as input of the RNN. After one epoch of training a loss of only 0.0606 was reached. The results based on training with five epochs, reaching a loss of 0.0388 are shown in Fig. 7 with (1) is the pitch angle, (2) is the pitch angle determined

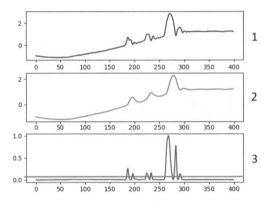

**Fig. 4.** Overrun of the cable channel, normalized values: (1) Vertical positions over ground; (2) Predicted signal; (3) Squared error – the anomaly measure. The horizontal gray line shows the threshold value.

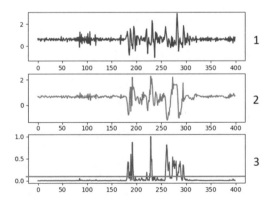

**Fig. 5.** Overrun of the cable channel, normalized values: (1) Vertical accelerations; (2) Predicted signal; (3) Squared error – the anomaly measure. The horizontal gray line shows the threshold value.

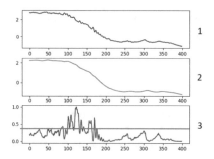

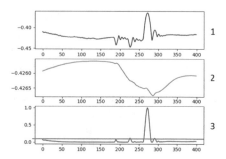

**Fig. 6.** Overrun of a bulb plate (between frame 67 and 210), normalized values: (1) Vertical positions over ground; (2) Predicted signal; (3) Squared error – the anomaly measure. The horizontal gray line shows the threshold value.

**Fig. 7.** Overrun of a bulb plate (between frame 67 and 210), normalized values: (1) Pitch angle determined from direction cosine matrix; (2) Predicted signal; (3) Squared error – the anomaly measure. The horizontal gray line shows the threshold value.

from the predicted nine timeseries and (3) is the normalized difference of the two. A clear detection of the robots takeoff was possible but no threshold value is found for detection of all three overruns.

## 7   Discussion

When an autonomous robot moves through an environment, even small anomalies, e.g. the only 9 mm high cable channel investigated in this project, can cause vibrations, finally leading to short takeoffs of the platform from the floor. That is why we think even over runnable obstacles should take ones attention and should be taken in account for a path quality score.

The takeoffs, shown in our concrete example, happens assumedly for the rear wheels as a product of the front and the drive wheels already past the cable channel and therefore pulling is more effectively. To avoid a damage of the platform or its cargo an online detection of the front wheels passing the cable channel could be used for dynamic adaptation: Whenever an anomaly is passed by the front wheels the platform should be slowed down before the rear wheels reach the cable channel.

We used the second derivative of the vertical position to simulate acceleration data, as it is typically available from inertial sensors. Training of the RNN with such data looks more difficult and even when the training is extended by more epochs it results in bigger loss values than with usage of the vertical position directly. This might be caused by the bigger complexity of the temporal structure of the signal and the increased noise (also for normal data).

Training the RNN with the nine dimensional data of the direction cosine matrix results in a score curve very similar to the one with the one dimensional data of the z position as input, if the pitch angle of a cardan decomposition is used as a score. So, the RNN looks robust against different count of input channels but domain knowledge was used to define a score. Furthermore the proportion of the peaks amplitudes (Fig. 7) has changed compared with Fig. 4, which makes the threshold definition difficult.

With the DL4J implementation we produced similar but not identical results (not shown). The small differences might be caused by implementation details: The GravesLSTM [16] layer is different from the one used in Keras and some further Keras specific details e.g. random_data_dup configuration, controlling the weights update by mini batch size or definition of a fix sequence length are not easy to implement identically based on the DL4J framework and have therefore been omitted.

# 8   Conclusion and Future Work

This work demonstrates the importance of real-time hump detection for dynamic adaptions and as a measure for quantifying path quality for path selection.

Deep learning based autoencoders allow a robust and easily expandable implementation of anomaly detection. We showed that autoencoders can be used for offline as well as online detection. The score determined by offline detection can easily be integrated over a path to determine an overall score characterizing path quality. The online detection based on DL4J enables dynamic adaptions.

One drawback of the implemented architecture is that the threshold value is not independent from the structure of the input data. Therefore the threshold has to be adjusted in order to detect other anomalies (here the bulb plate). To overcome this limitation future work should focus on VAE- or VRAE layers, which are promising RNN architectures. On the other side the adjustment of a threshold value is needed to define some kind of anomaly intensity, which is meaningful to the robot motion. This value should be independent of the terrain and should depends only on the robot and its charge.

**Acknowledgements.** The kinematic measurements were executed in the Gait- and Motionlab Heidelberg, University Clinics and were supported by the local laboratory team. The project is funded by the Federal Ministry for Economic Affairs and Energy of Germany.

# References

1. Meyer, J., Filliat, D.: Map-based navigation in mobile robots: II A review of map-learning and path-planning strategies. Cogn. Syst. Res. **4**, 283–317 (2003)
2. Hodge, V., Austin, J.: A survey of outlier detection methodologies. Artif. Intell. Rev. **22**(2), 85–126 (2004)
3. Edgeworth, F.Y.: On observations relating to several quantities. Hermathena **13**(6), 279–285 (1887)
4. Chandola, V., Banerjee, A., Kumar, V.: Anomaly detection: a survey. ACM Comput. Surv. (CSUR) **41**(3), 15 (2009)

5. Gamboa, J.C.B.: Deep Learning for Time-Series Analysis. arXiv preprint arXiv:1701.01887. (2017)
6. Tai, L., Liu, M.: Deep-learning in mobile robotics-from perception to control systems: A survey on why and why not. arXiv:1612.07139. (2016)
7. Neto, H.V., Nehmzow, U.: Real-time automated visual inspection using mobile robots. J. Intell. Rob. Syst. **49**(3), 293–307 (2007)
8. Sofman, B., Neuman, B., Stentz, A., Bagnell, J.A.: Anytime online novelty and change detection for mobile robots. J. Field Robot. **28**(4), 589–618 (2011)
9. Malhotra, P., Ramakrishnan, A., Anand, G., Vig, L., Agarwal, P., Shroff, G.: LSTM-based encoder-decoder for multi-sensor anomaly detection. arXiv preprint arXiv:1607.00148. (2016)
10. Kingma, D.P., Welling, M.: Auto-encoding variational bayes. arXiv preprint arXiv:1312.6114. (2013)
11. Rezende, D.J., Mohamed, S., Wierstra, D.: Stochastic backpropagation and approximate inference in deep generative models. arXiv preprint arXiv:1401.4082. (2014)
12. Sölch, M., Bayer, J., Ludersdorfer, M., van der Smagt, P.: Variational inference for on-line anomaly detection in high-dimensional timeseries. arXiv preprint arXiv:1602.07109. (2016)
13. An, J., Cho, S.: Variational autoencoder based anomaly detection using reconstruction probability. Technical report, SNU Data Mining Center (2015)
14. Fabius, O., van Amersfoort, J.R.: Variational recurrent auto-encoders. arXiv preprint arXiv:1412.6581 (2014)
15. Hochreiter, S., Schmidhuber, J.: Long short-term memory. Neural Comput. **9**(8), 1735–1780 (1997)
16. Graves, A.: Supervised Sequence Labelling With Recurrent Neural Networks. Studies in Computational Intelligence, vol. 385. Springer, Heidelberg (2012)
17. Bengio, Y.: Practical recommendations for gradient-based training of deep architectures. In: Neural Networks: Tricks of the Trade, pp. 437-478. Springer, Berlin, Heidelberg (2012)

# Efficient Coverage of Unstructured Environments

Elias Khsheibun[1], Norman Kohler[2], and Maren Bennewitz[1(✉)]

[1] Humanoid Robots Lab, University of Bonn, Bonn, Germany
`maren@cs.uni-bonn.de`
[2] Central Research and Development, Alfred Kärcher GmbH and Co. KG,
71364 Winnenden, Germany

**Abstract.** In this paper, we present a novel solution to mobile robot coverage of unstructured environments. We apply boustrophedon-based planning and use a heuristic to make the minimum sum of altitudes (MSA) decomposition, which computes an optimal exact cellular decomposition, applicable to more complex environments. Contrary to previous approaches, our technique explicitly takes into account different entry and exit points for the obtained cells and hence allows for minimizing inter-region distances in the corresponding traveling salesman problem (TSP) formulation. This is a highly important factor for unstructured environments, which heavily influences the quality of the final plan. Furthermore, we show how our method is applicable to coverage with finite resources. We implemented our planner in ROS and performed extensive experiments in a V-REP simulation environment in various scenarios. Comparisons with a state-of-the-art boustrophedon-based method show that our approach has a significantly lower total coverage time. Additionally, we demonstrate that our system is capable of performing online recharging and replanning in dynamic, crowded environments while obtaining a high coverage percentage. The results of this work are relevant for a variety of real-world applications such as autonomous floor cleaning.

## 1 Introduction

Coverage planning for planar environments has been frequently studied since it has various real-world applications that range from floor cleaning [13], lawn mowing [3], agricultural field operation [15] to surveillance, painting, and underwater mining. The efficiency of the task execution becomes a benchmark as soon as industrialized autonomous systems become available on economic markets as commercialized products. In the case of professional floor cleaning robots, real-world applications impose heavy constraints on resource capacities, thus requiring efficient solutions while the robots also need to be able to react to dynamic obstacles and temporary no-go areas. Furthermore, the planner has to comply to kinematic constraints as well as to task-specific motion strategies, such as the preference for so called "ox-plow" motions (i.e., going back and forth as can be

© Springer Nature Switzerland AG 2019
M. Strand et al. (Eds.): IAS 2018, AISC 867, pp. 111–126, 2019.
https://doi.org/10.1007/978-3-030-01370-7_10

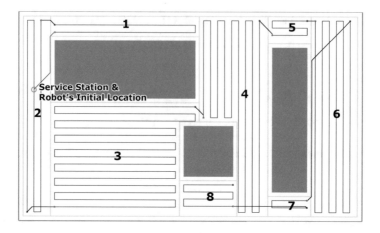

**Fig. 1.** Complete coverage plan including inter-region paths as generated by our algorithm (inter-region paths in black). Numbers indicate the order in which the regions from the MSA decomposition are processed according to our TSP solution. As can be inferred, inter-region distances play an important role and can seriously influence the total coverage time. Our aim is to generate a complete time-minimal coverage path. Note that even if the environment can be represented using only rectangular polygons in this example, our approach is generally applicable to environment representations consisting of arbitrary polygons.

seen in the path in Fig. 1). Such motions are known to be a good strategy for minimizing the number of turns [14], which are often time-consuming for larger robots.

In this paper, we propose an algorithm capable of both offline-planning and online-replanning for efficient coverage that overcomes limitations of state-of-the-art approaches as discussed in depth in the next section. Existing boustrophedon coverage techniques, which are based on ox-plow motion, try to find a near-optimal coverage path that visits all cells obtained from the decomposition. However, they commonly neglect inter-region paths, which are inherently also part of the final plan and play an important role as they can seriously influence the total coverage time.

Our boustrophedon path planning algorithm aims at generating a *complete* time-minimal coverage path. Given a grid map of the environment from which a polygonal representation is derived, our approach finds a segmentation using optimal exact cellular decomposition, calculates possible ox-plow paths for the obtained cells, and efficiently plans a TSP tour to solve the coverage problem. Hereby, the TSP tour is aware of all different entry and exit points for each considered cell. As an example, Fig. 1 shows a coverage plan for an unstructured environment. As can be seen, the regions are efficiently traversed in an order so as to minimize the overall coverage time.

Comparative experiments in a V-REP simulation environment demonstrate that the paths generated by our method have a significantly lower completion time than a state-of-the-art approach [23]. Additionally, we show that our framework can perform online recharging and replanning in dynamic scenes while still obtaining a high coverage percentage. Our ROS-based implementation can be easily integrated in service robots that are deployed in complex scenarios with finite-resource constraints to efficiently perform the coverage tasks.

## 2   Related Work

We first review general coverage planning algorithms and then present approaches that work with finite resources.

An extensive survey and background to the coverage path planning problem is found in the work of Choset [5] and Galceran et al. [11]. In the following, we address relevant works from the surveys, in addition to more recent results. Jimenez et al. [16] compute a trapezoidal decomposition and apply a genetic algorithm that generates a traversal order and constructs the path structure. Due to a non-optimal initial decomposition and a randomly initialized genetic algorithm, this approach might lead to a sub-optimal solution. Jin and Tang [17] use different types of robot motion turns to find the best coverage motion. However, this formulation does not provide the optimal positions for entering and exiting the regions nor for traveling between different regions. The work of Xu et al. [24] introduced an algorithm based on the Boustrophedon cellular decomposition [6]. The authors encode the cells as edges of a Reeb graph and use the Chinese Postman Problem (CPP) to calculate an Euler tour that accounts for the entry points of the robot. While this method provides a path that traverses each cell once, its construction phase ignores inter-cell distances. Choi [4] uses the MSA decomposition [14] and formulates the problem as a group traveling salesman problem, considering for each cell its entry and exit points. The author proposed to perform an exhaustive search using integer programming to provide the optimal solution. This method is only feasible for environments with few regions. The work of Pratama et al. [20] performs path planning to segment the map into an occupancy binary map. Afterwards the authors use a Morse decomposition and apply an efficient heuristic to find paths that minimize the number of turns in a plan. For getting the shortest coverage path, a depth-first search (DFS) is performed on the adjacency graph. This method achieves reasonable paths with a low number of turns. However, while DFS traversal is highly efficient to compute, it does not ensure an optimal solution.

Mei et al. [18] addressed the deployment of mobile robots with energy and time constraints. Their work provides a strategy for a group of robots for covering a given area but does not consider charging or receiving new resources as the problem is formulated without a service station. Aaron et al. [2] addressed the multi-robot, multi-depot dynamic coverage problem. The authors provided algorithms for an arbitrary fixed number of robots for a range of typologies, however, they do not consider full coverage of an arbitrary space. Their work rather concentrates on border coverage. Easton and Burdick [9] also presented a solution to the multi-robot, multi-depot coverage problem. They posed the coverage problem as the $k$-rural postman problem, provided a heuristic and used it to formulate the coverage routes. Although this approach incorporates a depot in the problem formulation, it is only used as a single dispatch and return point, and not for charging purposes to aid continuous planning. Shnaps and Rimon [22] addressed the coverage problem of a battery-constrained robot that uses a service station for recharging. Their work establishes a universal lower bound over all on-line battery powered coverage algorithms and provides a coverage algorithm that is based on circular motion paths. While the authors developed a close to optimal coverage algorithm, the circular movement used is disadvantageous for many practical robot tasks. Finally, Strimel and Veloso [23] presented an approximate solution to the coverage problem with finite resources. After decomposing the layout using a boustrophedon approach, the robot's service station is incorporated in a weighted full graph that represents the obtained cells. To find a reasonable tour, the authors apply an approximation of a reduction to the distance constrained vehicle routing problem [1]. The result is a list of sub-routes with the service station as start and end point. By following these routes the robot covers the entire environment. The proposed algorithm provides an elegant solution to the coverage problem but is based on a non-optimal decomposition. In contrast to our approach, their work does not consider different entry and exit points of cells and assumes that the cells are entered from a predetermined corner which is reasonable for office-like environments but not for unstructured environments where cells can be entered from different locations using multiple directions.

## 3    Problem Formulation

We consider the following problem: A mobile service robot is required to cover a known, unstructured environment with static and dynamic obstacles (e.g., moving people). The static obstacles in the environment are represented by a grid map from which a polygonal representation is created. The robot is equipped with finite resources (e.g., battery) that might not suffice to complete full coverage without being recharged. The environment contains a service station where the robot can renew its resources. We assume that the robot can continuously determine its pose and detect local obstacles. Given that the robot can travel from its starting point (which is typically a service station) to the farthest point in the environment and back to recharge, we aim at finding an efficient plan that minimizes the time needed to cover the whole environment. To do so, we consider as move cost between two locations the shortest distance computed from the grid map. The travel cost of a complete path is the sum of the distances of subsequent grid cells on this path and for recharging we use constant cost.

# 4   Our Approach to Coverage Planning

In this section, we first describe the applied technique to decompose the environment into disjoint segments, which can be efficiently covered with so-called boustrophedon (or ox-plow) motions minimizing the number of time-consuming turns. Afterwards, we introduce our approach to solve the coverage planning problem using a traveling salesman problem formulation.

## 4.1   Decomposition of the Environment

Given the polygonal representation derived from the grid map of the environment, we segment it using the MSA decomposition [14] and obtain a set of disjoint cells $\{c_1, \ldots, c_n\}$. Each corresponding region can be efficiently covered using a simple ox-plow (back-and-forth) motion. The MSA decomposition provides an optimal cellular decomposition that minimizes the sum of turns within the sub regions. The resulting decomposition consists of monotone polygons.[1] To efficiently find a good MSA decomposition for a complex environment, we implemented a heuristic. This heuristic is based on partitioning the cellular decomposition adjacency graph into sub-graphs and applying the MSA method for each. The sub-graphs are disjoint and their union is the original adjacency graph. Based on available computation resources, the maximum size of a sub-graph can be chosen. After applying the MSA on each sub-graph, we iteratively check whether adjacent regions can be merged, i.e., whether the merged polygons are still monotone wrt. the sweeping direction.

The problem we consider in the following is how to determine the best traversal order for the complete environment with the obtained regions. Here, we need to consider (1) the best order to process the cells and (2) the entry and exit point positions for all cells so that the total traveling distance between the cells is minimized. Given $n$ cells, the search space corresponding to (1) is $n!$. This, however, still does not take into account where to enter and exit the cells. If each cell permits two different start points to perform the boustrophedon motion, the size of the search space is increased to $2^n \cdot n!$.

## 4.2   Possible Boustrophedon Paths

The regions resulting from the MSA decomposition consist of monotone polygons with the associated optimal sweeping direction and for each cell we have two different starting corners and two different ending corners. In this work, we consider two possible optimal boustrophedon motions for each region, (1) a motion that starts at the bottom left corner and (2) a motion that starts at the upper left corner of the region (see Fig. 2 for an example in which the horizontal sweep direction is the optimal one).

---

[1] In geometry, a polygon P in the plane is called monotone with respect to a straight line L, if every line orthogonal to L intersects P at most twice.

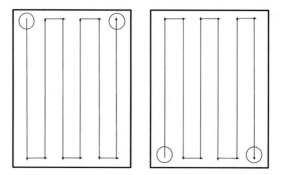

**Fig. 2.** Illustrative example. A simple rectangular cell with its two different optimal ox-plow motions given that the cell can be entered at the bottom/top left/right corner. As can be seen, the optimal sweep direction is horizontal since it minimizes the number of necessary turns. Note the different start and end locations for each path.

An ox-plow motion can then be represented as a list of $(x, y)$ points, that all reside in the cell. For a cell $c_i$ we define $Boust_i^1$ and $Boust_i^2$ for the above two paths respectively:

$$Boust_i^j = \left\langle \left(x_1^{i_j}, y_1^{i_j}\right), \ldots, \left(x_{n_i^j}^{i_j}, y_{n_i^j}^{i_j}\right)\right\rangle, j \in \{1, 2\}, i \in \{1, \ldots, n\} \qquad (1)$$

Here, $Boust_i$ contains $n_i$ grid map coordinates that represent the boustrophedon path.

A boustrophedon path is therefore guaranteed to pass over the entire region: if it starts at a *start* corner it will finish at an *end* corner. Since the notion of start and end is symmetric (i.e., a start point might be considered an end point and vice versa), we plan the motion for each cell for its two possible start corners and, thus, get two boustrophedon paths that correspond to its optimal sweep line. For cell $i$, let these points be denoted as $start_i^1, end_i^1, start_i^2, end_i^2$ corresponding to the start and end point for the bottom left and the top left corner of the cell, accordingly.

## 4.3   Formulation as Traveling Salesman Problem

To determine the best traversal order of the environment with the obtained cells, we formulate the problem as a graph traversal problem analogous to the traveling salesman problem (TSP). An exact solution of the TSP provides the optimal coverage tour for our problem. Strimel and Veloso [23] have already considered the reduction of the coverage tour problem as a TSP problem. However, in their proposed solution, the entry and exit points of the cell were neglected, even though they might form a major factor in the overall tour. Thus, we extend the graph formulation to account for the different entry and exit points and corresponding costs and aim at generating a TSP tour that computes the best traversal order while taking into account the possible entry and exit points of the cells.

We construct an undirected full graph $G = (V, E)$ between all cells including a special cell accounting for the service station $v_0^{station}$. We represent each cell $c_i, 0 < i \leq n$ as three vertices: $v_i^{start}, v_i^{middle}, v_i^{end}$ which account for the entry and exit points. Here, $v^{middle}$ plays a special role as it enforces the TSP solver to treat each cell atomically, i.e., once a route traverses through a start/end vertex, it has to go along the other two vertices, meaning that it traverses the complete cell before addressing other cells. A complete 2-Opt solver will compare every possible swapping of edges in this TSP formulation. Since between every two cells, there exists a route with a finite distance (by the way the graph is built), a solution is hence guaranteed. The complete graph formulation is as follows:

$$V = \left\{ v_0^{station}, v_1^{start}, v_1^{middle}, v_1^{end}, \ldots, v_n^{start}, v_n^{middle}, v_n^{end} \right\}$$
$$E = \left\{ \left(v_i^a, v_j^b\right) | v_i^a, v_j^b \in V, v_i^a \neq v_j^b \right\}$$
$$w(v_i^a, v_j^b) = \begin{cases} 0, & \text{if } i = j \text{ and} \\ & (a = \text{middle or } b = \text{middle}) \\ dist(v_i^a, v_j^b), & \text{if } i \neq j \text{ and} \\ & a \neq \text{middle and } b \neq \text{middle} \\ \infty & \text{otherwise} \end{cases} \tag{2}$$

where $dist(v_i^a, v_j^b)$ returns the shortest distance between the two points. In our implementation we used $A^*$ [12] to compute paths between the entry/exit points on a grid map. Recall that every cell can have up to two possible start points (with their corresponding end points). The above graph represents only one variation of a possible plan. To consider all possibilities, we need to compare the costs of all possible $2^n$ variations. We implemented an efficient binary representation to encode the multiple graphs. The encoding of the graph is a binary vector which consist of $n$ bits corresponding to $n$ cells, where each bit determines the corresponding cell's ox-plow coverage path. Depending on the value of each bit, a corresponding ox-plow motion will be chosen and thus different inter-cell distances for the whole graph will be calculated, while also considering the cost of the corresponding boustrophedon paths inside the cells. For each graph the TSP is solved and its travel costs are saved. All inter-cell distances are hashed to avoid extra calculations. In real-world environments, not all cells will have two entry/exit points (depending on the geometry of the monotone polygons) and the number of vector permutations will be low. Hence, the number of TSP problems that have to be considered is much smaller than $2^n$.

## 4.4    Solving the Traveling Salesman Problem

In general, assigning weights of $\infty$ and 0 in a TSP graph is not common. We use this technique to force the TSP solver once it "enters" a cell to remain in it. Our weight assignment in (2) breaks the triangular inequality and therefore our graph does not qualify for a metric space TSP solver. To address this, we use a TSP heuristic that still works outside the metric space. In our experiments, the 2-opt heuristic [10] always provided the best result in comparison to other heuristics. Our system computes the TSP heuristic for all the possible variations (up to $2^n$) and chooses the traversal tour with the lowest cost for execution.

# 5    Considering Capacity Constraints

The tour resulting from the solution of the TSP starts and ends at the station node. In this section, we now tackle the problem of finite resources, which means that we need to deal with cases in which the robot cannot traverse the complete tour without recharging. One solution is to apply an approximation to a reduction to the distance constrained vehicle routing problem (DCVRP) as proposed by Strimel and Veloso [23]. The idea here is to partition the TSP tour into sub-paths that start and end at the service station. While this method provides a solution to the imposed capacity constraint, it has two major downsides: (1) It does not account for live consumption of resources but rather computes an offline estimate and (2) the granularity of the sub-routes are in complete cells and cannot be broken into paths within cells.

To overcome these issues, we developed a method that applies replanning on the fly if needed and, thus, can react dynamically according to the availability of resources. Additionally, we allow for interrupting a path at any time within a cell. Thus, our method is independent from the dimensions of the cells. With our approach, the robot utilizes the on-board resources to the maximum. As a result, it tries to minimize the number of time-consuming round-trips needed to travel to the service station and back to the next uncovered region. In practice, this often provides significant performance gains as the service station is typically positioned in a corner of the environment.

During the execution of the coverage plan, we maintain for each cell its updated ox-plow path and the corresponding entry point. When a visit to the service station is needed, our system invokes replanning, i.e., it computes the best tour that takes into account the updated graph of the current status of the environment. For the current cell in which the robot operates, we continuously update the remainder of the path, which can be done in $\mathcal{O}(1)$ time since we only have to update the first coordinate of the cell's boustrophedon path $Boust$ (see Eq. 1). When a cell is completely covered, its corresponding vertices and the edges are removed from $G$.

# 6   Path Execution and Replanning

To execute the computed plan and deal with moving obstacles, we implemented a classical two-layered path planning architecture that consists of a planning and a collision avoidance layer. The planning layer provides the global coverage path according to the method described in Sect. 4. The collision avoidance layer then decides based on sensor data whether to execute the corresponding motion commands or not. To react to dynamic obstacles such as single humans or crowds of people on the coverage path, the robot stops, waits, and performs replanning if needed. Hereby, the robot uses a predefined threshold, that defines the maximum waiting time before recomputing the coverage plan.

Thus, the planning layer is called in two cases: (1) Resources are scarce, and hence a path to the service station and a new coverage plan for the remaining area is requested (see previous section) and (2) the collision-avoidance layer detects obstacles on the path and cannot proceed with the current plan. In both cases, a new coverage plan is computed for the remaining area.

# 7   Experimental Results

For testing our approach, we implemented our system in ROS [21] and simulated the execution of the coverage plan in V-REP [8]. For the implementation of the TSP heuristic, we used the LEMON library [7].

## 7.1   Comparative Experiments

**Alternative Planning Methods.** For a benchmark comparison, we implemented *BC Sweep*, the planner proposed by Strimel and Veloso [23] which can be considered as state of the art for coverage with boustrophedon paths. *BC Sweep* does not consider the costs of traversing inter-cell paths and segments the environment into rectangular cells. Nevertheless, *BC Sweep* can be used to compute a coverage plan of office-like environments with many rooms, such as the environment discussed in the work itself [23]. Note that most of the 53 rectangular cells that were generated in the experiments of this work, represent closed rectangular spaces with a single entry point (which is also the exit). For environments consisting of such self-contained cells, the coverage path of each cell can be calculated separately and then integrated in the entire plan as there are no different possible entry points that need to be considered. In our experiments, we focus on environments without such a clear room structure (e.g., airport halls or large hallways), where different entry/exit points of the cells highly influence the efficiency of the coverage path.

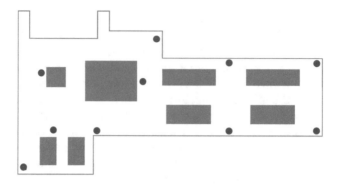

**Fig. 3.** Service station locations for experiments in the airport environment.

Standard *BC Sweep* does not consider the structure of the cells, uses only a horizontal sweep motion, and does not calculate the inter-cell paths to take into account the corresponding costs. Therefore, to have a fair comparison, we implemented two additional strategies based on *BC Sweep*:

- *BC Sweep* horizontal: based on the method proposed by Strimel and Veloso [23] with the Christofides TSP solver [19] using only horizontal motion.
- *BC Sweep* vertical: using only vertical motion.
- *BC Sweep* hybrid: using the best motion for each cell.

For all methods, we calculate the cost of the inter-cell paths in the TSP and choose the lower left corners of the cells resulting from our segmentation as the entry points. Additionally, to account for capacity constraints, we apply the technique presented in Sect. 5. Using this experimental design, our aim is to show the strength of our approach that considers different entry and exit points and the corresponding path costs in the TSP formulation. We performed the comparative evaluation in three different environments (see Figs. 3, 4 and Fig. 1). Note that even if in these test examples, the environments can be represented using only rectangular polygons, our approach is generally applicable to environment representations consisting of arbitrary polygons.

**Evaluation Criteria.** For comparison, we evaluate the total time the robot needs to cover the complete environment, which counts the time from the initial movement of the robot until the end of the coverage plan. Note that this also includes the time needed to recompute the path if necessary in case of recharging. We used different time constraints $\lambda$ on the capacity in the experimental evaluation. The initial planning time is excluded since the coverage path can

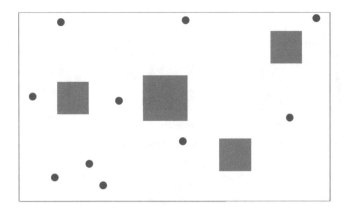

**Fig. 4.** Service station locations for the experiments (Environment 2).

be computed offline (for a complex environment as the one in Fig. 3, it took about five minutes to calculate). We neglected the time needed for acquiring new resource from the service station, since this is independent of the quality of a planner.

**Comparative Results.** We evaluated our method in comparison to the three *BC Sweep* methods described in Sect. 7.1 in three different environments. For each environment, we tested ten different starting positions which reflect the position of the service station and simulated for each environment/method, three different capacities $\lambda$. To be fair, we give all *BC Sweep* methods the possibility of issuing a recharge request at *any point*, and not only at the granularity of complete cells [23], i.e., at the end point of a cell's ox-plow path. The ten different locations for the service station were chosen randomly and are depicted in Figs. 3 and 4 for two environments.

Figures 5a–c show the obtained result of the different planning algorithms. As can be seen, our planner lowered the coverage time by at least 10% and significantly outperforms the *BC Sweep* planners as proven by a paired t-test (99.999%). Note that in many cases, our method has a lower total plan variance, which makes it more reliable and independent from the service station location. As a qualitative example, Fig. 8 shows a complete coverage plan and the resulting robot path after recharging and replanning.

## 7.2 Performance in Dynamic Scenarios

Finally, we tested our approach in a significantly more complex, dynamic environment corresponding to an airport scenario (Fig. 6). To achieve a close to real-world example, we simulated walking people, three different static queues,

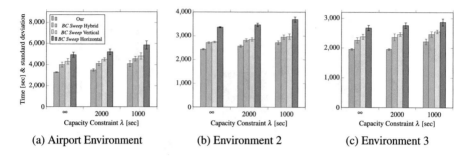

(a) Airport Environment     (b) Environment 2     (c) Environment 3

**Fig. 5.** Comparison of coverage time (including the time for replanning) for different capacity constraints in the three unstructured environments (see Figs. 3, 4 and Fig. 1). The results show the mean and standard deviation of the complete coverage time for 10 different robot starting positions. Note that this also includes the time needed to recompute the path if necessary in case of recharging. The initial planning time is excluded since the initial coverage path can be computed offline.

and two dynamic queues (which means that the corresponding people start moving at $t = 2000$ s) as can be seen in the figure. The behavior of a moving person is a repetition of the following: choose a random walking direction and follow it until a collision occurs.

Figures 7a–c show results for different capacity constraints and for different numbers of moving people that were added at the beginning of the simulation at random points in addition to the queues. The robot's starting pose was at the service station in the upper left in Fig. 3. The figure shows the mean values

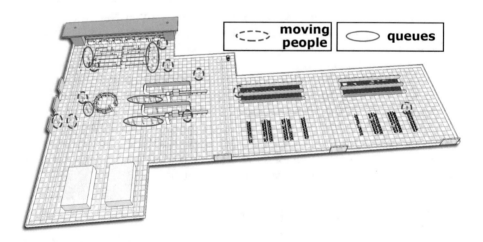

**Fig. 6.** Airport model in V-REP with simulated dynamic obstacles (moving people) that were added randomly at time $t = 0$. Additionally, there are three static queues of people and two dynamic queues (those are in the upper part of the image), in which people will start moving randomly at $t = 2000$ s.

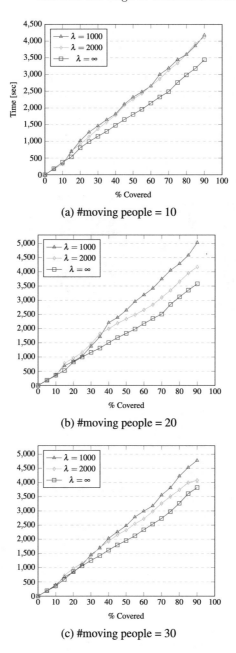

(a) #moving people = 10

(b) #moving people = 20

(c) #moving people = 30

**Fig. 7.** Coverage progress for the dynamic airport environment (Fig. 6): Each graph shows the percentage of the covered area with a number of randomly moving people in addition to two static and three dynamic queues. $\lambda$ refers to the capacity constraint, i.e., seconds until the next recharge. In order to account for the inherent randomness of the environment, we give the mean of five simulation runs. As can be seen, with our approach the robot is able to perform online recharging and replanning while achieving a high coverage percentage despite a high number of dynamic obstacles.

of five simulation runs to account for the randomness of the people's motion. In addition to the execution time, we evaluate the percentage of the environment covered by the robot. The results clearly demonstrate that our planner is able to perform online recharging and replanning while also obtaining a high coverage percentage despite the permanent presence of dynamic obstacles.

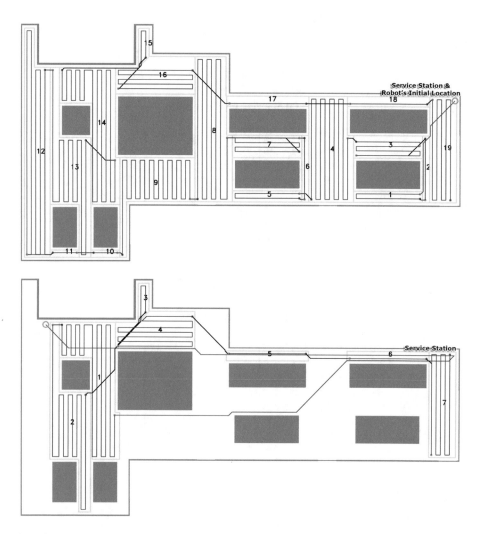

**Fig. 8.** Top: initial coverage plan. After covering Region 12, the robot had to recharge. Bottom: recomputed remainder of the plan after a recharge request at $\lambda = 2000$ s. The figure can be best viewed in color. Boustrophedon paths within the cells are drawn in blue, inter-region paths in black, and the path from the start to the first region as well as the path to the service station in red.

# 8   Conclusions

In this paper, we presented a novel technique for mobile robot coverage of unstructured environments that can deal with finite resources and dynamic obstacles. Our approach applies optimal exact cellular decomposition and formulates the coverage problem as a TSP on the obtained cells. In the TSP, we explicitly consider different inter-cell distances resulting from the possible ox-plow paths within the cells. Thus, the TSP tour is aware of different entry and exit points for the cells, which is especially important for environments without a clear room structure.

We thoroughly evaluated our coverage planner in environments of different complexity and with varying capacity constraints. In comparative experiments, we demonstrated that our algorithm provides significantly shorter paths than a state-of-the-art boustrophedon-based coverage method. Additionally, we showed that using our approach the robot can deal with dynamic obstacles and perform online replanning and recharging if necessary.

Another possibility would be to include multiple charging points and to consider recharging at different points in time and not only when resources are ceased. Such strategies might result in a shorter coverage time.

As future work, we will also evaluate how the performance of our heuristic on the MSA decomposition scales to even larger environments in comparison to applying the MSA on the complete graph which, however, significantly increases the run time.

# References

1. On the distance constrained vehicle routing problem. Oper. Res. **40**(4) (1992)
2. Aaron, E., Krizanc, D., Meyerson, E.: Multi-Robot Foremost Coverage of Time-Varying Graphs. Springer, Heidelberg (2015)
3. Bretl, T., Hutchinson, S.: Robust coverage by a mobile robot of a planar workspace. In: Proceedings of IEEE International Conference on Robotics and Automation (2013)
4. Choi, M.H.: Optimal underwater coverage of a cellular region by autonomous underwater vehicle using line sweep motion. J. Electr. Eng. Technol. **7**(6), 1023–1033 (2012)
5. Choset, H.: Coverage for robotics - a survey of recent results. Ann. Math. Artif. Intell. **31**(1), 113–126 (2001)
6. Choset, H., Pignon, P.: Coverage path planning: the boustrophedon decomposition. In: International Conference on Field and Service Robotics (1997)
7. Dezs, B., Jüttner, A., Kovács, P.: LEMON - an open source C++ graph template library. Electron. Notes Theor. Comput. Sci. **264**(5), 23–45 (2011)
8. Rohmer, E., Singh, S.P.N., Freese, M.: V-REP: a versatile and scalable robot simulation framework. In: Proceedings of The International Conference on Intelligent Robots and Systems (IROS) (2013)
9. Easton, K., Burdick, J.: A coverage algorithm for multi-robot boundary inspection. In: Proceedings of the IEEE International Conference on Robotics and Automation (2005)

10. Englert, M., Röglin, H., Vöcking, B.: Worst case and probabilistic analysis of the 2-Opt algorithm for the TSP. Algorithmica **68**(1), 190–264 (2014)
11. Galceran, E., Carreras, M.: A survey on coverage path planning for robotics. Robot. Auton. Syst. **61**(12), 1258–1276 (2013). https://doi.org/10.1016/j.robot.2013.09.004
12. Hart, P.E., Nilsson, N.J., Raphael, B.: A formal basis for the heuristic determination of minimum cost paths. IEEE Trans. Syst. Sci. Cybern. **4**(2), 100–107 (1968)
13. Hofner, C., Schmidt, G.: Path planning and guidance techniques for an autonomous mobile cleaning robot. In: Proceedings of IEEE/RSJ International Conference on Intelligent Robots and Systems (1994)
14. Huang, W.H.: Optimal line-sweep-based decompositions for coverage algorithms. In: Proceedings of IEEE International Conference on Robotics and Automation (2001)
15. Jensen, M.A.F.: Algorithms for operational planning of agricultural field operations. Technical Reports Mechanical Engineering, vol. 2, no. 3, p. 19 (2015)
16. Jimenez, P.A., Shirinzadeh, B., Nicholson, A., Alici, G.: Optimal area covering using genetic algorithms. In: Proceedings of IEEE/ASME International Conference on Advanced Intelligent Mechatronics (2007)
17. Jin, J., Tang, L.: Optimal coverage path planning for arable farming on 2D surfaces. Trans. Am. Soc. Agr. Biol. Eng. **53**(1), 283 (2010)
18. Mei, Y., Lu, Y.H., Hu, Y.C., Lee, C.S.G.: Deployment of mobile robots with energy and timing constraints. IEEE Trans. Robot. **22**(3), 507–522 (2006)
19. Papadimitriou, C.H.: The Euclidean traveling salesman problem is NP-complete. Theor. Comput. Sci. **4**(3), 237–244 (1977)
20. Pratama, P.S., Kim, J.W., Kim, H.K., Yoon, S.M., Yeu, T.K., Hong, S., Oh, S.H., KIm, S.B.: Path planning algorithm to minimize an overlapped path and turning number for an underwater mining robot. In: International Conference on Control, Automation and Systems (2015)
21. Quigley, M., Conley, K., Gerkey, B.P., Faust, J., Foote, T., Leibs, J., Wheeler, R., Ng, A.Y.: ROS: an open-source robot operating system. In: ICRA Workshop on Open Source Software (2009)
22. Shnaps, I., Rimon, E.: Online coverage of planar environments by a battery powered autonomous mobile robot. IEEE Trans. Autom. Sci. Eng. **13**, 32–42 (2016)
23. Strimel, G.P., Veloso, M.M.: Coverage planning with finite resources. In: IEEE/RSJ International Conference on Intelligent Robots and Systems (2014)
24. Xu, A., Viriyasuthee, C., Rekleitis, I.: Efficient complete coverage of a known arbitrary environment with applications to aerial operations. Autonom. Robots **36**, 365–381 (2013)

# Intelligent Systems

# Probabilistic Logic for Intelligent Systems

Thomas C. Henderson[1]([✉]), Robert Simmons[1], Bernard Serbinowski[1],
Xiuyi Fan[2], Amar Mitiche[3], and Michael Cline[4]

[1] University of Utah, Salt Lake City, USA
tch@cs.utah.edu
[2] University of Swansea, Swansea, Wales
[3] University of Montreal, Montreal, Canada
[4] Simon Fraser Unviersity, Burnaby, Canada

**Abstract.** Given a knowledge base in Conjunctive Normal Form for use
by an intelligent agent, with probabilities assigned to the conjuncts, the
probability of any new query sentence can be determined by solving the
Probabilistic Satisfiability Problem (PSAT). This involves finding a consistent probability distribution over the atoms (if they are independent)
or complete conjunction set of the atoms. We show how this problem can
be expressed and solved as a set of nonlinear equations derived from the
knowledge base sentences and standard probability of logical sentences.
Evidence is given that numerical gradient descent algorithms can be used
more effectively then other current methods to find PSAT solutions.

**Keywords:** PSAT · Probabilistic knowledge base · Nonlinear systems

## 1 Introduction

Given a logical knowledge base of $m$ sentences over $n$ atoms in Conjunctive Normal Form (CNF), and a probability associated with each sentence (conjunction),
the probabilistic satisfiability problem is to determine whether or not a probability distribution exists which assigns consistent probabilities to the complete
conjunction basis set of the $n$ atoms in the sentences. This means that for each
basis element, $\omega_k, k = 0 \ldots 2^n - 1$, $0 \leq \omega_k \leq 1$ and $\sum_k \omega_k = 1$, and, in addition,
the sentence probabilities follow from the complete conjunction probabilities.
Solutions for this problem were proposed by Boole (1857) and again by Nilsson
(1986); they set up a linear system relating the sentence probabilities to the
models of the sentences. The major drawback is that this system is exponential in the number of sentences. Georgakopoulos (1988) showed this problem is
NP-complete and exponential in $n$.

We have proposed that the individual conjuncts in the CNF be converted into
a system of nonlinear equations, that is, since each conjunct is a disjunction, it
can be re-written in terms of the probabilities of the atoms and conditional

This research supported in part by Dynamic Data Driven Application Systems AFOSR
grant FA9550-17-1-0077.

M. Strand et al. (Eds.): IAS 2018, AISC 867, pp. 129–141, 2019.
https://doi.org/10.1007/978-3-030-01370-7_11

probabilities of a smaller number of terms Henderson (2017a) and Henderson (2017b). [Note: although this expression itself is exponential in the number of literals in the disjunction, the CNF can always be converted to 3CNF form in polynomial time, and thus, limit the complexity to O(8).] Let the KB have $m$ sentences, and let the $k^{th}$ disjunction, $L_1 \vee R$, have probability, $p$, where $L_1$ is the first literal in the disjunction, and $R$ represents the remaining literals in the sentence, then the corresponding equation is:

$$F(k) = -p + P(L_1) + P(R) - P(L_1 \mid R)P(R)$$

If $R$ has more than one literal, then any $P(R)$ term will be recursively defined until there are only single literal probabilities, whereas the conditional probabilities will be left as is. Once these equations are developed, each unique probability expression will be replaced with an independent variable. At this point, there is a set of $m$ equations with $n + c$ unknowns, where $c$ is the number of unique conditional probabilities. A scalar error function can then be defined as the norm of the vector $F$ defined above. It is now possible to apply nonlinear solvers to this problem and determine variable assignments that are constrained to be in $[0, 1]$, and which result in the assigned sentence probabilities.

We describe here detailed results of experiments comparing the use of (1) Newton's method, and (2) gradient descent using the Jacobian of the error function. We show that excellent results can be achieved using these methods, and provide a statistical framework for medium-sized KBs (this is to allow the computation of the complete conjunction set probabilities in order to know ground truth), as well some results on a large KB (e.g., 80 variables and over 400 sentences). Although the answer found by the method described here is not guaranteed to be a PSAT solution, it does provide an approximate solution, and in some cases can be determined not to be a solution.

## 2    Related Work

Stated more technically, consider a logical knowledge base expressed in CNF with $m$ conjuncts over $n$ logical variables: $CNF \equiv C_1 \wedge C_2 \wedge \ldots \wedge C_m$. Each conjunct is a disjunction of literals: $C_k = L_{k,1} \vee L_{k,2} \vee \ldots \vee L_{k,n_k}$ and has an associated probability, $p_k$. Let $\bar{p}$ be the $m$-vector of sentence probabilities. Define the complete conjunction set (all combinations of atom truth value assignments): $\omega_k = L_1 \wedge L_2 \wedge \ldots \wedge L_n$, $k = 0 : 2^n - 1$, where the literal assignments correspond to the bit assignments in the $n$-digit binary number representing $k$; $\Omega$ is the set of all $\omega_k$. The *Probabilistic Satisfiability Problem* is to determine if there exists a probability distribution $\pi : \Omega \to [0, 1]$ such that $0 \leq \pi(\omega_k) \leq 1$ and $\sum_k P(\omega_k) = 1$, and

$$p_k = \sum_{\omega_j \vdash C_k} P(\omega_k)$$

We also let $\bar{\pi}$ represent the vector of probability assignments to the complete conjunctions.

Nilsson described a solution to this problem by creating a linear system based on the CNF sentences and a query sentence (Boole 1857 first proposed a very similar method). The semantic tree (see Kowalski 1983) over the sentence truth assignments is determined and the valid sentence truth assignments are used to form an $m \times 2^m - 1$ matrix, A. I.e., $A(i,j) = 1$ if $w_j$ satisfies $C_i$. Then he solves $\bar{p} = A\bar{\pi}$ for $\bar{\pi}$. For more detailed discussion of this approach, including the geometric approach, see Henderson (2017a). The problem with this method is the exponential size of the matrix $A$. For broader discussions of this problem and approach (see Adams 1998, Hailperin 1996, and Hunter 2013).

Others have explored inconsistencies in probabilistic knowledge bases (see Thimm 2009 and Thimm 2012). Another important approach is Markov Logic Networks (see Biba 2009, Domingos 2009 and Gogate 2016). All these methods suffer from exponential cost (e.g., Markov Logic Networks in the number of maximal cliques), or resort to sampling methods to estimate the solution while not guaranteeing the result solves PSAT.

## 3   Method

The method proposed here, called *Nonlinear Probabilistic Logic (NLPL)* involves conversion of the probabilistic CNF to a set of equations and the use of numerical solvers. First, consider the case when the variables are independent, i.e., $P(A \wedge B) = P(A)P(B)$. For this, the basic formula is:

$$P(A \vee B) = P(A) + P(B) - P(A)P(B)$$

where this is applied recursively if $B$ is not a single literal. Consider Nilsson's example problem (modus ponens):

$$[p_1 = 0.7]C_1 : P$$

$$[p_2 = 0.7]C_2 : \neg P \vee Q$$

with query: $P(Q)$?. This gives rise to the equations:

$$F(1) = -0.7 + P(P)$$

$$F(2) = -0.7 + P(\neg P) + P(Q) - P(\neg P)P(Q)$$

This requires no search (is not exponential) and can be solved:

$$P(P) = 0.7$$

$$P(Q) = \frac{0.7 - 0.3}{0.7} = 0.571$$

$P(Q)$ is in the solution range $[0.4, 0.7]$ and we provide an answer in this interval (note that MLNs produce a single answer as well: the maximal entropy solution).

A first question is: Does *NLPL* solve PSAT? Suppose that *NLPL* finds a probability assignment for the atoms that results in the sentence probabilities (using the nonlinear equations), where each atom probability is in the interval $[0, 1]$; i.e., the atom probability vector is in the unit hypercube in $n$-space. We need to show that the complete conjunction probabilities generated satisfy the properties described above.

**Theorem:** Given a set of atom probabilities, $a_k$, $k = 1 : n$, with $0 \leq a_k \leq 1$, then each complete conjunction probability is in the range $[0, 1]$ and the $\sum_k \omega_k = 1$.

**Proof:** First, note that since the variables are independent, the complete conjunction probabilities are the product of $n$ numbers each between 0 and 1; thus their product is between 0 and 1. Next, we will show that the sum of the complete conjunction probabilities is 1. Each $\omega \in \Omega$ is the product of $n$ distinct literals. Given variable independence, then:

$$P(\omega_k) = P(L_1 \wedge L_2 \wedge \ldots \wedge L_n)$$

$$= P(L_1)P(L_2) \ldots P(L_n)$$

where each $P(L_i)$ is either $P(a_i)$ or $1 - P(a_i)$. We show the theorem by induction on $n$.

Case $n = 1$: Then there are 2 complete conjunctions: $A$ and $\neg A$; the sum of their probabilities is: $P(A) + (1 - P(A)) = 1$.

Case $n$: Suppose the theorem holds for $n - 1$. For $n$ variables, there are $2^n$ summands which can be paired as follows:

$$P(L_1)P(L_2) \ldots \wedge P(L_k) \wedge \ldots \wedge P(L_n)$$

$$P(L_1)P(L_2) \ldots \wedge P(\neg L_k) \wedge \ldots \wedge P(L_n)$$

When these are summed, the result is:

$$(P(L_k) + (1 - P(L_k)))(P(L_1) \ldots \wedge P(L_{k-1})$$

$$\wedge P(L_{k+1}) \wedge \ldots \wedge P(L_n))$$

which reduces to an expression in $n-1$ variables. Applying this to all appropriate pairs results in a sum of elements of length $n - 1$. **QED**

Call $n$-modus ponens the problem with conjuncts $A_1, \neg A_1 \vee A_2, \ldots, \neg A_{n-1} \vee A_n$. Then the standard approach needs $2^n$ models (as does MLNs), whereas we solve it in linear time.

## 3.1     When Variables Are Not Independent

Suppose the independence of variables is not assumed. Then:

$$P(A \vee B) = P(A) + P(B) - P(A \mid B)P(B)$$

This equation is developed recursively for any non-singleton variable probability, but conditional probabilities are considered as unique unknowns. Solving the nonlinear system provides a set of atom and conditional probabilities. However, this may not be part of a PSAT solution; i.e., the space of solutions for this system contains the solutions (if they exist) for independent variables. In general, we propose to use the solution found by gradient descent.

Although we cannot guarantee that the result of the algorithm is a PSAT solution, there is a check to determine some cases when it is not. Given a conditional probability over two variables (i.e., one of $P(A \mid B)$, $P(\neg A \mid B)$, $P(A \mid \neg B)$, or $P(\neg A \mid \neg B)$) the other three can be determined from it. If any of these is not in the range $[0,1]$, then the result is not a PSAT solution. This same check can be applied to conditionals over three variables as well.

# 4  Experiments and Results

## 4.1  Independent Variables

We have tested this approach on sets of randomly generated knowledge bases. This involves selecting a number of variables ($n$), specifying a maximum number of sentences to generate, as well as the maximal length of any one sentence. A set of sentences is generated which satisfies these constraints, and then a set of probabilities are produced for the compete conjunction set, and from these the sentence probabilities are computed. This ensures that there is a solution, although it does not preclude the existence of other solutions (generally a non-zero measure subset of the unit hypercube).

A set of 100 KB's was generated this way, with $n = 5$, $m_{max} = 30$, and $len_{max} = 5$. Figure 1 shows the number of iterations required by Newton's method to solve PSAT; the blue trace shows when initial points are far from the known solution, and red when they are near (within 0.5 vector norm). The mean number of iterations is 4.26 when starting near, and 12.63 when starting far. The method fails on 6 of the 100 KB's. As for gradient descent, Fig. 2 shows the number of required iterations. Although a few KB's require over 1000 iterations, the mean number of iterations required when starting near a solution is 21.19, and when starting far is 171.58. Note that the search is terminated when a sentence error of less than 0.01 is reached.

## 4.2  Non-independent Variables

The equations must include variables for whatever conditional probabilities arise from the sentences, and are thus a bit more complicated. Figure 3 shows the number of iterations required for Newton's method on 100 random general KB's with the same parameters as above. In this case, solutions were found for 75 of the 100 KB's, and the mean number of iterations was 3.91 when starting near the known solution, and 10.27 when starting far from it. Figure 4 shows the results for gradient descent (which found solutions for all the KB's) and had mean number of iterations 662.17 for far starting points and mean number of iterations 638.61 for near points.

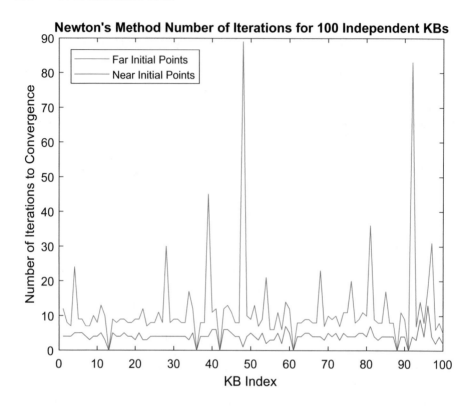

**Fig. 1.** Newton's method results for 100 KB's.

What these results indicate is that Newton's method should be tried first given the low iteration cost, and then gradient descent used if Newton's Method fails. Also, note that even though in the case of failure (i.e., local minimum found), the methods were allowed to re-start at new random initial locations. Gradient descent only re-started this way twice and then only tried 2 alternate points. When Newton's Method finds a solution is does so with the initial guess; when it failed, it did so for both near and far initial starting points.

Finally, Fig. 5 shows the maximal individual atom probability error comparing the atom probabilities from the actual 100 general KBs to the atom probabilities found by the numerical solver. The mean of the max atom probability error for near starting points is 0.09, while for far starting points is 0.10. This is very promising in that the discovered solutions are near the actual underlying solution for most KBs.

### 4.3    Trajectory Visualization and Finding Good Initial Guesses

As pointed out above, if the initial guess is too far from a solution, these methods may not converge. Thus, it would help to be able to identify good starting points. In order to get insight into the convergence sequence, we have developed

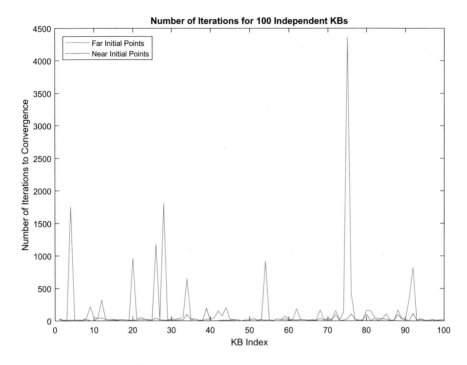

**Fig. 2.** Gradient descent results for 100 KB's.

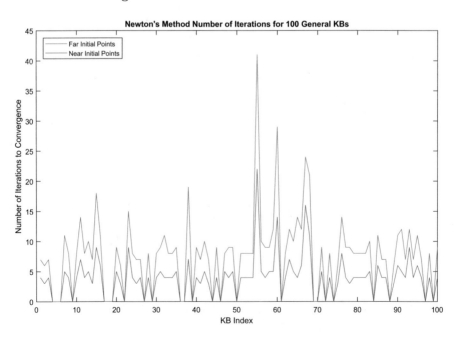

**Fig. 3.** Newton's method results for 100 KB's.

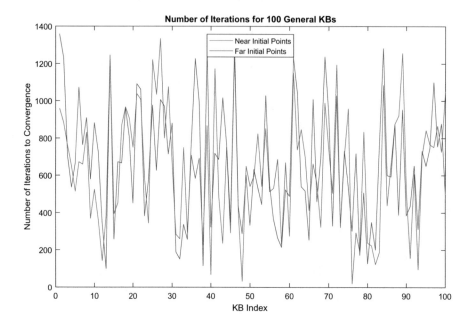

**Fig. 4.** Gradient descent results for 100 KB's.

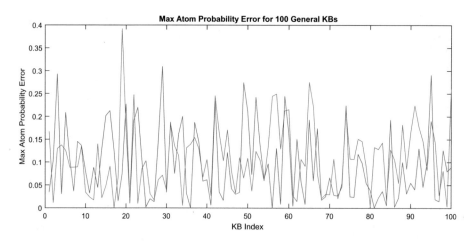

**Fig. 5.** Maximum atom probability error for 100 general KBs (near starting points in red and far starting points in blue).

a visualization method which maps $n$-D points to 2-D points. Given a point, $\bar{a}$, in $n$-D, define the corresponding 2-D coordinates as follows:

$$x = \sum_{i=1}^{n}(a_i cos(\frac{(i-1)\pi}{n})) \tag{1}$$

$$y = \sum_{i=1}^{n}(a_i sin(\frac{(i-1)\pi}{n})) \tag{2}$$

Figure 6 shows the convergence trajectories for four different initial points. The $q$-convergence of the method can be estimated by determining the $c_k$'s in the following equation:

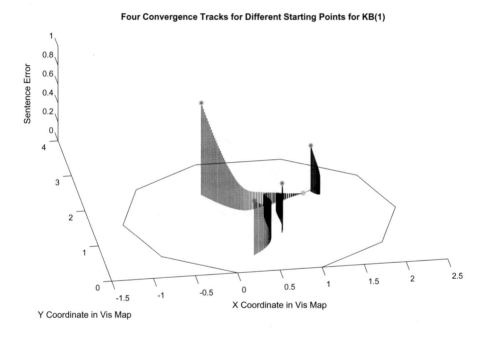

**Fig. 6.** Convergence tracks for 4 random starting points for 5-D problem; $x$ and $y$ values are projections of 5-D points to 2-D, and $z$ value is sentence error value.

$$|\bar{x}^{k+1} - \bar{x}^*| \leq c_k | \bar{x}^k - \bar{x}^*| \tag{3}$$

Figure 7 shows these values for the 100 tracks for gradient descent on the general KB's. The plots indicate that the method is $q$-superlinear/quadratic.

Another interesting aspect of this visualization method is its use to find good starting points. Given fixed $x$ and $y$ in the plane, we have developed a method to obtain a unique point in the pre-image of Eqs. (1) and (2). Each equation defines a hyperplane in $n$-space; taken together they represent a hyperplane of

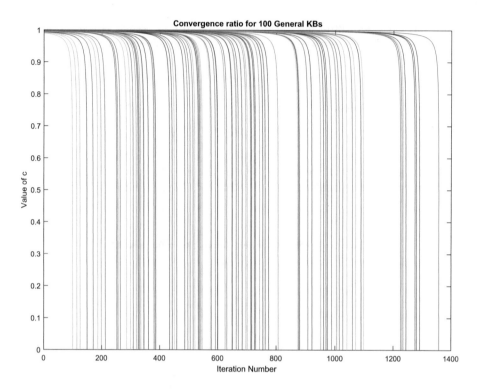

**Fig. 7.** Convergence step ratio for 100 general KB's using gradient descent.

dimension $n - 2$. One way to understand the map defined by Eqs. (1) and (2) is as an $n$-joint prismatic manipulator, where joint $k$ translates in the direction $\theta = \frac{(k-1)\pi}{n}$. The manipulator's workspace is a $2n$-gon (as shown in Fig. 6). By uniformly sampling this workspace, and then finding pre-image points in $n$-D, the sentence error can be found, and then the lowest such value used to pick the initial point. Of course, since there is an infinite number of potential pre-image points for each $x$ and $y$ location, other methods can be used to sample that subspace to find better starting points.

## 5    Conclusions and Future Work

We propose a novel approach to approximately solve PSAT (or at least as much of it as is useful) which avoids the computational complexity of previous methods as well as the error introduced using MC-SAT methods. Instead we solve a system of nolinear equations derived directly from the meaning of the probability of the logical sentences. The experiments reported here show that solving these systems is possible and not overly complex (evidence shows $q$-superlinear/quadratic convergence). The number of variables and sentences used in these experiments are modest as is required so that the ground truth can be ascertained (i.e., that there is a solution!); however, in previous work we have demonstrated that the method can solve problems with 80 variables and over 400 sentences (Henderson 2017a).

It is also possible to convert SAT instances into the form of PSAT by assigning every clause the probability of 1, and then apply numerical solvers to this problem. In particular, if we limit our input to 3-SAT instances, we can recursively expand the equation for a CNF equation to:

$$P(A \lor B \lor C) = P(A) + P(B) + P(C) - P(A \land B)$$
$$-P(A \land C) - P(B \land C) + P(A \land B \land C)$$

where we can treat each distinct probability as a separate variable. Note that we restrict input to 3-SAT instances because this recursive expansion is exponential in the number of variables in the clause. We can also force consistency constraints given, for any literals $L_i$, $L_j$, $L_k$:

$$P(L_i) = P(L_i \land L_j) + P(L_i \land \neg L_j)$$
$$P(L_i \land L_j) = P(L_i \land L_j \land L_k) + P(L_i \land L_j \land \neg L_k)$$

There are $O(n^3)$ such constraints. The system of equations that this creates is far too large to be a practical SAT solver, but it is nevertheless a polynomial number of linear equations, which can be solved in polynomial time. If some $\pi$ can be found that satisfies the new sentences, every $w_k$ such that $\pi(w_k) \geq 0$ is a solution to the original SAT equation. Every sentence has probability 1, and since the sum of all $\pi(w_k)$ must be equal to 1, only atom truth assignments that satisfy every clause may hold probability. As such, if it can be shown that nonindependent solutions can be used to construct valid probabilities for the complete conjunction set, then $P = NP$.

Other future work includes the investigation of:

1. The problem encountered with Newton's Method. It is possible that the Hessian as computed does not remain positive definite which can cause failure. It may be possible to address this with SVD methods.
2. The discovery of good initial starting points. For this, the trajectory visualization method will be studied; i.e., the inverse kinematics of the planar $n$-joint prismatic manipulator.
3. The exploitation of the method to support a knowledge base providing probabilistic logic and in the future, argumentation. such a capability will provide decision makers and analysts a robust estimate of the confidence of a statement or the consequences of an action. The current application domain for this is geospatial knowledge bases (see Henderson 2017b).

## 5.1  Numerical Solutions

Given a set of nonlinear equations resulting from a CNF KB and the associated sentence probabilities, it is necessary to create the sentence error function, find an intitial guess at a solution, and then apply Newton's method or some other technique. We have applied two methods: (1) Newton's method, and (2) gradient descent using the Jacobian.

### 5.1.1 Newton's Method

Given the vector function $F$ defined above (a vector function of $m$ elements), Newton's method iterates the following until within tolerance of a solution:

1. Produce next step vector
   $$H_{F(\bar{x}^k)}\bar{s} = \nabla F$$
2. Move toward solution
   $$\bar{x}^{k+1} = \bar{s} + \bar{x}^k$$

where $H_F$ is the Hessian matrix for $F$, $\bar{s}$ is the step vectorm and $\bar{x}^k$ is the $k_{th}$ solution estimate. The development of the Hessian is done symbolically then solved numerically in Matlab. This imposes constraints of the application of this method to larger problems. However, we give results below for moderate size KB's.

### 5.1.2 Gradient Descent Using the Jacobian

Gradient descent using the Jacobian should have q-quadratic convergence when starting not too far from a solution, but may hit a local (non-solution) minimum otherwise. The method iterates until within tolerance of a solution as follows:

1. Determine the Jacobian
   $$J = \nabla F$$
2. Move toward lower sentence error
   $$\bar{x}^{k+1} = \alpha * J(\bar{x}^k) + \bar{x}^k$$

where $\alpha$ is the step size.

# References

Adams, E.W.: A Primer of Probability Logic. CLSI Publications, Stanford (1998)

Biba, M.: Integrating logic and probability: algortihmic improvements in markov logic networks. Ph.D. thesis, University of Bari, Bari, Italy (2009)

Boole, G.: An Investigation of the Laws of Thought. Walton and Maberly, London (1854)

Domingos, P., Dowd, D.: Markov Logic: An Interface Layer for Artificial Intelligence. Morgan and Claypool, San Rafael (2009)

Georgakopoulos, G., Kavvadiass, D., Papadimitriou, C.H.: Probabilistic satisfiability. J. Complex. **4**, 1–11 (1988)

Gogate, V., Domingo, P.: Probabilistic theorem proving. Commun. ACM **59**(7), 107–115 (2016)

Hailperin, T.: Sentential Probability Logic. Lehigh University Press, Cranbury (1996)

Henderson, T.C., Mitiche, A., Simmons, R., Fan, X.: A preliminary study of probabilistic argumentation. Technical report UUCS-17-001, University of Utah, Salt Lake City, UT, February 2017a

Henderson, T.C., Simmons, R., Mitiche, A., Fan, X., Sacharny, D.: A probabilistic logic for multi-source heterogeneous information fusion. In: Proceedings of the IEEE Conference on Multisensor Fusion and Integration for Intelligent Systems, Daegu, South Korea, November 2017b

Hunter, A.: A probabilistic approach to modeling uncertain logical arguments. Int. J. Approximate Reasoning **54**, 47–81 (2013)

Kowalski, R., Hayes, P.J.: Semantic trees in automatic theorem proving. In: Automation of Reasoning, Berlin, pp. 217–232 (1983)

Nilsson, L.: Probabilistic Logic. Artif. Intell. J. **28**, 71–87 (1986)

Sacharny, D., Henderson, T.C., Simmons, R., Mitiche, A., Welker, T., Fan, X.: A novel multi-source fusion framework for dynamic geospatial data analysis. In: Proceedings of the IEEE Conference on Multisensor Musion and Integration, Daegu, South Korea, November 2017

Thimm, M.: Measuring inconsistency in probablilistic knowledge bases. In: Proceedings of the 25th Conference on Uncertainty in Artificial Intelligence, Montreal, Canada, pp. 530–537, June 2009

Thimm, M.: A probabilistic semantics for abstract argumentation. In: Proceedings of the 20th European Conference on Artificial Intelligence, Monpelier, France, August 2012

# Aggregating Models for Anomaly Detection in Space Systems: Results from the FCTMAS Study

Francesco Amigoni[1(✉)], Maurizio Ferrari Dacrema[1], Alessandro Donati[2],
Christian Laroque[3], Michèle Lavagna[1], and Alessandro Riva[1]

[1] Politecnico di Milano, 20133 Milan, Italy
{francesco.amigoni,maurizio.ferrari,michelle.lavagna,
alessandro.riva}@polimi.it
[2] European Space Agency (ESA), Advanced Mission Concepts
and Technologies Office, 64293 Darmstadt, Germany
Alessandro.Donati@esa.int
[3] Telespazio Vega Deutschland GmbH, 64293 Darmstadt, Germany
christian.laroque@telespazio-vega.de

**Abstract.** The *Flight Control Team Multi-Agent System (FCTMAS)*
study, funded by the European Space Agency (ESA), has investigated
the use of multiagent systems in supporting flight control teams in routine operations. One of the scientific challenges of the FCTMAS study
has been the detection of anomalies relative to a space system only on the
basis of identified deviations from the nominal trends of single measurable variables. In this paper, we discuss how we addressed this challenge
by looking for the best structure that aggregates a given set of models,
each one returning the anomaly probability of a single measurable variable, under the assumption that there is no *a priori* knowledge about the
structure of the space system nor about the relationships between the
variables. Experiments are conducted on data of the Cryosat-2 satellite
and their results are eventually summarized as a set of guidelines.

## 1 Introduction

Autonomous software agents have been successfully applied in several space
applications to improve timely decision making of space systems. Starting from
the Remote Agent (RAX) technology [16], several architectures for autonomy
have been developed by NASA (National Aeronautics and Space Administration), including IDEA (Intelligent Distributed Execution Architecture) [15],
CASPER (Continuous Activity Scheduling Planning Execution and Replanning system) [7], and CLARAty (Coupled Layer Autonomous Robot Architecture) [25]. In Europe, ESA (European Space Agency) has also investigated the
possibility of using software agents in a number of studies, including DAFA
(Distributed Agents For Autonomy) [3], with the objective of demonstrating
the advantages of using distributed autonomous agents in a space system, and

M. Strand et al. (Eds.): IAS 2018, AISC 867, pp. 142–160, 2019.
https://doi.org/10.1007/978-3-030-01370-7_12

MECA (Mission Execution Crew Assistant) [17], which makes use of distributed agents to amplify the cognitive abilities of human-machine teams during planetary exploration missions in order to cope with unexpected, complex, and potentially hazardous situations.

Against this background, the *Flight Control Team Multi-Agent System* (*FCT-MAS*) study has been funded by ESA in order to investigate the use of multiagent systems in supporting flight control teams in routine operations. The project has run from 2011 to 2015. In addition to the several technological aspects, one of the main scientific challenges addressed in the FCTMAS study has been the detection of anomalies relative to a space system starting from identified deviations from the nominal behaviors of single measurable variables. Not all the deviations detected at the level of single variables correspond to global anomalies for the space system (e.g., a variable could take a value larger than usual if, at the same time, another variable takes a value smaller than usual).

In this paper, we discuss how we addressed this problem in the FCTMAS study in order to find the best aggregation structure for a given set of models, each one returning the anomaly probability of a single measurable variable, under the assumption that we have no *a priori* knowledge about the structure of the space system nor about the relationships between the variables. To the best of our knowledge, this problem has not been addressed in the literature and, in this sense, we provide an original contribution.

The approach we propose represents an aggregation structure as a *tree* in which the leaves are the given models and the internal nodes are aggregation functions, each one "harmonizing" and "merging" the anomaly probabilities of its children and returning an aggregated anomaly probability to its parent. Overall, the root of the tree returns the global anomaly probability for the whole system, that can be provided to a human operator for decision making. The nodes are implemented as agents. Without any *a priori* knowledge on the relationships between variables, the best tree structure is searched in the space of possible trees in order to maximize the accuracy in identifying global anomalies in test data. We characterize this space according to different choices for aggregation functions (maximum, average, cooperative negotiation) and we propose some algorithms for searching the space. Experiments on data of the Cryosat-2 satellite show the pros and cons of different aggregation functions and of the algorithms we employ, which are eventually summarized in a set of guidelines.

## 2   The Application Context

The overall objective of the FCTMAS study has been to conduct a dedicated analysis to evaluate to which extent a multiagent system can support flight control teams in routine operations under nominal and contingency conditions. The description of all activities and results obtained in the FCTMAS study is out of the scope of this paper. Here, we focus on a specific aspect that emerged during the study and that motivates the relevance of the problem we address. The analysis of Mission Control System (MCS) events logs received at ground stations is performed by human operators in order to extract relevant information. This generates a problem of information overload for human operators: events in the logs are relative to variables measured onboard the space system and many of them are not relevant, even if they are triggered as errors, because they do not correspond to any anomaly for the space system. The FCTMAS study has developed an agent-based system called *MCS Subsystem* for the extraction and the analysis of the relevant information about unexpected events (anomalies) from log files containing events generated by a space system. For example, Fig. 1 shows a portion of an MCS events log file relative to the Cryosat-2 satellite. Each event corresponds to a line and has a timestamp (first two columns) and a description (last column). Basically, each event can be considered as an assignment to a variable, either numerical or categorical. (If an event involves multiple variables, like the third line of Fig. 1, it can be decomposed in multiple events, each one assigning a value to a single variable, with the same timestamp.) The former variables take numerical values, like fhp = 730 in the last line of Fig. 1 (referred to the telemetry packetizer of the MCS), while the latter variables take symbolic values, for example, in the second line of Fig. 1, the value PASSED is assigned to variable SSC09000 (that represents a telecommand sent to the satellite).

```
2013-08-03 07:08:13.553  20755 2   BEHVLimCPB    crymca  Information  Log     2013.215.05.08.50.190 DHT30304 VAL: ON STATE: ON STATUS
    limit is back to nominal
2013-08-03 07:08:11.275  13524 1   CMDHveri      crymca  Information  Log     TC: SSC09000, APID: 812, SSC: 13900,
    set stage: EV\_APP\_ACCEPT status to: PASSED
...
2013-08-22 23:32:20.754  23001 1   TPKT          crymca  Error        System  4 Missing Source Packets, APID = 68, VCID = 0,
    SSC = 14894, Time = 2013-08-22T23:32:18.710216
2013-08-22 23:32:19.511  10307 1   NCDUadmi      crymca  Warning      Log     NCDU:TM007 Data gap on TM link VC 250/0,
    Mode: Onl-TIM from KR14 . Reason: unk, Size: 15
2013-08-22 23:32:19.448  13842 1   CMDHmplx      crymca  Information  Log     Commanding link status set to: TC: NO RF, TM: GREEN
2013-08-22 23:32:19.443  14778 1   TPKT          crymca  Error        System  Unexpected spill-over data (fhp = 730)
```

**Fig. 1.** A portion of an MCS events log file relative to the Cryosat-2 satellite.

The MCS Subsystem developed in the FCTMAS study to detect anomalies in space systems works as follows.

– Learning of single models phase. The MCS Subsystem receives in input a nominal MCS log file like that of Fig. 1 (representing reference, ground truth, correct behavior of the satellite) and builds the nominal models for the single variables. This phase models the trends of the values that single variables take during nominal behavior of the satellite and the times at which they appear in logs. For numerical variables, the nominal trend of the values and the nominal time intervals between values are modeled using neural networks (NNs) or autoregressive moving-average (ARMA) models. For categorical variables, the nominal time intervals between values can be modeled using NNs or ARMA models and the nominal succession of values is modeled using Markov chains.
– Learning of aggregate models phase. The MCS Subsystem receives in input a *labeled* MCS log file in which, for each timestamp, a human expert has specified if the behavior of the satellite is nominal or not, and calculates the best structure to aggregate the anomaly identifications performed by the models for single variables.

– Operative phase. The MCS Subsystem receives a new MCS log file and feeds the data to the structure found above to actually identify global anomalies of the whole satellite.

In this paper, we consider the nominal models for single variables as given and we address the second phase of the learning process by calculating the best aggregation structure for anomaly detection according to the input labeled MCS log file. Our approach is agnostic about the models for single variables, meaning that they can be obtained in any way.

## 3   Related Work

Systems for anomaly detection have the goal to identify deviations from the nominal behavior of a target system by observing some measured variables. Usually, anomalies are associated to faults of the target system. Countless approaches have been proposed for anomaly detection and it is beyond the scope of this section to provide a comprehensive review; we just discuss some of their main principles. Approaches for anomaly detection can be classified in two broad classes. The first class refers to *model-based anomaly detection* and includes methods that exploit a model of the functioning of the target system. The model can be represented mathematically, for instance as a set of equations [24], or qualitatively, for instance using causal relationships [22]. The second class is composed of approaches that perform *data-driven anomaly detection*. They do not require any initial explicit model of the target system, but exploit a large amount of data about the functioning of the system to learn a representation of its behavior [23,26]. Compared to model-based anomaly detection, data-driven anomaly detection requires less initial domain knowledge, but usually builds approximate representations of the target systems that do not allow to identify the nature of detected anomalies.

In space applications, model-based anomaly detection has been largely investigated. The Advanced Fault Detection Isolation and Recovery (AFDIR) framework [11] includes several tools like Kalman filters, parity vectors, generalized likelihood test [9], and random sample consensus [10]. Another example of model-based anomaly detection technique for space systems is that of Smart-FDIR [14], which exploits fuzzy inductive reasoning in a mixed quantitative-qualitative approach. Other similar systems employ fuzzy sets [6] and neuro-fuzzy models [13]. The ARPHA (Anomaly Resolution and Prognostic Health management for Autonomy) system [8] can run onboard a rover for planetary exploration and uses Bayesian networks and fault trees [20] to model the target system.

From the data-driven anomaly detection side, the Anomaly Monitoring Inductive Software System (AMISS) [18] analyzes and clusters the data relative to the nominal behavior of a space system and populates a knowledge base with models that represent the nominal values for the variables of each subsystem. The anomaly detection approach we propose in this paper is data-driven and, unlike AMISS, is not focused on building models of nominal behaviors of single subsystems but, considering them as given, aggregates the anomalies detected by these models in a structure able to provide a global probability of anomaly. To the best of our knowledge, we are not aware of any data-driven anomaly detection method that aggregates anomaly probabilities that refer to different variables of a single space system without knowing how these variables are related.

Multiagent systems have been applied to anomaly detection in several fields, including wind turbines [27], patient monitoring [19], and surveillance [21]. In [27], anomaly detections performed by different agents on different data are all presented to the human operators, without attempting to integrate them. A similar approach is used by [19], where all alarms identified by different agents are sent to an healthcare center. In [21], knowledge on anomalies coming from different agents is integrated using domain-dependent rules. Differently from the above and other similar works, in this paper, we provide a general method for aggregating anomaly probabilities coming from different models (agents).

## 4    Problem Formulation

For the purposes of this paper, the anomaly models of single variables are given, fixed, and regarded as black boxes for their aggregation. They are represented as detection functions. Given a labeled log file $\mathcal{D}$ for which we know, at each timestamp $t \in \mathcal{T}$, if the behavior of the satellite is nominal or not, we define $I$ single-variable *detection functions*:

$$D_i : \mathcal{T} \to [0, 1]$$

each of which, given a $t \in \mathcal{T}$, returns a probability of anomaly referred to variable $i \in [1, I]$. We assume that $\mathcal{D}$ is such that, for each timestamp $t$ appearing in it, a value for every variable $i$ is present and, consequently, a probability $D_i(t)$ can be calculated. (This assumption can be satisfied by considering the latest values for variables that are not assigned new values at $t$.) Our approach is independent of how detection functions $D_i$ are built. In our experiments, we build them from data as follows. Consider the value $\hat{y}_i(t)$ predicted by a single-variable nominal model for the *numerical* variable $i$ at time $t$ (provided by NNs or ARMA models, see Sect. 2) and the actual value $y_i(t)$ measured at time $t$ (which is read from $\mathcal{D}$), the probability of anomaly returned by $D_i$ is $D_i(t) = 1 - \exp\left[-\frac{1}{2} \cdot \left(\frac{\epsilon_i(t) - \bar{\epsilon}_i}{\sigma_i}\right)^2\right]$, where $\epsilon_i(t) = |\hat{y}_i(t) - y_i(t)|$ is the prediction error, $\bar{\epsilon}_i$ is the mean of the prediction error, and $\sigma_i$ is its standard deviation (both $\bar{\epsilon}_i$ and $\sigma_i$ are calculated during the training of the nominal model for $i$). Note that $\sigma_i$ accounts for the imprecision of the nominal model for the variable $i$ when calculating probability $D_i(t)$. For *categorical* variables, the probability of anomaly is calculated according to the regularity of the values. For instance, if a variable has a nominal behavior modeled by a Markov chain in which the probability of transition from the current value $y_i(t-1) = $ OFF to the same value $\hat{y}_i(t) = $ OFF is close to 1, then any measured value different from $y_i = $ OFF generates a high probability of anomaly. Moreover, in general, the anomaly probability for a categorical variable grows with the length of the sequence of consecutive anomalous values.

The problem we tackle in this paper is to find the aggregation of detection functions $D_i$ that provides the best detection of the satellite anomalies given $\mathcal{D}$. Aggregation is subject to the constraint that there is no *a priori* domain knowledge about the structure of the satellite and, consequently, there is no domain knowledge about the relationships between the variables (e.g., there is no knowledge about correlations between multiple variables, for instance because the data in the available log files are not enough to learn a model of such correlations). Consider now *aggregation functions*:

$$A_j : [0,1]^{k_j} \rightarrow [0,1]$$

such that function $A_j$ returns a probability of anomaly that "aggregates" $k_j$ probabilities of anomaly (which can come from detection functions or from other aggregation functions). We assume $k_j \geq 2$.

We consider tree structures $T$ such that:

- the leaves are the $I$ detection functions (a detection function corresponds to one and only one leaf, so the number of leaves is equal to $I$),
- internal nodes are aggregation functions,
- children of an aggregation function $A_j$ are $k_j$ detection or aggregation functions providing input probabilities to $A_j$.

Note that, in order to have a tree, there must be at least an aggregation function, the root of the tree. The idea is that the aggregation function node at the root of tree $T$ returns the global probability of anomaly when, for each $t$, data from $\mathcal{D}$ are fed to all the leaves (detection functions) of $T$. For instance, consider the tree of Fig. 2a, in which 6 detection functions are aggregated using 2 aggregation functions. In the example, for each $t$ present in $\mathcal{D}$, a global probability of anomaly is returned by $A_1$. This global probability can be used to provide the human operator a single information about anomalies in the space system. Note also that the tree structures we consider do not represent probability distributions and are different from the trees used in the context of Bayesian networks to represent conditional probability distributions (see, e.g., [12]).

Our goal is to find the tree structure $T^*$ that optimizes a *performance function* $P(\mathcal{D}; T)$, which measures the ability of the tree structure $T$ to identify global anomalies in a labeled log file $\mathcal{D}$. In Sect. 6 we describe a specific instance of the performance function $P(\mathcal{D}; T)$. For the moment, we assume that $P(\mathcal{D}; T)$ is a cost function that does not contain any term relative to the complexity of $T$, so it only depends on how well the anomalies in $\mathcal{D}$ are detected.

In the MCS Subsystem developed in the FCTMAS study, detection and aggregation functions are implemented as independent agents (see also Sect. 6).

# 5   Problem Analysis and Solution

## 5.1   Analysis of the State Space

The state space that should be explored to find the optimal tree structure $T^*$ is the set of all possible trees having as leaves (only and all) the $I$ detection functions. Given $n$ (labeled) nodes, the number of trees that can be built using these nodes is given by the Cayley's formula: $n^{n-2}$. If we assume to have $I$ detection functions, in our case we have that $n \geq I + 1$ (at least, the tree includes all the $I$ detection function nodes and the root) and $n \leq 2I - 1$ (the maximum number of internal nodes of a tree with $I$ leaves is $I - 1$, assuming that each internal node has at least two children, $k_j \geq 2$). It is clear that the size of state space quickly becomes huge. However, according to the specific form of the aggregation functions, many of these trees are equivalent in the sense that they return the same global probability for anomaly for any $t$. In the following, we investigate this issue for better characterizing the state space when all the aggregation functions are of a given form.

**Maximum Aggregation Functions.** In this case, each aggregation function $A_j$ returns the maximum of its arguments:

$$A_j(p_1, p_2, \ldots, p_{k_j}) = \max\{p_1, p_2, \ldots, p_{k_j}\} \tag{1}$$

where $p_1, p_2, \ldots, p_{k_j}$ are probability values returned by detection functions or aggregation functions that are children of the node corresponding to $A_j$. For instance, in Fig. 2a, $A_1(p_1, p_2, p_3, P_2) = \max\{p_1, p_2, p_3, P_2\}$, where, for any $t$, $p_1 = D_1(t)$, $p_2 = D_2(t)$, $p_3 = D_3(t)$, and $P_2 = A_2(D_4(t), D_5(t), D_6(t))$. If we consider the space of trees in which all the aggregation functions are of the form (1), then it is easy to see that all these trees are equivalent and return at the root the largest anomaly probability returned by the detection functions at the leaves $D_i$ ($i \in [1, I]$). In this case, the actual structure of a tree has no influence on the returned global probability of anomaly. As a consequence, if aggregation functions are limited to the form (1), the state space is composed of equivalent trees and $T^*$ can be found trivially by building a random tree.

**Average Aggregation Functions.** Now consider the case in which each aggregation function $A_j$ returns the average of its arguments:

$$A_j(p_1, p_2, \ldots, p_{k_j}) = \frac{p_1 + p_2 + \ldots + p_{k_j}}{k_j} \tag{2}$$

where, as before, $p_1, p_2, \ldots, p_{k_j}$ are the probability values returned by detection functions or aggregation functions that are children of the node corresponding to $A_j$. If all the aggregation functions in a tree are of the form (2), we can "flatten" the tree, eliminating the aggregation functions (except the root) and assigning a weight to each detection function. For example, for the tree of

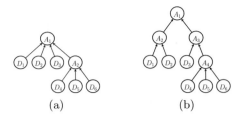

$$\text{(a)} \qquad\qquad\qquad\qquad \text{(b)}$$

**Fig. 2.** A tree structure (a) and an equivalent tree structure (b), when the aggregation functions perform the average of their inputs.

Fig. 2a: $A_1(D_1(t), D_2(t), D_3(t), A_2(D_4(t), D_5(t), D_6(t))) = \frac{p_1+p_2+p_3+\frac{p_4+p_5+p_6}{3}}{4} = \frac{p_1}{4} + \frac{p_2}{4} + \frac{p_3}{4} + \frac{p_4}{12} + \frac{p_5}{12} + \frac{p_6}{12}$.

Thus, the global anomaly probability returned by a tree in which all the aggregation functions are of the form (2) is fully specified by a weight vector $\mathbf{w} = [w_1, w_2, \ldots, w_I]$, where $I$ is the number of detection functions. In the above example: $\mathbf{w} = [\frac{1}{4}, \frac{1}{4}, \frac{1}{4}, \frac{1}{12}, \frac{1}{12}, \frac{1}{12}]$. The weight $w_i$ of a detection function $D_i$ is the inverse of the product of the number of children of its parent (including itself) and of the numbers of children of all its ancestors, up to the root. In the above example, $w_6 = \frac{1}{12} = \frac{1}{3 \times 4}$ because $D_6$ corresponds to a node that is a child of a node ($A_2$) with 3 children and the parent node of $A_2$ (which is the root) has 4 children. From this observation, it easily derives that multiple tree structures can have the same weight vector $\mathbf{w}$, and can thus provide the same global probability of anomaly starting from the same $\mathcal{D}$. For instance, Fig. 2b shows another tree associated to $\mathbf{w} = [\frac{1}{4}, \frac{1}{4}, \frac{1}{4}, \frac{1}{12}, \frac{1}{12}, \frac{1}{12}]$. Hence, when all the aggregation functions are of the form (2), the state space can be partitioned in classes of equivalent trees.

Actually finding the partitions of the state space is not trivial because there could be an arbitrary number of aggregation functions, up to $I - 1$. If we fix the number of aggregation functions in the tree, then the trees belonging to the equivalence class corresponding to a given $\mathbf{w}$ can be found by formulating an assignment problem that "allocates" the detection functions to the aggregation functions. Due to space constraints, we do not further discuss this issue here, but we note that the above analysis can be generalized to the case in which the aggregation functions perform weighted averages of their inputs.

**Cooperative Negotiation Aggregation Functions.** We now consider the case in which an aggregation function node acts as a mediator to manage an iterative negotiation process performed by its children in order to reach an agreement over the value of the anomaly probability. An agreed-upon value of the probability of an anomaly is decided using *cooperative negotiation* as defined in [4] (to which the reader is referred for further information beyond the general description we provide here). In the MCS Subsystem developed in the FCTMAS study, each aggregation function acts as a *mediator agent* and its children are *negotiating agents*.

---

**Algorithm 1.** Negotiating Agent $i$

1: $p_i^0 \leftarrow D_i(t)$ (or $p_i^0 \leftarrow A_i(\ldots)$)
2: SendToMediator($p_i^0$)
3: **loop**
4:      status, $a^\tau \leftarrow$ WaitForMediatorAnswer()
5:      **if** status $=$ *agreement_found* **then return** $a^\tau$
6:      **else**
7:          $p_i^{\tau+1} \leftarrow \mathcal{F}_i(p_i^\tau, a^\tau) = p_i^\tau + \alpha_i \cdot (a^\tau - p_i^\tau)$
8:          SendToMediator($p_i^{\tau+1}$)
9:          $\tau \leftarrow \tau + 1$
10:     **end if**
11: **end loop**

---

**Algorithm 2.** Mediator Agent $j$

1: **repeat**
2:      $\{p_1^\tau, p_2^\tau, \ldots, p_{k_j}^\tau\} \leftarrow$ CollectAllProposalsFromAgents()
3:      $a^\tau \leftarrow \mathcal{A}(\{p_1^\tau, p_2^\tau, \ldots, p_{k_j}^\tau\}) = \left(\sum_{i=1}^{k_j} p_i^\tau\right)/k_j$
4:      **if** all $p_i^\tau$ are equal **then** status $\leftarrow$ *agreement_found*
5:      **end if**
6:      SendToAgents(status, $a^\tau$)
7: **until** status $=$ *agreement_found*

---

The negotiation process evolves in steps (rounds). At each step $\tau$, first, the negotiating agents send their proposals to the mediator and, then, the mediator sends its counter-proposal (current agreement) to the negotiating agents. The process continues until convergence. Note that a negotiation process is performed for each aggregation function of a tree $T$, in sequence starting from the bottom and eventually obtaining probability values at the root of $T$, at each time instant $t$ (for which data in $\mathcal{D}$ are available).

The algorithm for a negotiation agent $i$ is reported as Algorithm 1. In the algorithm, probabilities $p_i^0$, $p_i^\tau$, and $p_i^{\tau+1}$ are the *proposals* of agent $i$ at steps $0$, $\tau$, and $\tau + 1$, respectively; probability $a^\tau$ is the counter-proposal of mediator at step $\tau$; and $\mathcal{F}_i(p_i^\tau, a^\tau)$ is the *strategy function* of agent $i$. The parameter $\alpha_i \in (0, 1]$ is called *concession coefficient* and represents the rigidity of agent $i$ to move toward the counter-proposal $a^\tau$ received from the mediator ($\alpha_i \to 0$ means that agent $i$ is rigid, $\alpha_i \to 1$ means that it is concessive). The rationale is that agent $i$ proposes an initial anomaly probability (i.e., that returned by $D_i(t)$ when agent $i$ corresponds to a detection function or that returned by a previously completed negotiation when $i$ corresponds to an aggregation function) at the first negotiation step $\tau = 0$ and then revises its proposals at later steps to take into account the proposals of other agents.

The algorithm for the mediator agent $j$ is shown as Algorithm 2, where $\mathcal{A}(\{p_1^\tau, p_2^\tau, \ldots, p_{k_j}^\tau\})$ is the *agreement function* that calculates the current agreement as average between the proposed probabilities and, in a way, expresses the aggregated preferences of all agents.

When, at some $\tau$, all $p_i^\tau$ are equal to the same value $\bar{p}$, then the final agreement is reached, and the negotiation process ends. In this case, status $=$ *agreement_found* and $a^\tau = \bar{p}$. Under mild conditions (basically related to the fact that all $\alpha_i \neq 0$), the cooperative negotiation protocol is guaranteed to eventually converge to an agreement (see Theorem 5.9 of [4]). In our case, we set $\alpha_i = 1 - e^{-G(D_i(t))}$, where $G()$ is a non-negative decreasing function such that the agent is rigid in conceding ($\alpha_i \rightarrow 0$) when $D_i(t) \rightarrow 1$.

It has been shown that the above cooperative negotiation schema (performing iterated averages over proposals) can produce better results than the (one shot) average of Sect. 5.1 (see [1,2] for examples in other application domains). In principle, also cooperative negotiation partitions the state space of trees in equivalence classes. However, given the non-linearity of the cooperative negotiation process, the study of these partitions is not trivial and left for further investigation.

## 5.2   Solving Algorithms

In order to practically find $T^*$ in the general case (which can involve aggregation functions of different forms), we propose different algorithms.

**Enumeration Algorithm.** This algorithm naïvely lists and evaluates all the possible trees that can be built starting from the initial $I$ detection functions, in order to find out the best tree. Basically, it enumerates all the trees of the state space. This approach is viable only for small instances (i.e., for few detection functions). Given a tree structure with $J$ aggregation functions as generated by the enumeration, all the $J^m$ combinations are evaluated by the algorithm, where $m$ is the number of possible forms for aggregation functions ($m = 3$ is our case, corresponding to maximum, average, and cooperative negotiation). The same happens for the other algorithms.

**Simulated Annealing Algorithm.** This algorithm starts from a "flat" tree $T_{\text{init}}$, namely from a tree with a single aggregation function as the root and all the detection functions as children, then it creates neighboring trees, evaluates them in order to select the best one, considers the selected neighboring tree as the new current tree (with some probability, see below), and iterates until a fixed number of iterations is performed.

Starting from the current tree, like for example that of Fig. 3a, the neighboring trees are created by swapping two detection functions contributing to two different aggregation functions (Fig. 3b), by moving a detection function from one aggregation function to another aggregation function (Fig. 3c), by inserting a new aggregation function (Fig. 3d), and by removing an aggregation function (Fig. 3e). All the above actions are constrained to create legal trees, according to the constraints of Sect. 4. It is easy to see that the state space can be fully explored using sequences of the four actions.

As in a typical simulated annealing approach, a temperature $t_\ell$ at iteration $\ell$ is calculated as $t_\ell = t_{\text{MAX}}(1 - \beta)^\ell$ (we use $t_{\text{MAX}} = 1000$ and $\beta = 0.01$ in our experiments). A selected neighboring tree $T$ of the current tree $T_c^{\ell-1}$ becomes

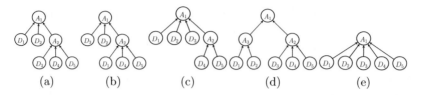

**Fig. 3.** Actions for the simulated annealing algorithm: initial tree structure (a), swap two detection functions ($D_1$ and $D_3$) (b), move a detection function ($D_3$) (c), insert a new aggregation function ($A_3$) (d), and remove an aggregation function ($A_2$) (e).

the new current tree $T_c^\ell$ if $T$ has a lower cost (calculated according to $P(\mathcal{D};T)$) than $T_c^{\ell-1}$ or, with probability $e^{\frac{P(\mathcal{D};T_c^{\ell-1})-P(\mathcal{D};T)}{t_\ell}}$, if $T$ has a larger cost than $T_c^{\ell-1}$.

**Greedy Algorithm.** This algorithm constructively builds a tree in the following way. It starts from a tree composed of a single aggregation function as the root and of a random number (from 1 to $z$) of detection functions as children (see, for instance, Fig. 4a). It then incrementally adds a random number (from 1 to $z$) of detection functions (not yet present in the tree) as children of a randomly-selected aggregation function node (Fig. 4b) or inserts an aggregation function as child of a randomly-chosen aggregation function node (Fig. 4c). The newly inserted aggregation function node has as children a random number (from 1 to $z$) of detection functions (not yet present in the tree). The two above operations are guaranteed to create legal trees, according to the constraints of Sect. 4. From all the applicable actions, the greedy algorithm selects the one that produces the best tree (again, according to $P(\mathcal{D};T)$), which is considered as the new current tree for calculating new actions. The process ends when all the detection functions are present in the tree.

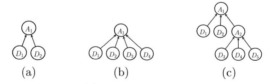

**Fig. 4.** Actions for the greedy algorithm: initial tree structure (a), add detection functions ($D_3$ and $D_4$) (b), add aggregation function ($A_2$, with children $D_3$, $D_4$, and $D_5$) (c).

This greedy algorithm is very fast, so it can be run several times with different randomly-chosen initial trees in order to obtain a set of trees from which the best one is selected. In experiments, given a problem instance, we restart the greedy algorithm about 300 times. The greedy algorithm does not explore the whole state space of trees. In particular, using the actions defined above it cannot obtain a tree in which an aggregation function node has only other aggregation functions as children (like that of Fig. 2b).

## 6   Experimental Evaluation

In this section, we' report the results of some of our experimental activities devoted to evaluate the performance of the algorithms introduced above, combined with the different forms of aggregation functions. All the algorithms have been implemented in Java (as the rest of the MCS Subsystem, which is built using the JADE framework [5]). As anticipated in Sect. 4, each detection or aggregation function is embedded in a different agent in order to have a highly flexible system that could be physically distributed. Considering detection and aggregation functions as agents is further motivated by the fact that the former ones embed different knowledge, while the latter ones can in principle embed arbitrary complex ways to merge anomaly probabilities. The tree structure is embedded in the communication patterns between agents.

Given a tree $T$, the global probability of anomaly returned at the root of $T$ at a time $t$ is considered to signal an anomaly when it is larger than a threshold, set to 0.2 in our experiments. We implement the performance function $P(\mathcal{D}; T)$ of Sect. 4 as follows. Given a labeled log file $\mathcal{D}$ (remember that $\mathcal{D}$, for each time instant $t$, indicates whether the satellite has an anomaly or not), $P(\mathcal{D}; T)$ at time instant $t$ is calculated according to the cost matrix of Table 1, which has been filled considering our application. True positives and true negatives (correct classifications) have cost 0, while we assume that false positives (false alarms) have cost 10 and are less serious than false negatives (unidentified anomalies) that have cost 50. Given $\mathcal{D}$, $P(\mathcal{D}; T)$ is the sum of the costs for each $t$ in $\mathcal{D}$. Note that $P(\mathcal{D}; T)$ is used to find the best aggregation tree (recall the three phases of Sect. 2).

Table 1. Cost matrix to evaluate tree $T$ at time instant $t$.

|  | $\mathcal{D}$ is nominal | $\mathcal{D}$ is anomalous |
|---|---|---|
| $T$ returns nominal | TN: 0 | FN: 50 |
| $T$ returns anomaly | FP: 10 | TP: 0 |

As labeled log file $\mathcal{D}$ for our experiments, we start from a one month log (from July 23, 2013 to August 22, 2013) of the Cryosat-2 satellite, composed of $336,549$ timestamps all corresponding to a nominal behavior of the satellite. The log is then manually injected with artificial events, namely with timestamps for which the values of variables are specified, which can represent anomalies or nominal behavior of the satellite. Note that an injected event at timestamp $t$ produces a probability value for each $D_i(t)$. Using injected anomalies is typical in testing space systems and allows the experimenters to have full control over the settings (e.g., decide the number and the types of injected anomalies).

In the experiments presented in the following, we consider $I = 5$ detection functions, which have been built starting from (nominal behavior) logs of the Cryosat-2 satellite relative to February 2013 and which we consider as fixed.

They represent both numerical and categorical variables whose nominal models are ARMA and Markov models, respectively (see Sect. 4). Our approach is then applied to organise these detection functions in the tree that maximizes anomaly detection on the (injected) log file $\mathcal{D}$.

To the best of our knowledge, there is no other approach that performs anomaly detection in space systems aggregating probabilities relative to single variables. Hence, we can only compare against manual identification of anomalies represented by the labels of events injected in $\mathcal{D}$ performed by a human expert (namely, to the ground truth). Accordingly, the smaller $P(\mathcal{D};T)$, the more similar the identification of anomalies performed by our system to the manual identification of anomalies performed by human operators who labeled data in $\mathcal{D}$ (and the better our approach can support human operators in identifying anomalies).

**In the first experiment**, we inject 20 anomalies in $\mathcal{D}$ according to Table 2 (left), which shows, for each injected event, the probability of anomaly returned by the 5 detection functions we consider (empty cells mean zero). In this case, each anomaly is recognized by just one detection function that returns an anomaly probability of at least 0.3. Intuitively, the best tree structure is the "flat" one, with all the detection functions as children of the root, which is associated to a maximum aggregation function.

Using cooperative negotiation for aggregation functions, there are two optimal solutions (Figs. 5a and 5b) that both produce two false negatives (so cost $P(\mathcal{D};T) = 100$). These solutions have been found by all the algorithms. When allowing the aggregation functions to be either maximum or cooperative negotiation, the algorithms find several optimal tree structures with cost 0, which, as expected, have the root associated to a maximum aggregation function, as for example that of Fig. 5c. It is interesting to note that the children of $A_2$ are all the detection functions that present a probability of 0.3 (at least) in presence of anomalies (see Table 2 (left)). The other detection functions (that return higher

**Table 2.** Injected events for the first, second, and third experiment, respectively (from left to right, A/N means Anomaly/Nominal).

| # | $D_1$ | $D_2$ | $D_3$ | $D_4$ | $D_5$ | A/N |
|----|------|------|------|------|------|-----|
| 01 | 0.5 | | | | | A |
| 02 | 0.5 | | | | | A |
| 03 | | 0.3 | | | | A |
| 04 | | 0.4 | | | | A |
| 05 | | 0.7 | | | | A |
| 06 | | 0.5 | | | | A |
| 07 | | | 0.3 | | | A |
| 08 | | | 0.8 | | | A |
| 09 | | | 0.9 | | | A |
| 10 | | | | 0.7 | | A |
| 11 | | | | 0.7 | | A |
| 12 | | | | | 0.8 | A |
| 13 | | | | | 0.4 | A |
| 14 | | | | | 0.4 | A |
| 15 | | | | | 0.3 | A |
| 16 | 0.4 | | | | | A |
| 17 | 0.8 | | | | | A |
| 18 | 0.5 | | | | | A |
| 19 | | 0.5 | | | | A |
| 20 | | 0.5 | | | | A |

| # | $D_1$ | $D_2$ | $D_3$ | $D_4$ | $D_5$ | A/N |
|----|------|------|------|------|------|-----|
| 01 | | | | 0.60 | | A |
| 02 | | 0.20 | | | | N |
| 03 | 0.10 | 0.15 | 0.35 | | | N |
| 04 | 0.10 | 0.12 | 0.35 | | | N |
| 05 | 0.10 | 0.10 | 0.35 | 0.70 | | A |
| 06 | | | 0.20 | | | N |
| 07 | | | | 0.50 | | A |
| 08 | 0.10 | 0.09 | 0.25 | | | N |
| 09 | 0.20 | 0.17 | 0.35 | | | N |
| 10 | 0.25 | 0.18 | 0.35 | | | A |
| 11 | | | 0.20 | | | N |
| 12 | | | | 0.40 | | A |
| 13 | | | 0.10 | 0.50 | | A |
| 14 | 0.10 | 0.10 | 0.25 | 0.60 | | A |
| 15 | 0.10 | 0.08 | 0.25 | | | N |
| 16 | 0.05 | 0.07 | 0.20 | | | N |
| 17 | 0.30 | 0.28 | 0.45 | | | A |
| 18 | 0.35 | 0.30 | 0.55 | | | A |
| 19 | 0.25 | 0.10 | 0.35 | | | A |
| 20 | 0.10 | 0.12 | 0.25 | | | N |

| # | $D_1$ | $D_2$ | $D_3$ | $D_4$ | $D_5$ | A/N |
|----|------|------|------|------|------|-----|
| 01 | 0.15 | 0.20 | 0.25 | | | N |
| 02 | 0.10 | 0.15 | 0.20 | | | N |
| 03 | 0.25 | 0.28 | 0.32 | | | N |
| 04 | 0.30 | 0.32 | 0.35 | | | A |
| 05 | 0.25 | 0.35 | 0.40 | | | A |
| 06 | 0.35 | 0.40 | 0.50 | | | A |
| 07 | 0.45 | 0.55 | 0.60 | | | A |
| 08 | 0.10 | 0.09 | 0.25 | | | N |
| 09 | 0.20 | 0.17 | 0.35 | | | N |
| 10 | 0.25 | 0.18 | 0.35 | | | N |
| 11 | | | | 0.30 | 0.20 | N |
| 12 | | | | 0.40 | 0.30 | N |
| 13 | | | | 0.50 | 0.40 | N |
| 14 | | | | 0.60 | 0.45 | A |
| 15 | | | | 0.65 | 0.55 | A |
| 16 | | | | 0.70 | 0.40 | A |
| 17 | | | | 0.80 | 0.70 | A |
| 18 | 0.05 | 0.10 | 0.15 | 0.20 | 0.30 | N |
| 19 | 0.15 | 0.20 | 0.25 | 0.30 | 0.40 | N |
| 20 | 0.20 | 0.25 | 0.30 | 0.40 | 0.50 | N |

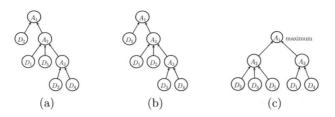

**Fig. 5.** Optimal tree structures for the first experiment with cooperative negotiation aggregation functions ((a) and (b)) and with cooperative negotiation ($A_2$ and $A_3$) and maximum ($A_1$) aggregation functions (c).

probabilities in presence of anomalies) are children of $A_3$. In this sense, we say that our approach can create trees that capture the structure of the problem.

Now we consider the case in which all aggregation functions perform average. In this case, the best solutions found by the algorithms are definitely worse than those found before and correspond to 13 false negatives. The problem is that the weights $w_i$ associated to the detection functions are large and so, being in this first experiment an anomaly identified by a single detection function, it is difficult for the probability returned by that function "to emerge" over the zero probabilities of the other detection functions.

**In the second experiment**, we inject in $\mathcal{D}$ the events shown in Table 2 (center). In this case, we have three detection functions ($D_1$, $D_2$, and $D_3$) whose behavior is similar (i.e., they are correlated in identifying the same anomalies). However, $D_3$ returns large anomaly probabilities also in case of nominal behavior. With this experiment, we would like to check if our approach is able to find tree structures that model the correlation between $D_1$, $D_2$, and $D_3$ and that do not produce false positives. Indeed, our algorithms are able to find several optimal solutions $T^*$ with cost 0. One of them is reported in Fig. 6a and displays the expected structure in which $D_1$, $D_2$, and $D_3$ are children of the same aggregation function (all aggregation functions perform cooperative negotiation). It is also interesting that no false positive is returned despite the fact that $D_3$ returns probabilities of anomaly up to 0.35 in presence of nominal behavior.

Figure 7 (left) compares the algorithms. The y-axis shows the cost $P(\mathcal{D}; T)$ of the trees, while the x-axis shows the solutions (trees) generated by the

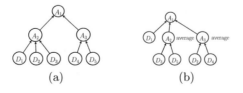

**Fig. 6.** Optimal tree structure for the second experiment with cooperative negotiation aggregation functions (a) and for the third experiment with cooperative negotiation ($A_1$) and average ($A_2$ and $A_3$) aggregation functions (b).

algorithms ordered according to their increasing costs. For the enumeration algorithm, all the $N$ generated trees are reported; while for the simulated annealing and greedy algorithms, the best $N$ trees they build are reported. From the graph, it seems that simulated annealing algorithm finds more optimal solutions (with cost 0) than the enumeration algorithm. However, some of the trees built by the simulated annealing algorithm are repeated (the same happens for the greedy algorithm). The greedy algorithm performs worse than the other two algorithms. The results of the greedy algorithm shown in the figure are obtained considering $z = 2$. If we set $z = 4$, the greedy algorithm is not able to find any optimal solution.

In this second experiment, there are no advantages in employing the maximum aggregation functions (because they introduce false positives) nor in using the average aggregation functions (because they generate false negatives, as in the first experiment).

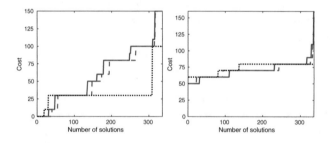

**Fig. 7.** Number of solutions for the second (left) and third (right) experiment and corresponding costs generated by the enumeration algorithm (continuous blue curve), by the simulated annealing algorithm (dashed red line), and by the greedy algorithm (dotted black line). Results of simulated annealing and greedy algorithms are averages over 10 runs. Aggregation functions perform cooperative negotiation.

**In the third experiment**, we inject in $\mathcal{D}$ some events in order to create two subsets of correlated detection functions (Table 2 (right)). Specifically, detection functions $D_1$, $D_2$, and $D_3$ are correlated and trigger an anomaly when one of them returns a probability larger than 0.3 and, similarly, detection functions $D_4$ and $D_5$ are correlated and trigger an anomaly when at least one of them returns a probability larger than 0.5. Note that the injected events 19 and 20 are particularly hard to correctly classify as nominal behavior, since the probabilities returned by the detection functions are high.

Using cooperative negotiation for all aggregation functions, the best tree structures the algorithms can find produce 5 false positives (cost $P(\mathcal{D}; T) = 50$, according to our cost matrix). The performance of the algorithms is compared in Fig. 7 (right), which shows that simulated annealing and greedy algorithms perform worse than the enumeration algorithm. This can be explained by noticing that the best solutions correspond to deep and unbalanced trees, which are generated less frequently by the simulated annealing and greedy algorithms.

Considering maximum aggregation functions does not improve the results, because they are not able to solve the problem of false positives. On the contrary, considering average aggregation functions improves the quality of the tree structures found by the algorithms, because average tends to "uniform" the values of probability. Finally, allowing any combination of aggregation functions (maximum, average, and cooperative negotiation), the optimal solution is that of Fig. 6b with cost $P(\mathcal{D}; T) = 20$, due to 2 false positives.

We now discuss the **scalability** of the algorithms with the number $I$ of detection functions. We consider that all the aggregation functions are of the same form. (As said, if they can be of $m$ different forms, $m = 3$ in our case, we can have $J^m$ different configurations for each tree, where $J$ is the number of aggregation functions.) Assuming that aggregation functions can be calculated almost instantaneously (this is trivially true for maximum and average, but requires that all cooperative negotiations are run off-line for all time instants $t$ comprised in $\mathcal{D}$), Table 3 shows that the enumeration algorithm does not scale well with the number $I$ of detection functions. On the other hand, computing time of simulated annealing and greedy algorithms is controllable by the programmer, who can set the maximum number of iterations and the value of $z$, respectively. This flexibility comes at the cost of the impossibility to guarantee any bound on the quality of the solutions. On typical instances, the simulated annealing algorithm analyzes a number of trees per second comparable to that of the enumeration algorithm, while the greedy algorithm generates about 800 trees per second (using about 300 restarts and $z = 2$).

**Table 3.** Computing times ($^*$ are estimates) of the enumeration algorithm on an i7-4770K 3.50 GHz computer.

| $I$ | 5 | 6 | 7 | 8 | 9 | 10 | 11 | 12 |
|---|---|---|---|---|---|---|---|---|
| Trees | $3 \cdot 10^2$ | $6 \cdot 10^3$ | $1 \cdot 10^5$ | $4 \cdot 10^6$ | $2 \cdot 10^8$ | $8 \cdot 10^9$ | $4 \cdot 10^{11}$ | $3 \cdot 10^{13}$ |
| Time | 0 s | 0 s | 0 s | 15 s | 9 m | 8 h | 19 d$^*$ | 3.7 y$^*$ |

From our results, some **guidelines** for aggregating detection functions can be derived. If the global anomalies in $\mathcal{D}$ are strongly related to single detection functions (as in the first experiment), then a simple "flat" tree structure with a maximum aggregation function at the root is the best solution. If a group of detection functions are correlated in signalling global anomalies (like in the second experiment), then tree structures with cooperative negotiation aggregation functions seem to work well. Finally, when there are multiple groups of correlated detection functions (like in the third experiment), a combination of different types of aggregation functions is needed. In general, the maximum aggregation function increases the sensitivity to small anomaly probabilities, while the average aggregation function smooths large anomaly probabilities. Cooperative negotiation aggregation functions seem to balance the two extremes above. Turning to algorithms for finding the best aggregation tree, the simulated annealing algorithm shows the best trade-off between quality of solutions (which is often similar to that of solutions found by the enumeration algorithm, as Fig. 7 shows) and computing time (which is tunable by setting the number of iterations).

## 7    Conclusion

In this paper we have presented an approach that returns a global anomaly probability for a space system by aggregating the anomaly probabilities returned by models that identify deviations from nominal behaviors of single measurable variables of the system, which we have developed in the context of the FCTMAS study. Experiments show that the proposed approach can build aggregation trees able to identify global anomalies in space systems with performance comparable with that of human operators. Given the nature of our detection functions (which we consider as given), the global anomalies that our approach is able to identify well are those signalled by anomalous behavior of single variables or of groups of variables at *single* time instants. Global anomalies that can be detected by looking at the values of variables at *different* time instants are not currently identified.

Future work includes considering the complexity of a tree $T$ when evaluating $P(\mathcal{D}; T)$ in order to promote trees with a desired structure (e.g., balanced trees or trees with a given number of levels). While we do not consider any *a priori* domain knowledge, an interesting direction for future research is to include domain knowledge to direct the search of the best tree structure. Finally, other aggregation functions and algorithms could be considered.

**Acknowledgment.** The authors kindly acknowledge the contributions of Matteo Gallo and Matteo Garza to the development of the MCS Subsystem described in this paper.

## References

1. Amigoni, F., Beda, A., Gatti, N.: Multiagent systems for cardiac pacing simulation and control. AI Commun. **18**(3), 217–228 (2005)
2. Amigoni, F., Beda, A., Gatti, N.: Combining rate-adaptive cardiac pacing algorithms via multiagent negotiation. IEEE Trans. Inf. Technol. B **10**(1), 11–18 (2006)
3. Amigoni, F., Brambilla, A., Lavagna, M., Blake, R., le Duc, I., Page, J., Page, O., de la Rosa Steinz, S., Steel, R., Wijnands, Q.: Agent technologies for space applications: the DAFA experience. In: Proceedings of IAT, pp. 483–489 (2010)
4. Amigoni, F., Gatti, N.: A formal framework for connective stability of highly decentralized cooperative negotiations. Auton. Agent Multi Agent Syst. **15**(3), 253–279 (2007)
5. Bellifemine, F., Caire, G., Greenwood, D.: Developing Multi-agent Systems with JADE. Wiley, Hoboken (2007)
6. Cayrac, D., Dubois, D., Prade, H.: Handling uncertainty with possibility theory and fuzzy sets in a satellite fault diagnosis application. IEEE Trans. Fuzzy Syst. **4**(3), 251–269 (1996)

7. Chien, S., Sherwood, R., Tran, D., Cichy, B., Rabideau, G., Castano, R., Davies, A., Lee, R., Mandl, D., Frye, S., Trout, B., Hengemihle, J., D'Agostino, J., Shulman, S., Ungar, S., Brakke, T., Boyer, D., Gaasbeck, J.V., Greeley, R., Doggett, T., Baker, V., Dohm, J., Ip, F.: The EO-1 autonomous science agent. In: Proceedings AAMAS, pp. 420–427 (2004)
8. Codetta-Raiteri, D., Portinale, L., Guiotto, A., Yushstein, Y.: Evaluation of anomaly and failure scenarios involving an exploration rover: a Bayesian network approach. In: Proceedings of iSAIRAS (2012)
9. Daly, K., Gai, E., Harrison, J.: Generalized likelihood test for FDI in redundant sensor configurations. J. Guid. Control Dyn. **2**(1), 9–17 (1979)
10. Fischler, M., Bolles, R.: Random sample consensus: a paradigm for model fitting with applications to image analysis and automated cartography. Commun. ACM **24**(6), 381–395 (1981)
11. Holsti, N., Paakko, M.: Towards advanced FDIR components. In: Proceedings of DASIA (2001)
12. Koller, D., Friedman, N.: Probabilistic Graphical Models. The MIT Press, Cambridge (2009)
13. Lavagna, M., Sangiovanni, G., Da Costa, A.: Modelization, failures identification and high-level recovery in fast varying non-linear dynamical systems for space autonomy. In: Proceedings of DCSSS, pp. 451–550 (2004)
14. Massioni, P., Sangiovanni, G., Lavagna, M.: Innovative software for autonomous fault detection and diagnosis on space systems. In: Proceedings of ASTRA (2006)
15. Muscettola, N., Dorais, G., Fry, C., Levinson, R., Plaunt, C.: IDEA: planning at the core of autonomous reactive agents. In: Proceedings of IWPSS (2002)
16. Muscettola, N., Nayak, P., Pell, B., Williams, B.: Remote agent: to boldly go where no AI system has gone before. Artif. Intell. **103**, 5–47 (1998)
17. Neerincx, M.: Situated cognitive engineering for crew support in space. Pers. Ubiquit. Comput. **15**(5), 445–456 (2011)
18. Smith, E., Korsmeyer, D.: Intelligent systems technologies for human space exploration mission operations. In: Proceedings of SMC-IT, pp. 169–176 (2011)
19. Sneha, S., Varshney, U.: Enabling ubiquitous patient monitoring: model, decision protocols, opportunities and challenges. Decision Support Syst. **46**(3), 606–619 (2009)
20. Ulerich, N., Powers, G.: On-line hazard aversion and fault diagnosis in chemical processes: the digraph+ fault-tree method. IEEE Trans. Reliab. **37**(2), 171–177 (1988)
21. Vallejo, D., Albusac, J., Castro-Schez, J., Glez-Morcillo, C., Jimenez, L.: A multi-agent architecture for supporting distributed normality-based intelligent surveillance. Eng. Appl. Artif. Intel. **24**(2), 325–340 (2011)
22. Venkatasubramanian, V., Rengaswamy, R., Kavuri, S.: A review of process fault detection and diagnosis: part II: qualitative models and search strategies. Comput. Chem. Eng. **27**(3), 313–326 (2003)
23. Venkatasubramanian, V., Rengaswamy, R., Kavuri, S., Yin, K.: A review of process fault detection and diagnosis: part III: process history based methods. Comput. Chem. Eng. **27**(3), 327–346 (2003)
24. Venkatasubramanian, V., Rengaswamy, R., Yin, K., Kavuri, S.: A review of process fault detection and diagnosis: part I: quantitative model-based methods. Comput. Chem. Eng. **27**(3), 293–311 (2003)
25. Volpe, R., Nesnas, I., Estlin, T., Mutz, D., Petras, R., Das, H.: The CLARAty architecture for robotic autonomy. In: Proceedings of IEEE Aerospace Conference, pp. 121–132 (2001)

26. Yin, S., Ding, S., Xie, X., Luo, H.: A review on basic data-driven approaches for industrial process monitoring. IEEE Trans. Ind. Electron. **61**(11), 6418–6428 (2014)
27. Zaher, A., McArthur, S., Infield, D., Patel, Y.: Online wind turbine fault detection through automated SCADA data analysis. Wind Energy **12**(6), 574–593 (2009)

# Detection of Motion Patterns and Transition Conditions for Automatic Flow Diagram Generation of Robotic Tasks

Guilherme de Campos Affonso[✉], Kei Okada, and Masayuki Inaba

JSK Laboratory, The University of Tokyo, Tokyo, Japan
affonso@jsk.imi.i.u-tokyo.ac.jp

**Abstract.** In this paper a method for detecting motion patterns and transitions between them in source code describing robotics tasks is proposed, being used to automatically generate flow diagrams of the corresponding task. This is done through the combination of both static and dynamic program analysis, first segmenting original code based on token matching and then using runtime data to judge importance of each segment and relations between them. Generation of flow diagrams is not only a simple way of visualizing overall procedure of complex tasks, but can also be used for current state identification and fail recovery systems. Proposed system is verified through experimentation using PR2 robot on the household task of cleaning up a table, being able to automatically generate a flow diagram with reasonable number of states.

**Keywords:** Flow diagram · Visualization of robotic tasks
Dynamic programming analysis

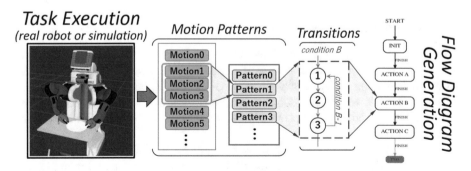

**Fig. 1.** System proposed extracts motion patterns and transitions between them from robotic tasks, through analysis of execution time data and source code itself. This information is then used for automatic generation of flow diagrams, having a simplified version shown at the right (expanded version is shown at the end of this paper).

M. Strand et al. (Eds.): IAS 2018, AISC 867, pp. 161–173, 2019.
https://doi.org/10.1007/978-3-030-01370-7_13

# 1   Introduction

Not anymore relying on simple iteration of repetitive tasks, current robotic development focuses on system's adaptability and functionality when placed under unstructured environments. One of the keys for realizing such adaptability is the capacity for the robot to recognize the environment and select optimal actions for each situation, using highly branched code. Under these circumstances, it is easy for the system to exponentially grow more and more complex, making it difficult to understand and deal with. Specially when facing situations in which the user is required to read code wrote by other developers, such as when dealing with larger projects that require collaborative work, performing software maintenance by multiple personal, or inheriting previously build libraries, having accessible ways of understanding the system directly affects productivity.

Visualization of the system's overall procedure by e.g. flow diagrams, is a known counteract for such problem, also proving to be a good documentation tool [1]. In the particular case of robotic tasks, it is also possible to use flow diagrams and information on detected motion patterns to determine current state during execution and establishing fail recovery systems [2]. This way, extraction of motion patterns and information regarding translations between them not only makes the system easy to be understood by humans, but can also be an effective method for improving task's adaptability and self-reliance of the code.

This paper explores the possibility of automatically generating flow diagrams describing robotic tasks through the extraction of common motion patterns and analysis of the conditions between them. In this work implementation of both the proposed system and robotic task itself is done in Lisp, being the task definition done within a subset of the language, which cannot include any *goto* statement or directly recursive procedures.

# 2   Visualization of Source Code

## 2.1   Schematic Representations

Schematic graphical representations are widely used in the computer science field, as a way of abstracting and allowing easier understanding of complex structures. Among such, binary decision diagrams [3], state machines [4] and flow diagrams [5] can be pointed out for special relevance. Such representations are commonly used in description of boolean functions, automata behavior and algorithms, respectively.

Not only being of easy visualization by humans, such data structures provide ways for machines to organize, interpret and execute source code, being widely used on compile time processing. Application examples for binary decision diagrams and finite state machines include automatic generation of test code [6] and development of design aid verification tools [7], specially for integrated circuits.

Creation of such diagrams from source code, however, faces a variety of problems such as the *halting* [8] and *aliasing* [9] determination, tending to be undecidable for Turing-complete languages. Under such circumstances, past approaches are divided between those dealing with subsets of higher level languages [10,11] and those which implement new languages, suitable for easy transition between literal and visual representations [12,13]. This work takes the former approach.

## 2.2   Our Method

Our aim is to automatically generate flow diagrams able to schematically and precisely express the overall procedure described on source code, being easily understandable by humans. Differently from past work centered on machine-readable results, generation of human-readable outcome demands for the final diagram to have reasonable size. In this work this is achieved by combining both static and dynamic program analysis, applying token matching based algorithms for identifying branching points and using runtime data to regroup blocks into common motion patterns, minimizing the total number of states. Finally, detection of transition conditions – edges in the final diagram – allows further grouping of related motion patterns into actions, which is done by recursive lexical analysis of contexts and also aided by execution data.

# 3   Detecting Motion Patterns in Robotic Tasks

Motion patterns are extracted by (i) Identifying and labeling all conditional clauses present in the source code, segmenting it into minimal indivisible blocks; (ii) Using runtime data to detect significant branching points, which are used to regroup segments into larger, more meaningful blocks – the motion patterns.

## 3.1   Code Segmentation

Segmentation of original code into minimal blocks that do not include any branching is done through the following procedure, which at this point does not support *goto* statements nor directly recursive processes. Overall task execution can be described as the combination of such segments, being further subsets of them at no time required.

1. **Code expansion**
   At first, macros and user defined functions are expanded into the original definition, generating output that is at the same time functionally equivalent to the original code and granted to be exclusively composed by built-in functions. This step is important to expose conditional clauses that might be hidden in the code, as illustrated in Fig. 2.
2. **Search for conditional clauses and labeling**
   During code expansion, tokens signalizing the function name are matched against a list containing the name of all built-in conditional special operators,

such as *if* and *while*. When matching succeeds, each possible outcome is given an unique label, used as the segment ID. A simple example of labeling is illustrated at Fig. 3.

```
1    (defun func1 (a) (if (> a 10) (print 1)))
2    (defun func2 (a b) (let ((m (* a b))) (if (> m 100) (func1 a))))
3
4    (expand '(func2 input 5))
5    ⇒
6    ((lambda (a b)
7      (let ((m (* a b)))
8        (if (> m 100)
9          ((lambda (a)
10            (if (> a 10) (print 1)))
11           a))))
12    input 5)
```

**Fig. 2.** Expansion of the function *func2* into its components, disclosing hidden *if* clauses present at both *func2* and *func1* definitions.

```
process_0          ←     LABEL 0
while condition_1 do
    process_1      ←     LABEL 1
    if condition_2 then
        process_2  ←     LABEL 2
    else
        process_3  ←     LABEL 3
    end if
    process_4      ←     LABEL 4
end while
process_5          ←     LABEL 5
```

**Fig. 3.** Example of labeling a simple program with *if* and *while* clauses. Every possible outcome of conditional clauses is given a different and unique label.

## 3.2   Pattern Detection and Regrouping

Number of labels obtained in segmentation can prove to be incredibly high, reaching order of hundreds even for relatively simple coding. In order to minimize the extracted number of blocks while preserving its indivisible property, this work proposes the use of dynamic program analysis methods, based on recursive solving of the longest common substring problem on gathered runtime data. Detailed procedure is given at the following.

1. **Execution**
   Expanded and labeled code is executed a few times, reproducing the original task under different conditions. Labels of every block called during execution are gathered, making possible to represent each trial as a sequence of labels.

---

**Algorithm 1.** Extract orthogonal common blocks from sequences

---

**Require:** *array* holding all of input sequences
**Ensure:** list of blocks present on *array* elements.
  $n \leftarrow$ {length of *array*}
  $res \leftarrow$ []
  **while** there are elements on *array* **do**
    *com* $\leftarrow$ {common longest substring present in at least $n$ elements of *array*}
    **if** *com* **then**
      $res \leftarrow append(com, res)$
      remove all occurrences of *com* from all elements of *array*
      remove all empty lists from *array*
    **else**
      $n \leftarrow n - 1$
    **end if**
  **end while**
  **if** *res* differs from input *array* **then**
    *array* $\leftarrow res$
    **apply** procedure from start
  **end if**
  **return** *res*

---

2. **Pattern detection**
   Common patterns are extracted by recursively solving the common longest substring problem [14] on the sequence of labels obtained during task execution. Pseudo code for this algorithm is shown in Algorithm 1, and example of execution is illustrated on Fig. 4.
3. **Importance evaluation and filtering**
   Since robotic tasks are fully described by the actual movements of the robot, it is not desired to recognize other kinds of operations as fundamental blocks. Here, information gathered from robot controllers is used to filter blocks that do not contain any command to the real robot, providing only meaningful patterns.

## 4  Detecting Transition Conditions

Transition conditions are extracted by (i) Identifying the triggering clause of each motion pattern by focusing on the first label of each block; (ii) Analyzing the relation between motion patterns, determining to which motion patterns should the triggering clauses be linked to.

### 4.1  Identifying Triggering Clauses

If sufficient execution data is given, sequence of motions extracted through the previous method can also be seen as minimal and indivisible blocks. This means that besides having several conditional statements (labels), they are called in

```
INPUT: array ← [[a b c], [a b c a b]]
length(array) = 2

common(2,array): [a b c]
;; Remove common from elements in array:
array = [[], [a b]]
;; Remove empty lists from array:
array = [[a b]]

common(2,array): none
common(1,array): [a b]
array =   []

;; [[a b c], [a b]] ≠ [[a b c], [a b c a b]]
INPUT: array ← [[a b c], [a b]]
;; Repeat while res differs from input array

;; Return common blocks:
OUTPUT ⇒ [[a b], [c]]
```

**Fig. 4.** Sample execution of algorithm for extracting common blocks from sequences. This process allows to minimize the number of segments while preserving it's indivisible property, under the executed conditions.

ways that always lead to the same outcome, without any kind of branching. The only valid conditional clause on such sequences is the one corresponding to the first element, whose evaluation determines if the motion pattern as a whole will be executed or not – being, therefore, the triggering clause of given block.

Triggering clauses can be extracted by previously associating each label with its corresponding condition, during the segmentation process. When considering the example given on Fig. 3, for instance, conditional statements relative to labels 1 to 3 would be: *condition_1*, *condition_2*, *NOT condition_2*, respectively. Furthermore, labels 0, 4 and 5 do not have any conditional statement matching, meaning that corresponding motions would be executed unconditionally after execution of previous blocks (or at the beginning, for *process_0*).

It is common for such clauses, however, to rely on locally bound variables, offering little information about its containing. For example, in Fig. 2, extraction of the first conditional statement results on *IF(> m 100)*, which gives no information about original input values *input* and 5.

In this paper, above problem is coped with by recursively analyzing lexical bindings on sequentially broader contexts, starting from the clause itself and being expanded through back search and matching of context delimiters. Detailed procedure is given at Algorithm 2, and example of execution for Fig. 2 code is illustrated at Fig. 5. Here, *substitute(new, old, src)* is a function that substitutes occurrences of *old* by *new* at *src*, and *context_of(seg, src, lvl)* one that outputs the block of code describing the context of segment *seg* at *src*. Integer variable *lvl* indicates how many additional times context should be expanded relative to original point. For instance, when dealing with the tree (a b (c (d (e) f))), point indicating letter 'e' is expanded as 'e'; (e); (d (e) f); (c (d (e) f)); (a b (c (d (e) f))), for levels 0 to 4.

---

**Algorithm 2.** Evaluate local bindings

---

**Require:**
  *source* being the original source code
  *clause* to be evaluated
**Ensure:** expansion of variables present at *clause* into original definitions
  **Procedure** *expand_symbol* (clause $c$, source $src$, int $lvl$)
  **if** $c$ is a function call **then**
    $fn \leftarrow$ function name
    $args \leftarrow$ function arguments
    $expanded\_args \leftarrow$ for term in args, collect $expand\_symbol$(term, $src$, 0)
    **return** $fn(expanded\_args)$
  **else**
    $ctx \leftarrow context\_of(ctx, src, lvl)$
    **if** $c$ is a symbol unbound in $ctx$ **then**
      **return** $expand\_symbol(c, src, lvl + 1)$
    **else**
      $eval \leftarrow$ evaluation value of $c$
      **if** $eval$ equals $c$ **then**
        **return** $c$
      **else**
        $new\_src \leftarrow substitute(eval, ctx, src)$
        **return** $expand\_symbol(eval, new\_src, 0)$
      **end if**
    **end if**
  **end if**
  $start\_ctx \leftarrow context\_of(clause, source, 1)$
  $start\_src \leftarrow substitute(clause, start\_ctx, source)$
  **RETURN** $expand\_symbol(clause, start\_src, 0)$

---

Because this method does not take in account variable assignments appearing at block's body, it is not suitable for cases including destructive operations. For instance, condition of code (let ((i 0)) (while (< i 10) (setq i (1+ i)))) would be mistakenly expanded as *WHILE(< 0 10)*, an endless loop. As a partial counteract for this problem, here runtime data is used to determine if the variable's value remains constant or not, outputting instantiating forms e.g. (initialize i 0) in the case it doesn't.

## 4.2   Determining Relations Between Motion Patterns

Since labels are given in order of expansion and, as long as there are no *goto* like jumping statements, execution, overall flow is assumed to be a linear progression from motion patterns with lower value labels to higher value ones. In fact, this is verified to be true when dealing with cases that do not include nested structures with multiple motion patterns, here denominated as actions. If the task does include such actions, representation using motion patterns becomes more complex and branched, possibly being of hard visualization. In such cases,

```
clause ← (> m 100)
start_src ⇒
  ((lambda (a b)
     (let ((m (* a b)))
       (> m 100)))
   input 5)

;;   expand_symbol(clause, start_src, 0)
 is_function(clause): true
fn ← >
args ← [m, 100]

;;   expand_symbol(m, start_src, 0)
ctx ← m
is_unbound(m, ctx): true

;;   expand_symbol(m, start_src, 1)
ctx ← (> m 100)
is_unbound(m, ctx): true

;;   expand_symbol(m, start_src, 2)
ctx ← (let ((m (* a b))) (> m 100))
eval ← (* a b)
equal(eval, m): false

new_src ← substitute((* a b), m, src) ⇒
  ((lambda (a b) (* a b)) input 5)

;;   expand_symbol((* a b), new_src, 0) → is_function
;;   expand_symbol(a, new_src, 0) → unbound
;;   expand_symbol(a, new_src, 1) → unbound
;;   expand_symbol(a, new_src, 2) → unbound
;;   expand_symbol(a, new_src, 3) → eval to input

;;   expand_symbol(input, input, 0)
eval ← input
equal(eval, m): true
⇒ input

expand_symbol(b, new_src, 0) ⇒ 5
expand_symbol(100, start_src, 0) ⇒ 100

OUTPUT ⇒ (> (* input 5) 100)
```

**Fig. 5.** Sample execution of algorithm for evaluating local bindings. This process allows the extraction of context independent clauses, more meaningful for generated diagram.

considering actions as the diagram nodes allows to represent the overall flow as a simple linear progression, as showed at the right side of Fig. 1, simplified version of the flow diagram given in Fig. 10.

Actions can be identified by focusing on the upper context of each triggering clause i.e. *context_of(clause, src, 1)*, listing all labels present on it and collecting motion patterns with corresponding labels. If the structure is not nested, this will return only one element – the motion itself.

## 5   Experiment on Flow Diagram Generation

Proposed system is verified through experimentation with PR2 robot, performing the domestic task of cleaning up a table, removing all plates and cutlery on it. This task is accomplished by consecutively piling up the dishes, placing all cutlery on top of it and then picking everything up, as shown in Fig. 6. In this experimentation, task is executed with both the real robot and simulation a total of six times, variating the number of plates from 1 to 3 and number of cutlery from 0 to 2, as shown at Table 1. Number of trials and conditions of each trial were decided in order to ensure the variety of combinations of the executed steps, shown at the bottom of Table 1. Since the system only recognizes branching points appearing at runtime data as relevant, executing the task under various conditions is crucial for ensuring the indivisible property of extracted motion patterns.

**Fig. 6.** Task of cleaning up a table, implemented on PR2 robot. Task is accomplished by piling up dishes, collecting cutlery and picking everything up, in order.

Automatic detection of motion patterns leads to the extraction of six blocks, given in Fig. 7. At this implementation, Block 0 always describes the initial state. Automatic identification of triggering clauses – clauses associated with the first label of each block – lead to the result described on Fig. 8. These statements are segments of the original source code, where *one-shot-subscribe* is a system function that receives messages from the recognition topic and *ri* is the robot interface, to where controller commands are send to. Interpreting the code, clause of label 1 translates to 'repeat (*number_of_plates* - 1) times', label 286 to 'repeat (*number_of_cutlery*) times', and label 300 to 'try to grasp with effort 0.007, repeating if final gripper position value is bigger than 25 mm'.

**Table 1.** Execution conditions.

| Trial | a | b | c | d | e | f |
|---|---|---|---|---|---|---|
| Number of plates | 1 | 1 | 1 | 2 | 2 | 3 |
| Number of cutlery | 0 | 1 | 2 | 0 | 1 | 0 |
| Pile plate executions | 0 | 0 | 0 | 1 | 1 | 2 |
| Collect cutlery exec. | 0 | 1 | 2 | 0 | 1 | 0 |
| Pick up executions | 1 | 1 | 1 | 1 | 1 | 1 |

```
BLOCK 0 ⇒ (0)
BLOCK 1 ⇒ (1 2 3 5 6 8 ... 234 235 236 237)
BLOCK 2 ⇒ (286 287 289 290 292 294 295 296
    298 299)
BLOCK 3 ⇒ (300)
BLOCK 4 ⇒ (301 302 303 305 306 307 308)
BLOCK 5 ⇒ (309 355 356 ... 429 430 432 433)
```

**Fig. 7.** Collected motion patterns.

Finally, analyzing the relations between motion patterns allows identification of four actions: Block 0, Block 1, Block 2 + Block 3 + Block 4, Block 5.

This information is then used to automatically generate the flow diagram shown in Fig. 10, drawn with the tool SMACH [15]. Furthermore, generated diagram is also used for real time visualization of the state being currently executed, as illustrated on Fig. 9. This allows to understand Action A as pilling up plates; Action B as collecting cutlery and Action C as picking everything up.

```
LABEL 0   ⇒ none
LABEL 1   ⇒ WHILE (< (initialize i 0) (1- (length (send (one-shot-subscribe
    "bounding_box_array/plate_boxes" jsk_recognition_msgs::boundingboxarray)
    :boxes))))
LABEL 286 ⇒ WHILE (< (initialize i 0) (length (send (one-shot-subscribe
    "bounding_box_array/cutlery_boxes" jsk_recognition_msgs::boundingboxarray
    :timeout 500) :boxes)))
LABEL 300 ⇒ WHILE (>= (send *ri* :start-grasp :rarm :gain 0.007) 25)
LABEL 301 ⇒ none
LABEL 309 ⇒ none
```

**Fig. 8.** Conditional clauses extracted from the first label of each motion pattern.

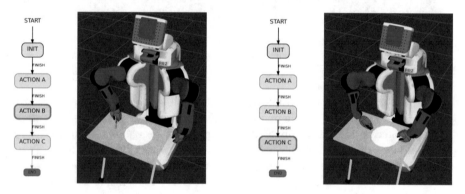

**Fig. 9.** Real time visualization of state in progress, using reduced flow diagram and simulation. Left side image shows Action B: collect cutlery and right side image shows Action C: pick everything up.

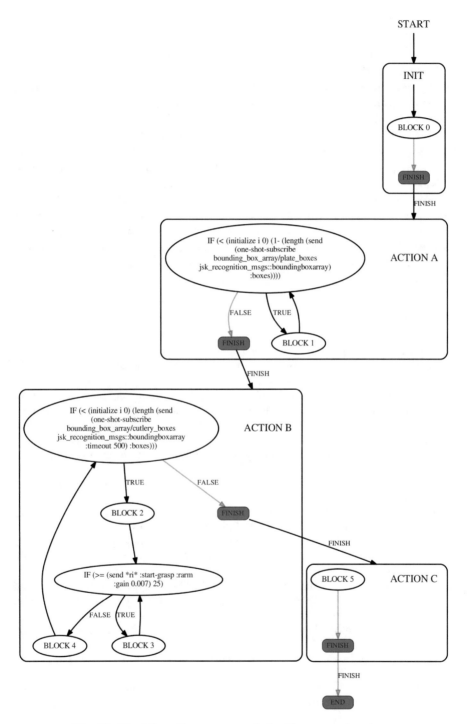

**Fig. 10.** Final flow diagram generated with proposed system.

# 6  Conclusion

Combination of both static and dynamic program analysis, by top down lexical evaluation of source code and comparison of runtime data, can prove to be useful for detecting common patterns in source code and relations between them. As one of this applications, automatic generation of flow diagrams with reasonable number of states is given, which has the merits of easing the understanding of complex systems and proving to be a good documentation tool. When dealing with robotic tasks, such diagrams also provide a compact data structure useful for e.g. improving the self-reliance of the code by implementing fail-recovery systems or displaying actions in progress at real time.

Since labels given to states do not hold any semantic meaning (e.g. BLOCK3, ACTION1), real time visualization is thought to be a significant aid to understanding what each block represent, improving the readability of the generated diagram. Implementation of more self-explaining naming for states and conditions in the flow diagram by e.g. performing semantic analysis of source code is left to future work.

# References

1. Protsko, L.B., Sorenson, P.G., Tremblay, J.P., Schaefer, D.A.: Towards the automatic generation of software diagrams. IEEE Trans. Softw. Eng. **17**(1), 10–21 (1991)
2. Furuta, Y., Inagaki, Y., Okada, K., Inaba, M.: Self-improving robot action management system with probabilistic graphical model based on task related memories. In: International Conference on Intelligent Autonomous Systems, pp. 811–823. Springer (2016)
3. Bryant, R.E.: Graph-based algorithms for Boolean function manipulation. IEEE Trans. Comput. **100**(8), 677–691 (1986)
4. Minsky, M.L.: Computation: Finite and Infinite Machines. Prentice-Hall Inc., Hoboken (1967)
5. Böhm, C., Jacopini, G.: Flow diagrams, turing machines and languages with only two formation rules. Commun. ACM **9**(5), 366–371 (1966)
6. Cheng, K.-T., Krishnakumar, A.S.: Automatic functional test generation using the extended finite state machine model. In: 30th Conference on Design Automation, pp. 86–91. IEEE (1993)
7. Burch, J.R., Clarke, E.M., McMillan, K.L., Dill, D.L.: Sequential circuit verification using symbolic model checking. In: Proceedings of the 27th ACM/IEEE Design Automation Conference, pp. 46–51. ACM (1991)
8. Boyer, R.S., Strother Moore, J.: A mechanical proof of the unsolvability of the halting problem. J. ACM (JACM) **31**(3), 441–458 (1984)
9. Landi, W.: Undecidability of static analysis. ACM Lett. Program. Lang. Syst. (LOPLAS) **1**(4), 323–337 (1992)
10. Cheng, K.-T., Krishnakumar, A.S.: Automatic generation of functional vectors using the extended finite state machine model. ACM Trans. Des. Autom. Electron. Syst. (TODAES) **1**(1), 57–79 (1996)

11. Jou, J.-Y., Rothweiler, S., Ernst, R., Sutarwala, S., Prabhu, A.: Bestmap: behavioral synthesis from c. In: Proceedings of the International Workshop on Logic Synthesis (1989)
12. Costagliola, G., Tortora, G., Orefice, S., De Lucia, A.: Automatic generation of visual programming environments. Computer **28**(3), 56–66 (1995)
13. Naebi, A., Khalegi, F., Hosseinpour, F., Zanjanab, A.G., Khoshravan, H., Kelishomi, A.E., Rahmatdoustbeilankouh, B.: A new flowchart and programming technique using bond graph for mechatronic systems. In: 2011 UkSim 13th International Conference on Computer Modelling and Simulation (UKSim), pp. 236–241. IEEE (2011)
14. Gusfield, D.: Algorithms on Strings, Trees and Sequences: Computer Science and Computational Biology. Cambridge University Press, New York (1997)
15. Open Source Robotics Foundation. smach - ROS Wiki. http://wiki.ros.org/smach

# Least Action Sequence Determination in the Planning of Non-prehensile Manipulation with Multiple Mobile Robots

Changxiang Fan[✉], Shouhei Shirafuji, and Jun Ota

The University of Tokyo, 5-1-5 Kashiwanoha, Kashiwa City, Chiba Prefecture, Japan
fan@race.u-tokyo.ac.jp

**Abstract.** To complete a non-prehensile manipulation task, using the lowest number of manipulation sequences with a well-determined number of robots is desirable to improve efficiency of the manipulation and to bring stability to it. Since many possible states exist in the manipulation, various manipulation sequences exist to finish the task. Furthermore, the number of robots should be determined according to the environment. In this work, a graph-based planning method was used to determine the lowest possible number of sequences in the non-prehensile manipulation of mobile robots with a determined number of robots for gravity closure. Based on all possible object-environment contacts, the required number of robots for gravity closure was determined to generate possible manipulation states in the contact configuration space. A state transition graph was created by representing the obtained states as nodes and then determining the least action sequences by searching for the shortest path in the graph.

**Keywords:** Multiple mobile robots · Non-prehensile manipulation
Manipulation planning · State transition graph

## 1 Introduction

In robotic manipulations, it is often seen that objects are manipulated by manipulators using prehensile method. However, such manipulation method is unavailable for big-sized objects in narrow space, as big industrial robots cannot enter such environment. Therefore, the small-sized mobile robots are widely adopted to such kind of manipulation tasks owing to their motion flexibility. As the small mobile robots cannot manipulate big objects with prehensile method, non-prehensile method is adopted, in which an object is operated by pushing, pulling, or other similar actions [1]. Manipulations via multiple-robot systems must be effectively designed to perform the task cooperatively and efficiently. As non-prehensile manipulation is often conducted by using the environment that the targeted object lies in, the interaction between the targeted object and other

© Springer Nature Switzerland AG 2019
M. Strand et al. (Eds.): IAS 2018, AISC 867, pp. 174–185, 2019.
https://doi.org/10.1007/978-3-030-01370-7_14

objects in the surroundings should be considered in the planning of the manipulation. Maeda and Arai [2] discretized the contact configuration space of objects and conducted manipulation planning for an object while comprehensively considering the contact on the object from the robots and their environment. In their method, the object-environment contacts were considered and the manipulation feasibility was measured to determine the most stable simultaneous contacts. When the robot contacts with an object become indefinite, the planning becomes very complicated. Comparatively, Kijimoto [3] and Lee [4] adopted a quite similar hierarchical planning method to separate the contacts on the object into those from environment and robots. After generating the topological contact states between the object and the environment, optimal motion sequences of object were selected to furtherly do the robot contact planning. They planned the manipulation for two robot contacts to operate an object while interacting with other objects in the surrounding. In the case of multiple mobile robots, the robots may not be able to contact any surface of the targeted object, due to the size limitation of the robots. As different types of robots exert different types of contacts on an object, the resulting constraints can also vary. Adopting the methods of [3,4] to a multiple mobile robot system with different types and an indefinite number of robots would lead to many possible manipulation states because of the indefinite combination of robots. If the number of robots is not properly determined according to the working conditions and the type of robots in certain manipulation states, the constraints on the object may not be stable enough to resist external perturbations. Therefore, it is necessary to determine the number of robots when planning non-prehensile manipulation with multiple mobile robots, which then allows for the creation of states with manipulation stability. The state transition graph can then be generated to conduct the manipulation planning (Fig. 1).

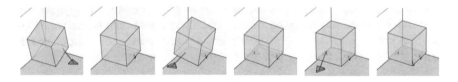

**Fig. 1.** Non-prehensile transportation adopting multiple mobile robots

Another important aspect of manipulation planning is efficiency. Obtaining high efficiency requires a series of compact action sequences with the least state transformations. Thus, manipulation planning will be conducted by searching for the shortest paths from the initial state node to the targeted state node in the state transition graph. Each of the obtained shortest paths contains the least action sequences to efficiently complete the task.

In this work, the hierarchical planning method used by Kijimoto and Lee was adopted while considering the necessary number of robots for non-prehensile manipulation in a discretized contact configuration space. The least number of

mobile robots needed to achieve force closure for manipulation stability according to the environment contact on object was determined. The possible object-
environment and object-robot contact states were determined, and the dimensions of the wrench spaces spanned by the contacts in each contact state were
calculated. According to the obtained dimensions of the wrench spaces, the possible robot combinations were matched with the object-environment contact
states to create manipulation states that are possible to achieve force closure.
The state transition graph was then generated using the obtained manipulation
states. Given the initial state of the object and the targeted state, the lowest
amount of manipulation sequences could then be computed by searching for the
shortest path in the state transition graph. As an example, the method was then
adopted for the manipulation of a cuboid lying against two walls.

## 2    Methodology

Each possible state comprises the targeted object and the robots manipulating
this object. Here, $s$ and $S$ were used to represent a single state and the state space
formed by all the possible states, respectively. If one state can transit into another
state through a certain manipulation without any intermediate states, the two
states are deemed transitable. When an initial state $s_I$ and a final state $s_G$ are
given, manipulation planning comprises searching for all feasible sequences of
states from $s_I$ to $s_G$ according the transition between the states [5].

Planning was conducted in the discretized contact configuration space by
adopting the hierarchical manipulation planning method. The manipulated
object can come into contact with robots and with fixed elements in its surroundings, such as the floor, walls, and other objects lying nearby in each state.
These fixed elements in the surroundings were defined as the environment. The
state of contact between the targeted object and the environment was defined as
the environment contact (EC) state, denoted as $C^E$. Correspondingly, the state
of contact between the targeted object and robots was defined as robot contact (RC) state, denoted as $C^R$. Therefore, a manipulation state in discretized
contact configuration space was denoted by combining these contacts as

$$s = \{C^E, C^R\}.$$

The transition among the manipulation states was determined by comprehensively considering the transformation of ECs and RCs.

### 2.1    Generation of Manipulation States

The generation of manipulation states in discretized contact configuration space
is discussed in this section. All possible ECs and RCs were first created. The
ECs and RCs were then combined to generate manipulation states, given that
the resulting contact set achieved force closure for manipulation stability. Considering the flexibility of contact with robots compared with contact with the
fixed environment, the manipulation states were created by matching the RCs

with each EC according to the minimum requirement for force closure. In this section, the procedure to generate the manipulation states will be introduced.

**Contacts with Environment.** The method proposed by Xiao and Ji [6,7] was used to determine the possible contact states between the object and the environment and the transitions between them. In this method, the possible contact states between objects were determined by conducting the compliant motion of the targeted object to extract the possible transitions of the contact states between objects.

To express the single fundamental EC, the principal contact (PC) [8] was defined by three types of contact primitives: vertex, edge, or face contact. Here, a PC referred to a pair of contact primitives on the object and a polyhedron of the environment, denoted as $c^e = (a, b)$, where $a$ is a contact primitive of the manipulated object and $b$ is a contact primitive on a polyhedron of the environment. In the case that an object was in multiple contacts with the environment, the set of PCs was used to express the corresponding EC, which is called contact formation (CF) in [8]. An EC is denoted as

$$C^E = \{c_1^e, \ldots, c_N^e\},$$

where $c_i^e$ is a PC between the object and the environment, and $N$ is the number of contacting parts between object and environment.

The CFs created by adopting the method in [6,7] kept the object in contact with the environment in all obtained states. However, the object does not contact the environment in certain states of non-prehensile manipulation, such as when the object is loaded onto mobile robots. Thus, this individual state was added with the denotation $C^E = \varnothing$.

**Contacts with Robot.** All possible RCs corresponding to each EC were then generated by taking different types and number of robots as combinations.

The robots used were divided into active and passive robots, where the active robots actively exert force on the object and the passive robots do not. If a robot has active joints and acts as the actuating part in the manipulation to exert driving force and change the state of the object, it is defined as an active robot. If a robot acts only as an auxiliary in the manipulation to support the object passively, it is defined as a passive robot. A single contact of either type of robot on the object was treated as a point contact.

Contact between a robot and an edge or vertex of the object could easily be broken; thus, robots were only allowed to contact the surfaces of the object. Additionally, as the robot can contact the object flexibly on various locations of various faces, the exact contacting face for each contact does not need to be considered when creating the possible RC states. In this way, a single robot contact was expressed as $c^r = (f, r)$, where $r$ is the adopted robot and $f$ is the contacting surface. To incorporate multiple robots of different types, the contact state was expressed as the combination of single robot contacts, denoted as

$$C^R = \{c_1^{ar}, \ldots, c_A^{ar}, c_1^{pr}, \ldots, c_B^{pr}\}$$

where $c_i^{ar}$ is an active robot contact, $A$ is the number of active robot contact, $c_j^{ar}$ is a passive robot contact, and $B$ is the number of passive robot contact.

Therefore, the possible RCs in the manipulation planning could be generated by making combinations of robots. However, in some states, the object itself was able to maintain a stable contact with environment while no robot was needed to support the object. In such a case, the RC was an empty set, given as $C^R = \varnothing$. The manipulation states were created by matching the generated RCs with each generated EC, according to the minimum requirement to achieve force closure.

**EC-RC Combination.** To guarantee the stability and Controllability of the object in manipulation, the object-robot system should be able to resist the external perturbation in every state by achieving force closure. Aiyama [9,10] extended the theory on force closure and defined gravity closure as the force closure formed by robot contacts, environment contacts and gravity, treating gravity as a virtual finger. This definition was adopted, and gravity was treated as an external, virtual, frictionless contact on the object, defined as gravity contact (GC). Thus, the ECs and RCs were combined to generate manipulation states together with the GC to achieve gravity closure.

For a given force exerted on the object, the resultant force wrench was defined as

$$w = \begin{bmatrix} f \\ p \times f \end{bmatrix} \tag{1}$$

where $f$ is the force applied to the object, and $p$ is the position of the contact point in the object coordinate frame.

The calculation of force wrenches varied with differing contact types. For frictionless contact, the generated wrench can be expressed as Eq. (1). For frictional contact in three-dimensional space, the frictional cone can be represented by approximating the cone as a pyramid generated by a finite set of vectors and each of the vectors can also be described by Eq. (1). Edge contacts and face contacts must take into consideration all the boundary points of the contacting part. Thus, the contact force formed a set of wrenches as

$$W = \{w_1, \dots, w_i\}. \tag{2}$$

In three-dimensional space, the force wrenches on object achieve force-closure when the wrench vectors positively span $\mathbb{R}^6$, described as

$$\text{pos}(W) = \mathbb{R}^6. \tag{3}$$

According to the Carathéodory Theorem, at least seven vector frames in six-dimensional space are needed to positively span $\mathbb{R}^6$. Thus, in the obtained set of wrenches from all the contacts exerted on the object, at least seven vector frames of wrench should exist to achieve gravity closure. Then when creating the manipulation states, as a minimum requirement for gravity closure, the wrench vector set resulted by all the contacts should contains at least seven vector frames.

ECs exert force wrenches on the object. The wrench vector set of all contained PCs were determined for each type of EC to form the wrench vector set of environmental contact, denoted as $W^{C^E}$. In most of the manipulation states, RC is needed to complement the gravity closure. The set of wrench vectors exerted by GC and RC are denoted by $W^{C^G}$ and $W^{C^R}$, respectively. As gravity can be viewed as a frictionless point contact, $\dim(W^{C^G}) = 1$. For each type of EC, there exists the possibility that

(i) $\dim(W^{C^E}) = 6$,
     (a) $\text{pos}(W^{C^E} \cup W^{C^G}) \neq \mathbb{R}^6$;
     (b) $\text{pos}(W^{C^E} \cup W^{C^G}) = \mathbb{R}^6$.
(ii) $\dim(W^{C^E}) < 6$.

In case i-a, the object cannot maintain stable contact with the environment as gravity closure cannot be achieved by EC and GC alone. RC is needed to complement the gravity closure.

In case i-b, gravity closure can be achieved by EC and GC; the object can maintain stable contact with the environment without RC. RC can still be exerted to the object to form the possible manipulation states without any necessity to complement the gravity closure.

In case ii, $W^{C^E}$ is not able to positively span $\mathbb{R}^6$ even if the $W^{C^G}$ is included, as the total number of wrench vector frames is less than seven. Thus, RCs are needed to complement the gravity closure for manipulation stability.

Each EC can be represented by $a$ vector basis if $\dim(W^{C^E}) = a$. Similarly, if $\dim(W^{C^R}) = b$ for each RC, it has $b$ vector basis. To ensure $W^{C^E} \cup W^{C^G} \cup W^{C^R}$ would be possible to positive span $\mathbb{R}^6$ for gravity closure, the corresponding RCs will be matched to an EC if

$$a + 1 + b \geq 7 \tag{4}$$

To determine the EC-RC combinations, the dimensions of $W^{C^E}$ and $W^{C^R}$ must be calculated.

The dimension of the wrench vector space $W^{C^E}$ is very difficult to calculate analytically because singular points that may result in the degeneration of the dimension exist in the wrench vector space. Therefore, a sampling method was adopted to check the degeneration of the wrench vector space. Three possible configurations of the object were sampled using the method proposed in [6] and [7] for the same EC. If any of the resultant rank from the sampled datas was smaller than the others, then the corresponding orientation was designated as the singularity of the EC. The singularity is caused in limited configurations of the object and it can be avoided in the detailed planning stage. Thus, the rank of the wrench vector space from the sampled datas for a EC was considered as the dimension of its wrench vector space.

For RCs, the dimension of wrench vector space $W^{C^R}$ was easier to consider since robot contacts are viewd as point contact, and the contact positon can be placed flexibly due to the motion of robot. Taking into account the difference of frictional and frictionless robot contact, only the maximum dimension of space that the robot contact can span was considered. In this way, the dimension of wrench vector space was determined for each kind of RC. According to the dimension of their wrench vector space, they were then matched with the ECs to generate manipulation states.

Although the type and the number of the robots can be determined with this method, appropriate robot contact placement to ensure gravity closure was not addressed. If the contacts are not properly placed, $W^{C^E} \cup W^{C^G} \cup W^{C^R}$ will not positively span the $\mathbb{R}^6$. The procedure to decide the placement of robots will be addressed in a later stage of the hierarchical planning.

## 2.2   Transition Between Manipulation States

Transition rules between the obtained manipulation states must be defined to generate the state transition graph. The cost for the transformation between two states was calculated by the number of sequences in our case. Given two states in the state space as $s_a$ and $s_b$, the cost function is defined as

$$l(s_a, s_b) = k, \tag{5}$$

where $k$ is the cost for $s_a$ to transform into $s_b$. If the two states can transit to each other without going through any intermediate state, the cost for this transition is 1. Thus, $k = 0$ when $s_a = s_b$; $k = 1$ when $s_a$ can transit to $s_b$ directly; and $k = +\infty$ when $s_a$ cannot transit to $s_b$.

In a state transition graph, each node represents a state and an arc is created to connect two nodes if the two states are transitable. The arc between two nodes $s_a$ and $s_b$ is defined as

$$E(s_a, s_b) = d \tag{6}$$

where $d = 1$ if the arc can be generated between the two nodes and $d = 0$ if no arc can be generated.

The state transition graph was created based on the goal-contact relaxation (GCR) graph proposed in [11], where the nodes are connected by arcs in the GCR graph if the corresponding states are transitable when the possible ECs are created. In our case, the state $C^E = \varnothing$ was added to the obtained set of ECs. Since any kind of EC is able to transit to the floating state, the node representing $C^E = \varnothing$ in the GCR graph could be connected to any other node. For two connected ECs in GCR graph, their transition ability is denoted as

$$G(C_a^E, C_b^E) = 1 \tag{7}$$

The transformation of EC and RC must be considered comprehensively when determining the rules for transition of manipulation states because robot contacts are also envolved in the manipulation. Since a passive robot can only exert contact to the object passively, its contact on object was viewed as a point in the environment. The subset of contacts formed by EC and passive robot contacts was defined as fixed contact set in the manipulation state, denoted as $C^{\mathrm{F}}$. The active robot contact and the passive robot contact were denoted as $c^{\mathrm{a}}$ and $c^{\mathrm{p}}$, respectively. For any two obtained manipulation states, given as $s_1 = \{C_1^{\mathrm{E}}, C_1^{\mathrm{R}}\}$ and $s_2 = \{C_2^{\mathrm{E}}, C_2^{\mathrm{R}}\}$, with their fixed contact subset as $C_1^{\mathrm{F}}$ and $C_2^{\mathrm{F}}$ respectively, $E(s_{\mathrm{a}}, s_{\mathrm{b}}) = 1$ in any of the following cases:

(i) if $C_1^{\mathrm{R}} \backslash C_2^{\mathrm{R}} \cup C_2^{\mathrm{R}} \backslash C_1^{\mathrm{R}} = \{c^{\mathrm{a}}\}$, and $C_1^{\mathrm{E}} = C_2^{\mathrm{E}}$;

(ii) if $C_1^{\mathrm{R}} \backslash C_2^{\mathrm{R}} \cup C_2^{\mathrm{R}} \backslash C_1^{\mathrm{R}} = \{c^{\mathrm{p}}\}$, $C_1^{\mathrm{E}} = C_2^{\mathrm{E}}$, $c^{\mathrm{a}} \in C_1^{\mathrm{R}}$, and $\min\{\dim(\mathrm{W}^{C_1^{\mathrm{F}}}),$ $\dim(W^{C_2^{\mathrm{F}}})\} < 6$;

(iii) if $C_1^{\mathrm{R}} = C_2^{\mathrm{R}}$, $G(C_1^{\mathrm{E}}, C_2^{\mathrm{E}}) = 1$, $c^{\mathrm{a}} \in C_1^{\mathrm{R}}$, and $\min\{\dim(\mathrm{W}^{C_1^{\mathrm{F}}}), \dim(\mathrm{W}^{C_2^{\mathrm{F}}})\} < 6$.

When the EC remains unchanged, the robot contact can only add to the contact state one by one. But the conditions to add the active robot and the passive robot for state transition are different.

Case i shows if the EC remains unchanged, and the difference between $C_1^{\mathrm{R}}$ and $C_2^{\mathrm{R}}$ is only one active robot, the states are transitable. This is because an active robot can actively exert contact to the object to make the whole system transform into another state.

Case ii shows that if the EC remains unchanged, and $\dim(W^{C^{\mathrm{F}}}) < 6$ in one state, the addition of a passive robot contact can change it to another state. This is because the passive robot cannot actively exert contact to the object. When the fixed contact subset does not constrain all the six DOFs of the object (or, $\dim(W^{C^{\mathrm{F}}}) < 6$), the object can be moved by the active robot to contact the passive robot. Similarly, the active robot can also move the object to lose contact with the passive robot. If the fixed contact is a two-point contact with friction, the dimension of the wrench vector space is five due to the redundant force.

Case iii shows if $\dim(W^{C^{\mathrm{F}}}) < 6$, the object can be moved by the active robot contact to change the EC to another EC connecting with it in GCR graph, with the RC maintains unchanged (or another state is needed to change the RC with the EC unchanged). This is because if $\dim(W^{C^{\mathrm{F}}}) < 6$, the fixed contact set does not constrain all the six DOFs of the object. It can be manipulated by the active robots and change the contact state with the environment. If $\dim(W^{C^{\mathrm{F}}}) \geq 6$, the object can be moved and lose certain parts of contact with the environment, and the $\dim(W^{C^{\mathrm{F}}})$ reduce to less than six. A state with $\dim(W^{C^{\mathrm{F}}})$ equal or over six is not able to directly transit to another state also with $\dim(W^{C^{\mathrm{F}}})$ equal or over six, since in both of the states the six DOFs of object are constrained.

The state transition graph was thus created. For the given initial and the final states, the manipulation strategy with least action sequences was obtained by searching for the path with the lowest cost in the state transition graph with an objective function of

$$\min \sum_{i=1}^{i=n-1} l(s_i, s_{i+1}), \qquad (8)$$

where $s_1$ is the initial state and $s_n$ is the final state.

## 3   Implementation

The proposed method was applied to the non-prehensile manipulation planning of a cuboid lying against two walls, as shown in Fig. 2, to be transported away by adopting a multiple mobile robot system. As the object lies against in a corner of two adjacent walls, it could not be moved away by directly pushing; rather, the strategy was to load the cuboid onto transporter robots to carry the cuboid away. The main task thus became loading the cuboid onto the transporter robot without grasping.

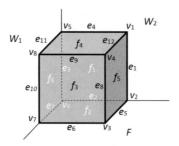

**Fig. 2.** Targeted object for manipulation labeled by its vertices $(v)$, edges $(e)$, and faces $(f)$. $W_1$, $W_2$, and $F$ represent the two walls and the floor, respectively, and were used to show the EC states.

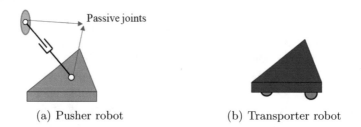

(a) Pusher robot                    (b) Transporter robot

**Fig. 3.** The mobile robots applied in the non-prehensile manipulation

To tilt the object, the pusher robot [12] with a manipulator containing a linear actuated joint that can actively push the object was adopted. Thus, the pusher robot worked as the active robot in the system, denoted as $r^a$. Because of the special mechanism adopting passive joints, the resultant force wrenches of the pusher robot could only span one-dimensional vector space, so the pusher robot contact was equivalent to a frictionless point contact. The transporter robot was only used to support the object during the loading manipulation and thus worked as a passive robot, denoted as $r^p$. The transporter robot exerted frictional contact on the object, and the resultant force wrenches could span three-dimensional vector space. Both of the robots are shown in Fig. 3.

The floor was represented by a frictional plane, denoted as $F$, whereas the walls were frictionless, denoted as $W_1$ and $W_2$. The proposed method was then used to generate the possible manipulation states in contact configuration space and search for optimal manipulation strategy to change the state of the object from the initial state to the targeted states from the floor with least number of action sequences.

The implementation was performed in Python. All possible object-environment contact states were first determined to form a set of ECs. The wrench vector space dimension was determined for each kind of EC. Meanwhile, the possible RCs were also created by making combinations of six pusher robots and three transporter robots. Their corresponding wrench vector spaces were also calculated. After combined the ECs and RCs, 3330 manipulation states were created in total.

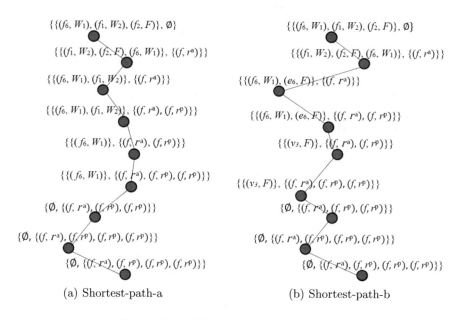

(a) Shortest-path-a          (b) Shortest-path-b

**Fig. 4.** Two of the obtained shortest paths

With the obtained manipulation states, the state transition graph for manipulation planning was generated. Given $s_I = \{\{(f_6, W_1), (f_2, F), (f_1, W_2)\}, \varnothing\}$ as the initial state, where $C_I^E = \{(f_6, W_1), (f_2, F), (f_1, W_2)\}$, and $C_I^R = \varnothing$; and $s_G = \{\varnothing, \{(f, r^P), (f, r^P), (f, r^P)\}\}$ as the final state, where $C_G^E = \varnothing$, and $C_G^R = \{(f, r^P), (f, r^P), (f, r^P)\}$, Dijkstra's Algorithm was used to search for the shortest path from the initial state to the final state. Twelve shortest paths were obtained, each contained nine nodes. Figure 4 shows two of the obtained shortest path in the state transition graph.

In the path shown in Fig. 4a, the object was pushed by a pusher robot and lifted up against the two walls, $W_1$ and $W_2$. A transporter robot was then inserted underneath it. The pusher robot then moved the object, so that only $f_6$ was in contact with $W_1$, and another transporter robot was inserted underneath the object. Finally, with two transporter robots on the bottom, the object lost contact with wall $W_1$ through the contact with the pusher robot, allowing the third transporter robot to enter underneath the object. Thus, the object was again loaded onto three transporter robots.

In the path shown in Fig. 4b, the object was tilted by a pusher robot to contact the floor $F$ by the edge $e_6$ and the wall $W_1$ by the face $f_6$. A transporter robot was then inserted under the object. The object was again tilted by a pusher robot to contact the floor $F$ via vertex $v_3$ and the inserted transporter robot, so that another transporter robot could be inserted under the object. Finally, a pusher robot pushed the object up so that a third transporter robot could be inserted under the object. Thus, the object was loaded onto three transporter robots and could then be carried away.

## 4  Conclusion

A methodology for the preliminary stage of the hierarchical planning for non-prehensile manipulation of a multiple mobile robot system was proposed. In the proposed planning method, the manipulation states of the robot-object system in contact configuration space were created with a sufficient number of robots to realize gravity closure for fully-constrained objects. The obtained manipulation states then allowed for the creation of the state transition graph. Given an initial and targeted state, the least number of action sequences to finish the manipulation task were obtained by searching for the shortest path in the graph. The proposed method was implemented on a specific non-prehensile manipulation in which pusher and transporter robots were adopted to load a cuboid onto transporter robots without grasping. The state transition graph for manipulation planning was generated, and the lowest amount of action sequences for manipulation was obtained by searching for paths with lowest cost in the graph using Dijkstra's Algorithm. Future work needs to address the feasibility of the obtained manipulation paths and the placement of the robot contact.

## References

1. Mason, M.T.: Mechanics of Robotic Manipulation. MIT press, Cambridge (2001)
2. Maeda, Y., Arai, T.: Planning of graspless manipulation by a multifingered robot hand. Adv. Robot. **19**, 501–521 (2005)
3. Kijimoto, H., Arai, T., Aiyama, Y., Yamamoto, T.: Performance analysis and planning in graspless manipulation. In: Proceedings of the 1999 IEEE International Symposium on Assembly and Task Planning, (ISATP 1999), pp. 238–243. IEEE (1999)

4. Lee, G., Lozano-Pérez, T., Kaelbling, L.P.: Hierarchical planning for multi-contact non-prehensile manipulation. In: 2015 IEEE/RSJ International Conference on Intelligent Robots and Systems (IROS), pp. 264–271. IEEE (2015)
5. LaValle, S.M.: Planning Algorithms. Cambridge University Press, Cambridge (2006)
6. Xiao, J., Ji, X.: Automatic generation of high-level contact state space. Int. J. Robot. Res. **20**, 584–606 (2001)
7. Ji, X., Xiao, J.: Planning motions compliant to complex contact states. Int. J. Robot. Res. **20**, 446–465 (2001)
8. Xiao, J., Zhang, L.: A general strategy to determine geometrically valid contact formations from possible contact primitives. In: IEEE International Conference on Robotics And Automation, vol. 1, pp. 2728–2728 (1993)
9. Maeda, Y., Aiyama, Y., Arai, T., Ozawa, T.: Analysis of object-stability and internal force in robotic contact tasks. In: Proceedings of the 1996 IEEE/RSJ International Conference on Intelligent Robots and Systems, IROS 1996, vol. 2, pp. 751–756. IEEE (1996)
10. Aiyama, Y., Arai, T., Ota, J.: Dexterous assembly manipulation of a compact array of objects. CIRP Ann. **47**, 13–16 (1998)
11. Xiao, J.: Goal-contact relaxation graphs for contact-based fine motion planning. In: 1997 IEEE International Symposium on Assembly and Task Planning, ISATP 1997, pp. 25–30. IEEE (1997)
12. Shirafuji, S., Terada, Y., Ota, J.: Mechanism allowing a mobile robot to apply a large force to the environment. In: International Conference on Intelligent Autonomous Systems, pp. 795–808. Springer (2016)

# Convolutional Channel Features-Based Person Identification for Person Following Robots

Kenji Koide$^{(\boxtimes)}$ and Jun Miura

Toyohashi University of Technology, Toyohashi, Aichi, Japan
koide@aisl.cs.tut.ac.jp, jun.miura@tut.jp

**Abstract.** This paper describes a novel person identification framework for mobile robots. In this framework, we combine Convolutional Channel Features (CCF) and online boosting to construct a classifier of a target person to be followed. It allows us to take advantage of deep neural network-based feature representation and adapt the person classifier to the specific target person depending on circumstances. Through evaluations, we validated that the proposed method outperforms existing person identification methods for mobile robots. We applied the proposed method to a real person following robot, and it has been shown that CCF-based person identification realizes robust person following.

**Keywords:** Person tracking · Person identification · Mobile robot

## 1 Introduction

Person identification is one of the fundamental functions for person following robots. To keep following a person, they have to reliably localize the target person real-time. In cases where the target person is occluded by another person, robots would lose track of him/her, and they have to find the person among surrounding persons with a person model learned before the lost (i.e., re-identification) to resume following.

In this paper, we propose a Convolutional Channel Features (CCF) [21] based person identification framework for person following robots. Figure 1 shows our person following robot and an overview of the proposed system. The robot is equipped with Laser Range Finders (LRFs) and a camera. We first detect and track people using the LRFs, and then find people regions on images based on the people positions provided by the LRFs. In order to reliably identify a target person among surrounding persons, we construct a target person model by learning his/her appearance. We employ CCF, a set of convolution filters trained by a neural network, to extract appearance features of persons. It produces robust and discriminative features for person identification. Then, the robot

---

Video available at: https://www.youtube.com/watch?v=semX5Li0yxQ.

M. Strand et al. (Eds.): IAS 2018, AISC 867, pp. 186–198, 2019.
https://doi.org/10.1007/978-3-030-01370-7_15

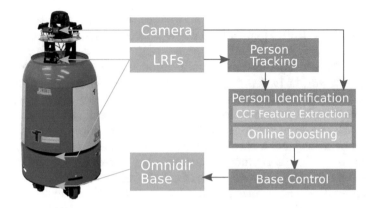

**Fig. 1.** Person following robot equipped with a camera and LRFs.

learns the extracted features of the target person using online boosting [8]. With this approach, the robot can adapt the person classifier to the specific target person. It is suitable for tasks where a specific person is important, like person following.

The contributions of this paper are two-fold. First, we introduce CCF, deep neural network-based representation, for person following robots. It significantly improves the identification accuracy, while keeping the processing cost low (it can be run real-time without GPU). To our knowledge, this is the first work to introduce it to online person re-identification. Secondly, we provide pieces of the framework, such as CCF related routines and parameters, as open source[1]. It can be re-used for person identification as well as other tasks, such as person detection and tracking.

The rest of the paper is organized as follows. Section 2 explains related works. Section 3 describes our LRF-based person tracking and visual person region detection methods. Section 4 describes the proposed CCF-based person identification method and its evaluation. Section 5 shows a person following experiment conducted to show that the proposed method can be applied to real person following robots. Section 6 concludes the paper.

## 2   Related Work

Person tracking is an essential function for person following robots. A lot of works proposed person tracking systems for mobile robots. In particular, LRFs [3,10] and depth cameras [16,19] have been widely used for person following robots. They provide a person's position accurately as long as he/she is visible from the sensors and promise reliable person following capabilities. However, once the systems lose track of a person to be followed due to occlusion, they cannot find the person even he/she re-appear in the sensor view, and robots are not able

---

[1] https://github.com/koide3/ccf_feature_extraction.

to continue to follow the person. In such cases, robots have to re-identify the person based on a target person model learned before the occlusion.

Several features, such as gait [14], height [4], and skeletal information [15], have been proposed for person re-identification. In cases of mobile robots, the most popular way is to use appearance features, such as color and texture of cloths, and learn them online [2,6,13] to construct a target person model. Appearance is one of the most discriminative features, and online learning methods adapt the person model to a specific target person. For instance, when there are persons wearing similar sheets and dissimilar trousers, online learning methods can focus on the discriminative part, trousers in this case, to re-identify the target person robustly. However, most of existing methods for mobile robots use naive hand-crafted appearance features, such as Haar-like features [13], Local Binary Patterns (LBP) [6], edge features [2] on color and depth images. They are not dedicated features for person re-identification, and they may not be discriminative when persons are wearing similar cloths.

Recently, deep neural networks have been applied to various vision applications. Person re-identification for surveillance is one of such applications, and Convolutional Neural Network (CNN) based methods outperform traditional systems [1,20] in terms of identification accuracy. However, a few works [5] applied such CNN-based methods to mobile robots due to the limitation of computation resource on mobile robots. We usually cannot use computers with GPUs for mobile robots, and it is hard to directly apply such CNN-based methods to person following robots. Moreover, in person following tasks, it is important to adapt the person model to the target person online. Although there are methods to update neural networks online [18], those methods are very costly.

Yang et al. proposed Convolutional Channel Features [21]. They take the first a few convolution layers from a trained deep CNN, and use the set of convolution layers as a feature extractor (called CCF). By training light-weight models, such as SVM and boosting, with the deep feature representations, they adapt the framework to several tasks without expensive tuning of the network. Following their work, we introduce CCF to person identification for mobile robots to take advantage of deep representation while keeping the processing cost low.

# 3    Person Tracking

## 3.1    LRF-Based Person Tracking

We first detect and track persons using LRFs placed at leg and torso heights. We detect leg/torso candidate clusters by finding local minimas in range data, and then we validate if they are real torsos/legs using classifiers based on cluster shapes. We use Arras's method (14 shape features and boosting) [3], and Zainudin's method (4 shape features and SVM) [22] to validate torsos and legs, respectively. We assume that the torso and at least one of the legs of a person must be detected, and aggregate the detected torsos and legs by considering torsos with no legs under it are false positives. We track the detected torsos using Kalman filter with a constant velocity model and global nearest neighbor data association [17].

**Fig. 2.** ROIs calculated from person positions provided by the LRFs (Red transparent regions), and detected upper body regions (Green rectangles).

## 3.2  Visual Person Region Detection

In order to find person regions on an image for appearance feature extraction, we project a cylinder at each person position into the image, and calculate a rectangle which surrounds the projected cylinder as an ROI. Then, we use HOG cascaded classifier [7] to accurately localize the upper body region of the person. Figure 2 shows an example of calculated ROIs and detected upper body regions. We double the height of each region so that it covers the whole body of the person, and then extract appearance features from the detected regions to train a target person classifier.

## 4  Person Identification

### 4.1  Convolutional Channel Features

To take advantage of deep CNN-based feature representation, we employ Convolutional Channel Features (CCF) [21] instead of traditional appearance features which have been used for mobile robots, such as color histograms [9], haar-like [3], and edge features [2]. CCF consists of a few convolutional layers taken from a trained deep CNN. It takes an input image and yields a set of response maps (i.e. feature maps) which is optimized for a specific task, such as person detection and classification.

In this work, we train Ahmed's network for person re-identification [1] as the base of CCF, and use the first two convolution layers of the network to extract appearance features for online person identification (see Fig. 3). Ahmed's network takes a pair of person images and then applies convolution filters to extract

Offline Feature Training (Ahmed's network[1])

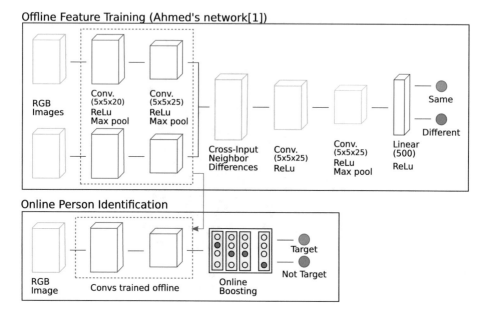

**Fig. 3.** Convolutional Channel Features-based person identification framework. We take the first two layers of a network for person re-identification and use them to extract features for online person identification.

feature maps of each input image. The extracted feature maps are compared together by taking the difference between each pixel of a feature map and the neighbor pixels of the corresponding pixel of the other map. Then, it applies convolution filters again to the differences map, and through a linear layer, the network judges whether the input images are the same person or not. The numbers of filters of the first and the second convolution layers in the network are 20 and 25, and thus, they yield 25 feature maps. Since it may be costly for mobile systems to directly use the network, we also trained a tiny version of the network, where the numbers of convolution filters of both the first and the second layers are 10. We trained both the networks with a dataset consisting of CUHK01 [11] and CUHK03 [12]. The total number of identities in the dataset is about 2300, and the number of images is about 17000. We used nine tenths of the dataset for training and the rest for testing and confirmed that both the networks show over 98% of identification accuracies on the test set. In the rest of this paper, the CCFs taken from the original and the tiny version networks are denoted as CCF25 and CCF10, respectively.

Figure 4 shows example feature maps extracted by CCF10. We can see that each layer shows strong responses for different color properties. For instance, layer 2 shows higher values on darker or blue regions, while layer 8 strongly responds orange regions. We can obtain diverse feature representation using CCF, without hand-crafting, and such diverse features would contribute to identification performances.

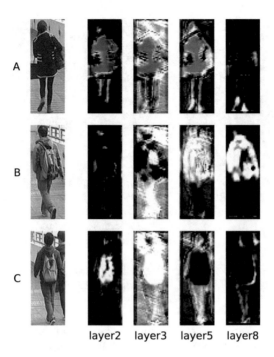

layer2   layer3   layer5   layer8

**Fig. 4.** Feature maps extracted by CCF10. Each layer shows strong responses for different color properties.

### 4.2   Online Boosting-Based Person Classifier

With the offline trained CCF, we extract feature maps from person images, and then train a target person classifier online. Following Luber's work [13], we employ online boosting [8] to construct the classifier. Online boosting constructs an ensemble of weak classifiers and uses it as a strong classifier. In this work, each weak classifier takes the sum of pixel values in a random rectangle region on a feature map and classifies images into the target and other persons using a naive Bayes classifier. Since online boosting selects weak classifiers with better accuracies, regions, which are effective to identify the target, are automatically chosen for identification. In this work, we use online boosting with 10 weak classifier selectors, and each selector contains 15 weak classifiers. Thus, the total number of weak classifiers is 150, and 10 of them are selected to construct an ensemble. Figure 5 shows an example of features selected by online boosting. We can see that online boosting automatically selects the discriminative regions, the upper body regions, in this case, to construct a classifier ensemble.

### 4.3   Evaluation

To evaluate the proposed CCF-based person identification method, we created a dataset consisting RGB image and LRF data sequences taken from a person

Target

Other

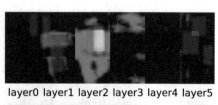

layer0 layer1 layer2 layer3 layer4 layer5

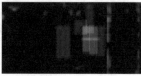    

layer6 layer7 layer8 layer9        all

(a) A snapshot of the environment.    (b) Features selected by online boosting. Heatmaps of selected regions are overlayed on feature maps.

**Fig. 5.** An example of features selected by online boosting. The discriminative regions, the upper body regions in this case, are automatically selected.

following robot (shown in Fig. 1). Figure 6 shows snapshots of the dataset. We controlled the robot manually and made it follow a target person in indoor and outdoor environments. We collected six sequences, and two of them are recorded in indoor, and the rest are recorded in outdoor environments. In each sequence, a target person to be followed stands in front of the robot for the first seconds so the robot can know and learn the appearance of the person, and then he/she starts walking. During the recording, the target person was often occluded by

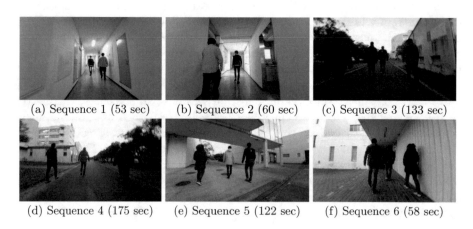

(a) Sequence 1 (53 sec)    (b) Sequence 2 (60 sec)    (c) Sequence 3 (133 sec)

(d) Sequence 4 (175 sec)    (e) Sequence 5 (122 sec)    (f) Sequence 6 (58 sec)

**Fig. 6.** Snapshots of the dataset for evaluation of person identification in person following tasks. The dataset consists of RGB images and LRF data recorded from a mobile robot. The robot was manually controlled and following a person in indoor and outdoor environments.

other persons so he/she become invisible from the robot, and the robot loses track of him/her.

We evaluated the proposed method with CCF25 and CCF10 on the dataset. We also evaluated online boosting with Haar-like features on intensity images and *Lab* color histograms. This is almost identical to [13] except that we didn't use Haar-like features on depth images.

Table 1 shows a summary of identification results. We categorized identification results in four states. CT (Correctly Tracked) means that the target was

**Table 1.** Person identification evaluation result. Bold indicates best results.

| | | Duration [sec] | | |
|---|---|---|---|---|
| | | Haar Lab | CCF10 | CCF25 |
| Seq. 1 | CT | 38.78 (73.23%) | **40.84 (77.11%)** | 37.96 (71.69%) |
| | CL | 6.62 (12.49%) | 6.78 (12.80%) | **7.37 (13.92%)** |
| | WT | 3.91 (7.38%) | 3.75 (7.08%) | **3.16 (5.96%)** |
| | WL | 3.65 (6.90%) | **1.59 (3.01%)** | 4.47 (8.44%) |
| Seq. 2 | CT | 43.76 (73.78%) | 43.86 (73.95%) | **43.87 (73.97%)** |
| | CL | **11.28 (19.02%)** | 10.76 (18.14%) | 10.90 (18.37%) |
| | WT | **2.52 (4.24%)** | 3.04 (5.12%) | 2.90 (4.89%) |
| | WL | 1.76 (2.96%) | 1.65 (2.79%) | **1.64 (2.77%)** |
| Seq. 3 | CT | 48.08 (36.11%) | **106.31 (79.84%)** | 88.60 (66.55%) |
| | CL | 7.67 (5.76%) | **20.18 (15.16%)** | 19.67 (14.77%) |
| | WT | 46.45 (34.89%) | **3.94 (2.96%)** | 6.47 (4.86%) |
| | WL | 30.94 (23.24%) | **2.71 (2.04%)** | 18.40 (13.82%) |
| Seq. 4 | CT | 37.89 (21.56%) | **141.19 (80.33%)** | 85.60 (48.70%) |
| | CL | **24.83 (14.13%)** | 23.18 (13.19%) | 21.57 (12.27%) |
| | WT | 12.08 (6.88%) | **5.83 (3.32%)** | 6.30 (3.58%) |
| | WL | 100.95 (57.44%) | **5.56 (3.16%)** | 62.29 (35.44%) |
| Seq. 5 | CT | 98.33 (80.38%) | 98.75 (80.73%) | **98.89 (80.84%)** |
| | CL | 16.66 (13.62%) | **18.39 (15.03%)** | 18.36 (15.00%) |
| | WT | 5.12 (4.19%) | **3.32 (2.71%)** | 3.38 (2.76%) |
| | WL | 2.22 (1.81%) | 1.88 (1.53%) | **1.70 (1.39%)** |
| Seq. 6 | CT | 33.10 (59.67%) | 41.90 (75.55%) | **43.67 (78.74%)** |
| | CL | 2.68 (4.84%) | **9.01 (16.24%)** | **9.01 (16.24%)** |
| | WT | 16.80 (30.28%) | **0.06 (0.11%)** | **0.06 (0.11%)** |
| | WL | 2.88 (5.20%) | 4.49 (8.10%) | **2.73 (4.91%)** |
| Total | CT | 299.94 (50.08%) | **472.86 (78.94%)** | 398.60 (66.55%) |
| | CL | 69.75 (11.64%) | **88.29 (14.74%)** | 86.87 (14.50%) |
| | WT | 86.89 (14.51%) | **19.94 (3.33%)** | 22.26 (3.72%) |
| | WL | 142.40 (23.77%) | **17.89 (2.99%)** | 91.24 (15.23%) |

visible from the robot and correctly identified. CL (Correctly Lost) means that the target was invisible from the robot due to occlusion, and the system judged that he/she is not in the view correctly. WT (Wrongly Tracked) means the robot identified a wrong person as the target while the target was invisible, and WL (Wrongly Lost) means the robot judged that the target is not visible, although he/she was actually visible from the robot.

CCF-based methods show better identification performance than the traditional appearance feature-based method thanks to their robust deep feature representations. Even in sequences where cloths of the target and others are similar, they correctly identified the target while the traditional one identified wrong persons as the target.

CCF10 and CCF25 show comparable results. However, in a few sequences, CCF25 failed to keep identifying the target person. For instance, it identified a wrong person as the target in sequence 3 and failed to re-identify the target after occlusion in sequence 4. We consider that this is due to the limitation of feature selection of online boosting. Online boosting selects better classifiers among a limited number of weak classifiers. When the feature space is vast, the set of weak classifiers cannot cover enough feature space, and thus, online boosting would fail to select good features. The performance of CCF25 could be improved by increasing the number of weak classifiers. However, it increases the processing cost, and it may lead to over-fitting. Although the feature space of CCF10 is smaller than CCF25, "average effectiveness" of CCF10 features might be better than CCF25 since it was optimized to identify persons with fewer filters. As a result, CCF10 showed a better result than CCF25 in this case.

Note that, we also tested the original Ahmed's network on this dataset, however, the results were very poor. In each sequence, we compared every person image with the target person images of the first ten seconds using the network, and classified the image into the target and others by majority-voting. However, it worked well on only easy situations (Sequence 1 and 2), and in the rest of sequences, it classified all similar persons as the target (Sequence 3, 4, and 6) or classified the target as other persons (Sequence 5). The results suggest that, even with deep feature representations, we cannot obtain a good identification result without online learning approaches. In addition to that, it took about 1 s for each frame and was far from real-time performance.

Table 2 shows average processing time of feature extraction and person classifier update on a computer with Core i7-6700K (without GPU). While the

**Table 2.** Processing time for each person image

|  | Method | Time [msec] |
|---|---|---|
| Feature extraction | Haar & Lab | 1.2 |
|  | CCF10 | 4.2 |
|  | CCF25 | 6.0 |
| Classifier update | All | 0.1 |

traditional feature extraction method takes 1.2 ms for each person image, CCF10 and CCF25 take 4.2 ms, and 6.0 ms, respectively. Although CCFs are more costly than the traditional one, they are still able to run real-time. Since the processing time of updating the person classifier depends on only the number of weak classifiers, every method takes the same time for updating (0.1 ms per person image).

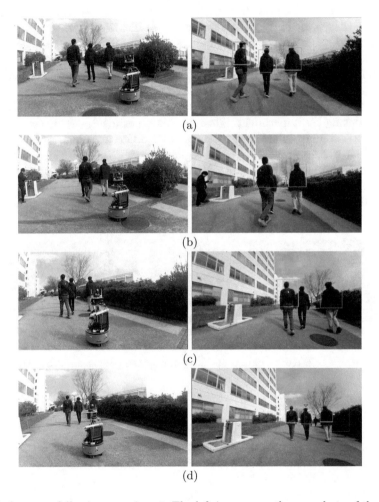

Fig. 7. A person following experiment. The left images are the snapshots of the experiment, and the right images are the detection and the identification results. The red triangles in the right images indicate the person identified as the target.

## 5   Person Following Experiment

To show that the proposed method can be applied to real person following tasks, we conducted a person following experiment. We implemented a simple person following strategy; the robot moves toward the target person, and when the robot loses track of the target, it stops and waits until the person re-appears.

Figure 7 shows snapshots of the experiment. The target person was occluded by another person (b), and the robot lost track of him (c). However, when the target person re-appeared, the robot correctly re-identified him with the boosting model trained before the occlusion and resumed to follow him (d). The duration of the experiment was 220 s. During the experiment, the robot lost the track of the target person seven times due to occlusion. Every time the target person re-appeared in the view of the robot, the robot correctly re-identified the target person and resumed to follow him. Figure 8 shows the features selected by online boosting during the experiment. Since, in this experiment, persons were wearing jackets with similar colors and trousers with different colors, online boosting selected features around the trousers region.

In this experiment, we used Intel NUC with Core i7-5557U for not only person identification but also other components required to drive the robot. Although the processor is not very powerful one, the system run real-time (about 10 Hz).

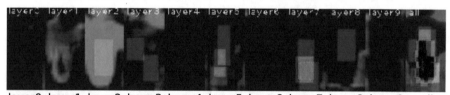

layer0  layer1  layer2  layer3  layer4  layer5  layer6  layer7  layer8  layer9      all

**Fig. 8.** Features selected by online boosting during the person following experiment.

## 6   Conclusion

We proposed a novel person identification framework for mobile robots. It is based on Convolutional Channel Features-based appearance features and online boosting to reliably identify the target person to be followed. We validated that the proposed method outperforms a traditional person identification method through evaluations. We also applied the proposed method to a real person following task. It has been shown that the robot with the proposed method is able to robustly follow a target person for a long time.

# References

1. Ahmed, E., Jones, M., Marks, T.K.: An improved deep learning architecture for person re-identification. In: IEEE Conference on Computer Vision and Pattern Recognition, pp. 3908–3916. IEEE (2015)
2. Alvarez-Santos, V., Pardo, X.M., Iglesias, R., Canedo-Rodriguez, A., Regueiro, C.V.: Feature analysis for human recognition and discrimination: application to a person-following behaviour in a mobile robot. Robot. Auton. Syst. **60**(8), 1021–1036 (2012)
3. Arras, K.O., Mozos, O.M., Burgard, W.: Using boosted features for the detection of people in 2D range data. In: IEEE International Conference on Robotics and Automation, pp. 3402–3407. IEEE (2007)
4. Berdugo, G., Soceanu, O., Moshe, Y., Rudoy, D., Dvir, I.: Object reidentification in real world scenarios across multiple non-overlapping cameras. In: European Signal Processing Conference, pp. 1806–1810 (2010)
5. Chen, B.X., Sahdev, R., Tsotsos, J.K.: Integrating stereo vision with a CNN tracker for a person-following robot. In: Lecture Notes in Computer Science, pp. 300–313. Springer (2017)
6. Chen, B.X., Sahdev, R., Tsotsos, J.K.: Person following robot using selected online ada-boosting with stereo camera. In: Conference on Computer and Robot Vision, pp. 48–55 (2017)
7. Dalal, N., Triggs, B.: Histograms of oriented gradients for human detection. In: IEEE Conference on Computer Vision and Pattern Recognition, vol. 1, pp. 886–893. IEEE (2005)
8. Grabner, H., Bischof, H.: On-line boosting and vision. In: IEEE Conference on Computer Vision and Pattern Recognition, vol. 1, pp. 260–267. IEEE (2006)
9. Koide, K., Miura, J.: Identification of a specific person using color, height, and gait features for a person following robot. Robot. Auton. Syst. **84**, 76–87 (2016)
10. Leigh, A., Pineau, J., Olmedo, N., Zhang, H.: Person tracking and following with 2D laser scanners. In: IEEE International Conference on Robotics and Automation, pp. 726–733 (2015)
11. Li, W., Zhao, R., Wang, X.: Human reidentification with transferred metric learning. In: Asian Conference on Computer Vision (2012)
12. Li, W., Zhao, R., Xiao, T., Wang, X.: DeepReID: deep filter pairing neural network for person re-identification. In: IEEE Conference on Computer Vision and Pattern Recognition (2014)
13. Luber, M., Spinello, L., Arras, K.O.: People tracking in RGB-d data with on-line boosted target models. In: IEEE/RSJ International Conference on Intelligent Robots and Systems, pp. 3844–3849. IEEE (2011)
14. Makihara, Y., Mannami, H., Yagi, Y.: Gait analysis of gender and age using a large-scale multi-view gait database. In: Asian Conference on Computer Vision, pp. 440–451. Springer (2011)
15. Munaro, M., Ghidoni, S., Dizmen, D.T., Menegatti, E.: A feature-based approach to people re-identification using skeleton keypoints. In: In: IEEE International Conference on Robotics and Automation, pp. 5644–5651. IEEE (2014)
16. Munaro, M., Menegatti, E.: Fast RGB-d people tracking for service robots. Auton. Robot. **37**(3), 227–242 (2014)
17. Radosavljevic, Z.: A study of a target tracking method using global nearest neighbor algorithm. Vojnotehnicki glasnik **2**, 160–167 (2006)

18. Sahoo, D., Pham, Q., Lu, J., Hoi, S.C.H.: Online deep learning: Learning deep neural networks on the fly. CoRR abs/1711.03705 (2017). arXiv:1711.03705
19. Satake, J., Chiba, M., Miura, J.: A SIFT-based person identification using a distance-dependent appearance model for a person following robot. In: IEEE International Conference on Robotics and Biomimetics, pp. 962–967. IEEE (2012)
20. Schumann, A., Stiefelhagen, R.: Person re-identification by deep learning attribute-complementary information. In: IEEE Conference on Computer Vision and Pattern Recognition Workshops. IEEE (2017)
21. Yang, B., Yan, J., Lei, Z., Li, S.Z.: Convolutional channel features. In: IEEE International Conference on Computer Vision. IEEE (2015)
22. Zainudin, Z., Kodagoda, S., Dissanayake, G.: Torso detection and tracking using a 2D laser range finder. In: Australasian Conference on Robotics and Automation, ARAA (2010)

# Service Robot Using Estimation of Body Direction Based on Gait for Human-Robot Interaction

Ayanori Yorozu[1(✉)] and Masaki Takahashi[2]

[1] Graduate School of Science and Technology, Keio University,
3-14-1 Hiyoshi, Kohoku-Ku, Yokohama 223-8522, Japan
ayanoriyorozu@keio.jp
[2] Department of System Design Engineering, Keio University,
3-14-1 Hiyoshi, Kohoku-Ku, Yokohama 223-8522, Japan
takahashi@sd.keio.ac.jp

**Abstract.** Recently, there have been several studies on the research and development of service robots, and experimental results in real environments have been reported. To realize socially acceptable human-robot interaction for service robots, human recognition, including not only position but also body direction, around the robot is important. Using an RGB-D camera, it is possible to detect the posture of a person. However, because the viewing angle of the camera is narrow, it is difficult to recognize the environment around the robot with a single device. This study proposes the estimation of the body direction based on the gait, that is, not only the position and velocity, but also the state of the legs (stance or swing phase), using laser range sensors installed at shin height. We verify the effectiveness of the proposed method for several patterns of movement, which are seen when a person interacts with the service robot.

**Keywords:** Service robots · Human-robot interaction · Gait measurement
Kalman filter

## 1 Introduction

As the society rapidly ages, there has been active research and development of service robots, and experimental results in real environments have been reported. For example, the Haneda Robotics Lab was established at the Haneda Airport, Japan, and verification such as the guidance and transportation of luggage has been carried out [1]. In addition, as shown in Fig. 1(a) and (b), demonstrations of a robot teahouse with multiple robots were carried out [2]. Figure 1(c) also shows the demonstration of a guidance robot at the open campus of a university. For these service robots, it is necessary to recognize the surrounding environment and human behavior, and to perform appropriate human-robot interactions (HRIs). The importance of the initial interaction with the human has been previously reported [3]. Therefore, in the situation shown in Fig. 1, the robot is required to detect and recognize people approaching the robot from various directions, and it then turns in the direction of the approaching persons and provides service instantly. However, with respect to the reactions of humans to the robot, there are some

© Springer Nature Switzerland AG 2019
M. Strand et al. (Eds.): IAS 2018, AISC 867, pp. 199–209, 2019.
https://doi.org/10.1007/978-3-030-01370-7_16

cases where he/she turns back and walks away, or requests another service after receiving service from the robot. Alternatively, as shown in Fig. 1(c), he/she steps backward and observes how other people are receiving robot services. Therefore, it is necessary for service robots to provide appropriate services based on their observations of the target person providing the service and people around the robot. Moreover, as shown in Fig. 1(a), when the robot guides people to a destination or attends to moving people, it may be necessary for the robot guides to move depending on their relative positions to the people and the surrounding environment [3–5].

(a) Pepper©Softbank responds to customers and guides them to table in a robot teahouse.

(b) HSR©TOYOTA delivers an object to     (c) Pepper guides visitors at an open campus of
a customer in a robot teahouse.                        a university.

**Fig. 1.** Demonstration of service robots in real environments.

Therefore, to successfully realize socially acceptable HRI, there is a need for human recognition that includes not only the position but also the body direction (posture). By using an RGB-D camera such as Kinect, it is possible to detect the posture of a person. However, because of the narrow viewing angle of the camera it is difficult to recognize the surroundings of the robot with a single device, and the sensor fusion of multiple devices is required. A method has been proposed for the detection of the position and body direction of a person using multi-layered laser range sensors (LRSs), which can measure a wide range of two-dimensional (2D) distance data [6]. However, depending on the mounting height of the LRS and the height of the person, it may be difficult to detect the body direction. Alternatively, we have proposed a gait-measurement method

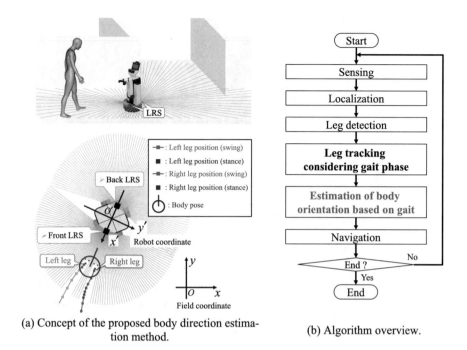

(a) Concept of the proposed body direction estimation method.

(b) Algorithm overview.

**Fig. 2.** Overview of proposed method.

that can measure not only the position and velocity of both legs, but also the state of the legs (stance or swing phase) using an LRS installed at shin height [7, 8]. This method may be applied to people with different body shapes because there are small differences in the thicknesses of individuals' legs. Although it is possible to obtain the movement of the center position and velocity of both legs using the method, it is difficult to estimate the body direction based on the movement of the center of both legs.

To deal with these problems, in this study, we focus on the gait and propose a method to estimate the body direction based on the gait, which is measured using the LRS installed at shin height. With the proposed method, it may be possible to measure the position and body direction of the person around the robot with fewer sensors. In addition, it is expected to improve the recognition of HRI by sensor fusion, where the robot points the RGB-D camera towards people approaching the robot based on the estimation result of the LRS-based proposed method.

## 2 System Overview

Figure 2 shows the overview of the proposed method. As shown in Fig. 2(a), we define two coordinate systems: a field $x - y - \theta$ coordinate system and a robot $x' - y' - \theta'$ coordinate system, where the prime symbol indicates the robot coordinates. This study

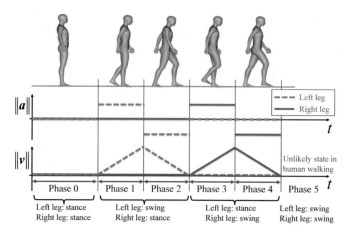

**Fig. 3.** Human walking model with six gait phases.

proposes the body-pose (position and direction) estimation method in field coordinates based on the gait (position, velocity, and states of both legs) using the shin height LRSs. Figure 2(b) gives an overview of the proposed algorithm. The robot estimates its own pose in field coordinates using the LRS scan data and its built-in map.

Then, the robot detects legs in the robot coordinate system, and it estimates the position, velocity, and states of both legs (gait) in the field coordinate system using the Kalman filter and the estimated coordinates of its own pose. Next, the body position and direction are calculated based on the gait, and these are estimated using the Kalman filter.

## 3   Gait Measurement Algorithm for Moving Robot

### 3.1   Human Walking Model

During normal human walking, one leg swings while pivoting on the other, alternating roles with a rhythmic pattern. We have proposed an accelerating walking model with six gait phases [7]. Figure 3 shows six gait phases. Phase 0 is the state where both legs are in the stance phase; phase 1 is the state where the left leg is accelerating in the swing phase and the right leg is in the stance phase; phase 2 is the state where the left leg is decelerating in the swing phase and the right leg is in the stance phase; phase 3 is the state where the left leg is in the stance phase and the right leg is accelerating in the swing phase; phase 4 is the state where the left leg is in the stance phase and the right leg is decelerating in the swing phase; and phase 5 is the unlikely state in human walking where both legs are in the swing phase.

## 3.2    State Equation of Each Leg

The leg position and velocity are estimated using the Kalman filter with an accelerating motion model that takes into account the gait phase. The discrete-time model of the leg motion at time step $k$ in field coordinates is given by

$$x_k^f = Ax_{k-1}^f + B_u u_{k-1}^f + B\Delta x_{k-1}^f \quad (f = L, R) \tag{1}$$

where
$$A = \begin{bmatrix} 1 & 0 & \Delta t & 0 \\ 0 & 1 & 0 & \Delta t \\ 0 & 0 & 1 & 0 \\ 0 & 0 & 0 & 1 \end{bmatrix}, B_u = B = \begin{bmatrix} \Delta t^2/2 & 0 \\ 0 & \Delta t^2/2 \\ \Delta t & 0 \\ 0 & \Delta t \end{bmatrix}$$
and

$x_k^f = \begin{bmatrix} p_{x,k}^f & p_{y,k}^f & v_{x,k}^f & v_{y,k}^f \end{bmatrix}^T$; $\left( p_{x,k}^f, p_{y,k}^f \right) := p_k^f$ is the estimated position and $\left( v_{x,k}^f, v_{y,k}^f \right) := v_k^f$ is the estimated velocity of the leg in field coordinates $(f = L, R,$ where $L$ and $R$ indicate the left and right legs, respectively); $u_k^f = \begin{bmatrix} u_{x,k}^f & u_{y,k}^f \end{bmatrix}^T$ is the acceleration input vector corresponding to the gait phase; $\Delta x_k^f = \begin{bmatrix} n^{v_{x,k}} & n^{v_{y,k}} \end{bmatrix}^T$ is the acceleration disturbance vector, which is assumed to be zero mean and has a white noise sequence with covariance $Q$; and $\Delta t$ is the sampling time. The leg measurement model $y_k^f = \begin{bmatrix} p_{x,k}^{\prime f} & p_{y,k}^{\prime f} \end{bmatrix}^T$ is defined using the estimated robot pose $p_k^{Robot} = \begin{bmatrix} p_{x,k}^{Robot} & p_{y,k}^{Robot} & \theta_k^{Robot} \end{bmatrix}^T = \begin{bmatrix} y_k^{Robot} & \theta_k^{Robot} \end{bmatrix}^T$ at time step $k$ in field coordinates,

$$y_k^{\prime f} = C_k x_k^f + C_k^\prime y_k^{Robot} + \Delta y_k^{\prime f} \tag{2}$$

where $C_k = \begin{bmatrix} \cos\theta_k^{Robot} & \sin\theta_k^{Robot} & 0 & 0 \\ -\sin\theta_k^{Robot} & \cos\theta_k^{Robot} & 0 & 0 \end{bmatrix}$, $C_k^\prime = \begin{bmatrix} -\cos\theta_k^{Robot} & -\sin\theta_k^{Robot} \\ \sin\theta_k^{Robot} & -\cos\theta_k^{Robot} \end{bmatrix}$, and $\Delta y_k^\prime = \begin{bmatrix} n^{p_{x,k}^\prime} & n^{p_{y,k}^\prime} \end{bmatrix}^T$ is the measurement noise, which is assumed to be zero mean and has a white noise sequence with covariance $R$.

## 3.3    Gait Phase Judgement

From the validation compared with the force plate [9], it is possible to identify the gait phase (in the stance phase or swing phase) considering the speed of both legs during human walking. The condition where the right leg is in the stance phase is:

$$\|v_k^R\| < \|v_k^L\| \quad \vee \quad \|v_k^R\| < v_{st\_th} \tag{3}$$

The condition where the right leg is in the swing phase is:

$$\|v_k^R\| > \|v_k^L\| \quad \vee \quad \|v_k^R\| > v_{sw\_th} \tag{4}$$

where $v_{st\_th}$ and $v_{sw\_th}$ are the thresholds of the maximum speed in the stance phase and the minimum speed in the swing phase, respectively. The gait phase of the left leg is identified in the same way. Then, the gait phase is identified considering the relative positional relationship of both legs and the velocity. First, if both legs are in the stance phase, the gait phase is identified as Phase 0. Second, if the left leg is in the swing phase and the right leg is in the stance phase, the gait phase is identified based on the inner product of the velocity vector of the left leg and the relative position vector of the right leg from the left leg:

$$\left(\boldsymbol{p}_k^R - \boldsymbol{p}_k^L\right) \cdot \boldsymbol{v}_k^L \tag{5}$$

If Eq. (5) has a positive value, the gait phase is identified as Phase 1; otherwise, the gait phase is identified as Phase 2. If the left leg is in the stance phase and the right leg is in the swing phase, the gait phase is identified in the same way. Finally, if both legs are in the swing phase, the gait phase is defined as Phase 5.

### 3.4  Leg Tracking Considering Gait Phase

The observed leg positions $\boldsymbol{y}'^j_k (j = 1, \cdots, J)$ in the robot coordinates are calculated based on five observed leg patterns [7]. To reduce the likelihood of false tracking and improve the measurement accuracy, an improved leg-tracking method using a data association (one-to-one matching of a tracked leg and an observation from LRS data) that considers the gait phase is used (see [7] for details on the leg-tracking algorithm).

## 4  Estimation of Body Direction Based on Gait

### 4.1  Observation of Body Pose Based on Gait

To estimate the body position, velocity, direction, and rotation velocity in field coordinates using the Kalman filter, observations of the body position and direction are obtained considering the position, velocity, and states of both legs. The observed body position $\left(p_{x,k}^B, p_{y,k}^B\right) := \boldsymbol{p}_k^B$ is calculated as the center of the estimated leg positions as follows,

$$\left(p_{x,k}^B, p_{y,k}^B\right) = \left(\frac{\hat{p}_{x,k}^L + \hat{p}_{x,k}^R}{2}, \frac{\hat{p}_{y,k}^L + \hat{p}_{y,k}^R}{2}\right) \tag{6}$$

where $\left(\hat{p}_{x,k}^f, \hat{p}_{y,k}^f\right) := \hat{\boldsymbol{p}}_k^f$ indicates the estimated position of the leg at time step, $k$. It is difficult to estimate the body direction based on the movement of the center of both legs. In this study, we focus on the gait, and the observation of the body direction $p_{\theta,k}^B$ is calculated according to the gait as follows.

### (a) Both Legs in Stance Phase (Gait Phase 0)

As shown in Fig. 4(a), the observation of the body direction is defined as the direction of the perpendicular of the line segment connecting both legs.

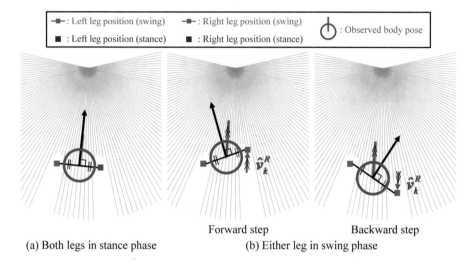

(a) Both legs in stance phase      Forward step      Backward step

(b) Either leg in swing phase

**Fig. 4.** Observation of body pose based on gait.

### (b) Either Leg in Swing Phase (Gait Phases 1 to 4)

As shown in Fig. 4(b), assuming that the body direction is oriented toward the direction of motion of the swing leg, the observation of the body direction is defined using the estimated velocity vector of the swing leg $\left( \hat{v}_{x,k}^{f}, \hat{v}_{y,k}^{f} \right) := \hat{v}_{k}^{f}$. However, the body direction is reversed in the case of backward steps. Therefore, using the perpendicular vector obtained in (a), it is determined whether there are forward or backward steps, and then the observation of the body direction is obtained, as shown in Fig. 4(b).

### 4.2  State Equation of Body

The body position, velocity, direction, and rotation velocity in field coordinates are estimated using the Kalman filter. The discrete-time model of the body motion at time step $k$ in field coordinates is given by

$$x_{k}^{B} = A^{B} x_{k-1}^{B} + B^{B} \Delta x_{k-1}^{B} \tag{7}$$

$$where \quad A^B = \begin{bmatrix} 1 & 0 & 0 & \Delta t & 0 & 0 \\ 0 & 1 & 0 & 0 & \Delta t & 0 \\ 0 & 0 & 1 & 0 & 0 & \Delta t \\ 0 & 0 & 0 & 1 & 0 & 0 \\ 0 & 0 & 0 & 0 & 1 & 0 \\ 0 & 0 & 0 & 0 & 0 & 1 \end{bmatrix}, B^B = \begin{bmatrix} \Delta t^2/2 & 0 & 0 \\ 0 & \Delta t^2/2 & 0 \\ 0 & 0 & \Delta t^2/2 \\ \Delta t & 0 & 0 \\ 0 & \Delta t & 0 \\ 0 & 0 & \Delta t \end{bmatrix}, \quad and$$

$x_k^B = \begin{bmatrix} p_{x,k}^B & p_{y,k}^B & p_{\theta,k}^B & v_{x,k}^B & v_{y,k}^B & v_{\theta,k}^B \end{bmatrix}^T$. $\Delta x_k^B = \begin{bmatrix} n^{v_{x,k}^B} & n^{v_{y,k}^B} & n^{v_{\theta,k}^B} \end{bmatrix}^T$ is the acceleration disturbance vector, which is assumed to be zero mean and has a white noise sequence with covariance $Q^B$. The measurement model $y_k^B = \begin{bmatrix} p_{x,k}^B & p_{y,k}^B & p_{\theta,k}^B \end{bmatrix}^T$ in field coordinates is given by

$$y_k^B = C^B x_k^B + \Delta y_k^B \tag{8}$$

where $C^B = \begin{bmatrix} 1 & 0 & 0 & 0 & 0 & 0 \\ 0 & 1 & 0 & 0 & 0 & 0 \\ 0 & 0 & 1 & 0 & 0 & 0 \end{bmatrix}$ and $\Delta y_k^B = \begin{bmatrix} n^{p_{x,k}^B} & n^{p_{y,k}^B} & n^{p_{\theta,k}^B} \end{bmatrix}^T$ is the measurement noise, which is assumed to be zero mean and has a white noise sequence with covariance $R^B$.

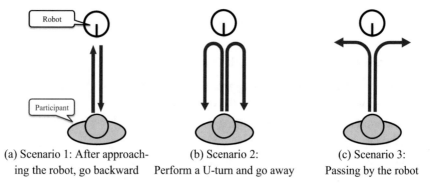

(a) Scenario 1: After approaching the robot, go backward

(b) Scenario 2: Perform a U-turn and go away

(c) Scenario 3: Passing by the robot

**Fig. 5.** Verification scenarios during experiments.

**Table 1.** System parameters used in experiments.

| $\Delta t$, Sampling time | 65 ms |
|---|---|
| $Q^B$, Motion covariance | $\text{diag}\left(2.5^2, 2.5^2, \left(\frac{2}{3}\pi\right)^2\right)$ |
| $R^B$, Observation covariance | $\text{diag}\left(0.07^2, 0.07^2, \left(\frac{\pi}{6}\right)^2\right)$ |

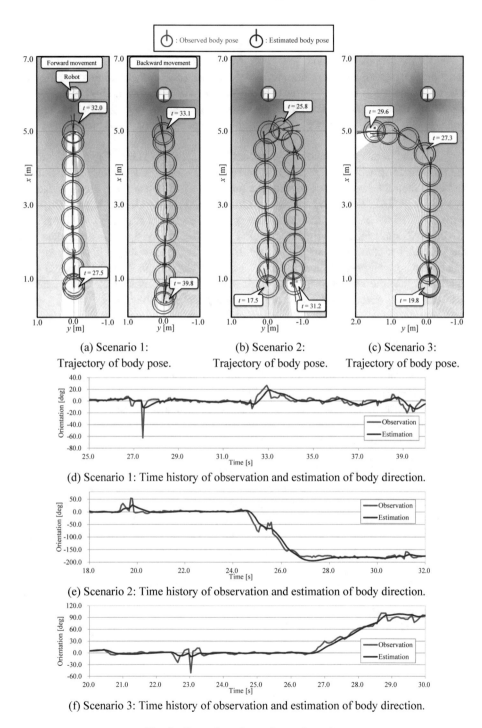

(a) Scenario 1:
Trajectory of body pose.

(b) Scenario 2:
Trajectory of body pose.

(c) Scenario 3:
Trajectory of body pose.

(d) Scenario 1: Time history of observation and estimation of body direction.

(e) Scenario 2: Time history of observation and estimation of body direction.

(f) Scenario 3: Time history of observation and estimation of body direction.

**Fig. 6.** Examples of experimental results.

## 5   Experiments

To verify the proposed method, two young volunteers (one male and one female) were recruited as participants for this study. As shown in Fig. 5, assuming the human interaction with the robot, we verified the proposed estimation method of the body direction in three scenarios. Figure 5(a) shows a movement in which the participant once gets close to the robot and then moves backwards. Figure 5(b) shows the movement,

where the participant makes a U-turn and leaves from the robot. Figure 5(c) shows the movement in which the participant passes by the robot. The robot was stationary and estimated the body direction with the proposed method using the distance data of two LRSs (UTM-30LX-FW, Hokuyo Automatic Co., Ltd., Osaka, Japan [10]) that were installed 0.3 m above the floor. The system parameters are shown in Table 1.

Figure 6 shows the experimental results. As shown in Fig. 6, it was confirmed that the proposed method is applicable to human forward, backward, and U-turn movements that are observed when a person interacts with a robot. In addition, it was confirmed that the proposed method could reduce the influence of the instantaneous deterioration of accuracy of the observation using Kalman filtering.

## 6   Conclusion

To realize socially acceptable human-robot interaction for service robots, it is important for human recognition to include not only position but also body direction around the robot. By using an RGB-D camera, it is possible to detect the posture of a person. However, because the viewing angle of the camera is narrow, it is difficult to measure the surroundings of the robot with a single device. This study proposes the estimation of the body direction based on the gait, which includes not only the position and velocity, but also the states of the legs (stance or swing phase) using laser-range sensors that are installed at shin height. From the experimental results, it was confirmed that the proposed method is applicable to human forward, backward, and U-turn movement, which are observed when a person interacts with the service robot.

Human-robot interaction and accompanying control for a person based on the estimated body pose by the proposed method is future work.

**Acknowledgement.** This study was supported by JSPS KAKENHI Grant Number 17K14619 and "A Framework PRINTEPS to Develop Practical Artificial Intelligence" of the Core Research for Evolutional Science and Technology (CREST) of the Japan Science and Technology Agency (JST) under Grant Number JPMJCR14E3.

# References

1. Haneda Robotics Lab. https://www.tokyo-airport-bldg.co.jp/hanedaroboticslab/en.html
2. Nakamura, K., Morita, T., Yamaguchi, T.: PRINTEPS for development integrated intelligent applications and application to robot teahouse. In: International Conference on Web Intelligence, pp. 1199–1206 (2017)
3. Satake, S., Kanda, T., Glas, D.F., Imai, M., Ishiguro, H., Hagita, N.: A robot that approaches pedestrians. IEEE Trans. Robot. **29**(2), 508–524 (2013)
4. Hu, J.S., Wang, J.J., Ho, D.M.: Design of sensing system and anticipative behavior for human following of mobile robots. IEEE Trans. Ind. Electron **61**(4), 1916–1927 (2014)
5. Karunarathne, D., Morales, Y., Kanda, T., Ishiguro, H.: Model of side-by-side walking without the robot knowing the goal. Int. J. Soc. Robot. **10**, 1–20 (2017)
6. Carballo, A., Ohya, A., Yuta, S.: Reliable people detection using range and intensity data from multiple layers of laser range finders on a mobile robot. Int. J. Soc. Robot. **3**(2), 167–186 (2011)
7. Yorozu, A., Moriguchi, T., Takahashi, M.: Improved leg tracking considering gait phase and spline-based interpolation during turning motion in walk tests. Sensors **15**(9), 22451–22472 (2015)
8. Yorozu, A., Takahashi, M.: Navigation for gait measurement robot evaluating dual-task performance considering following human in living space. In: Workshop on Assistance and Service Robotics in a Human Environment in Conjunction with IEEE/RSJ International Conference on Intelligent Robots and Systems (2016)
9. Matsumura, T., Moriguchi, T., Yamada, M., Uemura, K., Nishiguchi, S., Aoyama, T., Takahashi, M.: Development of measurement system for task oriented step tracking using laser range finder. J. NeuroEngineering Rehabil. **10**, 47 (2013)
10. Hokuyo Automatic Co., Ltd. http://www.hokuyo-aut.jp/

# Accurate Pouring with an Autonomous Robot Using an RGB-D Camera

Chau Do$^{(\boxtimes)}$ and Wolfram Burgard

University of Freiburg, Autonomous Intelligent Systems,
Georges-Köhler-Allee 80, 79110 Freiburg, Germany
{do,burgard}@informatik.uni-freiburg.de

**Abstract.** Robotic assistants in a home environment are expected to perform various complex tasks for their users. One particularly challenging task is pouring drinks into cups, which for successful completion, requires the detection and tracking of the liquid level during a pour to determine when to stop. In this paper, we present a novel approach to autonomous pouring that tracks the liquid level using an RGB-D camera and adapts the rate of pouring based on the liquid level feedback. We thoroughly evaluate our system on various types of liquids and under different conditions, conducting over 250 pours with a PR2 robot. The results demonstrate that our approach is able to pour liquids to a target height with an accuracy of a few millimeters.

**Keywords:** Liquid perception · Robot pouring · Household robotics

## 1  Introduction

A capable and effective domestic service robot must be able to handle everyday tasks involving liquids. Some examples are filling a cup or measuring out a certain amount of liquid for baking or cooking. This requires the ability to perceive the liquid level while pouring and using this information to decide when to stop pouring. This is a challenging task, considering the large selection of liquids available and their varying characteristics. In this paper, we consider the problem of tracking a liquid and determining when to stop pouring using depth data from a low-cost, widely available RGB-D camera.

Visually, liquids change their appearance depending on the environment, which makes it particularly challenging to detect the fill level of a container using vision. In this paper we therefore investigate the estimation of the fill level based on data provided by a depth camera such as the ASUS Xtion Pro or Microsoft Kinect. However, even for this modality the task at hand is complicated by the fact that liquids have different appearances depending on their transparency and index of refraction. For transparent liquids such as water and olive oil, the infrared light is *refracted* at the liquid boundary, causing the resulting liquid level to appear lower than it actually is. In the case of water, depending on the view angle, a full cup appears one-third to one-half full based on the depth

© Springer Nature Switzerland AG 2019
M. Strand et al. (Eds.): IAS 2018, AISC 867, pp. 210–221, 2019.
https://doi.org/10.1007/978-3-030-01370-7_17

(a) PR2 with cups and bottles used     (b) PR2 pouring water
in the experiments

**Fig. 1.** PR2 robot with bottles and cups used in the experiments (a) and close-up of a water pouring trial (b).

data. On the other hand, opaque liquids such as milk and orange juice, reflect the infrared light and the resulting depth measurement correctly represents the real liquid height. The approach presented in this paper is designed to deal with opaque and transparent liquids, by switching between the detection approaches depending on the type of liquid. We utilize a Kalman filter handling the uncertainty in the measurements and for tracking the liquid heights during a pour. For controlling the pour, we employ a variant of a PID controller that takes the perception feedback into account. We demonstrate the effectiveness of our approach through extensive experiments, which we conducted using a PR2 robot depicted in Fig. 1.

The two main contributions of this paper are a novel approach for pouring liquids into a container up to a user-defined height and the extensive analysis of the approach with respect to a large variety of parameters of the overall problem.

## 2   Related Work

A substantial amount of research has been carried out on various aspects of pouring. For example, Pan and Manocha [1] and Tamosiunaite et al. [2] focus on learning a pouring trajectory, but do not consider the problem of pouring to a specific height. In the area of learning from humans, Langsfeld et al. [3] and Rozo et al. [4] demonstrate how to pour a specific volume, and learn the pouring motion parameters for that volume. However, in their work no perception of the liquid is used. Regarding the area of liquid perception, Elbrechter et al. [5] focus on the problem of detecting liquid viscosity. Morris and Kutulakos [6] look at reconstructing a refractive surface, but this requires a pattern placed underneath the liquid surface. Mottaghi et al. [7] use deep learning to infer volume characteristics of containers. Neither of these papers consider the problem of tracking a liquid level and pouring to a specific height. Yamaguchi et al. [8] focus on the planning side of pouring. They detect the fill level using a plastic cup, modified such that the back half is colored and use color segmentation. Our approach does not require modified containers. Yamaguchi et al. [9] use a stereo camera and apply optical flow detection to determine the liquid flow. They do not apply their approach to the problem of pouring to a specific height. Furthermore, flow detection cannot be used to determine static liquid already present in a container, which our approach is capable of as shown by our experiments. Yoshitaka et al. [10] use an RGB-D camera to detect the presence and height of liquid in a cup. For transparent liquids, they outlined a relationship between the measured height from the depth data and the real liquid height. They did not consider the problem of pouring liquids and only dealt with the detection of static liquids.

In our previous work [11], we presented a probabilistic approach to liquid height detection, which makes use of both RGB and depth data. It assumes no knowledge of the liquid type and only considers a static liquid height (i.e., non-pouring) scenario. This previous approach uses multiple images and point clouds of the static liquid, taken from different viewing angles. Capturing data from different viewing angles is required to disambiguate between liquid types and accurately determine the height, since no prior knowledge of the liquid is given. In this paper, in contrast to our prior work, we determine the height of the liquid during the pour and without observing the cup from multiple viewpoints.

Schneck and Fox [12] demonstrate it is possible to segment and track liquids using deep learning. However, this was only shown with synthetic data. In follow-up work, Schneck and Fox [13] consider the problem of pouring to a specific height. They train a deep network to classify image pixels as either liquid or not liquid. For ground truth, they synchronized a thermal camera with an RGB-D camera and recorded pouring data using water heated to 93 °C. Using this detection they then track the liquid volume while pouring. For 30 pours, they reported a mean error of 38 ml. As we will demonstrate in our experimental evaluations, our approach is able to achieve a substantially lower mean error while only employing an RGB-D camera and not requiring an expensive thermal camera as well as the heating of the liquid for collecting the training data.

Compared to these previous approaches, the method presented in this paper allows accurate pouring of liquids to given levels using a low-cost RGB-D sensor. It can be applied to different types of liquids without any need for training and is usable in every environment in which an RGB-D camera such as the Kinect or Xtion can operate.

## 3   The Autonomous Pouring System

Figure 2 depicts our PR2 robot used for evaluating our pouring approach. We control the pouring by rotating the angle of the robot's wrist joint. As input the system receives the point clouds from the robot's RGB-D sensor. The cup should be positioned such that the camera can look into it and can see part of the cup bottom.

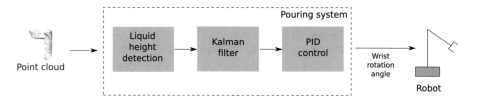

**Fig. 2.** Overview of the approach: the sensor inputs are point clouds from an RGB-D camera. The output is a rotation angle for the wrist joint of the robot to pour the liquid.

### 3.1  Liquid Height Detection

In our approach, the detection and tracking of the liquid level relies only on the depth information from an RGB-D camera such as a Kinect or an ASUS Xtion Pro. Throughout this paper, we assume that the robot knows which liquid is being poured and which target height is required. Typically, this information can easily be provided by the user, for example, by issuing a command such as "pour me a full cup of water" or following a recipe for baking and cooking.

There are two cases to consider with respect to the liquid type: opaque and transparent liquids. Opaque liquids, such as milk and orange juice, are accurately represented in the point cloud and the extracted height can be taken as the true liquid height. Transparent liquids such as water and olive oil, however, refract the light and the point cloud height is incorrect. It is possible to estimate the actual liquid height given a refracted depth measurement using a relationship based on the view angle and the index of refraction of the liquid (see Yoshitaka et al. [10] and Do et al. [11] for a full derivation). This relationship for finding the liquid height $h$ is given by:

$$h = \left( \frac{\sqrt{n_l^2 - 1 + \cos^2(\alpha)}}{\sqrt{n_l^2 - 1 + \cos^2(\alpha)} - \cos(\alpha)} \right) h_r. \tag{1}$$

Thus, we estimate the liquid height $h$ given the raw depth measurement height $h_r$, index of refraction of the liquid $n_l$ and the angle $\alpha$, where $h_r$ is the liquid height determined from the point cloud and $\alpha$ is the incidence angle of infrared light from the camera projector with respect to the normal of the liquid surface. The latter is also determined from the point cloud. The index of refraction $n_l$ is determined from the liquid type, which is provided by the user. It should be noted that Eq. (1) is an approximation and does not account for all physical effects.

To determine the liquid height from the depth data, we first need to detect the cup. We achieve this by finding the plane representing the table and extracting all points above the table. Then we use RANSAC [14] to determine the cylinder model for the cup. In order to avoid detecting the cup rim as part of the liquid height, we determine the diameter of the cup from the detected cylinder model and only consider a reduced diameter section for the liquid. In other words, we only search an area defined by the cup height and the reduced diameter. Finally we extract the points inside this area and average over them to get the raw measured height.

## 3.2   Kalman Filter

To track the liquid height and filter the typically noisy depth measurements from the RGB-D camera [15], we use the Kalman filter for tracking. Thereby we assume that the liquid height follows a constant velocity motion given by

$$x_k = F x_{k-1} + w_k, \tag{2}$$

where $w_k \sim \mathcal{N}(0, Q_k)$ is zero-mean Gaussian noise with covariance $Q_k$, which represents the system noise and is given by

$$Q_k = q \begin{bmatrix} \frac{1}{3}\Delta t^3 & \frac{1}{2}\Delta t^2 \\ \frac{1}{2}\Delta t^2 & \Delta t \end{bmatrix}. \tag{3}$$

Here, $x_k$ and $F$ are given by

$$x_k = \begin{bmatrix} h_k \\ \dot{h}_k \end{bmatrix} \text{ and } F = \begin{bmatrix} 1 & \Delta t \\ 0 & 1 \end{bmatrix}. \tag{4}$$

The term $h_k$ refers to the liquid height and $\dot{h}_k$ refers to the rate of change in liquid height at time $k$, while $\Delta t$ is the time interval. Note that we chose not to model a system input, as this term and its effect on the system depends on several factors such as the bottle opening, amount of liquid in the bottle and the tilt angle of the bottle. Modeling this term from recorded data would be unique for that situation only. We determined the measurement noise covariance matrix $R_k$ and $q$ from collected pouring data.

## 3.3   Pouring Control

To realize the pour, we need to determine the rotation angle of the wrist joint. In our approach we employ a variant of the PID controller that we augmented by additional control policies to ensure a smooth pour. In the event that no liquid height is detected in the point cloud, we slow down the pouring. If no liquid height update is received (i.e., the detection is too slow), we maintain the rotation angle so as not to continue pouring blindly. As soon as we detect that the required liquid height has been reached, we rotate the bottle back to its initial position.

## 4  Experiments

We implemented our approach on a PR2 robot equipped with an ASUS Xtion Pro camera and carried out a series of experiments to evaluate our approach in real-world settings. The different cups and bottles involved in the experiments can be seen in Fig. 1. For the experiments, we positioned the cup relative to the camera with its bottom center point at an approximate horizontal distance of 25 cm and an approximate vertical distance of 75 cm. The vertical angle of the camera was chosen so that the cup bottom could be detected in the depth data. In all experiments, we placed the bottle in the gripper of the PR2 and positioned it close to the cup. Before the experiments, we fine-tuned the parameters of our PID controller. We found that proportional control was sufficient for achieving accurate results. A pouring trial for milk can be seen in Fig. 3.

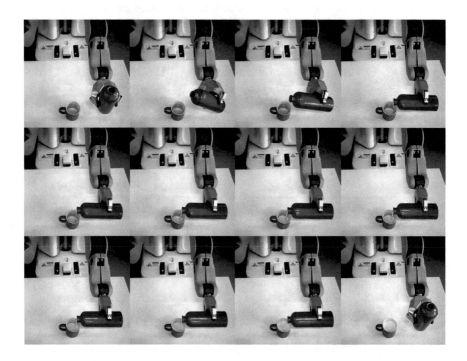

**Fig. 3.** Pouring sequence for a milk pouring trail (best viewed enlarged).

In total, we performed six different types of experiments, which we designed to analyze how changes in liquid type, initial liquid volume, target height, bottles and cups affect the pouring accuracy. In all of these experiments we used the same parameter values. Overall, we performed a total of 290 pours. To measure the ground truth data we hand measured the liquid height with a ruler.

## 4.1    Influence of Different Liquids

To analyze the impact of the liquid type, we performed experiments with water, carbonated water, olive oil, milk and orange juice. The first three are transparent liquids, while the latter two are opaque. Overall, they represent differences in properties such as viscosity, index of refraction and carbonation.

For each liquid we poured ten times. For each pour, the initial volume in the bottle was randomly chosen from the values of [200 ml, 250 ml, 300 ml, 350 ml, 400 ml, 450 ml, 500 ml] and the target height in the cup was randomly chosen from the range of [20 mm, 30 mm, 40 mm, 50 mm, 60 mm, 70 mm]. We have the criteria that the initial volume in the bottle should be at least 100 ml more than the target height, to make sure the robot does not pour all the liquid out. To focus only on the changes caused by the liquids, we used the blue bottle and blue cup shown on the rightmost side of Fig. 1 for all the pours in this experiment.

Figure 4 shows the absolute mean error and standard deviation for the ten pours of each liquid. The crosses show the maximum error that occurred. In almost all cases the robot over-pours. We believe this is due to the fact that we send the *STOP POUR* signal when we detect that the desired liquid height has been reached. However, this is minimized by the PID controller, which reduces the amount poured as the liquid height approaches the target height. The opaque liquid results show that this delay accounts for only around 2 mm additional liquid. Between the three transparent liquids there is not much variation and the same can be said of the two opaque liquids.

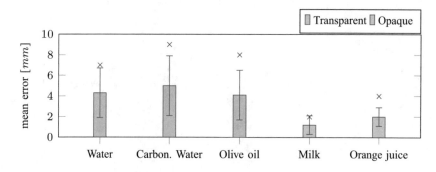

**Fig. 4.** Mean error and standard deviation for ten pours of different liquids with varying initial volume and target height. The crosses mark the maximum error.

As can be seen, the transparent liquids have higher mean errors and standard deviations versus the opaque ones. To investigate this further, we additionally poured 60 mm of water and milk into the same blue cup and recorded 200 depth measurements. The results can be seen in Table 1. In this table, "raw" refers to the raw depth measurement taken from the point cloud. In the case of water, this is before being transformed by Eq. (1). For the raw measurements, the standard deviation is about the same for both water and milk. The transformed values for water have a higher standard deviation. Looking at Eq. (1), it is clear that any noise in the raw measurement $h_r$ is magnified by the term before it. For the recorded data, the term before $h_r$ is around 4.03. A second observation, is that the transformed value for water underestimates the liquid height. We note that Eq. (1) is only an approximation and does not factor in all physical effects. Accordingly, we can expect a higher mean error and standard deviation for the transparent liquids versus the opaque ones.

**Table 1.** Static error - target height 60 mm

| Measurement | $\mu \pm \sigma$ [mm] |
|---|---|
| Water (raw) | $18.64 \pm 0.18$ |
| Water | $58.77 \pm 0.56$ |
| Milk (raw) | $60.37 \pm 0.12$ |

## 4.2   Influence of Varying Initial Volume

In this experiment, we investigate how robust our system is to changes in initial volumes in the bottle. We kept the target height fixed at 40 mm and varied the initial volume by changing it to one of [350 ml, 400 ml, 450 ml, 500 ml]. In other words, for each volume we poured ten times to a height of 40 mm. As before, we only used the blue bottle and blue cup. The results for transparent liquids depicted in Fig. 4 are very similar and the same can be said of the opaque liquids. Hence for this experiment and the following ones we only used water and milk.

The top figure in Fig. 5 shows the results of this experiment for the pours. There is a slight increase in mean error as the volume increases. For water, the difference in mean error between 500 ml and 350 ml is 2.4 mm and for milk, it is 1 mm, which are not very large differences in view of the overall results.

## 4.3   Influence of Varying Target Heights

Here we look at the influence of the target height. We kept the initial volume constant at 400 ml and varied the target height to one of [30 mm, 40 mm, 50 mm, 60 mm]. For each height, we poured ten times for both water and milk. The blue bottle and blue cup were used each time for consistency.

The results can be seen in the bottom figure of Fig. 5. The mean errors and standard deviations remain fairly consistent for both liquids. So it does not appear as if different target heights have a large influence on the pouring error.

## 4.4   Influence of Bottle Opening

In this experiment, we investigate the influence of the bottle opening. The silver bottle (referred to as the wide opening bottle) pictured on the left in Fig. 1 has an opening of 4.5 cm while the blue bottle (referred to as the small opening

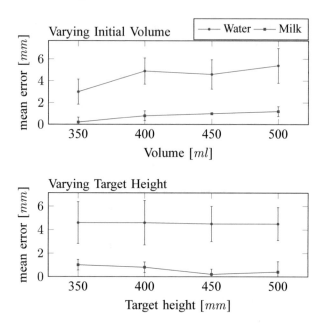

**Fig. 5.** Mean error and standard deviation over ten pours for varying initial liquid volumes in the bottle and constant target height (top). Mean error and standard deviation over ten pours for varying target height with constant initial liquid volumes (bottom).

bottle), pictured on the right, has an opening of 2.5 cm. We conducted ten pours each for water and milk, using the wide opening bottle. The initial volumes and target heights were varied in the same manner as in Sect. 4.1 for each pour.

The results can be seen in the top figure of Fig. 6. We include the results for the small opening bottle from Sect. 4.1 for reference. A small increase in mean error (0.3 mm for water and 0.2 mm for milk) for the wide opening bottle can be seen. This can be expected since a wider opening makes it more difficult to control the flow of the fluid as it leaves the bottle. But overall, this experiment demonstrates that our system is able to deal with different openings.

### 4.5    Influence of Different Cups

So far, each experiment was conducted with the blue cup, depicted on the right side in Fig. 1. In this section we investigate the influence of different cups, in particular whether the shape and cup bottom diameter affect the system. We refer to the leftmost cup in Fig. 1 as the *text cup*, the middle cup as the *patterned cup* and the rightmost cup as the *blue cup*. For both the text cup and the patterned cup, we performed ten pours each, using the blue bottle for water and milk. As in Sect. 4.1, we varied the initial volume of liquid in the bottle and the target height. We used the same approach for detecting the cup in each case.

The results can be seen in the middle figure of Fig. 6. For reference we include the results for the blue cup from Sect. 4.1. In general, the differences in mean

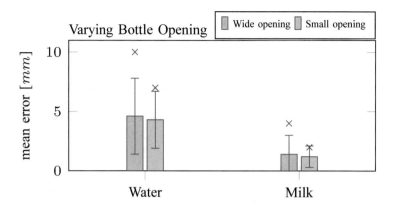

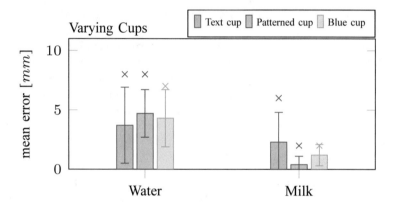

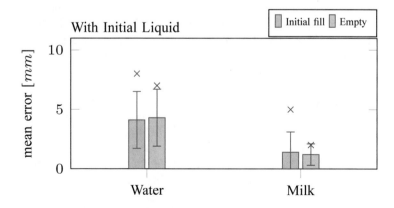

**Fig. 6.** Mean error and standard deviation over ten pouring experiments with a wide and a small opening bottle (top). Mean error and standard deviation over ten pours for each of the three different cups (middle). Mean error and standard deviation over ten pours for each liquid for pouring into a partly pre-filled cup (bottom). In all cases, the crosses represent the worst results over all ten pours.

error are minor between cups and there is no clear trend. The cup with the smallest bottom is the text cup, having a diameter of around 5 cm, compared to 6 cm for the patterned cup and 7.5 cm for the blue cup. In the case of water, where it is the refracted cup bottom that is being measured, one would expect a smaller cup bottom would be more problematic, but this does not appear to be the case. Furthermore, it was unclear beforehand whether the shape of the cups would affect Eq. (1) (due to reflections off the cup sides). But this does not appear to be an issue.

### 4.6   Influence of Initial Liquid in Cup

In this experiment, we look at how the system can deal with initial liquid levels in the cup. For both water and milk, we performed ten pours each. In each pour, we chose an initial amount of liquid in the cup ranging between 10–30 mm. We varied the initial volume in the bottle and ensured the target height was at least 10 mm more than the initial liquid amount. Once again, we used the blue bottle and blue cup.

The bottom figure in Fig. 6 shows the mean errors, standard deviations and maximum pour error. We also include the case of starting with an empty cup (see Sect. 4.1 for reference). The results show only minor differences between pouring into an empty cup and pouring into one with an initial amount of liquid. Accordingly our system is able to robustly deal with situations in which the cup is partly pre-filled.

### 4.7   Comparison of Pouring Accuracy

As noted before, Schneck and Fox [13] report a mean error of 38 ml for water over 30 pours, using a system that also controls the rotation angle of the wrist joint gripping the bottle. For each of our cups, we take the worst 10-pour trial, which all occurred while pouring water, and converted the height errors to volume errors. This resulted in mean volume errors of 23.9 ml for the blue cup, 13.2 ml for the text cup and 30.5 ml for the patterned cup. The patterned cup is also the widest cup of the three, meaning errors in height result in larger volume errors. Overall, our system was able to achieve better accuracy in pouring, while under more extreme testing conditions.

## 5   Conclusion

In this paper, we presented a novel approach for liquid pouring that uses data received from an RGB-D camera and allows for pouring to a specific, user-defined height. Our approach tracks the liquid height during the pour which allows it to accurately stop as soon as the desired height has been reached. We conducted extensive experiments with our PR2 robot, which show that we are able to accurately pour both a transparent and an opaque liquid under various conditions. In future work we will look at reducing the dependency on user input and improving the handling of delays between the perception and the controller so as to further improve the accuracy.

# References

1. Pan, Z., Manocha, D.: Motion planning for fluid manipulation using simplified dynamics. In: Intelligent Robots and Systems (IROS), pp. 4224–4231 (2016)
2. Tamosiunaite, M., Nemec, B., Ude, A., Wörgötter, F.: Learning to pour with a robot arm combining goal and shape learning for dynamic movement primitives. Robot. Auton. Syst. **59**, 910–922 (2011)
3. Langsfeld, J., Kaipa, K., Gentili, R., Reggia, J., Gupta, S.: Incorporating failure-to-success transitions in imitation learning for a dynamic pouring task. In: Workshop on Compliant Manipulation: Challenges and Control (2014)
4. Rozo, L., Jiménez, P., Torras, C.: Force-based robot learning of pouring skills using parametric hidden Markov models. Robot Motion and Control (RoMoCo), pp. 224–232 (2013)
5. Elbrechter, C., Maycock, J., Haschke, R., Ritter, H.: Discriminating liquids using a robotic kitchen assistant. In: Intelligent Robots and Systems (IROS), pp. 703–708 (2015)
6. Morris, N., Kutulakos, K.: Dynamic refraction stereo. IEEE Trans. Pattern Anal. Mach. Intell. **33**, 1518–1531 (2011)
7. Mottaghi, R., Schenck, C., Fox, D., Farhadi, A.: See the glass half full: reasoning about liquid containers, their volume and content. arXiv preprint arXiv:1701.02718 (2017)
8. Yamaguchi, A., Atkeson, C., Ogasawara, T.: Pouring skills with planning and learning modeled from human demonstrations. Int. J. Humanoid Rob. **12**, 1550030 (2015)
9. Yamaguchi, A., Atkeson, C.: Stereo vision of liquid and particle flow for robot pouring. In: Humanoid Robots (Humanoids), pp. 1173–1180 (2016)
10. Yoshitaka, H., Fuhito, H., Takashi, T., Akihisa, O.: Detection of liquids in cups based on the refraction of light with a depth camera using triangulation. In: Intelligent Robots and Systems (IROS), pp. 5049–5055 (2014)
11. Do, C., Schubert, T., Burgard, W.: A probabilistic approach to liquid level detection in cups using an RGB-D camera. In: Intelligent Robots and Systems (IROS), pp. 2075–2080 (2016)
12. Schenck, C., Fox, D.: Detection and tracking of liquids with fully convolutional networks. arXiv preprint arXiv:1606.06266 (2016)
13. Schenck, C., Fox, D.: Visual closed-loop control for pouring liquids. In: Robotics and Automation (ICRA), pp. 2629–2636 (2017)
14. Fischler, M., Bolles, R.: Random sample consensus: a paradigm for model fitting with applications to image analysis and automated cartography. Commun. ACM **24**, 381–395 (1981)
15. Andersen, M.R., Jensen, T., Lisouski, P., Mortensen, A.K., Hansen, M.K., Gregersen, T., Ahrendt, P.: Kinect depth sensor evaluation for computer vision applications. Technical report ECE-TR-6 (2012)

# MS3D: Mean-Shift Object Tracking Boosted by Joint Back Projection of Color and Depth

Yongheng Zhao$^{(\boxtimes)}$ and Emanuele Menegatti

Department of Information Engineering (DEI), University of Padova,
Via Gradenigo 6/B, 35131 Padova, Italy
{zhao,emg}@dei.unipd.it
http://robotics.dei.unipd.it/

**Abstract.** In this paper, we present MS3D tracker, which extends the mean-shift tracking algorithm in several ways when RGB-D data is available. We fuse color and depth distribution efficiently in the mean-shift tracking scheme. In addition, in order to improve the robustness of the description of the object to be tracked, we further process the pixels in the rectangular region of interest (ROI) returned by mean-shift. We apply depth distribution analysis to pixels of the ROI in order to separate background pixels from pixels belonging to the object to be tracked (i.e. the target region). Then, we use the color histogram of the target region and its surroundings to create a discriminative color model, which has the capability to distinguish the object from background. The proposed algorithm is evaluated on the RGB-D tracking dataset proposed by [1]. It ranked in the first position and it runs in real-time showing both accuracy and robustness in the challenge sequences of background clutter, occlusion, scale variation and shape deformation.

**Keywords:** Mean-shift · RGB-D object tracking
Fusion of color and depth

## 1  Introduction

Visual object tracking is one of the fundamental tasks for computer vision applications ranging from surveillance to human-computer interactions and augmented reality, from autonomous driving to robotics and industrial automation [2]. Although recent years have seen great progress regarding this field [3,4], this task is still far from to be solved because of several challenging factors e.g. background clutter, illumination variation, occlusion, shape deformation, scale variation and fast motion.

Over the last decade, although the object tracking algorithms based on colors have shifted from models based on color statistics (*e.g.* [5]) to complex features based descriptors (*e.g.* [6]), trackers based on the color statistics are still able to achieve excellent performances [7–9]. Algorithms based on color statistics work

© Springer Nature Switzerland AG 2019
M. Strand et al. (Eds.): IAS 2018, AISC 867, pp. 222–236, 2019.
https://doi.org/10.1007/978-3-030-01370-7_18

well with object's shape changes, while they suffer to discriminate the target from background in challenging and cluttered situations [10]. In order to deal with this problem, many authors [7,8] apply discriminative models, which use surrounding color distribution to weigh the target model. Since the appearance of the target and its surrounding background can vary significantly during a sequence, many state-of-the-art algorithms exploit the tracking results in past frames as training data to continuously update the object model. This is dangerous, because (even small) tracking errors can accumulate and cause model drift, especially when there is occlusion or fast shape changes.

The contribution of this paper is four-fold. First, we propose a depth distribution based target extraction method to prevent the background in ROI interfering the accuracy of initial color model. Then we present a background weight algorithm to obtain a discriminative color model. In order to deal with tracker drifting to self-similar background or objects, we propose the joint color and depth back projection to the original mean-shift tracking algorithm. At last we propose a image moments based occlusion handle mechanism which actively detects the target occlusion and recovery during tracking process.

The reminder of the paper is organized as follow. In Sect. 2, we review the work on mean-shift based trackers, other color distribution based trackers as well as RGBD data based trackers. Our contributions are presented in Sect. 3, while experimental results and discussions are provided in Sect. 4. Finally, we summarize our conclusion in Sect. 5.

## 2  Related Work

Mean-shift is a very popular nonparametric and fast matching object tracking algorithms due to its simplicity and efficiency. In the mean-shift tracking algorithm, the color histogram is used to represent the target because of its robustness to camera motion scaling, rotation and partial occlusion [5]. Tracking algorithms based on color distribution are more robust to spatial shape deformation, but less discriminative to similar colored background, compared to template based trackers which rely on spatial descriptions. In order to overcome this problem, Stolkin et al. [11] proposed an algorithm named Adaptive Background Camshift which applies a background model and on-line learning method to obtain the background with little computational expense. Ning et al. [12] presented a tracking algorithm based on the representation of joint color-texture histogram in which the texture information is extracted by means of local binary pattern (LBP) technique.

Other researchers have thought to resolve this issue by regarding the color distribution of object surroundings as a weight on target model to enhance its capability to distinguish the object from its surrounding background (e.g. [7,8,13]). This background weighting method is used a lot in the recent color distribution based trackers which achieved state-of-the-art performance. In DAT tracker [7], the authors employ Bayes rule on target and its surrounding color distribution to obtain the object likelihood in the search region. Pixels with the color that only shown in the target region will be assigned the maximum probability of 1.0 while Pixels which is unseen in the both target and surrounding region will be assigned the maximum entropy prior of 0.5. This background model is not able to describe the color probability density in detail, thus it is not suitable for mean-shift based trackers. Therefore, the author used tracking-by-detection principle to localize the object rather than mean-shift.

Because of the movement of the object and illumination variations, both target and surrounding background can vary significantly during a sequence. Several researchers proposed on-line learning methods to update the model in so as to maintain its discriminative ability towards background interference [7,13]. However the model will learn also from inaccurate predictions and this causes model drift, especially in case of fast motion, occlusion, and fast changes in shape. Hu et al. [14] consider Correlation Filter (CF) based trackers as an open-loop systems and propose a generic mechanism to use closed loop feedback to reduce or prevent tracking drift. Besides, Fan et al. [15] apply a verifier based on a VGG-net detector which works in parallel with the tracker. The object model will be updated if the tracking result passes verification. However the verifier works slower than the tracker (approximately 20 ms) so that the tracker needs to repeat the processing of the frames 20 ms ago, if the verification has negative result. With this feature, the tracker works well with the recorded sequences while may has a problem with on-line tracking. There are also some short-term trackers which do not update the model during tracking. Vojir et al. [8] adopt a regularization terms by considering two problem: scale expansion by background clutter and scale implosion on self-similar objects. In this paper, the color model update is also not exploited and those two problems are solved with depth distribution analysis.

Nowadays, with the advent of consumer RGB-D cameras which can achieve high-quality color and depth information in real time, many researchers have studied on combining the color and depth information to track the object. In a previous work [16], we proposed a fast and robust multiple object tracking algorithm based on a RGB-D version of mean-shift and exploited RGB-D camera networks when multiple RGB-D sensors are available. The algorithm is not generic enough and it has only discrete performances when the object moves fast toward the camera, because of the simple depth model update policy.

One of the most significant RGB-D tracking algorithm was proposed by Song and Xiao [17] who released a large RGB-D based general object tracking dataset (PTB) as well as a series of baseline trackers based on RGB and RGB-D data. The tracking performance of these trackers have shown that depth data convey a lot of meaningful information in object tracking, especially in challenging scenarios like occlusion and background clutters. In the last years there are a lot of relevant works evaluated on this dataset. DS-KCF [18] is build upon kernel correlation filter tracker and apply depth information to deal with occlusion, scale variation and shape deformation problems. OAPF tracker [19] is based on particle filter and applies a probabilistic method to deal with both partial and full occlusion. Xiao et al. [1] presented a RGB-D tracker, that exploits RGB and depth features independently to get two tracking results. The tracker composed by two layers (global and local layer) switches between each other by comparing the two tracking results. It is worth to mention that Xiao et al. noted that color images and depth images are not synchronized in some sequences of the PTB dataset. Thus, in [1] they proposed a new rich dataset with a large number of RGB-D sequences both from moving and stationary cameras. In this paper, we also use the dataset presented by Xiao et al.

# 3   Proposed Algorithm

In this section, we detail the building blocks composing the MS3D algorithm we are proposing. MS3D uses a discriminative color model to identify the foreground pixels belonging to the object to be tracked with respect to the background pixels (see Sect. 3.1). This model exploits the color Probability Density Function (PDF) to identify the tracked object in RGB-D sequences. Our approach is robust to shape deformation and complex cluttered background. Indeed, instead of updating the color model with on-line learning scheme, we use a mask which back projects of color and depth PDF into the image (see Sect. 3.2). This generates a joint probability distribution (JPD) image, which exploit differences in depth to handle complex backgrounds and interferences/occlusions of similar objects in subsequent frames. In order to achieve real-time performances, the mean-shift algorithm is applied on the JPD image to localize the target (Sect. 3.3). This provides accurate localization without exhaustive search. The moments of the pixels belonging to the foreground are used to detect occlusion and re-localize the target in a probabilistic manner.

## 3.1   Discriminative Color Model

**Target Region Extraction.** Most tracking algorithms build object models from a ROI (region of interest) containing the object to be tracked, see Fig. 1(a). This ROI commonly is a rectangle consisting of both target pixels and a small number of background pixels, see Fig. 1(b). If a model of the appearance of the object to be tracked is calculated from the pixels within the ROI, the background pixels will spoil the object description. This can make the tracker easier drift to the background. To solve this problem in [20] the combination of mean-shift segmentation and region growing was proposed for object localization. Instead, we use the depth distribution analysis to extract the pixels belonging to the target from ROI. The depth distribution of the ROI in Fig. 1(b) is displayed as histogram in Fig. 1(c). We exploit a peak searching method to find the peaks of depth histogram. Here, we make an assumption: the target object is not occluded in the initial frame of the sequence. This makes the pixels belonging to the target object closer to camera than other pixels in the ROI. With this assumption, the depth histogram can be divided into two parts: foreground peak with the smallest depth and background peaks with higher values of depth. Then we estimate the target depth interval around the foreground peak. This depth interval can be used to create a mask on the ROI to filter out the ROI pixels belonging to the background, as shown in Fig. 1(d). The pixels in the ROI, belonging to the background have been masked in black.

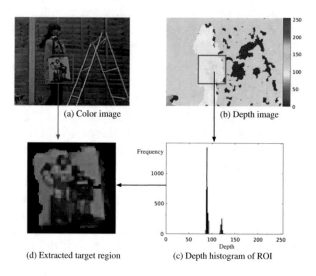

(a) Color image

(b) Depth image

(d) Extracted target region

(c) Depth histogram of ROI

**Fig. 1.** Target region extraction. The depth distribution of the ROI in (b) is displayed as histogram in (c). The extracted target region is shown in (d) with background masked in black.

This target extraction method performs well when the object and background vary a lot in depth, but, can perform poorly when the object is close to (or even touching) the background, as shown in Fig. 2 where the athlete stands on the floor. In this case the calculated foreground can easily include background pixels, failing to mask them out. To prevent this we apply the so called *surrounding background weighting* to filter out pixels erroneously assigned to the foreground.

**Background Weighting.** If pixels of the background can not be masked out using the depth information, like in Fig. 2(d). We propose to use the probabilistic description of the color of the background to filter them out. The intuition is to use the color histogram of a larger ROI surrounding the target ROI, and thus likely comprising a large portion of the background, to weight the color histogram of the ROI. The higher is the frequency (amount of pixels) of one color that is present in the surrounding ROI, the lower weight will be used to weigh the frequency of this color in the ROI. After this weighting, we can obtain a discriminative color model for the tracked object as a PDF that represents the likelihood of every color to represent a pixel of the foreground.

Let $H_C^\Omega$ denotes the $b$-th color frequency (amount of pixels) of the non-normalized color histogram $H$ calculated over region $\Omega$. As shown in Fig. 3, $O$ is the initial rectangular ROI which is also named bounding box, $T$ is the target

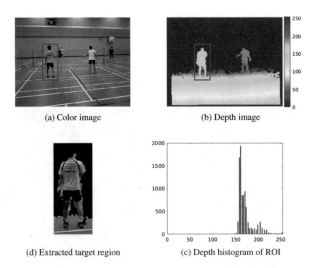

(a) Color image                (b) Depth image

(d) Extracted target region        (c) Depth histogram of ROI

**Fig. 2.** Inaccurate target region extraction. The depth histogram showed in (c) represent the depth distribution of athlete and floor. It is hard to be divide into different Gaussian components which caused the inaccurate target region extraction in (d)

region while $S$ is its surroundings. The object likelihood of $b$-th color $P(b)$ in PDF is calculated below:

$$
P(b) = \begin{cases} H_C^O(b) \times \dfrac{\min(H_C^S > 0)}{H_C^S(b)} & \text{if } H_C^S(b) > 0 \\ H_C^O(b) & \text{if } H_C^S(b) = 0 \end{cases}
\tag{1}
$$

where $\min(H_C^S)$ is the lowest frequency in the surrounding histogram and it is used to normalize the color PDF.

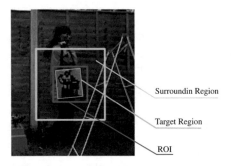

Surroundin Region

Target Region

ROI

**Fig. 3.** ROI, target and surrounding region

The object color PDF obtained from color image can be back projected to image in order to obtain probability distribution images which represents the object model likelihood in pixels. The probability distribution images obtained with original color PDF, DAT [7] weighted PDF, our proposed weighted PDF are shown in Fig. 4(b), (c), (d) respectively. Figure 4(a) represents the original color image.

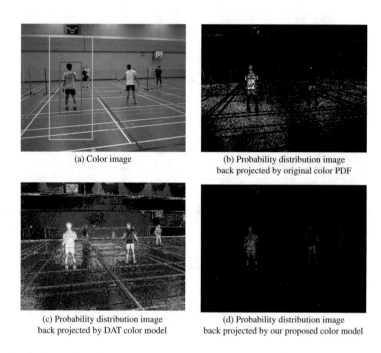

(a) Color image

(b) Probability distribution image
back projected by original color PDF

(c) Probability distribution image
back projected by DAT color model

(d) Probability distribution image
back projected by our proposed color model

**Fig. 4.** Probability distribution images back projected by different color model. The object color model (PDF) obtained from color image can be back projected to image in order to obtain probability distribution images which represents the object model likelihood in pixels.

As shown in Fig. 4, compared with other probability distribution images show in Fig. 4(a)–(c), the one (Fig. 4(d)) projected by our proposed color model is able to discriminate the target from background in the premise of reserving the original color distribution in detail.

## 3.2  Back Projection of Color and Depth PDF

**Color and Depth PDF.** With discriminative model described above, the tracker can works robustly against background interference if the background in the following frames is similar with initial object surroundings. However, the surrounding background can vary significantly during a sequence. Instead of using a model updating mechanism to deal with the continuously changing background clutter, we combine color and depth PDF with a joint back projection method and try to use the depth difference to discriminate the target object from the background and self-similar objects.

The color PDF is obtained in the initial frame and kept the same in the whole tracking procedure. The depth PDF represents the depth distribution of the target in current frame. We calculate the depth PDF by applying a mask to the ROI, in the depth image, containing the target object, similarly to the proposed technique for target region extraction in color image. In the initial frame, the depth PDF is calculated with the same mask that is used in color target region extraction. In the following frames, the extraction takes place in a search window which is an expansion of the last target region and composed by target, background, even occluders. We exploit a peak searching method to find the peaks in the depth histogram of this search window. By matching these candidate peaks with the depth distribution of the last target region, we can obtain the predicted depth interval of the current target. Then, we can apply it as a mask on the depth PDF of the search window to filter out non target pixels. Then the masked PDF will be used later with the color PDF to do the back projection.

The difference between the color and depth target extraction is that the color one only operates once in the initial frame to calculate the color PDF, while the depth is used in every frame to update the depth PDF, except when there is heavy occlusion. This depth PDF model update is necessary since the depth of target varies in all the frames.

**Joint Back Projection.** We apply the color PDF $P_C$ calculated with Eq. 1 and the updated depth PDF $P'_D$ in the joint back projection to generate the JPD image for mean-shift tracking. The intensity of JPD image $I_J$ which represents the target likelihood is calculated in (2)

$$I_J(x, y) = P_C\left(\frac{h(x,y)}{h\_bins}, \frac{s(x,y)}{s\_bins}, \frac{v(x,y)}{v\_bins}\right) * P'_D\left(\frac{d(x,y)}{d\_bins}\right) \qquad (2)$$

Here we use HSV color space and divide it into several bins: ($h\_bins \times s\_bins \times v\_bins$). The depth space is divided into $d\_bins$ bins.

Figure 5 shows an example of the discriminative power of the joint back projection. In Fig. 5a, one can notice that there is a colored object in background similar to the target object, which could easily cause tracking drift if only the color information is considered (see the Probability distribution image projected by color model only in Fig. 5b). At the same time, the depth between target and the person is similar which will also lead to tracking drift if only exploit depth distribution (see the probability distribution image projected by depth PDF only in Fig. 5c). Finally, the object model is more discriminative when a joint color and depth back projection is applied, see Fig. 5d.

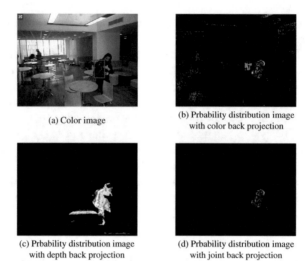

(a) Color image

(b) Prbability distribution image with color back projection

(c) Prbability distribution image with depth back projection

(d) Prbability distribution image with joint back projection

**Fig. 5.** Probability distribution image by different back projection. In (d), the proposed joint back projection shows the discriminative power on both self-color similar objects and self-depth similar objects.

### 3.3    Occlusion Aware Mean-Shift Tracking

**Mean-Shift for Localization.** In this paper, we exploit JPD image $I_J$ rather than color probability distribution (CPD) image which is used in the original mean-shift tracker.

**Occlusion Handle.** Being able to detect occlusions is of great importance to enhance the overall tracking quality. In the converged search window obtained from mean-shift iteration, the occlusion detection mechanism is based on the difference of depth and color probability distribution caused by occlusion. The statistical probability of search window can be represented by the JPD image moment and the difference can be measured by the ratio of initial and current statistical probability.

We present window quality $W_Q$ and density $W_D$ in Eq. 3 to represent the statistical probability of search window.

$$
\begin{cases}
W_Q = M_{00} = \sum_x \sum_y I_J(x,y) \\
W_D = \dfrac{M_{00}}{l \times w}
\end{cases}
\tag{3}
$$

Furthermore, by introducing $W_{Q_i}, W_{Q_c}$ as the statistical probability of the initial and current frame respectively, we ensure to detect occlusion by using a quality tolerance $Q_T$ in the statistical probability ratio (see Eq. 4).

$$
\frac{W_{Q_i}}{W_{Q_c}} < Q_T
\tag{4}
$$

Once an occlusion is detected, a global target re-localization procedure will generate an expanding search window centered in the last tracked window to recover the tracked object. The depth PDF will stop updating during occlusion since the changed depth data of the object can not be fetched. On the CPD (color probability distribution) image generated from the original color PDF, the re-localizer uses mean-shift iterations and a strict restriction to ensure an accurate re-localization after occlusion. This restriction is composed by two joint ratio of initial and current statistical probability: window quality and density. With the expanding search window, the request of window quality is easy to meet in a large window, while that of window density is much harder which can ensure a more accurate re-localization. When the object is partially recovered from full occlusion, the window quality can meet the request while the window density can not because of the expanded search window. During this partial occlusion, we will rescale (usually reduce) the target window by Cam-shift for scale estimation which will be discussed later.

The tracking result with and without occlusion handle during occlusion is shown in Fig. 6. The target is occluded from the third image. In the top row without occlusion handle, the tracker drift to the background where there is a self-similar colored object. However, in the bottom row, the occlusion is well detected and the global re-localizer start to expand the search window in the third frame. When the object is partial recovered from full occlusion in the fifth frame, the re-localizer start to rescale the search window until the object is fully recovered in the last frame and the request two joint ratio of statistical probability is met to finish the re-localization.

**Fig. 6.** Tracking without occlusion handle (top) and with occlusion handle (bottom)

**Scale Estimation.** In order to deal with scale variation problem, we apply the scale estimation mechanism in Cam-shift [21] to adjust the size of the converged search window by invariant moments. The adjusted window size $(l, w)$ is calculated as outlined in Eq. 5:

$$
\begin{cases}
l = \sqrt{\dfrac{(a+c) + \sqrt{b^2 + (a-c)^2}}{2}} \\
w = \sqrt{\dfrac{(a+c) - \sqrt{b^2 + (a-c)^2}}{2}}
\end{cases}
,
\quad
\begin{cases}
a = \dfrac{M_{20}}{M_{00}} - x_c^2 \\
b = 2 \times (\dfrac{M_{11}}{M_{00}} - x_c \times y_c) \\
c = \dfrac{M_{02}}{M_{00}} - y_c^2
\end{cases}
\tag{5}
$$

The tracking result is the resized search window which will be expanded into the search window of next frame.

## 4    Experiment Results

As mentioned above, we evaluated the proposed tracker on a public RGB-D object tracking dataset built by Xiao et al. [1] rather than the famous PTB benchmark [17]. In [1], the author outlined 4 problems of PTB. First, it contains some sequences where the RGB and depth images pairs are unsynchronized. Second, over a half the PTB dataset is devoted to people tracking which introduce a bias in the results of evaluating tracking of generic targets. Third, the majority of the benchmark videos are captured by a stationary camera. Fourth, many videos in PTB are from similar scenes.

We tested our tracker on this new dataset with a state-of-the-art evaluation methodologies named OTB [2] which runs the trackers from initial frame with a bounding box, without re-initialize trackers by any tracking failures.

We present the results of comparing the proposed tracker with two top-ranked RGB-D trackers in PTB benchmark (PT [17], DS-KCF1 [22], DS-KCF2 [18], OAPF [19]), the base-line tracker of the new dataset (STC [1]) and a mean-shift based tracker (ASMS [8]) which achieved state-of-the-art performance. The tracking result of DSKCF, OAPF, STC are obtained from the new benchmark presented in [1] (see [1] for a detailed description of the labels).

**Qualitative Evaluation.** As shown in Fig. 8, the proposed tracker (MS3D) outperforms others in various of challenge scenarios. From top to bottom, the challenge factors are scale variation and outdoor, fast motion, shape deformation, self-similar colored objects interference and outdoor in dark, shape deformation with self-similar background. In these challenge scenes, the objects are hold by a person which can lead into self-similar depth surrounding clutters. However, MS3D survived in all of them because of its capability to discriminate both color and depth difference.

**Quantitative Evaluation.** We exploit the AUC (area-under-the-curve) of region overlap ratio to evaluate all the trackers which is also the evaluation

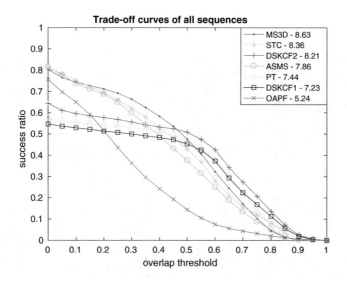

**Fig. 7.** Trade-off curves of all sequences

way used in the new dataset. In order to compare with all the trackers mentioned above, we use the evaluation tool of the new benchmark [1]. The trade-off curves of all sequences is shown in Fig. 7.

It is clear to see from the Fig. 7 that our algorithm (MS3D) outperforms others. The overall performance has been improved by 0.44 by comparing MS3D with STC which is the baseline algorithm of the new dataset. Table 1 shows the AUC performance of trackers in different attributes. Our tracker outperforms others in both stationary, moving cameras as well as most of attributes. However, it performs worse in IV, CDV, SCC since the color varies a lot between the initial frame and the rest in these sequence (*e.g.* turn off lights, changing clothes) and MS3D doesn't exploit color model update strategy. However, in our opinion, the great change of appearance should belongs to detection problem rather than tracking.

The frame rate of MS3D is tested on a desktop computer (Intel Core i7-3820 @3.60 GHz) with the image resolution of $640 * 480$. Since the proposed tracker depends on mean-shift localizer rather than sliding window, it has a high frame rate of over 100 FPS when there is no occlusion. The frame rate will decrease to 30–50 FPS because of the global searching based re-localizer which runs during fully occlusion.

**Fig. 8.** A visual review of the trackers' performance in sequences with various of challenges. The tracking results are shown in different colored bounding box with a demonstration legend on the top-right corner. The overlap ratio between MS3D tracking result and ground truth is shown at the top-left corner

**Table 1.** AUC of bounding box overlap, RED demotes best performing tracker and BLUE demotes the second.

| | MS3D (ours) | STC | DSKCF2 | ASMS | PT | DSKCF1 | OAPF |
|---|---|---|---|---|---|---|---|
| Overall | **8.63** | 8.36 | 8.21 | 7.86 | 7.44 | 7.23 | 5.24 |
| Stationary | **10.22** | 9.57 | 9.36 | 8.48 | 8.54 | 7.52 | 6.0 |
| Moving | **7.18** | 7.18 | 7.13 | 7.28 | 6.43 | 6.85 | 4.54 |
| IV | 5.72 | 5.78 | **6.10** | 6.99 | 4.15 | 5.50 | 3.18 |
| DV | 7.78 | 7.56 | **7.88** | 7.48 | 6.65 | 7.06 | 4.45 |
| SV | 5.52 | 5.07 | 4.39 | **5.66** | 2.81 | 3.36 | 3.07 |
| CDV | 3.07 | 5.12 | 0.94 | 6.58 | 0.43 | 1.43 | 3.22 |
| DDV | **8.26** | 7.66 | 5.26 | 6.96 | 3.49 | 4.16 | 3.71 |
| SDC | **8.17** | 8.01 | 7.93 | 7.37 | 6.80 | 7.90 | 5.00 |
| SCC | 9.01 | 9.53 | 9.81 | 7.41 | 8.23 | 8.25 | 6.13 |
| BCC | 7.05 | 6.67 | 5.66 | **6.67** | 5.73 | 4.82 | 3.79 |
| BSC | 7.10 | 7.17 | 6.50 | 7.28 | 5.76 | 5.16 | 4.82 |
| PO | **8.30** | 7.73 | 7.76 | 7.75 | 6.20 | 6.02 | 5.82 |

# 5   Conclusion

In this work, mean-shift tracker has been improved with a novel target representation which fuses color and depth statistics. The new representation can handle background clutter and self-similar objects interference which is the main challenge factor of original mean-shift tracker. We also provide a target extraction method from ROI with depth distribution analysis so as to obtain target model with less background information involved. At last we propose a joint color and depth probability distribution image moments based occlusion handle mechanism which actively detects the target occlusion and recovery during tracking process. et. The proposed tracker runs in real time and ranks in the first position on a new RGB-D tracking dataset. As a short term tracker without model update, it has inferior performance in long-term occlusion and great illumination change. All the code developed has been released as open-source in order to allow future comparisons and to provide benefit to the wide community.

# References

1. Xiao, J., Stolkin, R., Gao, Y., Leonardis, A.: Robust fusion of color and depth data for RGB-D target tracking using adaptive range-invariant depth models and spatio-temporal consistency constraints. IEEE Trans. Cybern. (2017)
2. Wu, Y., Lim, J., Yang, M.-H.: Object tracking benchmark. IEEE Trans. Pattern Anal. Mach. Intell. **37**(9), 1834–1848 (2015)
3. Henriques, J.F., Caseiro, R., Martins, P., Batista, J.: High-speed tracking with kernelized correlation filters. IEEE Trans. Pattern Anal. Mach. Intell. **37**(3), 583–596 (2015)

4. Kalal, Z., Mikolajczyk, K., Matas, J.: Tracking-learning-detection. IEEE Trans. Pattern Anal. Mach. Intell. **34**(7), 1409–1422 (2012)
5. Comaniciu, D., Ramesh, V., Meer, P.: Kernel-based object tracking. IEEE Trans. Pattern Anal. Mach. Intell. **25**(5), 564–577 (2003)
6. Hare, S., Golodetz, S., Saffari, A., Vineet, V., Cheng, M.-M., Hicks, S.L., Torr, P.H.: Struck: structured output tracking with kernels. IEEE Trans. Pattern Anal. Mach. Intell. **38**(10), 2096–2109 (2016)
7. Possegger, H., Mauthner, T., Bischof, H.: In defense of color-based model-free tracking. In: Proceedings of the IEEE Conference on Computer Vision and Pattern Recognition, pp. 2113–2120 (2015)
8. Vojir, T., Noskova, J., Matas, J.: Robust scale-adaptive mean-shift for tracking. Pattern Recognit. Lett. **49**, 250–258 (2014)
9. Kristan, M., Pflugfelder, R., Leonardis, A., Matas, J., Čehovin, L., Nebehay, G., Vojir, T., Fernandez, G., Lukežič, A., Dimitriev, A., et al.: The visual object tracking VOT2014 challenge results (2014)
10. Bertinetto, L., Valmadre, J., Golodetz, S., Miksik, O., Torr, P.H.: Staple: complementary learners for real-time tracking. In: Proceedings of the IEEE Conference on Computer Vision and Pattern Recognition, pp. 1401–1409 (2016)
11. Stolkin, R., Florescu, I., Kamberov, G.: An adaptive background model for CAMSHIFT tracking with a moving camera. In: Advances In Pattern Recognition, pp. 147–151. World Scientific (2007)
12. Ning, J., Zhang, L., Zhang, D., Wu, C.: Robust object tracking using joint color-texture histogram. Int. J. Pattern Recognit. Artif. Intell. **23**(07), 1245–1263 (2009)
13. Ning, J., Zhang, L., Zhang, D.: Robust mean-shift tracking with corrected background-weighted histogram. IET Comput. Vis. **6**(1), 62–69 (2012)
14. Hu, Q., Guo, Y., Chen, Y., Xiao, J., An, W.: Correlation filter tracking: beyond an open-loop system
15. Fan, H., Ling, H.: Parallel tracking and verifying: a framework for real-time and high accuracy visual tracking. arXiv preprint arXiv:1708.00153 (2017)
16. Zhao, Y., Carraro, M., Munaro, M., Menegatti, E.: Robust multiple object tracking in RGB-D camera networks. In: IEEE/RSJ International Conference on Intelligent Robots and Systems (IROS), pp. 6625–6632. IEEE (2017)
17. Song, S., Xiao, J.: Tracking revisited using RGBD camera: Unified benchmark and baselines. In: IEEE International Conference on Computer Vision (ICCV), pp. 233–240. IEEE (2013)
18. Hannuna, S., Camplani, M., Hall, J., Mirmehdi, M., Damen, D., Burghardt, T., Paiement, A., Tao, L.: DS-KCF: a real-time tracker for RGB-D data. J. Real Time Image Process. 1–20 (2016)
19. Meshgi, K., Maeda, S.-I., Oba, S., Skibbe, H., Li, Y.-Z., Ishii, S.: An occlusion-aware particle filter tracker to handle complex and persistent occlusions. Comput. Vis. Image Underst. **150**, 81–94 (2016)
20. Hidayatullah, P., Konik, H.: CAMSHIFT improvement on multi-hue object and multi-object tracking. In: 3rd European Workshop on Visual Information Processing (EUVIP), pp. 143–148. IEEE (2011)
21. Bradski, G.R.: Real time face and object tracking as a component of a perceptual user interface. In: Proceedings of the Fourth IEEE Workshop on Applications of Computer Vision, WACV 1998, pp. 214–219. IEEE (1998)
22. Camplani, M., Hannuna, S., Mirmehdi, M., Damen, D., Paiement, A., Tao, L., Burghardt, T.: Real-time RGB-D tracking with depth scaling kernelised correlation filters and occlusion handling (2015)

# Heterogeneous Multi-agent Routing Strategy for Robot-and-Picker-to-Good Order Fulfillment System

Hanfu Wang[1,2,3], Weidong Chen[1,2,3(✉)], and Jingchuan Wang[1,2,3]

[1] Key Laboratory of System Control and Information Processing,
Ministry of Education of China, Shanghai, China
wdchen@sjtu.edu.cn
[2] Department of Automation, Shanghai Jiao Tong University, Shanghai, China
[3] Shanghai Key Laboratory of Navigation and Location Based Services,
Shanghai, China

**Abstract.** In this research heterogeneous multi-agent routing strategy for a robot-and-picker-to-good order fulfillment system based on collaboration between mobile robots and human workers is proposed. As it is intractable to solve all agents' routing problems as a large global optimization problem, individual agent's routing is planned separately based on utility-based heuristics. Both the robot and the worker routing problems are formulated as graph optimization problems based on the graph representation of robot's task and dynamically induced subtasks of the worker. The total travel distance of the robot and the total waiting time criteria are optimized by genetic algorithm. The long-term performance of the proposed routing strategy is evaluated according to different indicators using hybrid event simulation. The results show that order fulfillment system with the proposed routing strategy can either extensively save manual labor or improve the system efficiency.

**Keywords:** Heterogeneous multi-agent system · Routing strategy
Order fulfillment system

## 1 Introduction

Robot armies are progressively invading humans' daily life and relieving them from boring, tedious, repetitive, burdensome and dangerous situations. In recent years innovative robotic order fulfillment systems (OFS) designed for material transportation in warehouses or factories, such as Kiva [14], Fetch and Freight [2], and Swisslog Click&Pick [1], gain their success in the tide of industrial automation transition. According to [10], there are "two representative solutions of using mobile robots to upgrade warehousing operations: the robotic G2P (good-to-picker) picking method and the R2G (robot-to-good) picking method" (p70). Nevertheless, until now most commercially available solutions for e-commerce

© Springer Nature Switzerland AG 2019
M. Strand et al. (Eds.): IAS 2018, AISC 867, pp. 237–249, 2019.
https://doi.org/10.1007/978-3-030-01370-7_19

warehouses are based on G2P solution, because the problem of fast, accurate and robust robotic manipulation of versatile and universal products remains unsolved. Robotic manipulation is inferior to humans' picking in rapidity, accuracy and robustness, which makes fully-automatic R2G solution commercially unavailable yet.

In order to take advantage of capabilities of robots and humans, robots are utilized to augment human workforce, rather than totally replace them. In this view, a robot-and-picker-to-good (RP2G) solution can be an commercially available alternative to the prior two solutions. In this solution, order picking tasks are accomplished by the collaboration between multiple autonomous mobile robots and human workers, with the robot acting as the transporter role and the worker as the picker role. This method is a combination of zone-based (for workers) and order-based picking with batching (for robots). This method is cost-effective, flexible, and highly adaptive to traditional warehouses.

Routing strategy for both agents arises when robots and workers collaborate to fulfill the order picking tasks. Routing strategy is one of the most important aspects for the order fulfillment system. The routing problems in RP2G are challenging because of complex relationship between environment, products, orders, robots and workers. For research on routing problems of P2G method for traditional manual-labor warehouses, readers can refer to [4,6–8]. For research on routing problems of G2P method for automatic warehouses, readers can refer to [5,14]. Although research on routing strategies for other solutions is instructive for RP2G solution, as far as the authors investigated, materials on the routing strategy for RP2G are seldom seen.

This research is a first step for the design and assessment of RP2G-OFS, and subject to several assumptions. Coordination level aspects, including worker safety, congestion, conflict and deadlock are neglected; the agent will execute its route plan exactly as the solver indicates; finally, all uncertainties are neglected. For simplicity all robots' travel velocities are the same, and all workers' walk velocities the same as well.

Two major contributions are highlighted. Firstly, the heterogeneous multi-agent routing strategy for the RG2P-OFS is proposed; secondly the long-term performance of the routing strategy is evaluated and analyzed in comparison with human labored warehouses, and the results can be instructive before deployment.

This paper is organized as follows. In Sect. 2 routing problems of robot and worker in RP2G method are formulated as graph optimization problems; in Sect. 3 genetic algorithms are developed to solve the graph optimization problems; in Sect. 4 the simulation settings, instances, algorithmic parameters, results and analysis are presented; finally in Sect. 5 conclusions and future work are summarized.

## 2    Routing Problems Formulation

Routing problems of RP2G-OFS can be defined in a 3-tuple $< L, O, A >$ category, as Table 1 shows. Terminology notation in [6] is partially adopted. According to [3], the travel time matters the most in a traditional manual labor warehouse and it is the largest component of the labor. Correspondingly, for robots in RP2G-OFS, time consumption is composed of two parts: travel time (traveling between stations) and idle time (staying adjacent to racks and waiting for workers' picking operation). Meanwhile, as it is impossible to solve all agents' routing problems as a large global optimization problem, each agent's routing is planned separately based on utility-based heuristics [14]. As a consequence, RP2G-OFS routing problems are divided into the robot routing problem and the worker routing problem, and both problems are formulated as graph optimization problems with distinct optimization objectives.

To define RP2G-OFS routing problems, an undirected and connected graph $\mathcal{G} = (\mathcal{V}, \mathcal{E})$ is constructed. A bijection from the location set to the set of vertices $\mathcal{V}$ can be constructed: $\mathcal{V} = \mathcal{V}(L) = \mathcal{V}(S) \cup \mathcal{V}(D)$. Edges in set $\mathcal{E}$ connect two neighbor vertices of $\mathcal{V}$. $\boldsymbol{d}_{l(i),l(j)} \geqslant 0$ represents the distance between vertices $i$ and $j$. For readability $\boldsymbol{d}_{l(i),l(j)}$ is abbreviated as $\boldsymbol{d}_{ij}$. $\boldsymbol{d}_{ij} = \boldsymbol{d}_{ji}$. The robot velocity is $v^r$ and the worker velocity is $v^w$.

**Table 1.** Terminology notations

| Term | Notation | Description |
|------|----------|-------------|
| Location | $L = S \cup D; s_a \in S, a = 1, 2, \cdots, |S|; d_b \in D, b = 1, 2, \cdots, |D|$ | Set of Locations is a union of set of stations and set of depots. Stations are locations from where robots can stay and workers can collect products. Depots are locations from where robots depart and to where they must return |
| Order batch | $o_d \in O, d = 1, 2, \cdots, |O|$ | Set of order batches to be collected. Each order batch contains a list of order lines, and each order line corresponds to a spatial storage location in station set |
| Agent | $A = R \cup W; r_i \in R, i = 1, 2, \cdots, |R|; w_j \in W, j = 1, 2, \cdots, |W|$ | Agents is a union of set of robots and works. The robot is the transporter of an order batch. The worker works in a designated zone and acts as a picker |

## 2.1  Robot Routing Problem Formulation

A solution to robot routing problem in $\mathcal{G}$ is a closed route for the robot. For robot $r_i$ and allocated order batch $o_i \subseteq O$, the route $p_i$ is associated with a subgraph $\mathcal{G}_i = (\mathcal{V}(L_i) \cup \mathcal{V}(d_i^s) \cup \mathcal{V}(d_i^e), \mathcal{E}_i)$ of $\mathcal{G}$, in which $\mathcal{V}(L_i)$ is the order batch mapped vertex set, $\mathcal{V}(d_i^s)$ is the start depot vertex and $\mathcal{V}(d_i^e)$ is the end depot vertex. Each route starts at $\mathcal{V}(d_i^s)$, traverses a set $\mathcal{E}_i \subseteq \mathcal{E}$ of selected edges and returns to $\mathcal{V}(d_i^e)$.

Robot routing problem of robot $r_i$ is formulated as a classic traveling salesman problem: finding a walk on subgraph $\mathcal{G}_i$ with designated end vertices, and the optimization objective is:

$$min: \sum_{(m,n)\in\mathcal{E}_i} d_{m,n} \tag{1}$$

## 2.2  Worker Routing Problem Formulation

Worker routing problem is influenced by robots' routing strategy. In order to improve the efficiency of the system, the optimization objective of worker routing problem is to minimize the total waiting time of robots in a certain zone. The optimization is conducted in a rolling way: whenever a worker finishes one picking, the local zone information is updated, the subgraph is dynamically generated, and the optimization process is triggered.

In a certain time for worker $w_j \subseteq W$ to be optimized, the arrived robot set is $R_j^a$, and the future arriving robot set (the robots which are heading worker $w_j$'s zone) is $R_j^f$. $R_j = R_j^a \cup R_j^f$ is all the robots taken into consideration in this optimization process. For each future arriving robot $r_{j,m}^f \in R_j^f, m = 1, 2, \cdots, |R_j^f|$, the future station is $l_{j,m}^f \in L_j^f$ and arriving time is $\tau_{j,m}^f$; for each arrived robot $r_{j,n}^a \in R_j^a, n = 1, 2, \cdots, |R_j^a|$, the location is $l_{j,n}^a \in L_j^a$, and the arriving time can be regarded as zero, $\tau_{j,n}^a = 0$. For each robot $r_{j,k} \in R_j, k = 1, 2, \cdots, |R_j|$, the required picking time of corresponding products is $\tau_{j,k}$.

The route $q_j$ of worker $w_j$ is associated with a dynamic induced subgraph $\mathcal{G}_j = (\mathcal{V}_j, \mathcal{E}_j)$ of $\mathcal{G}$ updated in accordance with robots' routes in real-time. $\mathcal{V}_j = \mathcal{V}(L_j^f) \cup \mathcal{V}(L_j^a) \cup \mathcal{V}(l_j^s)$, in which $\mathcal{V}(L_j^f)$ is the arriving robots' station mapped vertex set, $\mathcal{V}(L_j^a)$ is the arrived robots' station mapped vertex set, and $\mathcal{V}(l_j^s)$ is worker's current vertex.

Let $v_{j,k} \subseteq \mathcal{V}_j$, $t_{j,k}$ represents the whole processing time for robot that is staying or arriving at the mapped station $\mathcal{V}^{-1}(v_{j,k})$. Let $T_j^k$ represents the elapsed time after finishing the picking on station $\mathcal{V}^{-1}(v_{j,k})$ if worker $w_j$ executes route $q_j$. For worker $w_j$ and vertex $v_{j,k}$ of route $q_j$, the time transition is constrained by the simultaneity constraint between the robot and the worker. $T_j^k$ and $t_{j,k}$ can be computed iteratively, i.e.,

$$T_{j,0} = 0; \quad T_{j,k} = \sum_{l=1}^{k} t_{j,l}; \quad t_{j,l} = \tau_{j,l} + max\left(\frac{d_{l-1,l}}{v_w}, \tau_{j,l}^f - T_{j,l-1}\right)$$

The waiting time of robot in $v_{j,k}$'s is obtained:

$$t_{j,k} = \begin{cases} T_{j,k-1} + \frac{d_{k-1,k}}{v_w} - \tau_{j,k}^f + \tau_{j,k}, & T_{j,k-1} + \frac{d_{k-1,k}}{v_w} - \tau_{j,k}^f > 0 \\ \tau_{j,k}, & T_{j,k-1} + \frac{d_{k-1,k}}{v_w} - \tau_{j,k}^f \leqslant 0 \end{cases}$$

Then worker routing problem is formulated as a graph optimization problem: finding a walk on dynamic induced subgraph $\mathcal{G}_i$ and the optimization objective is:

$$min : \sum_{k=1}^{|\mathcal{V}_j|-1} t_{j,k} \tag{2}$$

# 3   Genetic Algorithm Based Optimization

Both agents' routing problems are formulated as graph optimization problems, and they can be solved by heuristic combinatorial algorithms. The genetic algorithm [9] is selected because GA is a population-based swarm intelligent algorithm, and has been successfully used to solve traveling salesman problem. The framework of the algorithm is based on [12] and summarized as follows:

- **Input:** category $< L, O, A >$; subgraph $\mathcal{G}_i$ for robot $r_i$'s routing problem or $\mathcal{G}_j$ for worker $w_j$'s routing problem;
- **Output:** optimal route plan $p_i$ for robot $r_i$ or route plan $q_j$ for worker $w_j$;
- **Encoding:** route plan $p_i$ or $q_j$ is represented naturally as an ordered set of vertices using path representation. A chromosome is a permutation vector of vertices in $\mathcal{G}_i$ or $\mathcal{G}_j$;
- **Fitness Evaluation:** fitness functions are evaluated based on Eq. (1) for robot or Eq. (2) for worker;
- **Initialization:** a random mechanism is used to initialize chromosomes, both to seed the first generation and to introduce new individuals into the population in subsequent generations to maintain diversity;
- **Selection:** elitism is adopted to ensure the convergence performance; a roulette method is adopted for the selection of chromosomes;
- **Crossover:** for random paired chromosomes with a certain crossover rate, a one-point crossover method is adopted; after each crossover operation, validation of the chromosomes is necessary;
- **Mutation:** for random selected chromosomes with a certain mutation rate, a two-point exchange method is adopted and followed by chromosome validation;
- **Termination Condition:** maximum generations, maximum stall generations, fitness limit are set to ensure the termination of the algorithm.

# 4   Simulations and Results

## 4.1   Description of Warehouse, Order Batches and Agents

The planar warehouse layout of the simulations is based on a real human labor e-commerce warehouse layout. The original layout is adjusted to accelerate the

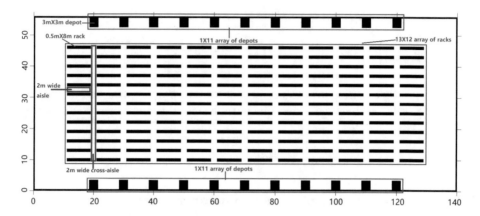

**Fig. 1.** Warehouse layout

simulation process according to layout design principles in [3], as well as some benchmark layout in [8]. As Fig. 1 shows, the warehouse is 56 m wide and 140 m long, and composed of two main constituents, the rack and the depot. In the center of this warehouse are 13 × 12 array of racks, and then a 12 × 12 array of aisles is obtained between these racks. The workspace is easily partitioned into different zones, as 12 is the lowest common multiple of 1, 2, 3, 4. Each rack is 0.5 m × 8 m and height is neglected; each aisle is 2 m wide and each cross-aisle is 2 m wide, thus providing sufficient space for robots to travel in parallel. On the upper and lower sides of the rack array are 22 packing and shipping depots, and each depot is a 3 m × 3 m area.

The prior layout is spatially discretized for hybrid dynamic system simulation: the layout is transformed into a 240 × 100 grid map and one grid represents a 0.5 m × 0.5 m space. Each grid has a value of 0 or 1, with 0 representing free space and 1 representing obstacles. Racks and depots are simplified as occupied space on the grid map. The dynamic system is evolved and all agents' states are updated every 0.5 s. Based on the discretized layout of the warehouse, stations are represented as unoccupied grids adjacent to racks in each aisle.

To simplify the simulations, each order line is directly represented as its corresponding station. Each order batch's length (order lines in one batch) is subject to a normal distribution of $N(40, 0.8)$; in each order batch, the order lines are uniformly generated. Sufficient order batches are generated to ensure that workers and robots are always busy in one day.

The robot or the worker is abstracted to a point on this grip map: the robot is represented as red circle and the worker as green square. In any discrete time, robot or human always occupies one grid. When one worker grabs products on racks, he always stay on the right side of the corresponding robot. Traffic rules are adopted, in which robots always remain on the right side of the aisle or cross-aisle when moving forward. Path planning algorithm for both agents is A*. The overlap between agents is neglected. Robots and workers either travel at speed

1 m/s or stay still, and the acceleration process is neglected; worker's pick time for an order line is set to 5 s, and depot's pack time is set to 20 s.

## 4.2   Simulation Settings and Algorithmic Parameters

Based on different combination of robot number and different human number in RP2G-OFS, instances for RP2G-OFS are generated with different configurations as Table 2 shows. For one particular worker number, there may be many zoning strategies. On performance evaluation of different zoning strategies, readers can refer to [13] for more information.

As far as the authors investigated, no publication is found on the performance evaluation of RP2G-OFS. For a comparative study, manual labor order fulfillment system (H-OFS) instances are generated based on different worker number: 100, 120, 140, 160, 180. Worker in H-OFS adopts the same routing algorithm and path planning method as robot in RP2G; instances for H-OFS and RP2G-OFS share the same mechanisms on other aspects.

**Table 2.** Different configurations of RP2G-OFS

| Worker number | Zoning setting | Robot number |
|---|---|---|
| 36 | 2 × 2 aisles/worker | 100, 140, 180, 220, 260, 300, 340, 380 |
| 48 | 1 × 3 aisles/worker | 100, 140, 180, 220, 260, 300, 340, 380 |
| 72 | 1 × 2 aisles/worker | 100, 140, 180, 220, 260, 300, 340, 380 |
| 144 | 1 × 1 aisle/worker | 100, 140, 180, 220, 260, 300, 340, 380 |

**Table 3.** Algorithmic parameters

| Parameter | Robot routing algorithm | Worker routing algorithm |
|---|---|---|
| Population Size | 100 | 50 |
| Elitism rate | 0.05 | 0.05 |
| Crossover rate | 0.8 | 0.8 |
| Mutation rate | 0.2 | 0.2 |
| Maximum generations | 2000 | 1000 |
| Maximum stall generations | 100 | 50 |
| Fitness limit | 0.1 | 0.1 |

In the simulations, each loop represents a discrete time 0.5 s in real world clock, and the routing optimization and path planning of agents is independent of their state update. This means that the all agents' states are updated as

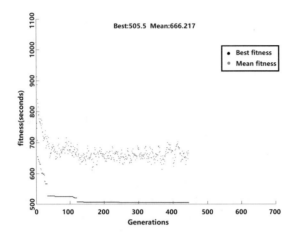

**Fig. 2.** Convergence of the routing optimization algorithm

long as all necessary computations are completed. To abate the influence of the stochastic routing algorithms, each instance runs for 10 days and each day for 8 hours continuously, and all the statistics are averaged on daily basis. Parameters for both routing optimization algorithms are adjusted separately to balance the convergence speed and optimality, as Table 3 shows. For each instance a single thread is performed on a computer with Intel Core(TM) i7-6700 3.40 GHz CPU and 16 GB RAM. Figure 3 shows a scene from one simulation instance. Figure 2 shows the convergence of the genetic algorithm for a particular worker with 5 arriving robots and 10 arrived robots. It is the same case for the robot and the convergence analysis is omitted here.

### 4.3   Results, Analysis and Solution Comparison

With the dynamic behavior of this complex system is certainly a demanding work to explore, in this performance evaluation the emphasis is laid on static statistical results. Based on some common key performance index (KPI) in [11], the performance criterion is defined as follows:

- *throughput:* total order lines collected in one day;
- *average order lines:* average order lines of each worker or each robot;
- *average walk/travel distance:* average meters in one day of each worker walks or each robot travels;
- *average effective time:* agents' average time spend on specialized roles, i.e., pick time of each worker, or travel time of each robot;
- *average idle time:* agents' average time spend on waiting or doing nothing.

Figures 4 and 5 show the results of H-OFS and RP2G-OFS.

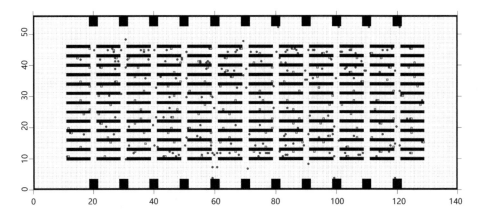

**Fig. 3.** A scene from an simulation instance

**Results and Analysis of H-OFS:** the throughput is roughly linear to the worker number, however, the average order lines decreases as the work number increases. Generally each worker has to walk over 20 km to achieve about 500 order lines every day. As a matter of fact, the mechanism of H-OFS is simple and the worker in H-OFS has to conduct both the transportation and picking operations.

**Results and Analysis of RP2G-OFS**

- *throughput:* for a large worker number, the throughput is roughly linear to the robot number; in contrast, if the worker number is small, the throughput growth slows down. This tendency can be explained along with other indicators;
- *average order lines:* in contrast to the throughput indicator, average order lines per robot increase as worker number increases, however, decrease as robot number increases. This implies that regardless of the worker number, adding more robots means a reduction of average robot utilization;
- *average walk/travel distance:* for the worker number of 36 and 48, the average walk distance of workers firstly increases as robot number increases, however, decreases as robot number increases. This phenomenon can be explained by the fact that as robot number increases, robot congestion always occurs. Workers need not walk a long way to pick. For other worker numbers, robot number is not large enough to reveal this phenomenon. For the robot, the travel distance is reduced as robot number increases because of limited manual labor;
- *average effective time:* for the worker number of 36, 48 and 72, the average effective time is relatively unaffected by robot number. Nevertheless, for the worker number of 144, the average effective time increases as robot number increases. For the robot, as a result of linear relation between time and

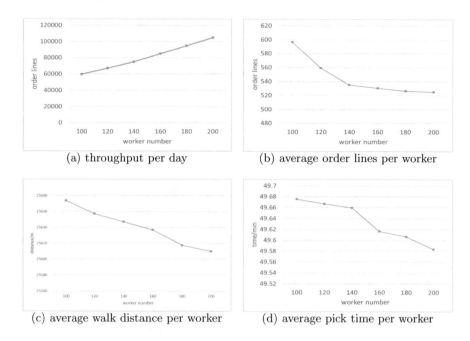

(a) throughput per day          (b) average order lines per worker

(c) average walk distance per worker    (d) average pick time per worker

**Fig. 4.** Statistical results of H-OFS

distance, this indicator appears as the same trend of average travel distance indicator;

– *average idle time:* for the worker, the average idle time decreases as robot number is getting larger and zones is getting larger. For the robot, idle time increases as robot number increases and worker number decreases.

**Solution Comparison:** based on the performance evaluation of H-OFS and RP2G-OFS, warehouse managers can make decisions according to their most concerning aspects, reduction of manual labor, improvement of throughput, or reduction of the global operation cost. For example, RP2G-OFS with 150 robots and 36 workers, performs equally compared with H-OFS with 100 workers. Roughly two third of the workers are saved without reducing the throughput. Although the realistic situation would be much more conservative, this estimation reveals the overall tendency for decision-makers to assess the performance before deployment.

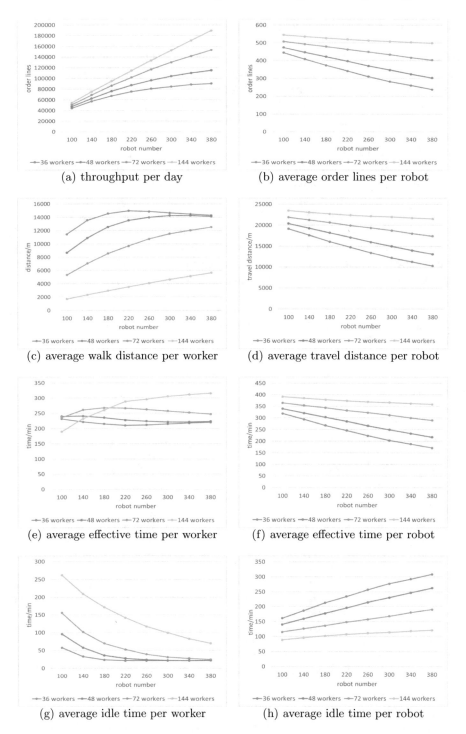

(a) throughput per day

(b) average order lines per robot

(c) average walk distance per worker

(d) average travel distance per robot

(e) average effective time per worker

(f) average effective time per robot

(g) average idle time per worker

(h) average idle time per robot

**Fig. 5.** Statistical results of RP2G-OFS

# 5   Conclusions and Future Work

In this research, routing strategy for RP2G-OFS is designed and its performance is tested using simulations. The results show that RP2G-OFS with proposed routing strategy can either extensively save manual labor, without reducing the throughput, or improve the efficiency with the same worker number. The proposed routing strategy provides a reference for practitioners to evaluate the balance before deployment.

This work is researched merely in a optimization framework and many challenging problems are open in future work. From a warehouse operation point of view, how warehouse managers combine strategic, tactical and operational levels and make decision jointly in designing routing strategy is the most necessary work before deployment; from a robot control point of view, further taking realistic robotic task level and control level coordination, uncertainties (i.e., agent's task execution discrepancies, agent failures, emergent situation and dynamic environment), stochastic online scheduling and planning, different human working habits, human-machine safety into consideration are possible future work.

**Acknowledgment.** This work is supported by the National Key R&D Program of China (Grant 2017YFB1303601); Natural Science Foundation of China (Grant 61773261 and 61573243); and the Innovation Action Plan of STCSM (Grant 16111106202).

# References

1. https://www.swisslog.com/
2. Ackerman, E.: Fetch robotics introduces fetch and freight: Your warehouse is now automated (2015). https://spectrum.ieee.org/automaton/robotics/industrial-robots/fetch-robotics-introduces-fetch-and-freight-your-warehouse-is-now-automated
3. Bartholdi, J.J., Hackman, S.T.: Warehouse & Distribution Science, Atlanta (2017)
4. De Koster, R., Le Duc, T., Jan Roodbergen, K.: Design and control of warehouse order picking: a literature review. Eur. J. Oper. Res. **182**, 481–501 (2006)
5. Enright, J., Wurman, P.: Optimization and coordinated autonomy in mobile fulfillment systems. In: Proceedings of the 9th AAAI Conference on Automated Action Planning for Autonomous Mobile Robots, pp. 33–38, January 2011
6. van Gils, T., Ramaekers, K., Caris, A., De Koster, R.: Designing efficient order picking systems by combining planning problems: state-of-the-art classification and review. Eur. J. Oper. Res. **267**, 1–15 (2017)
7. Gu, J., Goetschalckx, M., Mcginnis, L.: Research on warehouse operation: a comprehensive review. Eur. J. Oper. Res. **177**, 1–21 (2007)
8. Gu, J., Goetschalckx, M., Mcginnis, L.: Research on warehouse design and performance evaluation: a comprehensive review. Eur. J. Oper. Res. **203**, 539–549 (2010)
9. Holland, J.H.: Adaptation in Natural and Artificial Systems. MIT Press, Cambridge (1992)
10. Huang, G., Chen, Z.Q., Pan, M.: Robotics in ecommerce logistics. J. HKIE Trans. **22**, 1–10 (2015)

11. Lamballais, T., Roy, D., De Koster, R.: Estimating performance in a robotic mobile fulfillment system. Eur. J. Oper. Res. **256**, 976–990 (2016)
12. Larrañaga, P., Kuijpers, C., Murga, R., Inza, I., Dizdarevic, S.: Genetic algorithms for the travelling salesman problem: a review of representations and operators. Artif. Intell. Rev. **13**(2), 129–170 (1999)
13. Petersen, C.: Considerations in order picking zone configuration. Int. J. Oper. Prod. Manag. **22**, 793–805 (2002)
14. Wurman, P., D'Andrea, R., Mountz, M.: Coordinating hundreds of cooperative, autonomous vehicles in warehouses. AI Mag. **29**, 9–20 (2008)

# Multiple Mobile Robot Management System for Transportation Tasks in Automated Laboratories Environment

Ali A. Abdulla[1,2(✉)], Steffen Junginger[3], X. Gu[2], Norbert Stoll[3], and Kerstin Thurow[2]

[1] College of Engineering, University of Mosul, Mosul, Iraq
alialtaee2008@gmail.com
[2] Center for Life Science Automation (Celisca), University of Rostock, 18119 Rostock, Germany
[3] Institute of Automation, University of Rostock, 18119 Rostock, Germany

**Abstract.** This paper introduces a new multiple mobile robot management strategies for transportation tasks in automated laboratories environment. In this strategy, two aspects have been considered. First, the appropriate robot selection method which is either based on the highest charging level or based on the nearest robot to the transportation station. The other aspect is the robot-robot interaction which employs for robots collision avoidance especially in the narrow area (corridors). Server/client communication sockets with TCP/IP command protocol are employed for data transmission. A series of experiments have been performed to validate the performance of the presented strategy.

**Keywords:** Mobile robot management · Collision avoidance
Robot-Robot interaction

## 1 Introduction

Life sciences comprise various fields, including biochemistry, biotechnology, agriculture and medical technology. The automation of laboratories in these fields is being established in different areas, from applied research to production and drugs development. The automation of life science was brought to realize the demands for higher capacity, throughput and efficiency. Advanced automated laboratories integrate many sub-systems such as high-throughput screening, analytical and measurement systems, product line managers (PLMs), and robotic systems. Two kinds of robotic systems using stationary robots [1, 2] and mobile robots [3, 4] are commonly used in laboratory automation processes [5, 6]. At the Center for Life Science Automation (CELISCA), researchers are working to develop a fully automated life science laboratory of the future. For a facility such as a laboratory, a system consisting of several H20 mobile robots has been established which allows samples, and labwares to be transported between individual automated sub-systems.

Hui et al. presented a single floor transportation system based on the multiple H20 mobile robot [7]. This system deals with the robot remote center (RRC) and robot

© Springer Nature Switzerland AG 2019
M. Strand et al. (Eds.): IAS 2018, AISC 867, pp. 250–260, 2019.
https://doi.org/10.1007/978-3-030-01370-7_20

onboard control (RBC) control levels. The RRC receives the transportation task and forwards it to the appropriate robot which has the highest battery charge value without considering the robot in charging status. The final transportation results are reported back to the higher control level. The RBC was equipped with the necessary strategies to execute single floor transportation task. The path of the single floor transportation task starts from central charging stationary to the grasping position, then the mobile robot navigates to the placing position, and finally returns back to the same charging station.

An intelligent multi-floor transportation system based on H20 mobile robots was highly demanded to handle movements in a complex building structure with facilities distributed on different floors. Many issues should be considered in realizing an efficient navigation system, including questions of the representation of the working environment (mapping), knowledge of the mobile robot's current position within the working environment (localization), and a strategy for motion that connects the source point to the destination point (path planning). The multiple floor navigation system, the automated doors management system as well as the elevator handling system have been developed earlier [8–11]. The need to create new RRC strategies (Multiple Robot Management System, MRMS) appeared after developing the transportation system to transport multiple labware in multi-floor environments. The transportation time, paths and the elevator occupancy should be optimized in the new strategy. The chosen robot to perform the transportation task based on the new strategy either depends on battery value where the higher battery value robot with considering the charging status is chosen, or the nearest robot to the destination. The nearest robot working mode requires waypoints distributed near the transportation stations and robot power monitoring function. In addition to the working strategies, the robot-robot interaction has been innovative to avoid robot collisions in narrow areas. Mobile robots in automated laboratories are usually used for the transportation of samples and other material between sub-systems located in different places. Meanwhile automated laboratories typically lack space due to the large amount of equipment. Each sub-system in an automated laboratory has a limited workspace, and thus, an optimal solution for a mobile robot navigation system to work conveniently and safely will require an effective method to avoid collisions with the building structure, equipment, humans, and other mobile robots. There are many techniques that can be used for avoiding the robots collision [8, 12–15]. These techniques cannot be applied in life science laboratories due to lack of space which prevents multiple robots from using the same path. In this paper, a robot collision avoidance system has been developed by specifying safe points around the building which are used later as wait points for the nearest robot to free the path to other robots when collision is possible. The organization of this paper is as follow: Sect. 2 describes the architecture of the system. The communication between different control levels is clarified in Sect. 3. The two modes of robot selection strategy are presented in Sect. 4. The collision avoidance is described in Sect. 5. Experimental results are given in Sect. 6 followed by a conclusion in Sect. 7.

## 2  System Structure

The transportation system has three control layers as demonstrated in Fig. 1, the HWMS is the highest control layer. It manages the whole transportation by selecting either a robot to perform the required transportation task or calling human assistance based on the collected feedbacks from the connected systems. Next, the HWMS sends the transportation order with the labware number in the grasping station and the labware position in the placing station. The MRMS receives the transportation orders if the robot is chosen for performing the task, forwards it to the appropriate robot based on the selected working mode, and finally reports back a real-time transportation status to the HWMS. The multiple floor transportation system (MFTS) which is installed in the robot onboard computer is the third control level. It includes the mapping, indoor localization [8], path planning [9], automated door management system [9] and elevator handling strategies [10, 11]. A multi-floor mapping system was designed in which two kinds of maps are developed for autonomous mobile robots. Relative mapping provides a representation of the multi-floor global map, and path mapping is based on this to define the position of a set of waypoints which are then used as the basis for path planning and movement core. The indoor localization relies on the stargazer sensor with passive landmarks where the StarGazer Module (SGM) acts as a hex reader module [16]. Floyd searching algorithm is utilized for dynamically path planning, due to its efficiency and simplicity.

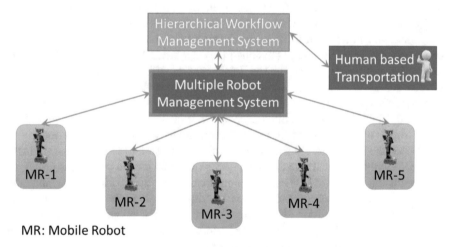

**Fig. 1.** System structure

## 3  Communication

In this application, a wireless IEEE 802.11g network is employed to realize the data transmission among the HWMS, MRMS and the mobile robots. To guarantee the reliability of the communication system, a TCI/IP command protocol with a client/server structure is utilized. Multiple routers were used to provide the wireless signal in a large area of the automated laboratories. During the robot movement, it is necessary to recognize the disconnection status especially in a weak signal area which can be caused by concurrent building or changing from router coverage area to another. Thus, a stable multiple socket communication between the MRMS and the distributed mobile robots was established. Each robot has a server socket while the MRMS has a client sockets. The socket management system was developed in the MRMS side to check each robot connection sequentially based on the heart beat monitor technique. Each connection is checked for a specific time and in the instant of signal missing, reconnection procedures are started in both sides to solve the detected connection problem. The developed sockets management process is demonstrated in Fig. 2.

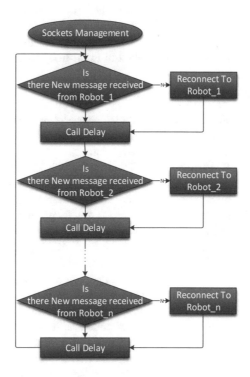

**Fig. 2.** Socket management between MRMS and mobile robots

## 4    Robot Selection Strategies

The MRMS selects the appropriate robot to perform the required transportation task either based on the robot battery or the robot position. Both strategies will be explained in detail in the following sections.

### 4.1   The Highest Charging Value

This strategy is based on the robot charging level and the HWMS transportation requests. The working procedures are as follows:

- collect the information about all connected mobile robots regarding the battery voltage since each robot has two power bank
- mark the robot status during transportation task as busy
- calculate the average power of the remaining robot
- compensate the power value in case the robot is on charging since the measured voltage when the robot in charging is not accurate.
- Finally, select the robot with higher voltage to perform the task.

The robot path in this strategy starts from the central charging station and ends at the same station after passing through the grasping and placing stations.

This strategy has a stable performance. The main disadvantage are long transportation times especially when it is applied in multi-floor large scale.

### 4.2   Nearest Robot Working Strategy

In this strategy the mobile robots are distributed near each working station if they have enough charging level to perform the transportation task. New waypoints are defined near each laboratory station as a robot wait point.

The HWMS, the MRMS, and the MFTS control levels are utilized in this strategy. When the HWMS selects the transportation based on mobile robot, it sends a transportation order to the MRMS including the start position with the labware label for grasping, and the end position with the labware position for placing. The MRMS initiates a parallel processing to find the distance between each robot to the first position in the task (grasping station). Each mobile robot computer is a cell of this parallel processing. These procedures start by enabling the available robots (see Fig. 3), then send an order to these robots to calculate the path from their current position to the required destination and repeat the distance request order in case of disconnection. If previous attempt has failed during 10 s, the disconnected robot is eliminated from the available robots for the task.

The MFTS finds the distance to the destination as follows: loading the complete path map, initializing a Floyd algorithm to find distance and sequence matrix, finding the nearest waypoint to the robot, extracting the distance to the goal from the Floyd distance matrix and finally replying the distance information to the MRMS. The MRMS collects the available robot replies and starts a comparison to find the shortest calculated distance. The robot with the shortest distance receives a transportation message to perform the required transportation. At the end of the transportation task, the selected robot waits near the placing station for a specific time to serve the possible new transportation task, which may start from the same station. After this period, the robot navigates to the station waitpoint to free a space for the labor staff to use this station. The robot battery level is monitored continuously. If the battery voltage becomes lower than the charging threshold, the navigation system controls the mobile robot until it reaches to the charging station. Finally, if the wait point near station was blocked by another robot, the robot returns to the central charging station (see Fig. 4).

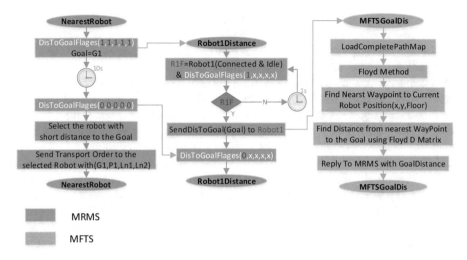

Fig. 3. Nearest mode robot selection strategy

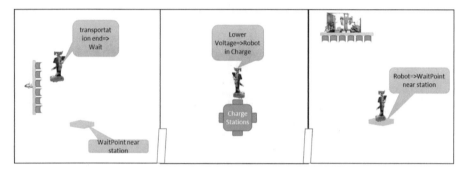

Fig. 4. Robot positions possibilities

This strategy optimizes the transportation time, the paths equipping, and minimizes the possibility of robot collisions.

## 5   Robot-Robot Interaction

In this paper, a new method is developed based on the robot-robot interaction to avoid the collision by specifying safe points around the building, which are used later as waitpoints for the nearest robot to free the path to other robot when collision is forecasting.

The first stage is to forecast the possible collision among mobile robots during transportation tasks. For this stage, four levels of nested loops are responsible to check the possible interference in future motions between all running mobile robots as shown in algorithm 1. Every robot's path waypoints sequence is compared with each running robot path sequence till finishing the task. If the path sequence of two robots moving in the opposite directions matches, the collision avoidance procedure is started to control the robots (see Fig. 5). These procedures start by sending a postpone request for both forecasted colliding robots. Then, each robot works separately to find the nearest safe point to its current position and calculates the required distance to reach it. The calculated distance is then reported back to the MRMS, which chooses the nearest robot to the safe point. Later the MRMS sends an order to the selected robot to navigate to the safe point and waits until this task is finished. Next, it asks the other robot to resume its movement. When the second robot passes the collision area, the MRMS sends an order to the robot in safe point to complete its transportation task. Finally, the forecasting function for the new possible interaction is enabled.

```
Algorithm 1 Collision forecasting between all running mobile robots

Input: RobotsNextWP matrix with the robots waypoints (current, next, next+1)
Initialize CollisionForcating fag;
Initialize robot1;
Initialize robot2;
NoOfRobot= RobotsNextWP.GetLength(0);
NoOfTrackedWP= RobotsNextWP.GetLength(1);

for (int i = 0; i < NoOfRobot - 1; i++) // First Robot
    {
            for (int j = i + 1; j < NoOfRobot - 1; j++) // Other Robot
            {
                for (int z = 0; z < NoOfTrackedWP - 1; z++)
                {
                    for (int k = 0; k < NoOfTrackedWP - 1; k++)
                    {

                        if ( ((RobotsNextWP[i, z] == RobotsNextWP[j, k + 1]) && (RobotsNextWP[i, 1 + z] == RobotsNextWP[j,
k])) && ((RobotsNextWP[i, z] != -1) && (RobotsNextWP[i, 1 + z] != -1)) )
                        {
                            CollisionForcating = true;
                            robot1 = i;
                            robot2 = j;
                            break;
                        }
                    }
                    if (CollisionForcating) break;

                }
                if (CollisionForcating) break;
            }
            if (CollisionForcating) break;
    }
```

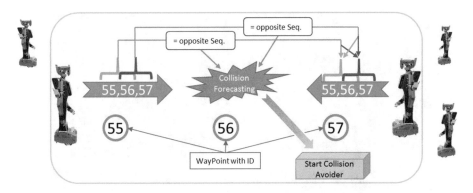

**Fig. 5.** Collision forecasting

## 6  Experiment

The first experiment has been established to check the developed sockets between the HWMS, the MRMS and the distributed mobile robots. A transportation task between laboratories distributed among different floors was executed twenty-five times in the CELISCA building. More than once, the connection has been deliberately lost in the HWMS, MRMS side, and the robot sides in addition to the robot movement in the weak signal locations to validate the real-time feedbacks to the HWMS and the re-connection management system. The developed system proved its efficiency to provide real-time feedbacks and its stability to manage the disconnection problems.

In the second experiment, the integration of the developed nearest robot strategy approach into a real transportation system was tested. In this experiment, the mobile robots have been distributed in different floors and different location around the building. The experiment has been repeated for 50 times and covered all robot status scenarios (in transportation, in charge, available). In the first 30 times, the repeatability of the system has been validated by requesting a new transportation operation each 5 times. While in the other experiments, the mobile robot position has been randomly changed before executing the transportation task. In each experiment, the MRMS successes to exclude both the busy and low charging level robots then call the nearest robot to perform the transportation task.

The last experiment has been done to validate the integration of the proposed collision avoidance strategy into the multi-floor transportation system. The robot-robot interaction based robots collision avoidance system is integrated directly into a real mobile robot multi-floor transportation system in life science automation. Two trans-portation tasks with opposite directions were assigned for two robots. Figure 6 shows the collision avoidance working procedures when both robots reached to a narrow area (corridor) at the same time. Figure contents explained as following: (a) the robot reaches to corridor area based on the earlier developed navigation system. (b) both robot pause their movement and calculate the distance to reach the safe point then send it to the MRMS which selects the robot with the shortest distance to the safe point. (c) the selected robot navigate till reaches the safe point. (d) The other robot resumes its

**Fig. 6.** Collision avoiding between two robots in narrow area.

movement to complete the transportation task. (e, f) after a specific time, the robot in safe point leaves it to complete its task.

## 7   Conclusion

In this paper, a multiple mobile robot management system for transportation tasks in automated laboratories environment is presented. A hierarchal work flow management (HWMS) has been developed earlier as a higher control level to manage the whole transportation system. The multiple robot management system selects the appropriate robot for each transportation task after receiving the transportation order either based on the higher charging level or on the nearest robot to the destination. The first transportation path strategy starts from the central charging station, passes through the grasping and placing stations and finally returns back to the central charging station. The nearest robot strategy starts from the nearest point to the required destination and ends at the waitpoint near the last transportation station, which may be the first station in the next transportation task. The nearest robot strategy optimizes the required transportation time and the path occupation. Connection sockets based server/client architecture and TCP/IP command protocol have been used to connect the higher control level with the MRMS and the last with the distributed mobile robots to grantee the reliability. A robot collision avoidance strategy based on robot-robot interaction is presented. Waypoints have been established as a safe point, where the robot waits until the other robot cleared the collision area. Thus, both robots can pass the narrow area and complete their transportation task. Three experiments to validate the connection sockets, the nearest robot selection strategy, and the robots collision avoidance system have been performed. The experimental results show an efficient performance for the presented systems.

**Acknowledgment.** The authors would like to thank the Mosul University in Iraq, and the German Federal Ministry of Education and Research for the financial support (FKZ:03Z1KN11, 03Z1KI1).

# References

1. Sakaki, K., et al.: Localized, macromolecular transport for thin, adherent, single cells via an automated, single cell electroporation biomanipulator. IEEE Trans. Biomed. Eng. **60**(11), 3113–3123 (2013)
2. Gecks, W., Pedersen, S.: Robotics-an efficient tool for laboratory automation. IEEE Trans. Ind. Appl. **28**(4), 938–944 (1992)
3. Miyata, N., Ota, J., Arai, T., Asama, H.: Cooperative transport by multiple mobile robots in unknown static environments associated with real-time task assignment. IEEE Trans. Robot. Autom. **18**(5), 769–780 (2002)
4. Montgomery, J.F., Fagg, A.H., Bekey, G.A.: The USC AFV-I: a behavior-based entry in the 1994 International Aerial Robotics Competition. IEEE Expert **10**(2), 16–22 (1995)
5. Petersen, J.G., Bowyer, S.A., y Baena, F.R.: Mass and friction optimization for natural motion in hands-on robotic surgery. IEEE Trans. Robot. **32**(1), 201–213 (2016)
6. Chapman, T.: Lab automation and robotics: automation on the move. Nature **421**(6923), 661, 663, 665–666 (2003)
7. Liu, H., Stoll, N., Junginger, S., Thurow, K.: Mobile robotic transportation in laboratory automation: multi-robot control, robot-door integration and robot-human interaction. In: 2014 IEEE International Conference on Robotics and Biomimetics (ROBIO), pp. 1033–1038 (2014)
8. Abdulla, A.A., Liu, H., Stoll, N., Thurow, K.: A new robust method for mobile robot multifloor navigation in distributed life science laboratories. J. Control Sci. Eng. **2016**, 2 (2016)
9. Abdulla, A.A., Liu, H., Stoll, N., Thurow, K.: A backbone-floyd hybrid path planning method for mobile robot transportation in multi-floor life science laboratories. In: IEEE International Conference on Multisensor Fusion and Integration for Intelligent Systems (MFI 2016), Baden-Baden, Germany, pp. 406–411 (2016)

10. Abdulla, A.A., Liu, H., Stoll, N., Thurow, K.: A robust method for elevator operation in semioutdoor environment for mobile robot transportation system in life science laboratories. In: 2016 IEEE International Conference on Intelligent Engineering Systems (INES), Budapest, Hungary, pp. 45–50 (2016)
11. Abdulla, A.A., Liu, H., Stoll, N., Thurow, K.: An automated elevator management and multi-floor estimation for indoor mobile robot transportation based on a pressure sensor. In: 2016 17th International Conference on Mechatronics - Mechatronika (ME), pp. 1–7 (2016)
12. Nilsson, N.J.: A mobius automation: an application of artificial intelligence techniques. In: Proceedings of the 1st International Joint Conference on Artificial Intelligence, San Francisco, CA, USA, pp. 509–520 (1969)
13. Ghandour, M., Liu, H., Stoll, N., Thurow, K.: A hybrid collision avoidance system for indoor mobile robots based on human-robot interaction. In: 2016 17th International Conference on Mechatronics (ME2016), Prague, Czech Republic (2016)
14. Kohtsuka, T., Onozato, T., Tamura, H., Katayama, S., Kambayashi, Y.: Design of a control system for robot shopping carts. In: König, A., Dengel, A., Hinkelmann, K., Kise, K., Howlett, R.J., Jain, L.C. (eds.) Knowledge-Based and Intelligent Information and Engineering Systems, pp. 280–288. Springer, Berlin Heidelberg (2011)
15. Nakahata, K., Dorronzoro, E., Imamoglu, N., Sekine, M., Kita, K., Yu, W.: Active sensing for human activity recognition by a home bio-monitoring robot in a home living environment. Presented at the 14th International Conference on Intelligent Autonomous Systems, Shanghai, China, vol. 531, p. 143 (2017)
16. Abdulla, A.A., Liu, H., Stoll, N., Thurow, K.: Multi-floor navigation method for mobile robot transportation based on StarGazer sensors in life science automation. In: 2015 IEEE International Instrumentation and Measurement Technology Conference (I2MTC), pp. 428–433 (2015)

# Robot Design

# A Rolling Contact Joint Lower Extremity Exoskeleton Knee

Jonas Beil$^{(\boxtimes)}$ and Tamim Asfour

High Performance Humanoid Technologies Lab (H$^2$T),
Institute for Anthropomatics and Robotics, Karlsruhe Institute of Technology (KIT),
Karlsruhe, Germany
{jonas.beil,asfour}@kit.edu

**Abstract.** This paper presents the design, kinematics modeling and experimental evaluation of a rolling contact joint for usage as a knee joint in lower limb exoskeletons. The goal of the design is to increase wearability comfort by exploiting the migrating instantaneous joint center of rotation which is characteristic for rolling contact joints. Two 3D-printed parts with convex surfaces form the mechanism, which is coupled by two steel cables and driven by a linear actuator. This coupling allows rotations around all axis as well as predefined translations. We conducted a kinematic simulation to optimize the shape of the convex joint surfaces and to estimate the expected misalignment between the subject's knee and exoskeleton joint. In our experimental evaluation we compared forces measured at the exoskeleton interface between subject and exoskeleton prototype with attached rolling contact or revolute joint. The results indicate a reduction of forces and therefore increased kinematic compatibility of the proposed joint design.

**Keywords:** Wearable robot · Joint mechanism
Exoskeleton knee joint

## 1 Introduction

Remarkable research efforts regarding the development of lower limbs exoskeletons have been made in recent years with the goal of improving the wearabilty and comfort of such devices. For exoskeletons designed for augmentation of human performance these are key requirements to increase user acceptance and allow the application of such exoskeletons in real world settings. Considering the human knee anatomy in the design process of a lower limb exoskeleton is therefore crucial to achieve high comfort and wearability as well as to prevent injuries caused by interaction forces at the interface during repetitive movements of the knee during walking. To this end, most commercial and many research lower limb exoskeletons use revolute joints to replicate the motion of the knee accepting macro and micro misalignments in order to keep the mechanical design as simple as possible ([1–4]).

© Springer Nature Switzerland AG 2019
M. Strand et al. (Eds.): IAS 2018, AISC 867, pp. 263–277, 2019.
https://doi.org/10.1007/978-3-030-01370-7_21

Other approaches align the instantaneous centers of rotation (ICR) of the human knee and the exoskeleton in the sagittal plane by utilizing additional joints. In [5], a cam mechanism for this alignment has been realized to follow the anatomical path of the knee joint axis during flexion motion while other systems use the four bar linkage mechanism [6], a Schmidt coupling [7] or a series of three revolute joints with parallel joint axis [8].

Since the knee is principally capable to perform rotations around all anatomical joint axis as well as translations in all spatial directions and the exoskeleton is coupled to the knee (parallel kinematics), exoskeletons with six degrees of freedom (DoF) were developed. The IT-knee uses a series of two articulated parallelograms, providing six degrees of freedom (DOF) to on the one hand self-align to the anatomy of different users and their knee articulation during motion and on the other hand to provide pure assistive torque to the flexion/extension axis [9]. In [10], a kinematic chain consisting of five revolute and one sliding joint is proposed to design an exoskeleton knee that self adjusts to the physiological knee movement. Both devices provide very good functionality but also require expanded space along the user's thigh and shank, impeding the implementation of hip or ankle joints.

**Fig. 1.** Prototype of the optimized rolling contact exoskeleton knee joint

In this paper, we present the conceptual design of a rolling contact joint (RCJ) for an knee exoskeleton for augmentation, which is capable of providing rotations around the anatomical joint axis as well as a prescribed translations of the ICR in the sagittal plane (see Fig. 1). The paper is organized as follows. In Sect. 2 the requirements for the system are derived from simulation of the human knee computing the translations projected to the assumed plane of the exoskeleton. The kinematic design and actuation of the device based on the requirements is described in Sect. 3. To gather indications of the kinematic compatibility a prototype was manufactured and used for experimental evaluation which is presented in Sect. 4. Section 5 concludes the paper.

## 2   Requirements

The knee joint as a condyloid hinge joint allows rotations around all anatomical axes and translations in all directions. However those rotations and translations are limited by the musculoskeletal system and are coupled to flexion/extension (F/E) motion during passive knee movement as previous studies have shown (see [11,12]). The instantaneous center of rotation (ICR) of the knee axis migrates on an evolute while the knee is flexed, rolling and sliding simultaneously [13]. Since the exoskeleton is placed parallel to the human knee and is coupled to the thigh and shank this could affect the required translations and rotations it has to perform.

Therefore, a kinematic simulation was set up which allows the projection of knee articulations to a parallel exoskeleton plane where the behavior of devices with different kinematic configurations could be investigated. Figure 2 presents the basic structure of this simulation. To gather the required rotations and translations in the exoskeleton plane $(EP)$, the knee joint is modeled as a four-link kinematic chain of cylindrical joints as described in [14], using the relations of Walker [11] to determine the translations $(S_1, S_2, S_3)$.

Knee joint rotations $(\theta_1, \theta_2, \theta_3)$ of a walking movement where gathered from [15], where the joint angles of five healthy subjects (mean age: 27 years, mean

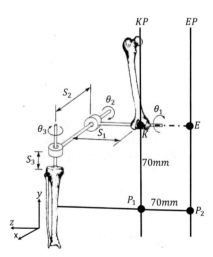

**Fig. 2.** Knee joint model labeled with the exoskeleton plane (EP), the knee plane (KP), the exoskeletons ICR (E) and the reference points on the user's leg (P1) and the exoskeleton (P2) used to determine the required trajectory of the exoskeleton joint (Figure adopted from [14])

height: 180.6 cm, mean body mass: 75.2 kg) were measured with markers fixed to the tibia and femur with intra-cortical traction pins. The walking movement was chosen because it is assumed to be the most repeated motion while wearing the exoskeleton.

The knee rotations and translations were projected from point $P_1$ in the sagittal plane $(KP)$ to point $P_2$ in the exoskeleton plane $(EP)$. Plane $EP$ is parallel to $KP$, has a distance of 70 mm in lateral direction to plane $KP$ and $P_1$ is placed 70 mm distal to the knee joint. $P_1$ is the location of the physical human robot interface (pHRI) connecting the human leg with the exoskeleton. The results of the simulation are presented in Fig. 3 showing the scatter-plot of the migrating point $P_1$ and its projection at $P_2$. The maximum translations in the XY-plane during a forward walking motion is 62.32 mm in x-direction and 39.33 mm in y-direction at a flexion angle of 65°.

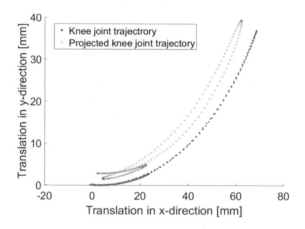

**Fig. 3.** Comparison of the knee joint trajectory at $P_1$ and its projection at $P_2$

The simultaneous appearance of flexion, abduction/adduction (ABD/ADD) and internal/external rotation (IR/ER) causes additional translations in the projection plane while the elliptic shape remains similar to the one observed in $KP$. Since the translations of $P_2$ will be used for further calculations, they will be denoted by $P_{refx,y,z}$.

The initial simulation is performed with a single revolute joint at point $E$ to compute a comparative value for later simulations with the proposed RCJ. Since this joint allows only rotations around $\theta_1$, misalignments and therefore unsolvable kinematic configurations occur during the simulation which are compensated with a 6DoF joint (three rotational and three translational DoFs) at point $K$. The 6 DoF joint should only deflect in the case of misalignment, so a spring stiffness of 9 N/m or 9 N/rad as well as a damping coefficient of 0.02 N/ms or 0.02 Nm/rads was added to all DoFs of the joint. All other joints have no stiffness or damping and are therefore preferred when solving the inverse kinematics.

While running the simulation the deflections of the 6 DoF joint and the spatial position of $P_2$ are recorded 280 times per one gait cycle and the translations of $P_2$ are compared to $P_{refx,y,z}$ (see Eq. 1).

$$F = \sum_{n} \left(6D_{x,y,z} + |P_{refx,y,z} - P_{2x,y,z}|\right) [m] \tag{1}$$

The variable $n$ denotes the number of recorded values and $6D_{x,y,z}$ are the translations of the 6 DoF joint. They are added to the absolute difference of the translations of the projected reference point $P_{refx,y,z}$ and the translations resulting from the motions of the revolute joint $P_{2x,y,z}$. The value of $F$ amounts to 49.009 m for the revolute joint while the ideal exoskeleton joint with six DoF would have an value of 0. Rotations of the 6 DoF joint are not taken into account in the equation because they correlate with the translations, meaning that a reduction of the observed translations at point $P_2$ and in the 6 DoF joint automatically leads to a reduction of the rotations.

The proposed exoskeleton knee joint should have a reduced value for $F$ and be capable of producing a similar shaped trajectory as presented in Fig. 3. Additionally, the range of motion (RoM) for the F/E axis is required to be equivalent to the human knee (i. e. 135°), while RoMs of the ABD/ADD and IR/ER axis should exceed the respective values arising in forward walking motions (ABD/ADD: 7.5°, IR/ER: 10°). The joint actuation is as important as the kinematic joint structure. Since the goal for our joint is to augment healthy people during working, the actuation should reduce peak torques around the F/E axis and if possible also around the ABD/ADD axis.

## 3   Design

As stated in Sect. 2, rolling and sliding motion influenced by the shape of femur and tibia in the human knee result in a migration of the ICR during flexion. Hence reproducing this behaviour would lead to a reduction on misalignments significantly. To achieve this we investigate the design of a system with two spheres forming a rolling contact joint. Similar system have already been proposed for robotic fingers ([16,17]), prosthesis and other technical applications ([18–20]).

### 3.1   Modelling of the Rolling Contact Joint

The kinematics of bodies rolling on each other are well described in the literature e.g. in [21]. Figure 4 presents the case where one cylinder is rolling on a non-moving second cylinder. The ICR migrates on a circle with a radius equivalent to the radius of the non-moving body while the resulting rotation $\alpha$ (corresponding to F/E) is the sum of $\alpha_1$ and $\alpha_2$. By providing two different radii for the bodies, the joint angle in relation to the joint translations can be changed (see Eq. 2). Additionally, the moving cylinder can rotate around an axis along $r_2$ (corresponding to IR/ER) without moving the ICR in Fig. 4.

$$\alpha = \alpha_1 + \alpha_2 = \alpha_1 + \frac{r_1 \cdot \alpha_1}{r_2} \tag{2}$$

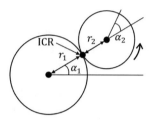

**Fig. 4.** Schematic drawing of a deflected RCJ in 2D

Replacing the two cylinders with two spherical bodies adds a third rotational DoF (corresponding to ABD/ADD) perpendicular to the aforementioned joint axis. This leads to the kinematic equivalent system shown in Fig. 5, which we used in our simulation. It consists of two revolute joints for the pitch (corresponding to F/E) and yaw (corresponding to ABD/ADD) axis each, as well as one revolute joint for the roll axis (corresponding to IR/ER) which are connected by links with a length equivalent to the radii $r_1$ and $r_2$. The rotation sequence is $Pitch_1$, $Yaw_1$, $Roll$, $Yaw_2$, $Pitch_2$. Using the simulation described in Sect. 2 with the RCJ model, the behavior during the gait cycle can be investigated. Up to this point the values for $r_1$ and $r_2$ are not defined. Since manual identification of the values is difficult and time consuming, an optimization process using the pattern search algorithm was included in the simulation. The algorithm minimizes the variable $F$ introduced in Sect. 2.

$$F_{opt} = min(F) \tag{3}$$

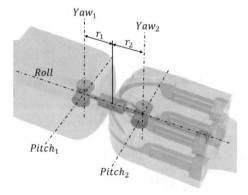

**Fig. 5.** Kinematic chain used to model the rolling contact joint (RCJ)

**Table 1.** Optimized parameters of the RCJ

| Parameter | Initial value [mm] | Lower/upper boundary [mm] | Final value [mm] |
|---|---|---|---|
| $r_1$ | 20 | 10/30 | 11.4 |
| $r_2$ | 20 | 10/30 | 13.1 |
| $IP_x$ | 0 | −50/50 | 1.7 |
| $IP_y$ | 0 | −50/50 | −4.2 |

This means, that the translations of the 6 DoF joint $(6D_{x,y,z})$ have to be minimized and that the RCJ should follow the reference trajectory $(P_{ref\,x,y,z})$ in the $EP$-plane as close as possible. Figure 6 shows the schematic setup of the simulation as well as the variables which are used in the optimization process. In addition to $r_1$ and $r_2$ the initial position of the RCJ in the XY-plane described by $IP_x$ and $IP_y$ is varied by the pattern search algorithm. The last two parameters were introduced because the knee joint is abducted at the beginning of the gait cycle causing initial translations at the exoskeleton knee joint due to parallel kinematics. Table 1 summarizes the initial values, the lower and upper bounds as well as the final values of the optimization process.

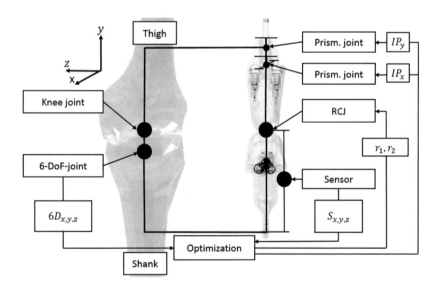

**Fig. 6.** Schematic of the simulation and optimization process

The optimization function $F_{opt}$ sums to 2.422 m using the final values meaning a reduction of 7.37 m compared with the initial values. Figure 7 presents the translations at the 6 DoF joint during one gait cycle, equivalent to the misalignment of the ICRs of exoskeleton and knee. Since no translations in z-direction were provided for the knee joint those values remain low (maximum translation of −0.95 mm in the swing phase). Translations in $x$ and $y$ direction show higher values especially during swing phase when the highest flexion occurs, since knee ICR migration is coupled to the flexion angle. Joint rotations of the 6 DOF joint are close zero ($\sim 10^{-5}$ rad) during the whole simulation. Since the initial position of the revolute joint was not optimized, the parameters $IP_x$ and $IP_y$ were set so zero leading to a value of 3.009 m. Comparison to the revolute joint value derived in Sect. 2, results in a reduction of 46.081 m.

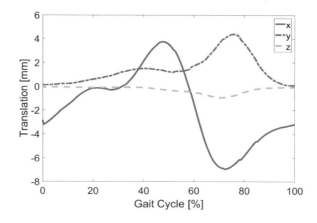

**Fig. 7.** Translations at the 6 DoF joint during one gait cycle

## 3.2   Mechanical Design

Two base components forming the RCJ (thigh and shank part) were designed using the final radii and width of the optimization process. They are connected by two steel cables with a diameter of 1.2 mm which are guided diagonally through the RCJ. Figure 8 presents the construction which is also equipped with four adjustment screws to change the initial cable length and three profiled cam rollers.

Cable 1 starts at the adjustment screw of the anterior thigh side and is guided through the grooved contact surfaces to the posterior shank side. After passing the shank part it rotates around the profiled cam roller and is lead back to the thigh. Basically the same guiding is also used for cable 2 with the difference that the starting and end point is at the posterior thigh side. Assuming that both cables are equally pretensioned, this arrangement leads to torque equilibrium around all joint axis if the joint is undeflected. Since both convex part surfaces should stay in contact, the grooves have a depth of 1.5 mm to incorporate the 1.2 mm steel cables.

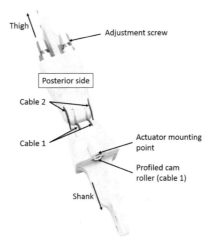

**Fig. 8.** Mechanical design of the RCJ

The rolling contact joint design prevents cable elongation (as well as arising joint torques) when the joint is deflected around the F/E axis. Joint deflections around the yaw or roll axis would lead to cable elongation. Therefore, the cables are guided over profiled cam rollers to compensate this elongation and only small torques occur caused by friction in the cable channels.

## 3.3 Actuation

As stated in Sect. 2 the actuation should reduce peak torques around the F/E axis and if possible also around the ABD/ADD axis. Therefore, a linear actuator mounted at the thigh part of the exoskeleton and the anterior actuator mounting point at the shank part (see Fig. 8) via ball bearings is proposed. Figure 9 presents the actuation principle for a configuration with no joint deflection (left) and a configuration with deflected yaw and roll joint (right). These two cases will be investigated to compute the torques for F/E (first case) and analyze parasitic torques for IR/ER and ABD/ADD (second case). The second case was selected because the knee is mainly abducted during the gait cycle (max angle of 7.5°) combined with both internal and external rotation.

The lever arm to the pitch axis $(l_p)$ amounts to 52.26 mm for an angle of 0° and reduces to 36.71 mm for a pitch angle of 90°. Using Eq. 1, the motion of the ICR and the actuator mounting point during Abd. and IR/ER motions is calculated. Assuming an actuator length of 263 mm the direction of the actuator force is obtained. The resulting torque arm of $F_{actuator}$ to the yaw axis $(l_y)$ amounts to 0.25 mm (0.6% of $l_p$) for an Abd. and IR. angle of 7.5°. For a combined motion around the Abd. and ER axis a lever arm $l_y$ of 1.29 mm (3.5% of $l_p$) is obtained.

The linear actuator introduced in our previous work (see [22]) has a maximum force of 900 N at 100 mm/s and was chosen to calculate the arising torques. With

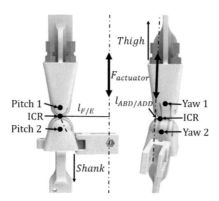

**Fig. 9.** Influence of joint deflections on the actuation

the aforementioned values a maximum torque of 47 Nm at 180 deg/s around the pitch axis is obtained. This parasitic torques around the yaw axis amount to 0.25 Nm for ABD. and IR. or 1.16 Nm for ABD. and ER. Since these values are low compared to the maximum torque in the pitch axis this result is considered acceptable.

## 4     Evaluation and Results

Two experiments were conducted to validate the simulation results. In the first experiment maximum joint angles of the prototype from Fig. 1 were measured. Therefore, the joint was manually deflected around all axis until a significant increase of joint stiffness was detected by the person that articulated the joint. In pitch direction the stiffness is consistently low until the rolling parts collide. Deflection around the yaw axis is initially possible at low stiffness, too. Increasing deflection causes collision of the cables with the edges of the grooves, guiding them. The same holds for a deflection around the roll axis. After collision, further joint deflection can only be obtained by elongating the cables, meaning a drastically increased joint stiffness. The values presented in Table 2 exceed the maximum joint angles occurring in our reference walking motion and the maximum angle in pitch direction is higher than the maximum active flexion of the human knee. Therefore, a prototype exoskeleton was manufactured to conduct further experiments.

**Table 2.** RoMs of the RCJ compared to the required joint angles

|          | Yaw    | Pitch | Roll |
|----------|--------|-------|------|
| RCJ      | 19.5°  | 138°  | 34°  |
| Required | 7.5°   | 135°  | 10°  |

## 4.1  Prototype

Our prototype, shown in Fig. 10 consists of two base components which hold the thigh and shank part of the RCJ and connect them to the user's leg via Velcro straps. To be able to adapt the RCJ as well as the revolute joint (17 B47 = 20, Otto Bock HealthCare GmbH, Duderstadt) to the subject's knee axes, four linear and two revolute joints are incorporated as well. These joints are adjusted during the donning process and are fixed during operation. It is possible to exchange the RCJ with the revolute joint while the user is wearing the device in order to provide equal conditions for the experiments (e.g. that the Velcro straps remain tightened while joints are exchanged).

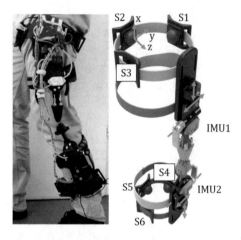

**Fig. 10.** Subject wearing the passive exoskeleton used during the experiments (left) and rendering presenting the sensor setup of the device (right)

The goal of the second experiment is to compare the kinematic compatibility of the RCJ to that of a revolute joint. Therefore, pressure and shear forces were measured with six 3D force sensors (Optoforce OMD-30-SE-100N, OptoForce Kft., Budapest) between the exoskeleton and the subject's leg, while the subjects is perfoming different motions wearing a passive exoskeleton. The sensors are mounted to 3D-printed interfaces and are positioned at posterior, medial and anterior side of the thigh and shank respectively. Due to the semi-spherical shape of the force sensors the maximum force, the resolution and the maximum dome deflection in compression direction (100 N, 6.25 mN, 3 mm) deviates from the aforementioned properties in shear direction (25 N, 7 mN, 2.5 mm). Two inertial measurement units (BNO055, Robert Bosch GmbH, Stuttgart) are utilized to compare joint deflections in the experiments.

## 4.2   Experiments

The experiments were conducted with three subjects with similar body characteristics (see Table 3) and the subject's knee axes was determined before donning the device. Then the exoskeleton with built-in RCJ was fixed to the user's leg by tightening two Velcro straps and the joint axis of RCJ and knee were adjusted using the aforementioned passive joints. Finally, the two Velcro straps with attached force sensors were tightened in a way that the forces were in between predefined intervals.

**Table 3.** Overview over basic body parameters of the three subjects

| Subjects | 3 male |
|---|---|
| Age | $27.33 \pm 5.03$ |
| Weight [$kg$] | $72.00 \pm 3.46$ |
| Height [$cm$] | $177.66 \pm 5.50$ |
| BMI [$kg/m^2$] | $22.81 \pm 0.79$ |

Each subject had to perform a predefined set of movements starting with standing relaxed in an upright position. The forces measured while standing relaxed are used to calculate an offset to the forces of all other movements later. Subsequently seven other movements were performed: Walking four steps forward (1), crouching (2), turn left (3), turn right (4), four sidesteps (5), sit down on a chair and stand up (6). Every movement was recorded four times including the relaxed standing. The same procedure was repeated after mounting the revolute joint. Figure 11 presents a comparison of the compression forces at the anterior thigh and the joint angles of the RCJ captured while executing one forward step. Negative values for $S6_z$ indicate compression forces since the sensor's z-axis is pointing away from it. In this trial, peak compression forces using the RCJ ($S6_z RCJ$) decreased by approximately 4 N while the progression of the force remains similar compared to forces when using the revolute joint. The highest forces occur during late stance (40–60 % of gait cycle) when the knee is extended. Reduced flexion angles (compared to literature) were observed for all subjects conducting the experiments which can be explained by the compression of the thigh muscles caused by the Velcro strap. Abduction and rotation of the RCJ is similar to the observed angles in the human knee.

To gather indications about the kinematic compatibility, median and maximum values of all forces were calculated. First the median of the forces gathered from the standing motion was subtracted from all other joint values recorded with the same joint. Then all four trials of every movement were combined to one dataset to calculate its median and peak value. Table 4 summarizes the peak and median forces between the RCJ and the revolute joint of all sensors for the walking forward movement. This movement was selected because it was used exemplary during the whole design process. The third and six row denote the

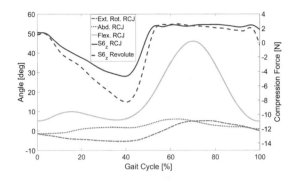

**Fig. 11.** Joint deflection and compression forces at the anterior thigh during one gait cycle

difference between the peak and median values of the two joints. Negative values indicate higher peak forces when using the revolute joint and vice versa. 24 out of 36 values show decreased peak forces ($S1_z \ldots S6_z$) during the trials, while the most significant reductions occur on the medial sensors likely emerging from the enhanced mobility of the RCJ around the roll axis. Values of the posterior thigh sensor are mainly increased by the RCJ, which can be explained by an increased stiffness of the RCJ compared to the revolute joint, affecting the forces during stance phase (closed kinematic chain). Peak pressure forces are generally reduced ($S_2 \ldots S_5$) or close to the value measured with the revolute joint ($S_1$). A similar force distribution was observed for the other movements as well. Combining all movements and peak forces leads to an average peak force reduction of 1.08 N.

**Table 4.** Comparison of peak and median force values in all spatial directions between RCJ and revolute joint (all values in $[N]$)

| | Thigh | | | | | | | | | Shank | | | | | | | | |
|---|---|---|---|---|---|---|---|---|---|---|---|---|---|---|---|---|---|---|
| | $S_1$ | | | $S_2$ | | | $S_3$ | | | $S_4$ | | | $S_5$ | | | $S_6$ | | |
| | x | y | z | x | y | z | x | y | z | x | y | z | x | y | z | x | y | z |
| Peak RCJ | −3.3 | −1.6 | −4.0 | −1.2 | −1.1 | −1.5 | −1.5 | −1.3 | −1.8 | −0.8 | −1.6 | −1.7 | −1.5 | −1.5 | −3.4 | −0.6 | −0.9 | −3.7 |
| Peak Rev. | −2.6 | −1.1 | −3.7 | −2.4 | −2.7 | −2.3 | −4 | −2.1 | −3.3 | −0.7 | −1.9 | −2 | −3.1 | −1.4 | −4.8 | −0.5 | −1.3 | −4.6 |
| Diff. Peaks | −0.7 | −0.5 | −0.3 | 1.2 | 1.6 | 0.8 | 2.5 | 0.8 | 1.5 | −0.1 | 0.3 | 0.3 | 1.6 | −0.1 | 1.4 | −0.1 | 0.4 | 0.9 |
| Med. RCJ | −1.3 | −0.8 | −1.1 | −0.6 | −0.6 | −0.6 | −0.5 | −0.4 | −0.4 | −0.3 | −0.6 | −0.6 | −0.4 | −0.7 | −0.9 | −0.3 | −0.5 | −1.6 |
| Med. Rev. | −1.0 | −0.5 | −0.9 | −1.7 | −1.7 | −1.3 | −3.0 | −1.2 | −2.0 | −0.3 | −0.7 | −0.8 | −0.7 | −0.6 | −1.2 | −0.2 | −0.5 | −2.2 |
| Diff. Med. | −0.3 | −0.3 | −0.2 | 1.1 | 1.1 | 0.9 | 2.5 | 0.8 | 1.6 | 0.0 | 0.1 | 0.2 | 0.3 | −0.1 | 0.3 | −0.1 | 0.0 | 0.6 |

To this end, the joint design has limitations regarding the joint angle sensing and the material wear. Since there are no distinct joint axis, joint angles were measured using IMUs which have a lower accuracy and tend to drift compared to other joint encoders. After the experiments significant wear of the 3D-printed surfaces rolling on each other was observed. It is not clear if this is caused by the assumed rolling motions or if there are slipping motions as well.

## 5    Conclusion and Future Work

We presented the conceptual design of a rolling contact joint intended to use as an exoskeleton knee mechanism, which is capable of performing three rotations and predefined translations in the sagittal plane to reduce macro- and micro-misalignments. Knee joint trajectories from literature served as simulation input to calculate the required rotations and translations in the plane where the device is assumed to be placed.

Our design consists of two parts with optimized shapes that are rolling on each other. The optimization was performed using a kinematic equivalent joint model coupled to a model of the human knee. The simulation results indicate increased alignment compared to a single revolute joint. The mechanical construction includes two cables to couple the rolling parts, which are guided over profiled cam rollers allowing joint deflection without cable elongation in a certain range around the Abd/Add. and IR/ER axes.

The resulting ranges of motions exceed knee angles while walking forward and the device provides a maximum flexion angle of 138°. Experimental evaluation to support the simulation results were also conducted. Indications regarding the kinematic compatibility were derived from the compression and shear forces captured with six 3D force sensors during multiple movements between the subject's leg and a prototype exoskeleton. To this end, forces with attached RCJ are compared to forces with attached revolute joint. The results indicate decreased compression and shear forces in five of the six sensors and therefore increased kinematic compatibility of the proposed joint design. Based on these promising results, we will conduct experiments using an actuated prototype to determine the forces between subject and exoskeleton for that case. The aforementioned limitations regarding joint angle sensing and wearability will be further addressed as well.

**Acknowledgment.** This work has been supported by the German Federal Ministry of Education and Research (BMBF) under the project INOPRO (16SV7665).

## References

1. CYBERDYNE Inc. HAL (2018). www.cyberdyne.jp
2. EKSO Bionics (2018). www.eksobionics.com
3. Wang, S., van der Kooij, H.: Modeling, design, and optimization of Mindwalker series elastic joint. In: IEEE International Conference on Rehabilitation Robotics (ICORR), pp. 1–8 (2013)
4. Hyon, S.-H.: et al.: Design of hybrid drive exoskeleton robot XoR2. In: IEEE/RSJ International Conference on Intelligent Robots and Systems (IROS), pp. 4642–4648 (2013)
5. Wang, D., et al.: Adaptive knee joint exoskeleton based on biological geometries. IEEE/ASME Trans. Mechatron. **19**(4), 1268–1278 (2014)
6. Kim, K., et al.: Development of the exoskeleton knee rehabilitation robot using the linear actuator. Int. J. Precis. Eng. Manuf. **13**(10), 1889–1895 (2012)

7. Celebi, B., Yalcin, M., Patoglu, V.: AssistOn-Knee: a self-aligning knee exoskeleton. In: IEEE/RSJ International Conference on Intelligent Robots and Systems (IROS), pp. 996–1002 (2013)
8. Choi, B., et al.: A self-aligning knee joint for walking assistance devices. In: International Conference of the Engineering in Medicine and Biology Society (EMBC), pp. 2222–2227 (2016)
9. Saccares, L., Sarakoglou, I., Tsagarakis, N.G.: iT-Knee: an exoskeleton with ideal torque transmission interface for ergonomic power augmentation. In: IEEE/RSJ International Conference on Intelligent Robots and Systems (IROS), pp. 780–786 (2016)
10. Cai, D., et al.: Self-adjusting, isostatic exoskeleton for the human knee joint. In: Annual International Conference of the IEEE Engineering in Medicine and Biology Society, pp. 612–618 (2011)
11. Walker, P., Rovick, J., Robertson, D.: The effects of knee brace hinge design and placement on joint mechanics. J. Biomech. **21**(11), 969–974 (1988)
12. Wilson, D., et al.: The components of passive knee movement are coupled to flexion angle. J. Biomech. **33**(4), 465–473 (2000)
13. Neumann, D.A.: Kinesiology of the Musculoskeletal System: Foundations for Rehabilitation. Elsevier Health Sciences, London (2013)
14. Grood, E.S., Suntay, W.J.: A joint coordinate system for the clinical description of three-dimensional motions: application to the knee. J. Biomech. Eng. **105**(2), 136–144 (1983)
15. Lafortune, M., et al.: Three-dimensional kinematics of the human knee during walking. J. Biomech. **25**(4), 347–357 (1992)
16. Collins, C.L.: Kinematics of robot fingers with circular rolling contact joints. J. Rob. Syst. **20**(6), 285–296 (2003)
17. Kim, S.H., et al.: Force characteristics of rolling contact joint for compact structure. In: IEEE International Conference on Biomedical Robotics and Biomechatronics (BioRob), pp. 1207–1212 (2016)
18. Mun, J.H., Lee, D.-W.: Three-dimensional contact dynamic model of the human knee joint during walking. KSME Int. J. **18**(2), 211–220 (2004)
19. Montierth, J.R., Todd, R.H., Howell, L.L.: Analysis of elliptical rolling contact joints in compression. J. Mech. Des. **133**(3), 031001 (2011)
20. Nelson, T.G., et al.: Curved-folding-inspired deployable compliant rolling-contact element (D-CORE). Mech. Mach. Theory **96**, 225–238 (2016)
21. Kuntz, J.P.: Rolling link mechanisms. Delft University of Technology (1995)
22. Beil, J., Perner, G., Asfour, T.: Design and control of the lower limb exoskeleton KIT-EXO-1. In: IEEE International Conference on Rehabilitation Robotics (ICORR), pp. 119–124 (2015)

# Towards a Stair Climbing Robot System Based on a Re-configurable Linkage Mechanism

Omar El-Farouk E. Labib[1], Sarah W. El-Safty[1], Silvan Mueller[1,2], Thomas Haalboom[2], and Marcus Strand[2(✉)]

[1] EMS (Engineering and Material Sciences), Mechatronics Department, German University in Cairo, New Cairo, Egypt

[2] Baden-Wuerttemberg Cooperative State University Karlsruhe, Karlsruhe, Germany
strand@dhbw-karlsruhe.de

**Abstract.** This paper contributes to a mechanical/kinematical design of a stair climbing robot. Based on the known Klann-mechanism, an extended re-configurable mechanism is proposed, in order to address the stair climbing problem. Steps offer a great variety of occurrences, since they differ in height, width or step length. Regarding staircases the variety is even higher since they differ in the number of steps per level, the size of the platform, the orientation etc. Due to these variations the concept proposes a re-configurable design, which is tested in simulation and in a real physical setup.

## 1 Introduction

There is a great variety of robots that are now used in industrial applications and special environments. These robots are either designed to perform a hand-picked task or a multitude of tasks. Albeit also service robots operating in human designed unstructured environments become more and more commercially available. There are many potential applications for human environment robots, such as home service robots, urban search and rescue missions, and Special Forces robots. Indeed there are already many systems available operating in household environments like vacuum cleaners, lawn mowers etc.

However current available systems cannot cope with heavily unstructured environments. Since there are many obstacles that face the mobility of these robots, such as narrow corridors, stairways, and steps. Especially step or even stair climbing face robotic systems in their current design (kinematic, sensor setup and algorithms) with huge problems. Stair climbing involves a lot of uncertainty and difficulty, having an accurate mathematical model to describe the interaction between the robot and the stair environment can be a very challenging task, as not all stairs have the same height, width, and step length. This paper presents an approach to kinematically overcome the addressed problems using a quadruped robotic system with reconfigurable leg mechanism, as the reconfigurable nature of the mechanism widens the movement possibilities.

M. Strand—This work was supported Baden-Wuerttemberg Cooperative State University Karlsruhe.

M. Strand et al. (Eds.): IAS 2018, AISC 867, pp. 278–288, 2019.
https://doi.org/10.1007/978-3-030-01370-7_22

# 2   Related Work

In this section, investigation of related research efforts is conducted. This section is divided into two parts, the first will cover the different robots that have been designed for stair climbing, and the second will be on re-configurable leg mechanisms.

## 2.1   Stair Climbing Robots

For robots that are deployed in an urban environment, the ability to ascend and descend stairs is essential. Robots that can perform this task are of many kinds, there are wheeled robots like the Rocker-Bogie, modular based wheeled robots, modified wheeled robots, and Segway robots. [1–4], Tracked robots [5–7], legged robots [7, 8]. Wheeled robots have an upper hand when it comes to speed in comparison to other stair climbing platforms, but speed is not always the measure of efficiency or practicality in this task. Spiral stair cases and wheel slippage are examples to the problems with these wheeled climbers. Tracked robots have design advantages over wheeled robots, tracked robots have a larger contact area and as a result they exert a much lower force per unit area on the ground compared to a wheeled robot of the same weight, which makes them suited for traversing uneven terrain, soft and low friction terrains such as snow. However, the primary disadvantage is that tracks are a more complex in design compared to wheels, and are more prone to failure modes such as derailed or snapped tracks, they also need the aid of a Flipper-stabilizer arm incorporated into the design to effectively climb stairs, which makes the control of such platforms complicated.

Legged platforms have advantages such as strategically choosing contact points on the ground which is a huge advantage over wheeled and tracked robots. This gives legged robots the ability to move smoothly over terrain, step over obstacles easily and climb stairs with their better reach. Another advantage is maintaining the direction the body of the robot is facing while changing its direction of motion. This ability is very useful in narrow spaces such that the robot has faster motion and still having a more natural motion. Legged robots are also kind to the surface they move on as they can distribute their weight and move their center of mass while maintaining the position of their legs [9]. This advantage is a useful in case of acceding or descending stairs where there is a long distance between supporting legs to step on.

Biped and quadruped robots are only two examples of legged robots. When dealing with biped robots they are similar to human beings, these robots have the characteristic of multi degree of freedom, and can achieve various motions through the different combination of its joints [7]. However having these many joints makes it challenging to establish a stable gait pattern and control when compared to a quadruped robot. Quadruped robots hold advantages when compared to biped robots, such as being inherently stable due to having the weight of the whole system distributed on each leg instead of only two. Leg design for quadruped robots also can have less degrees of freedom than bipedal, which makes them easier to control, this reduced complexity can be very useful for stair climbing.

Two examples of a simpler leg designs are the Theo Jansen mechanism and the Klann linkage mechanism. Both mechanisms are one degree of freedom mechanism, which are much simpler than other leg designs. The Theo Jansen mechanism shown in Fig. 1, is an eight bar mechanism that is constructed by an upper and lower four bar mechanism, a parallel mechanism, and two couplers. The crank that drives the system rotates in the opposite direction as the rotation as the contact point to the ground, and rotates 120° per stride. By configuring this mechanism in a certain configuration that mimics the human foot trajectory researchers were able to achieve stair climbing with this mechanism [10].

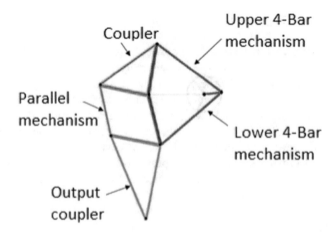

**Fig. 1.** Jansen mechanism

Klann linkage mechanism shown in Fig. 2, is a six bar mechanism. This mechanism is simpler than the Theo Jansen as it has fewer links, the crank rotation is the same direction as the contact point with ground and rotates 180° per stride. The main advantage of the Klann mechanism when compared to the Jansen mechanism is its longer reach, the leg in the Klann mechanism can go much higher than a Jansen mechanism of the same size.

## 2.2   Re-configurable Leg Mechanisms

Not all legged creatures in nature share the same locomotion, this is because they do not all have the same walking gaits and leg geometry. Each species will have its own walking patterns based on the morphology of their legs, and can only swap these gaits in a limited number of ways. Similarly, robots with fixed mechanical structures face the same issue. The ability to change leg geometry based on the given situation can be very beneficial for a robot in an urban environment, a reconfigurable design approach to legged robots moment that produces many different gait cycles, opening possibilities for new means of locomotion, such as cat walking like motion, jam avoidance motion, stair climbing motion, and drilling motion [11].

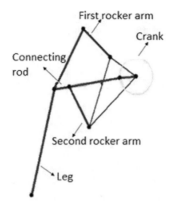

**Fig. 2.** Klann mechanism

A klann re-configurable mechanism was developed to be able to achieve a multi-tude of gait patterns and locomotion [12, 13]. However, designing a re-configurable mechanism with a lot of gait patterns has its issues, such as the increased number of actuators per leg, which can lead to power issues and can be complex to control. Weight is also an issue with such platforms as they has an increased number of components which can also lead to more possible modes of failure.

That being said, we believed that designing a reconfigurable Klann mechanism that can modify its gait pattern to be only able to climb stairs, thus resulting in a mini-malistic design that achieves only the design requirement. In the coming sections we will discuss a simulated model, designing a mechanism based on the simulation results, fabrication of the mechanical system, and experimental analysis of the prototype.

## 3   Simulated Model

In this section, the simulation of the re-configurable Klann mechanism will be dis-cussed. The simulation was run on a program called "Linkage", which is a 2D mechanical simulator. The simulated model tested the trajectory of the Klann mech-anism with one linear actuator attached instead of the second rocker arm. Replacing the second rocker arm with the linear actuator manipulates the amplitude of the mecha-nism's trajectory. The range of motion is greatly affected by the increase in the sec-ondary rocker arm. In Fig. 3, the simulated model is displayed, the dimensions of the links are as follows, the crank is 9.3 cm, the first rocker arm is 23.2 cm, the leg is 40.7 cm, and the secondary rocker arm has an initial length of 15.4 cm and at maxi-mum extension it is measured at 22.3 cm.

By controlling when the linear actuator extends and retracts, manipulation of the trajectory's amplitude can be achieved, in this case the initial height the mechanism can reach is about 20 cm, and at maximum displacement of the linear actuator it reaches 50 cm. By increasing the linear actuator through the increasing motion of the crank, then retracting the linear actuator through the decreasing motion of the crank a trajectory with maximum and minimum reach can be achieved shown in Fig. 4.

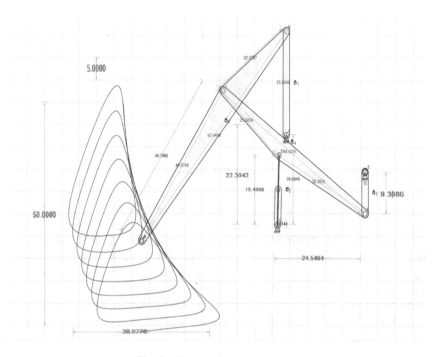

**Fig. 3.** Simulated trajectory range

## 4    Mechanical Design and Fabrication

In this section, the design process of the prototype to achieve our desired outcome will be discussed, which will be explained in details in the coming subsections. The design parameters and the fabrication will be explained.

### 4.1    Full Robot Design Concept

The end goal of the designed leg mechanism is a stair climbing platform with four reconfigurable Klann mechanism. There are two possible configuration for the system, a symmetric leg configuration where the front pair face a direction and the back pair face the opposite direction, and an asymmetric design configuration will all legs face the same direction. The symmetric layout is chosen as it is unidirectional and the robot can move forward and backwards in the same way, which would make the platform more stable than an asymmetric layout. The mass of the system would be distributed equally on all legs. The four legs would all be assembled on a box like chase which will act as the housing for all electronics and as a basket for any payload. The final CAD model is shown in Fig. 5.

## 4.2   CAD Model

The Leg mechanism was designed on "Solid works", the system contains 2 links (Crank and first rocker arm), the frame, the leg link, and a linear actuator taking the place of the second rocker arm. The measurements of the assembly's components are as follows: a crank of length 5.65 cm, a first rocker arm of length 12.6 cm, a triangle shaped frame of dimensions 24.31 cm × 14.20 cm × 10.79 cm, and a triangle shaped leg link of dimensions 312.2 cm × 111.94 cm × 201.35 cm. Both the leg link and the frame are hollowed from the center to decrease the weight of the leg. The spacing between the crank fixture and the first rocker arm is 4.2 cm in height × 9.75 cm in width, and the spacing between the linear actuator fixture and the crank fixture is 5.8 cm in height × 12.25 cm in width.

**Fig. 4.** Design concept

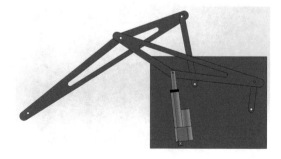

**Fig. 5.** CAD model

**4.3    Prototype Fabrication**

The crank, frame, first rocker arm, and the leg link were all manufactured from aluminum. A 3d printed box like structure was printed to act as a mounting body for the mechanism. The crank, the first rocker arm, and linear actuator were all mounted on the surface of the box with M4 screws, the frame is then connected to all three of them and finally the leg linkage is connected to the frame and first rocker arm. All links have one degree of freedom, they can only revolve around their mounting screws. The linear actuator was fastened in place with a zip lock as shown in Fig. 6. The crank is driven by a high torque 25 mm DC 6.0 V 210 rpm Geared motor, and the linear actuator is a heavy duty rated at a 1500 N/330lbs maxim Load at 12 V with 2 inch stroke length.

## 5    Experimental Results

A high speed camera was used to track the contact point of the leg link with the ground to track its trajectory. A white background is placed behind the mechanism and a spotlight is directed at the setup. The software used to track the motion of the contact point is "Tracker" a program by open source physics. The experiment conducted was running the DC motor connected to the crank at 6 V 210 rpm, while increasing the linear actuator till the tracking point reaches its maximum amplitude. Figure 7 shows the trajectory which the contact point moves, the contact point raises with the increase of the linear motors stroke and returns to the original trajectory when the linear actuator is fully retracted. Figure 8 shows the angular potion (Theta) in degrees with respect to time in seconds, the value of theta is between 60° and −60°.

**Fig. 6.** Mechanism setup

**Fig. 7.**  Actual setup

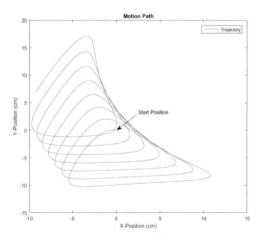

**Fig. 8.**  Trajectory

Figure 9 shows the angular velocity (Omega) in radian per second with respect to time in seconds. Figure 10 shows the angular acceleration (alpha) in radian per $seconds^2$ with respect to time in seconds.

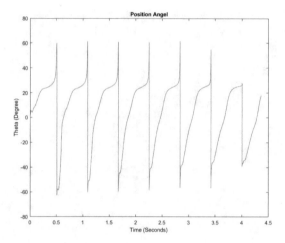

**Fig. 9.**  Theta

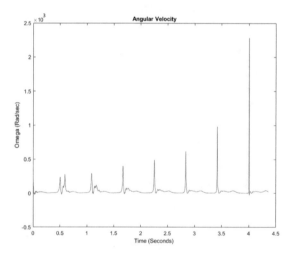

**Fig. 10.**  Omega

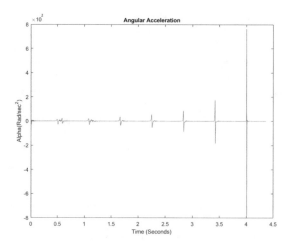

**Fig. 11.** Alpha

## 6   Conclusions and Future Work

In this paper, a re-configurable Klann linkage mechanism was proposed, focusing on stair climbing. The design, Fabrication, and experimental results are demonstrated. The aim of this mechanism was to approach stair climbing with a simple and straight forward designing and prototyping technique. A re-configurable Klann leg mechanism was designed with linear actuator instead of the secondary rocker arm, which increases the amplitude of the motion trajectory for the leg mechanism. A full four legged robotic system is currently being made based on the re-configurable design to experiment with the idea further (Fig. 11).

## References

1. Hong, H., Kim, D., Kim, J., Oh, J., Kim, H.S.: A locomotive strategy for a stair-climbing mobile platform based on a new contact angle estimation. In: IEEE International Conference on Robotics and Automation (ICRA) (2013)
2. Turlapati, S.H., Shah, M., Teja, S.P., Siravuru, A., Shah, S.V., Krishna, M.: Stair climbing using a compliant modular robot. In: 2015 IEEE/RSJ International Conference on Intelligent Robots and Systems (IROS) (2015)
3. Sun, Y., Yang, Y., Ma, S., Pu, H.: Modeling paddle-aided stairclimbing for a mobile robot based on eccentric paddle mechanism. In: 2015 IEEE/RSJ International Conference on Intelligent Robots and Systems (IROS) (2015)
4. Takaki, T., Aoyama, T., Ishii, I.: Development of inverted pendulum robot capable of climbing stairs using planetary wheel mechanism. In: 2013 IEEE International Conference on Robotics and Automation (2013)
5. Adiwahono, A.H., Saputra, B., Chang, T.W., Yong, Z.X.: Autonomous stair identification, climbing, and descending for tracked robots. In: 2014 13th International Conference on Control Automation Robotics & Vision (ICARCV) (2014)

6. Yu, S., Li, X., Wang, T., Yao, C.: Dynamic analysis for stairclimbing of a new-style wheelchair robot. In: 2010 Chinese Control and Decision Conference (2010)

7. Zhang, Q., Ge, S.S., Tao, P.Y.: Autonomous stair climbing for mobile tracked robot. In: 2011 IEEE International Symposium on Safety, Security, and Rescue Robotics (2011)

8. Moore, E.Z., Campbell, D., Grimminger, F., Buehler, M.: Reliable stair climbing in the simple hexapod 'Rhex'. In: IEEE International Conference on Robotics and Automation (ICRA) (2002)

9. Arikawa, K., Hirose, S.: Mechanical design of walking machines. Philos. Trans. R. Soc. A **365**, 171–183 (2007)

10. Liu, C., Su, N., Lin, M., Pai, T.: A multi-legged biomimetic stair climbing robot with human foot trajectory. In: 2015 IEEE International Conference on Robotics and Biomimetics (ROBIO) (2015)

11. Nansai, S., Rojas, N., Elara, M.R., Sosa, R.: Exploration of adaptive gait patterns with a reconfigurable linkage mechanism. In: 2013 IEEE/RSJ International Conference on Intelligent Robots and Systems (2013)

12. Sheba, J.K., Martinez-Garcia, E., Elara, M.R., Tan-Phuc, L.: Synchronization and stability analysis of quadruped based on reconfigurable Klann mechanism. In: 2014 13th International Conference on Control Automation Robotics & Vision (ICARCV) (2014)

13. Sheba, J.K., Martinez-Garcia, E., Elara, M.R., Tan-Phuc, L.: Design and evaluation of reconfigurable Klann mechanism based four legged walking robot. In: 2015 10th International Conference on Information, Communications and Signal Processing (ICICS) (2015)

# Can Walking Be Modeled in a Pure Mechanical Fashion

Antonio D'Angelo[(✉)]

Department of Mathematics and Computer Science, Udine University, Udine, Italy
antonio.dangelo@uniud.it

**Abstract.** The aim of this paper is to investigate the role of some mechanical quantities in the challenging task to make a robot walking or running. Because the upright posture of an humanoid is the main source of instability, the maintenance of the equilibrium during locomotion requires the gait-controller to deal with a number of constraints, such as ZMP, whose dynamical satisfactions prevent the humanoid from an harmful fall. Walking humanoids are open systems heavily interacting with a perturbing environment and the rapid loss of mechanical energy could be an hallmark of instability. In this paper we shall show how certain dimensionless parameters could be useful to design the walking gait of a bipedal robot.

## 1 Introduction

At the present bipedal walking is one of the most exciting challenge to build humanoid robots capable of supporting humans under their normal daily activities. This form of terrestrial locomotion, where an organism moves his two rear limbs, is an habitual method of locomotion only for a few modern species because the great majority of living terrestrial vertebrates are quadrupeds, with bipedal behavior occasionally exhibited in specific circumstances. Thus bipedal locomotion, rather than a structured behavior of a minor number of living systems, is an effective ability to explore and deal with the environment.

In the biological systems, the evolution has provided them with those tools which allow them to learn how to adjust their body during the motion against the force of gravity and any whatever disturbance. On the other side, to get the same results with humanoids we must build them to actively resist to the mechanical forces in a synergic fashion: this behavior requires to understand how all the involved mechanical quantities interfere or support the bipedal motion and, between them, the energy plays a central role.

### 1.1 Legged Locomotion Survey

The idea to stabilize dynamically legged locomotion in bipedal robots can be traced far back in the time [23]. Attempts have been made on one-legged robots [20] and on quadrupedal robots [8,15]. The first walking bipedal robots were

© Springer Nature Switzerland AG 2019
M. Strand et al. (Eds.): IAS 2018, AISC 867, pp. 289–301, 2019.
https://doi.org/10.1007/978-3-030-01370-7_23

developed in the 1970's; an example among all comes from Kato and Tsuiki [14]. More recently the problem to maintain bipedal gait has been addressed from a dynamical walking point of view. More specifically it has been handled either as direct control of the ZMP [3,17,25] or as stable walking generation through a CPG mechanism [12,27]. But also, the task of locomotion has been faced as a learning problem [5,26], compliance control [18] and so on. However, because the burden of providing humanoids with a reliable gait is far to be solved, much of work on bipedal robots is still focused on locomotion.

In general, the motion is divided into a single-support phase (with one foot on the ground) and a double-support phase. In ordinary human gait, the duration of the double-support phase covers about the 20% of the step cycle [11]. One of the simplest models of a walking robot is the 5-link biped introduced by Furusho and Masubuchi [9] so it has been used by several authors, such as Cheng and Lin [6], Pettersson [19] and many others. This simulated robot is constrained to move in the sagittal plane (Fig. 1), and has five degrees of freedom (DOF).

There exist many different formulations of the equations of motion for bipedal robots; for example, the Lagrangian formulation and the Newton-Euler formulation. For the rest of the paper, in the Sect. 2 we shall present and discuss the Lagrangian equations of motion for one of the leg of the simple five-link robot whereas in Sect. 3 we shall outline some consideration about how the gait control should be related to some meaningful mechanical parameters. At last, in Sect. 4 we shall draw some preliminary results and conclude with a suggestion for future improvements.

## 2   Problem Position

The Lagrangian formulation of the bipedal gait has the advantage that the constraining forces must not be made explicit to find out the motion of the robot. On the contrary, the Newton-Euler formulation is computationally the most efficient, with the computation time growing linearly with the number of degrees of freedom [24] and this property is generally suitable in simulations.

More advanced simulation models don't require the motion constrained in the sagittal plane and, moreover, they often include feet, arms, as well as further DOF in the hip (Hirai [10], Arakawa [4], Fujimoto [7]). In the follow we don't insist on the static walking of bipedal gaits (which requires the projection of the center of-mass of the robot to be kept within the foot supporting area, easier to implement but unacceptably slow) but, on the contrary, we focalize on dynamic walking which implies some form of *postural control* and it is generally based on the *zero-moment point* (ZMP), a dynamic generalization of the concept of center-of-mass[1] (Arakawa [4]) to generate stable bipedal gaits.

The ZMP, originally introduced by Vukobratovic [23] in 1969, is the point on the ground where the resulting torques around the (horizontal) x and y axes, generated by reaction forces and torques, are equal to zero. If the ZMP is contained within the convex hull of the feet support region, the gait is dynamically

---

[1] See *center of pression*.

balanced, namely, the robot can be prevented to fall down. Thus ZMP monitoring and control is the most popular method to maintain stable gaits within humanoids robots.

## 2.1   Spring-Loaded Inverted Pendulum Motion

Let's start by considering an humanoid robot walking on a roughly flat plain surface so that we can assume the motion to be confined on a plain where we introduce a cartesian frame of reference with X and Y the axes on this plain whereas Z is taken normal upward.

In the most general case the bipedal gait should be considered in a three-dimensional space because the instability sources stem from both the gravity force and the interactions with other human or humanoid bodies. However, because the main source of instability is the gravitational field, its perturbing action is better investigated in a simpler scenario where the humanoid motion is confined on its sagittal plane. By so doing we can concentrate on the motion of its center of mass with the aim to understand its mechanical properties along with those parameters whose viable values allow humanoids to maintain stable gaits. Hence, the reference model is a compass-like inverted pendulum framework where each leg is modeled as an extendable rod with a torque **K** applied at the bottom and a force **F** insisting at the top, directed along the rod upwards. Its current length can be triggered either as a rigid link or in a spring-fashion where its varying length results in a leg extension/contraction.

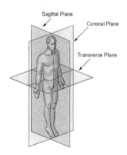

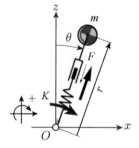

**Fig. 1.** General arrangement of a bipedal walker

**Fig. 2.** Spring-loaded model of walker based on the inverted Pendulum dynamics

Focusing on the SLIP model, sketched in Fig. 2, you can understand the named mechanical quantities as control variables whose values mainly depend on the inclination $\theta$ of the rod with the respect to vertical at the ground

$$F = mg\xi(\theta) \qquad K = mgr\eta(\theta) \qquad (1)$$

and the joint motors of the humanoid are continuosly adjusted by the feed-back provided to the gait controller. Its operation mode depends on a number of mechanical parameters whose values must be dinamically assigned and/or monitored. Thus, in the *linear inverted pendulum* (**LIP**) mode each leg is a rigid multi-link which doesn't allow to store and/or release mechanical energy, whereas in the *spring loaded inverted pendulum* (**SLIP**) mode the dynamics of the inverted pendulum pattern is augmented by the spring dynamics. Each leg is a variable length spring whose stiffness triggers leg performance by modulating the restore/delivering of mechanical energy. If the initial energy supplied to load the spring is rapidly released it causes a long step and/or jump.

In the follow we shall assume each leg working in a SLIP fashion where the energy supplied by the bipedal system balances the losses due to the interaction with the gravitational field (**balanced mode**) and the closed system includ-ing the gravity force becomes *conservative*. The Euler-Lagrange formulation of the mentioned arrangement is really straightforward: what we need is to make esplicit the *lagrangian* of the bipedal walker which in our case is built by the superposition of the inverted pendulum mechanics [13] with the linear spring motion

$$L = \frac{m}{2}(\dot{r}^2 + r^2\dot{\theta}^2) - mgr\cos\theta - \frac{k_e}{2}(r-a)^2 \qquad (2)$$

where the first term is called the *kinetic energy* $T$ of the system whereas the remaining terms are referred to as its *potential energy* $U$, the former $U_{gravity}$ due to the gravity and the latter $U_{spring}$ given by the spring-like behavior of the leg whose *stiffness* is denoted by $k_e$. The spring potential is taken null[2] when the leg is bent on its knee at length $a$, whereas the potential $U_{gravity}$ is assumed null at the ground. In this simplified model the center of mass of the humanoid is taken at the top of the leg. Remember that L is defined as $L = T - U$ so, by applying the generalized Euler-Lagrange equations

$$\frac{d}{dt}\frac{\partial L}{\partial \dot{q}_k} - \frac{\partial L}{\partial q_k} = Q_k \qquad (3)$$

for each generalized coordinate $q_k$ of the system, with $Q_k$ representing the exter-nal force/torque eventually insisting on it, it yields to

$$\ddot{r} - r\dot{\theta}^2 - \frac{k_e}{m}(r-a) = g[\xi(\theta) - \cos\theta]$$

$$\frac{d}{dt}(r^2\dot{\theta}) = gr[\sin\theta + \eta(\theta)] \qquad (4)$$

On the sagittal plane the center of mass covers a trajectory which can be described as a function of the *length* $r$ of the *stance leg* with the respect of *postural angle* $\theta$[3]. Now, Eq. (4) suggest us to apply the torque $\mathbf{K} = -mgr\sin\theta$ to the ankle so that the momentum for unity of mass is conserved.

---

[2] Squatting position.
[3] Again referred to the stance leg.

## 2.2 Energy-Based Motion

A more useful property for bipedal walking can be derived if we consider explicitly the role of energy. From the Euler-Lagrange formulation, by taking the total time derivative of the *Lagrangian L* and by substituting the terms as it is suggested by Eq. (3), it yields to

$$\frac{d}{dt}\left(\sum_k \frac{\partial L}{\partial \dot{q}_k}\dot{q}_k - L\right) = \sum_k Q_k \dot{q}_k \tag{5}$$

But the term under time derivative on the left side is the *Energy E* of the mechanical system and it is easily computed by summing up its kinetics and potential components[4] which in our case becomes

$$E = T + U_{spring} + U_{gravity} \tag{6}$$

The preceding Eq. (5) states that if we consider only internal forces and torques, the mechanical system is conservative. On the contrary, if we apply external forces and torques the system becomes dissipative. For example, if we assume that the applied force $F$ to the center of mass and the torque $K$ at the ankle take the form (1)

$$\mathbf{F} = mg\cos\theta \qquad K = -mgr\sin\theta \tag{7}$$

Equation (5) instantiates to

$$\frac{dE}{dt} = mg\dot{r}\cos\theta - mgr\sin\theta\dot{\theta} = \frac{dU_{gravity}}{dt}$$

which characterizes the bipedal gait as a purely mechanical motion through a conservative quantity derived from the mechanical energy. Let us term it the *balanced energy $E_{bal}$* which supplies the oscillatory motion of the stance leg. To compute its value we assume the humanoid to walk in a wadding fashion where each leg extends up to the length $l$, with $l > a$, which is the humanoid *standing position*. Since the *radial velocity* $\dot{r}$ at the point $r = l$ becomes null and the *total Momentum $M_0$* of the system is conserved, we shall write

$$E_{bal} = \frac{1}{2}k_e a^2 + \mu^2 k_e b^2 \tag{8}$$

where $b$ denotes the leg length such that $M_0 = m\omega_2 b^2$ with $\omega_2^2 = \frac{k_e}{m}$ the proper frequence of the spring-based behavior of the leg. The previous equation can be easily obtained converting the lagrangian (2) into the total energy of the leg by dropping out the gravity potential and changing the sign of the spring potential in according to Eq. (6) and using the parameters defined in the next sections. Thus, the parameter $\mu$ can be interpreted as a weight factor to accomodate the right quantity of energy to keep the gait up. By combining the above equation

---

[4] As it comes immediately from the definition.

with Eq. (6) and remembering that $E_{bal}$ is the total energy where the gravity potential has been dropped out, it yields to

$$r^2\dot{\theta} = w_2 b^2$$
$$\frac{m}{2}(\dot{r}^2 + r^2\dot{\theta}^2) + \frac{1}{2}k_e(r-a)^2 = \frac{1}{2}k_2^2 b^2 \overline{y}^2 + \frac{1}{2}k_e(l-a)^2 \tag{9}$$

by explicitly considering the conservation of the momentum for unit mass.

## 2.3    Balanced Mode

The only way to get an analytical solution to the problem of the dynamics of the center of mass of our walking biped comes from an explicit accounting of the energy balance given by Eq. (8). Thus, after some manipulations, we obtain the following relations for the polar components of the velocity of the center of mass

$$\frac{\dot{r}}{r} = w_2\sqrt{2\mu^2 y^2 + 2\lambda y - y^4 - 1}$$
$$\dot{\theta} = w_2 b^2 \frac{1}{r^2} \tag{10}$$

where the following parameters have been introduced

$$w_1^2 = \frac{g}{l} \quad \lambda = \frac{a}{b} \quad \rho^2 = \frac{al}{b^2} = \frac{\lambda}{\overline{y}} \quad y = \frac{b}{r}$$
$$w_2^2 = \frac{k_e}{m} \quad \overline{y} = \frac{b}{l} \quad \mu^2 = \frac{1+\overline{y}^4}{2\overline{y}^2} - \frac{\lambda}{\overline{y}} \quad \rho = \frac{w_1}{w_2} \tag{11}$$

and $y$ is the new variable introduced for convenience. Its dynamics is due to the superposition of the inverted pendulum oscillation, with proper frequence $w_1$, and the oscillation shown by the spring, with proper frequence $w_2$, so that $\rho$ is the parameter which makes explicit such a combination. It also affects the quantity $\mu$ which represents the amount of energy we need to supply to the leg besides that it necessary to load the *spring* which, in our model, provides the dynamical gait, as it appears from Eq. (8).

The motion of the center of mass results from Eq. (10); its trajectory can be reduced into a parametric form by the help of the elliptic functions. This circumstance is due to the quartic polynomial appearing under square root which shapes the length adjustment of the stance leg during a gait cycle. For more details you can consult the Abramowitz's handbook [1] which is also accessible via web [21]. For a good compendium of the main properties of this class of functions you can see also the reach bibliography on this topic, such as [2,16], just to name a few among the many (Fig. 4).

As it sketched in Fig. 3, the model outlined henceforth assumes that the roots of the quartic polynomial be all reals; let's term them $y_1$, $y_2$, $y_3$ and $y_4$. Now, because the coefficient of third degree is null, we can conclude that their sum is null, namely, $y_1 + y_2 + y_3 + y_4 = 0$ so that, by picking up two pairs at the time,

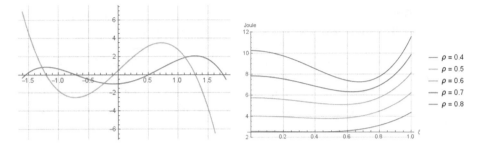

**Fig. 3.** Length adjustment of the stance leg during a gait cycle plotted with its derivative (red line)

**Fig. 4.** Course of Balanced Energy for different values of the $\rho$ parameter

we can find three real quantities $\sigma_1$, $\sigma_2$ and $\sigma_3$ such that the following relations hold

$$y_1 = \sigma_1 + \sigma_2 + \sigma_3 \quad y_2 = \sigma_1 - \sigma_2 - \sigma_3$$
$$y_3 = \sigma_2 - \sigma_1 - \sigma_3 \quad y_4 = \sigma_3 - \sigma_1 - \sigma_2 \tag{12}$$

and the quartic can be separated into the products of two 2nd order polynomials

$$\dot{r} = \omega_2 \frac{b}{y} \sqrt{[(\sigma_2 + \sigma_3)^2 - (y - \sigma_1)^2][(y + \sigma_1)^2 - (\sigma_2 - \sigma_3)^2]}$$

provided that the following identities are satisfied

$$\mu^2 = \sigma_1^2 + \sigma_2^2 + \sigma_3^2$$
$$\lambda = 4\sigma_1\sigma_2\sigma_3 \tag{13}$$
$$\sigma^3 - \bar{y}\sigma^2 + \tfrac{1}{2}(\bar{y}^2 - \mu^2)\sigma - \tfrac{\lambda}{4} = 0$$

with the unknown $\sigma$ indentifying one of $\{\sigma_1, \sigma_2, \sigma_3\}$ and $\bar{y}$ is one of the roots of the quartic polynomial. However, the analytic solution of this problem is not a simple task due to the specific variable substitution to be performed. If we take the inverse $y = \frac{b}{r}$ of the distance of the center of mass from the foot of the stance leg to have the form

$$y = \sigma_1 - (\sigma_2 + \sigma_3)\frac{1 + nsn^2 iz}{1 - nsn^2 iz} = y_1 - \frac{y_1 - y_2}{1 - nsn^2 iz} \tag{14}$$

the former of Eq. (10) simplifies to

$$\dot{r} = -\omega_2 \sqrt{\sigma_1^2 - \sigma_2^2} \frac{b}{y} \frac{dy}{dz} \tag{15}$$

which, combined with the latter of Eq. (10), provides the course of *postural values* $\theta$

$$d\theta = \frac{y_1}{\sqrt{\sigma_1^2 - \sigma_2^2}} dz - \frac{y_1 - y_2}{\sqrt{\sigma_1^2 - \sigma_2^2}} \frac{dz}{1 - nsn^2 iz} \tag{16}$$

whose shape comes by integration of the preceding differential form with the help of the incomplete elliptic integrals of first and third kind. As you can see the suggested variable substitution explicitly involves the *elliptic sinus sn* whose argument is purely immaginary. To understand how to compute this function and many other properties, please refer to the formulas explained in Chap. 22 of the already cited book [1]. The elliptic sinus also depends on the so called *modulus* **k** and, moreover, Eq. (14) includes the *elliptic characteristic* **n**. Their values are:

$$n = \frac{\sigma_1 - \sigma_3}{\sigma_1 + \sigma_2} \qquad k = \sqrt{\frac{\sigma_1^2 - \sigma_3^2}{\sigma_1^2 - \sigma_2^2}} \tag{17}$$

## 3   Gait Pattern

The course of postural values, provided by the stance leg during a gait cycle, results from the integration of Eq. (16) and it yields to

$$\theta = \frac{y_4 z + (y_2 - y_4)\Pi(n', k', z)}{\sqrt{\sigma_1^2 - \sigma_2^2}} \tag{18}$$

where we have also introduced the *complementary module* **k'** and the *complementary characteristic n'*, respectively defined by $k' = \sqrt{1 - k^2}$ and $n' = 1 - n$. Moreover, it appears the symbol $\Pi$ which notationally represents the incomplete elliptic integral of third kind. To get the correct form of the second term you need to perform appropriate substitutions on the original formula given by Eq. (16).

For sake of clarity, we consider now an example of bipedal walking. Let us suppose we have an humanoid having its center of mass *80* cm *height*, a *weight of 80* Kg and we choose the parameters $\mu = 1.25$, $\rho = 0.684$ and $\bar{y} = 0.5$. The bipedal model instantiates to

$$
\begin{array}{llll}
\omega_1 = 3.5\,\text{Hz} & k_e = 47\,\text{N/m} & a = 10\,\text{cm} \\
\omega_2 = 2.4\,\text{Hz} & P = 80\,\text{Kg} & b = 41\,\text{cm} \\
\rho = 0.684 & E_{bal} = 11\,\text{J} & l = 80\,\text{cm} \\
\lambda = 0.24
\end{array}
$$

where the computations of the $\sigma$-parameters give rise to the values reported below

$$\sigma_1 = 1.13206 \quad \sigma_2 = 0.103079 \quad \sigma_3 = 0.51509 \tag{19}$$

and the course of the postural values $\theta$ with the respect to the parameter $z$, argument of the elliptic sinus, is sketched in Fig. 6. As you can see, for the range of applicability of the proposed model, the *postural angle* $\theta$ given by Eq. (18), takes the form of a cubic parabola

$$\theta = n(\theta_1 - \theta_2)\frac{z^3}{3} + \theta_2 z \qquad \theta_1 = \frac{y_1}{\sqrt{\sigma_1^2 - \sigma_2^2}} \qquad \theta_2 = \frac{y_2}{\sqrt{\sigma_1^2 - \sigma_2^2}} \tag{20}$$

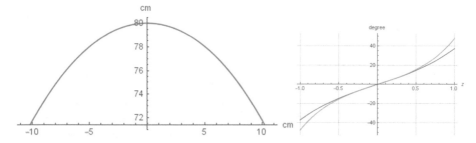

**Fig. 5.** Motion of the center of mass on the stance leg during a gait cycle

**Fig. 6.** Course of postural values for a walking biped during a gait cycle

where we have made explicit the dependence of the constants $\theta_1$ and $\theta_2$ on the set up parameters $\sigma_i$. With the help of Eq. (14) we can obtain an analitical standard form for the trajectory of the center of mass where the ordinary circular functions are substituted for the corresponding elliptical ones. After a lot of manipulations it yields to

$$r = b\,\frac{1 - n'\,sn^2(k',z)}{y_2 - n'y_4 sn^2(k',z)} \tag{21}$$

and the motion of the center of mass appears to be very similar to an arc of ellipse; moreover, when the real values $y_1$ and $y_2$ collapse, the arc of ellipse degenerates into an arc of circumpherence as one might expect. Thus, instead of using the elliptic functions we can approximate the trajectory covering of the center of mass with an ellipse that, in the sagittal plain XZ, takes the normal form

$$\frac{x^2}{c^2} + \frac{z^2}{l^2} = 1 \tag{22}$$

with the parameters $l$ and $c$, the latter chosen in such a way $c = lk' = l\sqrt{1-k^2}$ where $k$ is the modulus of the previous defined elliptic functions (Fig. 5).

## 3.1   Motivations for an Elliptic Model

The trigonometric approximation of the bipedal gait works very well if we want to trace a qualitative behavior of the motion of its center of mass. Nevertheless, for a quantitative analysis of those properties the key point is the elliptic approach where the relations among the involved parameters play their exact role. For example, the preceding discussion has stressed that the $\sigma$-parameters can characterize completely the behavior of the bipedal robot in its sagittal plain but we know those parameters are not indipendent, because their values come from the $\lambda$ and $\mu$ parameters.

Well, it can be shown that the model provides this feature because of the validity of the relations

$$\sigma_2 = \sigma_1 \, cn \, 2\alpha \qquad \sigma_3 = \sigma_1 \, dn \, 2\alpha \qquad sn \, \alpha = \sqrt{\frac{\sigma_1 - \sigma_2}{\sigma_1 + \sigma_3}}$$

where $\alpha$ is a quantity computed by inversion of the elliptic sinus, accordingly to the last equation. Remember that the elliptic functions also depend on the *modulus* **k** according to Eq. (17), whose different values result in different gait behaviors. For instance, if we go back to the previously cited example of walking humanoid, we find that

$$sn \, \alpha = 0.790382 \qquad k = 0.894205 \quad n = 0.499515$$

which yields to $\alpha = 1.02686$ and analogous risults come from a circular permutations of indexes. But this is not the only property for which this gait model has been chosen. In fact, all the parameters characterizing the bipedal walking can be easily expressed in term of elliptic functions. Hence, the gait pattern exhibits the character of cyclic motion that, accordingly to the suggestion of Siegwart in [22], is obtained through *elliptical rolling*: the *postural angle* $\theta$ and the distance $r$ of the center of mass from the *standing ankle* are linked together by the course of values of the argument of the elliptic sinus, running on the elliptic arc representing the trajectory of the center of mass.

## 3.2   Maintaining the Gait Cycle

The outlined *elliptic model* represents a good compromise between the behavior of a rigid multi-link leg and an arrangement performing in a spring fashion as it can be observed in the biological world. Here, only the knee and ankle joints should be explicitly dealt with whereas the hip joints and the articulation of the foot are implicitally referred by the features of its behavioral components.

While the bipedal robot moves forward, its center of mass covers a trajectory on the sagittal plane as a pure mechanical motion where the *balanced energy* doesn't change. However, this is not really true, because the total energy $E_{tot} = T + U_{spring} + U_{gravity}$ doesn't remain constant due to the already proved property, reported below,

$$\frac{dE_{tot}}{dt} = \frac{dU_{gravity}}{dt} \tag{23}$$

namely, during a *gait cycle* the total energy is dissipated so that it is required to be supplied in a *rolling fashion*, accordingly to Siegwart [22] that understands the bipedal gait just as a rolling[5] and the boost to maintain the motion is fed by the gravity force which triggers both the horizontal force and the torque at the ankle to avoid harmful falls down.

---

[5] The nature not invented the wheel but the walking biped is very similar to a rolling.

The pecularity of this bipedal walking is based on the non linear combination of two oscillations, one due to the inverted pendulum and the other provided by the spring feature of the leg; the *frequence ratio* $\rho$ and *spring stiffness* $k_e$ are the most meaningful parameters of the model and the former can be chosen so that $E_{bal}$ is minimized.

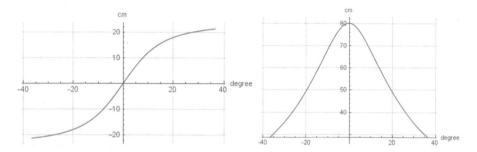

**Fig. 7.** Moving forward of the center of mass while the stance leg changes its inclination angle

**Fig. 8.** The position of the center of mass referred to the ground as the stance leg changes its inclination angle

Another interesting point of this model is that it considers the foot to be articulated in two parts, the rear one partecipating to the motion of the leg in a multi link fashion and the front part which is mainly responsible of the job to support the robot and also to control the swinging gait due to the action of the spring (Figs. 7 and 8).

## 4    Conclusions

In this paper we have presented a mechanical-based model of walking robot with the aim to investigate the role of some mechanical quantities to maintain the gait pattern. The problem has been fully analyzed from the mechanical point of view using the lagrangian approach to an extension of the inverted pendulum model where a rigid rod is substituted for a spring.

In this framework we have considered explicitly the role of the momentum whose component orthogonal to the sagittal plane (the only component) is conserved. To this aim we have assumed that an external force and an external torque are applied to the bipedal robot, the former acting on the center of mass to partially compensate the gravity force and the latter applied to the ankle in order to control the resulting momentum.

The apparent limitation of this simplified model is negligible because we were mainly interested on the gravity component which makes the motion potentially unstable but that a biped is able to turn to its own advantage. In the standard model of walking the inverted pendulum is supposed to counterbalance the gravity force by a fictional force applied along the leg upwards and whose horizontal component, parallel to the traveled path, provides the necessary forwarding boost. In the paper, on the contrary, we have supposed both a force and a torque balancing simultaneously the gravity effect.

The aim is to control the motion only through the mechanical properties of the spring-based leg so, because the gait should be generally considered to occur on an uneven surface, we have explicitly supposed a leg model based on a spring to be loaded/unloaded depending on the surface state. This is the reason why we have also considered a torque insisting on the ankle that, in this model, works actively to balance the biped against the gravity.

Thus, in future works, we want to explore these properties by refining the model and trying to implement it in more realistic simulations and even real bipeds besides the simple ones made until now as we only have simple small humanoids which don't allow the foot to be articulated.

**Acknowledgement.** This work was partially supported by a collaboration with the Intelligent Autonomous Systems Laboratory (IAS-Lab) of the University of Padua through a Grant of Consorzio Ethics, Abano Terme, Italy, for a three-years project (2014–2016) on a *Study on experimenting Exoskeletons in Medical Institutions*: we thank Enrico Pagello and Roberto Bortoletto et al. of IAS Lab for their valuable suggestions and discussions.

# References

1. Abramowitz, M., Stegun, I.A.: Handbook of Mathematical Functions, Applied Mathematics, vol. 55. National Bureau of Standards, Washington, D.C. (1972)
2. Akhiezer, N.I.: Elements of the Theory of Elliptic Functions. American Mathematical Society, Providence, Rhode Island (1990)
3. Alcaraz-Jiménez, J., Herrero-Pérez, D., Martínez-Barberá, H.: Robust feedback control of zmp-based gait for the humanoid robot nao. Int. J. Rob. Res. **32**(9–10), 1074–1088 (2013)
4. Arakawa, T., Fukuda, T.: Natural motion trajectory generation of biped locomotion robot using genetic algorithm through energy optimization. In: IEEE International Conference on Systems, Man and Cybernetics, p. 14951500 (1996)
5. Cardenas-Maciela, S.L., Castillo, O., Aguilar, L.T.: Generation of walking periodic motions for a biped robot via genetic algorithms. Appl. Soft Comput. **11**(8), 5306–5314 (2011)
6. Cheng, M., Lin, C.: Genetic algorithm for control design of biped locomotion. In: IEEE International Conference on Robotics and Automation, pp. 1315–1320. Nagoya (J), 21–27 May 1995
7. Fujimoto, Y., Kawamura, A.: Simulation of an autonomous biped walking robot including environmental force interaction. IEEE Robot. Autom. Mag. **5**(2), 33–42 (1998)
8. Furusho, J., Akihito, S., Masamichi, S., Eichi, K.: Realization of bounce gait in a quadruped robot with articular-joint-type legs. In: IEEE International Conference on Robotics and Automation, pp. 697–702. Nagoya (J), 21–27 May 1995
9. Furusho, J., Masubuchi, M.: Control of a dynamical biped locomotion system for steady walking. Dyn. Syst. Meas. Control **108**, 111–118 (1986)

10. Hirai, K., Hirose, M., Haikawa, Y., Takenaka, T.: The development of the Honda humanoid robot. In: IEEE International Conference on Robotics and Automation (1998)
11. Huang, Q., Yokoi, K., Kajita, S., Kaneko, K., Arai, H., Koyachi, N., Tanie, K.: Planning walking patterns for a biped robot. IEEE Trans. Robot. Autom. **17**(3), 280–289 (2001)
12. Ijspeert, A.J.: Central pattern generators for locomotion control in animals and robots: a review. Neural Networks **21**(4), 642–653 (2008)
13. Kajita, S., Kanehiro, F., Kaneko, K., Yokoi, K., Hirukawa, H.: The 3D linear inverted pendulum mode: a simple modeling for a biped walking pattern generation. In: IEEE/RSJ International Conference on Intelligent Robots and Systems, Maui (Hawaii), USA, pp. 239–246, 29 October–03 November 2001
14. Kato, I., Tsuiki, H.: The hydraulically powered biped walking machine with a high carrying capacity. In: Fourth Symposium on External Extremities. Dubrovnik (HR) (1972)
15. Kimura, H., Fukuoka, Y.: Adaptive dynamic walking of the quadruped on irregular terrain - autonomous adaptation using neural system model. In: IEEE International Conference on Robotics and Automation, San Francisco (CA), pp. 436–443, April 2000
16. Lawden, D.F.: Elliptic Functions and Applications, Applied Mathematical Sciences, vol. 80. Springer-Verlag, New York (1989)
17. Lee, S.H., Goswami, A.: A momentum-based balance controller for humanoid robots on non-level and non-stationary ground. Auton. Robots **33**(4), 116 (2012)
18. Ogino, M., Toyama, H., Asada, M.: Stabilizing biped walking on rough terrain based on the compliance control. In: IEEE/RSJ International Conference on Intelligent Robots and Systems (IROS 2007), San Diego (CA), pp. 4047–4052 (2007)
19. Pettersson, J., Sandholt, H., Wahde, M.: A flexible evolutionary method for the generation and implementation of behaviors in humanoid robots. In: IEEE-RAS International Conference on Humanoid Robots, pp. 279–286 (2001)
20. Raibert, M.: Legged Robots that Balance. The MIT Press, Cambridge (1986)
21. Reinhardt, W.P., Walker, P.L.: Jacobian Elliptic Functions, Digital Library of Mathematical Functions, vol. 22. NISTDigital Library of Mathematical Functions (2015). http://dlmf.nist.gov/22
22. Siegwart, R., Nourbakhsh, I.R.: Introduction to Autonomous Mobile Robots. The MIT Press, Cambridge (2004)
23. Vukobratović, M., Juricic, D.: Contribution to the synthesis of biped gait. IEEE Biomed. Eng. **16**(1), 1–6 (1969)
24. Walker, M., Orin, D.: Efficient dynamic computer simulation of robotic mechanisms. Dyn. Syst. Meas. Control **104**, 205–211 (1982)
25. Wieber, P.B.: Trajectory free linear model predictive control for stable walking in the presence of strong perturbations. In: 6th IEEE-RAS International Conference on Humanoid Robots, p. 137142 (2006)
26. Yi, S.J., Zhang, B.T., Hong, D., Lee, D.D.: Practical bipedal walking control on uneven terrain using surface learning and push recovery. In: IEEE/RSJ International Conference on Intelligent Robots and Systems, San Francisco (CA), pp. 3963–3968 (2011)
27. Yu, J., Tan, M., Chen, J., Zhang, J.: A survey on CPG-inspired control models and system implementation. IEEE Trans. Neural Netw. Learn. Syst. **25**(3), 441–56 (2014)

# Modular Robot that Modeled Cell Membrane Dynamics of a Cellular Slime Mold

Ryusuke Fuse, Masahiro Shimizu$^{(\boxtimes)}$, Shuhei Ikemoto, and Koh Hosoda

Osaka University, Suita, Japan
shimizu@sys.es.osaka-u.ac.jp

**Abstract.** Understanding of the design principles for implementing adaptive functions with respect to engineering currently remains stalled in the conceptual level. However, living organisms exhibit great adaptive function by skillfully relating shape and function in a spatio-temporal manner. In this study, we focus on amoeboid organisms because these organisms have a variable morphology that relates shape and function. Amoeboid organisms in the natural world (i.e., cellular slime molds) locomote through changing the cell membrane shape by inducing the internal protoplasmic streaming. Based on this mechanism, we developed modular robots that modeled the cell membrane dynamics of a cellar slime mold.

## 1 Introduction

In nature, we can see various phenomena in which many identical elements form some macroscopic structures by selfassembly. In the growth process of a crystal, for example, many atoms or molecules spontaneously organize regular lattice structure. Living organisms also show self-assembly process, where they grow by repeated cell division, form certain structures and develop their functions. These phenomena are very attractive not only from a scientific viewpoint but also from an engineering viewpoint. In general, a complex artifact is hard to manufacture. Picking up and assembling parts will not be a realistic process for constructing highly complex artifacts.

In this sense, it is important to consider how we can realize self-reconfigurable robots. Murata et al. [1] proposed a modular robotic system M-TRAN, which is a homogenous robot system and transforms to various forms such as snake-like, four-legged, and so on. Miyashita et al. [2] developed special floating pucks which organize structures utilizing the difference of surface tension. Origami robot is considered as a self-reconfiguring system based on the robot has foldable thin-plate-like shape [3]. Zhakypov *et al.* developed Tribot which is an origami robot designed with CAD [4]. This robot achieved jumping and crawl locomotions. Soft robots, that have bodies consisting of mechanically viscoelastic elements, recently are attracting a lot of attention [5]. Such a system is expected to adapt to the unstructured environment without complicated control systems.

© Springer Nature Switzerland AG 2019
M. Strand et al. (Eds.): IAS 2018, AISC 867, pp. 302–313, 2019.
https://doi.org/10.1007/978-3-030-01370-7_24

Umedachi *et al.* developed Softworm [6], which is a worm-shaped robot driven by SMA inside the body. The Softworm can generate various type of locomotions by exploiting flexibility of the body. These traditional self-reconfigurable robots, however, doesn't have self-reconfigurability and body softness, simultaneously.

On the other hand, in nature, we focus on amoeboid organisms (a cell of Cellular Slime Molds, physarum polycephalum, etc.) [7] because these organisms have a variable morphology that relates shape and function in the most simple and primitive manner. Amoeboid organisms in the natural world locomote through changing shape by inducing the internal flow of their protoplasm (protoplasmic streaming). It should be noted that the control of locomotion in slime molds is realized in an autonomous decentralized manner without being mediated by a central nervous system. Therefore, in this study, the locomotion of amoeboid organisms is employed as a practical example that enables adaptive locomotion with self-reconfiguration in autonomous decentralized manner. That adaptive function (i.e., amoeboid locomotion) could be modeled by exploiting a modular robot [8,9] that is capable of changing shape and locomotion by altering the relative positional relationship between mechanical modules according to environmental changes. Based on above consideration, in this study, we develop modular robots that modeled the self-reconfiguration mechanism of a cell of a cellular slime mold.

The main contribution of this paper is a development of the control scheme of self-reconfiguration based on cell membrane dynamics of a cellular slime mold. This paper is organized as follows: In Sects. 2, 3 and 4, the design strategies of the proposed model are explained; The experimental results are presented in Sect. 5; And, the paper is finalized with conclusions and the future works.

## 2 Self-reconfigurable Cellular Slime Mold

### 2.1 Cellular Slime Mold Deformation by Cell Membrane Dynamics

In nature, Amoeboid organisms world locomote through changing shape by inducing the internal flow of their protoplasm in an autonomous decentralized manner without being mediated by a central nervous system. So far, some researchers developed modular robots based on true slime molds, which is a kind of amoeboid organisms. We developed Slimebot [10] which modeled a true slime mold as a coupled oscillator system consisting of VDP oscillators. Umedachi *et al.* developed a true slime mold robot Slimy [11], which includes fluid circuit satisfies conservation of robot body volume against change of the entire shape.

Especially, a cell of a cellular slime mold exhibits intelligent behavior based on cell membrane dynamics of a cellular slime mold [7]. The mature cellular slime molds make a swarm state consisting of multi-cells. However, a cell of a cellular slime mold could be considered as a self-reconfigurable and motile living thing, which has the outer surface of the gel and inner protoplasm of sol. And, a cell of a cellular slime mold has a simpler inner structure of its body compared with a locomoting true slime mold. Therefore, we focused on such the great knowledge to embed self-reconfiguration function into modular robots system. A cell of a cellular slime mold elongates pseudopod to locomote. This elongation makes local deformation of its body, then, a cell of a cellular slime mold exhibits escape from enemies and/or exploring the environment. This mechanism is known as a result of chemical reaction inside the cell membrane. More specifically, polymerization/depolymerization of cell cytoskeleton inside the cell membrane governs the elongation of pseudopod [12]. The polymerization/depolymerization progress as autocatalytic reaction with other chemicals [7]. The cell membrane dynamics for self-reconfiguration of a cell of a cellular slime mold occurs based on above the chemical reaction system. Based on the knowledge, at first we modeled chemical reaction system of polymerization/depolymerization of cell cytoskeleton, then we connected the model with physical robotic body, finally we developed modular robots that modeled cell membrane dynamics of a cellular slime mold.

## 2.2   Cell Membrane Deformation by Pseudopod Elongation

Locomotion of a cell of a cellular slime mold occurs as chemotaxis for which it is necessary to detect concentration gradient of the chemical [13]. The phosphatidylinositol lipids signaling system is a trigger chemical reaction for such a chemotaxis [14,15]. This chemical reaction occurs at the micro-region inside cell membrane where the higher gradient of attractive chemical is detected. In such a micro-region of a cell of a cellular slime mold, phosphatidylinositol trisphosphate (PIP$_3$) is much produced [16]. The cell cytoskeleton polymerization is induced at the micro-region where PIP$_3$ is produced [17]. Pseudopod elongation occurs based on the cell cytoskeleton polymerization.

Based on the phosphatidylinositol lipids signaling system, we can model deformation mechanism by pseudopod elongation of a cell of a cellular slime mold. Here, we assume a micro-region of the cell membrane as a modular robot so that modular robots shape whole cell membrane of a cell. Therefore, we can consider cellular-slime-mold-like modular robots so that the cell membrane deformation (*i.e.*, pseudopod elongation) is controlled in a decentralized control manner. The movement of each modular robot, which determines how the cell membrane deform, is governed by the chemical reaction which describes PIP$_3$ increases and decreases. In order to embed the mechanism of a cellular slime mold, we simplify the cell shape into a 2-dimensional model (Fig. 1). Then, we divide the 2-dimensional cell membrane into discrete micro-region, and we also discretize protoplasm. Finally, the 2-dimensional discrete micro-region should be a modular robot. In what follows we consider a 2-dimensional space and discrete micro-region so that we can easily assume the modular robot system.

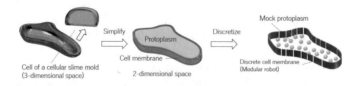

**Fig. 1.** Simplification of the cell shape into 2-dimensional model

# 3   Chemical Reaction Model for Deformation

## 3.1   The Phosphatidylinositol Lipids Signaling System

In the phosphatidylinositol lipids signaling system, lipidsPIP$_2$, PIP$_3$ are reversibly altered. A protein PI3K phosphorylate PIP$_2$, then works as a catalytic agent for PIP$_3$ production. This reaction works at the micro-region where a high concentration of an attractive chemical is detected. A protein PTEN phosphorylate PIP$_3$, then works as a catalytic agent for PIP$_2$ production. This reaction works at the micro-region where a low concentration of an attractive chemical is detected. These reactions are enzymatic catalytic reactions according to Michaelis-Menten equation. Therefore, the reaction rate at each micro-region is depending on the concentration of substrate (*i.e.*, PIP$_2$, and PIP$_3$). PIP$_2$ and PIP$_3$ also change the concentrations depending on the diffusion reaction of neighboring micro-region. As the result of above chemical reactions, PIP$_3$ is much produced at the micro-region inside the cell membrane, then polymerization of cell cytoskeleton is induced. After that, pseudopod elongation occurs, then chemotaxis could be realized.

In this study, we intend to deal with this chemical reaction model as control manner for each modular robot (*i.e.*, discrete micro-region inside the cell membrane). Shibata *et al.* modeled above chemical reactions as reaction diffusion systems consisting of three variables PIP$_3$, PIP$_2$, PTEN [14], followed by, added the effect of attractive chemicals, then chemotaxis was achieved [15]. The reaction diffusion model for each modular robot based on the model of Shibata *et al.* is indicated as follows:

$$\frac{\partial [\text{PIP}_3]}{\partial t} = -R_{\text{PTEN}} + A(\theta)R_{\text{PI3K}} - \lambda_{\text{PIP}_3}[\text{PIP}_3], \tag{1}$$

$$\frac{\partial [\text{PIP}_2]}{\partial t} = R_{\text{PTEN}} - A(\theta)R_{\text{PI3K}} + k_{\text{PIP}_2} - \lambda_{\text{PIP}_2}[\text{PIP}_2], \tag{2}$$

$$\frac{\partial [\text{PTEN}]}{\partial t} = k_{\text{PTEN}}[\text{PTEN}]_{\text{cyto}} \frac{K_{\text{PIP}_3} + \alpha[\text{PIP}_3]}{K_{\text{PIP}_3} + [\text{PIP}_3]} \frac{[\text{PIP}_2]}{K_{\text{PIP}_2} + [\text{PIP}_2]} - \lambda_{\text{PTEN}}[\text{PTEN}], \tag{3}$$

$$R_{\text{PTEN}} = V_{\text{PTEN}}[\text{PTEN}] \frac{[\text{PIP}_3]}{K_{\text{PTEN}} + [\text{PIP}_3]}, \tag{4}$$

$$R_{\text{PI3K}} = V_{\text{PI3K}} \frac{[\text{PIP}_2]}{K_{\text{PIP}_2} + [\text{PIP}_2]}, \tag{5}$$

$$A(\theta) = \frac{\Delta}{100} \frac{1}{2} \cos\theta. \tag{6}$$

Where, Eqs. (1)–(6) indicate equations at a discrete micro-region inside the cell membrane. And, the constants are listed in Table 1. Equation (1) shows time development of concentration of lipid PIP$_3$. In Eq. (1), the first term indicates that PIP$_3$ decreases, and PIP$_2$ increases, by PTEN; the second term indicates that PIP$_2$ decreases, and PIP$_3$ increases, by PI3K; the third term indicates that PIP$_3$ decreases by other chemical reactions. Equation (2) shows time development of concentration of lipid PIP$_2$. In Eq. (2), the first term indicates that

$PIP_3$ decreases, and $PIP_2$ increases, by PTEN; the second term indicates that $PIP_2$ decreases, and $PIP_3$ increases, by PI3K; the third term indicates that $PIP_2$ increases by other chemical reactions; the fourth term indicates that $PIP_2$ decreases by other chemical reactions. Equation (3) shows time development of concentration of protein PTEN. In Eq. (3), the first term indicates that PTEN inside the protoplasm transfer to the cell membrane; the second term indicates that PTEN inside the cell membrane transfer to protoplasm. Equation (6) shows activity against attractive chemical, which has $\Delta\%$ concentration, and the attractive chemical comes from the angle of $\theta$.

**Table 1.** Explanation of constants

| Constants | Mean |
|-----------|------|
| $k_{PIP_2}$ | Reaction coefficient of $PIP_2$ |
| $K_{PIP_3}$ | Negative feedback constant from $PIP_3$ to PTEN |
| $K_{PIP_2}$ | Positive feedback constant from $PIP_2$ to PTEN |
| $\lambda_{PTEN}$ | Dissociation constant of PTEN |
| $\lambda_{PIP_3}$ | Dissociation constant of $PIP_3$, which is not depending on PTEN |
| $\lambda_{PIP_2}$ | Dissociation constant of $PIP_2$, which is not depending on PI3K |
| $K_{PTEN}$ | Michaelis-Menten constant of PTEN dephosphorylation |
| $V_{PTEN}$ | Dephosphorylation rate of PTEN |
| $V_{PI3K}$ | Dephosphorylation rate of PI3K |
| $[X]_{total}$ | Total amount of X in the cell |
| $[X]_{cyto}$ | Total amount of X in the protoplasm |

In the chemical reaction inside the cell membrane, diffusion occurs between neighboring discrete micro-region (*i.e.*, neighboring modular robots). This is described as short distance interaction. Equations (7)–(9) indicate diffusion at time $t$:

$$\varepsilon^{PIP_3} = \frac{D}{\sqrt{3}} \frac{\Delta t}{\Delta x^2} \{[PIP_3]_{prev} - 2[PIP_3] + [PIP_3]_{next}\}, \tag{7}$$

$$\varepsilon^{PIP_2} = \frac{D}{\sqrt{3}} \frac{\Delta t}{\Delta x^2} \{[PIP_2]_{prev} - 2[PIP_2] + [PIP_2]_{next}\}, \tag{8}$$

$$\varepsilon^{PTEN} = \frac{D}{\sqrt{3}} \frac{\Delta t}{\Delta x^2} \{[PTEN]_{prev} - 2[PTEN] + [PTEN]_{next}\}. \tag{9}$$

Where, $[X]_{prev}$ and $[X]_{next}$ mean neighboring variables. $\varepsilon^X$ means the value after diffusion calculation, then this value will be the base value at time $t + 1$ (*i.e.*, the value before reaction diffusion at time $t + 1$).

Additionally, Protein PTEN satisfies law of conservation of mass. This is described as long distance interaction as follows:

$$[PTEN]_{cyto} = [PTEN]_{total} - \sum [PTEN].$$    (10)

## 3.2 Cell Cytoskeleton Polymerization Model

Cell cytoskeleton is a kind of protein which makes a mechanically stiff string like structure in the cell. This is built by cell cytoskeleton polymerization. The cell cytoskeleton produces the force in order to generate pseudopod elongation. The polymerization/depolymerization progress as autocatalytic reaction with other chemicals [7]. The cell membrane dynamics for self-reconfiguration of a cell of a cellular slime mold occurs based on the chemical reaction system explained above. We connect the model with the physical robotic body. Traditional studies show the cell cytoskeleton polymerization occurs much where $PIP_3$ increases inside the cell membrane [17]. Additionally, Nishimura $et$ $al.$ indicated that the cell cytoskeleton polymerization increases autocatalytically in simulations [7]. We determine the amount of cell cytoskeleton polymerization as the output of the sigmoid function of $PIP_3$ as follows:

$$F_{cytoskeleton} = \frac{1}{1 + e^{-f([PIP_3])}},$$    (11)

$$f([PIP_3]) = k_{cytoskeleton}[PIP_3].$$    (12)

Where, $k_{cytoskeleton}$ is a constant. $F_{cytoskeleton}$ indicates the force between modular robots ($i.e.$, discrete micro-region of the 2-dimensional cell membrane). This force outputs are assigned as follows: extension ($0.9 \leq F_{cytoskeleton}$), natural length ($0.7 \leq F_{cytoskeleton} < 0.9$), and contraction ($F_{cytoskeleton} < 0.7$).

# 4 Development of Real Physical Modular Robots

## 4.1 Overview of the Real Physical Modular Robots

Figure 2 indicates the overview of developed modular robots. The entire system is consisting of the mock protoplasm as many balls and the enclosing the cell membrane as connected modular robots. The neighboring modular robots are connected with spring and driven by SMA (Shape Memory Alloy), each of which is controlled by Raspberrypi 3. The number of mock protoplasm balls is constant so that the entire system satisfies law of conservation of mass. A modular robot is controlled based on the decentralized manner discussed in previous sections.

## 4.2 Actuation of Connected Modular Robots

The neighboring modular robots were connected, the spring and SMA were fabricated as Fig. 3. The length between neighboring modular robots was controlled by the cell cytoskeleton polymerization model. The structure composed of spring

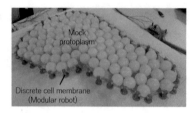

**Fig. 2.** Overview of the real physical modular robots

and SMA (BMX150, TOKI CORPORATION) produced the three states (extension, natural length, and contraction). The connector for spring and SMA, and each modular robot body were formed by 3D Printer (Object260 Connex, Stratasys). A thin teflon sheet was stuck on the bottom so that friction should be low between the modular robot and the ground. The entire system has 40 SMAs.

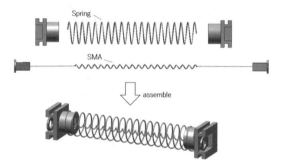

**Fig. 3.** Structure of a Spring and a SMA

### 4.3   Control System

Based on the phosphatidylinositol lipids signaling system, we modeled deformation mechanism by pseudopod elongation of a cell of a cellular slime mold. Here, we assume a micro-region of the cell membrane as a modular robot so that modular robots shape whole cell membrane of a cell. The movement of each modular robot, which determines how the cell membrane deform, is governed by the chemical reaction which describes $PIP_3$ increases and decreases. We connected the model with physical robotic body, finally we developed modular robots that modeled cell membrane dynamics of a cellular slime mold. Here, we explain the framework of the proposed entire system. In this study, the calculation of time development of $PIP_3$ and cell cytoskeleton polymerization were computed in host PC (Ubuntu). Based on the result of host PC, extension and/or contraction of SMA were controlled via GPIO ports of Raspberry pi 3. The modular robots (*i.e.*, raspberry pi 3) and host PC were organized with ROS. The cell cytoskeleton polymerization occurs much where $PIP_3$ increases inside the cell membrane. To do so, we have to produce the three states (extension, natural length, and contraction). In our modeling, this force output is assigned as Table 2. However, basically, we can only produce contractions in SMA. Therefore, we express the three states (extension, natural length, and contraction) by using two units (Fig. 4).

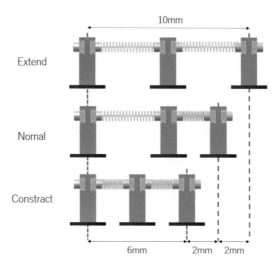

**Fig. 4.** Expression of extension, natural length, and contraction by two physical units

## 5   Verification of Self-reconfiguration

### 5.1   Experimental Setting

Based on above setting, we verified self-reconfiguration with developed modular robots. We modeled deformation mechanism by pseudopod elongation of a cell of a cellular slime mold based on the phosphatidylinositol lipids signaling system. The movement of each modular robot, which determines how the cell membrane deform, is governed by the chemical reaction which describes $PIP_3$ increases and decreases. We connected the model with physical robotic body, finally we developed modular robots that modeled cell membrane dynamics of a cellular slime mold. Here, we applied the virtual attractive chemical into certain modules (actually, here, we modified the parameter of the control system), then we observed deformation of cell membrane of modular robots with the setting of Fig. 5. After that, we applied the virtual attractive chemical again from opposite side. Through this process, we verified adaptability against environmental change.

**Fig. 5.** Experimental setting

## 5.2  Experimental Results

The experimental results were indicated in Figs. 6, 7, and 8. Here, we applied
the first virtual attractive chemical until 95 s, then we applied the second virtual attractive chemical from opposite side. As in the Fig. 6, we successfully
observed a local elongation of pseudopod (see the area around the red circle).
We observed the pseudopod-like elongation developed until 95 s. We started the
second attractive chemical from opposite side at 96 s, then we observed shrink
of pseudopod (see the figure at 105 s).

In order to understand how the phosphatidylinositol lipids signaling system
works, we analyzed time development of (a) $PIP_2$, (b) $PIP_3$, (c) amount of cell
cytoskeleton polymerization, (d) elongation of pseudopod (see Fig. 7) about the
focused modular robot (emphasized by the red circle in Fig. 6). As in the figure,
we can see $PIP_2$ increasing when $PIP_3$. On the other hand, we can also see $PIP_3$
increasing when $PIP_2$, then cell cytoskeleton polymerization and elongation of
pseudopod occurs according to $PIP_3$ increasing. After 96 s, we applied the second
virtual attractive chemical from opposite side. In this figure, we couldn't find
clear difference.

Under this circumstance, we observed the trajectory of the focused modular
robot (see Fig. 8). As in the figure, we observed the opposite side elongation of
pseudopod when we applied the second virtual attractive chemical.

**Fig. 6.** Local elongation of pseudopod

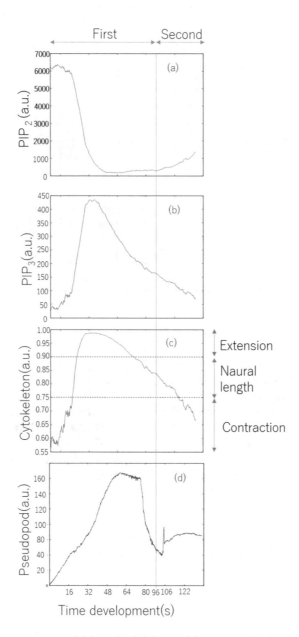

**Fig. 7.** Time development of (a) PIP$_2$, (b) PIP$_3$, (c) amount of cell cytoskeleton polymerization, (d) elongation of pseudopod.

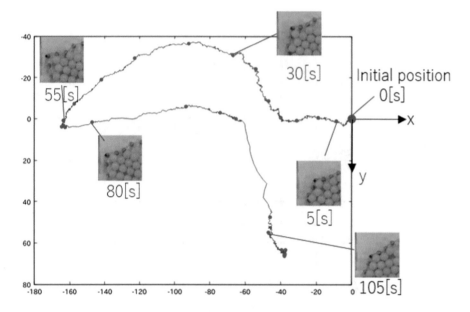

**Fig. 8.** The trajectory of the focused modular robot

# 6    Conclusion and Future Work

In this study, we developed modular robots that modeled the self-reconfiguration mechanism of a cell of a cellular slime mold. We focus on amoeboid organisms because these organisms have a variable morphology that relates shape and function. Amoeboid organisms in the natural world (i.e., cellular slime molds) locomote through changing the cell membrane shape by inducing the internal protoplasmic streaming. The main contribution of this paper is a development of the control scheme of self-reconfiguration based on cell membrane dynamics of a cellular slime mold. As future works, we are focusing on the relationship between elongation and motile velocity. We intend to consider how the pseudopod is used is related to exploring function.

**Acknowledgment.** This work was supported partially by Grant-in-Aid for Scientific Research on 15H02763, and 17K19978 from the Ministry of Education, Culture, Sports, Science and Technology of Japan.

# References

1. Murata, S., Yoshida, E., Kamimura, A., Kurokawa, H., Tomita, K., Kokaji, S.: M-tran: self-reconfigurable modular robotic system. IEEE/ASME Trans. Mechatron. **7**(4), 431–441 (2002)
2. Miyashita, S., Kessler, M., Lungarella, M.: How morphology affects self-assembly in a stochastic modular robot. In: IEEE International Conference on Robotics and Automation. ICRA 2008, pp. 3533–3538. IEEE (2008)

3. Firouzeh, A., Sun, Y., Lee, H., Paik, J.: Sensor and actuator integrated low-profile robotic origami. In: 2013 IEEE/RSJ International Conference on Intelligent Robots and Systems (IROS), pp. 4937–4944. IEEE (2013)

4. Zhakypov, Z., Falahi, M., Shah, M., Paik, J.: The design and control of the multimodal locomotion origami robot, tribot. In: 2015 IEEE/RSJ International Conference on Intelligent Robots and Systems (IROS), pp. 4349–4355. IEEE (2015)

5. Rus, D., Tolley, M.T.: Design, fabrication and control of soft robots. Nature **521**(7553), 467–475 (2015)

6. Umedachi, T., Vikas, V., Trimmer, B.A.: Softworms: the design and control of non-pneumatic, 3D-printed, deformable robots. Bioinspiration biomimetics **11**(2), 025001 (2016)

7. Nishimura, S.I., Ueda, M., Sasai, M.: Cortical factor feedback model for cellular locomotion and cytofission. PLoS Comput. Biol. **5**(3), e1000310 (2009)

8. Fukuda, T., Buss, M., Hosokai, H., Kawauchi, Y.: Cell structured robotic system cebot: control, planning and communication methods. Robot. Auton. Syst. **7**(2), 239–248 (1991)

9. Rubenstein, M., Ahler, C., Nagpal, R.: Kilobot: a low cost scalable robot system for collective behaviors. In: 2012 IEEE International Conference on Robotics and Automation (ICRA), pp. 3293–3298. IEEE (2012)

10. Shimizu, M., Ishiguro, A.: Amoeboid locomotion having high fluidity by a modular robot. IJUC **6**(2), 145–161 (2010)

11. Umedachi, T., Ito, K., Ishiguro, A.: Soft-bodied amoeba-inspired robot that switches between qualitatively different behaviors with decentralized stiffness control. Adapt. Behav. **23**(2), 97–108 (2015)

12. Pollard, T.D., Borisy, G.G.: Cellular motility driven by assembly and disassembly of actin filaments. Cell **112**(4), 453–465 (2003)

13. Bray, D.: Cell movements: from molecules to motility. Garland Science (2001)

14. Shibata, T., Nishikawa, M., Matsuoka, S., Ueda, M.: Modeling the self-organized phosphatidylinositol lipid signaling system in chemotactic cells using quantitative image analysis. J. Cell Sci. **125**(21), 5138–5150 (2012)

15. Shibata, T., Nishikawa, M., Matsuoka, S., Ueda, M.: Intracellular encoding of spatiotemporal guidance cues in a self-organizing signaling system for chemotaxis in dictyostelium cells. Biophys. J. **105**(9), 2199–2209 (2013)

16. Asano, Y., Nagasaki, A., Uyeda, T.Q.: Correlated waves of actin filaments and pip3 in dictyostelium cells. Cell Motil. Cytoskelet. **65**(12), 923 (2008)

17. Chen, C.-L., Wang, Y., Sesaki, H., Iijima, M.: Myosin i links pip3 signaling to remodeling of the actin cytoskeleton in chemotaxis. Sci. Signaling **5**(209), ra10 (2012)

# Locomotion of Hydraulic Amoeba-Like Robot Utilizing Transition of Mass Distribution

Takashi Takuma$^{(\boxtimes)}$ and Kyotaro Hamachi

Osaka Institute of Technology, Osaka 5358585, Japan
takashi.takuma@oit.ac.jp

**Abstract.** A soft robot constructed using a soft material and driven by a soft actuator is receiving increased attention because it is expected to passively change its shape upon making contact with the environment. This paper proposes novel design for a soft robot that changes not only its shape but also its mass distribution. The robot adopts liquid as a fluid, and realizes locomotion by changing the local friction that is generated by the amount of mass of the containing liquid. In order to observe the effects of liquid flow, we constructed a robot with two flexible chambers and observed that the robot succeeded in moving forward by arranging the material construction of the chamber and timing of supplying/releasing the liquid. Experimental results showed that the robot realized approximately 57 mm of locomotion per cycle. We conclude that the large deformation and movement of the mass distribution enables successful locomotion.

## 1 Introduction

A soft robot, which is supported by soft materials and driven by a soft actuator, is gathering increased attention. The robot deforms its shape or posture adaptively depending on what it comes into contact with, such as landforms [1] and grasping object [2,3]. One of the applications of the soft robot is passing through narrow space in order to enter collapsed buildings. It is generally constructed out of soft material, such as silicone rubber for a framework [4–7] and is driven by gas [1,4], shape-memory alloy (SMA) [5,6], strings [7], dielectric elastomer [8], or water [9,10]. Several studies have been conducted on soft robots [11,12]. Although they change their forms largely, the mass distribution of the robot is almost fixed because gas, SMA, strings, and dielectric elastomer are much lighter than the total weight of the robot. Therefore, it is very difficult for these soft robots to perform kicking motions and resist slipping, which require large friction generated by a large mass at the appropriate location.

This paper proposes a novel mechanism for the soft robot like an amoeba and explains how and why the robot realizes successful forward movement. The body is composed of elastic chambers, specifically water balloons in this paper. Instead of conventional air or electric material, we use water in this paper not

© Springer Nature Switzerland AG 2019
M. Strand et al. (Eds.): IAS 2018, AISC 867, pp. 314–324, 2019.
https://doi.org/10.1007/978-3-030-01370-7_25

only to deform the shape of the robot but also to change its mass distribution because water is heavy enough to influence the mass distribution of the robot. By utilizing the mass distribution and local higher or lower friction, the robot realizes sliding with small friction and resists slipping or kicking the ground with large friction. By considering the mass distribution, we deal with not only conventional kinematics such as posture and shaper of the robot but also dynamics. Thus, we'll discuss an acceleration and stability on the dynamic locomotion in future. A shift of a center of mass of the robot by deforming its shape like an underwater robot developed by Chen et al. [10] is one of the important factors to provide locomotion of the soft robot, and water is key material to achieve such locomotion. The proposed robot is equipped with two chambers, and water is supplied to and released from each chamber. When water is supplied to the chamber, the elastic chamber inflates and the mass increases and vice versa. By controlling the order of the supply or release of water for the two chambers, the robot realizes kicking, resisting, or sliding for forward locomotion. The robot is also equipped with a nylon net whose friction with the chamber is low and the one with the ground is moderate. By utilizing the features of the net, the robot realizes successful forward locomotion.

The rest of the paper is organized as follows: The configuration of the physical soft robot equipped with two elastic chambers is described in Sect. 2, and the strategy of the forward locomotion is explained in Sect. 3. The experimental results are explained in Sect. 4. Finally, the conclusions of this study are presented in Sect. 4

## 2   Configuration of the Robot

### 2.1   Configuration of the Robot

Figure 1 shows the proposed robot. Figure 1(a) shows the schematic design of the robot, and Fig. 1(b) shows the physical robot. The robot is equipped with two elastic chambers, specifically water balloon in this paper, and a connector (PISCO PV6) is attached to each chamber. The two connectors are interlocked by a plastic plate. Because the chamber is filled or emptied by the water, the weight and volume of the chamber varies. In this paper, we denote the front chamber as "fore-chamber", the rear chamber as "rear-chamber", and intermediate part including the connectors and the plastic plate as "trunk". The weight of the chamber is changed from approximately 0 to 400 g, and the diameter of the chamber is changed from approximately 0 to 91 mm. In the trial, we found that the trunk moved improper direction which prevents proper transition of the moving force from rear-chamber to fore-chamber when the trunk is light. Therefore, we placed a small weight on the trunk to provide proper transmission of the force from/to fore- and rear-chambers. The length, depth, height, and weight of the trunk are 140 mm, 40 mm, 35 mm, and 120 g, respectively.

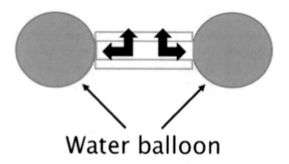

(a) Schematic design

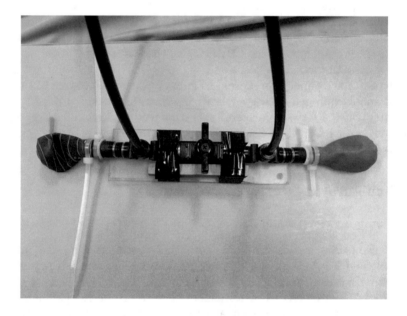

(b) Physical prototype

**Fig. 1.** Proposed hydraulic soft robot equipped with two elastic chambers

## 2.2 Pipe and Pump Arrangement

Figure 2 shows arrangement of pipes and a pump. We adopted the pipe made of urethane, whose inner diameter is 4 mm (PISCO UBT0640). Maximum pressure of the pump is 0.85 MPa and flow volume is 5 L/min. In the figure, the water on left side of the pump is pumped up, and it is discharged toward right side.

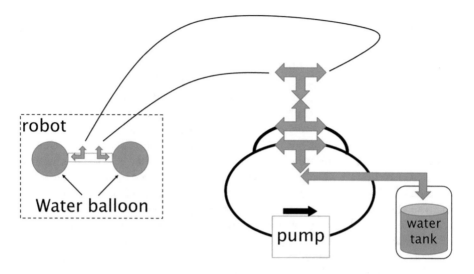

**Fig. 2.** Pipe and pump arrangement for supplying or releasing water

We adopted three three-way valves. By switching the handle of the valve, the connectors of left side and center are connected, and the ones of right side and center are connected.

Figure 3(a) shows the case when the water is supplied to an alternate chamber for inflation. In the figure, valve C connects the water tank to the left side of the pump, and valve B connects valve A and the right side of the pump. Valve A directs the water supply to either the fore- or rear- chamber. Figure 3(b) shows the case when the water is released from the chambers for deflation. Valve A determines whether water will be released from the fore- or rear-chamber. Valve B connects valve A and the left side of the pump. Valve C connects the right side of the pump and the water tank. By connecting these routes, the water in the fore- or rear-chamber is rapidly released using the pump.

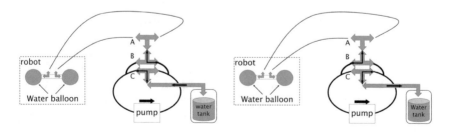

(a) Flow for supplying the water       (b) Flow for exhausting the water

**Fig. 3.** Procedure for supplying or releasing water

## 2.3    Slippery Nylon Net for Effective Inflation of Fore-Chamber

In order to move forward, it is important to extend the fore-chamber toward the direction of movement by supplying water to the fore-chamber. However, this operation raises two issues for forward extension. The first is that the chamber inflates in all directions, and it does not extend toward the direction of movement so effectively. The second is that the weight of the fore-chamber is increased by supplying water, and then the large friction caused by the large weight prevents the extension although the water is supplied in order that the fore-chamber extends. In order to extend the fore-chamber effectively and avoid such a contradictory phenomenas, we adopted the nylon net referring a construction of McKibben pneumatic actuator [13]. Figure 4(a) shows a schematic design of the net when it is fully extended. Shape of the net is trapezoidal, and it covers fore-chamber as shown in the figure. The net shrinks when the water is fully released from the fore-chamber. The shorter side of the net loops the connector, and the diameter of the longer side is 250 mm when the net is fully extended. When water is supplied to the fore-chamber, the nylon net is also extended at beginning. When the part of the net is fully extended, it prevents the fore-chamber from inflating. Because the friction coefficient between the net and fore-chamber is low, the fore-chamber slides within the net when water is continuously supplied to the fore-chamber. Figure 4(b) shows physical nylon net around the fore-chamber. As shown in the figure, the net shrinks when the inner fore-chamber does not inflate.

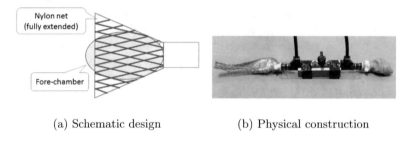

(a) Schematic design               (b) Physical construction

**Fig. 4.** Nylon net attached to the fore-chamber

Figure 5 shows the procedure for extending the fore-chamber. First, water is supplied to the rear-chamber. By increasing the mass, the friction between the robot and ground increases (Fig. 5A). Second, water is supplied to the fore-chamber (Fig. 5B). The chamber inflates toward all directions at the beginning. Accompanied with the inflation of the fore-chamber, the net also extends at the beginning, and then it stops extending. When the net stops extending, the direction of the inflation of the fore-chamber is limited because of the non-extensible nylon net. Instead of limited direction, the chamber extends along the longitudinal direction. Because the rear chamber with higher friction prevents backward slippage, the fore-chamber does not extend backward. Although the fore-chamber has a larger mass, the force to move backward is not occurred because the fore-chamber does not make contact with the ground and only the nylon net makes contact with the ground. Therefore, the fore-chamber internally extends in fully extended nylon sleeve (Fig. 5C), and then it achieves enough mass to make large friction to the ground.

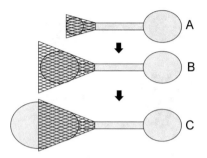

**Fig. 5.** Procedure for extending the fore-chamber by utilizing the nylon net

## 3   Realization of Locomotion

### 3.1   Procedure for Locomotion

Some researchers have realized locomotion by equipping their soft robot with legs. In this research, the proposed robot does not have limbs and deforms largely like an amoeba. The weight and volume are locally changed for locomotion. Figure 6 shows the open-loop control of the forward movement, including the procedure explained in the previous section. A force to move forward or backward is applied to the trunk when the chamber inflates or deflates, and the trunk is prevented from moving when the alternate or both chambers make contact with the ground and generate large friction even if the force is applied. First, water is supplied to the rear-chamber. The rear-chamber inflates, and the trunk moves forward according to the amount of inflation (Fig. 6 state 1 to 2). Second, as mentioned in the previous section, the fore-chamber extends toward the direction of movement (Fig. 6 state 2 to 3). Third, the water is released from the rear-chamber. In this process, the rear-chamber deflates and the force to move backward according to the amount of deflation is added to the trunk. However, because the friction between the fore-chamber (nylon net) and the ground is higher, the fore-chamber prevents backward slipping and the trunk does not move backward (Fig. 6 state 3 to 4). Finally, the water is released from the fore-chamber. Similar to the previous operation, the fore-chamber deflates and the force to move forward is added to the trunk. In contrast to the previous operation, the friction to prevent the trunk from moving forward was not generated on the rear-chamber because the weight of the rear-chamber is almost zero. Therefore, the trunk moves forward according to the amount of deflation of the fore-chamber.

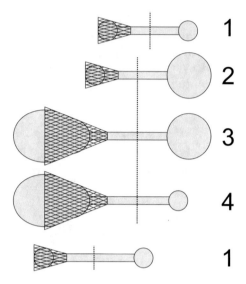

**Fig. 6.** Open-loop control of forward movement

## 3.2   Realization of the Locomotion

Figure 7 shows sequential pictures of the locomotion between the two states
shown in Fig. 6. Each row of pictures shows the sequence of inflating or deflating
the fore- and rear-chambers. The procedures between 1 to 2 are shown on the top
row, for example. The robot moved on the acrylic plate. The chambers inflate or
deflate following the procedure mentioned above. The period to supply water to
fore- and rear-chambers are both 20 s. In the procedures to release water, water
is fully released. As shown in Fig. 7(a), the rear-chamber is inflated by supplying
water to the rear-chamber, and the chamber makes contact with the ground
and the trunk is moved slightly forward by the inflation of the rear-chamber.
As shown in Fig. 7(b), the fore-chamber begins to inflate at time $t = 0.0$ s.
As explained earlier, the fore-chamber inflates, and the nylon net also inflates.
When some part of the nylon net is fully extended, the chamber is constrained
to inflate. Because the friction between the chamber and the nylon net is low,
the chamber slides to inflate toward longitudinal direction. By increasing the
mass of the fore-chamber, the friction between the nylon net and the ground is
increased. As shown in Fig. 7(c), the rear-chamber begins to deflate ($t = 0.0$ s).
While the rear-chamber deflates, the trunk does not move backward because the
fore-chamber generates enough friction to prevent backward sliding. As shown in
Fig. 7(d), while the fore-chamber deflates, the trunk is pulled forward in response
to the deformation of the fore-chamber because the mass-less rear-chamber does
not generates the friction to prevent forward movement. After one cycle of the
operation, the robot realized successful moving forward approximately 50 mm,

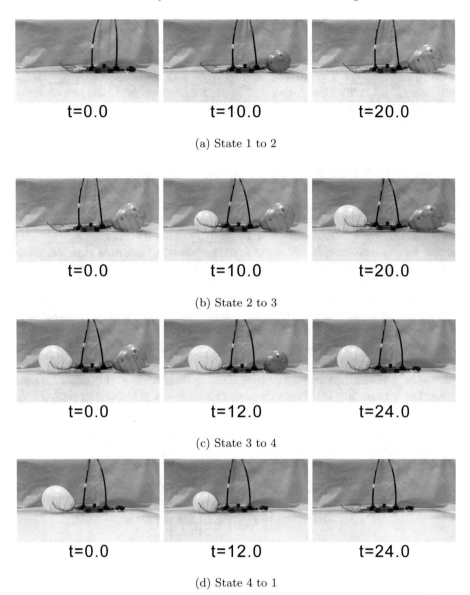

(a) State 1 to 2

(b) State 2 to 3

(c) State 3 to 4

(d) State 4 to 1

**Fig. 7.** Sequential pictures of physical robot that moves forward (unit of time: s)

approximately 35% of the length of the trunk. This experiment shows that the mass distribution was changed by supply and release of water, and the robot obtained sufficient local friction to prevent slipping and push the trunk. It also shows that the nylon net covering the fore-chamber provided successful forward extension.

## 4  Experiment

We observed relationships between periods of supplying water to both chambers and the locomotion performances. In the experiment, the period to supply water into rear-chamber $T_r$ (state 1 to 2 in Fig. 6) and one to supply into fore-chamber $T_f$ (state 2 to 3) are changed by 5.0, 10.0, 15.0 and 20.0 independently. Moving distance and the period of one cycle, i.e. total period of supplying/releasing water, are recorded. In this paper, water is fully released in the state 3 to 4 and 4 to 5(1), and the periods of releasing water are recorded. We had 3 trials for each set of supplying periods, and the average is calculated.

Table 1 shows total period for releasing water with respect to each set of supply periods. As shown in the table, the total releasing period is approximately 1.2 times longer than the total supplying period. For example, the total releasing period is 18.7 s when $T_r = 5.0$ s and $T_f = 10.0$ s. The extension of the releasing period will be examined in future. One of the possible reasons is that the pump should pump up the limited volume of water in the chamber when water is released from the chamber while the pump can pump up water from enough volume of water in the tank when water is supplied into the chamber, and it influences the inefficiency of releasing water. Table 2 shows average distance of one cycle with respect to the set of supplying periods. As shown in the table, the distance is longer when both $T_r$ and $T_s$ are longer. However, the period of one cycle is longer. Therefore, we evaluated the locomotion performance by calculating the velocity, average distance over average cycle in this paper. Table 3 shows the locomotion velocity with respect to the set of supplying periods. As

**Table 1.** Relationship between $T_r$ (top horizontal), $T_f$ (left vertical), and total period of releasing water: the unit of period is [s]

|            | $T_r =5.0$ | 10.0 | 15.0 | 20.0 |
|------------|------------|------|------|------|
| $T_f = 5.0$ | 12.6      |      | 18.7 | 24.8 | 30.6 |
| 10.0       | 18.7       |      | 25.0 | 30.8 | 37.5 |
| 15.0       | 24.8       |      | 31.3 | 36.8 | 42.5 |
| 20.0       | 31.2       |      | 37.0 | 42.9 | 49.4 |

**Table 2.** Relationship between $T_r$, $T_f$, and moving distance: the unit of distance is [mm]

|            | $T_r =5.0$ | 10.0 | 15.0 | 20.0 |
|------------|------------|------|------|------|
| $T_f =5.0$ | 14.7       | 21.3 | 20.3 | 22.7 |
| 10.0       | 20.7       | 24.7 | 24.7 | 28.0 |
| 15.0       | 24.7       | 33.0 | 40.0 | 41.0 |
| 20.0       | 33.3       | 43.0 | 51.3 | 57 |

**Table 3.** Relationship between $T_r$, $T_f$, and velocity: the unit of velocity is [mm/s]

|            | $T_r = 5.0$ | 10.0 | 15.0 | 20.0 |
|------------|-------------|------|------|------|
| $T_f = 5.0$ | 0.65       | 0.63 | 0.45 | 0.41 |
| 10.0       | 0.61        | 0.55 | 0.44 | 0.41 |
| 15.0       | 0.55        | 0.59 | 0.60 | 0.53 |
| 20.0       | 0.59        | 0.64 | 0.66 | 0.64 |

shown in the table, the velocity is lower than 0.5 mm /s when $T_f \leq 10.0\,\text{s}$ and $T_r \geq 15.0\,\text{s}$. In this case, the body moves backward by deflation of rear-chamber in the process from 2 to 3 in Fig. 5 because the friction of the fore-chamber is smaller due to short period of supplying water. When the supply and release periods are shorter as $(T_r, T_f) = (5.0, 5.0), (10.0, 5.0), (5.0, 10.0)$, the robot achieved faster movement over $0.6\,\text{mm/s}$. When these periods are longer as $(T_r, T_f) = (10.0, 20.0), (15.0, 20.0), (20.0, 20.0)$, the robot also achieved faster movement. The reason of former case is that the period of one cycle is shorter, and the velocity is higher though the moving distance is not so long. The reason of latter case is that, the moving distance is longer though the period of one cycle is not so short. From these results we conclude that an appropriate set of supply period provides faster locomotion.

## 5    Conclusion

This paper proposed a novel design for a soft robot whose mass distribution significantly changes. The robot has two elastic chambers at front and rear position, and water is supplied to or released from the chamber to inflate or deflate the chamber and increase the mass of the chamber. Fore-chamber is covered by the nylon net whose friction coefficient between the net and the chamber is low and the one between the net and the ground is not low. By shifting the mass in the two chambers placed in the front and rear position, the robot realizes successful movement by preventing slippage, pulling the trunk, and kicking the ground. In the experiment, the periods to supply water to fore- and rear-chamber are changed, and we found that the robot achieves effective movement by an appropriate combination of the supply periods. This is a first step toward realizing such a amoeba-like robot that deforms its shape and changes its mass distribution significantly. The developed robot does not have sensors to measure the state of the robot such as water pressure and shape of the chamber. We'll estimate the state of the robot by adopting appropriate sensors for more precise controlling in future.

## References

1. Drotman, D., Jadhav, S., Karimi, M., deZonia, P., Tolley, M.T.: 3D printed soft actuators for a legged robot capable of navigating unstructured terrain. In: International Conference on Robotics and Automation (ICRA), pp. 5532–5538 (2017)
2. Al-Abeach, L.A.T., Nefti-Meziani, S., Davis, S.: Design of a variable stiffness soft dexterous gripper. Soft Robot 4(3), 274–284 (2017)
3. Homberg, B.S., Katzschmann, R.K., Dogar, M.R., Rus, D.: Haptic identification of objects using a modular soft robotic gripper. In: 2015 IEEE/RSJ International Conference on Intelligent Robots and Systems (IROS), pp. 1698–1705 (2015)
4. Shepherda, R.F., Ilievskia, F., Choia, W., Morina, S.A., Stokesa, A.A., Mazzeoa, A.D., Chena, X., Wanga, M., Whitesides, G.M.: Multigait soft robot. In: Proceedings of the National Academy of Sciences of the United States of America, vol. 108, pp. 20400–20403 (2011)

5. Umedachi, T., Vikas, V., Trimmer, B.A.: Highly deformable 3-D printed soft robot generating inching and crawling locomotions with variable friction legs. In: 2013 IEEE/RSJ International Conference on Intelligent Robots and Systems (IROS), pp. 4590–4595 (2013)
6. Laschi, C., Cianchetti, M., Mazzolai, B., Margheri, L., Follador, M., Dario, P.: Soft robot arm inspired by the octopus. Adv. Robot. **26**(7), 709–727 (2012)
7. Calisti, M., Member, S., Arienti, A., Renda, F., Levy, G., Hochner, B., Mazzolai, B., Dario, P.: Design and development of a soft robot with crawling and grasping capabilities. In: 2012 IEEE International Conference on Robotics and Automation (ICRA), pp. 4950–4955 (2012)
8. Li, T., Li, G., Liang, Y., Cheng, T., Dai, J., Yang, X., Liu, B., Zeng, Z., Huang, Z., Luo, Y., Xie, T., Yang, W.: Fast-moving soft electronic fish. Sci. Adv. **3**(4), e1602045 (2017)
9. Katzschmann, R.K., Marchese, A.D., Rus, D.: Hydraulic autonomous soft robotic fish for 3D swimming. In: International Symposium on Experimental Robotics (ISER) (2014)
10. Chen, I.-M., Li, H.-S., Cathala, A.: Design and simulation of amoebot -0 a metamorphic underwater vehicle. In: Proceedings of the 1999 IEEE International Conference on Robotics and Automation (ICRA), pp. 90–95 (1999)
11. Polygerinos, P., Correll, N., Morin, S.A., Mosadegh, B., Onal, C.D., Petersen, K., Cianchetti, M., Tolley, M.T., Shepherd, R.F.: Soft robotics: review of fluid-driven intrinsically soft devices; manufacturing, sensing, control, and applications in human-robot interaction. In: Advanced Engineering Materials (2017)
12. Marchese, A.D., Katzschmann, R.K., Rus, D.: A recipe for soft fluidic elastomer robots. Soft Robot **2**(1), 7–25 (2015)
13. Chou, C.P., Hannaford, B.: Measurement and modeling of McKibben pneumatic artificial muscles. IEEE Trans. Robot. Autom. **12**(1), 90–102 (1996)

# Common Dimensional Autoencoder for Identifying Agonist-Antagonist Muscle Pairs in Musculoskeletal Robots

Hiroaki Masuda, Shuhei Ikemoto$^{(\boxtimes)}$, and Koh Hosoda

Osaka University, D538, 1-3, Machikaneyama, Toyonaka, Osaka, Japan
ikemoto@sys.es.osaka-u.ac.jp

**Abstract.** One of the distinctive features of musculoskeletal systems is the redundancy provided by agonist-antagonist muscle pairs. To identify agonist-antagonist muscle pairs in a musculoskeletal robot, however, is difficult as it requires complex structures to mimic human physiology. Thus, we propose a method to identify agonist-antagonist muscle pairs in a complex musculoskeletal robot using motor commands. Moreover, the common dimensional autoencoder, where the encoded feature has identical dimensions to the original input vector, is used to separate the image and the kernel spaces for each time period. Finally, we successfully confirmed the efficacy of our method by applying a 2-link planar manipulator to a 3-pairs-6-muscles configuration.

**Keywords:** Musculoskeletal robot · Redundancy · Autoencoder

## 1 Introduction

Musculoskeletal robots, which have driving systems inspired by the musculoskeletal systems of living organisms, are an intensively studied subfield of biorobotics [Hosoda2012, Marques2010]. While musculoskeletal systems have a variety of advantages, one of their most distinctive features is the redundancy of separately controlling the angles and stiffness of joints. The simplest model to explain this feature is the configuration of a one degree-of-freedom (DOF) manipulator driven by agonist-antagonist muscle pairs around a joint. Assuming that the agonist and antagonist muscles drive the joint with the same constant moment arm, if both muscles contract with the same force, the angle of the joint does not change regardless of what that force is. This demonstrates that there is redundancy in a mapping between torques and forces and living organisms adequately and adaptively solve it to realize different pairs of angles and the stiffness of joints. In studies on musculoskeletal robots, this phenomenon has been described as a fundamental advantage to employing a musculoskeletal structure. Where simple musculoskeletal structures (e.g. two-dimensional muscle configurations and constant moment arms) are adopted, extensive research based on their analytical kinematic models is available [Shirafuji2014, Ozawa2009, Sawada2012]. However,

© Springer Nature Switzerland AG 2019
M. Strand et al. (Eds.): IAS 2018, AISC 867, pp. 325–333, 2019.
https://doi.org/10.1007/978-3-030-01370-7_26

if more complex but more biologically plausible musculoskeletal structures are adopted, it is very difficult to identify where agonist-antagonist muscle pairs exist in the musculoskeletal robot as they are mostly state-dependent. Therefore, although there are many musculoskeletal robots that have biologically plausible complex musculoskeletal structures, this feature has been qualitatively explained but not qualitatively evaluated.

So far, many researchers have implicitly focused on complex kinematic problems derived from complex musculoskeletal robots. Because their kinematics are strongly nonlinear, machine learning techniques have been employed and their kinematic problems were often dealt with in parallel with solving dynamic problems. For example, Hartmann et al. focused on a 7 DOF musculoskeletal robot driven by 17 pneumatic artificial muscles, as shown in [Hosoda2012]. Furthermore, they demonstrated that their proposed method, which used an echo state network and Gaussian process regression, could learn strongly nonlinear kinematics and dynamics [Hartmann2012]. In addition, Diamond and Holland focused on realizing reaching control of an ECCERobot, as shown in [Marques2010], by finding out optimal combinations of muscle synergies using a reinforcement learning technique [Diamond2014]. These studies showed that machine learning techniques are sufficiently powerful for simultaneously dealing with the dynamic and kinematic problems of complex musculoskeletal robots. However, controllers exploiting the features of musculoskeletal robots where the angles and stiffness of joints could be separately controlled by exploiting agonist-antagonist pairs, could not be achieved owing to difficulties caused by the learning process. Because such controllers are not task-dependent, learning techniques based on evaluations of task achievements would not be suitable for this problem.

In this study, we aim to identify the state-dependent agonist-antagonist muscle pairs of a complex musculoskeletal robot and to use them as controllers. Mappings between states of muscles and joints cannot be described as linear transformations in complex musculoskeletal robots; therefore, their agonist-antagonist muscle pairs are state-dependent. This dependency further complicates the use of agonist-antagonist muscle pairs. For example, focusing on the relationships between muscle contraction forces and joint torques, joint stiffness cannot be simply increased/decreased by proportionally increasing/decreasing contraction forces of agonist-antagonist muscle pairs if they are state-dependent. Therefore, to identify "functional" agonist-antagonist muscle pairs based on analytical models of their hardware designs can be extremely difficult. Hence, we adopt neural networks to solve this problem. In particular, we employ autoencoders that have both forward and inverse mapping between given data and transformed data, as both forward and inverse problems are typically required to control robots.

In our proposed method(particularly for the common dimensional autoencoder), $N$-dimensional input vectors are first transformed into $N$-dimensional feature vectors; these feature vectors are then transformed back to the given $N$-dimensional input vectors. To train the network, we use not the original cost function to make the output vectors similar to the input vectors, and the additional cost function to make one portion of the elements of the feature vectors similar to one of low dimensional expressions obtained from the input vectors. Qualitatively speaking, this aims, in the rest of elements of the feature vectors, to obtain the kernel of the transformation from the input vectors to the part of elements of the feature vectors. For instance, some of the input vectors constitute elements of the feature vectors, and the remainder corresponds to muscle forces, joint torques, and bias tensions, respectively. In this paper, the validity of the proposed method is confirmed by applying a 2-link planar manipulator with a 3-pairs-6-muscles configuration.

## 2   Proposed Method

The core concept of this method becomes clear and simple when the analogy of linear algebra is introduced into this nonlinear problem. Thus, before explaining the proposed method, we first to introduce a method to extract a linear regression model equivalent to a neural network when the input vector is specified.

### 2.1   Linear Model Extraction

First, we explain the linear model extraction method in [Ikemoto2018] to easily illustrate the core concept of our proposed method. Assuming that a feedforward neural network employing the ReLU as the activation function receives the input vector $x \in \mathcal{R}^N$, the corresponding output vector $y \in \mathcal{R}^M$ can be obtained through the following calculation:

$$y = W_D h_D + b_{D+1}$$
$$h_i = \phi(W_{i-1}h_{i-1} + b_i) \tag{1}$$

where $D$, $\phi$, and $h_i$ indicate the number of hidden layers, the activation function (namely the ReLU), and the outputs of the hidden layers, respectively. Note that $i$ is an integer that $0 < i \leq D$ holds and $h_0$ is identical to the input $x$. $W_i$, $W_0$, $b_i$, and $b_{D+1}$ are weight matrices and biases. The ReLU is a function that outputs a value identical to the input; if the input is negative, this output will be zero. Assuming that $\forall i$, $\forall x$, $h_i > 0$ holds, the network is obviously identical to a linear regression. Therefore, if nonlinear mapping is required, some hidden units will output zero and the activation pattern will vary in response to changes in $x$. As the hidden units output zero when $x$ is given, they do not influence the output $y$ corresponding to $x$; thus, the weights and biases connected to these hidden units are also irrelevant. In other words, at each moment, the entire network is expressed by a subnetwork consisting of hidden units that output nonzero elements. Hence it is obvious that the subnetwork is expressed as an equivalent linear regression. Let $W_0'$, $W_i'$, $b_i'$, and $b_{D+1}'$ be weight matrices and biases, in which elements regarding inactive hidden units are removed. Thus, the equivalent linear regression is given as:

$$y = \left(\prod_{i=0}^{D} W_i'\right) \hat{x}(t) + \left(\sum_{i=1}^{D} \left(\prod_{j=i}^{D} W_j'\right) b_i'\right) + b_{D+1}'. \tag{2}$$

Linear model extraction helps us to understand the proposed method in terms of linear algebra, while contributing to determining its generality.

## 2.2   Common Dimensional Autoencoder

The common dimensional autoencoder is an encoder in which one hidden layer has the same dimensionality as the input and output; the remaining hidden layers have a larger number of elements. To train this network, we assume that there are two data types:

1. High dimensional data $\mathcal{D}_N$ whose dimensionality is $N$.
2. Low dimensional data $\mathcal{D}_I$ whose dimensionality is $I < N$ and that can be obtained by transforming the high dimensional data.

The initial data are used for the original cost function of the autoencoders:

$$L_{\mathrm{AE}} = \frac{1}{\|\mathcal{D}_N\|} \sum_{x \in \mathcal{D}_N} (y - x)^T Q (y - x) \tag{3}$$

where $y$, $\|\mathcal{D}_N\|$, and $Q$ indicate the output vectors of the common dimensional autoencoder, the size of the data, and a weighting matrix. Here, we denote a vector appearing on a hidden layer with $N$-dimensions as $h$. The remaining data are used for part of the elements of $h$ to express $h_I^* \in \mathcal{D}_I$ when $x \in \mathcal{D}_N$ is given. Denoting some of the elements of $h$ as $h_I$, the additional cost function is simply given as:

$$L_{\mathrm{I}} = \frac{1}{\|\mathcal{D}_I\|} \sum_{x \in \mathcal{D}_I} (h_I - h_I^*)^T R (h_I - h_I^*) \tag{4}$$

where $R$ indicates a weighting matrix which is involved in outputting $h_I$ and $\|\mathcal{D}_I\| = \|\mathcal{D}_N\|$ holds. The entire cost function for the common dimensional autoencoder is:

$$L = L_{\mathrm{AE}} + L_{\mathrm{I}}. \tag{5}$$

Regarding the training algorithm, our proposed method does not have any limitations.

Figure 1 shows the schematic of the common dimensional autoencoder; $h_K$ indicates the remaining elements of $h$. Qualitatively speaking, $h_K$ is required to retain information that would otherwise be lost by transforming $\mathcal{D}_N$ to $\mathcal{D}_I$, as the autoencoder must eventually reproduce $\mathcal{D}_N$ in the output layer. Considering that linear model extraction provides a linear regression model corresponding to an arbitrary $x$ (if the network is sufficiently trained), we can define Fig. 1 as *"the I-encoder is trained to be the transformation from $\mathcal{D}_N$ to $\mathcal{D}_I$ (namely the image); the K-encoder is trained to be the kernel"*. Because the kernel (zero-space) commonly explains how bias forces(stiffness, etc.) relate to the redundancy of simple musculoskeletal/tendon-driven robots with constant moment arms, the common dimensional autoencoder would expand the same viewpoints for complex musculoskeletal robots. To ensure this relationship, we use the ReLU as the activation function. However, it is not necessary for training and exploiting the common dimensional autoencoder.

Input layer          Hidden layer          Output layer

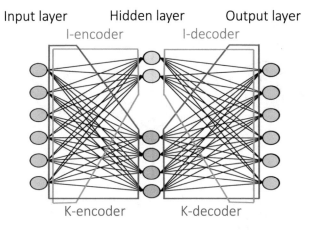

**Fig. 1.** Here, we expound the schematic of the common dimensional autoencoder proposed in this study, which consists of four parts: the I-encoder, the I-decoder, the K-encoder, and the K-decoder. Assuming that the learning process is a transformation from high-dimensional to low-dimensional data, the transformation will have zero space. By using an autoencoder that inputs/outputs high-dimensional data and adding a cost to make a part of the latent feature vectors low-dimensional data, the common dimensional autoencoder can express of the zero space (K-encoder) along with the transformation (I-encoder). This contributes Thus, the network can learn the causality that separates the image and the kernel in the transformation.

## 3   Simulation

### 3.1   Setup

To validate our proposed method, we conducted a simple simulation using a 2 DOF planar manipulator driven by the 3-pairs-6-muscles configuration. Figure 2 shows the parameters of the manipulator. In the manipulator, muscles are assumed to be series of elastic actuators, into which the displacement vector $d = [d_{f_1}, d_{f_2}, d_{f_3}, d_{e_1}, d_{e_2}, d_{e_3}]^T$ is fed as the input vector. In the simulation, we built $\mathcal{D}_N$ and $\mathcal{D}_N$ from a variety of displacements $d$ and the corresponding equilibrium positions of the end effector $p = [x, y]^T$.

In order to simply explain the statics and kinematics of the manipulator, let us define following vectors and matrix:

$$\theta = [\theta_1, \theta_2]^T \tag{6}$$

$$\tau = [\tau_1, \tau_2]^T \tag{7}$$

$$F = [F_{f_1}, F_{f_2}, F_{f_3}, F_{e_1}, F_{e_2}, F_{e_3}]^T \tag{8}$$

$$k = [k_{f_1}, k_{f_2}, k_{f_3}, k_{e_1}, k_{e_2}, k_{e_3}]^T \tag{9}$$

$$r = \begin{bmatrix} -r_1 & 0 & -r_1 & r_1 & 0 & r_1 \\ 0 & r_2 & r_2 & 0 & -r_2 & -r_2 \end{bmatrix}. \tag{10}$$

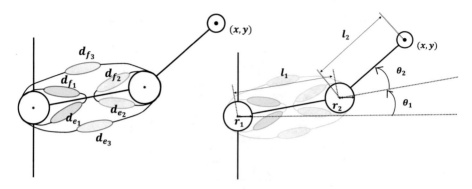

**Fig. 2.** The schematic of the 2 DOF planar manipulator driven by the 3-pairs-6-muscles configuration. Parameters are simply configured as $l_1 = l_2 = 0.2[m]$, $r_1 = r_2 = 0.02[m]$, and $k_{f_i} = k_{e_i} = 10$, $i \in 1, 2, 3[N/m]$.

When the manipulator takes a posture $p$ and the input $d$ is given, the contraction forces of muscles $F$ can be calculated as follows:

$$F = k \circ d - r^T \theta \tag{11}$$

where $\circ$ indicate Hadamard product. The contraction forces $F$ generate the joint torques as follows:

$$\tau = rF. \tag{12}$$

Because the $\tau$ must be zero at the equilibrium point, zeroing the left hand side of Eq. 12, we can obtain the equation to calculate the $\theta$ corresponding to $d$by substituting Eq. 11 into Eq. 12 and solving for $\theta$. As shown in Fig. 2, we greatly simplified the configuration of the parameters. Thus, they can be simply expressed as follows:

$$\theta = \frac{1}{6r_1} \begin{bmatrix} -2d_{f_1} + d_{f_2} - d_{f_3} + 2d_{e_1} - d_{e_2} + d_{e_3} \\ -d_{f_1} + 2d_{f_2} + d_{f_3} + d_{e_1} - 2d_{e_2} - d_{e_3} \end{bmatrix}. \tag{13}$$

To obtain $p$, following simple coordinate transformation was applied:

$$p = \begin{bmatrix} l_1 cos(\theta_1) + l_2 cos(\theta_1 + \theta_2) \\ l_1 sin(\theta_1) + l_2 sin(\theta_1 + \theta_2) \end{bmatrix}. \tag{14}$$

To obtain training data for the autoencoder, we randomly generated $d$; the corresponding position of the end effector was computed by: Eqs. 13 and 14. To define the network structure, we configured the number of units for each layer as 6, 180, 120, 6, 120, 180, and 6 from the input to the output. Following this, we trained the autoencoder in Eq. 3 x to correspond to d, and that in Eq. 4 $h_I$ to correspond to P, which is the hand position on the arm.

## 3.2   Results

To examine the trained autoencoder, we first gave periodical displacements for $d_{f_1}$ and $d_{e_1}$ by using two sinusoidal signals with a 45[deg] phase shift. The signal periods and their sampling frequencies were set to 4[sec] and 100[Hz], respectively. For remaining muscles, we assigned an adequate level of constant inputs to can maintain tension during movement.

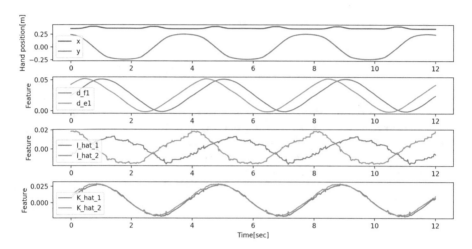

**Fig. 3.** The time-series signals for two muscles that were actively moved periodically. The top and the upper middle plots show changes in positions of the end effector caused by periodical actuations and the time series of decoded $d_{f_1}$ and $d_{e_1}$ that nearly identical to their actual values, respectively. The lower middle and the bottom plots show time-series of $d_{f_1}$ and $d_{e_1}$ decoded by the I-decoder and the K-decoder, respectively. Note that the manipulator takes the straight form on the x axis if $d_{f_1}$ and $d_{e_1}$ are the same.

Figure 3 shows the time series signals for these two muscles. As shown in the upper two subfigures, it is clear that the manipulator moved from the initial posture when $d_{f_1}$ and $d_{e_1}$ had different values. The important feature of the common dimensional autoencoder appears in the lower two subfigures. In contrast with that signals from the I-decoder, which showed alternating signals corresponding to the differences between the $d_{f_1}$ and $d_{e_1}$, signals from the K-decoder coincided with global changes within the $d_{f_1}$ and $d_{e_1}$. Because the differences between the $d_{f_1}$ and $d_{e_1}$ indicate changes in posture, and the features encoded by the I-encoder are the result of adding $L_I$, the alternating signals in the third subfigure were not surprising. However, it is interesting that these coincidental signals appeared despite the fact that the K-encoder and the K-decoder were not supervised. Thus, Fig. 3 shows that the common dimensional autoencoder worked well, as expected.

Following this, we generated $d$ data that realized ten postures on an arc trajectory in the task space with ten different bias forces. Figure 4 shows the outputs

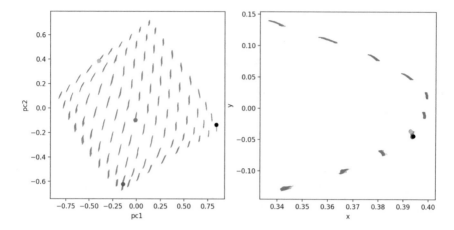

**Fig. 4.** The visualization of the mapping learned in the common dimensional autoencoder. The left and the right figures show outputs of the K-encoder and the I-encoder when the data of $d$ realized ten different postures on an arc trajectory in the task space, but ten different bias forces were given. Because the outputs of the K-encoder were four-dimensional, principal component analysis (PCA) was applied and the first and the second principal components were used in the left figure. Thus, similarly colored dots (in both figures) indicate that they were generated from the same $d$. Four kinds of $d$, corresponding to four colors, were generated with widely different bias forces.

of the I-encoder and the K-encoder where the data were fed. As intended by $L_I$, the outputs of the I-encoder successfully showed ten clusters corresponding to the ten postures. Therefore, ten different bias forces cannot be discriminated in the right figure. However, because we applied PCA, the ten different postures bias forces can be both discriminated (as shown in the right figure). To investigate the property of the outputs of the K-encoder, four points of $d$ (that resulted the same posture but widely different bias forces) were depicted on both figures. In contrast with the four dots appearing in the same cluster in the right figure, those in the left figure were widely distributed. Therefore, we concluded that the K-encoder actually extracted bias forces from $d$.

Based on these results, the validity of the proposed common dimensional autoencoder can be confirmed. In addition, based on the fact that the K-encoder successfully extracted bias forces that will directly contribute to the identification of agonist-antagonist pairs, we assert that the proposed common dimensional autoencoder contributes to the identification of these pairs in musculoskeletal robots.

## 4    Discussion and Conclusion

In this paper, we proposed a common dimensional autoencoder for constituting zero space while a network learns transformation from high-dimensional to low-dimensional data. The efficacy of this autoencoder validated by application to a simulated 2 DOF planar manipulator, which was driven by a 3-pairs-6-muscles system. Furthermore, preliminary experimental results showed that the K-encoder successfully extracted bias forces. However, clear identification of agonist-antagonist pairs could not been realized. Additional analysis of the kernel space of the common dimensional autoencoder, along with additional experiments, will be critical to validating the proposed method in future work. In addition, clearly demonstrating the relationships between agonist-antagonist pairs will be another important goal for future work in this domain. In this paper, we explained linear model extraction; however, it was used merely for better comprehension of the mechanism of the common dimensional autoencoder. We believe that linear model extraction will play the important role in future work as it easily shows relationships among elements of the control input vectors in the form of matrices, even where original mapping is nonlinear.

Finally, the proposed common dimensional autoencoder would also be useful in analyzing the real musculoskeletal systems of living organisms. Thus, expanding its application will also be important for future work in this domain.

# References

[Hosoda2012]  Hosoda, K., Sekimoto, S., Nishigori, Y., Takamuku, S., Ikemoto, S.: Anthropomorphic muscular-skeletal robotic upper limb for understanding embodied intelligence. Adv. Robot. **26**(7), 729–744 (2012)

[Marques2010]  Marques, H., Jantsch, M., Wittmeier, S., Holland, S., Alessandro, C., Diamond, A., Lungarella, M., Knight, R.: ECCE1: the first of a series of anthropomimetic musculoskeletal upper torsos. In: Proceedings of 10th IEEE-RAS International Conference on Humanoid Robots, pp. 391–396 (2010)

[Shirafuji2014]  Shirafuji, S., Ikemoto, S., Hosoda, K.: Development of a tendon-driven robotic finger for an anthropomorphic robotic hand. Int. J. Robot. Res. **33**, 677–693 (2014)

[Ozawa2009]  Ozawa, R., Hashirii, K., Kobayashi, H.: Design and control of underactuated tendon-driven mechanisms. In: Proceedings of IEEE International Conference on Robotics and Automation, pp. 1522–1527 (2009)

[Sawada2012]  Sawada, D., Ozawa, R.: Joint control of tendon-driven mechanisms with branching tendons. In: Proceedings of IEEE International Conference on Robotics and Automation, pp. 1501–1507 (2012)

[Hartmann2012]  Hartmann, C., Boedecker, J., Obst, O., Ikemoto, S., Asada, M.: Real-time inverse dynamics learning for musculoskeletal robots based on echo state Gaussian process regression. In: Proceedings of Robotics: Science and Systems (2012)

[Diamond2014]  Diamond, A., Holland, O.E.: Reaching control of a full-torso, modelled musculoskeletal robot using muscle synergies emergent under reinforcement learning. Bioinspiration Biomimetics **9**, 016015 (2014)

[Ikemoto2018]  Ikemoto, S., Duan, Y., Takahara, K., Kumi, T., Hosoda, K.: Robot control based on analytical models extracted from a neural network. In: The 1st International Symposium on Systems Intelligence Division (2018)

# Analysis of Variable-Stiffness Soft Finger Joints

Daniel Cardin-Catalan$^{(\boxtimes)}$, Angel P. del Pobil, and Antonio Morales

Robotic Intelligence Laboratory, Universitat Jaume I, Castellón de la Plana, Spain
cardin@uji.es

**Abstract.** This paper addresses the problem of designing an artificial finger with variable stiffness in its joints. Our approach is based on the principle of combining different means of actuation. Two different versions of variable-stiffness joints are presented and used in the design and manufacturing of three prototypes of gripper fingers. Diverse material configurations are used in order to determine which are the distinctive capabilities of each one and how they differ. In order to test the fingers we built a test bench that allows us to measure the movement of the tendon-driven actuation, the pressure of air actuation and the force that is deployed in the tip of the finger. Several tests are made to measure the relation between the actuation input and the force exerted by the fingertips. Our results suggest that the best mechanism to achieve variable stiffness in the joints is a soft-rigid hybrid finger.

**Keywords:** Variable stiffness · Soft robotics · Grippers

## 1 Introduction

Traditionally, grippers are meant to perform a single or low number of tasks. However, in the future robots will likely be required to change tasks regularly and this requires the development of multifunctional end effectors [1]. Soft robotics allow to create high-adaptable grippers in order to grasp the maximum possible number of objects. In most cases, these grippers are unable to change their stiffness. Although highly compliant fingers may be desirable for grasping some products, at other times stiffer fingers may be desirable.

Soft robotics is a subject of study that has been booming during the last years, under the assumption that bioinspired systems or robots may have better performance than classical robotics technologies [2,3]. This also includes soft manipulators and grippers [4]. Natural systems do not consist solely of soft materials, but they also contain rigid materials. For example, in the human body an 11% of the total volume is the skeleton and 42% are the muscles. Thus, there is a combination of hard and soft materials [5]. The combination of soft and rigid materials looks as a promising approach to compliant robotic systems [6]

Recent works [7–11] have developed grippers with the ability to change their own finger or joint stiffness in order to gain the capacity to grasp objects with different mechanical properties or having diverse grasp modes in the same gripper.

© Springer Nature Switzerland AG 2019
M. Strand et al. (Eds.): IAS 2018, AISC 867, pp. 334–345, 2019.
https://doi.org/10.1007/978-3-030-01370-7_27

Different mechanisms are possible [12]. For instance, some of them use materials whose stiffness can change depending on the temperature that is applied on it. This approach, however, needs some time to set up the desired temperature into the materials in the joints so that the gripper gets the programmed stiffness.

In natural systems, for example the human hand, various muscles with an agonist-antagonist function are used to achieve different finger stiffness while grasping. This kind of systems have been a source of inspiration to develop artificial variable-stiffness grippers [13]. Artificial systems made with an agonist-antagonist design tend to be quite complex [14]; typically using springs, pulleys and various types of elements.

In this article we present the design and evaluation of two types of variable-stiffness fingers. One design is based on the combination of agonistic/antagonistic tendons, and the second, on the combination of a tendon and a pneumatic valve. The aim of the research is to find designs which accomplish two desirable properties for variable-stiffness joints: namely, the joints must be able to change their stiffness almost instantly, and they must be as simple as possible.

To test these designs three different finger prototypes are built. Two of them are actuated with two tendons, but they differ in the type of material uses in the joints. The purpose of this is to test how the material can affect the finger performance without making changes to the design. The third finger prototype uses a combination of tendon and pneumatic valves.

## 2   Fingers

### 2.1   Finger Design

Finger design is inspired in the human finger, with three phalanxes and soft material in the joints that simulates the ligament between the phalanxes bones. In order to get a variable stiffness joint we also take inspiration of the hand, which uses a system of agonist and antagonist muscles. So, in one finger the agonist work is taken by a tendon and the antagonist work will be done by two elements: the own resilience of the soft material of the joints and the pressure air added into small chambers located in the finger joints. The pressure inside the joint chambers will change the stiffness of the joints, if we increase the pressure inside the chambers the stiffness of the fingers will rise too. In the other finger the agonist-antagonist work is done by two tendons, one that makes the flexion movement of the finger and the other the extension one.

The designs of the fingers are shown in Fig. 1. The parts shown in white were made of a rigid material and the black parts of soft material. The cut in the middle of the tendons will pass thought the holes in the rigid parts of the fingers, while there can also be seen the air cavities in the soft parts.

### 2.2   Prototype Manufacturing

The pieces of the fingers are made using fused deposition modelling (FDM), also known as 3D printing. FDM allows making fast prototyping models with

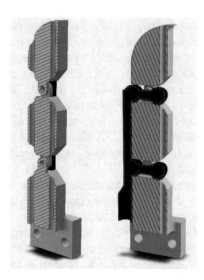

**Fig. 1.** Finger design

materials that can simulate the rigid and soft properties of natural elements. Rigid parts were made of PLA (Polylactic Acid) and soft parts of NinjaFlex (thermoplastic polyurethane). When all the pieces are printed, we assemble all the pieces in both finger models.

For the finger that is activated with tendons two options are manufactured. One is a material hybrid option, with the material distribution that is seen in Fig. 1. The other one is a simple monolithic soft material one. With this two options we can see how the finger performs depending on the material. The two options can be seen in Fig. 2.

The hybrid dual tendon will be HTFinger, the monolithic soft dual tendon will be STFinger, and the hybrid finger that is activated with pressure and tendon will be HBFinger.

## 3   Testing Bench

In order to measure the capabilities of the fingers a test bench was designed and manufactured. Its purpose was to measure the linear movement of the tendons, the force at the tip of the fingers and the pressure inside the air cavities. The final bench is shown in Fig. 3 with a finger mounted in it.

### 3.1   Making the Structure

To make the structure of the testing bench an aluminium frame was used, because it grants the main structure of the bench with resistance against the forces applied to it and also because it is easy to mount and unmount in different configurations.

**Fig. 2.** Tendon-driven fingers

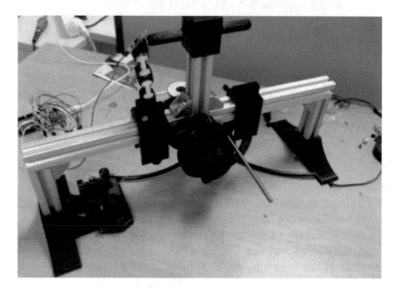

**Fig. 3.** Bench assembly

The rods are mounted with the help of structural elements as aluminium brackets that are manufactured to work with the frames used to make the core of the test bench. There are also pieces that have been 3D-printed to give more stability to the core.

The structure also provides the fingers with an opening-closing movement, so the grasp of an object can be simulated. This is made manually, thanks to a sort of pieces that transform the rotating movement into a lineal one. To grant this movement a carriage has been designed and 3D-printed, this carriage will be mounted bellow the fingers and the main rod will pass through its centre thus providing the finger with linear movement. This movement will simulate a parallel finger grasp. That mechanism is shown in Fig. 4.

**Fig. 4.** Open-close mechanism

The legs of the structure provides the testing bench with good stability and allows us to only use two feet instead of tree or more, giving more simplicity to the testing bench. The middle rod is used to mount on it the piece that will interact with a force sensor to provide us with the value of the fingertip force.

### 3.2 Sensors

In order to measure the performance of the fingers during the test, sensors have to be added in the structure. The data that we need to know is the force applied to a tendon that actuates the finger, along with the air pressure that can be applied inside the soft cavities -which can be used to actuate the finger or change the stiffness inside a cavity- and also the lineal movement of the tendon, as explained before. To embed the sensor in our system we need to design and manufacture some extra pieces. These pieces will be explained next.

**Linear Movement Measure.** The first thing that we need to know is how much the finger has moved when we apply the force, or how much force do we have for a known deformation. To be able to have these data we designed and make some parts that will be added to the main structure.

The first part is a pulley that will be the responsible to apply the movement to the finger, and to measure that movement we will measure radians that the pulley turns and transform them to mm.

$$r \times \theta = s \qquad (1)$$

Being r the radius of the pulley, $\theta$ the angle rotated in radians and s the linear displacement of the tendon. In one side of the pulley we make some indents that have two purposes. The first one is to measure the angular displacement of the finger, as there are 14 indents we can measure displacements of $1/7\pi$ radians, that knowing that the pulley has a radius of 10 mm, we can measure displacements in intervals of around 4,5 mm. The second function is to hold the little stick that will stop the pulley from moving when a force is applied to it.

**Force Sensor.** We need to know which is the force applied into the tendon that mainly activates the fingers that we are testing. The sensor will be a round Force Sensitive Sensor (FSR). These force sensors are used to measure the force that the finger applies in the tip. The sensor will be placed in the inner part of the distal phalanx of the fingers.

To interact with the sensor we build a piece whose function is to stop the finger movement where the sensor is, so that stopping force can be measured. That piece is shown in Fig. 5, this piece can also be moved up and down in the frame, so it can adapt to different fingers.

**Fig. 5.** Force measuring montage

**Sensor Integration.** To integrate all the sensors in the systems an Arduino Board [15] is used, because both electronic sensors work with it. To show the results in real time instead of using a PC we put an LCD screen in the Arduino Board, with that solution the sensor system is portable and it doesn't need any computer to work. Figure 6 shows how the LCD screen works, showing the analogic values for the Force (Fuerza) and the Pressure (Presión).

## 4    Experimental Testing of the Fingers

The next point of this research is to make tests with this bench and the three fingers.

**Fig. 6.** LCD screen

### 4.1  Test Description

The experiment consists in moving the finger into a known position that is almost touching the sensor with the piece added to the bench. Then we move the tendons to a specific position or inject pressure in the air cavities depending on the finger model. Finally, we measure the data given by the sensors.

The aim of this test is to observe how the force at the tip of the fingers varies when we change the way that the actuators affect the finger.

### 4.2  Test Implementation

First we make tests with the HTFinger. To test all the possibilities of the finger we firstly put the rear tendon in a position when it starts to make a force in the finger, and then we start to move the front tendon to different positions and see how the force value changes. Then we move the rear tendon and repeat the whole process. To the STFinger we follow the same steps as the ones followed with the hybrid fingers. Finally for the HBFinger we follow a process similar to the previous ones, but instead of changing the position of the rear tendon we change the pressure value inside the air chambers.

While doing the tests we move the stick that interacts with the sensor so that the sensor is always actuated in the same point. Due to the deformation of the finger while doing the tests, we need to correct the position of the stick in almost every measurement.

## 5  Results

### 5.1  Analogical Reading

The results are the analogical readings obtained from the sensors, which are later transformed in order to express the data in SI units. The tendon movements will be expressed as angular displacements; as it was said before, one angular displacement equals to more or less 4.5 mm of linear tendon displacement. All the data are the analogic input read in the Arduino screen.

## 5.2   Calculation of the Real Values

With the analogical values of the sensor readings we have an approximate idea of the relation between the inputs given the sensor readings. But to know better how these values really depend on each other is better to calculate the real value of each variable.

**Force Sensors.** The analogical force sensor is a sensor that gives a null value when there is no force actuating on it and a value of 1023 when there are 100 N force applying over it.

To calculate the force of the finger we have to make an interpolation of the values to get the actual force value in N. The next tables show the real values of the force that the fingers apply to get back to their original position. Also with the linear displacements instead of the angular ones. The linear displacement is in mm and the values of the force are in N (Tables 1, 2 and 3).

**Table 1.** Values of the fingertip force N for HTFinger

| Rear position | Front position | | | | | | | |
|---|---|---|---|---|---|---|---|---|
| Linear displacement (mm) | 0 | 4.5 | 8.9 | 13.5 | 18.0 | 22.4 | 26.9 | 31.4 |
| 0 mm | 0N | 19.5N | 27.8N | 42.6N | 48.3N | 50.4N | 52.0N | 54.2N |
| 4.588 mm | 0N | 0N | 14.7N | 26.6N | 42.0N | 46.9N | 48.6N | 50.3N |
| 8.976 mm | 0N | 0N | 10.5N | 20.8N | 39.7N | 44.7N | 46.1N | 48.9N |

**Table 2.** Values of the fingertip force for STFinger

| Rear position | Front position | | | | | | | |
|---|---|---|---|---|---|---|---|---|
| Linear displacement (mm) | 0 | 4.5 | 8.9 | 13.5 | 18.0 | 22.4 | 26.9 | 31.4 |
| 0 mm | 0N | 17.6N | 27.1N | 43.9N | 54.6N | 64.0N | 71.2N | 74.4N |
| 4.588 mm | 0N | 11.7N | 25.1N | 42.0N | 49.9N | 60.7N | 66.8N | 71.8N |
| 8.976 mm | 0N | 0N | 18.6N | 35.5N | 49.0N | 59.9N | 63.0N | 71.2N |

**Table 3.** Values of the fingertip force for HBFinger

| Rear pressure | Front position | | | | | | | |
|---|---|---|---|---|---|---|---|---|
| Linear displacement (mm) | 0 | 4.5 | 8.9 | 13.5 | 18.0 | 22.4 | 26.9 | 31.4 |
| 0 bar | 0N | 12.7N | 17.8N | 36.4N | 42.5N | 47.9N | 50.6N | 56.6N |
| 0.5 bar | 0N | 0N | 12.7N | 22.5N | 30.2N | 42.8N | 47.9N | 52.3N |
| 1 bar | 0N | 0N | 11.4N | 18.1N | 24.1N | 35.8N | 40.1N | 48.4N |

**Finger Equations.** The graphs shown in Fig. 7 are the force representations of the fingers depending on the linear position of the main activation tendon. The blue line represents the finger without the activation of the second actuator, the green one is when the secondary actuator is on the first position (4.5 mm in the tendon cases and 0.5 bar in the pressure one) and the yellow one is when the secondary activation is in the second position (8,9 mm in the tendon cases and 1bar in the pressure one).

For every one of the results we make a regression and see which kind of function fits better the data given in the results. While doing the statistics tests we will get an $r^2$ value, which shows the probability that the data is following the equation calculated by the software, the value of $r^2$ is between 0 and 1. We will only take the equations that surpass the value of 0.9 for $r^2$. In these equations the variable x is the linear displacement of the front tendon in mm.

The first finger (HTFinger) has the equations for the tree tests that are shown below. Equation 2 is the one that the fingers have when there is no rear activation, Eq. 3 is when the rear tendon has made a movement of 4,5 mm and Eq. 4 is when the rear tendon has made a movement of 8,9 mm. All the equations have an $r^2$ higher than 0.9.

$$F = 11.630x \tag{2}$$
$$F = 8.616x - 0.204 \tag{3}$$
$$F = 8.991x - 0.676 \tag{4}$$

The second finger (STFinger) has the equations for the tree tests that are shown below. Equation 5 is the one that the fingers have when there is no rear activation, Eq. 6 is when the rear tendon has made a movement of 4,5 mm and Eq. 7 is when the rear tendon has made a movement of 8,9 mm. All the equations have an $r^2$ higher than 0.9.

$$F = 13.546x + 1.842 \tag{5}$$
$$F = 14.872x + 1.269 \tag{6}$$
$$F = 12.810x - 0.757 \tag{7}$$

The last finger (HBFinger) has the equations for the tree tests that are shown below. Equation 8 is the one that the fingers have when there is no pressure in the air cavities, Eq. 9 is when there is a pressure of 0.5 bar in the cavities and Eq. 10 is when there is a pressure of 1 bar in the cavities. All the equations have an $r^2$ higher than 0.9.

$$F = 10.518x + 1.432 \tag{8}$$
$$F = 16.358x - 1.524 \tag{9}$$
$$F = 18.864x - 2.161 \tag{10}$$

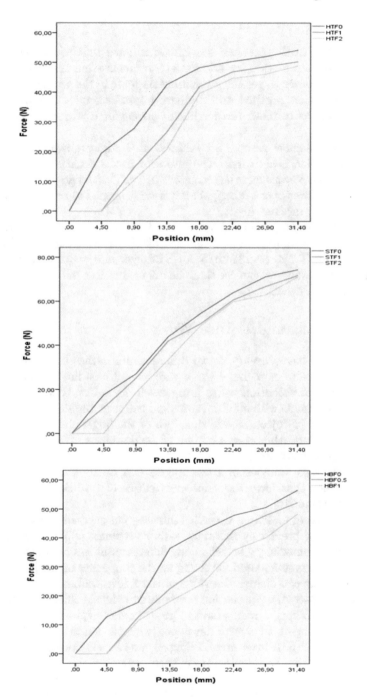

**Fig. 7.** Finger graphs

### 5.3    Discussion

From the results in the previous section we can see that when the secondary actuation is set into motion, an additional movement from the principal actuation is needed in order to get the same fingertip force in the sensor. So, when the secondary activation is called up the finger joints became stiffer and the principal actuation has to make further linear movement in order to get the same fingertip force.

The equations show us how the stiffness of the fingers augments when the secondary actuation is activated. The intercept in the equations becomes more negative as we keep actuating the secondary drive. That means that as the finger gets stiffer more force of the principal activation is needed to surpass the stiffness of the finger and start moving it.

The slope of the equations show us how the quantity of force that the fingers deploy with a millimetre of displacement of the principal actuation, being ~9 mm/N for HTFinger, ~13 mm/N for STFinger and ~15 mm/N for HBFinger. With these results we can see that the finger that in a medium term deploys more force is HBFinger.

## 6    Conclusions and Future Work

In this paper we have developed two different approaches to variable-stiffness actuation, that allow us to build three variable-stiffness fingers and test them in diverse forms of actuation usage using a self-made test bench. Out of these fingers, two were made with a hybrid soft-rigid use of materials and the other one as a soft monolithic object. In addition, two of the fingers have a dual tendon-driven actuation and the other a combined tendon-pressure actuation in order to get variable stiffness in the joints.

More developed versions of these fingers can be used to create variable-stiffness grippers that have the same capabilities of traditional grippers such as the Barret Hand [16].

Both methods of making a variable-stiffness joint succeed in overcoming the problems that had previously arisen for variable-stiffness joints. These problems were the lack of immediacy when changing different stiffness parameters, and the complexity of the systems used for changing the properties of the joints. Indeed, both our methods can change the value of the finger stiffness almost instantly, because of both secondary actuation methods can change their intensity almost instantly, and also both systems have a quite simple design.

For the two fingers that were developed with the same design but distinct material structure (HBFinger and STFinger), we can see from the results above that the soft solution can deploy more force per millimetre actuated than the hybrid one, but the slopes in the equations are more similar in the hybrid one if we do not count the inactivated one. Then, we can conclude that the soft material gives us more strength, but it turns out to be less reliable.

In future research we will implement the joints into a full gripper to test object grasping. The aim will be to be able to grasp the maximum number of objects with just one gripper by changing the stiffness of the joints in order to reproduce different kinds of grasps, such as power or precision grasps. We will mainly focus on the air cavities solution, mostly because of its design, that allows us to control different joints with a design and control structure that are simpler than in the case of the tendon-driven solution.

**Acknowledgement.** This paper describes research done at UJI RobInLab. Support for this research was provided by Ministerio de Economa y Competitividad (DPI2015-69041-R).

# References

1. Al Abeach, L.A.T., Nefti-Meziani, S., Davis S.: Design of a variable stiffness soft dexterous gripper. Soft Robot. J. **4**(3), 274–284 (2017)
2. Pfeifer, R., Lungarella, M., Iida, F.: The challenges ahead for bio-inspired soft robotics Commun. ACM **55**(11), 76 (2012)
3. Laschi, C., Rossiter, J., Iida, F., Cianchetti, M., Margheri, L.: Soft Robotics: Trends, Applications and Challenges Biosystems & Biorobotics, vol. 17. Springer (2017)
4. Hughes J., Culha U., Giardina F., Guenther F., Rosendo A., Iida, F.: Soft manipulators and grippers: a review. Front. Robot. AI **3**, 69 (2016)
5. Rus D., Tolley T.: Design, fabrication and control of soft robots. Nature **521**(7553), 467–475 (2015)
6. Culha, U., Hughes, J., Rosendo, A.L., Giardina, F., Iida, F.: Design principles for soft-rigid hybrid manipulators. Biosyst. Biorobotics **17**, 87–94 (2017)
7. Yufei, H., Tianmiao, W., Xi, F., Kang, Y., Ling, M., Juan, G., Li, W.: A variable stiffness soft robotic gripper with low-melting-point alloy. In: Proceedings of the 36th Chinese Control Conference, pp. 6781–6786 (2017)
8. Yang, Y., Chen, Y.: 3D printing of smart materials for robotics with variable stiffness and position feedback. In: IEEE/ASME International Conference Advanced Intelligent Mechatronics, AIM, pp. 418–423 (2017)
9. Memar, A.H., Mastronarde, N., Esfahani, E.T., Design of a novel variable stiffness gripper using permanent magnets. In: IEEE International Conference Robotics and Automation, pp. 2818–2823 (2017)
10. Li, Y., Chen, Y., Yang, Y., Wei, Y.: Passive particle jamming and its stiffening of soft robotic grippers. IEEE Trans. Robot. **33**(2), 446–455 (2017)
11. Firouzeh, A., Paik, J.: An under-actuated origami gripper with adjustable stiffness joints for multiple grasp modes. Smart Mater. Struct. **26**(5), 055035 (2017)
12. Manti, M., Cacucciolo, V., Cianchetti, M.: Stiffening in soft robotics: a review of the state of the art. IEEE Robot. Automat. Mag. **23**, 93–106 (2016)
13. Hogan, N.: Adaptive control of mechanical impedance by coactivation of antagonist muscles. IEEE Trans. Automat. Control **29**(8), 681–690 (1984)
14. Migliore, S., Brown, E., DeWeerth, S.: Biologically inspired joint stiffness control. In: IEEE International Conference on Robotics and Automation, pp. 4508–4513 (2005)
15. arduino.cc, December 2017
16. Townsend, W.T.: The Barrett Hand grasper-programmably flexible part handling and assembly. Ind. Robot Int. J. **10**(3), 181–188 (2010)

# Method for Robot to Create New Function by Uniting with Surrounding Objects

Yukio Morooka[✉] and Ikuo Mizuuchi

Tokyo University of Agriculture and Technology, Tokyo, Japan
{morooka,ikuo}@mizuuchi.lab.taut.ac.jp

**Abstract.** In this paper, we propose a robot that creates new functions by uniting with other objects. Such a robot can be applied to various situations by creating functions that fit to the situations. In this paper, we describe the elements of the proposed robot and development of two types of prototypes. The first prototype has an uniting function by gripping objects, and we conducted a demonstration of creating an automatic angle adjustment function on a projector using this prototype. The second prototype has an uniting function by using electromagnets and we conducted demonstrations of creating a function to handle object on a high place and creating automatic open and close functions for a door using this prototype.

**Keywords:** Function creation · Uniting · Modular robotics
Multipurpose robot

## 1 Introduction

When robots make an advance into general environments which are not prepared for robots from the industrial field, it is important that robots are able to deal with diverse situations. For example, studies on rescue robots that work in disaster sites are actively conducted. Such a rescue robot are required to do a various action such as getting over heaps of rubble and cracks in the ground and going forward in narrow ducts. Designing a robot to have all required functions is difficult and therefore a robot that can create functions extempore is quite useful. We set our ultimate goal to develop a robot which has high general versatility and is able to work in such environments and propose a robot that creates functions by uniting with surrounding objects.

Mounting a lot of functions and mechanisms on robots for getting over diverse situations is not seems to be better as a method for implementing the versatile robots because the number of functions and mechanisms necessary to deal with general environments is enormous. We propose a robot which has minimum function and creates function when it is faced with situations to get over. For the purposes of this paper, the term creating function will be taken to mean

© Springer Nature Switzerland AG 2019
M. Strand et al. (Eds.): IAS 2018, AISC 867, pp. 346–357, 2019.
https://doi.org/10.1007/978-3-030-01370-7_28

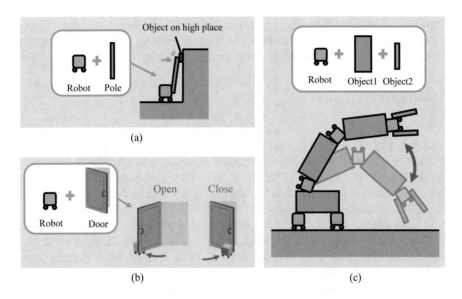

**Fig. 1.** Example of creating function by uniting

a situation in that the robots become to be able to do something that cannot be done by the robots originally. These robots utilize surrounding objects to create functions. Particularly, we insist that the robot is able to create more diverse functions by not only using but also uniting with objects and making them be a part of their own body. Figure 1 shows examples of the robot creating functions by uniting with surrounding objects. In an example of creating function of handling an object on a high place (Fig. 1a), the robot itself cannot reach the object. Ordinary, we have to design a robot that has function to reach the target places, and that robot may be too large to work in narrow places. By uniting with an object like a pole, the robot can create function of reaching and handling the object, and even cast off the function when it becomes useless. In an example of creating automatic open and close function for a door (Fig. 1b), the robot makes the door automatic by uniting with the door. We can automate the door without large-scale construction works. Figure 1c shows an example of creating a robot arm by uniting of multiple robots and objects. It is assumed that the uniting of multiple robots and objects can create more complicated functions such as Fig. 1c. In this paper, we propose a robot that uniting with surrounding objects and creating new functions and describe prototypes of the robot and demonstrations. We conducted demonstrations of creating automatic angle adjustment function on a projector, function for handling object on a high place and automatic open and close function for a door as cases in which the proposed robots are effective. In Sect. 2, we describe the related works and make the significance of our study clear. In Sect. 3, we describe examples of the application of the robot that we proposed and consider the necessary functions of the robot. In Sect. 4, we describe the first prototype we made and demonstration of creating a function by using prototype and show the effectiveness of method we proposed. In Sect. 5, we describe the second prototype and demonstrations of creating function. We mention the conclusion and future works in Sect. 6.

## 2  Related Works

A large number of previous works for robots that performs diverse tasks have conducted. Crawler robots [6] and legged robots [14] can move on diverse terrains and overcome obstacles. Amphibious robots [1] can move not only on the ground but also in underwater. Snake-like robots [15] and hyper-redundant robots [2] are robots that have a large number of degrees of freedom and can move variously according to situations. In researches of humanoid [7], it is insisted that robots which have human-like shapes are applicable to various situations in human life [4]. These studies have given the ability to get over diverse situations the robot congenitally, therefore it is hard for the robots to behave flexibly and get over the unexpected situations.

Some researchers insist the method that robots are adapted to situations by changing or creating its own function posteriorly. Reinforceable Muscle Humanoid [10] can change its power by changing number and position of actuators. In some studies, the method that robots extend its own functions by using tools has been insisted. These include research about humanoid using power tools [13] and research about humanoid that uses a box and board as tools to create steps and bridge [8] has been done. Modular robots [3,5,9,11,12] have been developed as robots that change their own functions posteriorly. Modular robots are robot systems consisted of robot modules that have comparatively simple mechanisms and functions and suggested to be versatile robot system since they can change their own shapes by changing combinations of modules.

In this paper, we propose robot that creates functions by uniting with surrounding objects as a robot that can be adapted to the various situations. Uniting not only with modules like modular robots, but also variable surrounding objects enable the robot to create more diverse functions and execute tasks.

## 3  Robot that Creates New Functions by Uniting with Surrounding Objects

### 3.1  Application

The method we propose depends on presence or absence of proper objects in the surroundings. Especially the method is effective in situations that have many objects. For example, disaster sites that are filled with rubbles and home where there are various goods. When rescue or search robots work in disaster sites, some obstacles such as steps that have various heights or cracks in the ground that have various widths obstacles in many rubbles. Our concept makes it possible that robots break through the obstacles by uniting with surrounding rubbles and extending its own body and functions. Although proposed methods like modular robots [3,12] effect similarly by constructing their own shapes from robot modules to be suitable to situations, the method that we propose is able to break through more various obstacles by using other objects. Furthermore, there are utilizes like that the robot units with a building that is damaged from disasters and reinforces it in order to prevent the collapse of it. Similarly, home robots have to manage various situations. There are many skills that robots should have such as skill to go upstairs, open doors or drawer and get objects on the top or under of a shelf when robots support human life. We consider that these can be solved by applying the method that we propose.

### 3.2   Necessary Factor of the Robot that Creates Function by Uniting with Surrounding Objects

The robot that creates functions by uniting with surrounding objects is able to supplement its poor ability. It means the robot itself does not need to have so many functions. And inconveniences may occur by increasing in size and mass of the robot result from implementations of extra functions. It seems to be suitable that we implement minimum functions for uniting with surrounding objects to the robot and create extra functions by uniting.

We have decided to implement three functions: a function for moving to objects, uniting with the objects and move the objects. We have decided to implement the function to move to objects by wire-driven method because it is applicable to various objects to make various movements. The advantage is due to flexibility and length adjustability of wire. Although we cannot push objects by wire, we can obtain same effects for example by using another robot and having it pull the objects.

## 4   First Prototype and Demonstration

### 4.1   First Prototype

In order to show a concrete instance and effectiveness of creating functions by uniting, we have made the first prototype. Figure 2 shows the appearance of the first prototype. The first prototype is composed of aluminum sheet metals and 0.8 kg in weight. We describe the implementations of the three functions that we have explained in Sect. 3.

#### Function to Unite with Surrounding Objects
We have decided to implement the function to unite with other objects by gripping a projecting part of objects. This made it possible that the first prototype units and parts with objects at will. Figure 3 shows the state and mechanism for gripping a projection. The first prototype moves the crawlers as shown in Fig. 3 (a), and grips a projection of an object and units with the object. Figure 3 (b) shows the mechanism for gripping by crawlers. The rotational motion of the RC servo module is divided into right and left by the gears, and move the crawlers via link mechanisms.

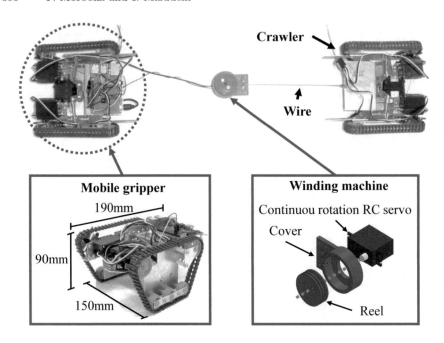

**Fig. 2.** The first prototype

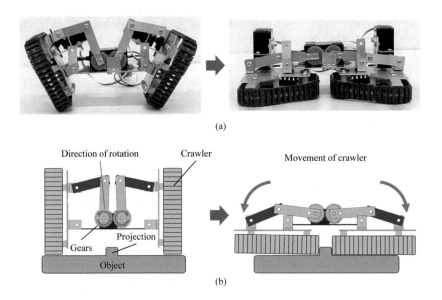

(a)

(b)

**Fig. 3.** Uniting by gripping protection

**Fig. 4.** Model of projector

**Function to Move Objects**

We implemented winding mechanism on the first prototype. This mechanism winds and can develop 2 kgf tension to the wire. We mounted a cover on the reel in order to prevent the wire from getting caught on the edge of the reel. The wire is made from ultra-high molecular weight polyethylene called Dyneema and has features of low friction, flexible and high intensity.

**Function to Move**

We have implemented the function to move by using crawlers. By rotating the left and right crawlers independently, the first prototype can move to forward and backward, and turn. The crawlers are driven by continuous rotation RC servo module, and its maximum drive force is about 2 kgf at one side.

We have conducted a demonstration that we describe in next section by remote operation.

## 4.2   Demonstration

For the purpose of showing a more concrete instance of our concept, we conducted a demonstration of creating automatic angle adjustment function on a projector using the first prototype. Figure 4 shows the model of the projector used for the demonstration.

Figure 5 shows the state of the demonstration. First, we set situation as shown in Fig. 5 (1). We controlled the first prototype and moved it to the projection of the floor as shown in Fig. 5 (2). We had the first prototype grip the projection and unite with the floor surface as shown in Fig. 5 (3). Next, we controlled the other mobile gripper and moved it to the top of the projector model as shown in Fig. 5 (4). And we had the mobile gripper grip the projection and unite with the projector model as shown in Fig. 5 (5). By winding the wire in this state, the first prototype adjusted the angle of the projector model as shown in Fig. 5 (6). Through the above process, we could demonstrate the creation of the automatic angle adjustment function on the projector by the first prototype. These processes take 59 s in total ((1)–(2) takes 11 s, (2)–(3) takes 8 s, (3)–(4) takes 23 s, (4)–(5) takes 6 s, (5)–(6) takes 11 s). When the first prototype winds the wire, the behavior of the projector depends on strength of the frictional force and the projector may only slide on the floor when frictional force is weak.

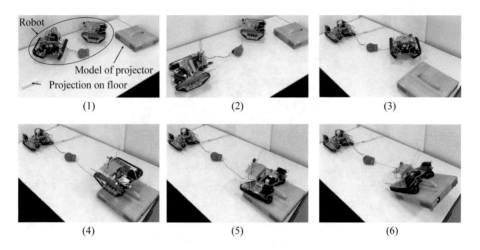

**Fig. 5.** Demonstration of creating automatic angle adjustment function on projector

## 5    Second Prototype and Demonstrations

We have made the second prototype of the robot that creates new functions by uniting with surrounding objects. Using the second prototype, we have conducted two demonstrations of creating functions in order to show the effectiveness of our concept. In this section, we describe the second prototype and state of the demonstrations.

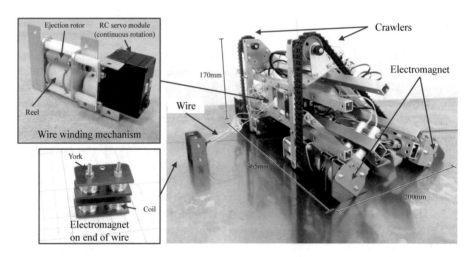

**Fig. 6.** The second prototype

## 5.1   The Second Prototype

Figure 6 shows the second prototype we have made. The second prototype is composed of aluminum frames and 2.5 kg in weight. We describe the implementation of the three functions on the second prototype.

### Function to unite with other objects

The second prototype units with objects by using electromagnets, unlike the first prototype. Using electromagnets has an advantage in uniting with various shapes because it does not need that objects have projections. Of course, electromagnets can unite only with irons or other magnets and we have to develop a method to unite with various object regardless of materials in the future. As Fig. 6 shows, electromagnets is implemented on under part of the second prototype and end of the wire in order to unite lower iron or magnet part of surrounding objects. The electromagnets have been made by rolling enameled wire around a pure iron core 1000 times and put them between yokes made of pure iron. The yokes make the loop of the magnetic line of force and make the adsorption power become stronger. The enameled wire's diameter is 0.32 mm, the iron core's diameter is 10 mm and length is 9 mm, and the yoke's thickness is 2 mm. Each of the electromagnets arises about 4kgf adsorption power by applying a voltage of 11 V. We have equipped 11 electromagnets on the second prototype.

### Function to Move Objects

We have implemented Dyneema wire and set a wire winding mechanism in front of the second prototype. By rotating the reel by a continuous rotation RC servo module and winding the wire, the mechanism can give 10 kgf tension

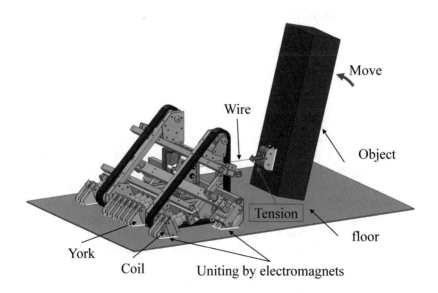

**Fig. 7.** The second prototype uniting with object.

to the wire. We have implemented another rotor to prevent tangling of wire when the wire winding mechanism ejects the wire. The second rotor rotates synchronize with the reel, and eject the wire without tangling of them.

**Function to Move**

We have implemented the function to move by using crawlers. By rotating the left and right crawlers independently, the second prototype can move to forward, backward and turn. Two continuous rotation RC servo modules move crawlers, and its maximum drive force is about 5 kgf at one side.

Figure 7 shows an example of uniting between the second prototype and objects and moving the object by the wire winding mechanism. In Fig. 7, yellow zone means parts of uniting by electromagnets and iron floor. We have conducted demonstrations that we describe in next section by remote operation.

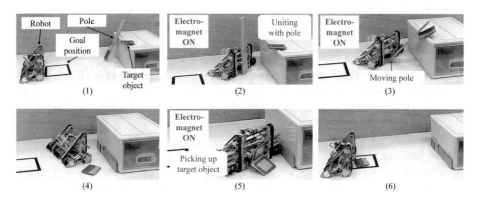

**Fig. 8.** Demonstration of creating function of handling object on high place.

## 5.2  Demonstration of Creating a Function to Handle Object on a High Place

The first demonstration shows creating function of handling objects on a high place like shown in Fig. 1(a). In this demonstration, we have set a pass case as a target object and set the goal to move the target object to the goal zone. Although the second prototype itself could not reach to the target object, by uniting with a pole, the second prototype gets the function of reaching to the target object. The target object and the pole have had iron parts and it was possible to unite with the second prototype by electromagnets.

Figure 8 shows the first demonstration. At first, we have set the situation as shown in Fig. 8(1). We have set the target object on the high place to which the second prototype itself can't reach. Figure 8(2) shows the uniting of the second prototype and the pole and then prototype has got the function of handling objects on the high place. The pole is 385 mm high and height of the second prototype has increased by 126% by uniting with it. In Fig. 8(3), the second prototype has moved the pole by ejecting the wire. Figure 8(4)–(6) shows the

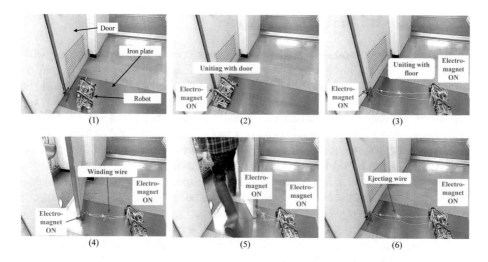

**Fig. 9.** Demonstration of creating automatic open and close function of door.

second prototype carrying the target object to the goal zone. Through the above process, the second prototype could create the function of handling objects on a high place. These processes take 152 s in total ((1)–(2) takes 42 s, (2)–(3) takes 15 s, (3)–(4) takes 30 s, (4)–(5) takes 40 s, (5)–(6) takes 25 s).

## 5.3   Demonstration of Creating Automatic Open and Close Function for Door

The second demonstration shows the uniting of a robot and a door and creation of the automatic open and close function of the door. We set iron plates on the door and the floor in order to unite with the second prototype.

Figure 9 shows the appearance of the demonstration. Figure 9(1) shows the setting at the start of the demonstration. At first, by the electromagnet on the end of the wire, the second prototype has united with the door as shown in Fig. 9(2). Then, the second prototype has ejected the wire and went backward to a suitable position. Then, the second prototype has united with the floor by electromagnets on under part of the body as shown in Fig. 9(3). Figure 9(4) shows that the second prototype opened the door by winding the wire.

By this, the second prototype was able to have opened the door as shown in Fig. 9(5) Then the second prototype has ejected the wire, and the door closed as shown in Fig. 9(6) work of the closer. Through these processes, the second prototype was able to create the automatic open and close function by uniting with the door. These processes take 201 s in total ((1)–(2) takes 13 s, (2)–(3) takes 52 s, (3)–(4) takes 50 s, (4)–(5) takes 48 s, (5)–(6) takes 38 s).

# 6    Conclusion and Future Works

In this paper, we proposed a robot that creates new functions by uniting with surrounding object. Such a robot can be applied to various situations by creating necessary functions. We made two prototypes of the concept and conducted the demonstrations of creating functions actually by using the prototypes. In the first demonstration, the first prototype has united with a model of a projector and created automatic angle adjustment function. In the second demonstration, the second prototype has united with a pole and created a function of handling objects on a high place. In the third demonstration, the second prototype has united with a door and created automatic open and close function of the door. Through these demonstrations, we have confirmed the effectiveness of our concept.

One of future works is the automation of the process in which robot unite with objects and creates function. In order to achieve this, it is necessary to develop a system that decides which object is best to create target function. Uniting between multiple robots and multiple objects is another future work. The robots would be able to create more diverse functions and be applicable to more diverse situations by achieving this. One example of this is to become a robot arm as shown in Fig. 1(c). In this paper, we describe the prototype that units with objects by using electromagnets, and the prototype can unite only with objects that have iron parts or magnet parts. In order to make the method that we proposed be effective, we have to develop the method to unite with more types of objects. For example, using vacuum suction or gluing might be effective to achieve this.

# References

1. Boxerbaum, A.S., Werk, P., Quinn, R.D., Vaidyanathan, R.: Design of an autonomous amphibious robot for surf zone operation: part i mechanical design for multi-mode mobility. In: Proceedings, 2005 IEEE/ASME International Conference on Advanced Intelligent Mechatronicsm, pp. 1459–1464, July 2005
2. Chirikjian, G.S., Burdick, J.W.: Design and experiments with a 30 DoF robot. In: Proceedings IEEE International Conference on Robotics and Automation, pp. 113–119, vol. 3, May 1993
3. Fukuda, T., Nakagawa, S.: Approach to the dynamically reconfigurable robotic system. J. Intell. Robot. Syst. **1**(1), 55–72 (1988)
4. Hirai, K., Hirose, M., Haikawa, Y., Takenaka, T.: The development of honda humanoid robot. In: Proceedings, 1998 IEEE International Conference on Robotics and Automation (Cat. No. 98CH36146). vol. 2, pp. 1321–1326, May 1998
5. Jorgensen, M.W., Ostergaard, E.H., Lund, H.H.: Modular ATRON: modules for a self-reconfigurable robot. In: Proceedings of IEEE/RSJ International Conference on Intelligent Robots and Systems(IROS 2004), vol. 2, pp. 2068–2073. IEEE (2004)
6. Kamimura, A., Kurokawa, H.: High-step climbing by a crawler robot DIR-2 - realization of automatic climbing motion. In: 2009 IEEE/RSJ International Conference on Intelligent Robots and Systems, pp. 618–624, October 2009
7. Kato, I.: Development of WABOT-1. Biomechanism **2**, 173–214 (1973)

8. Levihn, M., Nishiwaki, K., Kagami, S., Stilman, M.: Autonomous environment manipulation to assist humanoid locomotion. In: 2014 IEEE International Conference on Robotics and Automation (ICRA), pp. 4633–4638, May 2014
9. Lyder, A., Garcia, R.F.M., Stoy, K.: Mechanical design of odin, an extendable heterogeneous deformable modular robot. In: 2008 IEEE/RSJ International Conference on Intelligent Robots and Systems, pp. 883–888, September 2008
10. Mizuuchi, I., Waita, H., Nakanishi, Y., Yoshikai, T., Inaba, M., Inoue, H.: Design and implementation of reinforceable muscle humanoid. In: Proceedings of IEEE/RSJ International Conference on Intelligent Robots and Systems (IROS2004), vol. 1, pp. 823–833 (2004)
11. Mondada, F., Pettinaro, G.C., Guignard, A., Kwee, I.W., Floreano, D., Deneubourg, J.L., Nolfi, S., Gambardella, L.M., Dorigo, M.: SWARM-BOT: a new distributed robotic concept. Auton. Robots $17$(2–3), 193–221 (2004)
12. Murata, S., Yoshida, E., Kamimura, A., Kurokawa, H., Tomita, K., Kokaji, S.: M-TRAN: self-reconfigurable modular robotic system. IEEE/ASME Trans. Mechatron. $7$(4), 431–441 (2002)
13. O'Flaherty, R., Vieira, P., Grey, M.X., Oh, P., Bobick, A., Egerstedt, M., Stilman, M.: Humanoid robot teleoperation for tasks with power tools. In: Proceedings of IEEE International Conference on Technologies for Practical Robot Applications, pp. 1–6. IEEE (2013)
14. Waldron, K., McGhee, R.: The adaptive suspension vehicle. IEEE Control Syst. Mag. $6$(6), 7–12 (1986)
15. Wright, C., Buchan, A., Brown, B., Geist, J., Schwerin, M., Rollinson, D., Tesch, M., Choset, H.: Design and architecture of the unified modular snake robot. In: 2012 IEEE International Conference on Robotics and Automation, pp. 4347–4354, May 2012

# Sensing and Actuation

# Configuration Depending Crosstalk Torque Calibration for Robotic Manipulators with Deep Neural Regression Models

Adrian Zwiener[1(✉)], Sebastian Otte[2], Richard Hanten[1], and Andreas Zell[1]

[1] Cognitive Systems Group, University of Tübingen,
Sand 1, 72076 Tübingen, Germany
{adrian.zwiener,richard.hanten,andreas.zell}@uni-tuebingen.de
[2] Cognitive Modeling Group, University of Tübingen,
Sand 14, 72076 Tübingen, Germany
sebastian.otte@uni-tuebingen.de

**Abstract.** In this paper, an approach for articulated robotic manipulator which minimizes configuration depending crosstalk torques is presented. In particular, these crosstalk torques are an issue for the Kinova Jaco 2 manipulator. We can experimentally show that the presented approach leads to crosstalk minimization for the Kinova Jaco 2 manipulator. Crosstalk leads to a significant difference between sensor output and inverse dynamic models using CAD rigid body parameters. As a consequence, these disturbances lead to a hindered torque control and perception. Different machine learning techniques, namely Random Forests and various neural network architectures, are evaluated on this task. We show that particularly deep neural regression networks are able to learn the influence of the cross torques which improves perception.

**Keywords:** Crosstalk torques · Calibration · Machine learning

## 1 Introduction

Dexterous control of a robotic manipulator in human environments requires flexible joints to prevent injuries and damages. Flexible joints can be achieved by elastic joints with variable stiffness or by using torque control and software compliance control. In particular, torque control requires a precise dynamic model. Moreover, contact estimation can give more information about the environment.

Various phenomena, which cannot be modelled easily, lead to disturbances and an inaccurate dynamic model. These phenomena are friction, thermal effects, and crosstalk torques due to strains and unwanted loads which cannot be avoided in the robot assembly.

© Springer Nature Switzerland AG 2019
M. Strand et al. (Eds.): IAS 2018, AISC 867, pp. 361–373, 2019.
https://doi.org/10.1007/978-3-030-01370-7_29

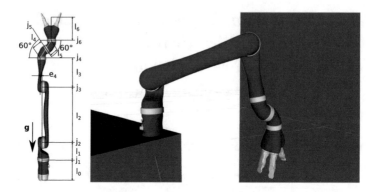

**Fig. 1.** Model of Jaco 2. Left: Jaco 2 in extracted configuration $\mathbf{q}_0 = (0, \pi, \pi, 0, 0, 0)^T$. The joints are called $j_i$ for $i \in 1, \ldots, 6$. Right: Planning scene in which the robot motion is planed. The red boxes are forbidden areas.

Unfortunately, crosstalk in the Jaco 2's torque sensors lead to inaccurate torque sensing. As a consequence, there is a configuration depending difference between the *inverse dynamic model* (IDM) and the sensed joint torque which is significant. This discrepancy cannot be cured by an optimization of the inertial parameters, link mass, and centre of masses.

Another consequence is that controller gains for torque control are hard to optimize. Usually, one introduces linear position and velocity feedback terms in the torque control law to compensate for inaccuracies in the dynamic model. Due to the unwanted torques, the gains of position and velocity gains have to be chosen to be the dominant therms in the control law. This causes unnecessarily stiff joints similar to a simpler velocity controller.

This problem may be solved by more rigorously obtained controller gains. For example, one could use Bayesian or *Gaussian Process* (GP) optimization [11,16], but for a simple trajectory controller for the Jaco 2 we had a 12 dimensional search space, which can be challenging. Furthermore, an optimal controller solves the problem just partially. It has been shown that tactile information enhances the success rate during grasping [5,14]. Besides, tactile information can also be inferred from joint torques [9,18,20]. However, if crosstalk torques are not minimized, a dynamic simulation of the Jaco 2 differs significantly from the real manipulator. Consequently, controller gains cannot be optimized in simulation, but can only be used as an initial guess for the real manipulator.

Thus, for our torque calibration we require three objectives: (i) the calibration can be used for tracking control, (ii) we can predict external contacts, and (iii) a realistic simulation can be obtained. If we improve our IDM in general, we can meet all objectives. For every configuration we have to predict the sensed joint torques as accurately as possible. Consequently, we discuss methods to predict the influence of the crosstalk torques for a given configuration leading to more accurate predictions on the joint torques.

The contents of this paper is organized as follows: In Sect. 2 related work is discussed. In Sect. 3 we briefly introduce dynamic models and define the problem. In Sect. 4 the used methodology is discussed. In Sect. 5 data acquisition and experiments are described. Finally, in Sect. 6 results are discussed and a conclusion can be found in Sect. 7.

## 2   Related Work

A control law can be designed so that disturbances and uncertainties can be overcome. But, since we are also interested in an accurate perception of the robot's dynamic state, we will not focus on controllers here. Instead, we want to discuss two other classes of approaches: One approach is calibration, the other is observers estimating the disturbances online. A review of different approaches determining rigid body parameters can be found in [17, Chap. 14].

Ishiguro et al. [4] used a neural network compensator to overcome the disturbances between an IDM and sensed torques. A *Multi-Layer-Perceptron* (MLP) is used to learn the error between torque sensors and the IDM. This MLP learns a function from sensed positions, velocities, and derived accelerations to a torque which compensates the error. Thus, not the full model has to be learned which simplifies the problem. In this paper we borrow this idea and learn the discrepancy between model and sensors instead of learning a full IDM.

A review of some ML algorithms, e.g. GP, learning the IDM can be found in [15, Chaps. 2.5 and 8.1].

Lee et al. [8] discuss a torque sensor calibration for manipulators with $n = 6$ joints minimizing crosstalk by introducing virtual loads. In a primary and secondary calibration, two $1 \times n$ matrices are determined with least squares to optimize the estimation of loads at the end effector. If no load is applied, the output will be zero. The approach is experimentally verified on a 3-DoF robotic wrist. While, this approach is designed to estimate external loads, we aim for a dynamic perception which can also be used for tracking control.

Similarly, Kim et al. [6] performed a crosstalk calibration for load estimation additionally optimizing the sensing frame. An optimization of the rotation and location of the sensor is considered since the actual sensor frame can differ from the joint frame. The approach is experimentally verified on a 6-DoF manipulator.

In Camorina et al. [1] an incremental semi-parametric method to learn the IDM is proposed. Firstly, the inertial parameters $\phi$ are determined using the *regressor-matrix-formulation* of Eq. (3). The first step is independent of the non-parametric kernel-based ML step. Secondly, *Recursive Regularized Least Squares* are used to learn non-linear effects, which are not modelled by the standard IDM Eq. (1). This online method was tested on the iCub humanoid robot.

A survey on collision detection, isolation, and identification is presented by Haddadin et al. in [3]. The best performing method is a generalized momentum based observer [10] depending on an accurate full dynamic model. In one of our previous works [20], we used this technique for contact point localization and showed that position depending crosstalk torques lead to an unsatisfactory localization for optimization.

Zelenak et al. [19] discuss an *Extended Kalman Filter* (EKF) for collision detection on a 7-DoF manipulator. While collision detection is done by thresholding, the EKF is used to enhance the accuracy of the dynamic model. Thus, the modelled joint torques are chosen as state variable, therefore the gradient of Eq. 1 for the joint torques has to be derived.

Mohammadi et al. [12] developed a Lyapunov stable torque disturbances observer for serial robots without any restrictions on the robot's degrees of freedom, configuration or joint types. The observer is validated in simulation and on the PHANToM Omni haptic device. However, external torques are merged together with all other sources of disturbances. Thus, external contacts cannot be accurately detected.

## 3   Problem Description

The inverse dynamic model of a serial manipulator with $n$ revolute joints maps joint positions $\mathbf{q} \in \mathbb{R}^n$, velocities $\dot{\mathbf{q}} \in \mathbb{R}^n$, and accelerations $\ddot{\mathbf{q}} \in \mathbb{R}^n$ to the modelled joint torques

$$\boldsymbol{\tau}_M = H(\mathbf{q})\ddot{\mathbf{q}} + \mathbf{c}(\mathbf{q},\dot{\mathbf{q}}) + \mathbf{g}(\mathbf{q}) + \boldsymbol{\tau}_{ext}. \tag{1}$$

Here, the $H(\mathbf{q}) \in \mathbb{R}^{n \times n}$ is the inertia moment matrix, $\mathbf{c}(\mathbf{q},\dot{\mathbf{q}}) \in \mathbb{R}^n$ corresponds to centripetal and Coriolis forces. Gravitational forces are given by $\mathbf{g}(\mathbf{q}) \in \mathbb{R}^n$. Besides, the influence of external torques/forces is introduced by $\boldsymbol{\tau}_{ext}$ [17, Chap. 2.4]. External torques $\boldsymbol{\tau}_{ext}$ are a priori only known if the robot actively provides a force/torque to an object. If the contact with the robot occurs passively, we have to estimate this contact torque from the torque disturbance.

In general, joint torque sensors can be effected by friction, strains or other effects which are hard to distinguish. Therefore, there will be a disturbance/inaccuracy between sensed torques $\boldsymbol{\tau}_S \in \mathbb{R}^n$ and the model

$$\boldsymbol{\Delta}_\tau = \boldsymbol{\tau}_M - \boldsymbol{\tau}_S. \tag{2}$$

The true nature of these disturbances $\boldsymbol{\Delta}_\tau$ is in general unknown.

Of course, inaccurate model parameters also contribute to the disturbance, but parameters can be optimized: Eq. (1) is linear in the rigid body parameters of the manipulator, which are the symmetric link inertia matrix $I_i \in \mathbb{R}^{3 \times 3}$, link mass $m_i \in \mathbb{R}$ and the links centre of mass $\mathbf{r}_{cm_i} \in \mathbb{R}^3$. One is able to reformulate Eq. (1) as matrix vector product [17, Chap. 14.3.4]:

$$\boldsymbol{\tau}_M = K(\mathbf{q},\dot{\mathbf{q}},\ddot{\mathbf{q}})\,\boldsymbol{\phi}, \tag{3}$$

where there are no external torques $\boldsymbol{\tau}_{ext} = 0$, the matrix $K \in \mathbb{R}^{n \times 10n}$ depends on the manipulator's structure and the state, and the vector

$$\boldsymbol{\phi} = \left(m_1,\, m_1 \cdot \mathbf{r}_{cm_1},\, \mathbf{l}(I_1),\, \ldots,\, m_n, m_n \cdot \mathbf{r}_{cm_n},\, \mathbf{l}(I_n)\right)^T \tag{4}$$

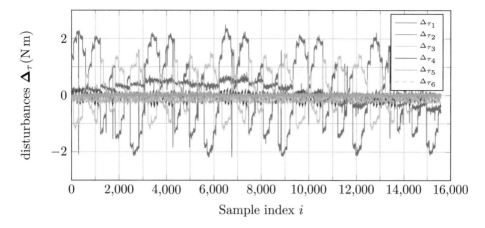

**Fig. 2.** Disturbance $\boldsymbol{\Delta}_\tau$ between joint torque sensors and a dynamic model using CAD inertial parameters. For sample index $i = 0$ we have the minimum configuration, $i = 5^6$ is the maximum. In between we have every possible configuration in the interval for 5 discrete steps. The mean RSME is 0.62 N m.

contains all the manipulator's rigid body parameters and where

$$\mathbf{l}(I_i) = \left(I_{i_{x,x}}, I_{i_{x,y}}, I_{i_{x,z}}, I_{i_{y,y}}, I_{i_{y,z}}, I_{i_{z,z}}\right)^T \tag{5}$$

is a vector containing the unique values of a symmetric inertia matrix $I_i$. The model parameters $\phi$ can be found by singular value decomposition and a dataset of exciting trajectories.

For modern manipulators designed with CAD programs and computer aided manufacturing, we can expect the real parameters $\phi$ to be close to ones used by CAD programs. In fact, using these parameters leads to already reasonable results: The mean RMSE of the disturbance on data set-1 (Sect. 5.1) is approximately 0.62N m. The disturbance is displayed in Fig. 2. Besides, in the left plane of Fig. 3, we plotted the normalized sensed torques against normalized modelled torques. In a perfect world, we would obtain the identity function $f(x) = x$. For joints 2, 3, 4, and 5, the sensed torques scatter around the identity function, thus the uncalibrated model is already reasonable. Nevertheless, this is not true for joints 1 and 6. In addition, the joint axis $z_1$ is parallel to gravity, see Fig. 1. Due to symmetry, joint torque $\tau_{S_1}$ should be close to zero for each sampled configuration. Similarly, sensed torques in the 6th joint should also be very small due to symmetry. On top of that, static torques in joint 6 are quite easy to calculate.

The control API provided by Kinova already has an option to optimize the static components of the inertia parameters (masses and centre of masses). Predefined trajectories are used to optimize inertial parameters internally. The mean RSME disturbance of this estimator is 0.26 N m on our dataset and already smaller than the mean RSME of the full IDM using CAD inertial parameters. Nevertheless, due to the odd sensor output, the crosstalk torque should be minimized first, otherwise inertial parameters compensate other disturbances which

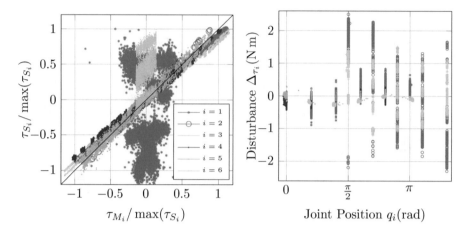

**Fig. 3.** Sampled configuration space with 5 steps per joint (same data as in Fig. 2). Left: Normalized sensed torques $\tau_{S_i}/\max(\tau_{S_i})$ plotted as a function of the normalized model torque $\tau_{M_i}/\max(\tau_{S_i})$. Ideally, we would obtain the identity function (black line). This is far off for joints 1 and 6. Right: Disturbance $\boldsymbol{\Delta}_{\tau_i}$ plotted against the joint position $q_i$. The vertical clustering indicates that a joint's position is not enough to cure the error.

do not originate from inaccurate parameters. This can be problematic, e.g. for admittance control. Wrongly predicted torques lead to virtual forces repelling the manipulator from valid configurations.

If we can neglect external torques/forces during a calibration phase, we might compensate inaccuracies due to friction and crosstalk torque. For the Jaco 2, we believe that crosstalk torques are the main source of disturbances because we can see huge differences even if the arm is not moving. Secondly, there are unreasonable measurements in joints 1 and 6. Thirdly, we can conclude from Fig. 2 and the left of Fig. 3 that there is an angle depending error which cannot be explained by the position of one joint alone. However, incorrect centre of masses may also lead to a joint angle depending error, although the centre of mass for link 6 cannot be that inaccurate since the joint torque in joint 6 is simple to calculate. In fact, optimizing inertia parameters leads to unphysical solutions, where the centre of masses locations differ from the theoretical centre by 1 m.

## 4    Methodology

In Sect. 3 we discussed the inverse dynamic model, the current accuracy of the model, and possible sources of the disturbance. The following strategies may minimize crosstalk and the disturbance between sensed torques and the inverse dynamic model. To determine the static disturbance $\boldsymbol{\Delta}_\tau$, regressors can be trained using suitable algorithms to find a mapping from the joint position $\mathbf{q}$ to the static $\boldsymbol{\Delta}_\tau$. To achieve this we have to present data of input values

$x_i \in \mathbb{R}^k; k, i \in \mathbb{N}$ and the corresponding desired output value $F(x)_i \in \mathbb{R}^l l, i \in \mathbb{N}$ to the regressor. Note that $l \neq k$ is possible. Thus, these regressors can be seen as functions $F : \mathbb{R}^n \mapsto \mathbb{R}^n$ with $n$ the number of joints, respectively $F(\mathbf{q}) = \boldsymbol{\Delta}_\tau$. The correction of model $\tilde{\boldsymbol{\tau}}_M$ or the sensors $\tilde{\boldsymbol{\tau}}_S$ is then

$$\tilde{\boldsymbol{\tau}}_M = \boldsymbol{\tau}_M - F \quad \text{or} \quad \tilde{\boldsymbol{\tau}}_S = \boldsymbol{\tau}_S + F, \tag{6}$$

where $F$ is the response of the regressor. In this paper, *Random Forests* (RFs) and *neural networks* (NNs) are used to find these regressions.

## 4.1  Random Forest Regression

A RF is an ensemble of decision trees which can be used for classification and regression. Actually, a RF predicts only mean values, but for an increasing tree depth, more and more splits of the input space can be made, leading to a mean response. Thus, a deep RF is able to learn non-linear functions. An increasing number of trees in the forest improves feature space coverage.

An advantage of RFs is that the importance of a variable or feature can be determined. Consequently, after training, one is able to draw a conclusion on which features are statistically relevant for a result or not.

In this work, we implemented the RF regressor with OpenCv 3.2.[1] However, only RF regression problems of type $RF : \mathbb{R}^n \mapsto \mathbb{R}$ are supported. Therefore, we train $n$ RFs, one for each joint separately.

## 4.2  Neural Network Regression

Artificial neural networks, which can be seen as trainable hierarchies of function compositions, have been shown to be powerful function approximators [2]. Particularly, deep neural architectures are of recent interest, as they can identify more and more high-level features from raw data with deeper layers, which results in state-of-the-art results on many (visual) pattern recognition tasks. Even though the non-visual learning problem of this paper does not fit in the usual application scenario of convolutional networks as the input dimension is only 6, a certain degree of abstraction might still be helpful to form a well-performing regression model. In order to investigate this hypothesis, different network architectures are studied in this paper. While we tested far more architectures (including variants of recurrent neural networks), we only present the most successful topologies in the results section, namely multi-layer feed-forward networks with *rectified linear units* (ReLU) in the hidden layers and a linear output layer with a bias term, and we applied *trainable biases* (b) in all stages. In our experiments, we used JANNLab [13] to train NNs. After training, the network's weights are stored to a text file. To load trained networks, we implemented a simple C++ library supporting feedforward networks with different activation functions. For training, we used Adam [7] with learning rate = 0.001 and 0.0001, $\beta_1 = 0.9$, and $\beta_2 = 0.999$ as well as different mini-batch sizes (1,16 and 64).

---

[1] https://github.com/itseez/opencv.

## 4.3   Lookup Table

The simplest solution could be to record a lookup table. In the calibration phase, we simply save the measured disturbance for the current configuration. Consequently, we would end up with a lookup table with $m^6$ dimensions for $m$ steps per joint, in which we have to store a 6-dimensional vector. Thus, if doubles are stored, we require $48 \cdot m^6$B to store our lookup table. If we assume each joint has a range of $360°$ and we want to have a resolution of $10°$, we already require $97\,\mathrm{GB}$ of storage. Therefore, we read and write the lookup table to a SSD because the required capacity is cheap and access is still fast, and we do not need a lot of RAM for such a simple task as calibration. In addition, since an even smaller resolution is not realistic, an incremental mean is used if we want to store data within the same bin multiple times.

---

**Algorithm 1.** Data Acquisition

---

1: **procedure** SAMPLE JOINT SPACE(Interval $Q \in \mathbb{R}^{6 \times 2}$, Steps $m$)
2:       $\delta\mathbf{q} \leftarrow$ steps size, $Q$ discretized in $m$ steps;
3:       $\mathbf{q}_g \leftarrow$ minimum of $Q$;
4:    **for** $i = 0 \rightarrow m - 1$ **do**
5:          plan collision free path to $\mathbf{q}_g + i\delta\mathbf{q}$;
6:          **if** collision free path found **then**
7:                move to $\mathbf{q}_g + i\delta\mathbf{q}$;
8:                **while** motion not finished & $||\dot{\mathbf{q}}||_2 >$ threshold  **do**
9:                      wait;
10:                Sample $\leftarrow$ record data for $1\,\mathrm{s}$;
11:                Save $mean$(Sample);

---

## 5   Experiments

### 5.1   Data Acquisition

To acquire datasets for the calibration process, a region of the joint space is sampled with $m$ steps. We believe that the main source of the disturbance is position depending. Velocity depending errors should be fixed by optimizing the inertial parameters. Therefore, we only sample different manipulator configurations in rest. The data acquisition is described by the Algorithm 1.

**Table 1.** RESULTS

| Method | Mean RMSE $\Delta_\tau$ (N m) | |
|---|---|---|
| | 10-fold CV set-1 | extra-test-set |
| LUT5 | $0.003 \pm 0.26^{a,b}$ | $0.460 \pm 0.290$ |
| LUT4 | $0.160 \pm 0.15^{a}$ | $0.380 \pm 0.240$ |
| RF10-16 | $0.080 \pm 0.0033$ | $0.230 \pm 0.120$ |
| 4×ReLU32b | $0.062 \pm 0.0011$ | $0.218 \pm 0.007$ |
| Kinova GC | $0.260 \pm 0.070$ | $0.200 \pm 0.060$ |

[a] Note, that this experiment is not very significant since only the stored data is recalled.
[b] For lookup tables, no CV was performed.

To conveniently record the data, we use *ROS MoveIt*[2], a motion planning frame work. Thus, we do not have to care about self-collisions or collisions with the environment since collision boxes are added around regions of the Cartesian space, where objects are e.g. a desk supporting the manipulator or walls, c.f. left panel of Fig. 1. The Jaco 2 moved for a maximum period of 10 h at a time without any collisions. We sample the configuration space with $m = 4$ and $m = 5$ steps and the following interval:

$$Q = \left( [0, \pi], \left[\pi, \tfrac{3\pi}{2}\right], \left[\pi, \tfrac{3\pi}{2}\right], [0, \pi], [0, \pi], [0, \pi] \right)^T \qquad (7)$$

In the following, the set with $m = 5$ steps is used for training and thus called **set-1**. The set with $m = 4$ is called **extra-test-set** since it is used to test how good the regressors generalize. Thus, **set-1** and **extra-test-set** have the same lower limit, but every other sample is slightly $(Q/20)$ shifted.

As an overall average, every 4 s, a sample is acquired. This amount of time is needed for planning, execution, and a small period of waiting until oscillations faded. Among these tasks, execution takes by far the most time. Recording set-1 took 4 h, whereas extra-test-set already required 17 h.

## 5.2   Evaluation of Lookup Tables

The evaluation of the lookup table is a bit problematic since a simple lookup table can only work well on sampled positions. Thus, we cannot simply perform a 10-fold cross validation because, for some samples, we had no data available, and the corrections were zero.

Recoding a sample takes 4 s. If we wanted to create a lookup table with 10° resolution, this solution took 207 years. Since the motion takes the most time, there is no realistic speed up of the data acquisition. Consequently, the use of lookup tables seems unrealistic. However, we still give some results of the RMSE as benchmarks: (i) A lookup table with 5 steps is recorded on the set-1 (LUT5)

---

[2] http://moveit.ros.org.

and evaluated. Results: Table 1. Please note that an evaluation of the lookup table on the recorded data is not fair. (ii) A lookup table with 4 bins is created and tested on the set-1 (LUT4), c.f. Table 1.

## 5.3    Evaluation of NN and RF Regression

For both ML approaches, 10-fold cross validation is performed. We present the results of the best NN architectures and their performance on the extra-test-set in Table 2. All network architectures have an input layer size of six and also six linear output neurons. In order to compare the architectures against a pure linear model, the best linear network is also listed in Table 2. The results of these experiments indicate that the problem has indeed non-linear facets, as the error of the linear model is significantly worse than error of the non-linear models.

**Table 2.** Results NN

| Network | Learning rate | Mini-batch size | Mean RMSE | (N m) |
|---|---|---|---|---|
| | | | 10-fold CV set-1 | extra-test-set |
| 4×ReLU32b | 0.001 | 1 | $0.06 \pm 0.011$ | $0.218 \pm 0.007$ |
| 2×ReLU32b | 0.0001 | 1 | $0.07 \pm 0.010$ | $0.202 \pm 0.006$ |
| 6×ReLU32b | 0.001 | 1 | $0.07 \pm 0.010$ | $0.223 \pm 0.008$ |
| 4×ReLU16b | 0.0001 | 64 | $0.07 \pm 0.004$ | $0.204 \pm 0.007$ |
| 2×ReLU32b | 0.001 | 1 | $0.07 \pm 0.019$ | $0.209 \pm 0.011$ |
| 2×ReLU32b | 0.001 | 16 | $0.07 \pm 0.011$ | $0.209 \pm 0.011$ |
| 2×ReLU16b | 0.0001 | 16 | $0.08 \pm 0.006$ | $0.203 \pm 0.008$ |
| 2×ReLU16b | 0.0001 | 64 | $0.08 \pm 0.007$ | $0.201 \pm 0.005$ |
| 2×ReLU16b | 0.001 | 16 | $0.08 \pm 0.013$ | $0.200 \pm 0.012$ |
| Linear | 0.0001 | 64 | $0.241 \pm 0.004$ | $0.285 \pm 0.003$ |

For RF, one has to make a trade-off between overfitting, depth, and accuracy. Deeper RFs tend to overfitting. However, this can checked with the cross validation. Moreover, for most problems, the RF validation error reduces asymptotically. The best RF is a forest with a depth of 10 and 16 trees (RF10-16).

The best NN architecture is an MLP with 4 ReLU-layers with bias and 32 neurons each and the last layer is a linear output layer with bias (4×ReLU32b). The results can be seen in Table 1. In addition, the resulting correction of the IDM is visible in Fig. 4. Besides, both regressors are tested on the extra-test-set (second column). Furthermore, in Table 1, we added the mean RSME and standard deviation for the calibrated *Kinova gravity compensator* (Kinova GC).

## 6   Result/Discussion

As one can conclude form Table 1, the ML regressors lead to an enhanced accuracy. Near sampled configurations, the error is minimal. With farther distance to the training set, the error is in the same order of magnitude as the gravity compensation proposed by Kinova. Overall, best results are achieved using feed-forward networks with rectified linear units with 4 hidden layers.

The design of the ML regressors guarantees that we can still use our ML compensator during motion. Experiments showed that during motion the RMSE is still reduced compared to an IDM using only inertial parameters, although due to noise in the sensor data RMSE is still higher than for static cases, but that can be expected.

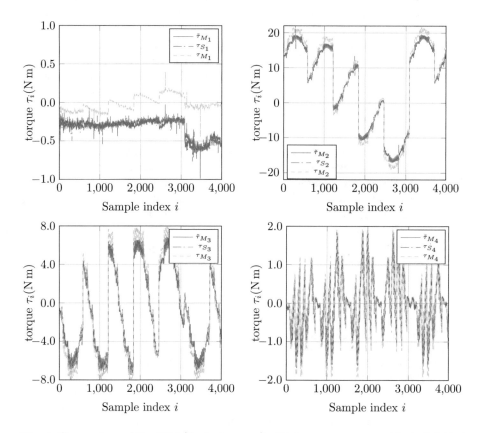

**Fig. 4.** Comparison of the IDM (dashed orange) with torque sensors $\tau_{S_i}$ (dashed dotted red) and the learned correction $\tilde{\tau}_{M_i} = \tau_{M_i} - F$ (solid blue) for joints 1 to 4 and set-1, where $F$ is the response of the 4×ReLU32b MLP. Note, the curves for joint 5 is similar to 4 and joint 6 to 1.

In addition, online experiments showed that, for configurations the regressors have not seen during training, the RMSE is still in the same order of magnitude as the one obtained by the IDM using CAD inertias, since the main disturbance is due to joints 2 and 3, and for these joints we sampled most of the joint space.

Lookup tables perform well if bins are sufficiently small and have been sampled. However, gathering the required data takes too long and usage of storage is not efficient, since lookup tables are very large and perform with similar or lower accuracy than RFs or MLPs. Moreover, the 4×ReLU32b has only 3,590 weights, the lookup tables store up to 875,000 values. If we consider our RF regressor to be a full binary tree, we have $2^{depth} \cdot N_{trees} \cdot 6 = 98,304$ leaves.

Without interpolation, the results of a lookup table cannot be used. As a consequence, it is far more efficient to use an MLP, since they are more efficient and also simpler to use because MLP implementations are very common. Lookup tables for a 6 dimensional problem interpolating between neighbours would most likely have to be implemented.

Finally, we can conclude that there are indeed other influences than inaccurate masses and centre of masses that disturb the torque sensors, due to the fact that the inertia calibration will not reduce the RSME below a certain error.

## 7   Conclusion

In this work, we showed an ML approach learning the disturbance between a IDM and real torque sensors. Here, the cross torque disturbance is mainly position depending. However, in general, additional disturbances can depend on velocities and accelerations as well. These effects, however, can also be learn by sampling the work space and using velocities and accelerations as input for the regressors.

Ongoing work is to incorporate an unsupervised online learning approach, so that disturbances can be minimized while the Jaco 2 is operated without running further calibration sequences.

## References

1. Camoriano, R., Traversaro, S., Rosasco, L., Metta, G., Nori, F.: Incremental semi-parametric inverse dynamics learning. In: 2016 IEEE International Conference on Robotics and Automation (ICRA), pp. 544–550. IEEE (2016)
2. Goodfellow, I., Bengio, Y., Courville, A.: Deep Learning. MIT Press (2016)
3. Haddadin, S., De Luca, A., Albu-Schäffer, A.: Robot collisions: a survey on detection, isolation, and identification. IEEE Trans. Robot. **33**(6), 1292–1312 (2017)
4. Ishiguro, A., Furuhashi, T., Okuma, S., Uchikawa, Y.: A neural network compensator for uncertainties of robotics manipulators. IEEE Trans. Ind. Electron. **39**(6), 565–570 (1992)
5. Jain, A., Killpack, M.D., Edsinger, A., Kemp, C.C.: Reaching in clutter with whole-arm tactile sensing. Int. J. Robot. Res. **32**(4), 458–482 (2013)
6. Kim, Y.L., Park, J.J., Song, J.B.: Crosstalk calibration for torque sensor using actual sensing frame. J. Mech. Sci. Technol. **24**(8), 1729–1735 (2010)

7. Kingma, D.P., Ba, J.L.: Adam: a method for stochastic optimization. ArXiv e-prints abs/1412.6980 (2014)
8. Lee, S.H., Kim, Y.L., Song, J.B.: Torque sensor calibration using virtual load for a manipulator. Int. J. Precis. Eng. Manuf. **11**(2), 219–225 (2010)
9. Likar, N., Zlajpah, L.: External joint torque-based estimation of contact information. Int. J. Adv. Robot. Syst. **11**(1), 107 (2014)
10. de Luca, A., Mattone, R.: Sensorless robot collision detection and hybrid force/motion control. In: Proceedings of the 2005 IEEE International Conference on Robotics and Automation, pp. 999–1004, April 2005
11. Marco, A., Hennig, P., Bohg, J., Schaal, S., Trimpe, S.: Automatic LQR tuning based on Gaussian process global optimization. In: 2016 IEEE International Conference on Robotics and Automation (ICRA), pp. 270–277 (2016)
12. Mohammadi, A., Tavakoli, M., Marquez, H., Hashemzadeh, F.: Nonlinear disturbance observer design for robotic manipulators. Control Eng. Pract. **21**(3), 253–267 (2013)
13. Otte, S., Krechel, D., Liwicki, M.: JANNLab Neural Network Framework for Java. In: Poster Proceedings of MLDM 2013, pp. 39–46. ibai-publishing, New York (2013)
14. Pastor, P., Righetti, L., Kalakrishnan, M., Schaal, S.: Online movement adaptation based on previous sensor experiences. In: 2011 IEEE/RSJ International Conference on Intelligent Robots and Systems (IROS), pp. 365–371. IEEE (2011)
15. Rasmussen, C.E., Williams, C.K.: Gaussian Processes for Machine Learning, vol. 1. MIT press, Cambridge (2006)
16. Shahriari, B., Swersky, K., Wang, Z., Adams, R.P., de Freitas, N.: Taking the human out of the loop: A review of bayesian optimization. Proc. IEEE **104**(1), 148–175 (2016)
17. Siciliano, B., Khatib, O.: Springer Handbook of Robotics. Springer, Heidelberg (2008)
18. Vorndamme, J., Schappler, M., Haddadin, S.: Collision detection, isolation and identification for humanoids. In: 2017 IEEE International Conference on Robotics and Automation (ICRA), pp. 4754–4761. IEEE (2017)
19. Zelenak, A., Pryor, M., Schroeder, K.: An extended Kalman filter for collision detection during manipulator contact tasks. In: ASME 2014 Dynamic Systems and Control Conference, p. V001T11A005. American Society of Mechanical Engineers (2014)
20. Zwiener, A., Geckeler, C., Zell, A.: contact point localization for articulated manipulators with proprioceptive sensors and machine learning. In: 2018 IEEE International Conference on Robotics and Automation (ICRA) (2018, accepted)

# Simulation of the SynTouch BioTac Sensor

Philipp Ruppel, Yannick Jonetzko, Michael Görner, Norman Hendrich$^{(\boxtimes)}$, and Jianwei Zhang

Informatics Department, University of Hamburg, 22527 Hamburg, Germany
{ruppel,jonetzko,goerner,hendrich,zhang}@informatik.uni-hamburg.de

**Abstract.** We present a data-driven approach to simulate the BioTac tactile fingertip sensor within physics engines. The behavior of the sensor is first captured in an experimental setup that records positions and external forces of contacts as well as the sensor output. This data is then used to fit a non-linear model that maps force-annotated mesh collisions of a simulator to sensor responses.

We discuss two deep network architectures that reproduce the BioTac data with high accuracy and demonstrate the simulation of simple grasps with the Shadow Dexterous Hand and five BioTac sensors. We present an open source plug-in for the simulator Gazebo and release the captured dataset alongside this paper.

**Keywords:** Tactile sensing · Physics simulation
Dexterous manipulation · Deep neural networks

## 1  Introduction

Simulation plays an important role in robotics: Whereas it often takes a lot of effort to run an experiment on a physical robot, a stable simulation of the same robot allows to develop and test new algorithms much faster and under reproducible conditions. At the same time it remains a challenge in itself to build a physics simulation that (1) is reasonably stable, (2) resembles the behavior of the actual robot, and (3) runs fast enough for online analysis.

The BioTac sensor, depicted in Fig. 1 and described in detail in Sect. 3 below, is approximately the size of a human fingertip and contains a thin layer of conductive liquid under a rubber cover. When contacting a surface, the pressure and distribution of the liquid inside the sensor changes and is measured as changes in voltage. Given the complex physical mechanisms, there is no simple relationship between the applied forces and the sensor output data.

Several authors have proposed models to estimate the contact locations and directed forces based on the raw sensor readings [1–4]. The resulting estimates

This research was funded by the German Research Foundation (DFG) and the National Science Foundation of China in project Crossmodal Learning, TRR-169.

© Springer Nature Switzerland AG 2019
M. Strand et al. (Eds.): IAS 2018, AISC 867, pp. 374–387, 2019.
https://doi.org/10.1007/978-3-030-01370-7_30

**Fig. 1.** Close-up of a single SynTouch BioTac sensor (left) [6]. The five-fingered Shadow Dexterous hand with BioTac tactile sensors (right) [7].

can be used in applications that require force feedback, including reactive grasping, see Fig. 2(a). Additionally, the raw data from the BioTac sensor can be used in end-to-end applications; for example [5] demonstrated the identification of a variety of surface attributes from raw sensor readings, see Fig. 2(b).

To test and develop BioTac applications in simulation, a BioTac simulation model is needed that reads forces and contacts from the physics engine and generates realistic raw sensor data; corresponding to use-cases (d) and (e) in Fig. 2. In this paper, a machine learning approach is proposed to build such a sensor model for the Gazebo framework [8] and ROS [9].

The rest of this paper is structured as follows. Section 3 describes the BioTac tactile sensor and summarizes key properties of the multi-modal raw output signals of the sensor. It also describes the experimental setup to record and process the raw data from the real sensors for machine learning. Section 4 then describes concept and architecture of the Gazebo simulation model for the BioTac sensor. Experimental results based on the recorded datasets are presented and analyzed in Sect. 5. Finally, Sect. 6 summarizes the proposed approach.

## 2   Related Work

Today, standard physics engines allow simulating many common robot scenarios with sufficient accuracy, including wheeled robots and industrial robot arms. For example, the Gazebo simulation framework supports pluggable physics engines, including ODE [10], Bullet game physics [11], Simbody [12] and DART [13]. Popular commercial frameworks offer similar functionality, for example Webots [14] and V-REP [15].

However, realistic physics simulation of robot grasping remains a challenge, because of the need to accurately generate and analyze multi-finger object contacts, see [16,17] for reviews. Given 3D mesh models of the fingers and the object to be grasped, a grasp is defined by the set of contact points between the fingers (and optionally the palm of the hand) and the object. Grasp stability can then be calculated from applied finger forces and the finger/object friction properties. This approach was implemented in the GraspIt! simulator [18] and combined with efficient grasp search and the idea of grasp synergies [19] to prepare a full database of grasps on nontrivial objects for several robot

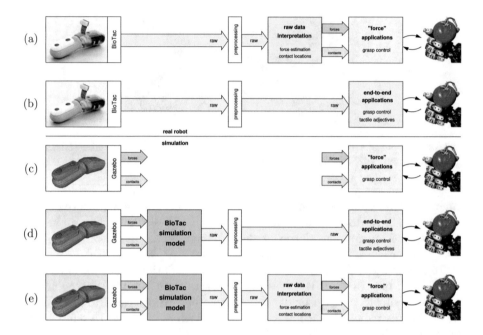

**Fig. 2.** Block diagram illustrating the five relevant BioTac sensor data processing options on the real robot and in Gazebo simulation. Traditionally, architecture (a) is the most common, while (b) is becoming popular due to recent advances in deep learning. All current Gazebo sensors directly use forces and contacts from the physics engine (c); raw data is not available. Architectures (d) and (e) show how the simulation model proposed in this paper can be used to test and verify architectures (b) and (a) in simulation. Preprocessing may be applied in (a), (b), (d), and (e).

hands [20]. However, GraspIt! is based on kinematics only, and is not optimized for dynamics simulation.

A second challenge in grasp simulation is due to the different size and weight of the components involved. Small inaccuracies and oscillations in base and arm motions can induce significant error in finger movements [21]. In our own earlier work, we presented a custom simulator based on the Bullet simulation engine, using high physics update rates (above $> 1\,\mathrm{kHz}$) to achieve stable grasping [22]. Still, some object slippage was observed. The OpenGRASP framework [23,24] is another example of a simulator optimized for grasping and dexterous manipulation with multi-fingered hands. OpenGrasp in turn is based on the Openrave motion planning and simulation framework [25].

For reactive grasping, feedback from force and tactile sensors must be included in the simulation. A variety of sensing principles and sensor types are available; see [26,27] for reviews. Most commercial force sensors are based on strain-gauges; typical lever deflections are in the order of ten micrometers so that the sensors can still be considered rigid without loss of accuracy for most applications. Several technologies have been proposed for tactile sensors (e.g.

capacitive, conductive polymers, optical fibers), but again the deflection under
load is usually small enough so that the sensors can be modeled as rigid objects.

In order to simulate such sensors, access to the forces and torques that act
on them is essential. These, however, are readily available in dynamics engines;
in order to calculate object acceleration and motion, the engine needs to detect
contacts between objects and has to accumulate the resulting forces anyway.

This in turn means that rigid tactile and force sensors are easy to implement;
for example the existing touch and bumper sensor plugins in Gazebo simply sub-
scribe to specific collision events to determine contact. The force/torque sensor
plugin just accumulates applied forces by calling the *GetForceTorque()* method
and publishing the result as a *WrenchStamped* message [28]. Finally, the *pressure sensor* accumulates the normal force and divides by the known area of the
sensor. All these sensors correspond to architecture (c) in Fig. 2.

A different group of force and tactile sensors rely on soft materials, where contacts result in significant sensor deflection. One example is the BioTac sensor discussed in this paper. Another example is provided by the Optoforce sensors [29],
where the deflection of a rubber dome is measured by the corresponding change
in light reflection to phototransistors inside the dome. Realistic simulation of
such sensors is possible using finite-element approaches [30] and multi-physics
frameworks (e.g. [31,32]), but such simulations are usually not real-time capable
and thus unsuited for rapid experiments. In order to avoid costly simulation of
sensor deformations, we exploit the spring-damper-based collision resolution of
rigid body physics engines to simulate soft contacts and utilize a rigid mesh of
the sensor.

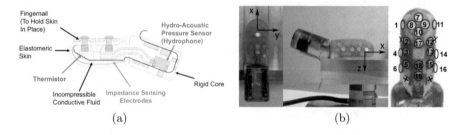

**Fig. 3.** (a) Cross section and main components of the BioTac sensor [6]. (b) Core of
the BioTac with coordinate system and layout of the electrodes [4].

## 3    The BioTac Sensor

The SynTouch BioTac sensors are one of the most advanced commercial biomimetic tactile sensors [6]. With a flexible rubber skin, the fingerprint-like surface
structure and the rigid core, the sensor mimics the human fingertip [2]. The
BioTac is equipped with three different sensor types: a hydro-acoustic-pressure
sensor, 19 electrodes, and a thermistor. The main components and the location
of the sensors within the BioTac are shown in Fig. 3.

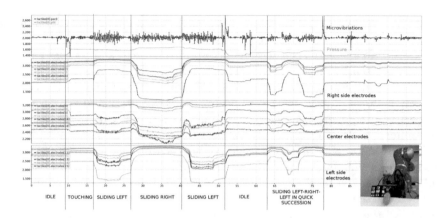

**Fig. 4.** Example of raw data from one BioTac sensor while sliding left and right along a surface. Thermistor data (tdc, tac) is not shown. See annotations for the different phases of the experiment. Note that shearing force can be estimated from the electrodes data.

### 3.1  Raw Sensor Data

See Fig. 4 for an example of the rich tactile data generated by the BioTac sensor, which can be used for tactile object exploration and surface classification, as well as for robot grasp control:

- a normal force applied to the rubber skin depresses the skin and increases the internal liquid pressure ($pdc$). Normal forces of up to $2.0\,$N relate to changes in pressure;
- changes in normal forces are recorded by the high-pass-filtered pressure value ($pac$); vibration patterns can be matched to surface roughness and contact slippage;
- deformation of the skin also changes the distribution of liquid, thus changing current from four excitation electrodes ($x_i$) to the measurement electrodes ($e_1, \ldots e_{19}$). Those values provide indirect information about contact location, contact size, and acting shearing forces;
- when contacting an object, heat is transferred from the sensor to the object; the resulting thermistor data ($tdc$) and high-pass filtered data ($tac$) are related to contact size, object temperature, and its thermal conductivity;
- deformation of the skin occurs at normal forces as low as $0.1\,$N; a sensitivity almost matching human fingertips;

Unfortunately, the deflection of the rubber skin is almost impossible to model and only indirectly related to the actual applied forces — and the effects of object contact on the thermistor readings are even more complicated. While Syn-Touch documents equations to relate raw 12-bit integer sensor readings (range $0\ldots 4095$) to physical quantities [6], those values depend on the skin deformation and require individual calibration. Not surprisingly then, no simulation model of the BioTac has been presented so far.

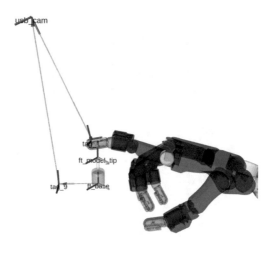

**Fig. 5.** Collecting ground truth and training data. Left: Experiment setup. The BioTac sensor touches a sample object (here, a cylinder with 4 mm diameter). Applied normal and shearing forces are recorded with a commercial six-axis F/T sensor (1). Contact point location is tracked optically via a calibrated webcam (4) and AprilTag fiducials (2,3). Right: Relevant coordinate systems for contact point location estimation in ROS.

## 3.2 Experimental Setup

We used the setup shown in Fig. 5 to collect the training data required for the simulation model. While recording data, the BioTac sensor is pressed repeatedly against a cylinder with 4 mm diameter and a spherical tip. The applied force is measured using a calibrated six axis force-torque sensor (ATi nano17e [33]) with a nominal force resolution better than 0.01 N. The contact location is reconstructed optically using a calibrated HD webcam and two AprilTag markers [34], one mounted on the BioTac and one attached to the probe object. The respective fixed affine transforms are modeled in a URDF robot model depicted in Fig. 5, right.

## 3.3 Captured Dataset and Postprocessing

The resulting dataset consists of more than 300.000 tactile readings. It encompasses the complete BioTac sensor data $(pdc, pac, tdc, tac, e_1, \ldots e_{19})$ and the applied reference forces $(F_x, F_y, F_z)$, both recorded at 100 Hz. Contact locations are estimated at roughly 10 Hz by the AprilTag detector and are converted into $(x, y, z)$ positions in the BioTac coordinate system.

Since we want to achieve submillimeter precision in our manually assembled setup, we applied a postprocessing step to calibrate systematic offsets in the estimated contact points. By analysis of the force-torque sensor data, we detect moments where there is only light contact between the measuring probe and the BioTac sensor (less than 0.3 N after a contact). In these cases the distance between the surface of the probe and the undeformed sensor has to be almost 0. Thus, we estimate calibration offsets for the rigid transforms of both AprilTags by minimizing the absolute distances between both bodies during these moments.

Starting from the initial transformation estimates, the optimization reliably converges to a stable solution, see Fig. 6.

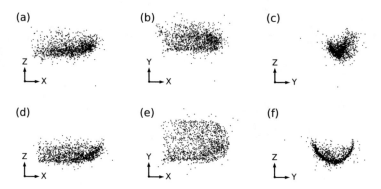

**Fig. 6.** Measured contact points, sampled at the end of each contact, before and after optimization-based calibration: side view unoptimized (a), top view unoptimized (b), front view unoptimized (c), side view optimized (d), top view optimized (e), front view optimized (f).

## 4   Simulation Model

To generate simulated sensor outputs for the BioTac sensors, we first extract contact points and contact forces from the physics simulation. If an object collides with the sensor model, a set of intersection points is computed. At each simulation step, the center of the current intersection points and the sum of the affecting simulated forces is selected.

To suppress the effects of simulation instabilities and to remove unrealistically sharp signal flanks which arise from the deformable surface of the BioTac sensor being approximated by a non-deformable rigid body, a configurable exponential decay low-pass filter is applied.

Gazebo resolves collisions through a dampened spring model. For the sensor collision models, the spring constants in the surface parameters are reduced to approximate the behavior of a soft surface. However, this does not affect the actual contact point and the unfiltered reported contact force may still be unstable, so filtering is still applied.

### 4.1   Neural Network Architecture

To model the non-linear relationship between contact points, applied forces, and output signals, we initially tested different machine learning approaches. Overall, artificial neural networks gave the most promising results. Several different network architectures were developed and compared, two of which will be presented in this paper.

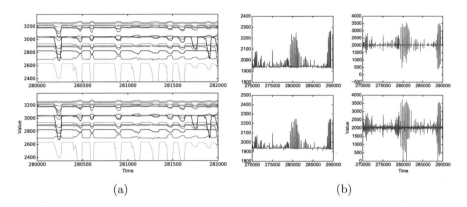

(a)                                                    (b)

**Fig. 7.** (a) Measured electrode signals ($e_1 \ldots e_{19}$) from the real BioTac sensor (bottom) and corresponding electrode signals re-constructed from forces and temperatures using the simulator (top). (b) pdc simulated (top left), pdc measured (bottom left), pdc simulated (top right), pdc measured (bottom right).

## 4.2 Network Inputs and Outputs

In our final architecture, the network receives a contact point, three force vector samples, and a temperature input. All signals are zero-mean and variance-one normalized according to the distributions observed in the captured data. The three force vector samples are captured at different time points 100 ms apart; shorter intervals resulted in overfitting. More force vector samples or shorter intervals can cause the network to react too strongly to high-frequency inputs which the physics simulator is unable to reproduce realistically.

The electrode values from the real BioTac sensor show a non-linear dependency on the device temperature. Since attempts to correct this dependency before feeding the data into the network performed worse, we now train the network to compensate the dependency by itself. During simulation, a constant temperate is assumed, which by default corresponds to the average temperature of the training set. The network generates simulated electrode and pressure signals as output (Fig. 7). Temperature outputs are currently not simulated.

We first developed a densely connected sequential model (hereafter referred to as Network A). After experimenting with different layer sizes, layer counts, activation functions, etc., we decided on a configuration with 5 hidden layers of sizes 1024, 4096, 512, 512, 512 and ReLU activation (Fig. 8).

In addition, we also designed a specialized network with the specific application in mind (Network B). The network first separates position inputs, force

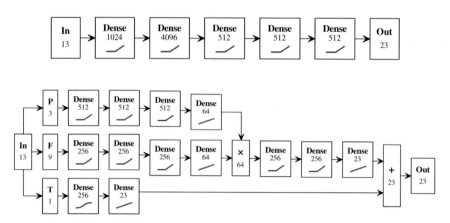

**Fig. 8.** Two deep neural network architectures for simulating BioTac sensor outputs. Dense network with five hidden layers (Network A, top). Specialized model for improved electrode output prediction (Network B, bottom). Network B splits inputs into positions (P), forces (F) and temperatures (T), and combines intermediate solutions through component-wise addition (+) and multiplication (×).

inputs, and temperature inputs. Position inputs and force inputs are processed by two separate network columns, each consisting of three densely connected layers, and both using ReLU activation. The position column uses 512 neurons per layer to be able to accurately discriminate between different areas on the sensor surface. The force column contains 256 neurons per layer with L1 bias regularization. Both columns have a smaller output layer of 64 neurons with linear activation. The outputs of the force and position columns are then mixed multiplicatively. The result is again processed by two densely connected layers with ReLU activation and 256 neurons each, and reduced to output size by a smaller densely connected layer with linear activation. A temperature correction vector is added, to obtain the final output vector. The temperature correction vector is generated by a third column, consisting of a densely connected 256-neuron layer with sigmoid activation for smooth temperature regression and a densely connected reduction layer at output size with linear activation.

### 4.3   Implementation

The sensor simulator was implemented in C++ as a Gazebo/ROS plug-in. For training, Keras and Tensorflow are used. To make installation and usage of the plug-in convenient, our own C++ implementations of the relevant layer types are added to the plug-in so that Keras and Tensorflow are only needed for training but not for running the plug-in.

## 5    Evaluation

The collected data was split into a training set and a validation set. The validation set contains 30000 samples and consists of 30 randomly selected contiguous ranges of 1000 samples each ($\sim$10% of the collected data). The training set contains the remaining samples. Model B outperformed model A for all output channels and is therefore selected as the default model in our plug-in. For Network B, the average electrode value error is $\sim$8.6% of the standard deviation, for Network A $\sim$11%. For all electrodes and for both networks, the average validation error is below 1% of the total range of output values found in the collected data set. (See Table 1 for results.)

To test the suitability of the sensor model for its intended applications, an existing geometrical contact estimation method [35] is applied to real sensor data and to data generated by the simulation model, and the results are compared. The contact point estimator behaves very similar, with a difference in average absolute estimation error of 0.04 mm. The average distance between contact points predicted from measured signals and contact points predicted from simulated signals is 0.87 mm.

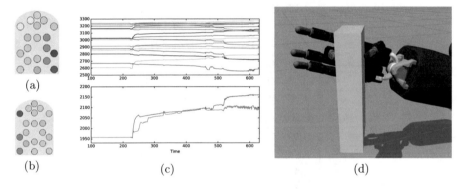

**Fig. 9.** Example grasp and corresponding simulated BioTac readings. (a), (b) electrode values for middle finger and thumb, respectively. Both images show the electrodes from behind the finger. (c) electrode readings of the middle finger and all pdc values during the applied grasp motion. (d) The applied two finger example grasp.

In Fig. 9 a two-finger grasp is applied to a box-shaped object using thumb and middle finger. The output of the simulated sensor during the motion is plotted for all pdc values and for the electrodes of the middle finger. With the pdc values, the increase of pressure on the object is clearly visible. The relative change of raw electrode values is visualized in (a) and (b), where yellow indicates decreasing and blue indicates increasing impedance measurements. In both images, the region of contact can be recognized.

**Table 1.** Validation errors for *pdc*, *pac* and electrodes $e_1 \ldots e_{19}$, as absolute values in the original scale of the sensor outputs (abs) and in multiples of the per-channel standard deviations (std), effective data ranges (eff. range), and average inference time ($t$) as well as number of trainable parameters (TP) for both networks.

| | Network A | | Network B | | |
|---|---|---|---|---|---|
| | abs | std | abs | std | eff. range |
| *pdc* | 10.812 | 0.13 | 10.145 | 0.122 | 578 |
| *pac* | 64.412 | 0.217 | 59.761 | 0.202 | 3958 |
| $e_1$ | 12.409 | 0.101 | 9.931 | 0.081 | 1803 |
| $e_2$ | 9.376 | 0.082 | 8.392 | 0.074 | 2147 |
| $e_3$ | 9.193 | 0.097 | 7.646 | 0.081 | 2076 |
| $e_4$ | 12.011 | 0.088 | 9.921 | 0.072 | 2467 |
| $e_5$ | 8.898 | 0.089 | 7.019 | 0.07 | 2778 |
| $e_6$ | 11.865 | 0.079 | 9.482 | 0.063 | 2193 |
| $e_7$ | 4.112 | 0.101 | 3.126 | 0.077 | 1137 |
| $e_8$ | 5.006 | 0.108 | 3.439 | 0.074 | 1170 |
| $e_9$ | 4.29 | 0.101 | 3.11 | 0.073 | 927 |
| $e_{10}$ | 4.434 | 0.114 | 3.274 | 0.084 | 610 |
| $e_{11}$ | 11.442 | 0.103 | 9.615 | 0.086 | 1981 |
| $e_{12}$ | 7.404 | 0.081 | 6.607 | 0.072 | 2123 |
| $e_{13}$ | 8.276 | 0.11 | 7.215 | 0.096 | 1841 |
| $e_{14}$ | 10.603 | 0.11 | 8.719 | 0.09 | 1917 |
| $e_{15}$ | 9.309 | 0.153 | 7.971 | 0.131 | 1370 |
| $e_{16}$ | 10.303 | 0.135 | 8.221 | 0.108 | 1899 |
| $e_{17}$ | 6.162 | 0.09 | 4.883 | 0.071 | 1393 |
| $e_{18}$ | 8.642 | 0.15 | 7.175 | 0.124 | 1618 |
| $e_{19}$ | 8.063 | 0.124 | 6.925 | 0.106 | 1287 |
| $e_{1..19}$ | 8.516 | 0.106 | 6.983 | 0.086 | 2828 |
| $t$ | 2 ms | | 0.2 ms | | |
| TP | 6,847,511 | | 805,550 | | |

## 5.1 Performance

On our test machine (Intel i5, 3.4 GHz) and with the current implementation, the average inference time is 0.2 ms for network B and 2 ms for network A. When simulating a full hand with 5 BioTac sensors at a physics update rate of 1000 Hz with network B running at an update frequency of 100 Hz (simulating the sampling rate of the real sensor), an average of 0.13 ms per simulation step is spent executing the update method of the plug-in.

# 6   Conclusions

In this paper, we introduced a simulation model for the tactile fingertip sensor BioTac. The model accurately reproduces sensor outputs that were measured on a real robot from measured contact forces and contact points, and produces plausible results for contact forces and contact points extracted from physics simulations.

A fast and portable C++ implementation with minimal third-party dependencies is provided as a plug-in for the robot simulation framework Gazebo. The implementation can be found at https://github.com/TAMS-Group/biotac_gazebo_plugin.

While the paper focused on the BioTac sensor, the proposed machine learning approach can easily be extended to provide simulation models for other complex sensors, whose output signals do not readily relate to physical properties.

# References

1. Fishel, J.A., Santos, V.J., Loeb, G.E.: A robust micro-vibration sensor for biomimetic fingertips. In: 2nd IEEE RAS & EMBS International Conference on Biomedical Robotics and Biomechatronics, pp. 659–663 (2008). https://doi.org/10.1109/BIOROB.2008.4762917
2. Wettels, N.B.: Biomimetic tactile sensor for object identification and grasp control. University of Southern California (2011)
3. Fishel, J.A.: Design and use of a biomimetic tactile microvibration sensor with human-like sensitivity and its application in texture discrimination using Bayesian exploration. University of Southern California (2012)
4. Loeb, G.E.: Estimating point of contact, force and torque in a biomimetic tactile sensor with deformable skin (2013). https://www.syntouchinc.com/wp-content/uploads/2016/12/2013_Lin_Analytical-1.pdf
5. Chu, V., et al.: Robotic learning of haptic adjectives through physical interaction. Rob. Auton. Syst. **63**, 279–292 (2015). https://doi.org/10.1016/j.robot.2014.09.021
6. Fishel, J.A., Lin, G., Matulevich, B., Loeb, G.: BioTac product manual. SynTouch LLC, V20 edn. (2015). https://www.syntouchinc.com/wp-content/uploads/2017/01/BioTac_Product_Manual.pdf
7. Shadow Robot Dextrous Hand. https://www.shadowrobot.com/
8. Koenig, N., Howard, A.: Design and use paradigms for gazebo, an open-source multi-robot simulator. In: IEEE/RSJ International Conference on Intelligent Robots and Systems, Proceedings, vol. 3, pp. 2149–2154. IEEE (2004). https://doi.org/10.1109/iros.2004.1389727
9. Quigley, M., Conley, K., Gerkey, B. P., Faust, J., Foote, T., Leibs, J., Wheeler, R., Ng., A. Y., ROS: an open-source robot operating system. In: ICRA Workshop on Open Source Software (2009)
10. Open Dynamics Engine. http://www.ode.org/
11. Bullet Physics Library. http://bulletphysics.org/wordpress/
12. Simbody Multibody Physics API. https://simtk.org/projects/simbody/
13. Dynamic Animation and Robotics Toolkit. http://dartsim.github.io/
14. Cyberbotics Inc., Webots robot simulator. https://www.cyberbotics.com/

15. Coppelia Robotics, V-REP Virtual Robot Experimentation Platform. http://www.coppeliarobotics.com/
16. Cutkosky, M.R.: On grasp choice, grasp models, and the design of hands for manufacturing tasks. IEEE Trans. Rob. Autom. **5**(3), 269–279 (1989). https://doi.org/10.1109/70.34763
17. Bicchi, A., Kumar, V.: Robotic grasping and contact: a review. In: IEEE International Conference on Robotics and Automation, Proceedings, ICRA 2000, vol. 1, pp. 348–353. IEEE (2000). https://doi.org/10.1109/ROBOT.2000.844081
18. Miller, A.T., Allen, P.K.: Graspit! a versatile simulator for robotic grasping. IEEE Rob. Autom. Mag. **11**(4), 110–122 (2004). https://doi.org/10.1109/MRA.2004.1371616
19. Ciocarlie, M.T., Allen, P.K.: Hand posture subspaces for dexterous robotic grasping. Int. J. Rob. Res. **28**(7), 851–867 (2009). https://doi.org/10.1177/0278364909105606
20. Goldfeder, C., Ciocarlie, M., Dang, H., Allen, P.K.: The columbia grasp database. In: IEEE International Conference on Robotics and Automation, ICRA 2009, pp. 1710–1716. IEEE (2009). https://doi.org/10.1109/ROBOT.2009.5152709
21. Taylor, J.R., Drumwright, E.M., Hsu, J.: Analysis of grasping failures in multi-rigid body simulations. In: IEEE International Conference on Simulation, Modeling, and Programming for Autonomous Robots (SIMPAR), pp. 295–301. IEEE (2016). https://doi.org/10.1109/SIMPAR.2016.7862410
22. Scharfe, H., Hendrich, N., Zhang, J.: Hybrid physics simulation of multi-fingered hands for dexterous in-hand manipulation. In: IEEE International Conference on Robotics and Automation (ICRA), pp. 3777–3783. IEEE (2012). https://doi.org/10.1109/ICRA.2012.6225156
23. León, B., et al.: OpenGRASP: a toolkit for robot grasping simulation. In: International Conference on Simulation, Modeling, and Programming for Autonomous Robots, pp. 109–120. Springer (2010). https://doi.org/10.1007/978-3-642-17319-6_13
24. OpenGRASP toolkit. http://opengrasp.sourceforge.net/
25. Diankov, R., Kuffner, J.: OpenRAVE: a planning architecture for autonomous robotics. Robotics Institute, Pittsburgh, PA, Technical report. CMU-RI-TR-08-34 79 (2008)
26. Dahiya, R.S., Metta, G., Valle, M., Sandini, G.: Tactile sensing-from humans to humanoids. IEEE Trans. Rob. **26**(1), 1–20 (2010). https://doi.org/10.1109/TRO.2009.2033627
27. Yousef, H., Boukallel, M., Althoefer, K.: Tactile sensing for dexterous in-hand manipulation in robotics-a review. Sens. Actuators A Phys. **167**(2), 171–187 (2011). https://doi.org/10.1016/j.sna.2011.02.038
28. Message and service publishers for interfacing with Gazebo through ROS. http://wiki.ros.org/gazebo_ros
29. Optoforce Ltd., 3D Force Sensor. https://optoforce.com/3d-force-sensor-omd
30. Grazioso, S., Sonneville, V., Di Gironimo, G., Bauchau, O., Siciliano, B.: A nonlinear finite element formalism for modelling flexible and soft manipulators. In: IEEE International Conference on Simulation, Modeling, and Programming for Autonomous Robots (SIMPAR), pp. 185–190. IEEE (2016). https://doi.org/10.1109/SIMPAR.2016.7862394
31. ANSYS multiphysics simulation. https://www.ansys.com/products/platform/multiphysics-simulation
32. Comsol Multiphysics suite. https://www.comsol.com/multiphysics

33. ATI Industrial Automation Inc.: F/T Sensor Nano17. http://www.ati-ia.com/products/ft/ft_models.aspx?id=Nano17
34. Olson, E.: AprilTag: a robust and flexible visual fiducial system. In: IEEE International Conference on Robotics and Automation (ICRA), pp. 3400–3407. IEEE (2011). https://doi.org/10.1109/ICRA.2011.5979561
35. Ciobanu, V., Popescu, D, Petrescu, A.: Point of contact location and normal force estimation using biomimetical tactile sensors. In: Eighth International Conference on Complex, Intelligent and Software Intensive Systems (CISIS), pp. 373–378. IEEE (2014). https://doi.org/10.1109/CISIS.2014.52

# Force Sensing for Multi-point Contact Using a Constrained, Passive Joint Based on the Moment-Equivalent Point

Shouhei Shirafuji[✉] and Jun Ota

Research into Artifacts, Center for Engineering, The University of Tokyo,
5-1-5 Kashiwanoha, Kashiwa-shi, Chiba 277-8568, Japan
shirafuji@race.u-tokyo.ac.jp

**Abstract.** In this paper, an analyzing method of a constrained joint using the Moment-Equivalent Point (MEP) is introduced that represents the balance between the torques exerted by two end joints in a robotic manipulator and the torque created by multiple-point contact of the flat surface that is in contact with the environment. By construction, the vector representing the summed reactive forces on the center of pressure (CoP) will always pass through the MEP. An important characteristic of the MEP is that it is fixed with respect to the link connecting the two joints if the ratio of the torques exerted at each joint is held constant. Therefore, if the robot has two passive joints that are mechanically constrained such that the ratio of the torques at each joint is constant, the MEP can be treated a single-contact point. Thus, we can model the robot's behavior as if contacts only with a point on MEP in the environment, even if the actual contact is over multiple points on the flat surface. Such mechanically constrained passive joints and the concept of the MEP result in an approach that is midway between the standard multi-point contact and standard single-point contact in terms of the contact kinematics. One advantage of considering the balance of forces between the robot and the environment based on the MEP is that the tangential force applied to the contact surface can be calculated just from the CoP position and the normal force at the CoP. Experimental results indicate that the tangential force at the foot of the robot can be estimated by measuring only the normal forces applied at the foot.

**Keywords:** Force sensing · Passive joint · Moment-equivalent point

## 1 Introduction

In many robotics investigations, the contact between the robot and the environment is assumed to be a point due to its ease of modeling. For example, the single-point assumption simplifies both the decision-making and the force location and direction measurement when attempting to stably grasp an item using a multi-fingered robot. The single-point-contact assumption is particularly

© Springer Nature Switzerland AG 2019
M. Strand et al. (Eds.): IAS 2018, AISC 867, pp. 388–400, 2019.
https://doi.org/10.1007/978-3-030-01370-7_31

effective in robotics applications that require fast feedback control such as fast manipulation or fast locomotion of a robot.

On the other hand, multi-point contact with a flat surface is more stable and enables more varied manipulations between the robot and the environment because it increases the frictional force between the two and allows the application of a torque to the environment. For example, if a robot is attempting bipedal humanoid walking motions, it is important that the multi-point contact between the robot's foot and the environment is maintained to prevent the robot from falling down. As another example, the multi-point contact at the end-link of a robotic hand stabilizes the pinching of an object. However, with multi-point contact, it is necessary to have accurate force sensing and feedback control is required because the application of the force can easily cause control instabilities. The differences between the single- and multiple-point contact types indicate that they have a trade-off relationship.

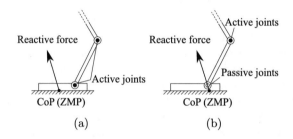

**Fig. 1.** CoP where the reactive force is applied for a surface with multi-point contact. (a) If the robot has no passive joints, the CoP location and the force direction depend on the forces at the active joints. (b) If one of the joints is passive, the reactive force at the CoP always passes through the axis of the passive joint.

When considering the sensing aspect of the problem, determining the applied force and its location is not easy for the multi-point contact case. Many tactile sensors have been investigated to facilitate the measurement of the applied force [1,2]. As an example, distributed force sensing is a standard approach for measuring the contact location (e.g., using a piezoresistive force sensor with very high resolution) [3]. In many cases, only knowing the center of pressure (CoP) caused by the multi-point contact is enough to realize the desired manipulation, as shown in Fig. 1a. Therefore, many sensors and methods that can identify the location of the CoP and the force applied at it have also been proposed [4–6]. CoP detection is particularly important in bipedal walking. In this application, the location of the CoP in the polygon formed by the robot's feet is called the Zero-Moment Point (ZMP) [7,8], and this representation of the force-equivalent point has been commonly used in bipedal walking control schemes for decades [9–11]. The measurement of the ZMP is crucial for the controller to follow the desired walking trajectory [12–14], and the location of the ZMP and force applied at it is usually measured by a six-directional force sensor embedded in the robot's foot. This method is employed in many humanoid robots including ASIMO [15], the HRP series [16–18], and HUBO [19]. However, the structural complexity of the sensors to identify the CoP location and force and the spatial limitations of the robot lead to problems during implementation.

If the contact is a single point, it is trivial to determine the CoP because it is always located at the contact point and the force direction is easily determined because the reactive force always passes through the point. If the end-joint of the robot is passive, the multi-point contact at the robot's end-link can be viewed as a single-point contact on the joint, as shown in Fig. 1b. In this case, the balance of forces guarantees that the reactive force at the CoP always passes through the axis of the passive joint, which simplifies the force measurement for the multi-point contact case. For example, the tangential force applied to the surface can be calculated by measuring only the location of the CoP and the normal force applied at it. Unfortunately, the passive joint prevents the robot from applying any torque onto the environment, which is one of the advantages of the multi-point contact setup. For example, it is difficult to prevent a humanoid robot from falling down without applying a rotational force from the foot when controlling a robot's posture with its leg.

In this paper, we focus on an interesting relationship between the forces applied on a constrained joint and the Moment-Equivalent Point (MEP). This point can be regarded as a virtual single-contact point if the two passive joints are mechanically constrained to exert the torques in a constant ratio, as shown in Fig. 2a. For the above-mentioned case in which the robot makes contact with the environment at multiple points via a passive joint, the force direction is easily determined because the reactive force must pass through the MEP. Additionally, the tangential force applied to the robot can be estimated by only measuring the normal force. Furthermore, although torques cannot be applied to the robot from the environment with a simple passive joint (as mentioned above), they can be applied with the constrained passive joints because the MEP is usually located at a specific distance along the line that connects these two joints, and this position changes based on the rotation of the joints. This approach can be regarded as an intermediate approach between the standard multi-point and single-point contact cases, as mentioned above, and it provides a new way of viewing the contact kinematics and realizing new mechanisms, such as the foot mechanism shown in Fig. 2a, or the fingertip of the robotic hand shown in Fig. 2b. This study aims to reveal the kinematic properties of this new type of passive joint and show its advantage as a mechanism in a robot.

The remainder of the paper is laid out as follows. First, the analytical proof that the reactive force from the environment always passes through the MEP is shown in Sect. 2. Secondly, the MEP-based method to determine the CoP location and the normal and tangential forces applied at the CoP by only measuring the normal forces on a few points of the surface is shown in Sect. 3. The proposed sensing method is validated experimentally in Sect. 4, and the paper is concluded in Sect. 5.

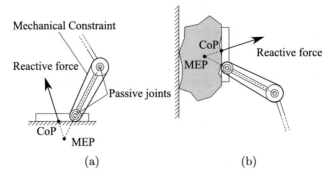

**Fig. 2.** The reactive force always passes through the MEP if the two end joints are mechanically constrained. (a) MEP in the foot of a robot in which the two end joints are constrained. (b) MEP in the fingertip of a robotic hand in which the two end joints are constrained.

## 2    Moment-Equivalent Point

In this section, it is shown that a manipulator that contacts the environment at multiple points is equivalent to a manipulator that contacts the environment at a single point if the manipulator has passively constrained joints. For simplicity, we consider a manipulator that has two rotational joints and moves in a plane as shown in Fig. 3. Let $\tau = (\tau_1, \tau_2) \in \mathbb{R}^2$ be the vector of torque exerted by the manipulator, where $\tau_1$ and $\tau_2$ are the torques at the first and second joints, ordered from lower to higher, respectively. $\theta_1$ and $\theta_2$ are the joint angles that are defined in the same order. The distance between the axes of these two joints is $l_1$. The kinematics of this manipulator are with respect to the spatial frame S, which is fixed on the ground.

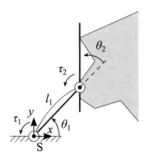

**Fig. 3.** Manipulator with two joints whose end-link contacts an object with multi-point contact.

Let the wrench caused by torque $\tau$ with respect to the spatial frame be denoted by $\boldsymbol{F}^{S} \in \mathbb{R}^{3}$. The relationship between the torque and this wrench is represented as follows [20]:

$$\tau = (\boldsymbol{J}^{S})^{T} \boldsymbol{F}^{S}, \tag{1}$$

where $\boldsymbol{J}^{S} \in \mathbb{R}^{3 \times 2}$ is the spatial manipulator Jacobian given by

$$\boldsymbol{J}^{S} = \begin{bmatrix} 0 & l_1 \sin \theta_1 \\ 0 & -l_1 \cos \theta_1 \\ 1 & 1 \end{bmatrix}. \tag{2}$$

For a given torque, the corresponding wrench is given by

$$\boldsymbol{F}^{S} = (\boldsymbol{J}^{S})^{+T} \tau + \boldsymbol{f}_{N}. \tag{3}$$

The mapping from the wrench space to the joint torque space in Eq. (1) is not injective, and its inverse matrix can not be defined. Therefore, the Moore–Penrose pseudoinverse of the manipulator Jacobian $(\boldsymbol{J}^{S})^{+T}$ is used in Eq. (3). This pseudoinverse can be obtained from Eq. (2) as follows:

$$(\boldsymbol{J}^{S})^{+T} = \begin{bmatrix} -\dfrac{\sin \theta_1}{l_1} & \dfrac{\sin \theta_1}{l_1} \\ \dfrac{\cos \theta_1}{l_1} & -\dfrac{\cos \theta_1}{l_1} \\ 1 & 0 \end{bmatrix}. \tag{4}$$

Here, $\boldsymbol{f}_{N}$ is the force vector in the kernel of the injective mapping from the wrench to the joint torque, that is $\boldsymbol{f}_{N} \in \ker(\boldsymbol{J}^{S})^{T}$, and is given by

$$\boldsymbol{f}_{N} = \begin{bmatrix} \cos \theta_1 \\ \sin \theta_1 \\ 0 \end{bmatrix} \xi, \tag{5}$$

where $\xi$ is any scalar. $\boldsymbol{f}_{N}$ represents the internal force, which is the force along the direction of the line connecting the axes of the two joints.

We next consider what external force the environment must apply to the manipulator in order to balance the joint torque. For simplicity, we assume that the external force is applied to the manipulator only through the end-link that has multi-point contact with the ground. When the flat surface has multi-point contact with the environment, the forces applied to the surface at the different points can be represented by summing the forces at the CoP; in other words, the forces applied from the multi-point contact can be represented by a single force being applied at the CoP. To balance the forces exerted by the joints and the force applied to the end-link, the force vector consisting of the tangential and normal forces applied at the CoP must balance with the wrench $\boldsymbol{F}^{S}$. The CoP is defined as the point where the sum of the moments from the forces applied at the contact points is zero. Therefore, the relationship between the force caused by the joint torque and the reactive force applied at the CoP is represented by

$$\boldsymbol{F}^{S} = - \begin{bmatrix} 1 & 0 & 0 \\ 0 & 1 & 0 \\ -p_y^S & p_x^S & 1 \end{bmatrix} \begin{bmatrix} f_x^S \\ f_y^S \\ 0 \end{bmatrix}, \tag{6}$$

where $p_x^S$ and $p_y^S$ are the $x$- and $y$-coordinates, respectively, of the ZMP in the spatial frame, and $f_x^S$ and $f_y^S$ are the $x$- and $y$-components, respectively, of the force applied at the ZMP in the spatial frame. The following is obtained from Eqs. (3) and (6):

$$f_x^S = \frac{\tau_1 - \tau_2}{l_1} \sin \theta_1 + \xi \cos \theta_1, \tag{7}$$

$$f_y^S = -\frac{\tau_1 - \tau_2}{l_1} \cos \theta_1 + \xi \sin \theta_1, \tag{8}$$

$$\tau_1 = f_x^S p_y^S - f_y^S p_x^S. \tag{9}$$

From Eqs. (7) and (8), Eq. (9) can be defined as the follows:

$$f_x^S \left( p_y^S - \left( \tau_1 \frac{l_1}{\tau_2 - \tau_1} \right) \sin \theta_1 \right)$$
$$= f_y^S \left( p_x^S - \left( \tau_1 \frac{l_1}{\tau_2 - \tau_1} \right) \cos \theta_1 \right). \tag{10}$$

If the position of the CoP, $p_x^S$ and $p_y^S$ are treated as variables, it can be seen that Eq. (10) represents a line with an angle $f_y/f_x$ that passes through the point $((\tau_1 l_1/(\tau_2 - \tau_1)) \cos \theta_1, (\tau_1 l_1/(\tau_2 - \tau_1)) \sin \theta_1)$. In other words, the vector of the reactive force at the CoP can be represented by a line that passes through this point, as shown in Fig. 4, and it is this point that we define as the MEP. The MEP is located on the line that connects the axes of the two end joints, and it can be seen from Eq. (10) that the distance from the axis of the first joint to the MEP is $l_m$ which is given by

$$l_m = \frac{\tau_1}{\tau_1 - \tau_2} l_1. \tag{11}$$

It is apparent from this equation that the distance between the first joint axis and the MEP is governed by the ratio between the two joint torques, $\tau_1$ and $\tau_2$. In the other words, if these joints were constrained so that the ratio between these torques remains constant, the MEP position is fixed with respect to the link connecting the two joints. In this case, we can model the manipulator as if it makes contact with the environment only at the MEP (i.e., it can be treated as a virtual single-point contact). This introduces a new concept to the mechanical design of a robot and its control, as mentioned in Sect. 1.

There are several methods to constrain the ratio of the torques of the two end joints that can be considered. One such method is to constrain the two passive joints by mechanical components, such as a timing belt or gears, as shown in Fig. 5a. In this case, these joints can then be regarded as passive joints with one degree of freedom and the end-link is assumed to make contact with the environment at the MEP instantaneously. The ratio of the joint torques is determined by the ratio of the radii of the timing pulleys. A second method is to use a wire-driven mechanism [21,22], as shown in Fig. 5b. The force is applied to the end-link by conveying the motor's tensile force to a point offset from the

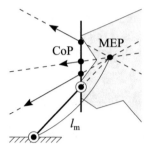

**Fig. 4.** The vector of the reactive force at the CoP always passes through a point specified by the ratio of the torques of the two end joints.

joints via a wire passing around pulleys placed on the passive joints. The ratio of the torques between the two joints is determined by the ratio of the pulley radii that the wire passes around. In this case, the system can be regarded as an under-actuated system whose end-link instantaneously makes contact with the environment at the MEP.

In the next section, the advantage of using the MEP in terms of force sensing is shown. Furthermore, we validate the method by constraining a robot's joints with a timing belt in Sect. 4.

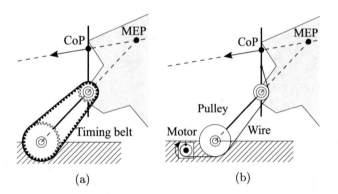

**Fig. 5.** Methods to constrain the joint torques so that the ratio between the two is constant. (a) Timing belt. (b) Wire-driven mechanism.

## 3   Force Sensing Based on MEP

As mentioned above, the force applied at the CoP of the end-link's surface, including the tangential force, can be identified by only measuring the normal force at the CoP if the MEP is fixed along the same line that connects the two joint axes. In this section, we present the method to calculate the tangential force at the CoP from the measured location of the CoP and the applied normal force for the simple case of two-dimensional space where the joints move in a plane. As an example, we consider a robotic leg whose flat foot contacts the floor and whose two end joints are constrained by a timing belt, as shown in Fig. 6a. In general, the position of the CoP and the normal force applied at it can be measured with simple methods. For example, as shown in Fig. 6a, the position and normal force could be calculated from the normal force applied at two protruding points. Let the positions of these two points with respect to the frame fixed on the axis of the end-joint on the end-link O be denoted by $p_1^O$ and $p_2^O$, and let the normal forces applied at the two points be respectively denoted by $f_1^n$ and $f_2^n$. Then, the position of the CoP $p_c^O$ and the normal force $f_c^n$ are obtained by

$$p_c^O = \frac{f_1^n p_1^O + f_2^n p_2^O}{f_1^n + f_2^n}, \tag{12}$$

$$f_c^n = f_1^n + f_2^n. \tag{13}$$

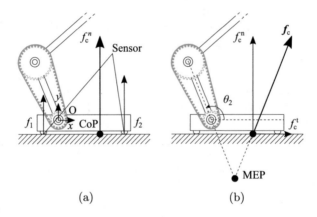

(a)                                    (b)

**Fig. 6.** Example of force sensing based on MEP. (a) Measurement of the CoP position and the normal force applied at the foot of a robotic leg. (b) Calculation of the tangential force from the measured CoP position and the normal force.

The tangential force applied at the CoP $f_c^t$ can be calculated from the relationship of the positions of the CoP and the MEP, as shown in Fig. 6b. The position of the MEP with respect to frame O is given by

$$p_m^O = -(l_m - l_1) \begin{bmatrix} \cos \theta_2 \\ \sin \theta_2 \end{bmatrix}. \tag{14}$$

Therefore, the tangential force is obtained by

$$f_c^t = f_c^n \frac{(p_c^O - p_m^O)_x}{(p_c^O - p_m^O)_y}, \tag{15}$$

where $()_x$ and $()_y$ represent the $x$- and $y$-component of a vector, respectively.

As shown in the above equations, the force applied on the flat surface including the tangential force can be obtained by measuring only the CoP position and the normal force if the MEP position is fixed by the constrained passive joints. We have demonstrated the method with an example of a foot moving in a plane, but the tangential force applied on the flat surface can also be calculated for a foot or manipulator that moves in a three-dimensional space by adding more constrained passive joints.

## 4    Experiment

In this section, the method to calculate the tangential force from the measured normal force based on the fixed MEP is validated experimentally.

### 4.1    Experimental Setup

We have developed the system shown in Fig. 7a for the experiment. The robotic leg consists of two passive joints that are mechanically constrained by a timing belt as discussed above. The length of the link connecting the joint axes was $l_1 = 100.0$ mm, and the pitch diameters of the timing pulleys for the first and second joints were 12.94 mm and 25.87 mm, respectively. Therefore, the ratio of the force exerted on the joints by the mechanical constraint is 12.94 : 25.87, and the distance from the first joint axis to the MEP is $l_m \approx 200.1$ mm.

The normal force applied at the flat foot was measured by the load cells (LMA-A-1KN, Kyowa Electronic Instruments Co., Ltd.) attached to the foot. Load cells were fixed on the lower plate of the foot, and the upper flat plate made contact with the load cells. The relative motion in the tangential direction between the upper and lower parts of the foot was restricted by linear bushings. The CoP position and the normal force applied at it were measured by these load cells using Eqs. (12) and (13).

The external force was applied to the system using a weight-induced gravitational force. The weight was connected to the robotic leg at a passive joint. The weight was also connected to a linear rail fixed in the environment, and the force direction was determined by the inclination angle of the rail. Let the mass of the weight be denoted by $m$ and let the inclining angle be given by $\phi$. Then, the external force applied to the robot from the weight is given by $mg \sin \phi$, where g is the gravitational acceleration. The ground truth of the reactive force that is applied on the system by the floor can be obtained by taking into account the gravitational forces caused by other components including links, joints, and gears in addition to the force applied by the weight. The CoP locations and the weights of the system components were measured before the experiment, and Table 1 shows the weights and the CoP positions in each frame for each system component described in Fig. 7a.

The joint angles $\theta_1$ and $\theta_2$ were measured by an angle sensor (CP-2HB, Midori Precisions Co., Ltd.), and the inclination angle of the linear rail $\phi$ was measured by a tilt angle sensor (D5R-L02-60, Midori Precisions Co., Ltd.).

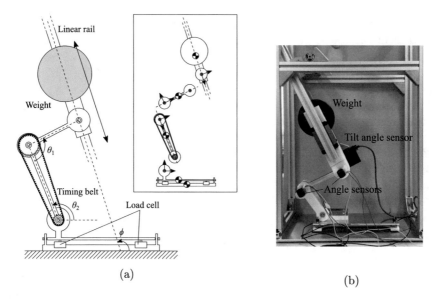

(a)

(b)

**Fig. 7.** Experimental setup for the validation of the proposed force measurement method using MEP, and the definition of the frame to represent the CoP of each part in the system.

**Table 1.** Weight (kg) and $x$ and $y$ positions of CoP (mm).

|        | Weighting | Link 1 | Link 2 | Upper foot | Lower foot |
|--------|-----------|--------|--------|------------|------------|
| Weight | 1.32      | 0.06   | 0.10   | 0.26       | 0.08       |
| $x$    | (112.50)  | 50.00  | 36.58  | 60.06      | 80.00      |
| $y$    | (0.0)     | 0.0    | 0.0    | 15.4       | 34.0       |

## 4.2 Experimental Procedure

Figure 7b shows the actual experimental setup, which was placed on a horizontal flat desk. The experiment was conducted by measuring the normal forces from the load cells for different applied external force conditions, which were determined by the inclination angle of the linear rail. The angle of the rail was changed manually to predetermined values that were selected such that the direction of the weight's force vector changed gradually. The direction represented by $\phi$ varied from 1.94 to 2.04 rad by 0.035 rad. The actual position of the CoP and the applied force on the foot were calculated from the rail posture and the other parameters mentioned above for each experimental condition. The CoP position and the normal and tangential forces estimated from the measured load cell forces using the proposed method were compared with their actual values for each condition.

The actual position of the CoP and the applied force on the foot was calculated from the rail posture and the other parameters mentioned above for every conditions. The CoP position and the normal and tangential force esti-mated from the forces measured by the load cells using the proposed method were compared with these actual valued for every conditions.

## 4.3   Results

Figure 8 shows the estimation results for the CoP position and the force applied at it for 16 external force conditions, along with their actual values. Figures 8a and b show the CoP position and the normal force calculated by Eqs. (12) and (13), respectively. These calculation methods are standard and are therefore reliable for estimating the CoP position and normal force; thus, it can be seen that the errors between the actual and estimated values are very small. The means of the errors of the two quantities are $1.44 \pm 0.86\,\text{mm}$ and $0.18 \pm 0.16\,\text{N}$, respectively.

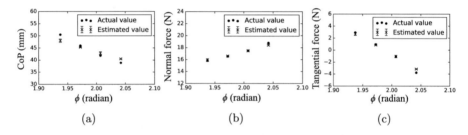

(a)                          (b)                          (c)

**Fig. 8.** Estimated and actual CoP position and force applied at the CoP for 16 exter-nal force conditions. (a) Position of the CoP. (b) Normal force applied at the CoP. (c) Tangential force applied at the CoP.

Figure 8c shows the tangential force estimated by the proposed method rep-resented by Eq. (15). The mean of the error between the estimated and actual values is $0.29 \pm 0.23\,\text{N}$. It can be seen that the error of the tangential force is small even when compared with the error of the estimated normal force. There-fore, we conclude that the proposed method is validated and can be used to measure the tangential force applied to the flat surface.

## 5   Conclusion

In this paper, a new concept was proposed to represent contact between a robot and multiple points on a flat surface. We showed analytically that if the two end passive joints are mechanically constrained such that the ratio of the torques at the joints is constant, the location of the MEP is fixed with respect to the link connecting the joints. This fact can be used to estimate the tangential force applied at the CoP from the measured CoP position and the normal force, and this was also validated experimentally. Therefore, the proposed method realizes a simple and reliable measurement of the force applied on the end-link. However, there are several problems from the viewpoint of application. One of them is the complexity of the mechanism. The constrained joints as shown in the experiment become large compared with the standard single joint. We can regard the proposed mechanism as a passive joint, and we also can realize the joint moved in the three-dimensional space by using several sets of this mechanism such as a universal joint. However, this makes the mechanism more complex. Besides, the reliability in the sensing is not high in the current implementation from the technical point of view. Therefore, we need to realize more compact and reliable implementation in our future work.

In this paper, we only discussed the application of the MEP concept to the force measurement, but it is also very interesting to consider the concept from a controls standpoint. One significant application is the fast operation of a bipedal robot, which requires fast feedback control of its posture in order to avoid it falling down. The foot mechanism with a constrained passive joint allows the foot to passively contact the floor, and a control scheme based on the MEP allows a moment to be applied to the floor to recover the robot's balance, in addition to the robust sensing of the reactive force. MEP-based control could also apply to the wire-driven mechanism as mentioned in Sect. 2 and shown in Fig. 5b, which would be useful for the manipulation of an under-actuated, wire-driven robotic hand such as the one investigated in our previous work [23]. To realize MEP-based control, extensions of the inverse pendulum control for the running bipedal robot and of the manipulation of a closed-loop system for the robotic fingers of an under-actuated, wire-driven robotic hand are needed. Such extensions in these control areas will also be part of our future work.

# References

1. Dahiya, R.S., Metta, G., Valle, M., Sandini, G.: Tactile sensing: from humans to humanoids. IEEE Trans. Rob. **26**(1), 1–20 (2010)
2. Stassi, S., Cauda, V., Canavese, G., Pirri, C.F.: Flexible tactile sensing based on piezoresistive composites: a review. Sensors **14**(3), 5296–5332 (2014)
3. Ahmed, M., Chitteboyina, M.M., Butler, D.P., Çelik-Butler, Z.: Mems force sensor in a flexible substrate using nichrome piezoresistors. IEEE Sens. J. **13**(10), 4081–4089 (2013)
4. Salisbury, J.: Interpretation of contact geometries from force measurements. In: IEEE International Conference on Robotics and Automation, Proceedings, vol. 1, pp. 240–247. IEEE (1984)
5. Bicchi, A., Salisbury, J.K., Brock, D.L.: Contact sensing from force measurements. Int. J. Rob. Res. **12**(3), 249–262 (1993)
6. Suzuki, Y.: Multilayered center-of-pressure sensors for robot fingertips and adaptive feedback control. IEEE Rob. Autom. Lett. **2**(4), 2180–2187 (2017)
7. Vukobratović, M., Stepanenko, J.: On the stability of anthropomorphic systems. Math. Biosci. **15**(1), 1–37 (1972)
8. Goswami, A.: Postural stability of biped robots and the foot-rotation indicator (FRI) point. Int. J. Rob. Res. **18**(6), 523–533 (1999)

9. Nagasaka, K., Inoue, H., Inaba, M.: Dynamic walking pattern generation for a humanoid robot based on optimal gradient method. In: IEEE International Conference on Systems, Man, and Cybernetics, IEEE SMC 1999 Conference Proceedings, vol. 6, pp. 908–913. IEEE (1999)

10. Sugihara, T., Nakamura, Y., Inoue, H.: Real-time humanoid motion generation through ZMP manipulation based on inverted pendulum control. In: IEEE International Conference on Robotics and Automation, Proceedings, ICRA 2002, vol. 2, pp. 1404–1409. IEEE (2002)

11. Kajita, S., Kanehiro, F., Kaneko, K., Fujiwara, K., Harada, K., Yokoi, K., Hirukawa, H.: Biped walking pattern generation by using preview control of zero-moment point. In: IEEE International Conference on Robotics and Automation, Proceedings, ICRA 2003, vol. 2, pp. 1620–1626. IEEE (2003)

12. Hirai, K., Hirose, M., Haikawa, Y., Takenaka, T.: The development of honda humanoid robot. In: IEEE International Conference on Robotics and Automation, Proceedings, vol. 2, pp. 1321–1326. IEEE (1998)

13. Choi, Y., Kim, D., You, B.-J.: On the walking control for humanoid robot based on the kinematic resolution of CoM Jacobian with embedded motion. In: IEEE International Conference on Robotics and Automation, ICRA 2006, Proceedings, pp. 2655–2660. IEEE (2006)

14. Kajita, S., Morisawa, M., Miura, K., Nakaoka, S., Harada, K., Kaneko, K., Kanehiro, F., Yokoi, K.: Biped walking stabilization based on linear inverted pendulum tracking. In: IEEE/RSJ International Conference on Intelligent Robots and Systems (IROS), pp. 4489–4496. IEEE (2010)

15. Hirose, M., Ogawa, K.: Honda humanoid robots development. Philos. Trans. R. Soc. London A: Mathe. Phys. Eng. Sci. **365**(1850), 11–19 (2007)

16. Hirukawa, H., Kanehiro, F., Kaneko, K., Kajita, S., Fujiwara, K., Kawai, Y., Tomita, F., Hirai, S., Tanie, K., Isozumi, T., et al.: Humanoid robotics platforms developed in HRP. Rob. Autonom. Syst. **48**(4), 165–175 (2004)

17. Akachi, K., Kaneko, K., Kanehira, N., Ota, S., Miyamori, G., Hirata, M., Kajita, S., Kanehiro, F.: Development of humanoid robot HRP-3P. In: 5th IEEE-RAS International Conference on Humanoid Robots, pp. 50–55. IEEE (2005)

18. Kajita, S., Kaneko, K., Kaneiro, F., Harada, K., Morisawa, M., Nakaoka, S., Miura, K., Fujiwara, K., Neo, E., Hara, I., et al.: Cybernetic human HRP-4C: a humanoid robot with human-like proportions. In: Robotics Research, pp. 301–314 (2011)

19. Zucker, M., Joo, S., Grey, M.X., Rasmussen, C., Huang, E., Stilman, M., Bobick, A.: A general-purpose system for teleoperation of the DRC-HUBO humanoid robot. J. Field Rob. **32**(3), 336–351 (2015)

20. Murray, R.M., Li, Z., Sastry, S.S.: A Mathematical Introduction to Robotic Manipulation. CRC Press, Boca Raton (1994)

21. Tsai, L.W.: Robot Analysis: The Mechanics of Serial and Parallel Manipulators. Wiley, New York (1999)

22. Ozawa, R., Kobayashi, H., Hashirii, K.: Analysis, classification, and design of tendon-driven mechanisms. IEEE Trans. Rob. **30**(2), 396–410 (2014)

23. Shirafuji, S., Ikemoto, S., Hosoda, K.: Development of a tendon-driven robotic finger for an anthropomorphic robotic hand. Int. J. Rob. Res. **33**(5), 677–693 (2014)

# Simulation and Transfer of Reinforcement Learning Algorithms for Autonomous Obstacle Avoidance

Max Lenk[1], Paula Hilsendegen[2(✉)], Silvan Michael Müller[2], Oliver Rettig[2], and Marcus Strand[2]

[1] SAP SE, Dietmar-Hopp-Allee 16, 69190 Walldorf, Germany
[2] Department for Computer Science, Duale Hochschule Baden-Württemberg, 76133 Karlsruhe, Germany
`hilsendegen.paula@student.dhbw-karlsruhe.de`

**Abstract.** The explicit programming of obstacle avoidance by an autonomous robot can be a computationally expensive undertaking. The application of reinforcement learning algorithms promises a reduction of programming effort. However, these algorithms build on iterative training processes and therefore are time-consuming. In order to overcome this drawback we propose to displace the training process to abstract simulation scenarios. In this study we trained four different reinforcement algorithms (Q-Learning, Deep-Q-Learning, Deep Deterministic Policy Gradient and Asynchronous Advantage-Actor-Critic) in different abstract simulation scenarios and transferred the learning results to an autonomous robot. Except for the Asynchronous Advantage-Actor-Critic we achieved good obstacle avoidance during the simulation. Without further real-world training the policies learned by Q-Learning and Deep-Q-Learning achieved immediately obstacle avoidance when transferred to an autonomous robot.

**Keywords:** Reinforcement learning · Machine learning
Obstacle avoidance · Collision avoidance · Simulation

## 1 Introduction

Successful navigation in autonomous mobile systems requires two key abilities: path planning and obstacle avoidance [1]. Both might be achieved easily in small, static and predefined environments. However, real-world scenarios share none of these characteristics. They are rather complex and dynamic, and partially or completely unknown to the moving agent. Reinforcement learning, a subtype of machine learning, offers promising solutions to the described challenges [2–4].

Reinforcement learning is inspired by the learning behavior of biological systems [5] and is based on the idea that direct interaction with the environment generates positive or negative outcomes. The thrive of the system for maximizing positive outcomes adjusts the probability of the occurrence of a particular

© Springer Nature Switzerland AG 2019
M. Strand et al. (Eds.): IAS 2018, AISC 867, pp. 401–413, 2019.
https://doi.org/10.1007/978-3-030-01370-7_32

behavior in the future [6]. In humans the acquisition of rather simple, e.g. grasping [7], as well as complex behavior, e.g. language [8], is ascribed to reinforcement learning. Albeit it is still under discussion to which extent reinforcement learning is accompanied or substituted by other learning mechanisms [9], at least some human learning processes seem to be guided by rewards as they can be mapped to neural systems specifically suited for reinforcement learning [10].

In mobile robotics the application of approaches using autonomous learning can overcome some of the limitations classical approaches (i.e. the detailed modeling of environment, robot and task) are bound to [11]. Especially generalization to new or dynamic environments or tasks is difficult to achieve with hand-designed robot controls. In this specifically lies the strength of autonomous learning and most notably reinforcement learning [5,12].

In contrast to other machine learning paradigms an agent learning by reinforcement has no information indicating which action to choose in a given state of the environment or which action is associated with positive long-term effects. Exclusively by interacting with the environment in a closed-loop manner the agent learns which actions can maximize the received reward [5,13]. According to [5] three elements are indispensable for this process: a policy, a reward signal and a value function. The policy defines which actions to choose in a given state of the environment. The reward signal divides actions in good or bad, according to the immediate reward received by the transition between states. The overall goal of the agent is to maximize the total reward and therefore the reward signal is the basis for adjusting the policy. Additionally, the value function evaluates which actions have positive long-term effects, by considering not only the immediate reward of a state but also the reward that is likely to follow in the long run.

Over the past decades a great deal of algorithms implementing reinforcement learning have been developed. In this paper we want to focus on four different reinforcement algorithms: Q-Learning [14], deep Q-Learning (DQN) [15], deep deterministic policy gradient (DDPG) [16,17] and asynchronous advantage actor-critic (A3C) [18,19].

All of these algorithms build upon the agent making experiences in the environment. This means in effect that the optimal policy can only be determined by training, i.e. exploration of the environment. However, real world exploration causes some problems that impair the application of reinforcement learning in robotics. First, expenditure of time for the training can be tremendous, especially when the movement velocity of the robot is low. Then, training needs to be supervised because in case of hardware or navigational problems the training could come to an unintentional halt [20]. Furthermore, training puts the robot in danger of getting damaged [2], causing thereby all kinds of other problems. Lastly, real-world training is to some extent inflexible. For example it is often difficult to vary the robot's sensor combination in a satisfactory amount of time. Taken together, this strongly suggests to simulate the training process.

Although simulations can overcome the described problems, they have one big disadvantage: the learned policy needs to be transferred to the real-world robot. For a successful transfer a careful design of the simulation is necessary. If the simulation is too realistic, the time-advantage is reduced and overfitting is much more likely to occur. However, if on the other hand the simulation is too abstract a real-world transfer might not work at all.

The objective of this study is to implement obstacle avoidance in an autonomous mobile robot by applying reinforcement learning. The major drawback of reinforcement learning is resolved by displacing the time-, labor- and potentially cost-consuming real-world training process to abstract simulation scenarios (see Fig. 1 for a visualization of our approach). Three simulation variants are tested, differing with regard to the agents' sensor combinations, the task to be solved and the start position being either constant or varying. Reinforcement learning is realized by the aforementioned algorithms. After successful training the learned policies are transferred to an autonomous mobile robot (Lego Mindstorms EV3) and tested in an office environment and on a small test field. We hypothesize that learning can be achieved by all algorithms. Furthermore, we expect that despite the abstractness of the simulation a real-world transfer is possible. Besides we are curious, whether the simulated training process needs to be extended by a short real-world training or whether the learned policies produce directly the desired outcome.

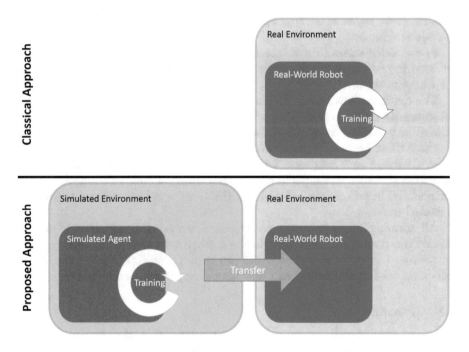

**Fig. 1.** On top of the figure the classical approach of reinforcement learning in robotics is visualized. At the bottom our proposed approach is visualized. The robot's real-world experience is preceded by a simulation, covering the complete training process.

## 2   Methods

### 2.1   Simulation

A simple abstract simulation was used to perform the training processes of the different reinforcement algorithms. Our simulation was integrated into the Open AI Gym [21], a framework specifically designed to test this type of algorithms. In this study we used three different scenarios. In all scenarios the agent's goal was to navigate through a square environment without colliding with the walls surrounding the environment or two obstacles placed into the area. The agent was punished for hitting an obstacle, received no reward for rotating at one position and was rewarded for driving straight. To ease or even eliminate the transfer process we paid particular attention to the simulation of the robot's control. Especially in continuous action spaces the design of the control can be challenging. In the first scenario the start position and initial rotation of the agent were held constant. The simulated agent had three ultrasonic sensors, mounted on the front whereby the outer sensors were rotated by 20° to the left or right respectively. Sensors were able to measure distances up to 255 cm.

The second scenario featured a robot that was equipped with two different sensors. One forward facing ultrasonic sensor measuring the distance to obstacles and a three-zone infrared sensor, which allowed to detect the position of an obstacle (left, right or ahead). Start position and initial rotation varied from episode to episode.

In the third scenario a target position was added to the environment. In addition to the aforementioned rewards the agent received a high reward when the target position was reached.

The reinforcement learning algorithms (Q-Learning, Deep Q-Learning, Deep deterministic policy gradient, asynchronous advantage actor-critic) were implemented in python. Except for Q-Learning we used the Tensorflow framework via the API of Keras. The simulated agent was implemented by Keras RL. The Open AI framework was used to construct the simulation environments. We used three different measures to assess the learning success of the agent: reward per episode, iterations per episode and the Q-value which denotes the average predicted action-value.

All training processes were performed unsupervised. At the end of each episode the simulation was reset. However, when there was no learning progress we aborted the training and adjusted the hyperparameters. When the applied algorithm used neural nets even the tuning of hyperparameters was automated.

### 2.2   Transfer

After successful simulation we transferred the learned models to our real-world robot. As a model system we used the Lego Mindstorms EV3 robot which offers despite its easy handling sufficient flexibility (Fig. 2). The robot was equipped with an ultrasonic sensor and a three-zone infrared sensor (both from Lego Mind-sensors) as described in simulation scenario two. Due to the limited computational power of our robot we exploited the node-like architecture of the robotic

**Fig. 2.** Lego Mindstorms EV3 robot

operating system (ROS). A sensor and an actor node were placed directly on the robot, whereas the agent's logic was outsourced to an external computer. The agent's logic was realized by python scripts which targeted the ROS node on the external computer and thereby communicated with the robot. Via the python scripts actions given a certain sensory input were selected on basis of e.g. the Q-table in case of Q-Learning or the trained neural nets in case of the other algorithms. Against our expectations we directly applied the models from the simulation and no further transfer process was needed.

## 3   Results

### 3.1   Simulation

**Q-Learning.** The Q-Learning algorithm was only tested in the first and second scenario. In the third scenario input was extended by two additional distance-sensors resulting in a non-linear increase of Q-table entries. Thus, training would become highly inefficient. Therefore we decided not to target the third scenario with Q-Learning. To apply Q-Learning we mapped the continuous input to discrete states.

Applied to the first scenario Q-Learning lead to good results (Fig. 3A). After a few training episodes the agent already learned to avoid obstacles. By following the edges of the simulated environment the agent maximized the reward and reached the maximal number of iterations. In the second scenario Q-learning also yielded good results. The desired behavior was learned quickly (Fig. 3B). Due to the varying start positions the reward was much more prone to outliers than in the first scenario. However, after about 180 training episodes, the reward stabilized. Nevertheless, due to the random start positions some episodes received low or negative reward. The number of iterations increased steadily and soon reached the maximum.

**Deep Q-Learning.** DQN builds upon Q-Learning. Because of the good results of Q-Learning in the first scenario we expected similar results for DQN and therefore tested it only in the second and third scenario. The neural net used

A                                          B

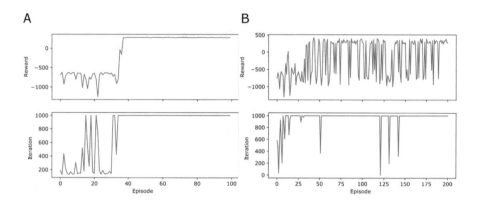

**Fig. 3.** Graphs depict reward and number of iterations per episode obtained by Q-Learning in A: the first scenario and B: the second scenario

to approximate the optimal policy consisted of one input layer with neurons for each sensor, three hidden layers each containing 8 neurons and an output layer with neurons for each action (go forward, turn left, turn right). First we trained the neural net for the second scenario. Due to the high number of training episodes reward and loss per episode was noisy (Fig. 4). After a positive start the reward was reduced between episodes 300 and 500. The agent's momentum might have been too strong and therefore the global maximum was exceeded. After that, the agent probably stepped into a local maximum which he couldn't leave until episode 500. This was also reflected by the loss which was quickly reduced in the beginning. While the agent was trapped in the local maximum the loss increased slightly before it finally decreased continuously. Learning success was also reflected by the average Q-Value, which increased after episode 500 continuously, reaching a stable level. In a second step we tested the learning success in the simulated environment. Except for one outlier the simulated agent received positive rewards in all episodes.

We then analyzed the behavior of the algorithm in the third scenario. Consequently we extended the neural net. It consisted of one input layer with eight neurons, four hidden layers with each 32 neurons and one output layer with one neuron per action. During the learning process reward increased only during the first episodes. After that it remained static (Fig. 5). The loss was quickly reduced and stagnated at its optimum. Up to episode 2100 the average Q-Value increased. However, the Q-Value was subject to fluctuations and collapsed after episode 2100, which is why we ended the training process. When we tested the trained policy in the simulated environment only a small learning success could be denoted. The simulated robot avoided obstacles but didn't target the goal position. Presumably, the delay of the reward for the goal position was too long, and thereby the effect on the learning process small.

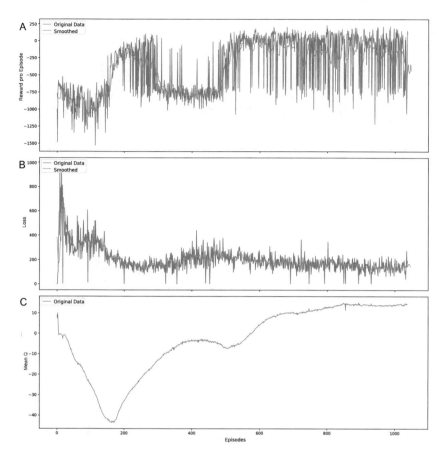

**Fig. 4.** Graphs depict A: reward B: loss and C: average Q-Value per episode obtained by DQN in the second scenario, original data is displayed in blue, orange curves denote smoothed data

**Deep Deterministic Policy Gradient.** DDPG was simulated in the first and second scenario. The continuous action space in DDPG necessitated an adaptation of the reward function: Rewards didn't increase when the robot went faster, rotations were slightly punished and moving backwards or colliding with an obstacle were highly punished. The applied actor net consisted of an input layer with three neurons, three hidden layers (32, 16 and 8 neurons) and three output neurons. The critic net consisted of four input neurons (three for state and one for the chosen action), 3 hidden layers with each 32 neurons and one output neuron. In the first scenario a positive reward is obtained after 100 episodes. In the following episodes the reward varies around zero. This implies that the simulated agent learned successfully. The decreasing loss and the increasing average Q-value were in the same line (Fig. 6).

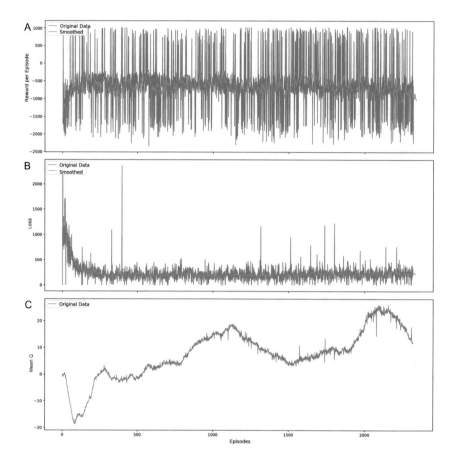

**Fig. 5.** Graphs depict A: reward B: loss and C: average Q-Value per episode obtained by DQN in the third scenario, original data is displayed in blue, orange curves denote smoothed data

We applied DDPG also in the second scenario and altered the configuration of the hidden layers of the neural nets. Then the actor net contained three layers each consisting of eight neurons and the critic net included also three layers each consisting of 16 neurons. As of the hundredth episode learning success was visible on the reward graph (Fig. 7). The continuously decreasing loss, as well as the average Q-Value, which reached its optimum asymptotically confirm this assumption.

**Asynchronous Advantage Actor-Critic.** Due to the similarity of A3C with DDPG and the good results of DDPG in the first scenario, we renounced the application of A3C in the first scenario and tested it only in the second scenario. The used neural net consisted of four hidden layers each containing 32 neurons. Eight agents were trained in parallel, by implementing multiple threads

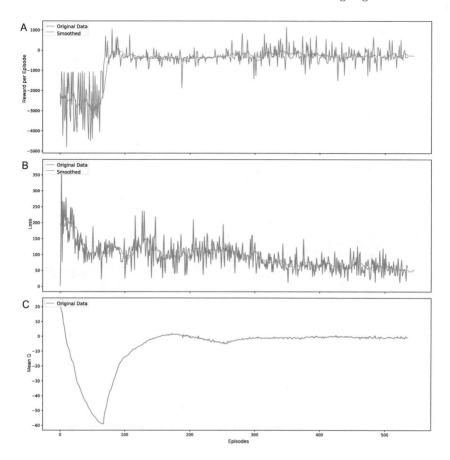

**Fig. 6.** Graphs depict A: reward B: loss and C: average Q-Value per episode obtained by DDPG in the first scenario, original data is displayed in blue, orange curves denote smoothed data

in Python. In all agents the reward increased for the first 500 episodes. After that reward decreased rapidly and stabilized at a negative reward of ca. −950. With our configuration of A3C we didn't reach any learning success. Therefore we neither applied A3C to the third scenario nor transferred it to the real-world robot.

## 3.2   Transfer

**Q-Learning.** We first tested the transfer of the policy learned by Q-learning in the second scenario to a real-world robot in an office environment. The input data of the robot was discretized and smoothed to assimilate it with the simulated sensor data. Without further adaption of the Q-Table avoidance of static and dynamic obstacles was reached, which can be seen by the consistently positive reward (Fig. 8A). We then tested success of transfer in a smaller test field.

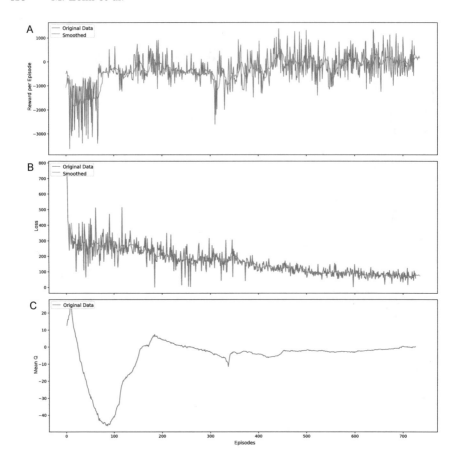

**Fig. 7.** Graphs depict A: reward B: loss and C: average Q-Value per episode obtained by DDPG in the second scenario, original data is displayed in blue, orange curves denote smoothed data

The robot generalized without further training, which is again depicted in a consistently high reward (Fig. 8B). Finally we placed the robot into a u-shaped environment. However, generalization of the learned policy was not high enough to avoid collision with the borders of this unknown environment.

**Deep Q-Learning.** Similar to the transfer of Q-Learning, obstacle avoidance was immediately obtained when the DQN policy learned in the simulation was transferred to the robot. However, the received reward was smaller than the reward obtained by Q-Learning (Fig. 8C). In an u-shaped environment the application of the policy learned by DQN was not successful.

**Deep Deterministic Policy Gradient.** First, we transferred only the actor net to the robot and tested its behavior without further training. Unfortunately with the model learned during the simulation we didn't obtain obstacle avoidance in the real robot (Fig. 8 D). The robot often got caught with its wheels or the mechanical threshold to turn the wheels was not reached. Building on this experience we also transferred the critic net and performed a short training. However, the robot's behavior was not improved.

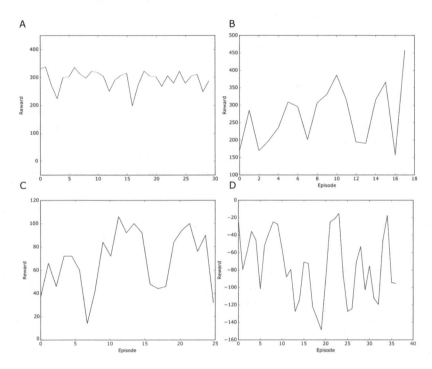

**Fig. 8.** Graphs depict the reward per episode obtained by the real-world robot with A: Q-Learning (office environment), B: Q-Learning (test field), C: DQN (test field), D: DDPG (scenario field)

## 4    Discussion

In this study we applied reinforcement algorithms to obtain obstacle avoidance in an autonomous mobile robot. By displacing the training processes to a simulation we were able to overcome the major drawback of reinforcement learning, i.e. the time-consuming training. Except for A3C all algorithms yielded successful obstacle avoidance during simulation. However, with our configurations, no algorithm allowed for the autonomous navigation to a target position. The transfer of the simulation results to an autonomous mobile robot worked effectively when learning was accomplished by Q-Learning or DQN.

The used simulation scenarios were highly abstract. Despite, or perhaps because of that, generalization of the learned policies was achieved. However, the abstractness of the simulation caused also some problems, especially with DDPG. In this approach the main reason for the failed transfer was that the simulation neglected the robot's dimensions and its mechanical details. For a successful application of DDPG it might be necessary to add an extended real-world training in which the agent acquires the necessary information directly from the environment.

We also didn't manage to apply A3C successfully. Although the hyperparameters of the algorithm were carefully selected, convergence was not achieved. Future research is needed to determine if this algorithm can be applied to solve the problem of obstacle avoidance. The other algorithms could also profit from a fine-tuning of the hyperparameters. Particularly, the possibilities to construct a neural net strive against infinity and so it's very likely that an altered configuration leads to similar or even better results.

In this study we only used model free algorithms, in order to circumvent the transfer of a specific internal model to a real-world robot. However, especially the delayed reward of a target position might necessitate the existence of such a model. How or if it is possible to transfer internal models acquired during a simulation is still an open question and needs to be determined in the future. In addition, the combination of model-free and model-based approaches is imaginable and has been proven useful [22].

The predominant advantage of displacing the training process to the simulation is the enormous time saving. In the simulated environment one iteration step during the training with Q-learning took about 2.5 msec. In the real world one iteration step would have taken 250 msec. The displacement of training reduced the training duration by a factor of 100. Due to the higher complexity of the other algorithms the temporal advantage was smaller (Factor of improvement in terms of time: DQN: 25, DDPG: 25, A3C: 40) but should not be despised.

Taken together, this study proves that reinforcement learning can be applied to solve obstacle avoidance. The usage of a simulation for training improves time- and cost-efficiency. The transfer to a real-world robot leads for Q-Learning and DQN to highly effective obstacle avoidance and turns out to be unproblematic. Research is needed to determine which algorithms other than the here tested can be applied to solve obstacle avoidance and what other problems in autonomous moving systems can be tackled by reinforcement learning.

# References

1. Siegwart, R., Nourbakhsh, I.R., Scaramuzza, D.: Introduction to Autonomous Mobile Robots. MIT Press, Cambridge (2011)
2. Kober, J., Bagnell, J.A., Peters, J.: Reinforcement learning in robotics: a survey. Int. J. Rob. Res. **32**(11), 1238–1274 (2013)
3. Azouaoui, O.: Reinforcement learning (RL) based collision avoidance approach for multiple Autonomous Robotic Systems (ARS). In: Proceedings 10th IEEE International Conference on Advanced Robotics, pp. 561–566 (2001)

4. Xie, L., Wang, S., Markham, A., Trigoni, N.: Towards monocular vision based obstacle avoidance through deep reinforcement learning. arXiv preprint arXiv:1706.09829 (2017)

5. Sutton, R.S., Barto, A.G.: Reinforcement Learning: An Introduction, vol. 1, No. 1. MIT Press, Cambridge (1998)

6. Thorndike, E.L.: Laws and hypotheses for behavior. In: Thorndike, E.L. (ed.) Animal Intelligence, pp. 241–281 (1970)

7. Needham, A., Barrett, T., Peterman, K.: A pick-me-up for infants' exploratory skills: early simulated experiences reaching for objects using 'sticky mittens' enhances young infants' object exploration skills. Infant Behav. Dev. **25**(3), 279–295 (2002)

8. Skinner, B.F.: The evolution of verbal behavior. J. Exp. Anal. Behav. **45**(1), 115–122 (1986)

9. Kuhl, P.K.: A new view of language acquisition. Proc. Nat. Acad. Sci. **97**(22), 11850–11857 (2000)

10. Holroyd, C.B., Coles, M.G.: The neural basis of human error processing: reinforcement learning, dopamine, and the error-related negativity. Psychol. Rev. **109**(4), 679 (2002)

11. Thrun, S.: An approach to learning mobile robot navigation. Rob. Autonom. Syst. **15**(4), 301–319 (1995)

12. Peters, J., Schaal, S.: Reinforcement learning of motor skills with policy gradients. Neural Netw. **21**(4), 682–697 (2008)

13. Szepesvári, C.: Reinforcement Learning Algorithms for MDPs. Morgan and Claypool Publishers, San Rafael (2010)

14. Watkins, C.J., Dayan, P.: Q-learning. Mach. Learn. **8**(3–4), 279–292 (1992)

15. Mnih, V., Kavukcuoglu, K., Silver, D., Graves, A., Antonoglou, I., Wierstra, D., Riedmiller, M.: Playing atari with deep reinforcement learning. Technical report arXiv:1312.5602 [cs.LG], Deepmind Technologies (2013)

16. Lillicrap, T.P., Hunt, J.J., Pritzel, A., Heess, N., Erez, T., Tassa, Y., Silver, D., Wierstra, D.: Continuous control with deep reinforcement learning. arXiv preprint arXiv:1509.02971 (2015)

17. Peters, J., Schaal, S.: Natural actor-critic. Neurocomputing **71**(7), 1180–1190 (2008)

18. Mnih, V., Badia, A.P., Mirza, M., Graves, A., Lillicrap, T.P., Harley, T., Silver, D., Kavukcuoglu, K.: Asynchronous methods for deep reinforcement learning. In International Conference on Machine Learning, pp. 1928–1937 (2016)

19. Palamuttam, R., Chen, W.: Vision enhanced asynchronous advantage actor-critic on racing games. Methods **4**, A3C (2017)

20. Mirowski, P., Pascanu, R., Viola, F., Soyer, H., Ballard, A., Banino, A., Denil, M., Goroshin, R., Sifre, L., Kavukcuoglu, K., Hadsell, R., Kumaran, D.: Learning to navigate in complex environments (2016). arXiv preprint arXiv:1611.03673

21. Brockman, G., Cheung, V., Pettersson, L., Schneider, J., Schulman, J., Tang, J., Zaremba, W.: OpenAI gym (2016). arXiv preprint arXiv:1606.01540

22. Racaniere, S., Weber, T., Reichert, D.P., Buesing, L., Guez, A., Rezende, D., Badia, A.P., Vinyals, O., Heess, N., Li, Y., Pascanu, R., Battaglia, P., Hassabis, D., Silver, D., Wierstra, D.: Imagination-augmented agents for deep reinforcement learning. In: Advances in Neural Information Processing Systems, pp. 5694–5705 (2017)

# HI-VAL: Iterative Learning of Hierarchical Value Functions for Policy Generation

Roberto Capobianco$^{(\boxtimes)}$, Francesco Riccio, and Daniele Nardi

Department of Computer, Control, and Management Engineering,
Sapienza University of Rome, via Ariosto 25, 00185 Rome, Italy
{capobianco,riccio,nardi}@diag.uniroma1.it

**Abstract.** Task decomposition is effective in various applications where
the global complexity of a problem makes planning and decision-making
too demanding. This is true, for example, in high-dimensional robotics
domains, where (1) unpredictabilities and modeling limitations typically
prevent the manual specification of robust behaviors, and (2) learning
an action policy is challenging due to the curse of dimensionality. In
this work, we borrow the concept of Hierarchical Task Networks (HTNs)
to decompose the learning procedure, and we exploit Upper Confidence
Tree (UCT) search to introduce HI-VAL, a novel iterative algorithm for
hierarchical optimistic planning with learned value functions. To obtain
better generalization and generate policies, HI-VAL simultaneously learns
and uses action values. These are used to formalize constraints within
the search space and to reduce the dimensionality of the problem. We
evaluate our algorithm both on a fetching task using a simulated 7-DOF
KUKA light weight arm and, on a pick and delivery task with a Pioneer
robot.

## 1 Introduction

Generating effective action policies is impractical in various applications that
are characterized by large state spaces and require strong generalization capabilities. Many techniques tackle the curse-of-dimensionality problem by using
expert demonstrations to initialize agents' behaviors and guide the learning process. However, this is not feasible when a reduced number of examples is available, and a direct mapping between the expert's and the agent's action space is
difficult to obtain (e.g. highly redundant robots). While generalization is typically achieved by means of function approximation (e.g., using neural networks),
solely relying on this and excluding prior knowledge can be inefficient, slow and
even dangerous in multiple applications, such as robotics. Hence, Monte-Carlo
tree search algorithms, and in particular the Upper Confidence Tree (UCT) algorithm [13], are widely used to exploit prior knowledge [21] in exploring search
space. Nonetheless, they show limitations in generalizing among related states.

We build upon these state-of-the-art techniques to directly learn action values from experience, and accordingly learn a policy by using UCT with focused
exploration. To this end, we introduce HI-VAL, a novel iterative algorithm for

© Springer Nature Switzerland AG 2019
M. Strand et al. (Eds.): IAS 2018, AISC 867, pp. 414–427, 2019.
https://doi.org/10.1007/978-3-030-01370-7_33

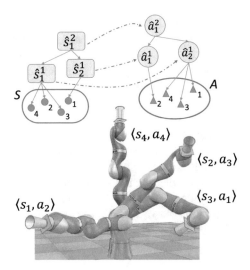

**Fig. 1.** HI-VAL generates high-level representations of actions, that are used to improve the exploration of the search space. In this figure, we show the action hierarchy generated for a fetching task using a redundant KUKA light weight arm.

learning hierarchical value functions that are used to (1) capture multi-layered action semantics, (2) generate policies by scaffolding the acquired knowledge, and (3) guide the exploration of the state space. HI-VAL improves the UCT algorithm and builds upon concepts from previous literature, such as Hierarchical Task Networks (HTNs) [8], semi-MDPs [23] and MAX-Q decompositions [7], to decompose the learning procedure and to generate both action abstractions and search space constraints. The action hierarchy formalized by HI-VAL is learned iteratively by evaluating state-actions pairs generated by UCT after each episode. Figure 1 shows an example of such a hierarchy where states and actions are associated at different layers of abstraction. HI-VAL assigns states and actions to different clusters $s_l^c$ and $a_l^c$ by evaluating the similarity of successor states that the agent can reach, by applying the actions in $a_l^c$ in the states contained in $s_l^c$. Intuitively, similar successor states have similar reward values and can be evaluated altogether when exploring the search space. Different layers provide different granularity of action semantics (the higher the more coarse) and help the learning process to evaluate states hierarchically. HI-VAL runs UCT to explore the environment by sampling the joint distribution of rewards and state-action pairs. Each sample is continuously aggregated into a dataset, that is used to estimate – by means of $Q$-learning – the value function $Q^\lambda$ of each layer $\lambda$ in the hierarchy. Specifically, at each layer, Monte Carlo search is ran for a subset of actions that are evaluated according to their $Q$-value, thus driving the node-expansion phase during episode simulation.

In this work, we aim at demonstrating that $Q$-values can be learned hierarchically to influence exploration, and to represent action semantics at different levels of abstraction, thus linking learning techniques to low-level agent controls. Our main contributions consist in (1) a novel integration of Monte-Carlo tree search, hierarchical planning and $Q$-learning, that enables good performance with selective state exploration and improved generalization capabilities, in (2) a two-sided extension of TD-search, that not only executes on multiple hierarchy layers, but also constructs upper confidence bounds on the value functions – and selects actions optimistically with respect to those – and in (3) a reduction of the curse-of-dimensionality that is obtained by means of focused exploration. We evaluated the Hi-Val performance in two different scenarios, an "object-fetching" task with a 7-DOF KUKA light weight arm and, a "pick and delivery" task in a simple environment with a Pioneer robot, where the agent has to collect an item and delivering it reduction in the number of states explored – which makes the method more suitable in robotic applications – and, the ability of Hi-Val to represent action semantics through its hierarchy in order to boost the learning process.

## 2   Related Work

Policy learning is widely adopted to generate practical behaviors in several applications. This is true for complex domains, such as robotics [4], where unstructured environments and uncertain dynamics through handcrafted policies – that typically fail or must be refined [14]. Although designing effective policies is impractical in most of these scenarios, and learning techniques are typically demanding and time consuming [11] for problems with large state spaces. The computational demand can be alleviated by initializing a policy with expert demonstrations, that restrict the learning process to a promising hypothesis space [12,20]. However, to apply imitation learning in complex domains, a large dataset of good-quality expert demonstrations is generally required, that can be efficiently mapped to the agent's action space. Unfortunately, this is not always possible due to the lack of (1) domain experts, (2) practical ways of providing demonstrations, and (3) action mappings from experts to agents (e.g. hyper-redundant robots).

To overcome these difficulties, we propose an approach that does not require expert demonstrations to initialize agent behaviors, and decomposes the learning procedure to generate action abstractions and search space constraints. In literature, multiple authors exploit the notions of skills and semi Markov Decision Processes (semi-MDPs), and define hierarchical representations such as *options* [23] and MAX-Q decompositions [7]. Unfortunately, applications of these methods in complex domains like robotics are limited, and prior knowledge has to be enforced in the learning process by means of expert demonstrations. In fact, although hierarchical learning and value function approximations techniques have been adopted in several applications, state-of-the-art techniques still show considerable margin of improvement. For example, [19] provide a better policy generalization by exploiting the concept of Generalized Value Functions, to improve value function approximation. In a different settings, [6] use expert demonstrations to learn high-level tasks as a combination of action-primitives. Unfortunately, these approaches only learn specific hierarchical structures, that poorly generalize and cannot profit from the expressiveness of value functions. Similarly, [22] apply hierarchical learning to sequences of motion primitives on a pick-and-place task with a hyper-redundant robotic arm. [15] initialize skill trees from human demonstrations, improving them over time. However, their representations use expert demonstrations and do not represent action on higher levels of abstractions. Conversely, [10] apply hierarchical policy learning to solve a 2-DOF stand-up task for a robotic arm. They exploit Q-learning and actor-critic methods to learn both task decompositions and local trajectories that solve specific sub-goals. Alternatively, [2,9] formalize action hierarchies to represent actions at different levels of abstraction. However, these procedures are not easily scalable to higher dimensionality problems.

Motivated by our discussion, we extend our previous work [16,17] to formalize action hierarchy by introducing Hi-Val, an iterative algorithm that learns hierarchical value functions to drive the policy search, and that achieves good generalization with a focused state exploration without the aid of expert demonstrations. Specifically, we enable Hi-Val to (1) learn a hierarchical value function directly from experience and (2) simultaneously use learned values during exploration, to generate a competitive policy. As in Hierarchical Task Networks, value functions are used to plan both over compound and primitive action spaces, whereas compound actions have specific implementations that depend on the state of the environment. Like [21], we exploit TD (temporal difference) methods to learn action values during a Monte-Carlo tree search. In this way, on the one hand, we are able to learn $Q$-values, and on the other hand, we adopt $Q$-values to support decision-theoretic planning for generalization and exploration at multiple levels of abstraction. Differently from [21], in fact, not only we improve our model by preserving the selective search of Monte-Carlo algorithms when bootstrapping is ongoing, but we also generate action and state abstractions. Specifically, Hi-Val extends TD-search by constructing upper confidence bounds on a hierarchy of value functions, and by selecting optimistically with respect to those. As the experimental evaluation shows in both scenarios, Hi-Val generates competitive policies – with the additional benefit of a reduction in (1) number of simulations (or expanded node), and (2) exploration of the search space – that alleviates the curse-of-dimensionality.

## 3   Hi-Val

*Formulation.* Hi-Val is an iterative algorithm that, at each iteration $i$, generates a new policy $\pi_i$ which improves $\pi_{i-1}$ [17]. To obtain an improved $\pi_i$, our algorithm leverages (1) data aggregation [18], and (2) Upper Confidence Bounds for Trees (UCT) [13], a variant of Monte-Carlo Tree Search that adopts an upper confidence bound strategy – UCB1 [3] – for balancing between exploration and exploitation on the tree. To describe Hi-Val we adopt the Markov Decision Process (MDP) notation, in which the decision-making problem is represented as a tuple $MDP = (\mathcal{S}, \mathcal{A}, \mathcal{T}, R, \gamma)$, where $\mathcal{S}$ is the set of discrete states of the environment, $\mathcal{A}$ represents the set of discrete actions, $\mathcal{T} : \mathcal{S} \times \mathcal{A} \times \mathcal{S} \to [0, 1]$ is a stochastic transition function that models the probabilities of transitioning from state $s \in \mathcal{S}$ to $s' \in \mathcal{S}$ when taking action $a \in \mathcal{A}$, $R : \mathcal{S} \times \mathcal{A} \to \mathbb{R}$ is the reward function, and $\gamma$ is a discount factor in $[0, 1)$.

*Function Approximation.* HI-VAL addresses the generalization problem by relying on previous literature [5]. We choose to approximate the $Q$ function using probability densities in the form of a mixture of $K$ Gaussians (i.e., Gaussian Mixture Models – GMMs). We integrate the approach in [1] with a data aggregation [18] procedure, where a dataset of samples is iteratively collected and aggregated. Specifically, at each iteration $i$, the values $Q^i$ are determined according to the $Q$-learning update rule

$$Q^i(s, a) = \hat{Q}'(s, a) + \alpha(r + \gamma \max_{a'} \hat{Q}'(s', a') - \hat{Q}'(s, a)), \tag{1}$$

where $\alpha$ is the learning rate, $\hat{Q}'$ is the function approximation learned at previous iteration, and $Q^0(s, a_j) = 0$. As we discuss later, the function approximation $\hat{Q}$ is learned over an aggregated dataset $\mathcal{D}^{0:i} = \{\cup \mathcal{D}^d | d = 0 \ldots i\}$.

## 3.1   Exploration and Sample Collection

At every step $h$, for $h = 1 \ldots H$, UCT simulates the execution of all the actions $\mathcal{A}_L \subseteq \mathcal{A}$ that are "admissible" in $s_h$, as detailed in next section. Specifically, each simulation executes an action $a \in \mathcal{A}_L$, followed by $K$ roll-outs, that run an $\epsilon$-greedy policy based on $\pi_{i-1}$ until a terminal state is reached. The best action $a_h$ is selected according to

$$e = C \cdot \sqrt{\frac{\log(\sum_a \eta(s_h, a))}{\eta(s_h, a)}}$$

$$a_h^* = \arg\max_a \hat{Q}(s_h, a) + e, \tag{2}$$

where $C$ is a constant that multiplies and controls the exploration term $e$, and $\eta(s_h, a)$ is the number of occurrences of $a$ in $s_h$. Since we assume a discrete state space $\mathcal{S}$, for continuous problems we define a similarity operator that informs the algorithm whether the difference of two states is smaller than a given threshold $\xi$ – thus discretizing the space.

During each roll-out:

- a dataset $\mathcal{D}^i$ of samples $x = (s, a, Q^i(s, a))$ is collected to improve our estimate $\hat{Q}$, as detailed in previous section;
- HI-VAL uses UCT as an expert and collects a dataset $\mathcal{D}_{\pi, uct}$ of $H$ samples $x = (s_h, a_h^*, s_h')$ that are selected by the tree search
- similarly to DAgger [18], $\mathcal{D}_{\pi, uct}$ is aggregated into a dataset $\mathcal{D}_{\pi, i} = \mathcal{D}_{\pi, uct} \cup \mathcal{D}_{\pi, i-1}$.

When $H$ UCT steps are run, $\mathcal{D}_{\pi, i}$ is used to generate hierarchy clusters – as detailed in the next section – and learn a new policy $\pi_i$. Our algorithm uses a discovery process that is supported by the UCT search and policy.

## 3.2   Hierarchical Action Selection

The hierarchical model adopted in Hi-Val builds upon the concepts of *High-Level Actions (HLAs)* and *Reachable Sets (RSs)* in HTNs [8].

- HLAs are defined recursively as a sequence of action primitives and/or other HLAs. When a HLA is composed by only primitives, such sequence is called "implementation".
- RSs model preconditions and effects of HLAs. They are defined as the union of possible states reachable by the different implementations of HLAs. A RS is bounded by a pessimistic $RS^-$ and optimistic $RS^+$ set. $RS^-$ represents the set of states that are reached independently from the chosen implementation, while $RS^+$ represents the set of states reached by all the possible implementations. Reachable sets describe interesting properties: (1) if $RS^-$ intersects the set of goal states, then a sequence of admissible actions leading to the goal is found; (2) if, instead, $RS^+$ intersects the set of goal states, a plan exists but its implementations do not reach the goal yet.

Using similar concepts, we allow Hi-Val to evaluate actions at multiple levels of abstraction. Specifically, a hierarchy of actions is obtained using an *agglomerative clustering* algorithm, which is ran on the set of next states $\{s'\}$ present in the $\mathcal{D}_{\pi,i}$. Our key assumption is that similar $s'$ encode information about actions with similar effects and thus can be clusterized altogether. Particularly, we refer to $\mathcal{H}$ as the set of layers in the action hierarchy generated by the clustering algorithm. The clustering routine organizes $\{s'\}$ in clusters $\hat{s}'$ over a predefined number of layers, which are then transferred into the action space, generating a set of action clusters $\hat{a}$ organized along the same structure. Such a mapping is realized by evaluating each element contained in the $\hat{s}'$ clusters and backpropagated to the original dataset $\mathcal{D}_{\pi,i}$ in order to retrieve the set of actions generating the transitions to the $\{s'\}$ states. In fact, dataset elements are a tuple of three components encoding a transition from the current state $s$ to the next state $s'$ by means of the action $a$. Figure 1 shows a simplistic example of such hierarchies.

In our algorithm, each action cluster $\hat{a}$ corresponds to a HLA in the layer $\lambda \in \mathcal{H}$, and each layer has an associated $Q^\lambda$ function, approximated as $\hat{Q}^\lambda$. The result of choosing an action $\hat{a}$ consists of selecting a cluster of lower level actions with a similar expectation to reach a desired cluster $\hat{s}'$. Noticeably, such a model intrinsically connects to the concept of reachable sets. Clusters $\hat{s}'$ are in fact an approximation of optimistic sets $RS^+$, and they evaluate actions $\hat{a}$ that lead to more rewarding states altogether.

Hi-Val uses each $Q^\lambda$ in $s$ to select the set $\mathcal{A}_L$ of "admissible" actions for UCT. Intuitively, a primitive action $a$ is admissible in $s$ if, for each layer $\lambda$, $a$ belongs to the cluster $\hat{a}^\lambda$ selected according to $Q^\lambda$. More formally:

$$\mathcal{A}_L = \bigcup_{a \in \mathcal{A}} \{a \mid \forall \lambda \in \mathcal{H} : a \in \hat{a}^\lambda\} \tag{3}$$

$$\hat{a}^\lambda = {}^{\arg\max}_{\hat{a}} Q^\lambda(s, \hat{a}) + \delta^\lambda_{s,\hat{a}}$$

$$\delta^\lambda_{s,\hat{a}} \sim \mathcal{N}(0, \sigma^2(Q^\lambda | s, \hat{a})), \tag{4}$$

where $\sigma^2$ is the standard deviation of the regression approximation [1].

Through $\delta^\lambda_{s,\hat{a}}$, the prediction error for each action abstraction is captured, leading to a more directed exploration of the action hierarchy. Action primitives are finally chosen and executed by UCT according to the lowest-level $\hat{Q}$, as detailed in previous section. To obtain a less biased exploration and avoid value function over-fitting, inadmissible actions are anyway expanded and selected by UCT with a 30% probability.

## 3.3   HI-VAL Algorithm

The goal of HI-VAL consists of iteratively updating each layer's value function approximation $\hat{Q}^\lambda$, to generate a policy $\pi_i$ that maximizes the expected reward of the agent. The underlying insight of HI-VAL is that, while exploring the search space, collected state-action pairs are used at each iteration $i$ to (1) update the approximated $Q$ functions for refining the policy $\pi_{i-1}$ into a policy $\pi_i$, and (2) use $Q$-values to influence UCT exploration in accordance with Eq. 3. The complete HI-VAL algorithm – described in Algorithm 1 – for each iteration proceeds as follows:

1. **Roll-in.** The agent follows the previous policy $\pi_{i-1}$ and generates a set of $s_t$ states for $T$ timesteps.
2. **UCT search.** For each of the generated states $s_t$, HI-VAL runs an UCT search with horizon $H$. At each step $h$, UCT simulates the execution of every "admissible" action in the set $\mathcal{A}_L$, computed according to Eq. 3. For each action $a \in \mathcal{A}_L$, a simulation consists of the execution of $a$, followed by $K$ $\epsilon$-greedy roll-outs based on $\pi_{i-1}$, which are used to estimate the $Q$-values of each visited state. Finally, for each step, the best action $a_h^*$ is (1) chosen according to Eq. 2 and (2) aggregated into a dataset $D_{uct}$ together with $s_h$ and $s_{h+1}$. It is worth remarking that a vanilla implementation of the UCT search evaluates all possible actions and explores a significant amount of states to generate an effective policy. Our approach, instead, leverages the hierarchical structure of $\mathcal{H}$ to generate a restricted subset of "admissible" actions, with high estimated $Q$-value. This efficiently reduces the exploration phase by guiding the algorithm to discard actions that are not expected to improve $\pi_{i-1}$.
3. **Hierarchical data aggregation and $\hat{Q}$ update.** After UCT, new data $D_{uct} = \{(s_{t+h}, a_{t+h}^*, s_{t+h+1}') \mid h = 1 \ldots H\}$ is available to be aggregated into a larger dataset $\mathcal{D}_{\pi,i}$. This dataset is used to generate clusters $\hat{s}$, $\hat{a}$, and $\hat{s}'$ in two steps: first, the sets of states $\{s\}$ and $\{s'\}$ in $D_{\pi,i}$ are separately clustered within $\lambda$ layers; then, the hierarchy of next state clusters $\hat{s}'$ is transferred into the action space to generate the action clusters $\hat{a}$, each of them corresponding to high-level actions. In order to correctly update the $\hat{Q}^\lambda$ estimation for every $\hat{a}$, samples of the form $\mathbf{x}^\lambda = (\hat{s}, \hat{a}, Q^\lambda(s, \hat{a}))$ are generated for each layer $\lambda$, with $Q^\lambda(s, \hat{a})$ determined as in Eq. (1). Such samples are then (1) aggregated into a dataset $\mathcal{D}^{\lambda,0:i}$ and (2) used to improve the estimate $\hat{Q}^\lambda$, as described in previous sections. Specifically, $\hat{Q}^\lambda$ is updated for each $(\hat{s}, \hat{a})$

---

**Algorithm 1.** Hi-Val

---

**Input**: $\mathcal{D}_0$ dataset of random state action pairs $\{(s, a, s')\}$.

**Output**: $\pi_N$ policy learned after N iterations of the algorithm.

**Data**: $\mathcal{A}$ set of primitive actions, $N$ number of iterations of the algorithm, $\Delta$ initial state distribution, $H$ UCT horizon, $K$ $\epsilon$-greedy roll-outs, $T$ policy execution timesteps, $\mathcal{H}$ set of layers.

**begin**

    Initialize $\hat{Q}^0$ to predict 0.

    Train classifier $\pi_0$ on $\mathcal{D}_{\pi,0}$.

    **for** $i = 1$ **to** $N$ **do**

        $s_0 \leftarrow$ random state from $\Delta$

        **for** $t = 1$ **to** $T$ **do**

            Get state $s_t$ by executing $\pi_{i-1}(s_{t-1})$.

            $\mathcal{D}_{uct} \leftarrow \text{UCT}(\mathcal{H}, s_t)$

            $\mathcal{D}_{\pi,i} \leftarrow \mathcal{D}_{\pi,i} \cup \mathcal{D}_{uct}$

            $\hat{s}, \hat{s}' \leftarrow \text{agglomerativeClustering}(\mathcal{D}_{\pi,i})$

            $\hat{a} \leftarrow \text{mapping}(\hat{s}')$

            **foreach** $\lambda \in \mathcal{H}$ **do**

                // update estimated Q-values

                $\bar{R}(\hat{s}, \hat{a}) \leftarrow \frac{1}{|(\hat{s},\hat{a})|} \sum_{s \in \hat{s}, a^* \in \hat{a}} R(s, a^*)$

                $\mathcal{D}^{\lambda,i} \leftarrow \text{getSamples}(\mathcal{D}_{uct}, \bar{R}(\hat{s}, \hat{a}))$

                $\mathcal{D}^{\lambda,0:i} \leftarrow \mathcal{D}^{\lambda,0:i} \cup \mathcal{D}^{\lambda,i}$

                Train $\hat{Q}^{\lambda,i}(\hat{s}, \hat{a})$ on $\mathcal{D}^{\lambda,0:i}$

            **end**

        **end**

        Train classifier $\pi_i$ on $\mathcal{D}_{\pi,i}$

    **end**

    **return** $\pi_N$

**end**

---

containing the state-action pairs $(s_{t+h}, a^*_{t+h}) \in \mathcal{D}_{uct}$, by using the averaged reward $R(s_{t+h}, a^*_{t+h})$ of the corresponding state-action pairs in the clusters $(\hat{s}, \hat{a})$.

4. **Training.** Once data aggregation has been performed, a new policy $\pi_i$ is trained from the dataset $\mathcal{D}_{\pi,i}$ (e.g., using a classifier).

## 4  Experimental Evaluation

We evaluate our approach in generating a policy for executing a "fetching" task, and a "pick and delivery" task in a simple environment [7], with a reduced number of state-action pairs explored by UCT. We compare our results with a random-UCT and vanilla-UCT implementations, the TD-search [21] algorithm and different configurations of the Hi-Val action hierarchy. We will refer to as Val as a basic implementation of the action hierarchy composed solely by a single layer of the primitive actions. Then, the random-UCT algorithm selects a random action at each $h$ step of the Monte Carlo search, while in the vanilla-UCT all actions are considered "admissible" at each step. In all algorithms, we implement a shaped reward function that computes reward values at each visited state. We deploy these algorithms within a simulated environment on a 7-DOF KUKA light weight arm and, on a Pioneer robot.

Experiments have been conducted within the V-REP simulation environment, using a single Intel i7-5700HQ core, with CPU@2.70GHz and 16 GB of RAM. For both the scenarios, the UCT search has been configured as follows: (1) search horizon $H = 4$; (2) exploration constant $C = 0.707$; (3) $K = 3$ rollouts. The number of components of the GMMs is evaluated according to the BIC criterion, which has been tested using up to 6 Gaussians. Q-values are updated with a learning rate $\alpha = 0.1$, and a discounted factor $\gamma = 0.8$. The algorithm is ran for $T = 15$ timesteps at each iteration. Moreover, stochastic actions are induced by randomizing the outcome of an action with a 5% probability.

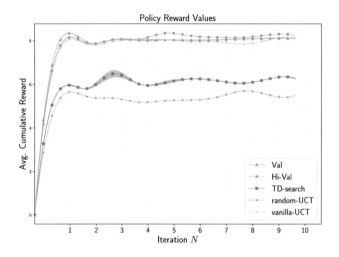

**Fig. 2.** Average cumulative reward obtained by the random-UCT, TD-search, vanilla-UCT, VAL and HI-VAL over 10 iterations.

## 4.1  Fetching Task

In this scenario the state of the problem is represented as a 7-feature vector, where 3 components represent the distance of the robot end-effector to the target, 3 components encode the distance to an obstacle introduced in the scene, and the last component is the angle difference between the end-effector and world axis Z. We include such component to bias the agent in planning to fetch an object with a preferable orientation. The reward function is a weighted sum of such components, and it is designed to promote states that are far from the

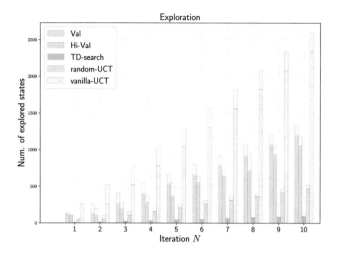

**Fig. 3.** Number of states expanded by the random-UCT, TD-search, vanilla-UCT, Val and Hi-Val.

obstacle, and close to the target position. Additionally, it penalizes states in which the end-effector does not point upwards, to simulate objects that have to be held with a preferred orientation (e.g. a glass full of water). The robot explores an action space composed by 13 actions: 6 translation actions to move the arm back and forth along the Cartesian axes, and 6 rotation actions to move the arm counter-/clockwise on the Roll, Pitch and Yaw angles. A *no-op* action is introduced to let the robot in its state. Figures 2 and 3 illustrate obtained results by reporting the average cumulative reward and the number of explored states obtained during 10 iterations. In detail, reward values are averaged over 10 simulated fetching trials for each of the iterations and for each algorithm, the continuous lines represent average cumulative rewards while the line width their standard deviation. In the explored state plots, the gray top part of each bar highlights the amount of states expanded during $i$ with respect to the total number of states explored until $i - 1$. While baseline algorithms perform worse in terms of obtained rewards (random-UCT, TD-search), only the vanilla-UCT shows results that are comparable to Hi-Val. However, the number of explored states of vanilla-UCT is significantly higher. Specifically, the naive implementation of UCT evaluates more than two times the number of states that Hi-Val ($\sim$55%). In fact, Hi-Val approximates the optimistic set of a HTN by evaluating only "admissible" actions that are expected to lead the search towards states with high reward. This is achieved by exploiting the action hierarchy updated at each iteration. Moreover, the results compare two different configurations of Hi-Val. The first, Val, is organized as a single layer structure, where the number of clusters within the layer is equal to the number of primitive actions, while the latter configuration, Hi-Val, is organized over 2 layers where the first layer also contains the set of primitive actions and, the second layer groups actions in 5 clusters. Again HI-VAL further reduces the number of explored states and confirms that a hierarchical evaluation of the search space improves the learning process. In this scenario, we do not notice a significant improvement between HI-VAL and VAL since, differently from the "pick and delivery" task, the structure of "fetching" is not hierarchical. However, we aim at showing that increasing the number of layers in the representation, even when that is not needed, does not damage the obtained performance and, still, slightly decreases the number of visited states.

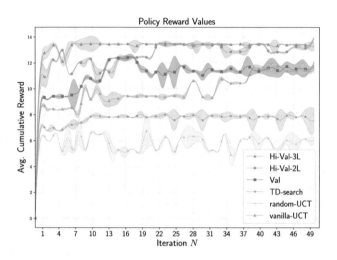

**Fig. 4.** Average cumulative reward obtained by the random-UCT, TD-search, vanilla-UCT, VAL, HI-VAL-2L and HI-VAL-3L over 49 iterations.

## 4.2 Pick and Delivery Task

Here the environment is represented as a $5 \times 5$ grid-world where the Pioneer has to collect an object at a random location and, carry it to an operator. The scenario resembles the one addressed by the "taxi-agent" in [7], however a comparison with max-Q would not be proper since our reward is implemented to be *shaped* and not *sparse*; and we implement our approach in a robotic context where the reduced number of samples and iterations are limiting constraints. Here, the state is a 9-feature vector where the first two components represent the position of the robot, the following two encode the current target of the robot (either the object station or the delivery one), the fifth component indicates whether the object is picked, and the last four components indicate whether there is an obstacle in one of the four possible directions (e.g. wall). The action space of the agent in composed by 6 actions, four to move through cells, and two to pick and drop the object. Moreover, we assume that a robot is helped to collect and drop the objects by an external operator. The reward function is a

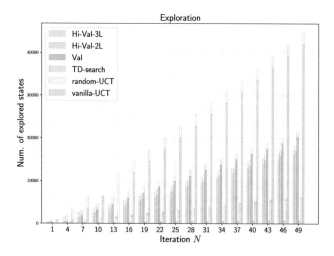

**Fig. 5.** Number of states expanded by the random-UCT, TD-search, vanilla-UCT, VAL, HI-VAL-2L and HI-VAL-3L over 49 iterations

weighted sum of two components encoding the distance of the robot to the target object and the distance of the object to its delivery station. This task represents a more complex scenario due to temporal constraints imposed by the status of the object, but as in the previous task, a similar analysis of the results can be observed. Figures 4 and 5 illustrate the average cumulative reward and the number of explored states obtained during 49 iterations. Also in this case, vanilla-UCT has comparable reward values, but the number of explored states is still significantly higher than each configuration of HI-VAL. The action hierarchy of HI-VAL, in fact, improves the overall performance showing the best results with a 3 layered structure. Particularly in these complex scenarios – where ordering constraints exist and the task can be decomposed – HI-VAL performs better and confirms that a multi-layered representation of action semantics improves the exploration of the search space.

## 5 Conclusion

In this paper we introduced HI-VAL, an iterative learning algorithm of hierarchical value functions for policy generation. We discussed its key features and described how it improves search space exploration in order to generate efficient policies. The results of our experimental evaluation show the efficacy of HI-VAL in enabling the agent to learn a good policy by evaluating a significant lower number of states. In fact, HI-VAL can be used to solve different tasks in multiple domains. Finally, we are investigating different directions to further improve HI-VAL, such as a proper formulation for continuous problems. Moreover, we want to explore the possibility of transferring hierarchical value functions among learning agents in order to take advantage of abstract actions and their semantics in different tasks.

# References

1. Agostini, A., Celaya, E.: Reinforcement learning with a Gaussian mixture model. In: The 2010 International Joint Conference on Neural Networks, pp. 1–8, July 2010

2. Anand, A., Grover, A., Mausam, M., Singla, P.: ASAP-UCT: abstraction of state-action pairs in UCT. In: Proceedings of the 24th International Conference on Artificial Intelligence, IJCAI 2015, pp. 1509–1515. AAAI Press (2015). http://dl.acm.org/citation.cfm?id=2832415.2832459

3. Auer, P., Cesa-Bianchi, N., Fischer, P.: Finite-time analysis of the multiarmed bandit problem. Mach. Learn. 47(2–3), 235–256 (2002)

4. Bagnell, J.A., Schneider, J.G.: Autonomous helicopter control using reinforcement learning policy search methods. In: 2001 IEEE International Conference on Robotics and Automation, vol. 2, pp. 1615–1620 (2001)

5. Chowdhary, G., Liu, M., Grande, R., Walsh, T., How, J., Carin, L.: Off-policy reinforcement learning with Gaussian processes. IEEE/CAA J. Autom. Sinica 1(3), 227–238 (2014)

6. Clair, A.S., Saldanha, C., Boteanu, A., Chernova, S.: Interactive hierarchical task learning via crowdsourcing for robot adaptability. In: Refereed Workshop Planning for Human-Robot Interaction: Shared Autonomy and Collaborative Robotics at Robotics: Science and Systems, Ann Arbor, Michigan. RSS (2016)

7. Dietterich, T.G.: Hierarchical reinforcement learning with the MAXQ value function decomposition. J. Artif. Intell. Res. (JAIR) 13, 227–303 (2000)

8. Erol, K., Hendler, J., Nau, D.S.: HTN planning: complexity and expressivity. In: the Twelfth National Conference on Artificial Intelligence, vol. 2, AAAI 1994, pp. 1123–1128. American Association for Artificial Intelligence, Menlo Park (1994). http://dl.acm.org/citation.cfm?id=199480.199459

9. Hostetler, J., Fern, A., Dietterich, T.G.: Sample-based tree search with fixed and adaptive state abstractions. J. Artif. Intell. Res. 60, 717–777 (2017). https://doi.org/10.1613/jair.5483

10. Jun, M., Kenji, D.: Acquisition of stand-up behavior by a real robot using hierarchical reinforcement learning. Rob. Autonom. Syst. 36(1), 37–51 (2001)

11. Kober, J., Bagnell, J.A., Peters, J.: Reinforcement learning in robotics: a survey. Int. J. Rob. Res. (2013)

12. Kober, J., Peters, J.R.: Policy search for motor primitives in robotics. In: Advances in Neural Information Processing Systems, pp. 849–856 (2009)

13. Kocsis, L., Szepesvári, C.: Bandit based monte-carlo planning, pp. 282–293. Springer, Heidelberg (2006). https://doi.org/10.1007/11871842_29

14. Kohl, N., Stone, P.: Policy gradient reinforcement learning for fast quadrupedal locomotion. In: 2004 IEEE International Conference on Robotics and Automation, vol. 3, pp. 2619–2624, April 2004

15. Konidaris, G., Kuindersma, S., Grupen, R., Barto, A.: Robot learning from demonstration by constructing skill trees. Int. J. Rob. Res. 31(3), 360–375 (2012)

16. Riccio, F., Capobianco, R., Nardi, D.: DOP: deep optimistic planning with approximate value function evaluation. In: Proceedings of the 2018 International Conference on Autonomous Agents and Multiagent Systems (AAMAS) (2018)

17. Riccio, F., Capobianco, R., Nardi, D.: Q-CP: learning action values for cooperative planning. In: 2018 IEEE International Conference on Robotics and Automation (ICRA) (2018)

18. Ross, S., Gordon, G.J., Bagnell, D.: A reduction of imitation learning and structured prediction to no-regret online learning. In: International Conference on Artificial Intelligence and Statistics, pp. 627–635 (2011)

19. Schaul, T., Ring, M.: Better generalization with forecasts. In: Proceedings of the Twenty-Third International Joint Conference on Artificial Intelligence, IJCAI 2013, pp. 1656–1662. AAAI Press (2013)

20. Silver, D., Huang, A., Maddison, C.J., Guez, A., Sifre, L., van den Driessche, G., Schrittwieser, J., Antonoglou, I., Panneershelvam, V., Lanctot, M., Dieleman, S., Grewe, D., Nham, J., Kalchbrenner, N., Sutskever, I., Lillicrap, T., Leach, M., Kavukcuoglu, K., Graepel, T., Hassabis, D.: Mastering the game of Go with deep neural networks and tree search. Nature **529**, 484–503 (2016)

21. Silver, D., Sutton, R.S., Müller, M.: Temporal-difference search in computer Go. Mach. Learn. **87**(2), 183–219 (2012)

22. Stulp, F., Schaal, S.: Hierarchical reinforcement learning with movement primitives. In: 2011 IEEE-RAS International Conference on Humanoid Robots, pp. 231–238, October 2011

23. Sutton, R.S., Precup, D., Singh, S.: Between MDPs and semi-MDPs: a framework for temporal abstraction in reinforcement learning. Artif. Intell. **112**(1–2), 181–211 (1999)

# Learning-Based Task Failure Prediction for Selective Dual-Arm Manipulation in Warehouse Stowing

Shingo Kitagawa[✉], Kentaro Wada, Kei Okada, and Masayuki Inaba

The University of Tokyo, 7-3-1, Hongo, Bunkyo-ku, Tokyo, Japan
s-kitagawa@jsk.imi.i.u-tokyo.ac.jp

**Abstract.** Stowing is one main task of warehouse automation, and manipulation with a vacuum gripper is recently known as a practical method. However, the gripper sticks an object from upper side, which causes task failures such as drop and protrusion by even small disturbance. In this paper, we aim to realize more stable stowing task and propose a stowing system which robot selectively stow an object by two arms in case the task failures may occur. For the selective stowing, we predict task failure occurrence by convolutional neural network (CNN) and select a proper motion from the prediction results. The network predicts probabilities of task failure occurrence for both single-arm and dual-arm stowing motion cases, and we design a motion select algorithm to evaluate the two motions and select optimal one. In experiment, we implemented our system in real stowing task and achieved higher success rate 58.0% than that of single-arm stowing system 49.0% in 100 trials.

**Keywords:** Dual-arm manipulation · Failure prediction
Motion select · Task-based learning · Warehouse automation
Stowing task

## 1 Introduction

Warehouse automation is recently common as robotics application [1], and manipulating various objects is one main challenge of it. Recent studies show that manipulation with a vacuum gripper is a practical method [4], because this approach can grasp and manipulate various objects except pre-known 3D object model. However, this approach results in unstable grasping because a vacuum gripper only sticks an object from upper side. Robot needs to avoid unstable grasping because it causes task failures by even small disturbance.

Thinking of human, can we always grasp an object stably at once? Even a human grasps an object unstably in cluttered environment, and the difference from robot is that we can observe unstable grasping condition and do extra manipulation to make the condition stable. Therefore, we suppose these human

© Springer Nature Switzerland AG 2019
M. Strand et al. (Eds.): IAS 2018, AISC 867, pp. 428–439, 2019.
https://doi.org/10.1007/978-3-030-01370-7_34

(a)                                    (b)

**Fig. 1.** (a) A robot sometimes fails to stow an object with single arm because of disturbance. (b) Our stowing system predicts task failure occurrence, and the robot selectively executes dual-arm stowing for avoiding the failures when necessary.

behaviors is essential for more stable manipulation, and we introduce this selective execution of stabilizing motion with the observation of grasping condition to warehouse automation.

In warehouse automation, we focus on stowing task. Stowing task is to grasp an target object, carry it and stow it in a shelf, and collision between an object and a shelf bin may results in a task failure as shown in Fig. 1(a) when robot grasp an object unstably. In this paper, we propose a stowing system which robot observe the grasping condition of object and selectively holds tight to it by two arms to prevent a task failure when necessary as shown in Fig. 1(b). For the observation of grasping condition, we propose learning-based failure prediction of the stowing task. We collect and annotate RGB image of grasping condition and task result of both single-arm and dual-arm stowing motion, and CNN is trained to predict the probabilities of task failure occurrence from the collected data. Finally, the predicted probabilities is used for the stability evaluation of single-arm and dual-arm stowing motion, and robot selects to execute the most stable motion.

In this paper, we propose a motion select system to determine which single-arm or dual-arm stowing motion is appropriate from grasping condition. First, we state the task failure definition and overview of our stowing system. Then, we introduce our system details, the dual-arm stowing motion and motion select system with task failure prediction. In the end, we demonstrate our stowing system is more stable for the stowing task comparing to single-arm one.

## 2   Related Work

**Warehouse Automation.** In warehouse automation, picking an object from a shelf and stowing it to a shelf are two main tasks, and a robot manipulates various type of objects in both tasks [1]. Both warehouse picking and stowing are industrial tasks and aimed to be done with no failure, and our approach has the same goal.

**Dual-arm Manipulation.** Since a human has two arms, a number of studies on dual-arm manipulation have been done with humanoids. Recent works show that dual-arm robots have advantages in pick-and-place task [2,3], and we implement these advantages in more complicated task, stowing task.

**Motion Select.** Contact states is a common index of motion strategy for object manipulation [7,8] because a robot physically interacts with objects and can detect reaction force. However, a vacuum gripper sticks an object by air flow, and it is difficult for a robot to observe the states. Therefore, we propose to observe the grasping condition using an RGB image and select a proper stowing motion. Our method first predicts the probabilities of task failures from an RGB image with learning method, evaluates task stability of two stowing motion, and determines which motion should be executed.

**Task Failure Prediction for Motion Select.** Deep learning is a common method for task probability prediction. Current methods predict success probability [6,9] from an RGB image. Although these methods predict success probability for each condition, there are several type of task failures in more complicated task, and the importance of task failures are not equal. For example, dropping an object is more critical than stowing an object incompletely in our task. Our method predicts the probabilities of each task failure for each single-arm and dual-arm stowing motion.

## 3    Overview

### 3.1    Stowing Task and Task Failures

Stowing task is to pick an object from tote and stow it in a shelf show in Fig. 2(a), and collision often occurs when robot stows an object in a narrow shelf bin. The collision may result in task failures when robot grasp an object unstably, and we define 2 task failures in stowing task as shown in Fig. 2(b)(c); drop and protrusion. Protrusion is defined as the condition that an object is stowed but some part is outside of the bin. Drop is mainly caused by collision between an object and a shelf bin floor, and protrusion occurs with flexible objects. If no task failure happens, the object will be successfully stowed as Fig. 2(d).

### 3.2    Selective Dual-Arm Stowing System

In this paper, we focus on stable stowing system, and stable stowing causes less task failures during the task. In order to realize more stable stowing, we argue that a robot needs to avoid task failure occurrence corresponding to the grasping condition, and we introduce selective execution of dual-arm stowing motion in single-arm stowing system. The overview of our selective dual-arm stowing system is shown in Fig. 3. In order to select proper stowing motion, we use CNN to predict the probabilities of task failure occurrence for both single-arm and dual-arm stowing cases. The network predicts the task failure probabilities from an RGB image of grasping condition. In the motion select

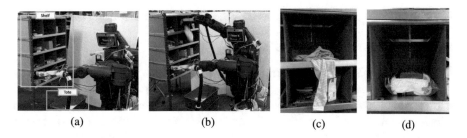

**Fig. 2.** (a) Stowing task is to pick an object and place it in a shelf bin. (b) A drop sometimes occur during the task caused by collision between an object and the shelf. (c) The object sometimes protrudes from the bin. (d) The task is completed with no failure.

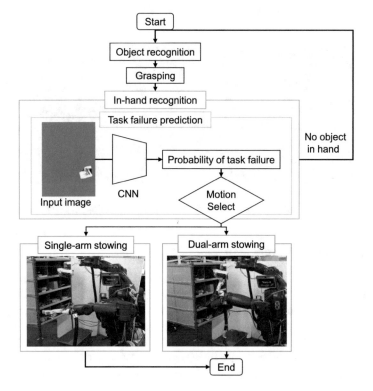

**Fig. 3.** We propose the stowing system, which the robot selectively executes dual-arm stowing motion corresponding to the grasping condition in hand

section, we design an algorithm which evaluates task stability of both stowing motion by the predicted probabilities and determine which motion should be executed. Therefore, we introduce stowing system which selects and execute proper stowing motion from by observing the grasping condition and predicting the task failure occurrence. We implement our system in real robot and evaluate its effectiveness by success rate and stability of the task.

## 4    Selective Dual-ARM Stowing with Failure Prediction

### 4.1    Dual-Arm Stowing Motion

We design a dual-arm stowing motion for stowing task. The dual-arm stowing motion is (1) pick an object by a single arm, (2) support the object from downside with another arm, (3) stow the object with supporting it. The main arm grasps an object, and another arm supports its center of mass from downside. This motion is shown in Fig. 4 and avoids the physical collision and draw-down of flexible objects, which suppresses task failure occurrence. Although dual-arm stowing motion is better than single-arm one in terms of robustness, single-arm stowing motion requires only 52 s compare to dual-arm one requires 67 s on average. Therefore, single-arm one needs less time so that robot should execute dual-arm one only when necessary in terms of task execution time.

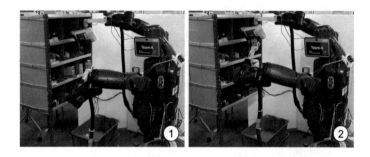

**Fig. 4.** The robot supports an object in hand and avoid its draw-down.

### 4.2    Task Failure Prediction

Then, we explain how to evaluate grasping condition and select optimal motion. For evaluation of task stability of grasping condition, we predict the probabilities of task failure occurrence for both single-arm and dual-arm stowing motion. We observe the grasping condition by an RGB image because the condition consists of object pose, grasping situation in hand and contact point position of the gripper, which robot can optically observe them. We use CNN for the prediction from an RGB image and train the network as follows.

**Dataset Preparation:** We execute the task with both single-arm and dual-arm stowing motions with a real robot and collect RGB images as input data and task failure labels as output data. When the robot grasps and holds an object, it observes an RGB image as Fig. 5(a), automatically generates a mask image from a depth image to extract the region of the object and its gripper and apply it on the original image as Fig. 5(b). The mean RGB pixel data of ImageNet dataset [5] is applied to the pixels in the masked region, and we collect $480 \times 640$ RGB images as input data. We also manually record binary label $y_i \in \{0, 1\}$ about 2 task failures, and label is annotated as 1 when it occurs and 0 when not.

We set $y_i = -1$ in labels about non-executed motion and collect $y$ (vector of length 4) as output data because robot can only execute either stowing motion in one trial.

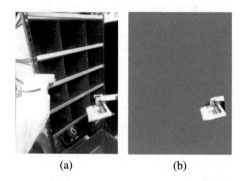

(a)                    (b)

**Fig. 5.** (a) An original RGB images are captured from RGB-D camera on another hand. (b) A mask image generated is applied to an RGB image in order to extract object and gripper region.

**Network Design:** Our network is based on AlexNet [5], and detailed structure is illustrated in Fig. 6. We design a network which can predict object class and the probabilities of task failure occurrence simultaneously, and, we only use the output about task failure prediction in this paper. Failure prediction has 4 output units because it returns probabilities about 2 task failures (drop and protrusion) for both stowing motion.

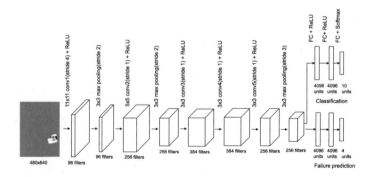

**Fig. 6.** We design a CNN based on AlexNet [5] to predict the probabilities of task failure occurrence. Activation function for all convolutional and fully connected layers except output layer is ReLU function. We use Softmax function for activation of output layer in order to get the probabilities of each task failure.

**Network Training:** In order to train the network properly with small dataset, initial weights of convolutional layers are taken from AlexNet [5] trained model as pre-training. Failure prediction is done with 4 labels; single-arm drop, single-arm protrusion, dual-arm drop and dual-arm protrusion, and the loss of failure prediction $L_{failure}$ is calculated with softmax cross entropy $SCE$ as follows:

$$L_{failure} = \frac{1}{N_{batch}} \sum_{n}^{N_{batch}} \sum_{i}^{N_{label}} \delta(y_i) SCE(y_i, \hat{y}_i) \tag{1}$$

where batch size $N_{batch}$ is batch size, number of labels is $N_{label}$, $y_i$ is one hot vector of $i$ th label,

$$\delta(y_i) = \begin{cases} 1 & (y_i \in \{0, 1\}) \\ 0 & (otherwise) \end{cases}$$

is a delta function for $y_i$ and $\hat{y}_i$ is the network output about $i$ th label. In this case, we set 4 failure labels so that number of labels $N_{label}$ is 4. With this $\delta$ function, 2 probabilistic values corresponding to input image are extracted because task failure labels about non-executed motion are annotated as $-1$, and the loss $L_{failure}$ can be treated as $SCE$ loss of the stowing motion, which is executed in data collection. Hence, only loss about stowing motion corresponding to input image is back-propagated when $N_{batch} = 1$, and the trained network can predict the probabilities about both single-arm and dual-arm stowing motions. The loss of the object classification $L_{class}$ is calculated with $SCE$ as normal object classification, and sum of these two losses are backpropagated.

We also do image augmentation of input RGB images for efficient network training, and random RGB modification and flip is added on the data.

### 4.3    Motion Select Algorithm

Our network predicts probabilities of task failure occurrence for both single-arm and dual-arm stowing motions. Using the predicted probabilities, robot calculates expected task scores $E(score)$ for both stowing motions, treats them as indices of task stability of each stowing motions, and select optimal one in terms of task stability. We set score penalty for each task failure and represent how we want to avoid each task failure. Drop is considered as fatal because robot sometimes cannot find object anymore without human help after drop occurs, and protrusion is regarded as less fatal failure than drop because the robot can recognize and fix the failure with another motion. The expected task score $E(score)$ is calculated as follows:

$$E(score) = \max(0, 10 - 10P_{drop} - 5P_{protrude}) \tag{2}$$

where $P_{failure}$ is the probability of *failure*.

With the calculated expected scores, we design stowing motion select algorithm described in Algorithm 1. Main rule of the algorithm is to select dual-arm stowing motion only if its expected score is higher than single-arm one.

If expected score of single-arm stowing is high enough and not different from that of dual-arm stowing, it is expected that no task failure happens with single-arm stowing, and robot selects to execute it.

---

**Algorithm 1.** Stowing motion select

---

$E(score)_{single}$: Expected score of single-arm stowing
$E(score)_{dual}$: Expected score of dual-arm stowing
*threshold*: Threshold for $E(score)$ comparison

**if** $E(score)_{dual} - E(score)_{single} > threshold$ **then**
    Execute Dual-arm Stowing Motion
**else if** $E(score)_{single} < 5$ **then**
    Execute Dual-arm Stowing Motion
**else**
    Execute Single-arm Stowing Motion
**end if**

---

## 4.4  Motion Select System

In this section, we summarize whole motion select system. We propose a motion select system for selective execution of dual-arm stowing motion, and whole system is described in Fig. 7. First, a robot predict the probabilities of task failure occurrence for both single-arm and dual-arm stowing with CNN. Then, the robot evaluate both stowing motion with their expected task scores calculated by predicted probabilities and select optimal one with the motion select algorithm using the calculated scores. Therefore, the robot observe the grasping condition and executes dual-arm stowing motion only when necessary, which improves task stability.

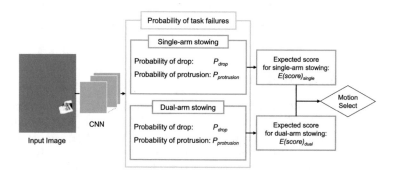

**Fig. 7.** Our network predicts the probabilities of task failure occurrence, and our system calculates expected scores from the predicted results, evaluates stowing motions and selects optimal one.

## 5   Experiments

### 5.1   Stowing Task Dataset

The stowing task dataset was collected over 10 objects shown in Fig. 8. We chose 10 objects from target objects set in Amazon Picking Challenge 2015–2017 [1,4], which includes flexible and unbalanced objects such as T-shirt, socks and book. We executed 200 trials of stowing task, and 1327 pairs of data were gathered for the training dataset. In data collection, a robot executed both single-arm and dual-arm stowing motions, and the summary of the collected dataset is in Table 1.

**Fig. 8.** We used 10 various objects; Book, Curtain, DentalTreats, ExpoEraser, IceTray, KidsBook, PaperTowels, Socks, Stems and T-shirt. The 10 objects are from Amazon Picking Challenge [1,4], and they include flexible and transformable objects such as cloth, books and bags.

**Table 1.** Stowing task dataset statistics

| Stowing motion | Success data | Drop data | Protrusion data | Total |
|---|---|---|---|---|
| Single-arm | 362 | 243 | 48 | 653 |
| Dual-arm | 398 | 130 | 146 | 674 |
| Total | 760 | 373 | 194 | 1327 |

### 5.2   Task Failure Prediction

We trained the prediction network with collected dataset. Since a number of data in dataset was not large, we did cross validation with 10% as validation ratio. Therefore, the 10% of the dataset were used as validation, and the rest for training, and training were iterated for 5000 times with $N_{batch} = 10$. For evaluation of the network, we set 0.5 as threshold for network outputs to calculate accuracy. If the output probability of a task failure is higher than the threshold, we set its output label as 1 and treat that network predict that the task failure occurs and vice versa. We regard network prediction as correct only when all output labels corresponding to the input image are correct. For comparison, we design a baseline method which return the average of task failures label in validation dataset as prediction output. As the result of validations, the prediction of our network results the average accuracy of 68.9%, which is higher than the baseline accuracy of 57.9%. Some examples of validation results are shown in Fig. 9. Hence, we can say that our method are generalized for the grasping condition and predicts the probability of task failures properly.

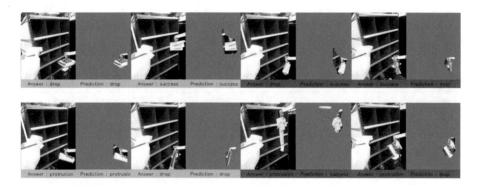

**Fig. 9.** Some examples of validation results of task failure prediction with single-arm (top) and dual-arm stowing motion (bottom) are shown with original and masked image. The green labeled results are successfully predicted, and the red ones show wrong prediction results

### 5.3   Selective Execution of Dual-Arm Support Motion

For the final experiment, we introduced our whole stowing system with the dual-arm humanoid Baxter. The target objects were same as ones used in the dataset collection, and we executed 10 trials for each object with task score threshold *threshold* in Algorithm 1 as 0.5. For comparison, we also executed the stowing task with non-selective system using only single-arm stowing motion. We used one network model trained in the prediction experiment for our system.

The results of the executed stowing task is shown in Tables 2 and 3. In Table 2, our system achieved higher success rate 58.0% and higher task score 6.90 than those of non-selective single-arm system 49.0% and 5.40, which means that selective execution of dual-arm stowing improves task success rate and task stability. Also, as you can see in Table 3, success rate and task score of single-arm stowing in our system are 56.8% and 6.91 and higher than non-selective single-arm system ones in Table 2. From this result, we can say that robot successfully select optimal stowing motion and avoid task failures by executing dual-arm motion instead of single-arm one. From these results, we can conclude that our system successfully selects to execute dual-arm stowing and optimizes the task in terms of task stability.

**Table 2.** Evaluation of Our Stowing System in 100 Trials

| Stowing system | Success times (Success rate) | Drop times (Drop rate) | Protrusion times (Protrusion rate) | Ave. score |
|---|---|---|---|---|
| Non-selective Single-arm | 49 (49.0%) | 41 (41.0%) | 10 (10.0%) | 5.40 |
| **Our method** | **58 (58.0%)** | 20 (20.0%) | 22 (22.0%) | **6.90** |

**Table 3.** Results of executed stowing motion with our system in 100 trials

| Executed stowing motion | Total times | Success times (Success rate) | Drop times (Drop Rate) | Protrusion times (Protrusion Rate) | Ave. score |
|---|---|---|---|---|---|
| Single-arm | 81 | **46 (56.8%)** | 15 (18.5%) | 20 (24.7%) | **6.91** |
| Dual-arm | 19 | **12 (63.2%)** | 5 (26.3%) | 2 (10.5%) | **6.84** |

# 6   Conclusion

For stowing task system, we propose to observe grasping condition and execute a dual-arm stowing when necessary. In order to determine whether single-arm or dual-arm stowing motion is proper for the grasping condition, we propose to predict the probabilities of task failure occurrence from an RGB image with CNN, and the network is trained by task results collected by executing both motions with real robot. We set task score and design an algorithm to evaluate two stowing motion with predicted probabilities and select optimal one. In the experiment, we executed the task with our system in real robot and achieved higher success rate 58.0% and higher task score 6.90 than those of non-selective single-arm system. The improvement of the success rate is 9%, which is effective for the stowing task, and the robot can successfully grasp more than one object in two with our method. From the experimental results, we conclude that task failure prediction by observing grasping condition is essential for proper motion select of dual-arm and single-arm stowing.

In this paper, we focus on stowing task, but the learning-based prediction of task failures can be applied to motion select system in various manipulation tasks. For further research, we want to generalize the concept of learning-based task failure prediction and implement it to various task systems. However, learning-based method have some difficulties. One main difficulty is that it takes huge time and human resources to collect data and label them. To solve the problem in task-based learning, self-supervised methods are invented [6,9], but they are only implemented to simple task such as grasping. For the implementation of our method to various tasks, we need to introduce self-supervised learning method to more complicated task system.

# References

1. Correll, N., Bekris, K.E., Berenson, D., Brock, O., Causo, A., Hauser, K., Okada, K., Rodriguez, A., Romano, J.M., Wurman, P.R.: Lessons from the Amazon Picking Challenge. CoRR abs/1601.05484 (2016)
2. Edsinger, A., Kemp, C.C.: Two arms are better than one: a behavior based control system for assistive bimanual manipulation. In: Recent Progress in Robotics: Viable Robotic Service to Human, pp. 345–355. Springer (2007)
3. Harada, K., Foissotte, T., Tsuji, T., Nagata, K., Yamanobe, N., Nakamura, A., Kawai, Y.: Pick and place planning for dual-arm manipulators. In: IEEE International Conference on Robotics and Automation (ICRA), pp. 2281–2286. IEEE (2012)
4. Hernandez, C., Bharatheesha, M., Ko, W., Gaiser, H., Tan, J., van Deurzen, K., de Vries, M., Van Mil, B., van Egmond, J., Burger, R., Morariu, M., Ju, J., Gerrmann, X., Ensing, R., van Frankenhuyzen, J., Wisse, M.: Team Delft's Robot Winner of the Amazon Picking Challenge 2016. CoRR, abs/1610.05514 (2016)
5. Krizhevsky, A., Sutskever, I., Hinton, G.E.: Imagenet classification with deep convolutional neural networks. In: Advances in Neural Information Processing Systems, pp. 1097–1105 (2012)
6. Levine, S., Pastor, P., Krizhevsky, A., Quillen, D.: Learning hand-eye coordination for robotic grasping with deep learning and large-scale data collection. CoRR, abs/1603.02199 (2016)
7. Murooka, M., Noda, S., Nozawa, S., Kakiuchi, Y., Okada, K., Inaba, M.: Manipulation strategy decision and execution based on strategy proving operation for carrying large and heavy objects. In: IEEE International Conference on Robotics and Automation (ICRA), pp. 3425–3432, May 2014
8. Nozawa, S., Murooka, M., Noda, S., Okada, K., Inaba, M.: Description and execution of humanoid's object manipulation based on object-environment-robot contact states. In: IEEE/RSJ International Conference on Intelligent Robots and Systems (IROS), pp. 2608–2615, November 2013
9. Pinto, L., Gupta, A.: Supersizing self-supervision: Learning to grasp from 50k tries and 700 robot hours. In: 2016 IEEE International Conference on Robotics and Automation (ICRA), pp. 3406–3413, May 2016

# Learning of Motion Primitives Using Reference-Point-Dependent GP-HSMM for Domestic Service Robots

Kensuke Iwata[✉], Tomoaki Nakamura, and Takayuki Nagai

The University of Electro-Communication, Chofu, Japan
datemitumasa@gmail.com

**Abstract.** In this paper, we propose a method for motion learning aimed at the execution of autonomous household chores by home robots in real environments. For robots to act autonomously in a real environment, it is necessary to define appropriate actions for the environment. However, it is difficult to define these actions manually. Therefore, body motions that are common to multiple actions are defined as motion primitives. Complex actions can then be learned by combining these motion primitives. For learning motion primitives, we propose reference-point-dependent Gaussian process hidden semi-Markov model (RPD-GP-HSMM). For verification, a robot is tele-operated in order to perform actions included in several domestic household chores. The robot then learned the associated motion primitives from the robot's body information and object information.

**Keywords:** Motion learning · Gaussian process
Hidden semi-Markov model · Reference-point

## 1 Introduction

In recent years, the development of service robots that support daily living has accelerated significantly. The introduction of such robots into real environments has begun. For a robot to support human life, both software and hardware must respond flexibly to environmental conditions and changes. However, service robots have yet to be introduced into ordinary households. If an environment is limited, it is possible for a robot to act autonomously by carrying out predefined actions. Robots already possess physical and recognition abilities that would facilitate work in ordinary households. However, in an environment that cannot be defined in advance, if a robot cannot recognize the meanings of actions and select the correct actions based on the circumstances, it cannot perform tasks autonomously. In order to select actions that are appropriate for an environment, it is necessary to define both the purpose and content of the action, which makes it difficult to define all possible actions manually. Therefore, robots must be able to learn and define appropriate actions for their environments autonomously.

© Springer Nature Switzerland AG 2019
M. Strand et al. (Eds.): IAS 2018, AISC 867, pp. 440–451, 2019.
https://doi.org/10.1007/978-3-030-01370-7_35

Although behavior varies depending on the environment, there are certain body motions that are common between environments. Physical control can be realized using existing technology. However, it is difficult to determine how a body should be moved to achieve a goal.

In this paper we take a motion learning approach [1]. Although many learning methods have been proposed in literature [2–5], it is still a difficult task to directly extract actions from body information during activity and define actions autonomously. The movements for actions differ based on positional relationships with objects, differences in posture, and the presence or absence of obstacles, even if the same action is performed repeatedly. This makes it difficult to learn behaviors as continuous body movements. Additionally, for similar reasons, it is difficult to learn standard activities. However, there are common body movements between different motions. Therefore, as shown in Fig. 6, we consider these common body movements as motion primitives. Activities are hierarchically represented by actions and motion primitives, and learning is performed iteratively. First, motion primitives are extracted from continuous body movements. The extracted motion primitives represent the basic information for expressing a change in the physical information associated with an activity. An action has a specific purpose, which is learned based on the environment and the target object. By combining motion primitives according to the environment and purpose, a robot can perform an activity. However, in the early stages of learning, it is difficult for robots to autonomously perform activities. Therefore, during the initial stage, activities in a home are performed through teleoperation by a human user and the learning of motion primitives is performed based on joint angles and object information. Then, using the extracted motion primitives, the learning of actions and tasks is performed. Through this method, it is possible to learn the elements necessary for autonomous activity.

In this study, we attempt to extract motion primitives from physical information, that the robot obtains through teleoperation, as a first step toward autonomy in real environments.

## 2   Discovery of Motion Primitives

### 2.1   Segmentation of Time-Series Data

Human beings perform recognition by dividing continuous information into meaningful segments. For example, by segmenting a speech waveform, it is possible to recognize words, which constitute meaningful segments. By segmenting continuous human actions, meaningful segmented motions can be recognized. For robots, the ability to segment time-series data flexibly is considered to be very important for learning languages, actions, gestures, etc. Here, we use Gaussian-process hidden semi-Markov model (GP-HSMM) as a method of segmenting time-series data and classifying it based on segment similarity [6]. In this model, each state in the hidden semi-Markov model (HSMM) has a Gaussian process, which represents a continuous unit time-series. Each class is represented by a Gaussian process and a sequence is generated by combining the

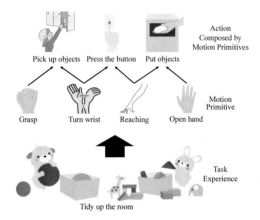

**Fig. 1.** Hierarchical representation of activities

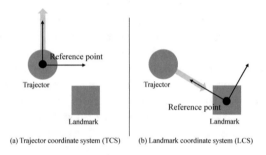

(a) Trajector coordinate system (TCS)     (b) Landmark coordinate system (LCS)

**Fig. 2.** Examples of reference points; (a) raising, and (b) reaching actions

segments generated from each class. By learning model parameters using only observed time-series, it is possible to estimate segmentation points and classify each segment autonomously. For model training, the forward filtering-backward sampling method can be utilized [6]. Learning is accomplished by simultaneously sampling boundaries and the class of each segment. Using this model, it has been shown that the segmentation of karate-type motion capture data is possible [6]. Because karate movements contain clear motion trajectories, one can segment the movements into primitive motions using only end effector trajectories with fixed coordinates as an origin. This is not the case for the domestic service since some motions act on objects, which makes the coordinate system more complex. For example, the movement trajectory for the motion of grasping an object changes based on the position of the object (Fig. 1).

In [2], Sugiura *et al.* proposed a method for learning motion based on the position of such an object. In this method, the learning of motion based on a reference point is accomplished by setting the object to be moved as a "trajector" and the object on which the movement is performed as a reference point to estimate the origin of the motion. For example, as shown in Fig. 2(a), because

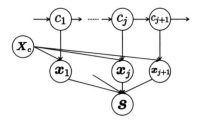

**Fig. 3.** Graphical model of GP-HSMM

the reference point is not important in the operation of "raising", the operation is expressed by a trajector coordinate system with the trajector as a reference point. In contrast, as shown in Fig. 2(b), because the position of the reference point to be approached is important in the operation of the "reaching action", the action is expressed in the landmark coordinate system (LCS), where the landmark is the reference point. In this manner, by using a coordinate system that emphasizes important elements for each motion, it is possible to learn motion trajectories that can better express the features of actions. However, in [2], motion and classes of motion are provided in advance and the parameters for the hidden Markov model to express the coordinate system and motion are estimated, which means that no unsupervised segmentation/classification of motion has been performed. In this paper, we combine the GP-HSMM with the idea of reference point to learn motion trajectories for domestic service robots.

## 2.2   Reference-Point-Dependent Hidden Semi-Markov Model

Figure 3 illustrates the graphical model of GP-HSMM. Conventional GP-HSMM segments raw time series data. In Fig. 3, $c_j (j = 1, 2, \cdots, J)$ denotes classes of segments and each segment $x_j$ is generated by a Gaussian process $\mathcal{GP}$ with parameters $X_c$ as following generative process:

$$c_j \sim P(c|c_{j-1}), \quad x_j \sim \mathcal{GP}(x|X_{c_j}), \tag{1}$$

where $X_c$ and $s$ denote a set of segments classified into class $c$ and the entire observed time series.

On the other hand, the proposed RPD GP-HSMM learns motion primitives with an appropriate reference-point (coordinate systems). Each primitive motion has a corresponding reference-point. Therefore, each segment of raw time series data $x_j$ is coordinate transformed into $\bar{x}_j$ based on the selected reference-point $\ell$ in order to inder the parameters of the $\mathcal{GP}$.

$$\bar{x}_j = T(x_j, \ell), \quad \bar{x}_j \sim \mathcal{GP}(x|\bar{X}_{c_j}), \tag{2}$$

where $T(x, \ell)$ represents the function of coordinate transform defined by the selected reference-point $\ell$. $\bar{X}_c$ denotes a set of coordinate transformed segments classified into class $c$.

**Algorithm 1.** Forward filtering-backward sampling
---
1: // Forward filtering
2: **for** $t = 1$ to $T$ **do**
3:    **for** $k = 1$ to $K$ **do**
4:      **for** $c = 1$ to $C$ **do**
5:        // Coordinate transformation according to reference point
6:        **for** $n = 1$ to $\text{len}(s_{t-k:k})$ **do**
7:          **if** cordinate[c]==TRAJECTOR **then**
8:            $p'_n = p_n - p_{t-k}$
9:          **else**
10:            **for** $j = 1$ to $\text{len}(reference)$ **do**
11:              $p^\ell_{n:t,k,j} = R(\theta)(p_n - p_{\ell[t][k][j]})$
12:            **end for**
13:            $l[t][k][c] = \text{argmax}_\ell \mathcal{GP}(p^\ell_{n:t,k,j}|c)$
14:          **end if**
15:        **end for**
16:        Calculate $\alpha[t][k][c]$ according to formula 4)
17:      **end for**
18:    **end for**
19: **end for**
20:
21: // Backward sampling
22: $t = T, j = 1$
23: **while** $t > 0$ **do**
24:    $k, c \sim p(x_j|s_{t-k:t})\alpha[t][k][c]$
25:    $x_j = s_{t-k:t}$
26:    $c_j = c$
27:    $l_j = \ell_{all}[t][k][c]$
28:    $t = t - k$
29:    $j = j + 1$
30: **end while**
31: return $(x_{J_n}, x_{J_n-1}, \cdots, x_1), (c_{J_n}, c_{J_n-1}, \cdots, c_1), (l_{J_n}, l_{J_n-1}, \cdots, l_1)$

**Forward Filtering-Backward Sampling.** In this paper, as in [7], both the segmentation points and the segment classes are regarded as hidden variables and sampled simultaneously via forward filtering-backward sampling. In forward filtering, the sample $s_{t-k:k}$ of length $k$, whose termination is the time step $t$ of a particular observation sequence $s$, is a segment. The probability that its class is $c$ is calculated as follows. First, the segment $s_{t-k:k} = \{p_{t-k}, p_{t-k+1}, \cdots, p_k\}$, where $p_t$ represents a vector of the segment at $t$, is transformed into the reference coordinate system according to the reference point of the class $c$. There are a plurality of coordinates, i.e. objects, that are candidates for reference points for each segment and the most appropriate coordinates are selected from among the candidates. In order to determine the most appropriate reference point $l$ from the candidates, the values of Gaussian process for the coordinate-transformed segments according to all candidates are compared.

$$\ell[t][k][c] = \underset{\ell}{\mathrm{argmax}}\, \mathcal{GP}(\boldsymbol{p}^\ell_{n:t,k,j}|c), \tag{3}$$

where $\boldsymbol{p}^\ell_*$ represents the coordinate-transformed segment according to the candidate $\ell$. Let $\boldsymbol{s}'_{t-k:k} = \{\boldsymbol{p}^\ell_{t-k}, \boldsymbol{p}^\ell_{t-k+1}, \cdots, \boldsymbol{p}^\ell_{k}\}$ be the segment transformed according to the selected reference point. The probability that this segment will be class $c$ is defined as follows:

$$\alpha[t][k][c] = P(\boldsymbol{s}'_{t-k:k}|\boldsymbol{X}_c) \sum_{k'=1}^{K} \sum_{c'=0}^{C} p(c|c')\alpha[t-k][k'][c'], \tag{4}$$

$$p(c|c') = \frac{N_{c'c} + \alpha}{N_{c'}L + C\alpha}, \tag{5}$$

where $C$ is the number of classes and $K$ is the maximum length of the segment. $P(\boldsymbol{s}'_{t-k:k}|\boldsymbol{X}_c)$ is the probability that $\boldsymbol{s}'_{t-k:k}$ is generated from class $c$ and is defined as follows:

$$P(\boldsymbol{s}'_{t-k:k}|\boldsymbol{X}_c) = \mathcal{GP}(\boldsymbol{s}'_{t-k:k,l[t][k][c]}|\boldsymbol{X}_c)P_{len}(k|\lambda). \tag{6}$$

$P_{len}(k|\lambda)$ is a Poisson distribution with an average of $\lambda$, which is a probability distribution of segment length. $p(c|c')$ in Eq. (4) represents the transition probability of the class, which is computed using Eq. (5). $N_{c'}$ and $N_{c'c}$ denote the number of segments whose classes are $c$ and the number of transitions from $c'$ to $c$, respectively. $k'$ and $c'$ denote the length and the class, respectively, in Eq. (4). Finally, based on the forward probability, we can extract the segment $\boldsymbol{x}_j$ and its class $c_j$ from the observation sequence by sampling the segment length and class backwards. The above algorithm is shown in Algorithm 1.

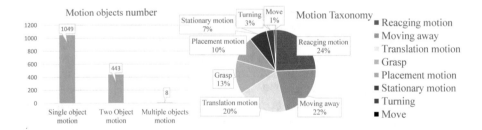

Fig. 4. 1,500 motions in the database

## 3   Design of Reference Point

### 3.1   Daily Action in the Home

The database called elderly physical function database summarizes activities performed in the home [8]. They summarize information regarding the physical functions of the elderly in nine different fields, such as cooking activities performed in daily life and sleeping. Each activity is organized into units, such as cooking and serving. Furthermore, these activities are divided into actions, such as carrying dishes. From the activities described in this database, 21 activities (excluding unnecessary and impossible activities for robots, such as bathing, undressing and eating meals), were selected for the extraction of primitive motions, including reaching for objects and holding hands. These primitive motions were selected as supposed units of motion with no changes in the target object during the motion. We selected 1,500 motions in total. These motions were manually classified into eight classes of motion based on the shape of their motion trajectories.

1. Reaching motion: The motion to move the end effector to an object or a predetermined position.
2. Moving away: The motion to move the end effector away from an object or a predetermined position.
3. Translation motion: The motion to move the end effector parallel to the axis of an object or the end effector itself.
4. Grasp: The motion to open and close the end effector.
5. Placement motion: The motion to place an object or the end effector on a desk or floor.
6. Stationary motion: A motion in which the start position and end position of the motion are the same.
7. Turning: The motion to draw an arcing trajectory around an object or a predetermined position.
8. Moving: The motion that expresses (robotic) base movement, rather than movement of the end effector.

Some motions are not only motions that are completed by oneself or motions that act on a single object, but instead make it important for an object to act on another object. As shown in Fig. 4, most of the motions are a motion targeting a single object or two objects. Motions targeting more than two objects account for less than 1% of the motions. In order to learn the motion trajectories of motions with multiple objects, an advanced concept for objects is required. Because this method targets the learning of motion primitives using the body of the robot itself, it is not possible to learn sophisticated motions targeting multiple objects. Therefore, the target motions in this paper are motions that act on a single object. These targets include trajectors. Movement based on a trajector is included motions targeting a single object. Among the eight types of classification based on motion trajectory, the robotic base movements are independent of movements of the end effector. For this reason, we design reference points that can express trajectories for the remaining seven types of motion.

## 3.2   Reference Points for Expressing Trajectory

Consider a coordinate system for expressing the seven types of motion trajectories mentioned above. "Reaching motion" and "moving away" are motions in which the distance from the reference point becomes important. Therefore, a

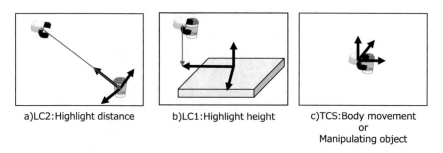

**Fig. 5.** Landmarks: (a) reference point coordinate system with two-axis rotation considering distance, (b) landmark coordinate system with two-axis rotation considering height, and (c) trajector coordinate system

coordinate system that emphasizes the distance between the end effector and reference point is chosen for these motions. A "placement motion" is a motion in which it is important to maintain an object and end effector in the same plane. The trajectory is very similar to that of a "reaching motion", but unlike the "reaching motion", it is important that the height difference to the target is zero, meaning it is important not to bring the end effector to a specific position, but to a specific plane. Therefore, it is important to choose a coordinate system that can emphasize height. "Translation motion" and "turning" are operations for manipulating an object. These motions include lifting an object, rotating an object, and pushing a door open. These motions manipulate objects using the end effector according to the coordinate system unique to each object. However, when the end effector grasps the object, the object and the end effector are often in a restrained relationship. Therefore, we consider that the motion trajectory of the end effector is very similar to the motion trajectory in the coordinate system unique to the object. In order to calculate a coordinate system unique to an object, recognition involving advanced knowledge of the object is required. Because the robot does not initially possess knowledge to that extent, we assume that the representation of the end effector, which can express a very similar motion trajectory, is appropriate. "Stationary motion" and "grasp" are operations in which the coordinates of the end effector at the start point and end point of motion are the same. The motion trajectory can be expressed by the posture of the end effector at the start point of motion. Based on these observations, we consider three reference point coordinate systems shown in Fig. 5. We can express all motion trajectories using these three coordinate systems.

## 4   Experiment

In order to show that motions in the home can be represented by the proposed method and designed coordinate system, we performed housekeeping household by teleoperating the robot, and extracted motion primitives from the time-series data that the robot obtained. The Human Support Robot (HSR), which was

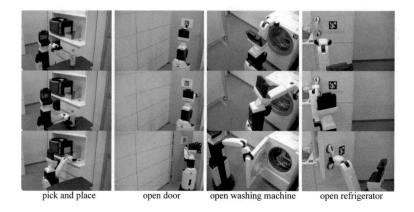

pick and place        open door    open washing machine    open refrigerator

**Fig. 6.** Four tasks in the experiment.

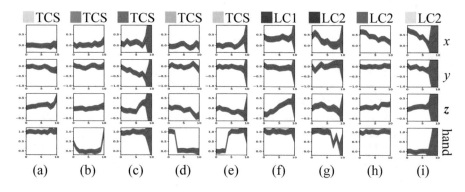

**Fig. 7.** Learned motion primitives (learned GPs); (a) move forward, (b) stay, (c) open door, (d) hand close, (e) hand open, (f) raise hand, (g) reaching above, (h) reaching front, and (i) place object

developed by Toyota Motor Corporation, was used for this experiment. The entire system is composed in ROS. The acquisition of information from sensors and object recognition results were performed using this system. For hand tip, 8-dimensional information, including 7-dimensional posture from the end effector and 1-dimensional opening/closing information of the hand gripper, was obtained. Object recognition via AR marker was performed using object information as a reference point. The robot performed the four types of activities presented in Fig. 6 three times and extracted motion primitives from the physical information during all 12 activities.

## 4.1  Discovered Motion Primitives

The discovered motion primitives are presented in Figs. 7 and 8. Each motion primitive in Fig. 7 represents the value of a Gaussian process acquired during

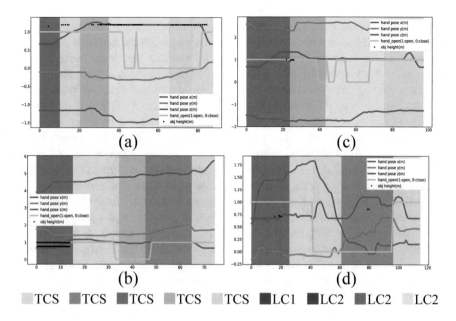

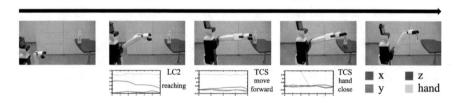

**Fig. 8.** Results of the proposed segmentation; (a) open refrigerator, (b) open door, (c) open washing machine, and (d) pick and place

**Fig. 9.** An example of generated action; grasp the object on a table

learning. The blue lines represent the averages of the Gaussian processes and the red areas surrounding the lines represent the variances of the Gaussian processes. In this method, it is necessary to give the number of classes in advance. Here, this value was determined by hand in an ad hoc manner. The labels on top of the trajectories for each Gaussian process were determined from the information regarding which portion of the physical information at that time during the Gaussian process represented which action. The segment indicated by the TCS is a motion segmented based on the coordinate system of its end effector reference. The $x$, $y$, and $z$ coordinates represent up-and-down, right-and-left, and front-to-back displacements, respectively. "Move forward" represents a motion to move the end effector forward. It is expressed as a motion to bring the hand closer to the target or to push a door open. "Stay" represents the motion immediately after grasping an object. In the case where the length of the segment is short, it represents the motion of remaining in the grabbed state. In the case where the length of the segment is long, it represents the operation of turning the knob on the door. "Open door" is an operation to open the refrigerator or washing machine. "Move forward" was also learned as opening the door, but a different motion trajectory was utilized. "Hand close" and "Hand open" are the motions used for grasping. These motions open and close the end effector in its current location. However, by combining these motions with a motion for moving the end effector closer to an object, they can be used as the action to grasp an object. "Raise hand" was learned as the motion of raising the hand at a position away from an object. At this point, it was desirable to classify the motion of placing an object. This motion was classified as a preliminary motion for opening the washing machine. Originally, it was desirable to classify this motion as a TCS standard motion. However, because it was an operation dealing with height not seen elsewhere in the executed activities, it was learned as motion in the coordinate system with a height emphasis. "Reaching above" is a motion to extend the end effector to grasp a shape, such as the rim of a washing machine handle. "Reaching front" is the motion of moving the hand to approach grasping an object. "Place object" is a motion for putting an object on a plane. These three motions were learned as motions in the coordinate system that emphasizes distance. It was desirable to classify "place object" as a motion in the coordinate system with a height emphasis. However, this trajectory was very similar to the trajectory of the coordinate system that emphasizes distance. In order to differentiate this placing trajectory different from the movement emphasizing distance, height was emphasized as a motion by repeating the placing motion against a wide plane.

These discovered motion primitives were modeled using Gaussian processes. Therefore, by reproducing their average trajectory, they can be reproduced as real robot motions. As shown in Fig. 9, we combined the learned motion trajectories and performed the action of grasping objects. However, in the case of actions requiring a more precise trajectory, such as opening and closing of the door or opening and closing the washing machine, it was impossible to execute them by simply executing the average trajectory. The average trajectory was able to bring the end effector into the vicinity of the target, but because of the vibration of the end effector, the motion easily failed. We believe this problem was caused by the fact that information on dispersion in the learned Gaussian process and the force applied to the hand cannot be considered. Therefore, it is important to generate a trajectory considering the existence of obstacles at the time of action and the information of dispersion.

## 5    Conclusion

In this study, we proposed a method for learning motion primitives from multiple housework activities and attempted to perform real actions based on the primitives by a robot. Regarding the learned motion primitives, motions common to each activity were extracted and a plurality of actions could be expressed by combining the motions. However, several actions could not be reproduced by mere execution of the average learned Gaussian processes. This is because the robot learned common motions from multiple activities containing motion variations. The key to solving this problem lies in the variance learned during the Gaussian process. By generating a motion trajectory using the position of a target object in combination with obstacle information and variance, we believe it will be possible to perform complicated object-manipulation actions that could not be executed in this study. As a future task, we aim to generate smoother trajectories by utilizing the learned distribution.

# References

1. Calinon, S.: A tutorial on task-parameterized movement learning and retrieval. Intell. Serv. Rob. **9**(1), 1–29 (2016)
2. Sugiura, K., Iwahashi, N., Kashioka, H., Nakamura, S.: Learning, generation and recognition of motions by reference-point-dependent probabilistic models. Adv. Rob. **25**(6–7), 825–848 (2011)
3. Sanzari, M., et al.: Human motion primitive discovery and recognition. arXiv preprint arXiv:1709.10494 (2017)
4. Kuniyoshi, Y., et al.: Learning by watching: extracting reusable task knowledge from visual observation of human
5. Chen, J., Zelinsky, A.: Programing by demonstration: coping with suboptimal teaching actions. Int. J. Rob. Res. **22**(5), 299–319 (2003)
6. Nakamura, T., Nagai, T., Mochihashi, D., Kobayashi, I., Asoh, H., Kaneko, M.: Segmenting continuous motions with hidden semi-Markov models and Gaussian processes. Frontiers Neurorob. **11** (2017)
7. Uchiumi, K., Tsukahara, H., Mochihashi, D.: Inducing word and part-of-speech with pitman-yor hidden semi-Markov models. ACL **1**, 1774–1782 (2015)
8. Fukazawa, N., Kariya, Y.: Database of elderly people's physical functions. J. Inf. Process. Manage. **42**(7), 583–590 (1999)

# Nonlinear Model Predictive Control for Two-Wheeled Service Robots

Shunichi Sekiguchi[1(✉)], Ayanori Yorozu[2], Kazuhiro Kuno[3],
Masaki Okada[3], Yutaka Watanabe[3], and Masaki Takahashi[1]

[1] Department of System Design Engineering, Keio University, 3-14-1 Hiyoshi,
Kohokuku, Yokohama 223-8522, Japan
quicklst97@keio.jp, takahashi@sd.keio.ac.jp
[2] Graduate School of Science and Technology, Keio University, 3-14-1 Hiyoshi,
Kohoku-ku, Yokohama 223-8522, Japan
ayanoriyorozu@keio.jp
[3] EQUOS RESEARCH Co., Ltd., 1-11-4, Mikawa-Anjo Minamimachi,
Anjo, Aichi 446-0058, Japan
{173066_Kuno,140121_OKADA,
126760_WATANABE}@aisin-aw.co.jp

**Abstract.** Two-wheeled service robots capable of following a person are an active research topic. These robots will support and follow customers like tourists to many places. It is necessary that an appropriate distance be maintained between the human and the robot. In addition, robots need to not only approach but also turn toward the human to provide services when the human stops walking. Therefore, the control system should change its property depending on the situation. However, many of the previous researches report an algorithm having only one property. Thus, this research proposed a motion control system of a two-wheeled service robot that could turn around while simultaneously following a human. To achieve this, we use nonlinear model predictive control (NMPC) and evaluation function weights depending on the relative distance between robots and human.

**Keywords:** Nonlinear Model Predictive Control · Two-wheeled robots
Service robots · Non-holonomic system

## 1 Introduction

Service robots have attracted considerable attention and have been actively developed to help people in many places, for example, at a shopping mall and sightseeing spots [1]. Conventionally, most robots provide guidance services in these places, and they lead people to pre-decided destinations, as demonstrated in Fig. 1. However, if the human changes their destination on their way, e.g., for impromptu shopping, then the robots cannot provide further support to the human. Hence, we suggest "accompanying robots", which remain close to the human at all times and move together, as depicted in Fig. 2. By following the human, the robots can carry baggage instead of the human doing so. Additionally, by turning their monitor toward the human quickly, the robots provide information of the shops where the human who is stopping. As can be seen from these examples, such robots can provide many services and therefore, it is very important for the accompanying robots to follow the human not only for straight movement but also turning movement. To follow the human, the robots need to follow the goal depending on the human position and posture. In addition to following the goal, the robots have to turn their monitor quickly toward the human stopping. Therefore, the robots should vary their movement property smoothly (Figs. 3 and 4).

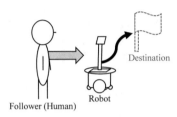

**Fig. 1.** Leading robots

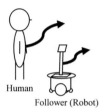

**Fig. 2.** Following robots

**Fig. 3.** Walking situation

**Fig. 4.** Stopping situation

Many conventional researches [2–5] used two-wheel robots. It is because two-wheel robots can get over small steps and be good at preventing from slipping. Hence, we also use two-wheeled robots.

The mathematical model of two-wheeled robots is expressed by a nonlinear and non-holonomic system. Therefore, previous researches reported nonlinear algorithm to control the robots [2, 3]. One of the popular nonlinear algorithms is nonlinear model predictive control (NMPC). NMPC [6, 7] is a control method for solving the finite-optimization problem at each sampling time. By solving the optimization problem at each sampling time, NMPC can not only control nonlinear systems but also consider constraints. Otsuka et al. applied NMPC to two-wheeled robots [2, 6, 8]. They showed that the robots can move to their origin state from any initial conditions except at equilibrium points. They used stable evaluation function weights of the finite-optimization problem which consists of tracking goal angel and tracking goal position weights. If we have greater tracking goal angle weights than tracking goal position weights, the robots make their posture turn to goal posture quickly and maintain it. This sometimes results in a useless Y-turn. However, if the position weights are bigger than the angle weights, the robots move goal position quickly, and the robots turn to goal position slowly. In this way, stable evaluation function weights can have only one property of systems nevertheless, service robots need to change their property of movement according to the situation. Therefore, we propose evaluation function weights of NMPC that depend on the relative distance between robots and the goal position. The robots can the change their movement property according to the situation. In addition, a consideration of the stability of the system is crucial. We demonstrate the stability of our method by confirming to the conventional methods [6, 8–11].

## 2   Two-Wheeled Service Robots

Figure 5 shows the appearance of the proposed two-wheeled robot and LRSs position. We set three LRSs. Two of them (UTM-30LX-FW, Hokuyo Automatic Co., Ltd., Osaka, Japan [13]) are installed at the front and back of the robot at 0.30 m height from the floor for the purpose of recognizing the human position [12], and the other (UST-10LX, Hokuyo Automatic Co., Ltd., Osaka, Japan [13]) is installed at 1.25 m height from the floor for the purpose of recognizing the human posture [12]. The wheel rotation data are obtained by encoders. Moreover, we just set acceleration sensors for the purpose of recognizing the posture of the robot. The computational flow is shown in Fig. 6. Firstly, we get the positions and angles of the human and the robot. By using those data, we set the goal position and angle. In addition, we calculate the evaluate function weights. Finally, we optimized the inputs. To control the system, we execute this flow each sampling time.

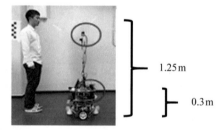

**Fig. 5.** Configuration and sensor position

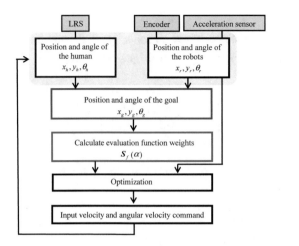

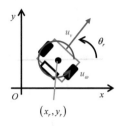

**Fig. 6.** Computational flows

**Fig. 7.** Configuration of wheels

## 3 Control System Design for Two-Wheeled Service Robots

### 3.1 Model

The mathematical model of two-wheeled robots is given by

$$\dot{x}_R = Bu \tag{1}$$

and as shown in Fig. 7. where, $B = \begin{bmatrix} \cos\theta_r & 0 \\ \sin\theta_r & 0 \\ 0 & 1 \end{bmatrix}$, and $x_r = \begin{bmatrix} x_r & y_r & \theta_r \end{bmatrix}^T$ is the

estimated position and angle (in radians) of the robot. Additionally, $u = \begin{bmatrix} u_v & u_\omega \end{bmatrix}^T$ is the estimated input command of the velocity and angular velocity of the robot.

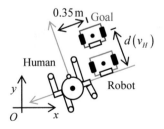

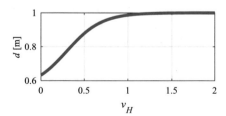

**Fig. 8.** Goal position and angle

**Fig. 9.** Function

## 3.2  Setting up the Goal

To follow a human, we have to estimate goal position $x_g, y_g$ and angle $\theta_g$ depending on the human position and posture as shown in Fig. 8. The goal state vector $\mathbf{x}_G = [x_g \quad y_g \quad \theta_g]^T$ given by

$$
\begin{bmatrix} x_g \\ y_g \\ \theta_g \end{bmatrix} = \begin{bmatrix} \cos(\theta_h) & -\sin(\theta_h) & 0 \\ \sin(\theta_h) & \cos(\theta_h) & 0 \\ 0 & 0 & 1 \end{bmatrix} \begin{bmatrix} d\left(v_{x_h}, v_{y_h}, v_{\theta_h}\right) \\ -0.35 \\ 0 \end{bmatrix} + \begin{bmatrix} x_h \\ y_h \\ \theta_h \end{bmatrix} \tag{2}
$$

and $d\left(v_{x_h}, v_{y_h}, v_{\theta_h}\right)$ is given by a sigmoid function, as shown in Fig. 9, and by

$$
d(v_H) = \frac{d_{Max} - d_{min}}{1 + \exp(-d_{Grad}(v_H + d_{Point}))} + d_{min}
$$
$$
d_{Max} = 1, d_{min} = 0.55, d_{Grad} = 5, d_{Point} = 0.3 \tag{3}
$$

where, $v_H = \sqrt{\left(v_{x_h}\right)^2 + \left(v_{y_h}\right)^2 + \left(v_{\theta_h}\right)^2}$ is a norm of the human velocity. Because of the estimated $d(v_H)$, which depends on the human velocity, the robot can approach the human when the latter stops. Additionally, the robot is able to keep the proper distance from the human when the latter is walking.

## 3.3  Nonlinear Model Predictive Control Problem

The information given in this section is about the NMPC. By using the robot state vector $\mathbf{x}_R$ and goal $\mathbf{x}_G$, we obtain the evaluation function $J^{Robot}$ of finite-optimization problems given by

$$
\begin{aligned}
J^{Robot} = &\frac{1}{2}(\mathbf{x}_R(t_0 + T) - \mathbf{x}_G(t_0))^T S_f(\alpha)(\mathbf{x}_R(t_0 + T) - \mathbf{x}_G(t_0)) \\
&+ \int_{t_0}^{t_0 + T} (\mathbf{u})^T(t)\mathbf{R}\mathbf{u}(t)dt
\end{aligned} \tag{4}
$$

where $\mathbf{R} = \mathrm{diag}[0.05 \quad 0.1]$, and $S_f(\alpha) = \mathrm{diag}[1 \quad 1 \quad S_{33}(\alpha)]$. Both are evaluation function weights. We describe the details of $S_f(\alpha)$ in Sect. 3.4. By using a new time variable $T$, the NMPC can be converted to a general optimal problem parameterized by the actual time $t_0$ as follows:

$$
\begin{aligned}
\text{minimize } J^{Robot} = &\frac{1}{2}(\mathbf{x}_R(t_0 + T) - \mathbf{x}_G(t_0))^T S_f(\alpha)(\mathbf{x}_R(t_0 + T) - \mathbf{x}_G(t_0)) \\
&+ \int_{t_0}^{t_0 + T} (\mathbf{u})^T(t)\mathbf{R}\mathbf{u}(t)dt
\end{aligned} \tag{5}
$$

$$
\text{subject to } \mathbf{x}_R = \mathbf{B}\mathbf{u}
$$

and actual input command $\mathbf{u}(t_0)$ is given by

$$u(t_0) = u^*(0, t_0) \tag{6}$$

where $u^*(0, t_0)$ indicates the optimized input. Equation (6) can convert the problem into a two-point boundary-value problem (TBPVP) by defining a Hamiltonian $H$ and variable $\lambda$. The TBPVP is given by

$$\frac{d}{dt}x_R = Bu$$

$$\frac{d}{dt}\lambda = -\left(\frac{\partial H}{\partial x_R}\right)^T = -\begin{bmatrix} 0 \\ 0 \\ -\lambda_1 \sin(\theta_r)u_v + \lambda_1 \cos(\theta_r)u_v \end{bmatrix} \tag{7}$$

$$\lambda(t_0 + T) = S_f(\alpha)(x_R(t_0 + T) - x_G(t_0))$$

$$\frac{\partial H}{\partial u} = \begin{bmatrix} u_v R_{11} + \lambda_1 \cos(\theta_r) + \lambda_2 \sin(\theta_r) \\ u_\omega R_{22} + \lambda_3 \end{bmatrix} = 0$$

where $H = (u)^T Ru + \lambda^T Bu$, and $\lambda = [\lambda_1 \quad \lambda_2 \quad \lambda_3]^T$. We can then obtain the optimized input command by calculating the TBPVP for each sampling time.

### 3.4 Setting the Evaluation Function Weight

To change the property of control system, we estimate the evaluation function weights $S_f(\alpha)$. $S_f(\alpha)$ given by

$$S_f(\alpha) = \begin{bmatrix} S_{11} & 0 & 0 \\ 0 & S_{22} & 0 \\ 0 & 0 & S_{33}(\alpha) \end{bmatrix} \tag{8}$$

$$S_{11} = 1, S_{22} = 1$$

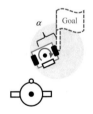

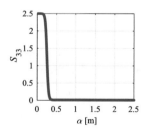

**Fig. 10.** Near the goal          **Fig. 11.** Far from goal          **Fig. 12.** Sigmoid function ($S_{33}(\alpha)$)

where $\alpha$ is the relative distance is given by

$$\alpha = \sqrt{(x_r(t_0) - x_G(t_0))^2 + (y_r(t_0) - y_G(t_0))^2} \tag{9}$$

If $S_{33}(\alpha)$ is bigger, the robot follows the goal angle more. Conversely, if $S_{33}(\alpha)$ is smaller, the robot follows the goal position more. By using these properties, we are able to change the property of the robot movement. Two scenarios are assumed, as shown in Figs. 10 and 11. When the robot is nearer the goal, it should follow the goal angle to enable it turn toward the human quickly. Therefore, we should set $S_{33}(\alpha)$ sufficiently bigger. On the contrary, when the robot is far from the goal, we should set $S_{33}(\alpha)$ sufficiently smaller because we want the robot to follow the goal position. We therefore express $S_{33}(\alpha)$ by

$$S_{33}(\alpha) = \frac{(s_{max} - s_{min})}{1 + \exp(-s_{Grad}(\alpha - s_{point}))} + s_{min} \tag{10}$$

$$s_{max} = 2.5, s_{min} = 0.01, s_{point} = 0.25, s_{Grad} = -50$$

Additionally, Fig. 12 shows a graph of $S_{33}(\alpha)$. By using a sigmoid function, we can change the property of the system smoothly.

### 3.5    Stability of Proposed System

To prove stability of system, we introduce Hamilton-Jacobian-Bellman equation (HJBE). Generally, the HJBE is given by $-\frac{\partial x_R}{\partial t} = \min_u H(x_R, u, \lambda, t)$. From the HJBE, we can obtain the optimized unique feedback input, which minimizes the evaluation function. The unique feedback is given by

$$u^*(t) = -R^{-1}B^T(S(\alpha)(x_R(t) - x_G(t_0)))^T \quad (t_0 \le t \le t_0 + T) \tag{11}$$

According to [6], if there is an input satisfying Eq. (13), then the system is stable.

$$((x_R(t_0 + T) - x_G(t_0))^T S_f(\alpha)(x_R(t_0 + T) - x_G(t_0))Bu) \le -u^T Ru \tag{12}$$

In this study, the optimized unique input of the system satisfied Eq. (13) all the time expect at some situations. The unique feedback control will be 0 when we have $x_R(t) - x_G(t_0) = [\gamma \quad \phi \quad 0]^T$ and as follows:

$$u^* = -R^{-1}B^T(S(\alpha)(x_R(t) - x_G(t_0)))^T$$

$$= \begin{bmatrix} R_{11} & 0 \\ 0 & R_{22} \end{bmatrix}^{-1} \begin{bmatrix} \cos(\theta_r) & \sin(\theta_r) & 0 \\ 0 & 0 & 1 \end{bmatrix} \begin{bmatrix} S_{11} & 0 & 0 \\ 0 & S_{22} & 0 \\ 0 & 0 & S_{33}(\alpha) \end{bmatrix} [\gamma \quad \phi \quad 0]^T \tag{13}$$

$$= \begin{bmatrix} S_{11}R_{11}^{-1}\cos(\theta_r)\gamma + S_{22}R_{11}^{-1}\sin(\theta_r)\phi \\ 0 \end{bmatrix}$$

Therefore, if $S_{11}R_{11}^{-1}\cos(\theta_r)\gamma + S_{22}R_{22}^{-1}\sin(\theta_r)\phi$ is equal 0, the optimized input vanishes. This implies that the robot is next to the goal. To summarize this section, we usually have an optimized input to minimize the evaluation function. Also, if we don't have the optimized input, the input is 0. We can therefore prove the stability of the system.

## 4  Simulation

### 4.1  Simulation Conditions

We carried out a simulation in order to verify that the proposed control system is more effective than the conventional methods. Table 1 shows the system parameters, along with the parameters of the conventional methods. We applied the proposed and conventional methods to the human walking data (U-turn), which were obtained from the LRSs of the robots, as shown in Fig. 13. To solve the finite-optimization problem, we used gradient descent methods [6]. Their parameters are shown in Table 2.

**Table 1.** System parameters.

| | |
|---|---|
| $\Delta t$, sampling time | 65 ms |
| $T$, prediction time | 650 ms |
| $R$, evaluation function weight | diag( 0.05    0.1 ) |
| $S_f$, evaluation function weight (Proposed) | diag( 1    1    $S_{33}(\alpha)$ ) |
| $S_{CON}^1$, evaluation function weight (Conventional 1) | diag( 1    1    2.5 ) |
| $S_{CON}^2$, evaluation function weight (Conventional 2) | diag( 1    1    0.01 ) |

**Table 2.** Optimization parameters.

| | |
|---|---|
| $\Delta t_{op}$, sampling time of optimization | 65 ms |
| $h$, amount of step width for each sampling time $\Delta t_{op}$ | 0.65 |
| Maximum number of iterations | 300 |
| Terminal conditions | $\partial H/\partial \mathbf{u} \le 10^{-3}$ |

**Fig. 13.** How to get human walking data (U-turn)

We considered the abilities of the robots after the optimization. For comparing the proposed method and the conventional methods, we set up evaluation indexes as follows:

- Angle settling time (start count when the human is stopping, stop count at $u_\omega < 10^{-3}$)
- Deviation of position
- Number of times of Y-turn

Initial conditions of the robot are given as $x_R(0) = x_G(0) = [\,-2.2 \quad -0.85 \quad -0.08\,]^T$.

## 4.2   Simulation Results

The trajectory of the $x$-$y$ position is shown in Fig. 14, and the time history of $\theta^{rel}$ and $S_{33}$ are presented in Figs. 15 and 16. Here, $\theta^{rel}$ is given in degrees and it is the relative angle based on the goal angle $\theta_g$. Table 3 shows the results of the evaluation indexes. The proposed method did not have a Y-turn and took a short time to settle the angles. On the contrary, the conventional 1 method had a Y-turn and the conventional 2 method settle their angle in 40.1 s. From 7.5 s to 10 s, the evaluation function weight was increasing. That is why the proposed method can settle their angle quickly. Additionally, from 2 s to 7 s, the weight was smaller. Thereafter, the robot avoided the useless Y-turn.

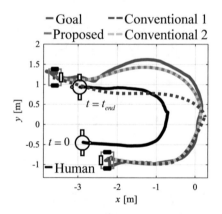

**Fig. 14.** Trajectory of $xy$ position

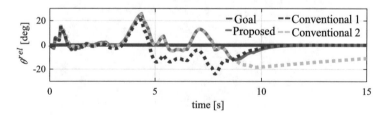

**Fig. 15** Time history of angle $\theta^{rel}$

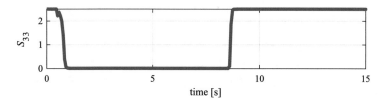

**Fig. 16** Time history of evaluation function weight $S_{33}$

**Table 3.** Results of evaluation indexes

|  | Proposed | Conventional 1 | Conventional 2 |
|---|---|---|---|
| Angle settling time [s] | 3.1 | 2.3 | 40.1 |
| Deviation of position [m] | 0.05 | 0.07 | 0.05 |
| Y-turn | No | Yes | No |

**Fig. 17.** Series of pictures

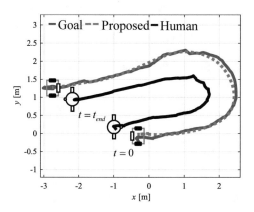

**Fig. 18.** Trajectory of $xy$ position

## 5   Experiments

We applied the proposed method to the robot to confirm actual movements of the robot. The position data of the robot was obtained from encoders. Figure 17 shows the series of pictures obtained and Fig. 18 shows the trajectory of the $x$-$y$ position. The robot was successfully controlled.

## 6   Conclusion

We proposed accompanying robots to provide support to a walking human. However, the conventional method cannot change their property according to the situations at hand, which is the reason for the occurrence of useless Y-turns and slow angle settlement. To rectify this, we suggested the use of NMPC with evaluation function weights depending on the relative distance between the robot and the goal. In simulation, the proposed methods could control robots correctly. The robot had no Y-turn and took a short time to settle their angles. Moreover, we verified the robot movement by experiment. In future work, we should consider avoiding obstacles, especially collision with the human using the robot.

## References

1. Haneda Robotics Lab Homepage. https://www.tokyo-airport-bldg.co.jp/hanedaroboticslab/. Accessed Jan 2018
2. Ohtsuka, T., Fujii, H.: Real-time optimization algorithm for nonlinear receding-horizon control. Automatica **33**(6), 1147–1154 (1997)
3. Gu, D., Hu, H.: Receding horizon tracking control of wheeled mobile robots. IEEE Trans. Control Syst. Technol. **14**(4), 743–749 (2006)
4. Indiveri, G.: Kinematic time-invariant control of a 2D nonholonomic vehicle. In: Proceedings of the 38th IEEE Conference on Decision and Control, vol. 3, pp. 2112–2117 (1999)
5. Aicardi, M., Casalino, G., Bicchi, A., Balestrino, A.: Closed loop steering of unicycle-like vehicle via Lyapunov techniques. IEEE Robot. Autom. Mag. **2**(1), 27–35 (1995)
6. Ohtsuka, T.: Introduction to Nonlinear Optimal Control. Corona Co., Ltd., Tokyo (2011). (in Japanese)
7. Maciejowski, J., Adachi, S.: Predictive Control with Constraints. Tokyo Electric University Press, Tokyo (2005). (in Japanese)
8. Ohtsuka, T.: Control of Distributed Parameter Systems and Nonlinear Systems in Aerospace Engineering. Doctoral Dissertation, Tokyo Metropolitan Institute of Technology (1994)
9. Chen, H., Allgower, F.: A quasi-infinite horizon nonlinear model predictive control scheme with guaranteed stability. Automatica **33**(10), 1205–1217 (1998)
10. Jadbabaie, A., Yu, J., Hauser, J.: Unconstrained receding-horizon control of nonlinear systems. IEEE Trans. Autom. Control **46**(4), 776–783 (2001)
11. Mayne, D.Q., Rawlings, J.B., Rao, C.V., Scokaert, P.O.M.: Constrained model predictive control: stability and optimality. Automatica **36**(6), 789–814 (2000)

12. Yorozu, A., Takahashi, M.: Navigation for gait measurement robot evaluating dual-task performance considering following human in living space. In: Workshop on Assistance and Service Robotics in a Human Environment (ASROB-2016) in Conjunction with IEEE/RSJ International Conference on Intelligent Robots and Systems (2016–2014), Daejeon, Korea (2016)
13. Hokuyo Automatic Co., Ltd., Homepage. http://www.hokuyo-aut.jp/. Accessed Jan 2018

# A Generalised Method for Adaptive Longitudinal Control Using Reinforcement Learning

Shashank Pathak[1(✉)], Suvam Bag[2], and Vijay Nadkarni[2]

[1] Visteon Electronics GmbH, An der RaumFabrik 33b, 76227 Karlsruhe, Germany
shashank.pathak@visteon.com
[2] Visteon Corporation, 2901 Tasman Drive, Santa Clara, CA 95054, USA
{suvam.bag,vijay.nadkarni}@visteon.com

**Abstract.** Adaptive cruise control (ACC) seeks intelligent and adaptive methods for longitudinal control of the cars. Since more than a decade, high-end cars have been equipped with ACC typically through carefully designed model-based controllers. Unlike the traditional ACC, we propose a reinforcement learning based approach – `RL-ACC`. We present the `RL-ACC` and its experimental results from the automotive-grade car simulators. Thus, we obtain a controller which requires minimal domain knowledge, is intuitive in its design, can accommodate uncertainties, can mimic human-like behaviour and may enable human-trust in the automated system. All these aspects are crucial for a fully autonomous car and we believe reinforcement learning based ACC is a step towards that direction.

## 1 Introduction

Recent advancements in machine learning has fuelled numerous applications – both in the academia and in the industries. In particular, supervised learning problems such image classification and object detection have matured to a product-level performance. On the downside, these approaches are limited by the requirements of well specified labelled data, which are either too expensive or too complex to obtain. Reinforcement learning [19], unlike supervised learning, is not limited to classification or regression problems, but can be applied to any learning problem under uncertainty and lack of knowledge of the dynamics. The approach indeed has been applied to numerous such cases where the environment model is unknown e.g - humanoids [18], in games [14], in financial markets [15] and many others.

Adaptive Cruise Control (ACC) is an important feature of autonomous driving where the car is designed to cruise at a speed both efficiently and safely over a wide range of scenarios and contexts. Numerous approaches for ACC have been proposed over the years. They can be divided in two major classes - one which primarily considers the physical signals of the environment and the other

© Springer Nature Switzerland AG 2019
M. Strand et al. (Eds.): IAS 2018, AISC 867, pp. 464–479, 2019.
https://doi.org/10.1007/978-3-030-01370-7_37

which considers the human perspective as the main design principle. We shall call them *environment-centric* and *driver-centric* ACC models.

**Environment-Centric Models**: One of the first environment-centric models proposed was Gazis-Herman-Rothery (GHR) model [7]. This was a linear-proportional control where the desired acceleration was some factor of the velocity difference between the ego car and the car ahead. Many variations of this simple model were suggested later on to address the limitation such as asymmetry between acceleration and deceleration, the non-zero reaction time of the driver, and interaction with multiple vehicles. We refer the reader to [6] for a thorough discussion. In order to address the fact that each driver may have a notion of safe-distance (that might vary with the speed of the ego car), a different class of models was proposed such as [10]. The most well-known model of this category is the Intelligent-driver model (IDM) [20]. Another approach in this class of models is where the driver reacts to the distance between the cars rather than the relative speed, such as in [8]. In yet another approach, an assumption of optimal safe velocity is made such as in [4]. Each of these three sub-classes have seen various – some even fairly complex –enhancements being proposed. Consequently, there have also been models that prioritised simplicity, such as [17] that identified that ACC should only consider two *modes* of driving; one in congestion and the other when the ego car is free of any obstacles ahead. A related idea is to consider ACC behaviour as an automaton where the switching happens at desired contexts; this stream of approaches can be traced back to the seminal work of Nagel and Schreckenberg [16].

**Driver-Centric Models**: One of the biggest flaws with the environment-centric models is the pre-assumption that the driver is perfectly rational, time-invariant, and can perceive all signals without failure and always seeks to optimise the control. None of these is actually true in the real-world. See [5] for the detailed criticism. Thus, there are classes of ACC design where the driver's perspective is central such as when a threshold to sensory perception is assumed [21] or the size of the objects [13] or visual angles [9].

*Contributions:* In this work, we propose a general approach to ACC which is significantly different from these two classes of ACC design. We identify ACC design as a highly context-dependent control problem. We incorporate changes due to both the environment and the driving styles. The key idea here is that instead of aiming for an optimal control for ACC, we strive for a human-like cruise driving that could adapt over time. Incorporating such sources of uncertainty increases the model complexity, hence in order to address that, we harness model-free approaches of the artificial intelligence research where exact modelling of the environment is skirted through efficient use of data. Additionally, the approach allows for truly adaptive cruise control that adapts not just to the environment but also the changing preferences of the driver over time.

## 2   Problem Formulation

The longitudinal control of an autonomous car has to cope with two major conflicting goals, viz., to maintain as close to the set speed as possible and to have as much safe distance from the preceding car as possible. There are other albeit less significant objectives besides this such as maintaining an optimally 'smooth' trajectory. Note that in actual deployment of ACC in the car, these additional objectives are very crucial too; ACC is after all a comfort feature. Also, in a real-world scenario, longitudinal control of the car is seldom, if ever, de-coupled from other aspects such as lateral control, traffic situations and behaviour of the other cars. However, these factors when considered together make the problem insurmountable.

Let us denote the ego car with $\mathcal{E}$ while the one ahead of the ego car as $\mathcal{A}$. Also, the position of the car is denoted by $x$ whereas velocity and acceleration by $\dot{x}$ and $\ddot{x}$ respectively. Furthermore, separation between these cars is denoted by $\Delta x$ i.e., $\Delta x := x_{\mathcal{A}} - x_{\mathcal{E}}$. The ego car would typically be controlled by some action; let $u_{\mathcal{E}}$ be such a control action. For example, in the simplest case, this control action could be the new position of the car. Since in the real world, longitudinal control is done through acceleration (gas pedal) and deceleration (brakes), we would assume the control action to be acceleration i.e., $u_{\mathcal{E}} \in [-b_{max}, a_{max}]_{\mathbb{Q}}$ where $b_{max}$ is the maximum possible deceleration and $a_{max}$ is the maximum acceleration that could be achieved given the physics of the car. There are couple of things to note here. First, the precision of actual control of acceleration (or deceleration) is governed by the dynamics of the car; we make this fact explicit through defining the control action range over rational numbers i.e., $u_{\mathcal{E}} \in \mathbb{Q}$. Secondly, the comfortable acceleration and deceleration limits are usually a strict subset of this range. We assume that the function $f^{\theta}(\Delta x) : \mathbb{R} \mapsto \mathbb{R}$ denotes the objective of maintaining a safe distance. Usually such a function would be parametrised by requirements like what is considered to be a safe distance; these parameters are denoted by $\theta$. Similarly, we let the function $g^{\theta}(\dot{x}_{\mathcal{E}}) : \mathbb{R} \mapsto \mathbb{R}$ stand for the objective of maintaining the set speed. Typically, a given situation or driver preference would determine such a set speed; this is subsumed in the parameters $\theta$. Finally, we assume that the dynamics of the environment is a black-box function $D(t)$ which at any instant $t > 0$ provides all necessary kinematics values i.e., $x_{\mathcal{E}}$, $\dot{x}_{\mathcal{E}}$, $\ddot{x}_{\mathcal{E}}$, $x_{\mathcal{A}}$, $\dot{x}_{\mathcal{A}}$ and $\ddot{x}_{\mathcal{A}}$; we shall denote it by $\mathbf{X}$. Equipped with these symbols, we are ready to define precisely what kind of control problem ACC is.

**Definition 1.** *Adaptive Cruise Control: Given the positions of the cars and possibly their time derivatives,* `classical-ACC` *is defined as a control problem of finding the optimal acceleration action that maintains a desired safe distance and also obeys a set speed. Mathematically,* `classical-ACC` *is equivalent to this problem:*

$$\begin{aligned} \underset{u_{\mathcal{E}}}{\text{minimize}} \quad & f^{\theta}(\Delta x) + \lambda g^{\theta}(\dot{x}_{\mathcal{E}}) \\ \text{subject to} \quad & u_{\mathcal{E}} \in [-b_{max}, a_{max}] \\ & \forall t > 0, \ \mathbf{X}_t = D(t) \end{aligned} \tag{1}$$

Here, $\lambda \in \mathbb{R}$ is a factor to facilitate convex combination of the conflicting objectives of maintaining a set-speed and a safe distance. In general, such a combination could also be some non-linear function; we however consider only linear combination for sake of simplicity. Note that classical-ACC does not consider the driver preferences explicitly. In fact, as is known already (e.g., see [5]), a crucial limitation of such models is that the parameters do not have intuitive meaning when actual traffic data are considered. In other words, the parameters of these models can not be set through observing actual traffic data and human behaviour.

# 3   Approach

## 3.1   Generalised Approach to Longitudinal Control

Like the approaches mentioned in the previous section, we consider the longitudinal control as a problem separate from other factors such as lateral control, traffic conditions and behaviours of the other drivers. The objective to be minimised in Eq. 1 is therefore a function of only longitudinal variables.

In the context of this work, the control problem involves controlling the ego car in the longitudinal direction when the position of the car ahead is known through some sensory input. Unlike the model-based approach, we do not consider the physical model of the problem and this position may even be provided in non-metric terms such as through viewing angle or the image itself or a laser scan. In reality, we would use the radar measurements as the inputs. To make this point explicit, we consider that the input variables are represented by the vector $\mathbf{Z}$. In the simple scenario, this would be all kinematics variables provided through ideal sensors i.e., $\mathbf{Z} = \mathbf{X}$. As is common in the learning community, we shall call the control strategy as *policy* which maps a given observation to the desirable action under that behaviour i.e., $\pi(\mathbf{Z}) = u_{\mathcal{E}}$.

Another aspect that we consider in our work is regarding intuitive behaviour model of the controller. This is similar to those car following models (see the previous Sect. 2) which consider human aspects of driving. However, unlike those approaches, we seek to model all the human-like preferences and biases under a single framework of *reward schema*. For example, a conservative driver would put a higher negative reward for an unsafe close distance than a usual driver. Similarly, in a highway of high-speed cars, faster speeds would carry bigger rewards.

As has been noted recently, longitudinal (and lateral) control of the car by human drivers are typically in the order of few seconds. E.g., once a driver commits to accelerating towards the car ahead, he is not likely to change it within the next few milliseconds. Considering the abstract-level of human-like control of the car greatly signifies the learning problem. This is similar to hybrid control strategy where the high-level controller is responsible for discrete jumps between low-level continuous state controllers. Similar to such approaches, we assume that such a low-level controller is provided to us. In the industrial setting, such controllers are designed based on specifics of the car and its numerous physical parameters.

In order to obtain a generalised longitudinal controller, in this work we have applied an artificial intelligence approach to the controller design. In particular, we use reinforcement learning to learn the controller through various active simulations. On one hand, this allows us to harness the reward schema mentioned before while on the other hand, it is a general approach which degenerates into known model-based controllers under suitable conditions. We shall now delve briefly into the learning approach.

To summarise, our approach is to generalise the classical-ACC problem into one where control is learned through the interaction with the world. To be more precise, we re-formulate the problem as:

**Definition 2.** *Learned Adaptive Cruise Control: Given the observations of the world* $\mathbf{Z}$*, RL-ACC is defined as a learning problem of finding the near-optimal policy* $\pi$ *(of acceleration control) under a given set of rewards specified for different scenes and actions i.e.,* $R^\pi(\mathbf{Z}_0)$*; here* $\mathbf{Z}_0 \in \mathcal{Z}$ *is the initial state at time* $t = 0$ *where after the policy* $\pi$ *is followed.* $\mathcal{Z}$ *is set of all possible initial states. Mathematically, RL-ACC is equivalent to this problem:*

$$
\begin{aligned}
\underset{u_\mathcal{E}}{maximise} \quad & \underset{\mathbf{Z}_0 \in \mathcal{Z}}{\mathbb{E}} \{R^\pi(\mathbf{Z}_0)\} \\
subject\ to \quad & u_\mathcal{E} \in [-b_{max}, a_{max}] \\
& \forall t > 0,\ \mathbf{Z}_t = \hat{D}(t)
\end{aligned}
\tag{2}
$$

Note that this approach does not require to set the functions $f^\theta(\Delta x)$ and $g^\theta(\dot{x}_\mathcal{E})$ explicitly. In the situations where these functions are known, such as from the previous models, we can readily incorporate them in our rewarding schema.

## 3.2   Least Square Policy Iteration

Markovian Decision Problem (MDP) involves a probabilistic system represented by the tuple $(\mathcal{S}, \mathcal{U}, \mathcal{T}, \mathcal{R}, \gamma)$. Here, $\mathcal{S}$ is the countable set of states of the system while $\mathcal{U}$ is the action space. The uncertain nature of the environment is encoded in the transition system $\mathcal{T} : \mathcal{S} \times \mathcal{S} \mapsto \mathbb{R}_{[0,1]}$. The rewards in each state (and action) are determined by $\mathcal{R} : \mathcal{S} \mapsto \mathbb{R}$ and $\gamma \in \mathbb{R}_{[0,1)}$ is the discounting factor. MDP is an NP-hard problem [12] of finding an optimal policy $\pi^* : \mathcal{S} \mapsto \mathcal{U}$ which maximises the total discounted reward. One of the approaches to achieve this approximately is to define *Q-value* as:

$$
Q^\pi(\hat{s}, \hat{u}) = \underset{u_t \sim \pi, s_t \sim \mathcal{T}}{\mathbb{E}} (\sum_{t=0}^{\infty} \gamma^t r_t | s_0 = \hat{s}, u_o = \hat{u})
$$

By applying the Bellman optimality criteria – i.e., Q-value at any step $t$ with the state $s$ and the action $u$ returned by an optimal policy is same as the optimal reward obtained at this step and following the optimal policy for all subsequent steps $(t+1$ onward) – we can compute the Q-value under the current policy. This is *policy evaluation*. In order to improve a policy, we may simply select those actions that maximises the Q-value thus computed. This step is *policy improvement*. When performed iteratively over the space of deterministic policies, policy evaluation and policy improvement result in optimal policy. This is the essence of *policy iteration*. However, under this algorithm, convergence is guaranteed only when the representation of the Q-values as well as the policy is exact such as with tabular form. For large (or infinite) state space, both these representations need to be parametrised and hence only approximate.

Least Square Policy Iteration – or LSPI – is a class of approximate policy iteration reinforcement learning algorithms [11]. The crucial idea here is to project the actual policy to an approximate space such that the iterative operator would still yield Q-values that are near-optimal (w.r.t. actual fixed point) in the sense of $L_2$-norm. We refer the interested reader to the seminal paper work on LSPI [11].

### 3.3   Learning the ACC Through LSPI

We are now ready with all ingredients to describe our approach for learning the ACC. We first generate the data of the interaction of the car with the environment. This is done through the third-party simulators described in Sect. 4. Once the required data of interaction is available, we apply the RL algorithm as described in Algorithm 1. Here, we devise rewarding schema based on the control and the environment. Another domain specific aspect is the choice of gaussians. Typically, we would like the gaussians to approximate the actual Q-value as close as possible hence it is always better to choose the centres of these gaussians to be at the interesting parts of the state-space. Based on the number of these gaussians, we construct a random weight vector. Note that once the problem design is accomplished, the purpose of the RL-ACC is to obtain optimum weight vector. Rest of the algorithm is simply the LSPI.

---

**Algorithm 1.** RL-ACC

---

**Require:** reliable simulator sim for the environment, rewarding schema reward, $\phi$ set of gaussians, discounting factor $\gamma$

1: Construct sample set $D \Leftarrow (\text{sim,reward})$
2: Initialise $\phi$ and weights $w$ randomly
3: $A = 0,\, b = 0$
4: **while** no convergence **do**
5:     **for** $(z, u, r, z') \in D$ **do**
6:         $A = A + \phi(z, u)(\phi(z, u) - \gamma\phi(z', \pi(u')))$
7:         $b = b + \phi(z, u)r$
8:     **end for**
9:     $w = A^{-1}b$
10: **end while**

---

### 3.4   Cruise Control Mode

A cruise control functionality is designed for the model in order to maintain a constant velocity when the ego vehicle can drive freely. This feature is primarily based on a proportional-integral-derivative (PID) controller. Through multiple simulations, the values of the proportional tuning constant ($K_p$) and the derivative constant ($K_d$) were set to **10.0** each. This value was found to be stable for the optimal smoothness at different velocities of the ego vehicle. For practical purposes, the output of the PID controller is limited to the maximum and minimum acceleration values. The desired acceleration is calculated through the Eq. 3.

$$\Delta^{cte} := \dot{x}_{cruise} - \dot{x}_{curr}, \quad \ddot{x}_{PID} = K_p \times CTE + K_d \times \frac{\Delta^{cte}_{i+1} - \Delta^{cte}_i}{dt} \qquad (3)$$

where subscripts *cruise* and *curr* stand for kinematics of the ego car for cruising and current state respectively. Note that the acceleration control is obtained via a PID controller. $\Delta^{cte}$ stands for *cross track error* while the time-step is $dt = 40\,\mathrm{ms}$

### 3.5   ACC and Cruise Control Modes Combined

Our model has the ability to switch between the cruise mode and the ACC mode based on the threshold ACC distance (TAD) and the ACC mode and the safety mode based on the threshold safety distance (TSD) between the ego vehicle and the lead vehicle set by the user (Fig. 1a) This inter vehicle distance (IVD) is a function of the long-range radar output from the ego vehicle.

## 4   Implementation

For sake of simplicity, we first consider the classical-ACC. Here, we have the ego vehicle equipped with the radar.[1] Recall that in this setting, ACC is a purely longitudinal control problem and we assume that the vehicle is already controlled appropriately in the lateral direction so as to maintain the lane it is driving in. Another aspect to notice is that the cut-ins and cut-outs are allowed to happen which may result in the separation between the vehicles vary discontinuously as is evident in the Figs. 1c and d.

### 4.1   Experimental Setup

In this section, we describe the extensive experiments that were carried out to evaluate RL-ACC empirically. For example, we first consider a simple scenario

---

[1] The vehicle is equipped with both long and short range radars with limits [80, 240] and [0.2 100] m respectively. Unlike the problems like parking, the ACC does not require very close range detection. Here, sensor fusion is used to homogenise the readings from both the radars.

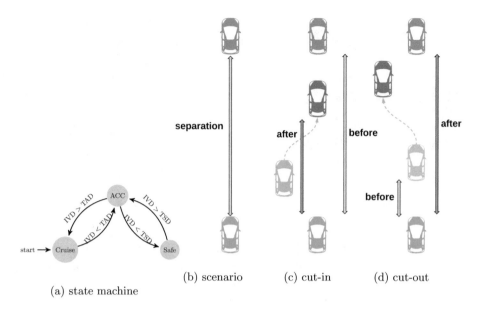

(a) state machine   (b) scenario   (c) cut-in   (d) cut-out

**Fig. 1.** Finite state machine regarding the safe control. Cut-in and cut-outs are shown in the scenario of ACC.

where there is only one leading vehicle with certain well-defined trajectory. This is named as `single-car`. While later on, we consider more realistic scenarios where there are multiple vehicles in various lanes, all of which exhibit some sort of intelligent and rule-following behaviour. This is denoted as `multi-car`. Thus, we have considered two automotive-grade simulators, albeit for different scenarios of evaluations. In this section, we describe very briefly, each of these simulators and their importance. We also describe the overall software and hardware pipeline utilised for this work. In the later section to follow, we shall describe how the overall architecture and these components were used to perform the experiments.

**Simulators**: The development phase experiments are conducted primarily in two simulators besides other mathematical software tests and closed loop tests. The simulators used are the Oktal SCANeR$^{\text{TM}}$studio [2] and Vires [3]. A model test vehicle with similar dynamics to the real vehicle is created for these tests. In order to test different functionalities of the vehicle including more sophisticated algorithms, multiple scenarios are created with different number of vehicles, trucks etc. as well as lanes. Sensor fusion from radars on the ego vehicle provide the desired input for the ACC. The set-up is capable of running the entire test vehicle's distributed system in simulation. Hence, ACC being a closed-loop feature can be tested on top of other components like domain controller, sensor fusion etc.

**Software:** The learning algorithm itself is implemented in MATLAB for ease of prototyping whereas during the actual deployment the learned policy is in native C++ code on a distributed system set-up on ROS. The software stack is

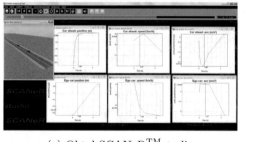

(a) Oktal SCANeR$^{\text{TM}}$studio

(b) DriveCore$^{\text{TM}}$ Studio

**Fig. 2.** Simulation and the actual DriveCore Studio platform

composed of several ROS nodes inter-communicating with each other at different levels based on their individual functionalities. Unfortunately, the detailed description of the system architecture is out of scope of this paper. The ROS nodes communicate to the simulators through a TCP/IP network configuration.

**Hardware:** The entire software stack runs on the DriveCore$^{\text{TM}}$ platform (Fig. 2b. It was designed as a centralised domain controller consisting of highly scalable hardware, in-vehicle middle-ware and PC-based software toolset all on a single platform handled by its three primary components - Compute, Runtime and the Studio. The detailed description of the platform is out of scope at the time of writing this paper, as the product will be released in CES 2018 [1].

**Overall Architecture.** We have considered both the cases of ACC applications – on highway with high set-speed and in urban environment with low set-speed. Even in this simple set-up of one leading car, many interesting situations arise for RL-ACC. We call these as *scenarios*; e.g., we may start with a close-by car ahead which then cuts-out increasing the separation immensely or with different maximum-allowed speeds in the lane. There are other features like driver behaviour – a conservative driver in the car ahead might accelerate (and decelerate) slower than a sporty one – which we encapsulate in the module *environment*. This module along with the given scenario is sufficient for implementing a theoretical MATLAB model on top of which learning can be performed. Environment module also affects how the Oktal/Vires simulator generates the next instance of the world. The result of RL algorithm applied on these models is a policy which can then be deployed in the Oktal/Vires simulator for visualisation. In order to ensure safety of the learned policy, it is rigorously checked against desirable specification using the simulator and the scenarios. Once found satisfactorily safe, the policy is deployed in the real car, but not before passing a few other levels of software and closed loop checks as per the safety guidelines implemented by the Systems and the Validation teams. Overall pipeline is shown in Fig. 3.

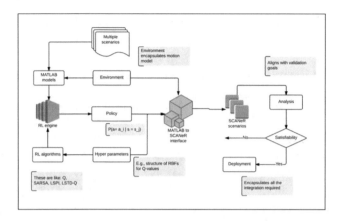

**Fig. 3.** Overall pipeline of learning-based ACC using SCANeR. Use of Vires is analogous.

## 5   Experiments

We conduct a variety of experiments to test `RL-ACC` in different scenarios. Like any other model of ACC, our primary criterion of success is for the ego vehicle to follow the lead vehicle at a safe distance while maintaining a smooth velocity profile. Besides that, we also compare velocity profiles with empirical models of ACC like the `IDM`. Finally, we try to explain our argument of the dynamic nature of `RL-ACC` through comparison of aggressive vs conservative behaviours of the controller in the same scenario.

### 5.1   Single-Car Scenario: `single-car`

The first set of experiments are conducted in Oktal in simple scenarios with relatively ideal conditions. There are 3 lanes (only 2 lanes shown in Fig. 2a) designed in the shape of a racetrack, with each non-curvy side greater than 3000 m, where the `RL-ACC` is primarily tested. The ego vehicle is modelled after the Lincoln MKZ (used in US road testing by Visteon ADAS) and the single lead vehicle is generated by the simulator. We leverage the powerful feature of setting checkpoints on the road provided by Oktal. These checkpoints are useful for altering the environment agents and other parameters.

*Lead Vehicle With Constant Velocity.* In this experiment the lead vehicle maintains a constant velocity throughout its journey. The main objective here is to test the very basic expectation of any ACC model i.e - for the ego vehicle to be able to adjust its velocity based on the preceding vehicle in the simplest possible scenario with no other agents involved. From the velocity profile in Fig. 4a, we can observe the drop in velocity from 11.11 m/s to 8.33 m/s between steps 3000–4000. Although we consider this outcome successful to an extent, it should be noted here that the model was still learning and hence the ACC velocity is not

perfectly smooth after stabilising. For the exact parameter values, please refer
to Table 1.

**Table 1.** Agent/parameter values in `single-car` experiments (see Sect. 5.1)

| Experiment | Parameter | Initial conditions | Checkpoint A | Checkpoint B |
|---|---|---|---|---|
| Exp. 1 | Ego vehicle | 5.55 m/s | N/A | N/A |
| | Lead vehicle | 8.33 m/s | N/A | N/A |
| | Distance between vehicles | >200 m | N/A | N/A |
| | Cruise speed | 11.11 m/s | N/A | N/A |
| Exp. 2 | Ego vehicle | 5.55 m/s | N/A | N/A |
| | Lead vehicle | 8.33 m/s | 13.89 m/s | N/A |
| | Distance between vehicles | <150 m | N/A | N/A |
| | Cruise speed | 11.11 m/s | 13.89 m/s | N/A |
| Exp. 3 | Ego vehicle | 11.11 m/s | N/A | N/A |
| | Lead vehicle | 8.33 m/s | 13.89 m/s | 11.11 m/s |
| | Distance between vehicles | <150 m | N/A | N/A |
| | Cruise speed | 11.11 m/s | 16.67 m/s | N/A |

*Lead Vehicle with Variable Velocity.* By this time we already know that RL-ACC
is capable of following vehicles at a constant velocity while maintaining a safe
distance. Hence, we step up a notch in the scenario complexity in the second
experiment. The lead vehicle here changes its velocity after crossing a preset
checkpoint on the simulator track, resulting in the ACC model to adapt to
multiple velocities over the course of the journey. We can see from Fig. 4b that the
ego vehicle initially accelerates to reach the cruise speed (cruise mode), followed
by decelerating once the ACC mode gets triggered and then accelerating again
after the lead vehicle has changed its velocity post checkpoint A. Compared to
Fig. 4a the velocity profile is smoother irrespective of the drop or rise in the
velocity. In Fig. 4b, we can also observe that the ego vehicle decelerates over
several steps in the first phase, but also has the ability to accelerate faster in a
short span of time in order to match the velocity of the vehicle ahead. This is
necessary as the lead vehicle changes its velocity to 13.89 m/s immediately after
crossing checkpoint A due to simulator constraints. However it is a testament
of the RL-ACC s flexibility to replicate human like driving, which always prefers
smooth deceleration over hard breaking but is less conservative when it comes
to accelerating.

*Test of Uncertainty - Sharp Rise/Drop in Velocity of the Lead Vehicle in Short
Ranges.* We often witness vehicles breaking away from their steady motion and
either accelerating/decelerating abruptly in both highways and country roads.
It is imperative that our model be able to adapt to these situations. Obviously,
the model will jump back and forth between the cruise and the ACC modes

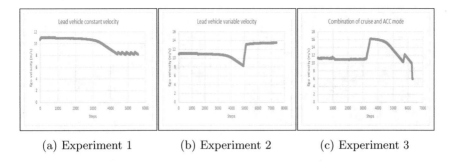

(a) Experiment 1          (b) Experiment 2          (c) Experiment 3

**Fig. 4.** Velocity profile in `single-car` scenarios

frequently. Here, the lead vehicle changes its velocity to 13.89 m/s after crossing checkpoint A, moving away from the ego vehicle rapidly. As the inter-vehicle distance becomes greater than 140 m, the ego vehicle switches back to the cruise mode like initially and starts accelerating. Post checkpoint B, as the lead vehicle slows down, the ACC starts decelerating. The important test here is whether the model can handle a sharp drop in velocity (5.55 m/s) over a very short distance while maintaining a safe distance with the lead vehicle. The Fig. 4c shows that it does adjust pretty well in this situation between steps 3500–5800. The final drop in velocity is at the end of the test track where the ego vehicle hard breaks in order to gradually slow down and stop eventually. This proves that our model can handle uncertain conditions well without manual human intervention, a key requirement in autonomous vehicles.

## 5.2   Multi-cars Scenario: `multi-car`

We use Vires for conducting experiments in more complex scenarios like realistic highway and residential roads with congested traffic while comparing with empirical models of ACC like the IDM. We designed two different scenarios for these experiments - (1) Crowded highway in the German Autobahn with a relatively higher congestion of vehicles (Fig. 5a) and (2) Country side roads where the average speed is much lower besides other traffic restrictions (Fig. 5b). The former helps us in testing our model on higher velocities as well as its adaptability to lane changes or other vehicles cut in/out, whereas the later creates many corner cases automatically because of the shorter inter-vehicle distances. We also test the conservative vs aggressive behaviour of our model on the country side roads. The ego vehicle is modelled on an Audi-A6 (used in German road testing by Visteon ADAS) here, whereas the other vehicles are generated from the simulator. There are no checkpoints unlike the experiments in Oktal, but instead we opt for smooth traffic flow with some hard-coded intelligent behaviour from other vehicles while following traffic rules of the respective scenarios.

*Crowded Highway with Lane Changes.* While it is relatively easier for human drivers to react to lane changes/merges with other vehicles cutting in/out, it

(a) highway scenario                    (b) residential country side scenario

**Fig. 5.** Two contexts of validating ACC

might be more challenging for the ACC to adapt in real time, being dependent sensor fusion of outputs of both radars. Instead of reacting to this output directly, we consider it over consecutive time steps which enables smoother control and also makes the learner context-aware. Hence, when the lead vehicle changes or is absent in the new lane, the ACC resets to either the cruise mode (lead vehicle absent) or adjusts its velocity based on the new lead vehicle's velocity in a short span of time. We test our model on a long stretch of the Autobahn while manually changing lanes a number of times. In Fig. 6, we can see that the ego vehicle adjusts well, even in situations where the lead vehicle is out of range of the radar (blue line hitting zero). Our approach being model-free, `RL-ACC` can act in a more aggressive or conservative manner depending on the inter vehicle distance or average speed of the other vehicles in the respective lane and tuning of the rewards.

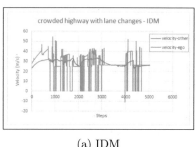
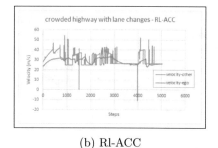

(a) IDM                                 (b) Rl-ACC

**Fig. 6.** RL-ACC vs IDM in crowded highway with lane changes

*Conservative vs Aggressive RL-ACC on Country Side Roads.* As discussed in Sects. 1 and 3, the primary advantage of the `RL-ACC` over `IDM` is its ability to adapt to different environments and driving conditions. Ideally, ACC should behave differently in highway scenarios and city roads. In the former, it should behave more conservatively i.e - the distance between vehicles should always be relatively large because of the average speed, whereas this distance would be a lot less in the later, resulting in a more aggressive behaviour of the ACC. In this experiment, we present our model behaving in a conservative vs aggressive manner on country side roads. The two sub-plots in Fig. 7 are produced from different episodes of the journey, thus having different velocity trajectories. Looking carefully at the plots, we can see that the conservative RL-ACC is able to maintain a smoother velocity profile than the aggressive model. The aggressive behaviour of the car here is the result of impatient driving style. Since our controller is designed for highway auto pilot we have set a high reward for optimal separation. When penalised appropriately for acceleration and deceleration, a smoother conservative behaviour is obtained.

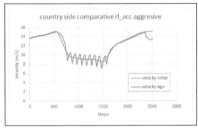 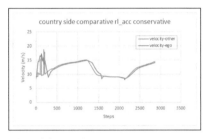

(a) RL-ACC aggressive                    (b) RL-ACC conservative

**Fig. 7.** RL-ACC aggressive vs conservative behaviour

## 6   Conclusion

Typically, adaptive cruise control (ACC) is approached as a model-based controller design. In this paper, we have presented our approach to the control problem of ACC by using artificial intelligence techniques; in particular, we used reinforcement learning and termed it as RL-ACC. While the classical approach – IDM– which is widely used as the go-to ACC model in the automotive industry performs decently on the highways, we argue that due to its model-based approach, it lacks the ability to adapt to the environments or driving preferences. Since the RL-ACC does not require domain knowledge, it can be trained irrespective of the environment. Moreover this approach also makes the controller design intuitive from the perspective of human-driving. Complex control *arises* out of rather simple rewarding strategies. This enhances both the explain-ability as well as trust of human passengers when the autonomous car is utilising ACC. We believe this is one concrete step towards human-like autonomous driving.

This work opens several avenues of scientific research and application. The two most prominent ones are ensuring the safety of learned control and devising a general human-like controller for autonomous driving. In order to achieve the former, systematic verification of the controller can be performed using state-of-the-art formal methods. For achieving the latter goal, one of the approaches is to invoke the deep learning methods that can ingest the input images and learn ACC as some sort of optimal behaviour in different contexts. In future, we shall be pursuing both these avenues.

# References

1. Drivecore - ces2018. http://wardsauto.com/technology/visteon-looks-play-big-role-autonomous-vehicles-drivecore/. Accessed Feb 2018
2. Oktal - simulation in motion. http://www.oktal.fr/en/automotive/range-of-simulators/software. Accessed Feb 2018
3. Vtd - virtual test drive. https://vires.com/vtd-vires-virtual-test-drive/. Accessed Feb 2018
4. Bando, M., Hasebe, K., Nakayama, A., Shibata, A., Sugiyama, Y.: Dynamical model of traffic congestion and numerical simulation. Phys. Rev. E **51**(2), 1035 (1995)
5. Boer, E.R.: Car following from the drivers perspective. Transp. Res. Part F: Traffic Psychol. Behav. **2**(4), 201–206 (1999)
6. Brackstone, M., McDonald, M.: Car-following: a historical review. Transp. Res. Part F: Traffic Psychol. Behav. **2**(4), 181–196 (1999)
7. Gazis, D.C., Herman, R., Rothery, R.W.: Nonlinear follow-the-leader models of traffic flow. Oper. Res. **9**(4), 545–567 (1961)
8. Gipps, P.G.: A behavioural car-following model for computer simulation. Transp. Res. Part B: Methodol. **15**(2), 105–111 (1981)
9. Gray, R., Regan, D.: Accuracy of estimating time to collision using binocular and monocular information. Vision Res. **38**(4), 499–512 (1998)
10. Helly, W.: Simulation of bottlenecks in single lane traffic flow, presentation at the symposium on theory of traffic flow. Research laboratories, General Motors, New York, pp. 207–238 (1959)
11. Lagoudakis, M.G., Parr, R.: Least-squares policy iteration. J. Mach. Learn. Res. **4**(Dec), 1107–1149 (2003)
12. Littman, M.L., Dean, T.L., Leslie, P.K.: On the complexity of solving markov decision problems, pp. 394–402. Morgan Kaufmann Publishers Inc. (1995)
13. Michaels, R.M.: Perceptual factors in car following. In: Proceedings of the 2nd International Symposium on the Theory of Road Traffic Flow, London, England, OECD (1963)
14. Mnih, V., Kavukcuoglu, K., Silver, D., Graves, A., Antonoglou, I., Wierstra, D., Riedmiller, M.: Playing atari with deep reinforcement learning. *arXiv preprint*arXiv:1312.5602 (2013)
15. Moody, J., Saffell, M.: Learning to trade via direct reinforcement. IEEE Trans. Neural Netw. **12**(4), 875–889 (2001)
16. Nagel, K., Schreckenberg, M.: A cellular automaton model for freeway traffic. Journal de physique I **2**(12), 2221–2229 (1992)
17. Newell, G.F.: A simplified car-following theory: a lower order model. Transp. Res. Part B: Methodol. **36**(3), 195–205 (2002)
18. Peters, J., Vijayakumar, S., Schaal, S.: Reinforcement learning for humanoid robotics. In: Proceedings of the Third IEEE-RAS International Conference on Humanoid Robots, pp. 1–20 (2003)

19. Sutton, R.S., Barto, A.G.: Reinforcement Learning: An Introduction, vol. 1. MIT Press Cambridge, Cambridge (1998)
20. Treiber, M., Hennecke, A., Helbing, D.: Congested traffic states in empirical observations and microscopic simulations. Phys. Rev. E **62**(2), 1805 (2000)
21. Wiedemann, R.: Simulation des straßenverkehrsflusses. schriftenreihe heft 8. Institute for Transportation Science, University of Karlsruhe, Germany (1994)

# Reconstructing State-Space from Movie Using Convolutional Autoencoder for Robot Control

Kazuma Takahara, Shuhei Ikemoto$^{(\boxtimes)}$, and Koh Hosoda

Osaka University, D538, 1-3, Machikaneyama, Toyonaka, Osaka, Japan
ikemoto@sys.es.osaka-u.ac.jp

**Abstract.** In contrast with intensive studies for hardware development in soft robotics, approaches to construct a controller for soft robots has been relying on heuristics. One of the biggest reasons of this issue is that even reconstructing the state-space to describe the behavior is difficult. In this study, we propose a method to reconstruct state-space from movies using a convolutional autoencoder for robot control. In the proposed method, the process that reduces the number of dimensions of each frame in movies is regulated by additional losses making latent variables orthogonal each other and apt to model the forward dynamics. The proposed method was successfully validated through a simulation where a two links planar manipulator is modeled using the movie and controlled based on the forward model.

**Keywords:** State-space reconstruction · Convolutional autoencoder

## 1  Introduction

In recent years, bioinspired and soft robotics has been attracting many researchers' interest as a new approach to develop robots. In contrast with conventional robots, which have been developed to be easy to be analytically modeled, the new approach focuses on designs that make the aimed tasks easier to achieve without complex controllers. Thus far, the development of many successful robots was reported and their number keeps increasing [Trimmer2014,Hosoda2012]. In these studies, advantages of new hardware designs have been central issues and their control was done by heuristics because analytical modeling becomes very difficult by adopting the new design policies. It is obvious that a theory and a method that enable to design their controllers in a systematic way will be necessary for further development.

Machine learning is one of the most promising theories and methods to design controllers of bioinspired and soft robots because it does not require to model the robot dynamics in advance. For instance, there are many studies about robotic applications of reinforcement learning, neural network, and so on [Peters2006,Akrour2016,Soloway1996]. In addition, the recent advent of deep

© Springer Nature Switzerland AG 2019
M. Strand et al. (Eds.): IAS 2018, AISC 867, pp. 480–489, 2019.
https://doi.org/10.1007/978-3-030-01370-7_38

neural network is opening possibility to drastically increase their applicability. For instance, Wahlstr et al. focused on convolutional neural networks to create an autoregression model of a motion sequence of a one link manipulator based on the movie [Wahlstrom2015]. On the other hand, Watter et al. approximated the forward model by using convolution neural networks based on the movie. In the latter study, a three links arm, which has more complex dynamics, was focused and successfully modeled [Watter2015]. This showed that the optimal control input at the moment can be calculated by the Iterative Linear Quadrature Regulator algorithm (iLQR). Lenz et al. focused on obtaining a forward model in a latent space created by using a convolution autoencoder, and used the model in the model predictive control. The validity was shown by using motion control problems of PR2 [Lenz2015]. Furthermore, many studies have been reported in recent years about reducing a state-space using an autoencoder [Berniker2015, Wang2016, Kashima2016] to figure out the latent space. In fact, the advantage of this approach is currently being investigated in the field of soft robotics [Soter2017].

Bioinspired and soft robots typically have complex body structures compared to conventional robots. In addition, because of the complex body, robots often cannot employ the conventional state-space definition, namely the joint space. Therefore, it will become particularly important to reconstruct a state-space of a bioinspired/soft robot from sensor data that can be easy obtained in many cases, such as movies. In order to estimate and control the movement of the robot using the state-space, it has to do not only capture sufficient information about the robot configuration but also allow for accurate prediction of the time evolution of the latent state. In [Watter2015] for that, Watter et al. regulated the learning to reconstruct the state-space by referring the error of the modeling of the forward dynamics, which is also a neural network and is trained in parallel with the autoencoder. However, the regularization was not fully investigated and was left on heuristic parameter tunings. Therefore, to reveal useful regulation terms in training of an autoencoder to model a forward dynamics in the latent space will be important to systematically design the controller of the bioinspired/soft robot.

In this study, we focus on reconstructing a state-space from a movie in which a robot is moving. To this end, the convolutional autoencoder was employed. The basic learning objective of autoencoders is to optimize the parameters to make the network's output identical to the given input. In addition to the regularization terms introduced by [Watter2015], we show another important term that decreases correlations between each axes of the latent space. Because autoencoders typically concentrate to figure out the most influential latent variable for reducing the reproduction error, it is difficult for the net dimensionality that appears on the feature layer to smoothly increase. As the result, modeling of the dynamics gets delayed and careful regulation of the learning speeds of the state-space reconstruction and the dynamics modeling becomes necessary. In this paper, the proposed method for reconstructing a state-space was evaluated by a preliminary simulation of controlling a two rigid links manipulator. As the control method, we employed the linear model extraction [Ikemoto2018]. As the result of the simulation, the validity of the proposed method was successfully confirmed.

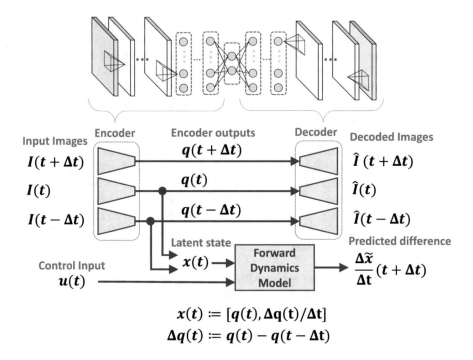

$$x(t) := [q(t), \Delta q(t)/\Delta t]$$
$$\Delta q(t) := q(t) - q(t - \Delta t)$$

**Fig. 1.** The overview of the network which we proposed in this paper. On the top of the figure, a schematic of the convolutional autoencoder is depicted. In the schematic, input images are fed onto the blue layer and corresponding reproduced images appear on the red layer. On the lower half, the entire structure consisting of the convolutional autoencoder is depicted. To retrieve velocity information, which is necessary to reconstruct the state-space, the same convolutional autoencoders parallely transform images at time $t - \Delta t$ and $t$ so that difference approximation is applied. The forward dynamics model, which structure is a common feedforward neural network with the ReLU activation, is trained in the reconstructed state-space. Here, the input vector is directly fed to the forward model.

## 2   Proposed Method

### 2.1   State-Space Reconstruction

Figure 1 shows the overview of the network structure which we proposed in this study. In the figure, a 8-bit grayscale image of a robot at time $t$ is denoted as $I(t) = [I_{i,j} \mid 0 \le I_{i,j} \le 255] \in \mathcal{N}^{H \times W}$. $I(t - \Delta t)$ and $I(t)$ are used to constitute the state vector at time $t$ on the reconstructed state-space. Let us denote the latent feature vector obtained by the convolutional autoencoder as $q(t) \in \mathcal{R}^N$ so that the state vector at time $t$ is defined as $x(t) = [q(t), (q(t) - q(t - \Delta t))/\Delta t] \in \mathcal{R}^{2N}$. Note that the batch normalization is applied to the layer that exhibits the latent feature vector so that $q(t)$ is standardized. In addition, we denote the reproduced image obtained by decoding $q(t)$ by the convolutional autoencoder as $\hat{I}(t)$.

The dynamics of the robot is learned on the latent space reconstructed by convolutional autoencoders. For this purpose, a common feedforward neural network is employed and the ReLU activation is employed to enable the linear model extraction [Ikemoto2018]. In contrast with the state vector, the input vector $u(t) \in \mathcal{R}^M$ is directly fed to the feedforward network. Therefore, the numbers of dimensions of the input and the output of the neural network are $2N + M$ and $2N$, respectively. Because the feedforward network outputs $\Delta \tilde{x}(t + \Delta t)/\Delta t$ that approximates $\Delta x(t + \Delta t)/\Delta t = (x(t + \Delta t) - x(t))/\Delta t$ made from images, the image at time $t + \Delta t$ is also encoded to feed as the supervised data.

The convolutional autoencoder is trained to reduce the following cost function in each batch in each epoch:

$$L = \alpha L_{\mathrm{AE}} + \beta L_{\mathrm{FM}} + \gamma L_{\mathrm{DC}} \tag{1}$$

where $\alpha$, $\beta$, $\gamma$ are coefficients to scale three terms adequately. With $\mathcal{D}$ denoting the batch of supervised data and expressing a possible set $\{I(t - \Delta t), I(t), I(t + \Delta t)\}$ in the given batch $\mathcal{D}$ as $\mathcal{I}$, the three terms on the right hand are defined as follows:

$$L_{\mathrm{AE}} = \frac{1}{N_D} \sum_{\mathcal{I} \in \mathcal{D}} \sum_{dt \in \{-\Delta t, 0, \Delta t\}} \|I(t + dt) - \hat{I}(t + dt)\|^2 \tag{2}$$

$$L_{\mathrm{FM}} = \frac{1}{N_D} \sum_{\mathcal{I} \in \mathcal{D}} \|\Delta x(t + \Delta t) - \Delta \tilde{x}(t + \Delta t)\|^2 \tag{3}$$

$$L_{\mathrm{DC}} = \frac{1}{C(N, 2)} \sum_{(x,y) \in \mathcal{P}_2(\{n \in \mathcal{N} \mid n < N\})} \left\| \frac{\sigma_{x,y}}{\sigma_x \sigma_y} \right\|^2 \tag{4}$$

where $N_D$ and $P_k(\cdot)$ are the size of the set consisting of possible $\mathcal{I}$ in $\mathcal{D}$ and a set of $k$-combination, respectively. In Eq. 4, $\sigma_{x,y}$, $\sigma_x$, and $\sigma_y$ indicate the covariance between x-th and y-th elements of $q(t)$ in $\mathcal{D}$, the standard deviation of x-th element of $q(t)$ in $\mathcal{D}$, and the standard deviation of y-th element of $q(t)$ in $\mathcal{D}$, respectively. For the training of the feedforward network approximating the dynamics of the robot, only $L_{\mathrm{FM}}$ is employed as usual.

The first $L_{\mathrm{AE}}$ is the original cost function for training autoencoder. By reducing the $L_{\mathrm{AE}}$, the latent feature vector $q(t)$ becomes to accurately express images in $\mathcal{D}$. However, it does not consider whether the reconstructed state-space is adequately smooth to model the forward dynamics of the robot. The second term $L_{\mathrm{FM}}$ regulates the training process of the convolutional autoencoder to make the state-space appropriate for modeling of the forward dynamics. Unfortunately, only $L_{\mathrm{AE}}$ and $L_{\mathrm{FM}}$ are not enough because the most influential element typically takes almost efforts in the learning process of the autoencoder. As a result, the dimensionality of the latent space sticks at one for a long period during the training, the forward model learning does not proceed in that period, and training the entire network becomes seriously time consuming. The third term $L_{\mathrm{DC}}$ is introduced to reduce correlations between two elements in the latent feature to overcome the problem. Thanks to the $L_{\mathrm{DC}}$, the learning speed and the stability of training of the entire network are drastically improved.

## 2.2   Linear Model Extraction and Control

Let us briefly explain about the linear model extraction and control [Ikemoto2018]. If the training of the feedforward network in the system shown in Fig. 1 is successfully done, the network stores the nonlinear state equation of the robot. Because of the simplicity of the ReLU, the nonlinearity indicates that some hidden units have to output zero and the pattern of which hidden units are still active and outputting non-zero values, must vary. The ReLU is a function that outputs zero if the input is negative otherwise outputs the input itself. Therefore, if the pattern does not vary, this indicates that the mapping is just a linear regression. The idea of the linear model extraction exploits this feature. By checking the pattern and ignoring inactive hidden units that output zero, we can obtain the linear regression equivalent the neural network, and linear state equations at each time can be obtained if the network stores the nonlinear state equation.

Specifically, the output vector of the neural network $\Delta \tilde{x}(t + \Delta t)/\Delta t$ is computed as follows:

$$\frac{\Delta \tilde{x}}{\Delta t}(t + \Delta t) = W_{N_L} h_{N_L} + b_{N_L+1} \tag{5}$$

$$h_i = \phi(W_{i-1} h_{i-1} + b_i)$$

where $N_L$, $\phi$, and $h_i$ indicate the number of hidden layers, the ReLU function, and outputs of hidden layers, respectively. In addition, $i$ is an integer that $0 < i \le D$ holds and $h_0$ is identical to the input of the network which a vector consists of $x(t)$ and $u(t)$. $W_i$, $W_0$, $b_i$, and $b_{N_L+1}$ are weight matrices and biases. Let $W'_0$, $W'_i$, $b'_i$, and $b'_{N_L+1}$ be weight matrices and biases, where elements relevant to inactive hidden units are removed, the linear state equation at time $t$ is extracted as follows:

$$\frac{\Delta x}{\Delta t}(t + \Delta t) = Ax(t) + Bu(t) + C \tag{6}$$

$$[A|B] = \left( \prod_{i=0}^{N_L} W'_i \right) \tag{7}$$

$$C = \left( \sum_{i=1}^{N_L} \left( \prod_{j=i}^{N_L} W'_j \right) b'_i \right) + b'_{N_L+1} \tag{8}$$

where $A$, $B$, and $C$ are matrices shaped $2N \times 2N$, $2N \times M$, and $2N \times 1$, respectively. Equation 7 means that A and B are obtained as submatrices of the right side. Based on Eqs. 6–8, we can design the control system using the well-established control theory every time new linear state equation is extracted. In this paper, we employed the linear quadratic regulator.

## 3   Simulation

We validated the proposed method by using a 2 DOF planer manipulator. The dynamics is written as follows:

$$\dot{x} = \frac{\mathrm{d}}{\mathrm{d}t}\begin{bmatrix}\theta_1\\\theta_2\\\dot{\theta}_1\\\dot{\theta}_2\end{bmatrix} = \begin{bmatrix}\dot{\theta}\\M^{-1}(\theta)\{-h(\theta,\dot{\theta})-g(\theta)+t\}\end{bmatrix} \tag{9}$$

where $M(\theta)$, $h(\theta,\dot{\theta})$, and $g(\theta)$ are the inertia matrix, the Coriolis and centrifugal force term, and the gravity term. In particular, they are written down as follows:

$$M(\theta) = \begin{bmatrix}M_1 & M_2\\M_2 & M_3\end{bmatrix} \tag{10}$$

$$M_1 = m_1 l_{g1}^2 + m_2 l_1^2 + m_2 l_{g2}^2 + I_1 + I_2 + 2m_2 l_1 l_{g2}\cos\theta_2$$
$$M_2 = m_2 l_{g2}^2 + I_2 + m_2 l_1 l_{g2}\cos\theta_2$$
$$M_3 = m_2 l_{g2}^2 + I_2,$$

$$h(\theta,\dot{\theta}) = \begin{bmatrix}-m_2 l_1 l_{g2}(2\dot{\theta}_1 + \dot{\theta}_2)\dot{\theta}_2\sin(\theta_2)\\m_2 l_1 l_{g2}\dot{\theta}_1^2\sin(\theta_2)\end{bmatrix} \tag{11}$$

$$g(\theta) = \begin{bmatrix}(gl_1 m_2 + gl_{g1}m_1)\cos(\theta_1) + m_2 gl_{g2}\cos(\theta_1 + \theta_2)\\m_2 gl_{g2}\cos(\theta_1 + \theta_2)\end{bmatrix} \tag{12}$$

where $I_i$ and $lg_i$ indicate the moment of inertia of the i-th link and the distance from the i-th joint to the center of gravity of the i-th link, respectively. We assumed that linkages are uniform rods. Masses and lengths of linkages were configured as 1.5 [Kg] and 0.35 [m], respectively.

Data for training were sampled to cover the entire joint space of the robot as dense as possible. In total, 3,000,000 pictures, specifically 1,000,000 independent tuples of $\{I(t-\Delta t), I(t), I(t+\Delta t)\}$, were prepared. They are all 8-bit grayscale image with the same size $(H, W) = (64, 64)$. For training, single batch consisted of 300 pictures (i.e. 100 tuples of $\{I(t-\Delta t), I(t), I(t+\Delta t)\}$). Figure 2 shows a part of samples used for training.

Figure 3 shows the visualization of the latent space reconstructed by the proposed method. The coordination is nonlinear but smooth. In addition, two variables constituting the latent space successfully expressed different joint angles. As explained in the previous section, the part of the cost function $L_{DC}$ was introduced to overcome the problem where hidden units of the autoencoder, which express feature vectors, are all attracted by the most influential element of the latent space and the increase of the dimensionality of the latent space is inhibited. Figure 4 shows the visualization of the latent space in the case where $L_{DC}$ is not employed. As expected, in Fig. 4, two variables of the latent space tended to express the same joint angle.

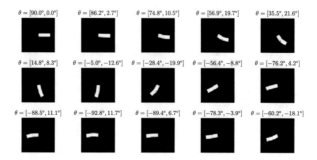

**Fig. 2.** Examples of 8-bit grayscale images used for training. They have the same size $64 \times 64$. To generate images, we simulated movements of the robot, whose dynamics was defined in Eq. 9, and used OpenGL to generate them.

By using the forward model obtained on the reconstructed state-space based on the result shown in Fig. 3, we controlled the robot using the movie. For the control, linear quadratic regulators were designed based on the linear state equations extracted from the feedforward neural network every time. The results are shown in Figs. 5 and 6. Figures 5 and 6 depict time-series data of two variables of the latent space on the true joint space and the latent space, respectively. In both figures, it can be seen that the robot successfully reached to the desired posture. However, in Fig. 5, vibratory moments appeared in the transient response. It is thought of as that the error in transforming an image to a two dimensional feature vector still remained and imposed inevitable error in the control. In fact, in Fig. 6 that shows the similar result but in the latent space, vibratory moments did not intensively appear. Based on these results, we would like to advocate that the proposed system was successfully validated and useful for controlling soft robots that requires to reconstruct state-spaces.

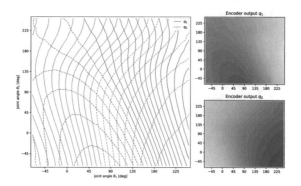

**Fig. 3.** Visualization result of the latent space obtained by the proposed system. The horizontal axis and the vertical axis of figures are $\theta_1$ and $\theta_2$ of the 2 DOF planar manipulator, respectively. The left figure visualizes how two latent feature variables encodes joint angles. It can be seen that contour lines are curved to make each vertex orthogonal. The right two figures separately visualize each variables for the further comprehension.

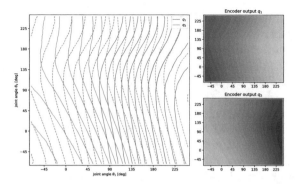
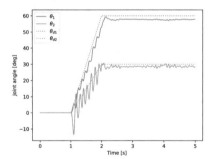

**Fig. 4.** Visualization result of the latent space obtained by the proposed system without using the $L_{DC}$ term. Comparing with Fig. 3, only difference is the absence of the $L_{DC}$ term. In this case, contour lines are almost parallely allocated. This means that both variables mainly express $\theta_1$ but not $\theta_2$ which is a clear drawback in modeling the dynamics of the robot.

**Fig. 5.** The control result in the joint space. Because the controller was designed in the latent space, the actual movement of the robot was influenced by how accurately the latent space was reconstructed. Although fluctuations are present, the reaching was successfully achieved.

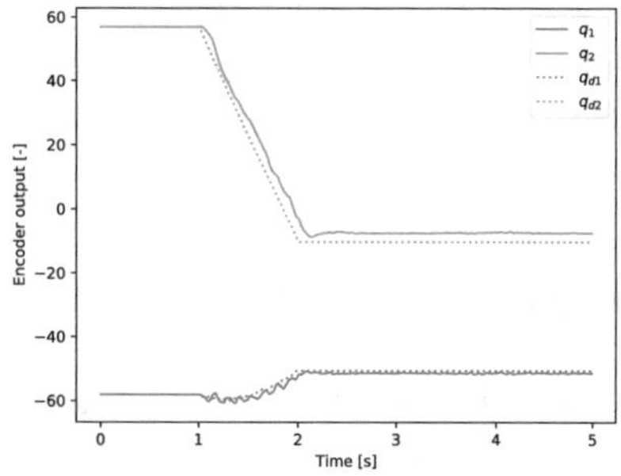

**Fig. 6.** The control result in the latent space. As clearly shown, the reaching was successfully achieved. The controller was designed in the latent space, the reaching performance looks better in the latent space than that in the joint space.

## 4 Discussion and Conclusion

In this study, we proposed a system which reconstructs a state-space from a movie of a robot by convolutional autoencoders, and models by a feedforward neural network. For modeling the dynamics of the robot in the reconstructed state-space, we introduced two costs in addition to the original cost that aims to make the network output identically with the given input. They focus on to make variables of the latent feature smoothly vary and decorrelated each other.

Through the simulation of the 2 DOF planer manipulator, it was confirmed that these costs actually contributed to reconstruct the state-space for modeling the forward model of the robot. In addition, by using the linear model extraction, we confirmed that the robot can be controlled by using only the movie.

As shown in Figs. 5 and 6, errors during control were directly imposed by errors caused in obtaining latent feature vectors from images. Aside from just trying to reduce the transformation error in convolutional autoencoders, in the current form of the proposed system, there is no way to reduce such errors during the control, such as constituting a feedback loop. To realize this functionality, it would be beneficial to use not only encoded features but also decoded images. To improve the proposed system to address such errors during control will be necessary for the practical use that will be focused in the future.

Applying the proposed system to real robots has to be focused in the future. We evaluated the proposed method as a method for reconstructing a latent space by simulating using the model of the rigid robot. This is still preliminary as an experiment for applying this method to control soft robots and it is desirable to simulate using a model of a soft robot. Moreover, in this paper we generated images for training by using OpenGL so that unnecessary information is not contained. However, in the realistic environments where robots work, to obtain such images is typically very difficult. To tackle this problem from the viewpoint of regulating the process of training will be one of important future works of this study.

# References

[Trimmer2014] Trimmer, B.: A journal of soft robotics: why now? Soft Rob. **1**(1), 1–4 (2014)

[Hosoda2012] Hosoda, K., Sekimoto, S., Nishigori, Y., Takamuku, S., Ikemoto, S.: Anthropomorphic muscular-skeletal robotic upper limb for understanding embodied intelligence. Adv. Rob. **26**(7), 729–744 (2012)

[Peters2006] Peters, J., Schaal, S.: Policy gradient methods for robotics. In: IEEE/RSJ International Conference on Intelligent Robots and Systems, pp. 2219–2225 (2006)

[Akrour2016] Akrour, R., Neumann, G., Abdulsamad, H., et al.: Model-free trajectory optimization for reinforcement learning. In: International Conference on Machine Learning, pp. 2961–2970 (2016)

[Soloway1996] Soloway, D., Pamela, J.H.: Neural generalized predictive control. Intell. Control, 277–282 (1996)

[Wahlstrom2015] Wahlstrom, N., Schon, T.B., Deisenroth, M.P.: Learning deep dynamical models from image pixels. IFAC-PapersOnLine **48**(28), 1059–1064 (2015)

[Watter2015] Watter, M., Springenberg, J., Boedecker, J., et al.: Embed to control: a locally linear latent dynamics model for control from raw images. In: Advances in Neural Information Processing Systems, pp. 2746–2754 (2015)

[Lenz2015] Lenz, I., Knepper, R.A., Saxena, A.: DeepMPC: learning deep latent features for model predictive control. Rob. Sci. Syst. (2015)

[Berniker2015] Berniker, M., Kording, K.P.: Deep networks for motor control functions. Frontiers Comput. Neurosci. **9** (2015)

[Wang2016] Wang, M., Li, H.X., Shen, W.: Deep auto-encoder in model reduction of lage-scale spatiotemporal dynamics. In: International Joint Conference on Neural Networks (IJCNN), pp. 3180–3186. IEEE (2016)

[Kashima2016] Kashima, K.: Nonlinear model reduction by deep autoencoder of noise response data. In: IEEE 55th Conference on Decision and Control (CDC), pp. 5750–5755. IEEE (2016)

[Soter2017] Soter, G., Conn, A., Hauser, H., Rossiter, J.: Bodily aware soft robots: integration of proprioceptive and exteroceptive sensors. arXiv:1710.05419v2 [cs.RO], 8 November 2017

[Ikemoto2018] Ikemoto, S., Duan, Y., Takahara, K., Kumi, T., Hosoda, K.: Robot control based on analytical models extracted from a neural network. In: The 1st International Symposium on Systems Intelligence Division (2018)

# BSplines Properties with Interval Analysis for Constraint Satisfaction Problem: Application in Robotics

Rawan Kalawoun[1,2]([⊠]), Sébastien Lengagne[1,3], François Bouchon[4], and Youcef Mezouar[1,2]

[1] CNRS, UMR 6602, Pascal Institute, 63171 Aubiére, France
rawankalawoun@hotmail.com
[2] SIGMA Clermont, Pascal Institute, BP10448, 63000 Clermont-Ferrand, France
[3] Clermont-Auvergne University, Pascal Institute, BP10448,
63000 Clermont-Ferrand, France
[4] Clermont-Auvergne University, Laboratory of Mathematics Blaise Pascal,
BP10448, 63000 Clermont-Ferrand, France

**Abstract.** Interval Analysis is a mathematical tool that could be used to solve Constraint Satisfaction Problem. It guarantees solutions, and deals with uncertainties. However, Interval Analysis suffers from an overestimation of the solutions, i.e. the pessimism. In this paper, we initiate a new method to reduce the pessimism based on the convex hull properties of BSplines and the Kronecker product. To assess our method, we compute the feasible workspace of a 2D manipulator taking into account joint limits, stability and reachability constraints: a classical Constraint Satisfaction Problem in robotics.

**Keywords:** Interval analysis · Pessimism · BSplines
Kronecker product · Constraint Satisfaction Problem · Robot
Feasible workspace

## 1 Introduction

Solving Constraint Satisfaction Problem (CSP) comes down to compute the feasible space of a set of variables that ensures a set of constraints. CSP has many applications in robotics: they are used in planning [1], sequential manipulation planning [2], robot control [3] and are applied to several tasks such as grasping, painting, stripping, etc. In robotics, those constraints may be the robot kinematic and dynamic limits, or the ones imposed by the desired task. Interval Analysis (IA) is a powerful mathematical tool that solve CSP while dealing with uncertainties. In motion generation, it has been shown that finding a new posture in the feasible space is faster than generating it explicitly: IA is an efficient tool to find the feasible space [4]. In parameters estimations, IA ensures the consistency of the results while dealing with some uncertainties [5].

© Springer Nature Switzerland AG 2019
M. Strand et al. (Eds.): IAS 2018, AISC 867, pp. 490–503, 2019.
https://doi.org/10.1007/978-3-030-01370-7_39

The main drawback of IA is the pessimism, i.e. an overestimation of the solutions. BSplines properties were already used to manage the pessimism in one dimensional case [6]. This paper proposes a new inclusion function for multi-dimensional systems that decreases the pessimism and the computation time using the convex hull property of BSplines and Kronecker product. We assess our method on a reaching task of 2D robot systems with 2, 3, 4 and 5 degrees of freedom while ensuring the system balance.

CSP is presented in Sect. 2. Section 3 introduces IA, its application in CSP, and its drawback, i.e. the pessimism. Section 4 details the proposed approach to reduce pessimism based on BSplines properties and the Kronecker product. Section 5 details the technical implementation of our method. The performance of the proposed method is evaluated using a 2D robot with 2, 3, 4, and 5 degrees of freedom in Sect. 6.

## 2   Problem Statement

CSP is a mathematical problem searching for a set of states or objects satisfying a certain number of constraints or criteria. CSP could be defined as:

$$
\begin{aligned}
\text{Find all} \quad & q \in Q \\
\text{such as } \forall j \in \{1, \dots, m\} \; & g_j(q) \le 0
\end{aligned}
\tag{1}
$$

where:

- $q = \{q_1, ..., q_n\}$ : the set of n variables.
- $Q = \{[\underline{q}_1 : \overline{q}_1], ..., [\underline{q}_n : \overline{q}_n]\} \subset \mathbb{R}^n$: the set of variable domains defined by the minimum and maximum values of each variable.
- $g_j(q)$: the $j$-th constraint equation.

In robotics, $q$ could be the set of joint angles, or the joint trajectory parameters [7], or even the feasible space with bounding errors [4], etc. Moreover, $g_j(q)$ could be non-linear and a piecewise-defined function. It may refer to joint limits, the collision avoidance constraint, the balance constraint, the reachability of the end-effector or manipulability criteria to avoid singularities.

To solve this CSP, four major algorithms can be used: backtracking, iterative improvement, consistency, and IA [8,9]. Generally, the backtracking algorithm is implemented using its recursive formulation. It can be related to a path in depth of a tree with a constraint on the nodes: as soon as the condition is no longer filled on the current node, the descent on this node is stopped. Iterative improvement algorithms start with a random configuration of the problems and modify it to find the solution. It does not assign an empty solution as an input to the algorithm. However, the consistency in CSP is used to reduce the problem complexity by reducing the search space: it changes the problem formulation in such a way the solutions remain the same. IA is used in CSP to deal with uncertainties, it guarantees finding solutions and avoids the lack of them.

In robotics, CSP is facing inherent uncertainties in the mechanical structure of the robot. Hence, IA is chosen to solve CSP since it deals with uncertainties and it ensures the reliability [7].

## 3    Interval Analysis

### 3.1    Presentation

IA was initially developed to take into account the quantification errors introduced by the floating point representation of real numbers with computers [10–12]. IA methods have been also used to solve optimization problems. Several works showed that IA is competitive compared to the classic optimization solvers since it provides guaranteed solutions respecting the constraints [13–15]. Nowadays, IA is largely used in robotics [16,17]. Recent works use IA to compute robot trajectories and the guaranteed explored zone by a robot [18–20].

Let us define an interval $[a] = [\underline{a}, \bar{a}]$ as a connected and closed subset of $\mathbb{R}$, with $\underline{a} = Inf([a])$, $\bar{a} = Sup([a])$ and $Mid([a]) = \frac{\underline{a}+\bar{a}}{2}$. The set of all real intervals of $\mathbb{R}$ is denoted by $\mathbb{IR}$. Real arithmetic operations are extended to intervals. Consider an operator $\circ \in \{+, -, *, \div\}$ and $[a]$ and $[b]$ two intervals. Then:

$$[a] \circ [b] = [inf_{u \in [a], v \in [b]} \, u \circ v, \quad sup_{u \in [a], v \in [b]} \, u \circ v] \tag{2}$$

Consider a function $\mathbf{m} : \mathbb{R}^n \longmapsto \mathbb{R}^m$; the range of this function over an interval vector $[\mathbf{a}]$ is given by:

$$\mathbf{m}([\mathbf{a}]) = \{\mathbf{m}(\mathbf{u}) \mid \mathbf{u} \in [\mathbf{a}]\} \tag{3}$$

The interval function $[\mathbf{m}] : \mathbb{IR}^n \longmapsto \mathbb{IR}^m$ is an inclusion function for $\mathbf{m}$ if:

$$\forall [\mathbf{a}] \in \mathbb{IR}^n, \; \mathbf{m}([\mathbf{a}]) \subseteq [\mathbf{m}]([\mathbf{a}]) \tag{4}$$

An inclusion function of $\mathbf{m}$ is evaluated by replacing each occurrence of a real variable by the corresponding interval and each standard function by its interval counterpart. The resulting function is called the natural inclusion function.

### 3.2    Solving CSP Using Interval Analysis

IA can be used to solve CSP as defined in Eq. 1. Given input bounds, the algorithm produces a subset of the input space that satisfies all the constraints. This subset is defined as a set of small boxes. A combination between bisection and contraction algorithms solves a CSP using IA. Those two operations are presented hereafter.

**Bisection.** Bisection is an iterative process that decomposes the set of inputs into smaller sets. It can be described as:

1. starting from the initial set $q = Q$,
2. the set of constraints $g(q)$ is computed,
3. $q$ is considered as a feasible box if: $\forall j, \; Sup(g_j(q)) \leq 0$,
4. $q$ is an infeasible box if: $\exists j$ such as $Inf(g_j(q)) \geq 0$,
5. in the other case, i.e. $\exists j$ such as $0 \in g_j(q)$, the current set $q$ is considered as a maybe feasible box and it is split into several sub-boxes $q_k$. Step 2 is processed considering each $q_k$.

Those operations are repeated until the size of the sub-boxes is inferior to a defined threshold.

**Contraction.** Contraction is based on the filtering algorithm concept. It manipulates the equation in order to propagate the constraints in two ways: from inputs to outputs and from outputs to inputs. Thus, contraction defines a smaller subset of input boxes respecting the different constraints. For instance, given three variables $x \in [-\infty, 5]$, $y \in [-\infty, 4]$ and $z \in [6, \infty]$, and the constraint $z = x + y$, find the intervals of x, y, z respecting this constraint [21]. One can process as follow:

$$z = x + y \Rightarrow z \in [z \cap (x + y)] \Rightarrow \quad z \in [6, \infty] \cap ([-\infty, 5] + [-\infty, 4]) = [6, 9]$$
$$x = z - y \Rightarrow x \in [x \cap (z - y)] \Rightarrow \quad x \in [-\infty, 5] \cap ([6, 9] - [-\infty, 4]) = [2, 5]$$
$$y = z - x \Rightarrow y \in [y \cap (z - x)] \Rightarrow \quad y \in [-\infty, 4] \cap ([6, 9] - [2, 5]) = [1, 4]$$

Thus, by using contractors, the intervals of the variables become tighter: $x \in [2, 5]$, $y \in [1, 4]$ and $z \in [6, 9]$. Contractions generate three types of boxes exactly as bisections: feasible box, infeasible box, and maybe feasible box.

Generally, a combination of contractions and bisections is used to solve CSP. Firstly, contraction reduces the input boxes size. Then, bisection is applied and contraction is called again until the dimension of the generated boxes is smaller than a threshold. This method is the classic contraction/bisection. It uses the inclusion function **m** during contraction and bisection. Hence, it suffers from pessimism explained in Sect. 3.3. In Sect. 4, we propose a new method to decrease pessimism.

### 3.3   The Pessimism

The pessimism overestimates intervals: it produces intervals larger than the real ones. For instance, consider the equation $y = (x + 1)^2$ with $x \in [-1, 1]$. Using this formulation $y = [0, 4]$, but using the following expression $y = x^2 + 2x + 1$, $y = [-1, 4]$. Both results are correct, but the solution range may be larger: this phenomena is called pessimism. It is clear that the inclusion function performance depends on the mathematical expression of **m**. Briefly, pessimism is an overestimation of the actual result. It is mainly caused by the multi-occurrence of variables in equations [22,23]. Each occurrence of the same variable is considered as a different variable relying on the same interval. This article uses BSplines and Kronecker product for the evaluation of the inclusion function in order to reduce pessimism. We must note that decreasing pessimism means finding tighter intervals, though, decreasing the total bisection number. The proposed method to evaluate the inclusion function is shown in Sect. 4.

## 4   BSplines Identification

BSplines properties were already used to tackle pessimism in one dimension: they are tested on continuous constraints used to optimize robot motion [6]. In this paper, we propose a generalization of this concept to multi-dimensional problems.

## 4.1 Definition and Convex Hull Property

BSplines function is the weighted sum of several basis functions. It is defined by $m$ control points $P_i$ and basis functions $B_i$. $K$ is the order of the basis function $B_i$.

$$F(q) = \sum_{i=1}^{m} B_i^K(q) P_i \tag{5}$$

A BSplines curve is totally inside the convex hull of its control poly-line [6]. This property is obtained quite easily from the following definition of a basis function:

$$\forall q \in [\underline{q}, \overline{q}] \quad \sum_{i=1}^{m} B_i^K(q) = 1 \tag{6}$$

This immediately yields:

$$\forall i \in [1, m] \quad \underline{F} \leq P_i \leq \overline{F} \Rightarrow \forall q \in [\underline{q}, \overline{q}] \quad \underline{F} \leq F(q) \leq \overline{F} \tag{7}$$

A conservative estimation of the bounds of $F(q)$ is made based on the minimum and the maximum of the control points. Thus, a bounding box of the considered function can be computed. However, the control points of this function must be identified.

## 4.2 Multi-dimension BSplines

N-dimensional BSplines are functions defined as:

$$F(q) = \sum_{i=0}^{m_1} \sum_{j=0}^{m_2} \dots \sum_{z=0}^{m_n} \left( B_i^{m_1}(q_1) B_j^{m_2}(q_2) \dots B_z^{m_n}(q_n) \right) \times P_{i,j,\dots,z} \tag{8}$$

where:

- $q = \{q_1, q_2, \dots, q_n\} \in [Q] \subset \mathbb{R}^n$,
- $m_i$ is the degree of the input $q_i$,
- $B^{m_i}(q_i)$ is the BSpline Basis function of degree $m_i$ relied to input $q_i$,
- $P_{i,j,\dots,z}$ are the Control Points grouped into vector $P$.

As in the one-dimensional case, the BSpline curve is entirely in the convex hull of its control poly-line:

$$\forall q \in [Q] \quad F(q) \in [\min(P), \max(P)] \tag{9}$$

## 4.3 Control Point Identification

**One Dimensional Case.** Considering that all the functions can be described as a polynomial expression of the input as done in [6], we have:

$$\forall q \in [\underline{q}, \overline{q}] \; F(q) \in \sum_{i=0}^{n} a_i \times q^i \tag{10}$$

where $\{a_0, a_1, \ldots, a_n\} \in \mathbb{R}^{n+1}$ are the coefficients of the polynomial. This equation can be written as :

$$F(q) = [1, q, \ldots, q^n] \times [a_0, a_1, \ldots, a_n]^T \tag{11}$$

and knowing the coefficients $a_i$, the coefficients $p_i$ of the equivalent control point are computed, such as:

$$F(q) = [1, q, \ldots, q^n] \times \mathbf{B} \times [p_0, p_1, \ldots, p_n]^T \tag{12}$$

where $\mathbf{B}$ is a matrix that contains the polynomial parameters of the BSplines basis functions. Therefore, we can compute the corresponding BSplines parameters as:

$$[p_0, p_1, \ldots, p_n]^T = \mathbf{B}^{-1} \times [a_0, a_1, \ldots, a_n]^T \tag{13}$$

**N-Dimensional Case.** The same procedure can be applied to the N-dimensional case. Let us note $P$ the vector of the control points and $X$ the vector of the polynomial coefficients. $P$ and $X$ are related through the following equation:

$$P = \mathbb{B}^{-1} X \tag{14}$$

Where $\mathbb{B}$ can be written as:

$$\mathbb{B} = \mathbf{B}_1 \otimes \mathbf{B}_2 \otimes \ldots \otimes \mathbf{B}_i \otimes \ldots \otimes \mathbf{B}_n \tag{15}$$

$\mathbf{B}_i$ are matrices linking the control point to the coefficient of the polynomial expression of the basis functions of input $q_i$ and $\otimes$ is the Kronecker product as defined hereafter.

## 4.4  Kronecker Product

**Definition.** The Kronecker product was firstly studied in the nineteenth century [24]. The Kronecker product of two matrices $A \in \mathbb{R}^{n_1, m_1}$ and $B \in \mathbb{R}^{n_2, m_2}$ is the matrix $A \otimes B \in \mathbb{R}^{n_1 \times n_2, m_1 \times m_2}$ defined as follows:

$$A = \begin{pmatrix} a_{1,1} & \cdots & a_{1,n_1} \\ \vdots & \vdots & \vdots \\ a_{m_1,1} & \cdots & a_{m_1,n_1} \end{pmatrix} \quad A \otimes B = \begin{pmatrix} a_{1,1} \times B & a_{1,2} \times B & \cdots & a_{1,n_1} \times B \\ a_{2,1} \times B & a_{2,2} \times B & \cdots & a_{2,n_1} \times B \\ \vdots & \vdots & \vdots & \vdots \\ a_{m_1,1} \times B & a_{m_1,2} \times B & \cdots & a_{m_1,n_1} \times B \end{pmatrix} \tag{16}$$

## 4.5  Properties of the Kronecker Product

More properties of the Kronecker product can be found in [25]. Here we focus on the associative (Eq. 17) and the invertible (Eq. 18) properties :

$$A \otimes (B \otimes C) = (A \otimes B) \otimes C \tag{17}$$
$$(A \otimes B)^{-1} = A^{-1} \otimes B^{-1} \tag{18}$$

Using Eq. (18), Eqs. (14) and (15) can be turned into :

$$P = \left(\mathbf{B}_1^{-1} \otimes \mathbf{B}_2^{-1} \otimes \ldots \otimes \mathbf{B}_n^{-1}\right) X \tag{19}$$

Assuming $\mathbb{B} \in \mathbb{R}^{x,x}$, with $x = \prod_{k=1}^{n} m_i$, the complexity of the inversion of the matrix $\mathbb{B}$ in Eq. (14) is $\mathcal{O}(x^3)$. Using the invertible property, the complexity decreases to $\mathcal{O}(x \sum_n m_i^2)$ where $x = \prod_n m_i$, that will allow faster computation. Hence, Eq. (19) is used to compute control points.

One can consider the following property relating matrix/vector multiplication and the Kronecker product:

$$A.X = (I \otimes A)X \tag{20}$$

where $A$ is a matrix and $X$ is a vector. Equation (20) is used to reduce computation time as it will be explained hereafter.

### 4.6    Recursive Inverse Kronecker Product

Using Eq. (20), Eq. (19) could be written as following:

$$\begin{aligned}
P &= \left(\mathbf{B}_1^{-1} \otimes \mathbf{B}_2^{-1} \otimes \ldots \otimes \mathbf{B}_n^{-1}\right) X_n \\
&= \left(\mathbf{B}_1^{-1} \otimes \ldots \otimes \mathbf{B}_{n-1}^{-1}\right)\left(I \otimes \mathbf{B}_n^{-1}\right) X_n \\
&= \left(\mathbf{B}_1^{-1} \otimes \ldots \otimes \mathbf{B}_{n-1}^{-1}\right) X_{n-1}
\end{aligned} \tag{21}$$
$$\text{With: } X_{n-1} = \left(I \otimes \mathbf{B}_n^{-1}\right) X_n$$

This equation can be used recursively to reduce the computation time. Regarding the low dimension of the matrix $\mathbf{B}_i$, the inversion is done numerically. However, future works will address this issue as presented in Sect. 7. As an example, consider two matrices $A \in \mathbb{R}^{2\times2}$ and $B \in \mathbb{R}^{2\times2}$, and $X = [X_1, X_2] \in \mathbb{R}^4$ with $\{X_1, X_2\} \subset \mathbb{R}^2$, the following product can be computed such as:

$$(A \otimes B)X = \begin{bmatrix} a_{1,1}.B.X_1 + a_{1,2}.B.X_2 \\ a_{2,1}.B.X_1 + a_{2,2}.B.X_2 \end{bmatrix} \tag{22}$$

Despite the multi-occurrence of $B.X_1$ and $B.X_2$, $B.X_1$ and $B.X_2$ are calculated only once. Though, the computation time is decreased.

### 4.7    Example

As an example of the use of BSplines to reduce pessimism, let us consider

$$f(q_1, q_2) = 1 - 3q_1 - 2q_2 + 4q_1q_2$$

with $q_1, q_2 \in [-10, 10]$. Thus, we have $X = [1, -3, -2, 4]^T$ and $\mathbf{B}_1 = \mathbf{B}_2 = \begin{bmatrix} 0.5 & 0.5 \\ -0.05 & 0.05 \end{bmatrix}$. We can compute the equivalent control point using:

$$P = \left(\mathbf{B}_1^{-1} \otimes \mathbf{B}_2^{-1}\right) X = [451, -389, -409, 351]^T \tag{23}$$

Thus using our method, we can deduce that $f \in [-409, 451]$ whereas using the natural inclusion functions we obtain $f \in [-449, 451]$. Using a 3D plot of $f$, we can deduce that $f$ is limited between $-409$ and $451$. Once can deduce that our method reduces pessimism in this example. Hence, the proposed method allows to avoid or at least reduce pessimism for n=2. Moreover, if the value of n increases, then the pessimism will be more reduced: the multi-occurrence increases once n increases.

## 5    BSplines and Kronecker Product in the Contraction

In this section, we present how to use the convex hull properties of BSplines and the Recursive Inverse Kronecker Product in the contraction process. Starting from a given box $[q]$, the algorithm will perform the contraction on each constraint $g_j(q)$ while it does not produce an empty set on the constraint or on one input. For each contraction, the minimum and the maximum of each variable $q_i$ are evaluated, and tighter $q_i$ intervals are found.

### 5.1    One Constraint Contraction

Let us consider a constraint $g_j(q) \leq 0$, our goal is to find the set of $q$ intervals respecting this constraint. Let $g_j(q)$ be the expression of the constraint $j$ and $v_j$ be a new variable initialized to $[-\infty, 0]$ (since $g_j(q) \leq 0$). We create an extended equation of this constraint $f_j(q, v_j)$ such as:

$$f_j(q, v_j) = g_j(q) - v_j \tag{24}$$

The contraction will be applied on the new polynomial $f_j(q, v_j)$. This polynomial is a sum of many monomials $\mu_i$.

If we consider that the bounds of $g_j(q)$ can be found, using $\mathbb{B}$ and $X$, through the minimum and maximum value of the vector $P$ (Eq. 14), thus the bounds of $f_j(q, v_j)$ can be found by the minimum and maximum value of $P_e$ such as:

$$P_e = \mathbb{B}_e^{-1} X_e \tag{25}$$

With :

$$\mathbb{B}_e = \mathbf{B}_v \otimes \mathbb{B} \tag{26}$$
$$X_e = [X^T, -1, 0, \ldots, 0]^T \tag{27}$$

With $\mathbf{B}_v$ a $2 \times 2$ matrix since $v_j$ is only of order 1 in Eq. (24).

Considering $x_i$ the $i-th$ element of vector $X_e$ that corresponds to the coefficient of the monomial $\mu_i$ of Eq. 24, the bounds of the monomial $\mu_i$ can be computed, for all $x_i \neq 0$ using:

$$\mu_i \in [Inf(\rho_i); Sup(\rho_i)] \tag{28}$$

$$\text{with } \rho_i = -\frac{1}{x_i} \left( \mathbb{B}_e^{-1} X_{e,i} \right) \tag{29}$$

With $X_{e,i}$ equals to $X_e$ except for the $i-th$ component that equals to zero.

## 5.2   Monomial Contraction

Considering a monomial $\mu_i = \prod_{\forall k} \sigma_{i,k}$ composed of the multiplication of several input intervals $[\sigma_{i,k}]$, if no input interval contains zero, the input interval can be contracted thanks to:

$$[\sigma_{i,k}] = [\sigma_{i,k}] \cap \frac{[Inf(\rho_i); Sup(\rho_i)]}{\prod_{\forall k \neq i} \sigma_{i,k}} \tag{30}$$

The current input set will be considered as infeasible if this intersection returns an empty interval.

## 5.3   Implementation Trick

The Recursive Inverse Kronecker Product requires a non negligible computation time, and hence must be used in a smart way. Hence, instead of Eq. 29, we rather implement the following Equation:

$$\rho_i = -\frac{1}{x_i}\left(\mathbb{B}_e^{-1} X_e + \mathbb{B}_e^{-1} X_i\right) \tag{31}$$

Considering $X_i + X_e = X_{e,i}$, the first part of the Equation ($\mathbb{B}_e^{-1} X_e$) is computed once for all the monomials of the considered constraint, whereas the structure of $X_i$ (only one non-zero value) makes possible faster computation of $\mathbb{B}_e^{-1} X_i$.

## 5.4   How to Deal with Non Linear Functions?

The proposed method assumes that the constraints can be formulated as polynomial equations of the inputs. Nevertheless, robotics equations are generally nonlinear (using sine and cosine). Here we proposed to create intermediate inputs for $\cos(q_i)$ and $\sin(q_i)$. Each time those intermediate inputs are contracted, they propagate the modification to the input $q_i$ that re-propagate on $\cos(q_i)$ and $\sin(q_i)$.

# 6   Simulations and Results

We tested our BSplines IA algorithm on 2D-robots with n degrees of freedom. The feasible space of the joint values is computed using our method. Hence, a set of joints boxes, i.e. intervals, of the robot respecting a set of constraints is defined. In this application, the end-effector must be inside a square surface while the quasi-static balance of the robot is guaranteed though the projection of the center of mass. We show the results of the implementation of the proposed method for three different positions of the square surface as presented in Fig. 1, with a robot of 2, 3, 4 and 5 degrees of freedom for each position. We consider the 2D robot with a total length of 2, with all the segment of equal values $(2/n)$ and equal mass, a root position in $(0,0)$ and with all joint limits of $[-1.5, 1.5]$.

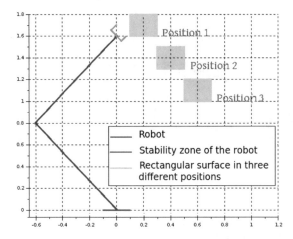

**Fig. 1.** Two dof Robot with a square surface in three different positions and a robot stability margin

The size of the desired square surface is 0.2 with two feasible positions (0.2; 1.7), (0.4; 1.4) and one infeasible position (0.6; 1.1) (due to balance constraint). The balance of the robot is considered by ensuring that the projection of its center of mass remains in the interval $[-0.1, 0.1]$ (considering that the center of mass of each segment is at the middle of the segment).

The surface reachability decision is based on the position of the end-effector. This position is computed using a composition of transformation matrices. The transformation matrix of $q_i$ is defined by: $T_{q_i} = \begin{pmatrix} cos(q_i) & -sin(q_i) & \beta_i \\ sin(q_i) & cos(q_i) & 0 \\ 0 & 0 & 1 \end{pmatrix}$ Where $\forall i > 1, \beta_i = \frac{l}{n}$ and $\beta_0 = 0$, since the initial robot position is along y axis. Hence the position of the end-effector $P_f = (T_f(0,2), T_f(1,2))$ Where $T_f = T_{\pi/2}T_{q_1}...T_{q_i}...T_{q_n}$.

We assign 0.01 to the box threshold during the bisection process. We compare three methods: the classic contraction with the bisection (Solver 1), the BSplines contraction of the monomial that are composed only of input variables with the bisection (Solver 2), and the BSplines contraction of all the monomials in the constraints equations with the bisection (Solver 3).

The number of iterations and the computation time required to find the feasible workspace are shown in Figs. 3a and b respectively, for a threshold of 0.01 and in Figs. 2a and b respectively, for a threshold of 0.1. The computing time required to find the feasible workspace for a robot of 5 dof using a precision of 0.01 was very huge: more than one week. Hence, we show the results of a robot with 1, 2, 3, and 4 dof for a precision of 0.01.

For both threshold, the number of iterations using the BSplines contraction with the bisection (Solver 2, 3) is smaller than the number of iterations in the state-of-the-art method (Solver 1). Though, BSplines IA uses less number of

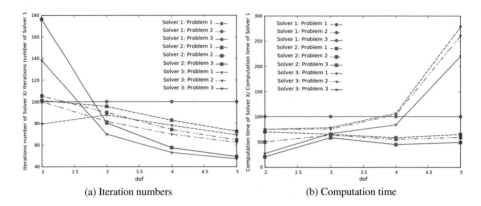

(a) Iteration numbers                    (b) Computation time

**Fig. 2.** Iteration numbers and computation time of the three different solvers compared to Solver 1 for a precision of 0.1

bisections to find the robot feasible workspace. Hence, our method decreases pessimism.

A comparison between the two solvers using the BSplines contraction (Solver 2 and 3) shows that (Solver 3) has lower number of iterations than (Solver 2). Though, the contraction of each monomials of the constraint equation helps to reduce pessimism. However, (Solver 3) is slower than (Solver 2), since doing better contractions requires more computation time (see Figs. 3b, 2b). It is clear that while increasing the accuracy (decreasing the threshold), our method (Solver 2) becomes faster than the classic method (Solver 1). Hence the BSplines contraction/bisection reduces pessimism, and it is faster than the contraction/bisection for high accuracy. We proposed two solvers using BSplines contraction: Solver 2 is faster than Solver 3 and 1, but Solver 3 reduces the pessimism more than Solver 2.

**Table 1.** The number of incrementation required to solve Position 1 problem using Solver 1, Solver 2 and Solver 3/PRECISION 0.01

| Position | Solver | DOF | | |
|---|---|---|---|---|
| | | 2 | 3 | 4 |
| 1 | 1 | 1975 | 2288339 | 1356748167 |
| | 2 | 1871 | 2032347 | 1086632883 |
| | 3 | 1555 | 1771783 | 999374305 |
| 2 | 1 | 705 | 466451 | 737351907 |
| | 2 | 681 | 397469 | 536515897 |
| | 3 | 477 | 326615 | 488179609 |
| 3 | 1 | 21 | 74895 | 231273567 |
| | 2 | 37 | 63129 | 152641419 |
| | 3 | 29 | 52201 | 132694819 |

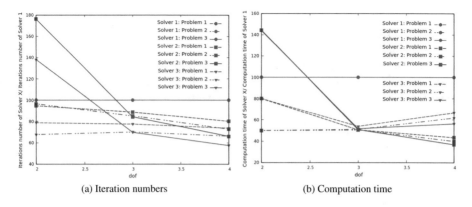

(a) Iteration numbers                    (b) Computation time

**Fig. 3.** Iteration numbers and computation time of the three different solvers compared to Solver 1 for a precision of 0.01

**Table 2.** The time (in days/HH:MM:SS) required to solve Position 1 problem using Solver 1, Solver 2 and Solver 3/PRECISION 0.01

| Position | Solver | DOF | | |
|---|---|---|---|---|
| | | 2 | 3 | 4 |
| 1 | 1 | 0/00:00:0.05 | 0/00:03:39 | 4/14:39:57 |
| | 2 | 0/00:00:0.04 | 0/00:01:51 | 2/00:02:09 |
| | 3 | 0/00:00:0.04 | 0/00:01:58 | 3/01:50:30 |
| 2 | 1 | 0/00:00:0.02 | 0/00:00:43 | 2/12:11:18 |
| | 2 | 0/00:00:0.01 | 0/00:00:21 | 1/00:02:35 |
| | 3 | 0/00:00:0.01 | 0/00:00:22 | 1/13:14:06 |
| 3 | 1 | 0/00:00:0.0005 | 0/00:00:07 | 0/19:08:45 |
| | 2 | 0/00:00:0.0007 | 0/00:00:03 | 0/07:02:04 |
| | 3 | 0/00:00:0.0007 | 0/00:00:035 | 0/10:47:06 |

Tables 1 and 2 show the number of iterations and the computing time required to find the feasible space of our different problems for a precision of 0.01.

## 7   Conclusion and Perspectives

In robotics, Constraint Satisfaction Problem is used to compute the feasible space of the set of robot joint angles: this computation is still an open research domain. In this paper, Interval Analysis is used to solve Constraint Satisfaction Problem based on bisection and contraction concept. This technique suffers from pessimism. Hence, we propose a new way to represent the inclusion function that decreases pessimism. Our method is based on the convex hull properties of BSplines functions and the resolution of an iterative Kronecker product. We assessed our method on different 2D cases and emphasize that it decreases the number of iterations and it may decrease the computation time. In future works, we do believe that the computation time can be reduced by solving the Kronecker inverse recursive product without inverting all the matrices. In the current implementation used in this paper, the bisection process is performed by splitting the largest input interval into two equal intervals. We plan to implement more efficient bisection process by bisecting the input interval that has the most significant impact on unsatisfied constraint. Eventually, the proposed method can be applied to 3D robot taking into account kinematic constraints such as collision and self-collision avoidance and also singularity avoidance or more dynamic constraints such balance or torque limits.

# References

1. Tay, N.N.W., Saputra, A.A., Botzheim, J., Kubota, N.: Service robot planning via solving constraint satisfaction problem. ROBOMECH J. **3**(1), 17 (2016)
2. Lozano-Pérez, T., Kaelbling, L.P.: A constraint-based method for solving sequential manipulation planning problems. In: 2014 IEEE/RSJ International Conference on Intelligent Robots and Systems, pp. 3684–3691, September 2014
3. Fromherz, M.P., Hogg, T., Shang, Y., Jackson, W.B.: Modular robot control and continuous constraint satisfaction. In: Proceedings of IJCAI-01 Workshop on Modelling and Solving Problems with Constraints, pp. 47–56 (2001)
4. Lengagne, S., Ramdani, N., Fraisse, P.: Planning and fast replanning safe motions for humanoid robots. IEEE Trans. Robot. **27**(6), 1095–1106 (2011)
5. Jaulin, L., Kieffer, M., Didrit, O., Walter, E.: Applied Interval Analysis with Examples in Parameter and State Estimation, Robust Control and Robotics. Springer, London (2001)
6. Lengagne, S., Vaillant, J., Yoshida, E., Kheddar, A.: Generation of whole-body optimal dynamic multi-contact motions. Int. J. Robot. Res. 17 (2013)
7. Merlet, J.-P.: Interval analysis and reliability in robotics. Int. J. Reliab. Saf. **3**(1–3), 104–130 (2009)
8. Yokoo, M., Hirayama, K.: Algorithms for distributed constraint satisfaction: a review. Auton. Agents Multi-Agent Syst. **3**(2), 185–207 (2000)
9. Gelle, E., Faltings, B.: Solving mixed and conditional constraint satisfaction problems. Constraints **8**(2), 107–141 (2003)
10. Sunaga, T.: Theory of interval algebra and its application to numerical analysis. RAAG Mem. Ggujutsu Bunken Fukuy-kai **2**, 547–564 (1958)
11. Moore, R.E., Bierbaum, F.: Methods and Applications of Interval Analysis (SIAM Studies in Applied and Numerical Mathematics (Siam Studies in Applied Mathematics). Siam Studies in Applied Mathematics, vol. 2. Soc for Industrial & Applied Math, Philadelphia (1979)
12. Neumaier, A.: Interval Methods for Systems of Equations. Cambridge University Press, Cambridge (1990)
13. Pérez-Galván, C., Bogle, I.D.L.: Global optimisation for dynamic systems using interval analysis. Comput. Chem. Eng. **107**, 343–356 (2017)
14. Jiang, C., Han, X., Guan, F., Li, Y.: An uncertain structural optimization method based on nonlinear interval number programming and interval analysis method. Eng. Struct. **29**(11), 3168–3177 (2007)
15. Ma, H., Xu, S., Liang, Y.: Global optimization of fuel consumption in J2 rendezvous using interval analysis. Adv. Space Res. **59**(6), 1577–1598 (2017)

16. Merlet, J.P.: Interval Analysis and Robotics, pp. 147–156. Springer, Heidelberg (2011)
17. Jaulin, L.: Interval analysis and robotics. In: SCAN 2012, Russia, Novosibirsk (2012)
18. Rohou, S., Jaulin, L., Mihaylova, L., Le Bars, F., Veres, S.M.: Guaranteed computation of robot trajectories. Robot. Auton. Syst. **93**, 76–84 (2017)
19. Desrochers, B., Jaulin, L.: Minkowski operations of sets with application to robot localization. In: SNR 2017, Uppsala (2017)
20. Benoît, D., Luc, J.: Computing a guaranteed approximation of the zone explored by a robot. IEEE Trans. Autom. Control **62**(1), 425–430 (2017)
21. Le Bars, F., Bertholom, A., Sliwka, J., Jaulin, L.: Interval SLAM for underwater robots; a new experiment. In: NOLCOS 2010, France, p. XX, September 2010
22. Wu, J.: Uncertainty analysis and optimization by using the orthogonal polynomials, Ph.D. dissertation (2015)
23. Netz, L.: Using horner schemes to improve the efficiency and precision of interval constraint propagation (2015)
24. Schäcke, K.: On the kronecker product (2013)
25. Loan, C.F.: The ubiquitous kronecker product. J. Comput. Appl. Math. **123**(1), 85–100 (2000). Numerical Analysis 2000. Vol. III: Linear Algebra

# Human Detection and Interaction

# Robot Vision System for Real-Time Human Detection and Action Recognition

Satoshi Hoshino[✉] and Kyohei Niimura

Department of Mechanical and Intelligent Engineering, Utsunomiya University,
7-1-2 Yoto, Utsunomiya, Tochigi 321-8585, Japan
hosino@utsunomiya-u.ac.jp

**Abstract.** Mobile robots equipped with camera sensors are required to
perceive surrounding humans and their actions for safe autonomous navigation. These are so-called human detection and action recognition. In
this paper, moving humans are target objects. Compared to computer
vision, the real-time performance of robot vision is more important. For
this challenge, we propose a robot vision system. In this system, images
described by the optical flow are used as an input. For the classification
of humans and actions in the input images, we use Convolutional Neural Network, CNN, rather than coding invariant features. Moreover, we
present a novel detector, local search window, for clipping partial images
around target objects. Through the experiment, finally, we show that the
robot vision system is able to detect the moving human and recognize
the action in real time.

**Keywords:** Robot vision · Real-time image processing · CNN
Optical flow

## 1 Introduction

For mobile robots, a perception of surrounding humans and their actions is
an essential capability in terms of safe autonomous navigation. For instance,
a mobile robot exhibits different behavior depending on whether someone is
moving or not around the robot. In many cases, mobile robots are equipped
with camera sensors. Since the camera plays a role for visual perception of the
robot, the sensor is called robot vision. For robot vision, it is required to achieve
the following challenges in real time:

- object (human) detection in the image; and
- action recognition of the detected object (human).

In this paper, moving humans are target objects. For the smooth processing of
the moving targets, we aim at speeding up the frame rate more than 30[fps].
This frame rate is a criterion for discussing the real-time performance.

In the field of computer vision, generic object recognition based on invariant
features is one of the main topics. As the invariant features in the images, Local

© Springer Nature Switzerland AG 2019
M. Strand et al. (Eds.): IAS 2018, AISC 867, pp. 507–519, 2019.
https://doi.org/10.1007/978-3-030-01370-7_40

Binary Patterns, LBP [1], Scale-Invariant Feature Transform, SIFT [2], and Bag of Features, BoF [3], have been proposed. Especially, Histogram of Oriented Gradients, HOG [4], is an efficient feature for human recognition. However, it is necessary to code different features from one image for simultaneous human recognition/detection and action recognition. This process might increase the computational cost. For human detection, moreover, this process is repeated for exhaustively searching the whole image. This is a serious problem for the robot vision.

Dollar *et al.* have shown the capability of the BoF in recognizing both the objects and actions [5]. However, the real-time performance was not discussed. Compared to computer vision, the real-time performance of robot vision is more important. For real-time human detection and action recognition, we propose a robot vision system as outlined in Fig. 1.

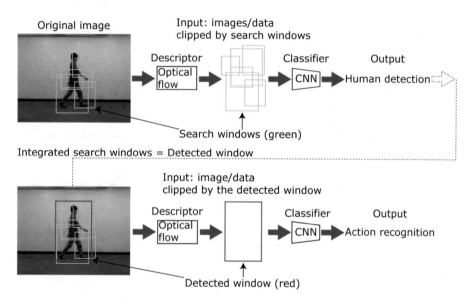

**Fig. 1.** Proposed robot vision system for human detection and action recognition

For moving humans, we focus on information about their movements contained in original images, i.e., optical flow. The optical flow is used as an input. For simultaneous human detection and action recognition, the effectiveness of Convolutional Neural Network, CNN [8], for extracting spatiotemporal features in the image, rather than coding invariant features, has been shown. Ren *et al.* have further achieved fast object detection by using a CNN classifier [9]. Therefore, CNN classifiers are used for human detection and action recognition. In consideration of the computational cost, a local window detector, e.g. selective window [6,7] based on multiple invariant color spaces, is more effective than the exhaustive search using a sliding window detector for clipping partial images around target objects. Therefore, we present a novel detector based on the optical flow for human detection. This object detector is named as local search window. The local search window(s) is/are also used as a detected window for action recognition.

In Sect. 2, a local descriptor based on the optical flow is explained. In Sect. 3, the CNN classifiers and the learning phase are described. In Sect. 4, the local search window is presented. Through the experiment in Sect. 5, finally, we show that the robot vision system is able to detect the moving human and recognize the action in real time.

## 2   Input Image Based on Optical Flow

Optical flow of a moving object in continuous images is represented by a vector form. Given that a point of an object located at a pixel, $(i, j)$, shifts toward another pixel, $(i + u, j + v)$, between two image frames at times $t - 1$ and $t$, the optical flow of the corresponding point is represented by $\boldsymbol{v} = [u, v]$. In this paper, we use Gunnar-Farneback method [10] to estimate the optical flow at every pixel.

In previous works, the effectiveness of the optical flow for action recognition has been shown [11,12]. On the other hand, it is not obvious whether the optical flow has a shape feature. Therefore, we investigate that the CNNs in Fig. 1 learn the convolutional filters for extracting spatiotemporal features from input images based on the optical flow. Since the CNN has a similar structure to a visual cortex, the optical flow is converted to HSV.

Given a set of the optical flow $\boldsymbol{v_x}$ at every pixel $\boldsymbol{x}$ in the image, the Euclidean distance of the corresponding points $d_x$ is defined as $d_x = \|\boldsymbol{v_x}\| = \sqrt{u_x^2 + v_x^2}$. Since $\boldsymbol{v_x}$ is composed of $u_x$ and $v_x$, the moving direction of an object in the image, $\theta_x$, is calculated as follows:

$$\theta_x = \begin{cases} \text{atan2}(v_x, u_x) & (v_x \geq 0) \\ \text{atan2}(v_x, u_x) + 2\pi & (v_x < 0). \end{cases} \tag{1}$$

The moving direction is converted to the hue, $H_x$, as follows:

$$H_x = \frac{\theta_x}{2\pi} \times 360. \tag{2}$$

As a result, the pixels with the optical flow are colored depending on the directions of the moving object.

The lightness value, $V_x$, of the hue derived from Eq. (2) is determined as follows:

$$V_x = \frac{d_x - d_{min}}{d_{max}} \times 100, \tag{3}$$

where $d_{max}$ and $d_{min}$ represent the maximum and minimum values of $d_x$ in the image. Equation (3) increases the brightness of the pixels as the object moves faster.

(a) Original image with optical flow (green lines)          (b) Flow image

**Fig. 2.** Examples of input images

The percentage of saturation, $S_x$, is constantly given as 100[%]. Figure 2 shows examples of the two input images.

In Fig. 2(a), the estimated optical flow, $v_x$, at every pixel in the original image is visualized by green lines. To be exact, the optical flow $v_x$ in the image is used as an input data. In Fig. 2(b), the optical flow is converted by HSV ($H_x$, $S_x$, and $V_x$) and visualized by RGB color ($R_x$, $G_x$, and $B_x$). This local descriptor is named as flow image.

## 3    Convolutional Neural Network

A Convolutional Neural Network, abbreviated as CNN, is one of feed-forward artificial neural networks. A CNN has a powerful role in extracting features of objects and classifying them in images [15]. The CNN classifiers used in the robot vision system are illustrated in Fig. 3. These are based on the CNN presented in [8]. In total, four classifiers depending on the combination of the optical flow and flow image for human detection and action recognition are developed.

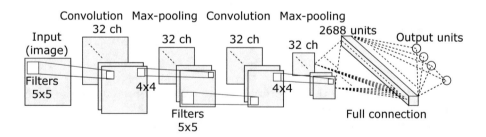

**Fig. 3.** Feature extraction and image classification based on CNN

The input layer is composed of two channels $u$ and $v$ for the optical flow and three color channels $R$, $G$, and $B$ for the flow image. For illustrative purposes, the input layer in this figure is composed of one channel. In each convolution layer, 32 convolutional filters are used. These filters have as many channels as the upper connected layer. The filter size is $5 \times 5$[pix] and the stride is 1[pix]. In the pooling layers, the pooling size is $4 \times 4$[pix] and the stride is 2[pix]. Therefore, the convolution and max-pooling layers are all composed of 32 channels. After that, the extracted features in these layers are processed through the fully-connected neural network layers. The first layer is composed of 2688 units. For human detection, the second layer is composed of two units. For action recognition, the units vary with the number of target actions. Finally, the output values are changed to probabilities through the softmax activation function. The unit with the highest probability yields the classification result.

During the learning phase, the convolutional filters and the connection weights in the fully-connected layers are modified by using backpropagation. In order to minimize output errors and optimize the above filters and weights, a stochastic gradient descent algorithm, Adam, is used. In this paper, cross entropy is used to define the loss function. The detailed process of Adam is described in [13]. For reducing overfitting, a dropout method [14] is used. In one backpropagation, each unit is dropped out with a probability of 0.5.

## 4   Local Search Window

If a target object is in an image, the CNN classifier in Fig. 3 is able to recognize the presence in the image. However, it is difficult to identify the position in the image. This is a problem for human detection. Furthermore, if there are multiple objects in the image, it is impossible to recognize the individual actions. In this paper, we present an object detector called local search window. The local search windows are generated around the target object locally in the image, and partial images clipped by the windows are inputted into the CNN classifier.

The accuracy of detectors depends on the computational cost. In this paper, a Region Of Interest, ROI, based on the optical flow is used for generating the local search windows. Figure 4 shows the process for calculating the ROI from an original image.

(a) Grayscaling          (b) Smoothing          (c) Binarization

**Fig. 4.** Original image processing for ROI

Figure 4(a) is a grayscale image of the optical flow shown in Fig. 2(b). Figure 4(b) is a result of applying a smoothing filter to the grayscale image.

Figure 4(c) is a binary image. As a result of using Otsu's method [16], the white moving body becomes segmented. The bounding rectangle of the moving body is defined as the ROI. On the basis of the ROI, the local search windows are generated as shown in Fig. 5.

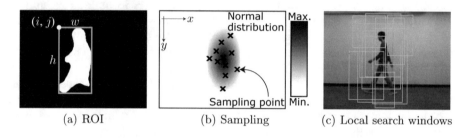

(a) ROI                          (b) Sampling                   (c) Local search windows

**Fig. 5.** Generation of local search windows based on ROI

In Fig. 5(a), the ROI is indicated by the blue rectangle. The size of the ROI is expressed by $h \times w$[pix]. The top-left coordinate of the ROI is $(i, j)$. In Fig. 5(b), a 2D normal distribution $f(x, y)$ based on the ROI is provided in the image. This distribution is used as a probability density function. Black marks, $\times$, indicate sampling points, $x$ and $y$, based on the probability density function. Finally, in Fig. 5(c), green local search windows are generated as many as the sampling points. It is notable that windows in different sizes are generated around the target object and even outside the ROI in some cases.

The 2D normal distribution, $f(x, y)$, is based on the mean and standard deviation of $x$ and $y$, which are represented by $\mu_x$, $\sigma_x$, $\mu_y$, and $\sigma_y$. The center of the distribution is $(\mu_x, \mu_y)$, where $\mu_x = i + \frac{1}{2}w$ and $\mu_y = j + \frac{1}{2}h$, as can be seen from Fig. 5(a). The standard deviations are $\sigma_x = \frac{1}{4}w$ and $\sigma_y = \frac{1}{4}h$ in consideration of the size of the ROI.

In order to generate the local search windows around the target object regardless of the posture and action, the aspect ratios are given as $1 : 2$, $2 : 3$, or $3 : 4$. In this regard, the minimum and maximum heights of the local search windows are given by $h_{min}$ and $h_{max}$. For one ROI, $N$ local search windows are generated. The aspect ratios and heights of the $N$ windows are randomly selected.

The $N$ images clipped by the local search windows are inputted into the CNN classifier. If an image is classified as human, the result, detection, is returned. The window corresponding to the image is also called as a detected window. If some of the images are classified as human, these are integrated together as the detected window. The position of the target object is identified as the center of the detected window. After the human detection, again, another image clipped by the detected window is inputted into the CNN classifier for action recognition.

## 5   Experiment

### 5.1   Settings

The environments in this experiment are illustrated in Fig. 6. A target person is moving between A and B in a camera view. In the horizontal and vertical directions to the stationary camera, he/she takes three and two actions, respectively.

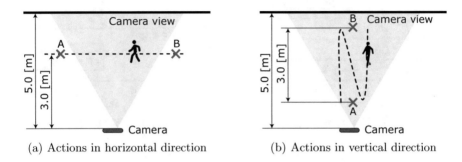

(a) Actions in horizontal direction          (b) Actions in vertical direction

**Fig. 6.** Top view of two environments

In Fig. 6(a), the target actions are walk, jump, and side step. The Bhattacharyya distance [17] of these actions based on the HOG feature coded from the flow image were up to 0.993 (walk and jump) and 0.990 (walk and side step). In other words, these are similar actions. In Fig. 6(b), the target actions are walk and jump. Please note that the optical flow of actions in the vertical direction is relatively low compared to actions in the horizontal direction.

Original images are captured with the use of the ZED stereo camera[1]. The image size is $1280 \times 720$[pix]. The captured image is then resized $320 \times 240$[pix] by using the bilinear interpolation algorithm. Beforehand, all the images are captured and resized. These images are also used in the learning phase in addition to the human detection and action recognition. The laptop computer, equipped with CPU Intel Core i7-7700K 4.20[GHz] and GPU GeForce GTX 1080, is used. The parameters used for generating the local search windows are given as $h_{min} = 80$[pix] and $h_{max} = 180$[pix]. For one ROI, eight points are sampled. Thus $N = 8$.

### 5.2   Classifier for Humans

During the learning phase of the classifier for humans, in total 60,000 images are used. In the half of the images, a target person is shown. 13 students joined this experiment. The input image size is $50 \times 80$[pix]. From the image data set, 54,000 images are randomly selected and used as the training data. The remaining 6,000 images are used as the test data.

---

[1] Note that the right image was only used in this experiment. In future works, we will use the 3D information obtained from both the images.

The learning progress is evaluated on the basis of the loss for the training data and classification accuracy for the test data. The phase is terminated when the train loss increases two times in a row. Figure 7 shows the transitions of the train loss and classification accuracy for the optical flow and flow image.

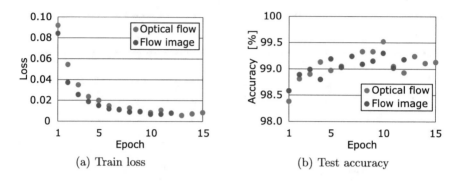

(a) Train loss                                        (b) Test accuracy

**Fig. 7.** Learning phase of CNN classifier for human

In Fig. 7(a), the train loss approached 0. This result indicates that the CNN classifier increasingly extracted the shape feature of humans from the optical flow and flow image. As a result, in Fig. 7(b), the classification accuracy for the test data increased to more than 99[%] at the end of the learning phase.

## 5.3    Classifier for Actions

In this learning phase of the classifier for actions, every image in the data set shows a target person taking any one of the five actions. The same 13 students joined this experiment. Original images with each of them are first inputted into the classifier in Sect. 5.2. If the result of human detection is true; then, partial images are clipped by the detected windows. The clipped images are resized 50 × 80[pix] and stored in the data set. For each action, 10,000 images are provided. In total, 50,000 images are used. From the data set, 45,000 images are randomly selected and used as the training data. The remaining 5,000 images are used as the test data. Figure 8 shows the transitions of the train loss and classification accuracy.

In Fig. 8(a), the train loss approached 0. This result indicates that the CNN classifier increasingly extracted the movement feature of humans from the optical flow and flow image. From the results shown in Figs. 7(a) and 8(a), it became obvious that a set of the optical flow in the image has spatiotemporal features. As a result, in Fig. 8(b), the classification accuracy for the test data increased to more than 92[%] at the end of the learning phase.

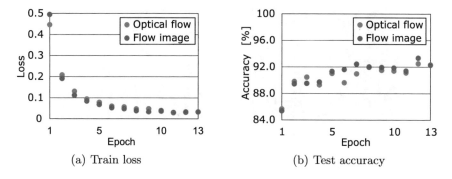

(a) Train loss     (b) Test accuracy

**Fig. 8.** Learning phase of CNN classifier for actions

## 5.4  Effectiveness of Local Search Window

In the process of generating local search windows, the size and the number of ROIs heavily depend on the threshold value given for binary image processing (cf. Fig. 4). In some cases, the moving body is divided into several parts as shown in Fig. 9.

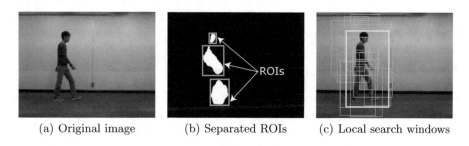

(a) Original image     (b) Separated ROIs     (c) Local search windows

**Fig. 9.** Robustness against image processing

In the original image as shown in Fig. 9(a), the target person is walking from left to right. Since the moving body was divided into the three parts as a result of the image processing, the ROIs are also separated as shown in Fig. 9(b). In this case, it is impossible for the ROIs to surround the whole body. Consequently, a part of the body, such as the head, torso, or legs, is shown in each input image clipped by the ROIs. For these images, it might be difficult for the CNN classifiers to recognize him. On the other hand, as shown in Fig. 9(c), $3 \times N(=8) = 24$ local search windows are generated; moreover, one of the windows colored by yellow is surrounding the whole body successfully. This result indicates the robustness of the local search windows against the image processing.

We further investigate the detector performance. Figure 10 shows a result of the human detection with the use of the sliding windows and local search

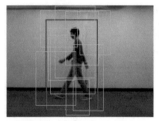
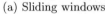

(a) Sliding windows                    (b) Local search windows

**Fig. 10.** Comparison of detectors: sliding window vs. local search window

windows. These windows are indicated by the green rectangles. The detected window is indicated by the red rectangle.

In Fig. 10(a), the sliding window detector exhaustively searches the image. On the other hand, in Fig. 10(b), local search windows are generated around the target object. As a result, these two detected windows are of similar size and located at the same position in the images.

The average detection rates for the five actions were 96.4[%] (sliding windows) and 97.5[%] (local search windows[2]). The average frame rates were 9.2[fps] (sliding windows[3]) and 50.6[fps] (local search windows). This result indicates that the local search window has the same detector performance as the sliding window with lower computational cost.

## 5.5 Real-Time Human Detection and Action Recognition

Four students joined this experiment. The CNN classifiers after the learning phases in the Sects. 5.2 and 5.3 are used for their images taking the five actions. These images were not used in the data set. In Fig. 11, example images of the detection and recognition results are shown. The detected window is indicated by the red rectangle and the recognized action is displayed.

Figure 11(a), (b), (c) are the horizontal actions. The walk and jump actions are represented as "c_walk" and "c_jump." Fig. 11(d) and (e) are the vertical actions. In these images, the detected window surrounding the whole body is generated, and the recognized actions are all correct. Furthermore, both of the results are simultaneously displayed. In Table 1, the average detection rates for each target person based on the optical flow and flow image are compared.

The CNN classifiers based on the optical flow and flow image resulted in more than 97[%] on average. In a related paper [19], the detection rate was about 96.5[%]. In comparison to this result, our classifiers provided sufficient performance. For the input images based on the optical flow and flow image, the CNN classifiers resulted in almost the same detection rates. In order to derive an advantage from the flow image, additional 3D information is necessary.

---

[2] The optical flow was used as an input to the CNN classifier.

[3] The mean shift clustering [18] was used for integrating the windows.

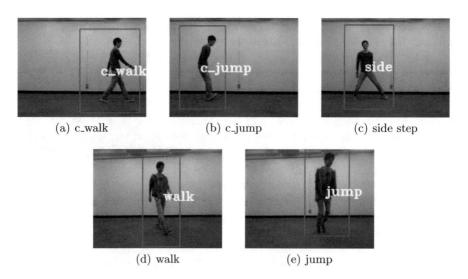

(a) c_walk                    (b) c_jump                    (c) side step

(d) walk                    (e) jump

**Fig. 11.** Examples of human detection and action recognition

**Table 1.** Comparison of human detection rate [%] depending on input images

| Input images | Person | | | | Av. |
|---|---|---|---|---|---|
| | 1 | 2 | 3 | 4 | |
| Optical flow | 98.0 | 96.1 | 97.8 | 97.9 | 97.5 |
| Flow image | 97.7 | 95.5 | 98.5 | 97.4 | 97.8 |

In Table 2, the average recognition rates for each action based on the optical flow and flow image are compared.

**Table 2.** Comparison of action recognition rate [%] depending on input images

| Input images | Action | | | | | Av. |
|---|---|---|---|---|---|---|
| | c_walk | c_jump | side step | walk | jump | |
| Optical flow | 97.3 | 93.9 | 69.0 | 94.4 | 85.1 | 87.9 |
| Flow image | 95.6 | 92.5 | 78.0 | 86.1 | 88.4 | 88.1 |

The CNN classifiers based on the optical flow and flow image resulted in more than 87[%] on average. In a related paper [20], the recognition rate was about 85.9[%]. In comparison to this result, our classifiers provided sufficient performance.

The average frame rates for four target persons taking the five actions were 47.0[fps] (optical flow) and 46.6[fps] (flow image). As mentioned in Sect. 1, these results were more than 30[fps]; therefore, the robot vision system successfully achieved the real-time human detection and action recognition.

As long as the whole bodies are in the image, the robot vision system is able to detect two or more persons and recognize their actions simulta-neously. Figure 12 shows an example for two target persons. For reference, the displayed actions in the image, "walk," indicate the walking action in the horizontal direction, i.e., "c_walk."

**Fig. 12.** Real-time human detection and action recogni-tion for two persons

In this experiment, original images were all captured and resized beforehand. As the robot vision, however, online processes through the cam-era should be performed in the system. In this regard, the processing rate for the image capture and resize was more than 30[fps] even if the laptop computer with GPU was used. Therefore, the robot vision system has the potential to perform a sequence of processes from the camera to the classification of the human and actions in real time.

## 6    Conclusions and Future Works

In this paper, a robot vision system for real-time human detection and action recognition was proposed. The system was composed of the following key techniques:

- input images described by the optical flow;
- CNN classifiers for simultaneous human detection and action recognition; and
- local search windows for clipping a target object in the image.

In the experiment, we showed that a set of the optical flow in the image has spatiotemporal features. Finally, the robot vision system was able to detect the moving human (97[%]) and recognize the action (87[%]) in real time (47.0[fps] and 46.6[fps]).

In future works, we will apply this system to actual mobile robots. The optical flow is caused by the relative motion between moving humans and robots. For autonomous navigation, therefore, we will work to correctly use the optical flow for human detection and action recognition in the robot vision system.

## References

1. Ojala, T., et al.: Performance evaluation of texture measures with classification based on Kullback discrimination of distributions. In: International Conference on Pattern Recognition, vol. 1, pp. 582–585 (1994)
2. Lowe, D.G.: Object recognition from local scale-invariant features. In: International Conference on Computer Vision, pp. 1150–1157 (1999)

3. Csurka, G., et al.: Visual categorization with bags of keypoints. In: International Workshop on Statistical Learning in Computer Vision, pp. 59–74 (2004)
4. Dalal, N., Triggs, B.: Histograms of oriented gradients for human detection. In: IEEE Conference on Computer Vision and Pattern Recognition, vol. 1, pp. 886–893 (2005)
5. Dollar, P., et al.: Behavior recognition via sparse spatio-temporal features. In: International Workshop on Visual Surveillance and Performance Evaluation of Tracking and Surveillance, pp. 65–72 (2005)
6. van de Sande, K.E.A., et al.: Segmentation as selective search for object recognition. In: IEEE International Conference on Computer Vision, pp. 1879–1886 (2011)
7. Uijlings, J.R.R., et al.: Selective search for object recognition. In: International Journal of Computer Vision, vol. 104, pp. 154–171 (2013)
8. LeCun, Y., et al.: Gradient-based learning applied to document recognition. Proc. IEEE $86(11)$, 2278–2324 (1998)
9. Ren, S., et al.: Faster R-CNN: towards real-time object detection with region proposal networks. IEEE Trans. Pattern Anal. Mach. Intell. $36(6)$, 1137–1149 (2016)
10. Farnebäck, G.: Two-frame motion estimation based on polynomial expansion. In: Scandinavian Conference on Image Analysis, vol. 2749, pp. 363–370 (2003)
11. Fathi, A., Mori, G.: Action recognition by learning mid-level motion features. In: IEEE Conference on Computer Vision and Pattern Recognition, vol. 2749, pp. 1–8 (2008)
12. Jain, M., et al.: Better exploiting motion for better action recognition. In: IEEE Conference on Computer Vision and Pattern Recognition, pp. 2555–2562 (2013)
13. Kingma, D., Ba, J.: Adam: a method for stochastic optimization. In: International Conference for Learning Representations (2015)
14. Srivastava, N., et al.: Dropout: a simple way to prevent neural networks from overfitting. J. Mach. Learn. Res. $15(1)$, 1929–1958 (2014)
15. LeCun, Y., et al.: Deep learning. Nature $521(7553)$, 436–444 (2015)
16. Otsu, N.: A threshold selection method from gray-level histograms. IEEE Trans. Syst. Man Cybern. $9(1)$, 62–66 (1979)
17. Goudail, F., et al.: Bhattacharyya distance as a contrast parameter for statistical processing of noisy optical images. J. Opt. Soc. Am. A $21(7)$, 1231–1240 (2004)
18. Comaniciu, D., et al.: Mean shift: a robust approach toward feature space analysis. IEEE Trans. Pattern Anal. Mach. Intell. $24(5)$, 603–619 (2002)
19. Oliveira, L., et al.: On exploration of classifier ensemble synergism in pedestrian detection. IEEE Trans. Intell. Transp. Syst. $11(1)$, 16–27 (2010)
20. Wang, H., Schmid, C.: LEAR-INRIA submission for the THUMOS workshop. In: ICCV Workshop on Action Recognition with a Large Number of Classes, vol. 2, no. 7 (2013)

# Movement Based Classification of People with Stroke Through Automated Analysis of Three-Dimensional Motion Data

John W. Kelly[1], Steve Leigh[2], Carol Giuliani[2], Rachael Brady[3], Martin J. McKeown[4], and Edward Grant[1(✉)]

[1] Center for Robotics and Intelligent Machines, Department of Electrical and Computer Engineering, North Carolina State University, Raleigh, NC, USA
egrant@ncsu.edu
[2] Center for Human Movement Science, Division of Physical Therapy, The University of North Carolina at Chapel Hill, Chapel Hill, NC, USA
[3] Visualization Technology Group, Pratt School of Engineering, Duke University, Durham, NC, USA
[4] Pacific Parkinson's Research Centre, Department of Medicine (Neurology), The University of British Columbia, Vancouver, BC, Canada

**Abstract.** Active recovery of motor function following a stroke requires intense and repetitive physical therapy. The effectiveness of such therapy can be greatly improved if it is fully customized for each patient. Motion tracking and machine learning algorithms can assist therapists in designing the therapy regimens, thereby saving valuable time. In this study, three-dimensional upper body movements both of people who had suffered a stroke and of healthy subjects were recorded as they performed a reaching task. A support vector machine with a five-dimensional feature space was used to automatically distinguish between the movements of people with stroke and those of healthy subjects. The success rate for this task peaked at over 95%. While this specific task is trivial for a clinician, it provides proof of concept, and a foundation for further work in developing classifiers that can locate more specific problems. Such a classifier may help clinicians treat the root cause of a complicated movement deficiency with a patient specific rehabilitation program. The results indicate that using machine learning approaches for analyzing stroke patient movement data shows potential for reducing clinician's workloads while providing improved treatment to specific subjects.

**Keywords:** Rehabilitation · Stroke · Physical therapy · Upper extremity
Pattern recognition

## 1 Introduction

Stroke is a leading cause of death and serious, long-term disability in the United States. Approximately 800,000 people suffer a stroke each year. On average, someone in the United States has a stroke every 40 s. The number of stroke related deaths has been falling since 1996, though. In 2006 stroke accounted for one out of every 18 deaths, but

© Springer Nature Switzerland AG 2019
M. Strand et al. (Eds.): IAS 2018, AISC 867, pp. 520–533, 2019.
https://doi.org/10.1007/978-3-030-01370-7_41

in that same year over 83% of people who suffered a stroke survived [1]. Many of these survivors, however, are left with severe physical impairments. Not only do these ailments sharply decrease the quality of life of individuals, they also significantly increase the cost of patient care and place a tremendous burden on the health care system [2–4]. The total economic cost of a stroke includes acute care costs, long term and ambulatory care costs, nursing home care costs; as well as a cost to the wider economy (through lost economic output, for example). It has been estimated that the lifetime cost associated with stroke is over $80 billion per year in the United States [5, 6]. Rehabilitation programs, therefore, are essential for improving patients' quality of life and reducing the economic load of stroke. Physical therapy is normally the best method to rehabilitate the motor function of a person with stroke. Only 20–25% of people who have had a stroke recover arm function, however, compared with approximately 75% of people who regain walking ability [7]. Recovery of arm function requires intense and repetitive task practice developed and overseen by therapists, which is labor intensive and costly [3, 4]. There is a need, therefore, to develop cost-effective and efficient methods of physical rehabilitation, particularly for the upper extremity.

To be maximally effective, stroke rehabilitation programs must address the unique needs of individual patients. Due to the large number of people who have suffered a stroke and the variance in their impairments, many clinics are left with three options: (1) use a generalized physical therapy regimen that changes little between patients, (2) devote a large amount of resources and clinicians' time to develop custom regimens for each patient, or (3) return the patient to their own home. The first option will not provide the most effective rehabilitation, because it is not tailored to each individual's needs. Ineffective therapy programs increases the burden on the healthcare system, due to the increased amount of time for which each patient will need therapy. The second option is considerably better for the patient, but creates an enormous cost up front to develop the custom therapy program. Option three, given many health plans, is adopted after only a few weeks of in-hospital care and treatment provided by insurance. Oftentimes it is a family member that bears the financial, physical, and mental burden. It is important then to develop an efficient, cost-effective way to create physical therapy regimens that specifically target the needs of each stroke patient. To contribute to that end goal, this study explores classification of motor impairments from automated analysis of motion capture data. Automated classification removes the cost from the therapist and places it on the machine. To be effective the proposed method for evaluating patients and subsequently designing individualized treatment programs must consume less time and resources than current practice. It must be noted that such a system is not intended to replace a clinician, but will be a valuable tool in assisting and augmenting the diagnosis. By decreasing evaluation time clinicians will have more time available to design patient specific therapy sessions and to spend working on rehabilitation instead of assessment. A sophisticated classification system may be able to accurately locate general underlying causes of complicated movement deficiencies, similar to a principal component analysis in statistics. This will allow the treatment sessions to be patient specific and straightforward, which may reduce recovery time. In addition to assisting in patient-specific diagnoses and therapy regimens, an effective classification system could help in the development of rehabilitative devices such as exoskeletons [8], functional electrical stimulation systems [9], neural prosthetics [10], and virtual fixtures [11] by providing a model for healthy movement.

The purpose of this study was to design a movement based automatic classification system, to test its accuracy in distinguishing between the motion patterns of people with stroke and of healthy subjects. In [12], a similar classification was performed using surface electromyographic recordings. The development of this simple classifier will help determine those parameters of movement that are most affected by stroke and the normal healthy range of those parameters, and could also provide the foundation for more specific classifiers. These classifiers will save valuable time when making diagnoses and determining underlying impairments of motion, which could assist therapists in developing customized rehabilitation regimens that takes into account the individual needs and limitations of each patient. It is important to note that such classifiers would not directly determine the best rehabilitation, but rather would provide valuable and readily available information when the therapist is developing the rehabilitation strategy.

## 2 Methods

### 2.1 Subjects

Eleven people who had suffered a stroke and twelve healthy subjects were recruited from the Chapel Hill and Durham areas in North Carolina. All potential subjects were examined before completing any test tasks to ensure they would be able to complete the study safely and to gather information about the specific nature of their stroke and physical limitations. All research was approved by the Duke University institutional review boards, and informed consent was obtained from all subjects prior to initiation of this study.

### 2.2 Tasks

Subjects sat in a chair facing a screen 2 m in front of them that filled their entire field of view. A virtual environment was projected onto the screen from a projector above them. Head tracking was used to provide a sense of depth perception. Stimuli were presented to subjects in the virtual environment that would engage them and keep their attention throughout the testing. Subjects were able to interact with the virtual environment by controlling a paddle that replicated the movement of their hand. Virtual flying balls roughly the size of a baseball were launched towards the subjects so that their trajectory would pass just laterally to the shoulder of the arm being tested at the time. Two types of balls were defined: distracter balls, which were entirely white; and target balls, which were marked with a large blue letter "T". Subjects were asked to move the paddle to block/hit target balls, but to ignore distracter balls. Collision detection was used to provide feedback to the subject. If the paddle contacted a ball it would drop to the floor. The virtual environment was custom written in C++ by engineers at Duke University; the program was named Syzergy.

During testing, seven unique conditions were presented to the subjects (Table 1). For conditions 1–5, the balls were released one after the other. In conditions 6 and 7, sets of five balls were released at once and traveled towards the subject together. In condition 1, the release speed, time between releases, and release height were varied until the subject was comfortable with the stimuli. In condition 2, a random mix of target balls and distracter balls were released and the release speed was varied until the subject was able to identify and block 20 target balls and they were deemed familiar with the task. Condition 3 used the same parameters as condition 2, but the subjects were asked to swat balls instead of blocking them. In condition 4, 20 target balls were released one after the other and the target balls were immediately revealed as targets. This served as a control condition. In condition 5, 50 total balls were released, but only 20 of the 50 were target balls. The target balls were randomly dispersed, and were only revealed to be target balls when they were one second from the subject. In condition 6, the sets of balls were released with varying release speeds, times between releases, and release heights until the subject was comfortable with the new stimuli. A target ball was revealed as one of the set of five when the set was one second from the subject. In condition 7, the parameters determined in condition 6 were used for 20 releases of sets of five balls. Subjects completed all conditions with both arms in a random order.

**Table 1.** Description of experiment conditions

| Condition number | Target balls | Total balls | Ball release and target identification methods |
| --- | --- | --- | --- |
| 1 | Varies | Varies | Individual release, varying ball speeds and heights |
| 2 | Varies | Varies | Individual release, varying ball speed |
| 3 | Varies | Varies | Individual release, varying ball speed, subjects swatted balls |
| 4 | 20 | 20 | Individual release, immediate identification |
| 5 | 20 | 50 | Individual release, targets randomly chosen, identified immediately |
| 6 | Varies | Varies | 5 balls released at once, 1 of the 5 a target, identified after a delay, varying ball speed |
| 7 | 20 | 100 | 5 balls released at once, 1 of the 5 a target, identified after a delay |

## 2.3   Data Collection

A global reference frame was established for the virtual environment and the subjects' location in such a way that the X axis was pointing to the right of the subject, the Y axis was pointing vertically upwards, and the Z axis was pointing out of the screen towards the subject. The origin of the global reference frame was set to be directly underneath the center of the subject's chair. During all trials, six degree of freedom electromagnetic position and orientation FASTRAK (Polhemus) sensors were placed on the subject's head, trunk, upper arm, forearm, and hand. The output of the sensors was controlled with Trackd (Inition), which streamed sensor data to Syzergy so that head, trunk, and hand movement could be fed back to the virtual environment. Syzergy recorded positions and orientations of the subject's trunk, head, and hand, as well as coordinates of the virtual balls, information about when the ball was released, when it was revealed to be a target ball (if it was a target ball), and when, if ever, it was contacted by the subject's virtual paddle. Motion Monitor (Innovative Sports Training, Inc.) software was used to record the positions and orientations of the trunk, upper arm, and forearm during conditions 4, 5, and 7 only.

## 2.4    Data Reduction

Trunk segment angles relative to the global reference frame, and shoulder and elbow joint angles were calculated using the Motion Monitor (Innovative Sports Training, Inc). The shoulder joint angle was calculated as the angle between the longitudinal axis of the trunk and the longitudinal axis of the upper arm, and does not represent a specific anatomical angle. The elbow joint angle was calculated as the inclination angle between the forearm and upper arm. Trunk and hand segment angles relative to the global reference frame were also calculated using MATLAB from the Syzergy data. The angle data calculated using Motion Monitor was synchronized with the angle data calculated using MATLAB by matching the trunk angles and adjusting the zero time of the data files. The synchronization was made using custom written MATLAB code.

The most important data relates to the subject's hand when it was actually moving towards a target ball in an attempt to make contact. The first step in reducing the data was then to determine the windows when this hand motion was occurring. This was a difficult process, especially people with stroke. The movement could be erratic, and in many cases there was some false movement that occurred before the true movement towards the target ball. There was also a chance that the hand was still in motion from reaching for the previous target ball. To determine the movement time as best as possible, the following steps were taken:

1. The end of movement was defined as the time at which contact with the ball was made, or if no contact was made then it was defined as the time at which the ball disappears from view.
2. A point X was defined as either 0.3 s prior to the time of contact, or if no contact was made then it was defined as 0.3 s prior to the time of peak hand velocity during the second half of the ball's flight. This determination was based on the assumptions that true hand movement occurred during the second half of the ball's flight, and peak velocity should occur close to the time when subjects attempted to make contact with the ball. The start of movement must occur before this point.
3. The start of movement was defined as the first point in time before point X when the hand velocity in the longitudinal direction (towards the ball) was greater than 0 and the hand displacement, i.e., a straight line with respect to the base frame of reference, in the longitudinal direction was greater than 5% above the minimum displacement that occurred during the ball's flight. This eliminated many false movements while retaining true movements that were erratic.

This method for determining movement time was verified graphically by plotting the hand movement and marking the calculated movement start and end points (Fig. 1).

It was determined through visual inspection that movement time was correctly deter-mined for 46 out of 50 randomly selected trials. In each of the 4 incorrect instances, it appeared that the hand never really moved at all.

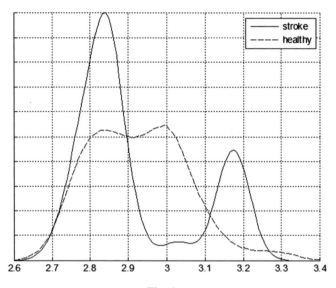

**Fig. 1.**

Once movement time was determined, it was possible to extract specific variables describing the movement. The initial variables of interest were suggested by an expe-rienced clinician and researcher, with additional variables added after a thorough examination of the data. The means and standard deviations of these variables over all conditions for both stroke subjects and healthy subjects are displayed in (Tables 2 and 3), and the definitions are as follows. Hit percentage is the percentage of target balls con-tacted and response time is the amount of time between the target ball's appearance and the start of movement. All remaining variables are measured only over the movement time, as defined earlier. Peak velocity is maximum velocity reached by the hand, and time to peak velocity is the time it takes to reach that speed. Range is the difference between the minimum and maximum, while excursion is the total change that occurred (i.e. OLE-object where P is a joint angle or the displacement of a sensor from the base frame of reference). Yaw is rotation with respect to (wrt) the global Y axis (longitudinal), pitch is rotation wrt the global X axis (frontal), and roll is rotation wrt the global Z axis (sagittal). Hand velocity smoothness was defined as the number of samples during movement time where longitudinal acceleration is not positive. The rationale behind this measure is once the hand begins moving towards the ball then it should continue accelerating towards the ball until contact. Any deceleration or hesitation is penalized, so lower numbers corre-spond to smoother velocities.

**Table 2.** Means of movement variables for healthy arms (mean ± standard deviation).

| | All | 4 | 5 | 7 |
|---|---|---|---|---|
| Hit percentage | 78.8 ± 20.2 | 89.3 ± 12.1 | 90.9 ± 15.7 | 77.9 ± 19.1 |
| Movement time (s) | 1.198 ± 0.406 | 1.028 ± 0.373 | 0.937 ± 0.377 | 1.320 ± 0.392 |
| Peak velocity (m/s) | 1.1 ± 0.4 | 1.1 ± 0.4 | 1.2 ± 0.5 | 1.2 ± 0.4 |
| Time to peak velocity (s) | 0.293 ± 0.133 | 0.298 ± 0.110 | 0.278 ± 0.109 | 0.294 ± 0.133 |
| Velocity smoothness | 36.4 ± 11.1 | 32.1 ± 11.1 | 30.4 ± 11.3 | 39.4 ± 9.9 |
| Response time (s) | 1.013 ± 0.569 | 1.042 ± 0.540 | 1.138 ± 0.491 | 0.478 ± 0.388 |
| Min shoulder angle (°) | 58 ± 27 | 60 ± 26 | 60 ± 30 | 55 ± 26 |
| Max shoulder angle (°) | 85 ± 28 | 86 ± 29 | 83 ± 29 | 87 ± 25 |
| Shoulder range (°) | 27 ± 13 | 26 ± 12 | 24 ± 15 | 31 ± 12 |
| Shoulder excursion (°) | 36 ± 17 | 33 ± 15 | 30 ± 17 | 44 ± 17 |
| Min elbow angle (°) | 44 ± 12 | 44 ± 12 | 46 ± 12 | 43 ± 11 |
| Max elbow angle (°) | 85 ± 16 | 85 ± 15 | 81 ± 16 | 89 ± 16 |
| Elbow range (°) | 46 ± 23 | 46 ± 19 | 40 ± 19 | 52 ± 28 |
| Elbow excursion (°) | 57 ± 47 | 54 ± 35 | 47 ± 31 | 70 ± 64 |
| Trunk excursion (m) | 0.022 ± 0.070 | 0.018 ± 0.064 | 0.017 ± 0.064 | 0.022 ± 0.075 |
| Total hand yaw (°) | 9 ± 6 | 8 ± 5 | 7 ± 4 | 11 ± 7 |
| Total hand roll (°) | 12 ± 10 | 11 ± 10 | 11 ± 10 | 14 ± 9 |
| Total hand pitch (°) | 24 ± 13 | 21 ± 11 | 19 ± 12 | 27 ± 11 |
| Total trunk yaw (°) | 3 ± 4 | 3 ± 3 | 3 ± 3 | 4 ± 3 |
| Total trunk roll (°) | 5 ± 6 | 4 ± 5 | 3 ± 3 | 5 ± 7 |
| Total trunk pitch (°) | 22 ± 9 | 18 ± 7 | 17 ± 8 | 24 ± 8 |
| Total head yaw (°) | 2 ± 1 | 2 ± 1 | 2 ± 1 | 2 ± 2 |
| Total head roll (°) | 3 ± 2 | 2 ± 2 | 2 ± 2 | 3 ± 2 |
| Total head pitch (°) | 21 ± 7 | 18 ± 6 | 17 ± 6 | 23 ± 8 |

## 2.5   Feature Extraction

The following rules were employed to extract optimal features for classification of healthy movements versus those affected by stroke: (1) glean as much information as possible from simple observations of the data, (2) combine this information with clinical knowledge, (3) extract any feature that has potential, and then examine the separation between the probability density function of the feature for healthy subjects and that of stroke subjects, and (4) conduct a quick performance test with a k nearest-neighbor classifier [13]. In this way well over a hundred possible features were examined.

None of the features differentiated well enough between subjects to create a reliable one-dimensional classifier, but when combined correctly they were effective. Some features were formed from linear combinations of some of the basic, related variables, such as hand movement in all 3 directions. This provided the benefit of giving the classifier more information without increasing the dimensionality of the feature space and incurring the curse of dimensionality [14]. Algorithms exist to do this

**Table 3.** Means of movement variables for stroke-afflicted arms (mean ± standard deviation).

| | All | 4 | 5 | 7 |
|---|---|---|---|---|
| Hit percentage | 61.9 ± 21.6 | 70.6 ± 22.8 | 77.3 ± 19.1 | 55.8 ± 21.3 |
| Movement time (s) | 1.602 ± 0.464 | 1.709 ± 0.690 | 1.386 ± 0.424 | 1.696 ± 0.484 |
| Peak velocity (m/s) | 0.9 ± 0.4 | 0.8 ± 0.3 | 0.9 ± 0.3 | 0.9 ± 0.3 |
| Time to peak velocity (s) | 0.298 ± 0.143 | 0.298 ± 0.135 | 0.291 ± 0.163 | 0.358 ± 0.157 |
| Velocity smoothness | 46.3 ± 13.1 | 49.4 ± 18.4 | 40.5 ± 12.8 | 47.7 ± 11.6 |
| Response time (s) | 1.049 ± 0.740 | 1.188 ± 0.392 | 1.480 ± 0.640 | 0.153 ± 0.453 |
| Min shoulder angle (°) | 66 ± 30 | 66 ± 32 | 70 ± 32 | 63 ± 30 |
| Max shoulder angle (°) | 86 ± 31 | 87 ± 33 | 89 ± 32 | 82 ± 33 |
| Shoulder range (°) | 20 ± 10 | 21 ± 11 | 19 ± 11 | 19 ± 7 |
| Shoulder excursion (°) | 26 ± 12 | 27 ± 14 | 23 ± 12 | 26 ± 9 |
| Min elbow angle (°) | 38 ± 11 | 36 ± 11 | 40 ± 12 | 38 ± 10 |
| Max elbow angle (°) | 81 ± 22 | 79 ± 22 | 80 ± 27 | 84 ± 18 |
| Elbow range (°) | 44 ± 16 | 45 ± 19 | 41 ± 16 | 47 ± 15 |
| Elbow excursion (°) | 53 ± 25 | 57 ± 35 | 45 ± 14 | 57 ± 22 |
| Trunk excursion (m) | 0.013 ± 0.006 | 0.014 ± 0.008 | 0.012 ± 0.005 | 0.013 ± 0.006 |
| Total hand yaw (°) | 16 ± 10 | 18 ± 16 | 13 ± 8 | 15 ± 8 |
| Total hand roll (°) | 29 ± 22 | 32 ± 25 | 25 ± 15 | 29 ± 27 |
| Total hand pitch (°) | 32 ± 19 | 34 ± 28 | 27 ± 19 | 33 ± 15 |
| Total trunk yaw (°) | 7 ± 5 | 7 ± 7 | 5 ± 5 | 7 ± 6 |
| Total trunk roll (°) | 4 ± 3 | 4 ± 3 | 4 ± 3 | 5 ± 4 |
| Total trunk pitch (°) | 23 ± 11 | 26 ± 15 | 20 ± 9 | 24 ± 10 |
| Total head yaw (°) | 2 ± 1 | 2 ± 2 | 2 ± 1 | 2 ± 1 |
| Total head roll (°) | 5 ± 4 | 5 ± 6 | 4 ± 4 | 4 ± 4 |
| Total head pitch (°) | 29 ± 9 | 30 ± 14 | 24 ± 7 | 30 ± 10 |

automatically, but the resulting features often have little clinical relevance. For example, principal components analysis (PCA) can reduce dimensionality in the direction of maximum variance in the data and linear discriminant analysis (LDA) can reduce dimensionality in the most discriminant direction [15], but for our purposes the resulting features should be interpretable and meaningful. Sparse logistic regression is a method with the potential to automatically select clinically relevant features [16]. This technique is rapidly gaining interest in the machine learning community and should be investigated in a future study.

Through this process we determined 5 features that were extremely effective at differentiating between the movements of people with stroke and healthy subjects. These features were, in order of effectiveness: lunge, linear combinations of the rotations of the hand, trunk, and head, and movement time. Lunge is a measure of how much the subject's body moves towards the ball, and is defined here as a linear combination of the maximum trunk displacement (wrt the base of the reference frame), the maximum head displacement, and the maximum trunk movement in the sagittal direction. Formulas for these features are shown below.

## 2.6   Data Classification

Data classification was performed using a 3rd order polynomial soft-margin support vector machine (SVM) [17, 18]. All features were normalized by subtracting out the mean and dividing by the standard deviation of the training samples. Normalization allows the different features to interact with each other on the same scale. The main difficulty in classification was in finding a method of separating the data set into a sample space that was large enough for classification, but still valid for the purposes of our study. The eventual goal of this classifier was to diagnose the condition of a subject's arm based on a reaching task, meaning each trial should be treated as a separate sample. Dividing the data in this way produced 250 samples across all testing conditions, but from only 43 distinct arms.

This sample space definition presents a problem when verifying the classifier with standard cross-validation techniques [19]. Ideally, if one trial from a particular arm is in the training set then all trials from that arm should be in the training set, since including trials from a particular arm in both the training and the validation data might artificially inflate results. To account for this, the sample space definition could use all trials for each arm averaged together to produce one sample per arm (43 resulting samples). This is normally too few samples to produce reliable results, though. To sidestep this problem we used an additional cross-validation technique in which we used the original 250 samples, but placed all samples from each arm in separate subsets. This created a 43-fold cross-validation test, but each fold was pre-determined rather than randomly selected. In doing so, the larger number of samples was maintained while preventing samples from the same arm to be in both the training and validation set. This also simulates the situation that the system would be intended for, in which it classifies the motions of an arm it has never seen before. To keep the discussion compact, this cross-validation method will from here on be referred to as arm-fold cross-validation.

Last, it would be interesting to see the results of the classifier on a few of the specific trial conditions given in Table 1. As mentioned before, conditions 4, 5, and 7 were the most consistent and important of the conditions tested. The problem with individually classifying the trials of these conditions is that there would only be just over 40 samples in each sample space. In order to simulate a more diverse and pop-ulated sample space, every 3 movements towards a target ball was considered 1 sample. So each trial was divided into about 6 samples, which produced between 250 and 255 samples depending on the condition.

## 3   Results

The classification results are shown below in Table 4, where the accuracy is the per-centage of samples where the condition of the arm (stroke/healthy) was correctly identified. For the first two sample space definitions, the results shown are from per-forming standard leave-one-out cross-validation (LOOCV). 10-fold cross-validation and random sub-sampling with 50 splits of half the samples produced nearly identical results. Using arm-fold cross-validation, 33 out of the 43 arms had 100% classification accuracy (Fig. 3). Note that data for subject e8 was not available and data was only available for the right arm of subject e6. Only 2 arms had the majority of their samples classified incorrectly. So if the classification results for each sample "voted" on the condition of the arm, over 95% accuracy is achieved. Results for the individual test conditions 4, 5, and 7 using LOOCV are also shown in Table 4.

**Table 4.** Overall classification results

| Conditions | Sample definition | Validation method | Accuracy |
|---|---|---|---|
| All | 1 trial | LOOCV | 96% |
| All | 1 arm | LOOCV | 83.7% |
| All | 1 trial | Arm-fold | 92.2% |
| All | 1 arm | Arm-fold, voting | 95.2% |
| 4 | 3 target balls | LOOCV | 89.9% |
| 5 | 3 target balls | LOOCV | 86.4% |
| 7 | 3 target balls | LOOCV | 81.7% |

## 4 Discussion

The purpose of this study was to design and test a movement based automatic classification system. The goal of making an effective classifier for distinguishing between the motion of healthy subjects and people with stroke was met, with a classification accuracy of over 95% (Table 4). This effectiveness would only increase with more unique test subjects and training. The 92.2% success rate of the arm-fold cross-validation show that the system is capable of effectively classifying a new arm that it has never seen before, which is the entire purpose of the classifier. The method of allowing multiple trials from the same arm to vote on the condition of the arm improved the classification results to 95.2%, most likely by smoothing out noise from bad or inconsistent trials. With enough data this technique could even be used as a crude measure of the severity of the subject's condition.

Additionally, the features used for the classifier isolate some of the details of how stroke affects different areas of motor control. If classifications were performed using each feature individually, then the results could help identify the corresponding variables of movement that are most deficient and need focused rehabilitation. For example, if the lunge of a person with stroke was classified as healthy while their hand rotation was not, this could assist a clinician in the development of more effective rehabilitation techniques that specifically address the needs of the individual patient. In turn, this will simultaneously increase the quality of life for many patients and reduce the cost of long-term patient care.

Of all the features used, lunge was not only the most effective, it was also one of the most interesting. The majority of potential features that were examined appeared to be Gaussian in nature, however, for lunge the probability distribution function for stroke subjects is bimodal (Fig. 2). Some stroke subjects have more lunge than normal, which seems to indicate an attempt to compensate for an impaired arm by lunging forward to reach the targets. Others seem to demonstrate an impairment that leaves their trunk more rigid than normal. This difference could be a result of the strokes occurring in different regions of the brain.

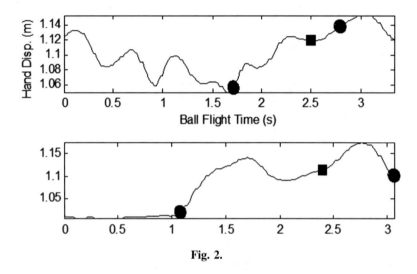

**Fig. 2.**

The 2nd, 3rd, and 4th features were, respectively, linear combinations of the integrals of hand, trunk, and head rotations. When reaching to swat a ball not much rotation should occur other than some hand pitch, but in stroke subjects more rotation occurred. This was even true in hand pitch, which is most likely due to the stroke subjects not being as quick and steady with their reaching motions. The only rotation for which healthy subjects exceeded stroke subjects was trunk roll, which indicates that some stroke subjects were unable to make the natural motion of leaning to the left or the right to reach the ball.

The last feature was movement time which was longer for stroke subjects, as expected. This was the least effective of the 5 features used, however, it should be noted that movement time indirectly plays a factor in calculating each of the other 4 features. This is because the other 4 features involve integrating over some variable of motion from the movement start to the movement end, so any increase in movement time will also naturally increase the other features slightly.

The results from the different individual testing conditions could also help to demonstrate what types of tests are most effective in diagnosing problem areas from a stroke. The accuracies can't be directly compared to the results from testing all conditions together since the sample space definition wasn't the same, but in general they aren't quite as good. This is probably mostly a result of there not being as many unique trials in the sample space. The results got worse as the testing condition got more complicated, though, decreasing from 89.9% to 86.4% to 81.7% from condition 4 to 5 to 7. This is most likely due to the adverse effect that mental complexity has on physical performance, especially in elderly subjects. A person who is perfectly healthy physically might still struggle with a complicated task. Nothing other than the person's physical condition should affect the classifier features. So the extra complexity in conditions 5 and 7 most likely just adds noise to the feature space.

There are several areas for improvement in both the data collection and the classifier. The first thing that should be done is to expand the data set. This research needs more subjects, and only data from truly stroke-afflicted subjects and truly healthy subjects should be used for classifier training. It severely harms the classifier when a subject who is deemed a person with stroke but has healthy motion (or vice versa) is used for training. This seemed to be the case for a few subjects in our dataset. As shown by Fig. 3, all but 10 arms actually had 100% classification accuracy for their trials, indicating that maybe the remaining 10 arms weren't good standards for their class. Obviously testing these tough cases would be the very purpose of the established classifier, but we will need further data and more information on our test subjects before we can validate the classifier in such a way.

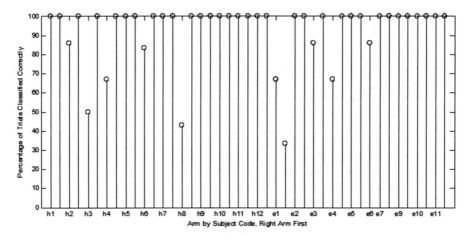

**Fig. 3.**

For further improvement, it would be good if a few extra parameters are recorded, such as a neutral resting position, and definitive start or stop times. With these data set improvements, the feature space and some of the techniques developed in this analysis could help create an effective tool for diagnosing physical problems caused by stroke.

# 5    Conclusion

The intention of most of the work completed here was not to be an end in itself, but to provide a foundation for future work. This study provides a set of features that are important in stroke diagnosis and recovery, and a classifier to distinguish an arm affected by stroke from a healthy one. This is a distinction that can easily be made by a clinician, but we hope that the concept can be expanded upon to help clinicians make more difficult diagnoses, and to provide physical therapy that is more accurately tuned to the needs of each patient. This could help improve recovery, decrease costs, and provide doctors with valuable time to treat more patients. It is important to note that this is not meant to create a fully automated system; it is meant to be a tool similar to AI-based systems that help doctors diagnosis other medical conditions.

**Acknowledgements and Funding.** Funding for this work was provided by the Park Foundation, the Tau Beta Pi Association, and the Whitaker Foundation (MJM). Thanks to David Zielinski for technical expertise with the virtual reality systems.

# References

1. Roger, V.L., et al.: Heart disease and stroke statistics—2011 update: a report from the american heart association. Circulation (2010)
2. Economic Impact of Acute Ischemic Stroke. Medical News Today, 26 February 2006
3. Krebs, H.I., Volpe, B.T., Aisen, M.L., Hogan, N.: Increasing productivity and quality of care: robot-aided neuro-rehabilitation. J. Rehabil. Res. Dev. **37**(6), 639–652 (2000)
4. Krebs, H.I., Hogan, N., Aisen, M.L., Volpe, B.T.: Robot-aided neurorehabilitation. IEEE Trans. Rehabil. Eng. **6**(1), 75–87 (1998)
5. Wolfe, C.: The burden of stroke. In: Wolfe, C., Rudd, T., Beech, R. (eds.) Stroke Services and Research. The Stroke Association, London (1996)
6. Taylor, T.N., Davis, P.H., Torner, J.C., Holmes, J., Meyer, J.W., Jacobson, M.F.: Lifetime cost of stroke in the United States. Stroke **27**(9), 1459–1466 (1996)
7. Richards, L., Pohl, P.: Theraputic interventions to improve upper extremity recovery and function. Clin. Geriatr. Med. **5**, 819–832 (1999)
8. Merritt, C.: A pneumatically actuated brace designed for upper extremity stroke rehabilitation. M.S. thesis, Department of Electrical and Computer Engineering, North Carolina State University, Raleigh, NC, USA (2003)
9. Davoodi, R., Brown, I.E., Loeb, G.E.: Advanced modeling environment for developing & testing FES control systems. Med. Eng. Phys. **25**, 3–9 (2003)
10. Davoodi, R., Urata, C., Hauschild, M., Khachani, M., Loeb, G.: Model-based development of neural prostheses for movement. IEEE Trans. Bio. Eng. **54**(11), 1909–1918 (2007)
11. Abbott, J.J., Hager, G.D., Okamura, A.M.: Steady-hand teleoperation with virtual fixtures. In: Proceedings of 12th IEEE International Workshop Robot and Human Interactive Communication (2003)
12. Li, J., Wang, Z.J., Eng, J.J., McKeown, M.J.: Bayesian network modeling for discovering "dependent synergies" among muscles in reaching movements. IEEE Trans. Biomed. Eng. **55**(1), 298–310 (2008)
13. Mitchell, T.: Machine Learning. McGraw-Hill, New York (1997)
14. Duda, R., Hart, P., Stork, D.: Pattern Classification, 2nd edn. Wiley, New York (2001)

15. Bishop, C.M.: Pattern Recognition and Machine Learning. Springer, New York (2006)
16. Tibshirani, R.: Regression shrinkage and selection via the lasso. J. R. Stat. Soc. Ser. B (Methodol.) **58**(1), 267–288 (1996)
17. Cristianini, N., Shawe-Taylor, J.: An Introduction to Support Vector Machines and Other Kernel-based Learning Methods, 1st edn. Cambridge University Press, Cambridge (2000)
18. Cortes, C., Vapnik, V.: Support-vector networks. Mach. Learn. **20**(3), 273–297 (1995)
19. Kohavi, R.: A study of cross-validation and bootstrap for accuracy estimation and model selection. In: Proceedings of the Fourteenth International Joint Conference on Artificial Intelligence, San Francisco, pp. 1137–1143 (1995)

# Real-Time Marker-Less Multi-person 3D Pose Estimation in RGB-Depth Camera Networks

Marco Carraro[1(✉)], Matteo Munaro[1], Jeff Burke[2], and Emanuele Menegatti[1]

[1] Department of Information Engineering, University of Padova,
Via Gradenigo 6/A, 35131 Padova, Italy
{marco.carraro,matteo.munaro,emg}@dei.unipd.it
[2] REMAP, School of Theater, Film and Television,
UCLA, Los Angeles, California 90095, USA
jburke@remap.ucla.edu

**Abstract.** This paper proposes a novel system to estimate and track the 3D poses of multiple persons in calibrated RGB-Depth camera networks. The multi-view 3D pose of each person is computed by a central node which receives the single-view outcomes from each camera of the network. Each single-view outcome is computed by using a CNN for 2D pose estimation and extending the resulting skeletons to 3D by means of the sensor depth. The proposed system is marker-less, multi-person, independent of background and does not make any assumption on people appearance and initial pose. The system provides real-time outcomes, thus being perfectly suited for applications requiring user interaction. Experimental results show the effectiveness of this work with respect to a baseline multi-view approach in different scenarios. To foster research and applications based on this work, we released the source code in OpenPTrack, an open source project for RGB-D people tracking.

## 1 Introduction

The human body pose is rich of information. Many algorithms and applications, such as Action Recognition [1–3], People Re-identification [4], Human-Computer-Interaction (HCI) [5] and Industrial Robotics [6–8] rely on this type of data. The recent availability of smart cameras [9–11] and affordable RGB-Depth sensors as the first and second generation Microsoft Kinect, allow to estimate and track body poses in a cost-efficient way. However, using a single sensor is often not reliable enough because of occlusions and Field-of-View (FOV) limitations. For this reason, a common solution is to take advantage of camera networks. Nowadays, the most reliable way to perform human Body Pose Estimation (BPE) is to use marker-based motion capture systems. These systems show great results in terms of accuracy (less than 1mm), but they are very expensive and require the users to wear many markers, thus imposing heavy limitations to their diffusion. Moreover, these systems usually require offline computations in complicated

© Springer Nature Switzerland AG 2019
M. Strand et al. (Eds.): IAS 2018, AISC 867, pp. 534–545, 2019.
https://doi.org/10.1007/978-3-030-01370-7_42

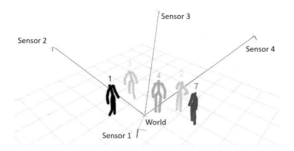

**Fig. 1.** The output provided by the system we are proposing. In this example, five persons are seen from a network composed of four Microsoft Kinect v2.

scenarios with many markers and people, while the system we propose provides immediate results. A real-time response is usually needed in security applications, where person actions should be detected in time, or in industrial applications, where human motion is predicted to prevent collisions with robots in shared workspaces. Aimed by those reasons, the research on marker-less motion capture systems has been particularly active in recent years.

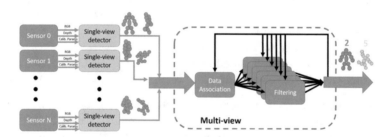

**Fig. 2.** The system overview. The camera network is composed of several RGB-D sensors (from 1 to N). Each single-view detector takes the RGB and Depth images as input and computes the 3D skeletons of the people in the scene as the output using the calibration parameters K. The information is then sent to the multi-view central node which is in charge of computing the final pose estimation for each person in the scene. First, a data association is performed to determine which pose detection is belonging to which pose track, then a filtering step is performed to update the pose track given the detection.

In this work, we propose a novel system to estimate the 3D human body pose in real-time. To the best of our knowledge, this is the first open-source and real-time solution to the multi-view, multi-person 3D body pose estimation problem. Figure 1 depicts our system output. The system relies on the feed of multiple RGB-D sensors (from 1 to N) placed in the scene and on an extrinsic calibration of the network. in this work, this calibration is performed with the `calibration_toolkit` [12][1]. The multi-view poses are obtained by fusing the single view outcomes of each detector, that runs a state-of-the-art 2D body pose estimator [13,14] and extend it to 3D by means of the sensor depth. The contribution of the paper is two-fold: (i) we propose a novel system to fuse and update 3D body poses of multiple persons in the scene and (ii) we enriched a state-of-the-art single-view 2D pose estimation algorithm to provide 3D poses. As a further contribution, the code of the project has been released as open-source as part of the OpenPTrack [15,16] repository. The proposed system is:

- *multi-view:* The fused poses are computed taking into account the different poses of the single-view detectors;
- *asynchronous:* The fusion algorithm does not require the different sensors to be synchronous or have the same frame rate. This allows the user to choose the detector computing node accordingly to his needs and possibilities;
- *multi-person:* The system does not make any assumption on the number of persons in the scene. The overhead due to the different number of persons is negligible;
- *scalable:* No assumptions are made on the number or positions of the cameras. The only request is an offline one-time extrinsic calibration of the network;
- *real-time:* The final pose framerate is linear to the number of cameras in the network. In our experiments, a single-camera network can provide from 5 fps to 15 fps depending on the Graphical Processing Unit (GPU) exploited by the detector. The final framerate of a camera network composed of $k$ nodes is the sum of their single-view framerate;
- *low-cost:* The system relies on affordable low-cost RGB-D sensors controlled by consumer GPU-enabled computers. No specific hardware is required.

The remainder of the paper is organized as follows: in Sect. 2 we review the literature regarding human BPE from single and multiple views, while Sect. 3 describes our system and the approach used to solve the problem. In Sect. 4 experimental results are presented, and, finally in Sect. 5 we present our final conclusions.

## 2   Related Work

### 2.1   Single-View Body Pose Estimation

Since a long time, there have been a great interest about single-view human BPE, in particular for gaming purposes or avatar animation. Recently, the advent of affordable RGB-D sensors boosted the research in this and other Computer Vision fields. Shotton et al. [17] proposed the skeletal tracking system licensed by Microsoft used by the XBOX console with the first-generation Kinect. This approach used a random forest classifier to classify the different pixels as belonging to the different body parts. This work inspired an open-source approach that was released by Buys et al. [18]. This same work was then improved by adding the

---

[1] https://github.com/iaslab-unipd/calibration_toolkit

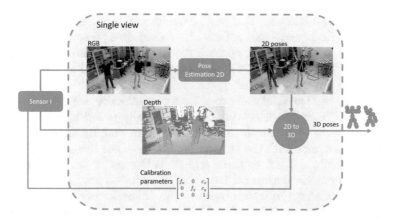

**Fig. 3.** The single-view pipeline followed for each sensor. At each new frame composed of a color image (RGB), a depth image and the calibration parameters, the 3D pose of each person in the scene is computed from the 2D one. Then, the results are sent to the central computer which will compute the multi-view result.

OpenPTrack people detector module as a preprocessing step [19]. Still, the performance of the detector remained very poor for non frontal persons. In these last years, many challenging Computer Vision problems have been finally resolved by using *Convolutional Neural Networks* (CNNs) solutions. Also single-view BPE has seen a great benefit from these techniques [14, 20–22]. The impressive pose estimation quality provided by those solution is usually paid in terms of computational time. Nevertheless, this limitation is going to be leveraged with newer network layouts and Graphical Processing Units (GPU) architectures, as proved by some recent works [14, 22]. In particular, the work of Cao et al. [14] was one of the first to implement a CNN solution to solve people BPE in real-time using a bottom-up approach. The authors were able to compute 2D poses for all the people in the scene with a single forward pass of their CNN. This work has been adopted here as part of our single-view detectors.

## 2.2   Multi-view Body Pose Estimation

Multiple views can be exploited to be more robust against occlusions, self-occlusions and FOV limitations. In [23] a Convolutional Neural Network (CNN) approach is proposed to estimate the body poses of people by using a low number of cameras also in outdoor scenarios. The solution combines a generative and discriminative approach, since they use a CNN to compute the poses which are driven by an underlying model. For this reason, the collaboration of the users is required for the initialization phase. In our previous work [19], we solved the single-person human BPE by fusing the data of the different sensors and by applying an improved version of [18] to a virtual depth image of the frontalized person. In this way, the skeletonization is only performed once, on the virtual depth map of the person in frontal pose. In [24], a 3D model is registered to the point clouds of two Kinects. The work provides very accurate results, but it is computationally expensive and not scalable to multiple persons. The authors of [25] proposed a pure geometric approach to infer the multi-view pose from a synchronous set of 2D single-view skeletons obtained using [26]. The third dimension is computed by imposing a set of algebraic constraints from the triangulation of the multiple views. The final skeleton is then computed by solving a least square error method. While the method is computationally promising (skeleton computed in 1s per set of synchronized images with an unoptimized version of the code), it does not scale with the number of persons in the scene. In [27] a system composed of common RGB cameras and RGB-D sensors are used together to record a dance motion performed by a user. The fusion method is obtained by selecting the best skeleton match between the different ones obtained by using a probabilistic approach with a particle filter. The system performs well enough for its goal, but it does not scale to multiple people and requires an expensive setup. In [28] the skeletons obtained from the single images are enriched with a 3D model computed with the visual hull technique. In [29] two orthogonal Kinects are used to improve the single-view outcome of both sensors. They used a constrained optimization framework with the bone lengths as hard constraints. While the work provides a real-time solution and there are no hard assumption on the Kinect positions, it was tested just with one person and two orthogonal Kinect sensors. Similarly to many recent works [25,27,28], we use a single-view state-of-the-art body pose estimator, but we augment this result with 3D data and we then combine the multiple views to improve the overall quality.

## 3 System Design

Figure 2 shows an overview of the proposed system. It can be split into two parts: (i) the single view, which is the same for each sensor and it is executed locally and (ii) the multi-view part which is executed just by the master computer. In the single-view part (see Fig. 3), each detector estimates the 2D body pose of each person in the scene using an open-source state-of-the-art single-view body pose estimator. In this work, we use the OpenPose[2] [13,14] library, but the overall system is totally independent of the single-view algorithm used. The last operation made by the detector is to compute the 3D positions of each joint returned by OpenPose. This fusion is done by exploiting the depth information coming from the RGB-D sensor used. The 3D skeleton is then sent to the master computer for the fusion phase. This is done by means of multiple Unscented Kalman Filters used on the detection feeds, as explained in Sect. 3.3.

### 3.1 Camera Network Setup

The camera network can be composed of several RGB-D sensors. In order to know the relative position of each camera, we calibrate the system using a solution similar to our previous works [15,16]. From this passage we fix a common

---

[2] https://github.com/CMU-Perceptual-Computing-Lab/openpose.

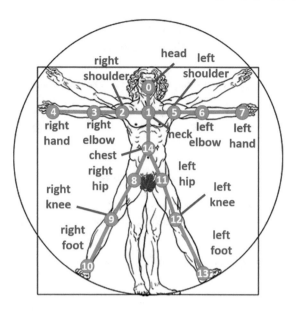

**Fig. 4.** The human model used in this work.

*world* reference frame $\mathcal{W}$ and we obtain a transformation $T_C^{\mathcal{W}}$, for each camera $C$ in the network, which transforms points in the camera coordinate system to the *world* reference system.

## 3.2 Single-View Estimation of 3D Poses

Each node in the network is composed of an RGB-D sensor and a computer to elaborate the images. Let $^R\mathfrak{F} = \{^R C, ^R D\}$ be a frame captured by the detector $R$ and composed of the color image $C$ and the depth image $D$ all in the $R$ reference frame. The color and depth images in $\mathfrak{F}$ are considered as synchronized. We then apply *OpenPose* to $^R C$ obtaining the raw two dimensional skeletons $\overline{\mathfrak{S}} = \{\overline{S_0}, \overline{S_1}, ..., \overline{S_k}\}$. Each $S = \{j_i \,|\, 0 \le i \le m\} \in \overline{\mathfrak{S}}$ is a set of 2D joints which follows the human model depicted in Fig. 4. The goal of the single-view detector is to transform $\overline{\mathfrak{S}}$ in the set of skeletons $\widehat{\mathfrak{S}} = \{\widehat{S_0}, \widehat{S_1}, ..., \widehat{S_k}\}$ where each $\widehat{S} \in \mathfrak{S}$ is a three dimensional skeleton. Given the RGB image $I$, let's consider a point $p = (x_p, y_p) \in I$ and its corresponding depth $d = proj(x_p, y_p)$. Considering $(f_x, f_y)$ and $(c_x, c_y)$ respectively the focal point and the optical center of the sensor, the relationship to compute the 3D point $P_R = (X_R, Y_R, Z_R)$ in the camera reference system R is explained in Eq. 1.

$$p = \begin{bmatrix} x_p \\ y_p \\ d \end{bmatrix} = \begin{bmatrix} f_x & 0 & c_x \\ 0 & f_y & c_y \\ 0 & 0 & 1 \end{bmatrix} \begin{bmatrix} X_R \\ Y_R \\ Z_R \end{bmatrix} = K P_R \tag{1}$$

**Algorithm 1.** The algorithm performed by the master computer to decide the association between the different skeletons in a detection and the current tracks.

**INPUT:**

- $^{W}\widehat{\mathfrak{S}}_i = \{S_0, S_1, ..., S_{k-1}\}$ - a new detection set from sensor $i$ in the world reference frame
- $\mathfrak{T} = \{T_0, T_1, ..., T_{l-1}\}$ - the current set of tracked persons pose.
- $\epsilon$ - maximum distance for a detection to be considered for the association

**OUTPUT:**

- $\mathcal{M} = \{(S_i, T_j) \in^{W} \widehat{\mathfrak{S}}_i \times \mathfrak{T}\}$ - the association between the pose tracked and the new observations
- $\mathcal{N} \subseteq^{W} \widehat{\mathfrak{S}}_i$ - the detections without an association. They will initialize a new track.
- $\mathfrak{T}_o \subseteq \mathfrak{T}$ - the tracks without an associated observations. They will be considered for removal

```
1: procedure DATA_ASSOCIATION($^{W}\widehat{\mathfrak{S}}_i$, $\mathfrak{T}$, $\epsilon$)
2:      $\mathfrak{T}_o \leftarrow \emptyset$
3:      $C \leftarrow \mathbf{0}_{k \times l}$
4:      for each $T_i \in \mathfrak{T}$ do
5:          for each $S_j \in^{W} \widehat{\mathfrak{S}}_i$ do
6:              $x_t(j) \leftarrow centroid(S_j)$
7:              $z_t(i, j) \leftarrow$ *v that $T_i$ would have if $S_j$ were associated to it*
8:              $\widehat{z}_{t|t-1}(i) \leftarrow$ *prediction step of $\mathcal{K}_{im}$*
9:              $\Sigma_t(i) \leftarrow \Sigma_t(\mathcal{K}_{im})$
10:             $\tilde{z}_t(i, j) = z_k(i, j) - \widehat{z}_{t|t-1}(i)$
11:             $C_{ij} \leftarrow \tilde{z}_t^T(i, j) \cdot \Sigma_t(i)^{-1} \cdot \tilde{z}_t(i, j)$
12:     $X \leftarrow solve\_Munkres(C)$
13:     for $i \in [0, l - 1]$ do
14:         for $j \in [i + 1, k - 1]$ do
15:             if $X_{ij} == 1$ and $C_{ij} < \epsilon$ then
16:                 $\mathcal{M} \leftarrow \mathcal{M} \cup \{(S_j, T_i)\}$
17:                 * update $\mathcal{K}_{im}$ with $S_j$ *
18:     $\mathcal{N} \leftarrow \{S_i \mid \nexists T_j, (S_i, T_j) \in \mathcal{M}\}$
19:     $\mathfrak{T}_o \leftarrow \{T_i \mid \nexists S_j, (S_j, T_i) \in \mathcal{M}\}$
20:     return $\mathcal{M}$, $\mathcal{N}$, $\mathfrak{T}_o$
```

Since the depth data is potentially noisy or missing, we compute the depth $d$ associated to the point $p = (x_p, y_p)$ by applying a median to the set $\mathfrak{D}(p)$, as shown in Eqs. 2 and 3.

$$\mathfrak{D}(p = (x_p, y_p)) = \{(x, y) \mid ||(x, y) - (x_p, y_p)|| < \epsilon\} \tag{2}$$

$$d = \phi(p) = \text{median}\{proj(x, y) \mid (x, y) \in \mathfrak{D}(p)\} \tag{3}$$

Given $\overline{\mathfrak{S}}$, we then proceed to the calculation of $\widehat{\mathfrak{S}}$ as shown in Eq. 4.

$$\forall 0 \leq j < k, \quad \overline{S_j} = \{\overline{p_i} = (x_i, y_i) \,|\, 0 \leq i < m\} \in \overline{\mathfrak{S}},$$

$$\widehat{S_j} = \left\{ \widehat{p_i} = \begin{bmatrix} |K^{-1}(\overline{p_i})|_x \\ |K^{-1}(\overline{p_i})|_y \\ \phi(\overline{p_i}) \end{bmatrix}, 0 \leq i < m \right\} \in \widehat{\mathfrak{S}} \qquad (4)$$

### 3.3    Multi-view Fusion of 3D Poses

The master computer is in charge of fusing the different information it is receiving from the single-view detectors in the network. One of the common limitations in motion capture systems is the necessity to have synchronized cameras. Moreover, off-the-shelves RGB-D sensors, such as the Microsoft Kinect v2, do not have the possibility to trigger the image acquisition. In order to overcome this limitation, our solution merges the different data streams asynchronously. This allows the system to work also with other RGB-D sensors or other low-cost embedded machine. At time $t$, the master computer maintains a set of tracks $\mathfrak{T} = \{T_0, T_1, ..., T_l\}$ where each pose tracked $T_i$ is composed of the set of states of $m$ different Kalman Filters, one per each joint, i.e: $T_i = \{\mathcal{S}(\mathcal{K}_{i0}), \mathcal{S}(\mathcal{K}_{i1}), ..., \mathcal{S}(\mathcal{K}_{im})\}$. The additional Kalman Filter $\mathcal{K}_{im}$ is maintained for the data association algorithm. At time $t+1$, it may arrive a detection $\widehat{\mathfrak{S}}_i = \{\widehat{S_0}, \widehat{S_1}, ..., \widehat{S_k}\}$ from the sensor $i$ of the network. The master computer first refers the detection to the common *world* coordinate system $\mathcal{W}$ (see Sect. 3.1):

$$^{\mathcal{W}}\widehat{\mathfrak{S}}_i = T_i^{\mathcal{W}} \cdot \widehat{\mathfrak{S}}_i = \{T_i^{\mathcal{W}} \cdot S_j \,|\, \forall S_j \in \widehat{\mathfrak{S}}_i\}$$

Then, it associates the different skeletons in $^{\mathcal{W}}\widehat{\mathfrak{S}}_i$ as new observations for the different tracks in $\mathfrak{T}$ if they belong to them or initializes new tracks if some of the skeletons do not belong to any $T_i \in \mathfrak{T}$. At this stage, the system also decides if a track is old and has to be removed from $\mathfrak{T}$. This step is important to prevent $\mathfrak{T}$ to grow big causing time computing problems with systems which are running for hours. We refer to this phase as *data association*. Algorithm 1 shows how it is performed. The data association is done by considering the centroid of each skeleton $S$ contained in the detection $^{\mathcal{W}}\mathfrak{S}_i$. The centroid is calculated as the chest joint $j_{14} \in S$, if this is valid, otherwise it is replaced with a weighted mean of the neighbor joints. Lines [6–9] of Algorithm 1 refers to the calculation of a cost associated to the case if the detection pose $S_j$ would be associated to the track $T_i$. To calculate this, we consider the Mahalanobis distance between the likelihood vector at time $t$ $\tilde{z}_t(i, j)$ and $\Sigma_t(\mathcal{K}_{i,x_t})$: the covariance matrix of the Kalman filter associated to the centroid of $T_i$. At this point, computing the optimal association between tracks and detections is the same as solving the Hungarian algorithm associated to the cost matrix $C$; Line 11 refers to the use of the Munkres algorithm which efficiently computes the optimal matrix $X$ with a 1 on the associated couples. Nevertheless, this algorithm does not consider a maximum distance between tracks and detections. Thus, it may happen that a couple is wrongly associated in the optimal assignment. For this reason, when inserting the couples in $\mathcal{M}$, we check also if the cost of the couple in the initial cost matrix $C$ is below a threshold. Once solved the data association problem, we can assign the tracks ID to the different skeletons. Indeed, we know which are the detection at the current time $t$ belonging to the tracks in the system and, additionally, we know also which tracks need to be created (i.e. new detections with no associated track) and the tracks to consider for the removal. Let $n$ be the number of people in the scene, we used a set of Unscented Kalman Filters $\mathfrak{K} = \{\mathcal{K}_{ij}, 0 \leq i < n, 0 \leq j \leq m\}$ where the generic $\mathcal{K}_{ij} \in \mathfrak{K}$ is in charge of computing the new position of the joint $j$ of the person $i$ at time $t$, given the new detection received from one of the detectors at time $t$ and the prediction of the filter $\mathcal{K}_{ij}$ computed from the previous position at time $t-1$ of the same joint $j$. The state of each Kalman Filter $\mathcal{K}_{ij}$ is dimensioned with the three dimensional position of the joint $j$. We used as motion model a constant velocity model, since it is good to predict joint movements in the small temporal space between two good detections of that joint.

## 4    Experiments

The algorithm described in this paper does not require any synchronization between the cameras in the networks. This fact makes particularly difficult to find a fair comparison between our proposed system and other state-of-the-art works. Thus, in order to provide useful indication on how our system performs, we recorded and manually annotated a set of RGB-D frames while a person was freely moving in the field-of-view of a 4-sensors camera network. We compare our algorithm with a baseline method called MAF (Moving Average Filter), in which the outcome of the generic joint $i$ at time $t$ is computed as an average of the last $k$ frames. In order to be as fair as possible, we fixed $k \geq 30$ to provide comparable results in terms of smoothness. We also demonstrated the effectiveness of the multi-view fusion by comparing our results with the poses obtained by considering just one and two cameras of the same network. In this comparison, we report the average reprojection error with respect to one of the cameras, $C_0$. Equation 5 shows how this error is calculated with $^W P$ as the generic joint expressed in the world reference system and $p^*$ as the corresponding ground truth:

$$e_{\text{repr}} = |p^* - K \cdot \mathcal{T}_W^{C_0} \cdot {}^W P| \tag{5}$$

Table 1 shows the results we achieved. As depicted, the proposed method outperforms the baseline in all the cases: single-view, 2-camera network and 4-camera network. In the first two cases (single and 2-camera network) the improvement is from 50% to 60%, while, when multiple views are available, it is from 18% to 32%. It is also interesting to note that the most noisy joints are the ones relative to the legs as confirmed by other state-of-the-art works [14, 20, 21].

**Table 1.** The results of the experiments. Each number represents the mean and the standard deviation of the re-projection error on the reference camera expressed in Eq. 5.

|  |  | r-shoulder | r-elbow | r-wrist | r-hip | r-knee | r-ankle |
|---|---|---|---|---|---|---|---|
| Single-camera network | $MAF_{30}$ | >100 | >100 | >100 | >100 | >100 | >100 |
|  | $MAF_{40}$ | >100 | >100 | >100 | >100 | >100 | >100 |
|  | **Ours** | **54.9 ± 58.6** | **42.4 ± 47.4** | **42.4 ± 40.0** | **51.7 ± 43.7** | **54.5 ± 31.0** | **63.3 ± 34.2** |
| 2-camera network | $MAF_{30}$ | 62.0 ± 33.0 | 62.9 ± 32.0 | 63.1 ± 34.5 | 76.4 ± 30.6 | 75.9 ± 27.4 | 88.3 ± 35.6 |
|  | $MAF_{40}$ | 83.7 ± 41.8 | 84.0 ± 40.9 | 83.1 ± 43.7 | 99.2 ± 40.4 | 96.3 ± 38.0 | >100 |
|  | **Ours** | **20.7 ± 17.2** | **21.0 ± 17.5** | **24.3 ± 17.5** | **22.4 ± 16.7** | **42.8 ± 17.2** | **59.7 ± 28.6** |
| 4-camera network | $MAF_{30}$ | 28.7 ± 16.4 | 31.0 ± 16.9 | 32.2 ± 22.5 | 40.2 ± 15.0 | 48.7 ± 12.8 | 58.6 ± 21.2 |
|  | $MAF_{40}$ | 38.4 ± 21.2 | 40.8 ± 21.7 | 41.6 ± 26.3 | 50.7 ± 19.4 | 56.2 ± 16.7 | 66.0 ± 24.5 |
|  | **Ours** | **22.7 ± 18.9** | **21.3 ± 18.5** | **26.3 ± 19.9** | **23.9 ± 18.0** | **46.5 ± 19.7** | **55.9 ± 25.1** |
|  |  | l-shoulder | l-elbow | l-wrist | l-hip | l-knee | l-ankle |
| single-camera network | $MAF_{30}$ | >100 | >100 | >100 | >100 | >100 | >100 |
|  | $MAF_{40}$ | >100 | >100 | >100 | >100 | >100 | >100 |
|  | **Ours** | **77.7 ± 74.4** | **79.1 ± 82.7** | **70.0 ± 61.8** | **97.8 ± 30.3** | **57.5 ± 38.9** | **69.2 ± 37.6** |
| 2-camera network | $MAF_{30}$ | 83.3 ± 33.4 | 85.8 ± 37.8 | 94.8 ± 45.4 | >100 | 85.4 ± 35.5 | 93.3 ± 37.0 |
|  | $MAF_{40}$ | >100 | >100 | >100 | >100 | >100 | >100 |
|  | **Ours** | **32.1 ± 23.0** | **33.4 ± 26.3** | **39.8 ± 35.1** | **98.3 ± 21.2** | **39.9 ± 18.3** | **58.6 ± 27.1** |
| 4-camera network | $MAF_{30}$ | 41.5 ± 17.9 | 39.9 ± 19.6 | 44.7 ± 29.5 | 94.1 ± 26.1 | 52.1 ± 17.8 | 57.8 ± 27.9 |
|  | $MAF_{40}$ | 53.0 ± 23.2 | 52.7 ± 24.6 | 57.6 ± 33.1 | 96.6 ± 30.8 | 61.2 ± 23.1 | 67.5 ± 31.6 |
|  | **Ours** | **22.5 ± 22.1** | **26.7 ± 25.9** | **31.8 ± 29.7** | **95.4 ± 22.0** | **45.1 ± 20.5** | **49.1 ± 25.2** |

### 4.1 Implementation Details

The system has been implemented and tested with Ubuntu 14.04 and Ubuntu 16.04 operating system using the Robot Operating System (ROS) [30] middleware. The code is entirely written in C++ using the Eigen, OpenCV and PCL libraries.

## 5 Conclusions and Future Works

In this paper we presented a framework to compute the 3D body pose of each person in a RGB-D camera network using only its extrinsic calibration as a prior. The system does not make any assumption on the number of cameras, on the number of persons in the scene, on their initial poses or clothes and does not require the cameras to be synchronous. In our experimental setup we demonstrated the validity of our system over both single-view and multi-view approaches. In order to provide the best service to the Computer Vision community and to provide also a future baseline method to other researchers, we released the source code under the BSD license as part of the OpenPTrack library[3]. As future works, we plan to add a human dynamic model to guide the prediction of the Kalman Filters to further improve the performance achievable by our system (in particular for the lower joints) and to further validate the proposed system on a new RGB-Depth dataset annotated with the ground truth of

---

[3] https://github.com/openptrack/open_ptrack_v2.

the single links of the persons' body pose. The ground truth will be provided by a marker based commercial motion capture system.

**Acknowledgement.** This work was partially supported by U.S. National Science Foundation award IIS-1629302.

# References

1. Han, F., Yang, X., Reardon, C., Zhang, Y., Zhang, H.: Simultaneous feature and body-part learning for real-time robot awareness of human behaviors, pp. 2621–2628 (2017)
2. Zanfir, M., Leordeanu, M., Sminchisescu, C.: The moving pose: an efficient 3D kinematics descriptor for low-latency action recognition and detection. In: Proceedings of the IEEE International Conference on Computer Vision, pp. 2752–2759 (2013)
3. Wang, C., Wang, Y., Yuille, A.L. : An approach to pose-based action recognition. In: Proceedings of the IEEE Conference on Computer Vision and Pattern Recognition, pp. 915–922 (2013)
4. Ghidoni, S., Munaro, M.: A multi-viewpoint feature-based re-identification system driven by skeleton keypoints. Robot. Autonom. Syst. **90**, 45–54 (2017)
5. Jaimes, A., Sebe, N.: Multimodal human-computer interaction: a survey. Comput. Vis. Image Underst. **108**(1), 116–134 (2007)
6. Morato, C., Kaipa, K.N., Zhao, B., Gupta, S.K.: Toward safe human robot collaboration by using multiple kinects based real-time human tracking. J. Comput. Inf. Sci. Eng. **14**(1), 011006 (2014)
7. Michieletto, S., Stival, F., Castelli, F., Khosravi, M., Landini, A., Ellero, S., Landš, R., Boscolo, N., Tonello, S., Varaticeanu, B., Nicolescu, C., Pagello, E.: Flexicoil: flexible robotized coils winding for electric machines manufacturing industry. In: ICRA Workshop on Industry of the Future: Collaborative, Connected, Cognitive (2017)
8. Stival, F., Michieletto, S., Pagello, E.: How to deploy a wire with a robotic platform: learning from human visual demonstrations. In: FAIM 2017 (2017)
9. Zivkovic, Z.: Wireless smart camera network for real-time human 3D pose reconstruction. Comput. Vis. Image Underst. **114**(11), 1215–1222 (2010)
10. Carraro, M., Munaro, M., Menegatti, E.: A powerful and cost-efficient human perception system for camera networks and mobile robotics. In: International Conference on Intelligent Autonomous Systems, pp. 485–497. Springer, Cham (2016)
11. Carraro, M., Munaro, M., Menegatti, E.: Cost-efficient rgb-d smart camera for people detection and tracking. J. Electr. Imaging **25**(4), 041007–041007 (2016)
12. Basso, F., Levorato, R., Menegatti, E.: Online calibration for networks of cameras and depth sensors. In: OMNIVIS: The 12th Workshop on Non-classical Cameras, Camera Networks and Omnidirectional Vision-2014 IEEE International Conference on Robotics and Automation (ICRA 2014) (2014)
13. Wei, S.-E., Ramakrishna, V., Kanade, T., Sheikh, Y.: Convolutional pose machines. In: CVPR (2016)
14. Cao, Z., Simon, T., Wei, S.-E., Sheikh, Y.: Realtime multi-person 2D pose estimation using part affinity fields. In: 2017 IEEE Conference on Computer Vision and Pattern Recognition (CVPR), pp. 1302–1310 (2017)

15. Munaro, M., Horn, A., Illum, R., Burke, J., Rusu, R.B.: OpenPTrack: people tracking for heterogeneous networks of color-depth cameras. In: IAS-13 Workshop Proceedings: 1st International Workshop on 3D Robot Perception with Point Cloud Library, pp. 235–247 (2014)

16. Munaro, M., Basso, F., Menegatti, E.: OpenPTrack: open source multi-camera calibration and people tracking for RGB-D camera networks. Robot. Autonom. Syst. **75**, 525–538 (2016)

17. Shotton, J., Sharp, T., Kipman, A., Fitzgibbon, A., Finocchio, M., Blake, A., Cook, M., Moore, R.: Real-time human pose recognition in parts from single depth images. Commun. ACM **56**(1), 116–124 (2013)

18. Buys, K., Cagniart, C., Baksheev, A., De Laet, T., De Schutter, J., Pantofaru, C.: An adaptable system for RGB-D based human body detection and pose estimation. J. Vis. Commun. Image Representation **25**(1), 39–52 (2014)

19. Carraro, M., Munaro, M., Roitberg, A., Menegatti, E.: Improved skeleton estimation by means of depth data fusion from multiple depth cameras. In: International Conference on Intelligent Autonomous Systems, pp. 1155–1167. Springer, Cham (2016)

20. Insafutdinov, E., Pishchulin, L., Andres, B., Andriluka, M., Schiele, B.: DeeperCut: a deeper, stronger, and faster multi-person pose estimation model. In: European Conference on Computer Vision, pp. 34–50. Springer (2016)

21. Pishchulin, L., Insafutdinov, E., Tang, S., Andres, B., Andriluka, M., Gehler, P.V., Schiele, B.: DeepCut: joint subset partition and labeling for multi person pose estimation. In: Proceedings of the IEEE Conference on Computer Vision and Pattern Recognition, pp. 4929–4937 (2016)

22. Carreira, J., Agrawal, P., Fragkiadaki, K., Malik, J.: Human pose estimation with iterative error feedback. In: The IEEE Conference on Computer Vision and Pattern Recognition (CVPR), June 2016

23. Elhayek, A., de Aguiar, E., Jain, A., Thompson, J., Pishchulin, L., Andriluka, M., Bregler, C., Schiele, B., Theobalt, C.: Marconi-convnet-based marker-less motion capture in outdoor and indoor scenes. IEEE Trans. Patt. Anal. Mach. Intell. **39**, 501–514 (2017)

24. Gao, Z., Yu, Y., Zhou, Y., Du, S.: Leveraging two kinect sensors for accurate full-body motion capture. Sensors **15**(9), 24297–24317 (2015)

25. Lora, M., Ghidoni, S., Munaro, M., Menegatti, E.: A geometric approach to multiple viewpoint human body pose estimation. In: 2015 European Conference on Mobile Robots (ECMR), pp. 1–6. IEEE (2015)

26. Yang, Y., Ramanan, D.: Articulated human detection with flexible mixtures of parts. IEEE Trans. Patt. Anal. Mach. Intell. **35**(12), 2878–2890 (2013)

27. Kim, Y.: Dance motion capture and composition using multiple RGB and depth sensors. Int. J. Distrib. Sens. Netw. **13**(2), 1550147717696083 (2017)

28. Kanaujia, A., Haering, N., Taylor, G., Bregler, C.: 3D human pose and shape estimation from multi-view imagery. In: 2011 IEEE Computer Society Conference on Computer Vision and Pattern Recognition Workshops (CVPRW), pp. 49–56. IEEE (2011)

29. Yeung, K.-Y., Kwok, T.-H., Wang, C.C.: Improved skeleton tracking by duplex kinects: a practical approach for real-time applications. J. Comput. Inf. Sci. Eng. **13**(4), 041007 (2013)

30. Quigley, M., Conley, K., Gerkey, B., Faust, J., Foote, T., Leibs, J., Wheeler, R., Ng, A.Y.: ROS: an open-source robot operating system. In: ICRA Workshop on Open Source Software, vol. 3, p. 5. Kobe (2009)

# People Finding Under Visibility Constraints Using Graph-Based Motion Prediction

AbdElMoniem Bayoumi$^{(\boxtimes)}$, Philipp Karkowski, and Maren Bennewitz

Humanoid Robots Lab, University of Bonn, Bonn, Germany
{abayoumi,philkark,maren}@cs.uni-bonn.de

**Abstract.** An autonomous service robot often first has to search for a user to carry out a desired task. This is a challenging problem, especially when this person moves around since the robot's field of view is constrained and the environment structure typically poses further visibility constraints that influence the perception of the user. In this paper, we propose a novel method that computes the likelihood of the user's observability at each possible location in the environment based on Monte Carlo simulations. As the robot needs time to reach the possible search locations, we take this time as well as the visibility constraints into account when computing effective search locations. In this way, the robot can choose the next search location that has the maximum expected observability of the user. Our experiments in various simulated environments demonstrate that our approach leads to a significantly shorter search time compared to a greedy approach with background information. Using our proposed technique the robot can find the user with a search time reduction of 20% compared to the informed greedy method.

**Keywords:** People tracking · Monte Carlo simulations
Particle filter-based prediction

## 1 Introduction

Finding a person is an essential functionality that is needed by several applications of mobile service robots. Typically, users do not stay at a fixed position but move along common paths between places where they remain for a while, e.g., to discuss work with a colleague, grab some material, or get a coffee. The robot needs a good strategy to find the user as fast as possible also in these situations to carry out its task.

All authors are with the Humanoid Robots Lab, University of Bonn, Germany. This work has been supported by the German Academic Exchange Service (DAAD) and the Egyptian Ministry for Higher Education as well as by the European Commission under contract number FP7-610532-SQUIRREL and by the DFG Research Unit FOR 2535 Anticipating Human Behavior.

© Springer Nature Switzerland AG 2019
M. Strand et al. (Eds.): IAS 2018, AISC 867, pp. 546–557, 2019.
https://doi.org/10.1007/978-3-030-01370-7_43

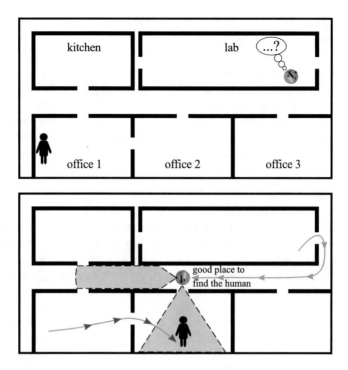

**Fig. 1.** Top: The robot needs to find a user whose current location is unknown. The user may walk toward any of a set of predefined destinations, known by the robot. Bottom: The robot needs to select a good search location that covers most of the expected paths of the user. Our approach selects a search location with maximum observability of the user at the time the robot reaches it.

One possible solution to the search problem is to apply techniques that try to maximally cover the visible area of the environment [1–3]. However, these approaches often lead to long search times and high navigation costs as they aim at covering the whole environment. Moreover, the maximum coverage techniques will not necessarily revisit already covered regions, which might be necessary since the user is assumed to move across the environment and the robot might miss him during the search.

In this paper, we make use of prior knowledge about frequently visited destinations of the user and their connecting paths. We developed an approach that determines good search locations using a particle filter based prediction model on a graph representation of paths the user typically takes. Our novel approach computes the likelihood of the observability of the user at each possible location based on Monte Carlo simulations. We hereby take into account the time needed by the robot to reach the search locations from its current position as well as the visibility constrains that arise from the robot's limited field of view and obstacles. Figure 1 highlights the strength of our approach. The location of the user is initially unknown. Our approach leads to the selection of an effective search location that provides the highest expected observability, i.e., the robot can observe the corridor and the entrances to multiple rooms and, thus, locate the user.

We show in extensive simulated experiments and in various environments that our technique generates search locations that significantly reduce the time to find the user compared to a greedy solution that is provided with background information about the possible destinations between which the user moves. In the experiments, we model noisy observations and dynamic obstacles to show the robustness of our approach.

## 2   Related Work

The problem of finding a moving person in an environment was early studied as a coverage problem based on the robot's visibility polygon [1–5]. Stiffler *et al.* [6] additionally considered the problem of unreliable sensors in this context. The authors developed a visibility-based geometric formulation to place the surveillance robot at specific environment locations that maximize the path of the intruder through the robot's visible region to increase the likelihood of observing the intruder. All these solutions to the coverage problem do not predict motion of the person and thus cannot provide any pose estimate. They lead to long search times and high navigation costs as they aim at covering the whole environment.

On the other hand, several approaches that predict motions and aim at minimizing the searching time for a mobile robot have been presented. Tipaldi and Arras [7] proposed to learn a spatial affordance map and apply a Poisson process to relate space, time, and occurrence probability of activity events. Afterwards, the spatio-temporal model can be used to generate an optimal path on a grid map of the environment for a mobile robot to encounter specific humans. This approach does not make use of any sensor modalities to update the belief about the location of a user but considers just the encounter probability of grid cells. Schwenk *et al.* [8] developed a search approach that uses a highly abstract topological representation of the environment and learns about the user's behaviors in order to estimate the likelihood of the user's current room. Here, it is assumed that the robot detects people when they are within a range of 1.8 m around the robot. Kulich *et al.* [9] introduced a model that learns the temporal likelihood of possible desired interactions to actively search for humans in order to interact with them in public space. Krajník *et al.* [10] presented a method based on spatio-temporal models to enable the robot finding non-stationary objects in an office environment. The authors represent the environment as an abstract topological map and combine it with periodic functions in order to compute the likelihood of existence of the objects at any node of the map with respect to the time. All these approaches, however, ignore the visibility constraints resulting from the environment layout.

Other approaches considered the frequency of human existence at specific locations. The idea here is to construct a probability distribution for every hour of the day. For example, in the work of Volkhardt and Gross [11], the robot searches for the human at pre-defined locations, where each location is assigned a probability relative to the frequency of observing the human there. Accordingly, the robot selects the location with the highest probability. Mehdi and Berns [12] presented a technique that generates a minimum set of view points that ensure maximum coverage of the environment with the robot's constrained field of view. The authors proposed to construct a probability distribution about the human's observability at these destinations during each hour of the day and take the navigation cost into account for deciding which of the view points to choose as search location. These methods do not model the human's motion and therefore cannot predict the expected position at a certain intermediate time step.

Goldhoorn *et al.* [13] proposed using particle filters to estimate the most likely location of the user at the current time step. The robot moves toward that location for few time steps then updates its estimate about the user's position and recomputes the robot's movement. As opposed to our method, this technique does not take into account the time needed by the robot to reach search locations from its current place. Moreover, moving the robot just for few time steps and then selecting another search location often leads to oscillating navigation behavior as the estimation jumps across the map as we realized in our experiments.

In contrast to all the mentioned search methods, our system models the human's motion and provides a probability distribution about his/her position at each time step. We consider the robot's limited field of view and visibility constraints when computing the likelihood of observing the user at a certain place and also take into account the time needed by the robot to reach the search locations.

## 3    Problem Formulation

The task of the robot is to find a non-stationary user as fast as possible. The environment is hereby known to the robot and it has prior knowledge about locations where the user frequently stays and his/her typical paths between these locations. We refer to these locations as *destinations*. After reaching such a destination, the user might stay there or move to another destination after a certain waiting time.

We represent the environment as a grid map with an overlaid topo-metric graph as shown in Fig. 2, where each cell in that grid is mapped onto its closest graph node [14]. The connections shown between neighboring nodes correspond to valid paths between these nodes. However, some of the paths are only passable by humans, e.g., due to the size of the robot or any other potential constraints of the searching environment.

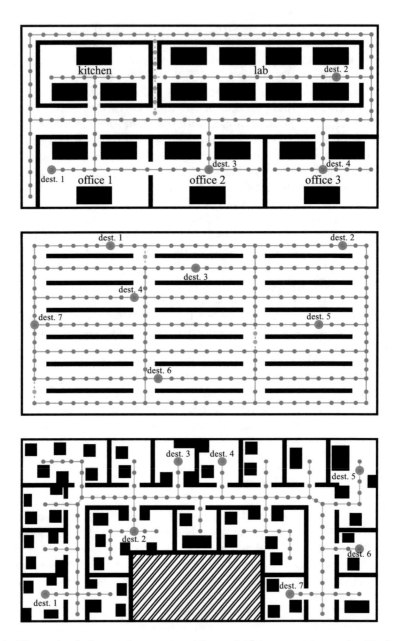

**Fig. 2.** Three simulation environments with overlaid topo-metric graphs. Each environment is represented as a grid map with an overlaid graph, where each grid cell is mapped to the closest graph node (green dots) in the same room. The orange dots represent paths that are only passable by the user but not by the robot. The bold green dots represent the predefined destinations between which the user moves.

The location of the user is initially unknown to the robot as well as his/her intended destinations when moving. After reaching a destination, the user might stay there or start moving to another destination after some time. We assume the moving velocity of the user to be within a certain range, however, the exact velocity of the user is unknown to the robot. Dynamic obstacles, e.g., other humans, can appear in the environment and temporarily constrain the robot's field of view. The task is considered as successful when the robot observes the user within its field of view.

## 4    Graph-Based People Tracking

To represent the belief about the location of the user and track its motion on the graph between the destinations, we apply a particle filter, inspired by the work of Liao *et al.* [15].

We use the information about the typical paths between the destinations and the times the user stays at the destinations to find the average time that the user occupies each node. We sample the pose of the particles according to this occupation likelihood. For each particle, we independently sample one of the destinations as the next "goal" based on the typical paths that lead through its node.

Each particle then moves to a graph node along the path to its destination according to a Gaussian motion model, taking into account the velocity range of the user. Whenever a particle reaches its destination (or is initialized at a destination), it remains there for a sampled time interval that corresponds to the typical waiting behavior of the user. Finally, we select another destination for the particle as its next goal according to the transition probability:

$$p(Dest_i = D_b | n_i = D_a) = p(D_b | D_a), \tag{1}$$

where $Dest_i$ is the chosen destination of particle $i$, $n_i$ is the graph node of particle $i$, $D_a$ and $D_b$ are two destinations, and $p(D_b | D_a)$ is the known probability of the user to move from $D_a$ to $D_b$.

The particles are weighted proportional to the observation likelihood. The weights are initially set to the same value and then updated at each time step as follows. For particles that fall within the robot's field of view while the user is not currently detected, the weights are reduced:

$$w_i = \begin{cases} \gamma w_i, & \text{if } (n_i \in \mathcal{FOV}) \wedge \textit{user not detected} \\ w_i & \text{otherwise} \end{cases}, \tag{2}$$

where $w_i$ is the weight of particle $i$, $n_i$ is again the current graph node of particle $i$, $\mathcal{FOV}$ is the area covered by the robot's visual sensors and $\gamma \in [0, 1)$ is a reduction factor. Since the likelihood of false negative observations increases with the distance of the user to the robot, $\gamma$ decreases with this distance.

As we assume a proper identification system, we do not model false positive observations. Note, however, that we can deal with false positive observations for a short time by requiring a minimum number of subsequent time steps where the human is detected before the search is assumed to be successful.

# 5   Selecting Search Locations via Monte Carlo Simulations

In this section, we describe our approach to selecting effective search locations for the robot to find the human. Relying only on the estimated most likely location of the user at each time step leads to an oscillating navigation behavior as the estimation might jump across the map. We, therefore, propose a method based on Monte Carlo simulations that takes into account the time needed by the robot to reach the possible search locations from its current place.

We first perform Monte Carlo simulations to compute the positions of the particles at future time steps according to the motion model. In particular, we simulate the particle propagation along the graph according to the motion model as many future time steps as needed by the robot to reach the furthest graph node relative to the robot's current node. We then compute the likelihood of the user's observability at each graph node, while considering the time needed to reach this node. For example, if a node lies ten time steps away, we consider the simulated particle distribution ten time steps into the future when computing likelihood of the user's observability at this node. The weights of the simulated particles stay unaffected during the Monte Carlo simulations.

We compute the observability likelihood $l_j$ of the user at each node $j$ as follows

$$l_j = \sum_{i \in \mathcal{O}_j^t} w_i, \quad \forall j \in \mathcal{R}^t, \quad 1 \le t \le T, \tag{3}$$

where $\mathcal{R}^t$ is the set of graph nodes that can be reached from the robot's current node $n_r$ within exactly $t$ future time steps, $T$ is the number of future time steps needed to reach the furthest graph node from $n_r$, and $\mathcal{O}_j^t$ is the group of simulated particles at future time step $t$ that can be observed from node $j$.

After computing $l_j$ for every $j$, we select the graph node with the highest observability likelihood $s$ as the next search location[1], i.e.,

$$s = \operatorname*{argmax}_{j} l_j. \tag{4}$$

The pseudo-code of our search goal selection algorithm is listed in Algorithm 1. As can be inferred from the example shown in Fig. 3, the robot selects the next search goal as the location that provides highest observability at the time the robot reaches it.

The robot then navigates to the selected node along the shortest path in the graph and does an observation action by performing a full rotation. If the user cannot be found anywhere on the way to the current search location nor while performing the observation action, a new search location is selected as previously and so forth. Performing the particle simulations in the described way and including them in the calculation of the observation likelihood provides an effective method for selecting a good search location that takes into account the dynamic behavior of the user.

---

[1] Note that the time is inherently considered in the computation of the $l_j$, such that $s$ does not need to have a time index.

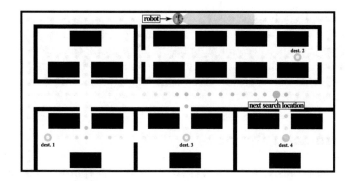

**Fig. 3.** This figure shows the selected search location according to Algorithm 1. The graph nodes are drawn with a color intensity corresponding to the observability likelihood of the user at the time the robot reaches this search location. The robot selects the node that provides the highest observability likelihood as next search location.

---

**Algorithm 1.** Selection of the next search location using Monte Carlo simulations

---

**Input**   : *particles* and *robotPose*
**Output**: next search location
*likelihood* ← {};
**for** $t \leftarrow 1$ **to** $T$ **do**
    *particles* ← simulate *particles* one step ahead acc. to the motion model;
    *reachableNodes* ← nodes that can be reached by the robot in exactly $t$ time steps;
    *nodesWeights* ← {};
    // calculate collective weight for each node at time step $t$;
    **for** $i \leftarrow 1$ **to** *particles.size* **do**
        *node* ← *particles[i].node*;
        *weight* ← *particles[i].weight*;
        *nodesWeights[node]* ←
          *nodesWeights[node]* + *weight*;
    **end**
    // calculate observability likelihood for each node;
    **for** $r \in$ *reachableNodes* **do**
        *visibleNodes* ← visible nodes from $r$ (incl. $r$) at time step $t$;
        **for** $v \in$ *visibleNodes* **do**
          *likelihood[r]* ←
            *likelihood[r]* + *nodesWeights[v]*;
        **end**
    **end**
**end**
**return** argmax$_{node}$ *likelihood[node]*;

---

While computing the next search location, we do not consider waiting actions or non-shortest paths, as this results in infinite possibilities to reach any node. Neither do we take into account the observability along the intermediate nodes to the considered search location. As we have found out in our experiments, this leads to a selection of search locations with longer paths and does not decrease the search time.

# 6   Experimental Results

We carried out extensive experiments to evaluate our approach and compare it to alternative methods.

## 6.1   Experimental Setup

We performed the experiments in three different, challenging simulation environments (see Fig. 2), each of size 41 m × 20.5 m with a grid map resolution of 0.25 m and a node distance of 1.5 m. In the first two environments, multiple paths exist between the destinations, among which the user chooses one based on a certain known probability distribution and the transition probabilities from one destination to the others are equally likely (see Eq. 1). Note, however, that some passages are impassible to the robot, i.e., the dotted line with orange nodes for the first two environments to make the search problem even more challenging.

In each experiment, the position of the user is initialized according to the occupation likelihood (see Sect. 4) and the user moves in the environment between the predefined destinations. The user does not necessarily move on the shortest path but might take detours. When the user reaches his/her destination, he/she waits there for a certain period of time. The user repeats this behavior until he/she reaches his/her fourth destination and remains there. The velocity of the user is sampled from a certain interval. At each time step, the position of the user is mapped onto the closest graph node given its grid map position. The initial location of the user is unknown to the robot and is outside its field of view.

We use 150 particles to represent the belief about the position of the user and track it on the graph representation of the environment. The particles are initialized and updated as described in Sect. 4.

The number of dynamic obstacles that constrain the robot's field of view ranges from three to five and their velocities are sampled from the same interval as the velocity of the user.

The search task is considered successful when the robot observes the user. The robot's field of view has a horizontal opening angle of 58°, which corresponds to that of an *ASUS Xtion Pro Live*, and a 10 m view distance. We set the probability of false negatives between 0.05 and 0.15 linearly increasing with the distance between the robot and the user. We do not consider false positive observations in the simulation experiments.

## 6.2   Evaluation and Results

We performed 5,000 experiments in each of the three environments. In order to evaluate the performance of our approach, we considered the search time and compared it to the time needed by two different approaches. The strategy of the first alternative approach is to visit all destinations in a greedy fashion using background information, i.e., the knowledge about the destinations of the user. The greedy approach does not consider any prediction about the user's location; it keeps selecting the closest unvisited destination as a search location until the user is found. After visiting all destinations it starts the search process all over again.

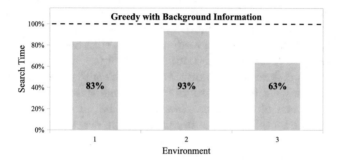

**Fig. 4.** Average relative search time achieved by our approach with respect to the greedy approach with background information. The times are normalized so that the greedy approach equals 100%.

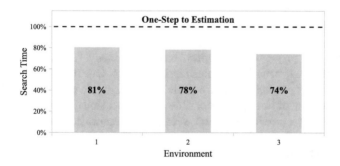

**Fig. 5.** Average relative search time achieved by our approach with respect to the one-step to estimation method. The times are normalized so that the one-step to estimation approach equals 100%.

We additionally compared our approach to a method that uses the particle filter representation to infer the currently most likely location of the user and moves the robot toward that location for one time step, then updates the estimation and so on. This method is similar to the approach of Goldhoorn *et al.* [13] and we refer to this method as the *one-step to estimation* method.

**Table 1.** Percentage of switching to the greedy approach.

|        | Our approach | One-step to estimation |
|--------|--------------|------------------------|
| Env. 1 | 1.64%        | 6.06%                  |
| Env. 2 | 0.91%        | 4%                     |
| Env. 3 | 5.96%        | 12.04%                 |

We evaluated the statistical significance of our comparative experiments with a *two-tailed paired t-test*. The experimental results show that our method performing Monte Carlo simulations significantly outperforms each of the other two approaches with a statistical significance of 99%. Figures 4 and 5 show the average relative search times achieved by our approach for each of the three environments with respect to the greedy approach with background information and the one-step to estimation method, respectively.

As our approach does not guarantee to cover the entire map and, thus, might not find a search location close to the user's final destination, we switch to the greedy approach after a given time limit. The maximum time limit was determined experimentally such as to minimize the overall search time. Table 1 shows the percentage of experimental runs using our approach and the "one-step to estimation" approach that exceeded this time limit and switched to the greedy method. As shown, our technique based on Monte Carlo simulations outperforms the one-step to estimation method for all the environments.

A video showing the advantages of our approach for an example run can be downloaded from https://www.hrl.uni-bonn.de/ias18bayoumi.mp4.

## 7  Conclusion

In this paper, we presented an approach that enables a mobile robot to quickly find a non-stationary user in complex environments. Our method selects the next search location by predicting future paths of the user. To compute the likelihood of the observability of the user at possible search locations, we apply Monte Carlo simulations using a particle filter on a graph representation of possible paths in the environment. We hereby take into account the time needed by the robot to reach the search locations as well as visibility constraints.

As our simulation experiments demonstrate, our approach enables the robot to select effective search locations to find the user within a short amount of time. We showed in extensive experiments that our proposed method significantly outperforms two other common search methods. The experiments included runs where occlusions caused by dynamic obstacles as well as false negative detection occurred, which will be the case for real-world scenarios.

# References

1. Suzuki, I., Yamashita, M.: Searching for a mobile intruder in a polygonal region. SIAM J. Comput. **21**(5), 863–888 (1992)
2. Guibas, L., Latombe, J., LaValle, S., Lin, D., Motwani, R.: A visibility-based pursuit-evasion problem. Int. J. Comput. Geom. Appl. **9**(04n05), 471–494 (1996)
3. Choset, H.: Coverage for robotics - a survey of recent results. Ann. Math. Artif. Intell. **31**(1), 113–126 (2001)
4. Moors, M., Rohling, T., Schulz, D.: A probabilistic approach to coordinated multi-robot indoor surveillance. In: Proceedings of the IEEE/RSJ International Conference on Intelligent Robots and Systems (IROS), pp. 3447–3452 (2005)
5. Kolling, A., Carpin, S.: Multi-robot surveillance: an improved algorithm for the graph-clear problem. In: Proceedings of the IEEE International Conference on Robotics & Automation (ICRA), pp. 2360–2365 (2008)
6. Stiffler, N., Kolling, A., O'Kane, J.: Persistent pursuit-evasion: the case of preoccupied pursuer. In: Proceedings of the IEEE International Conference on Robotics & Automation (ICRA), pp. 5027–5034 (2017)
7. Tipaldi, G., Arras, K.: I want my coffee hot! Learning to find people under spatio-temporal constraints. In: Proceedings of the IEEE International Conference on Robotics & Automation (ICRA), pp. 1217–1222 (2011)
8. Schwenk, M., Vaquero, T., Nejat, G., Arras, K.: Schedule-based robotic search for multiple residents in a retirement home environment. In: Proceedings of the National Conference on Artificial Intelligence (AAAI), pp. 2571–2577 (2014)
9. Kulich, M., Krajník, T., Přeučil, L., Duckett, T.: To explore or to exploit? Learning humans' behaviour to maximize interactions with them. In: Hodicky, J. (ed.) Modelling and Simulation for Autonomous Systems: Third International Workshop, MESAS 2016, Revised Selected Papers, pp. 48–63. Springer International Publishing (2016)
10. Krajník, T., Kulich, M., Mudrová, L., Ambrus, R., Duckett, T.: Where's Waldo at time t? Using spatio-temporal models for mobile robot search. In: Proceedings of the IEEE International Conference on Robotics & Automation (ICRA), 2140–2146 (2015)
11. Volkhardt, M., Gross, H.: Finding people in home environments with a mobile robot. In: Proceedings of the European Conference on Mobile Robots (ECMR), 282–287 (2013)
12. Mehdi, S., Berns, K.: Behavior-based search of human by an autonomous indoor mobile robot in simulation. Univ. Access Inf. Soc. **13**(1), 45–58 (2014)
13. Goldhoorn, A., Garrell, A., Alquézar, R., Sanfeliu, A.: Searching and tracking people in urban environments with static and dynamic obstacles. Robot. Auton. Syst. **98**, 147–157 (2017)
14. Bayoumi, A., Karkowski, P., Bennewitz, M.: Learning foresighted people following under occlusions. In: Proceedings of the IEEE/RSJ International Conference on Intelligent Robots and Systems (IROS) (2017)
15. Liao, L., Fox, D., Hightower, J., Kautz, H., Schulz, D.: Voronoi tracking: location estimation using sparse and noisy sensor data. In: Proceedings of the IEEE/RSJ International Conference on Intelligent Robots and Systems (IROS), vol. 1, pp. 723–728 (2003)

# Proposal and Validation of an Index for the Operator's Haptic Sensitivity in a Master-Slave System

Dongbo Zhou[1(✉)] and Kotaro Tadano[2]

[1] Department of Mechano-Micro Engineering, Tokyo Institute of Technology, R2-46, Nagatsutacho 4259, Yokohama 226-8503, Japan
zhou.d.aa@m.titech.ac.jp
[2] Institute of Innovative Research, Tokyo Institute of Technology, R2-46, Nagatsutacho 4259, Yokohama 226-8503, Japan
tadano.k.aa@m.titech.ac.jp

**Abstract.** In this paper, the authors propose an index for evaluating the haptic sensitivity of an operator when a slave device contacts the environment during operating a master–slave system. The index is derived from the velocity contrast of the master device before and after contact, which is hypothesized to represent the operator's sensation of the contact. By means of psychophysics experiment, how the change in the operator's haptic sensitivity being reflected in the index value is studied. The index is then validated by another psychophysics experiment. The experimental results show that the change of operator's haptic sensitivity can be represented correctly by the change of index value. This index is expected to be used in the parameter design of bilateral control systems.

**Keywords:** Force feedback · Haptic sensitivity · Master-slave system
Parameters design

## 1 Introduction

When operating a master–slave robot system, the effect of the force-feedback function has been demonstrated in many studies. For example, in robot-assisted surgery, a force-feedback function can reduce unwanted damage to the tissue [1]. In tasks that beyond a human's capability or access, such as micromanipulation, a force-feedback function can enhance the performance and efficiency of tasks [2].

A common method of providing a force-feedback function is impedance-adjusting bilateral control [3]. When using impedance-adjusting bilateral control, the systematic impedance, which is determined by system parameters, can be adjusted according to the requirements of different tasks or personal preferences.

In the operation of master-slave systems with force feedback function, when the slave devices contact the environment, a feeling of the contact is fed back to the operator. During this process, the haptic sensitivity of the operator to perceive the contact is an important factor that should be considered.

© Springer Nature Switzerland AG 2019
M. Strand et al. (Eds.): IAS 2018, AISC 867, pp. 558–572, 2019.
https://doi.org/10.1007/978-3-030-01370-7_44

Obviously the sensation for the contact is affected by various system impedance settings. Parameters that determines the system impedance should be designed properly to avoid unwanted effects upon the operator's haptic sensitivity. Some qualitative effects of adjusting parameters are summarized as a guideline for parameters design. For example, increasing the damping parameter of the master device can decrease hand vibration; decreasing the impedance between the master and the slave can reduce the impact of the slave device into the environment, but a too low system impedance between the master and slave devices will mask some delicate forces arising from the interaction with the environment, and thus reduce operator's haptic sensitivity for the contact [4].

However, knowing the qualitative effects alone is insufficient. A more precise guideline is needed for practical applications of master–slave systems, which is expected to represent the quantitative relationship between system parameters and an operator's haptic sensitivity.

The transparency is a widely used feature that guides the system parameter design. However, sometimes the absolute transparency (although nonexistent) is not always an ideal operation condition because some excitation signals from the environment are not expected [5]. Son et al. proposed a perception-based method for haptic teleoperation systems [6], but this index did not represent the relationship between parameters and operator's haptic sensitivity, and therefore cannot be directly used in parameter design. Christiansson et al. quantified the influence of stiffness and damping parameters on shape discrimination during grasping task [7]. However, the contact sensitivity was not studied.

To the best of our knowledge, no other study has addressed the lack of a simple and quantitative guideline for system parameters design. In fact, in practical applications of master–slave systems, these parameters are often adjusted by trial and error.

Therefore, we aimed at building a new guideline for system parameters design based on the operator's haptic sensitivity. As the first stage of this research, we intend to propose an index to represent the operator's haptic sensitivity, and validate the correctness of expressing the operator's haptic sensitivity by its value.

## 2  Proposal of the Haptic Sensitivity Index

When the slave device contacts the environment, the dynamic motions of both the master and slave devices will change due to the bilateral control. During operation of the master–slave system, we assume that the operator's fingers are always clinging to the master device; hence, it is obvious to consider that the contact sensation is determined by the contrast of master's dynamic factors before and after contact (i.e., changes in force, velocity, momentum, etc.). Therefore, we call this index that represents the contact sensation as: Dynamic Contrast (abbreviation: C), its value can be calculated from the dynamic factors.

## 2.1  Dynamic Factors in Perception

When the stimuli are weak, the contact sensation is tactile sensation that comes from the cutaneous receptors under the finger pad. And, the firing rate of cutaneous receptors (FA and SA receptors) are functions of the skin curvature's changing speed [8]. When the stimuli are intense, the kinesthetic sensation will become predominant in the contact sensation. From literature of kinesthetic sensation study [9], we can get that kinesthetic sensation is caused by the primary spindle receptors in the arm muscles, which code the velocity and acceleration to generate a sensation.

Haptic-feedback devices are therefore usually designed to combine these two types of sensations [10]. If we wish to measure a dynamic factor of the master-slave system that most easily tracks the operator's haptic sensation, that factor should be existed in both cutaneous and kinesthetic sensing. Based on the analysis above, we hypothesize that the change extent on master device velocity before and after contact to be the dynamic factor.

Figure 1 presents some examples of the master device's velocity profile before and after contact. As shown in the figure, when the stimulus that gives rise to a sensation is weak (the green line), under which condition the contact to environment is difficult to be detected, the velocity of the master device after contact changes slowly; when the stimulus becomes intense, the velocity of the master device after contact will change faster (the brown and red lines), under which condition the contact sensation is strong, which means the operator's haptic sensitivity for the contact is high.

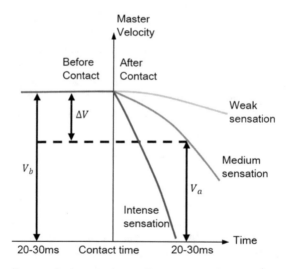

**Fig. 1.** Examples of master device velocity profiles. Operator's haptic sensitivity for the stimuli can be quantified by the extent of change in velocity ($\Delta V$).

## 2.2   Definition of the Index

Using the hypothesized dynamic factor of master velocity, we propose the index for representing operator's sensation of device contact can be defined as

$$C = 1 - V_a/V_b = (V_b - V_a)/V_b \tag{1}$$

In (1), $V_b$ and $V_a$ are the velocity of the master device before and after contact, respectively. According to (1), when the master device's velocity does not change, there is no contact sensation ($C = 0$); when the master device stops immediately after contact, the contact sensation is extremely intense ($C = 1$). In normal cases, numerical range of the index value is [0, 1].

When calculating the $C$ value, the average velocity in 50 ms before and after contact are used. The reasons are: first, the active muscle motion of the operator should not be involved in calculation. According to [11], to preclude volitional control, the time durations was set as 30 ms; second, according to the mechanism of human haptics, the contact sensation in the tactile modality is generated by the FA1 receptor, of which the temporal resolution is 15 to 50 ms [12]. And the time resolution of contact in the kinesthetic modality is measured to be 17–35 ms [13]. According to the above analysis, we choose the average velocity in 50 ms to represent the velocity level within 20–30 ms.

# 3   Properties of the Haptic Sensitivity Index

After proposing the definition of the index: Dynamic Contrast (C), properties of how the strengthening and weakening effects in haptic sensitivity is reflected in the index value should be studied, for which a psychophysics experiment was implemented.

## 3.1   Experimental Apparatus

We built a master–slave system with impedance adjusting bilateral control, as shown in Fig. 2. The master side is the Phantom Desktop haptic device (SensAble Technologies). To make an ideal system without interference from mechanical factors, the slave side and the operational environment are made in the virtual world; the slave side is a virtual ball and the operation environment is a virtual wall. The update rate of the system is fixed at 1 kHz.

According to the control model in Fig. 2, the equations of motion of the system are

$$f_h = M_m \ddot{r}_m + B_m \dot{r}_m + K_s(r_m - r_s) + B_s(\dot{r}_m - \dot{r}_s) \tag{2}$$

$$f_{en} = K_{en}(r_s - r_{wall}) = M_s \ddot{r}_s + K_s(r_s - r_m) + B_s(\dot{r}_s - \dot{r}_m) \tag{3}$$

In (2) and (3), $r_m$ is the position of the master device, $r_s$ is the position of the slave ball, and $r_{wall}$ is the position of the virtual wall, which is constant. For simplicity, the properties of the wall are only represented by its stiffness: $K_{en}$. Other definitions of each parameter are as follows: $B_m$: damping of the master device; $K_s$: stiffness between

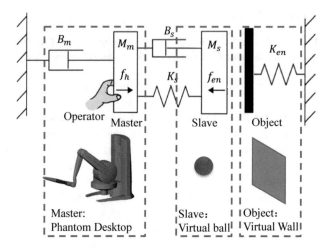

**Fig. 2.** Control model and device of experiment system.

master and slave; $B_s$: damping between master and slave; $f_h$: Hand force of the operator; $f_{en}$: Environment (virtual wall) force, $M_s$, $M_m$: inertia of master and slave respectively.

When the motion track between the master and slave was guaranteed, we found that the damping of the master device: $B_m$, and the stiffness of the virtual wall: $K_{en}$, affected the contact sensation remarkably more than the other parameters. Hence, in this paper, operator's sensitivities for the contact were measured by adjusting the parameters of $B_m$ and $K_{en}$; the other parameters were constant.

### 3.2   Subject and Motion

We enrolled 10 subjects all between 22 and 35 years old. Motion in the experiment was as follows; the subjects held the stylus of the master device and moved forward at any casual speed. When the force of contact between slave ball and the environment was sensed, withdraw the stylus just as knocking a door, the appearance of the experiment is shown in Fig. 3. During the whole experiment, subjects could not see the motion on the slave side; all motions were judged solely by haptic sensitivity.

### 3.3   Experiment Introduction

#### 3.3.1   Objective

While adjusting system parameters for some requirement, how the operator's haptic sensitivity for the contact is changed after adjusting the parameters should be considered. Among many kinds of changes in haptic sensitivity, the simplest and most basic are strengthening and weakening effects. Therefore, we studied the extent to which a change in index value can lead to a strengthening or weakening of the operator's haptic sensitivity, which is represented by the difference in contact sensation.

**Fig. 3.** Appearance of the experiment.

A psychophysics experiment was conducted, in which we measured the just noticeable difference (JND) for designated reference sensations, in both strengthen and weaken direction.

In directions of strengthening and weakening, JND is divided into upper threshold (UT: the reference sensation is strengthened until the difference can be just noticed) and lower threshold (LT: the reference sensation is weakened until the difference can be noticed). Therefore, for ten types of references sensations, twenty types of JND are measured in total.

### 3.3.2   Experimental Procedure

The experiment is designed based on the up-and-down method, as a variation of the method of limits, it is used extensively in studies of haptics to obtain the JND [14].

Two trials (reference and comparison) were conducted as one set. Both the two trials were shown to the subject, and the subject was instructed to answer whether 'ease of detecting the virtual wall' in the two trials was different. To prevent time error, we showed the reference and comparison trials to subjects in a random sequence. Here, we use the process of obtaining the UT of parameter setting $K_{en}$: 0.1 N/mm, $B_m$: 0.001 Ns/mm to illustrate the experimental process in details. For concrete explanation, this process is also illustrated in Fig. 4.

In the reference trials, the reference $K_{en}$ parameter was constant as 0.1 N/mm. In comparison trials, we kept $B_m$ constant and $K_{en}$ began at 0.3, which is far higher than 0.1 in the reference trial, thus allowing the difference between the reference and comparison trials to be easily discriminated. Hence, the subject should answer "Yes".

When the subjects answered "Yes", we reduced $K_{en}$ in the comparison trials by every 0.01 to approach the reference $K_{en}$ (0.1) gradually, showed the reference and comparison trials set to the subject at each decrement and asked him whether the ease of detecting the virtual wall was different. This was the descending series, which

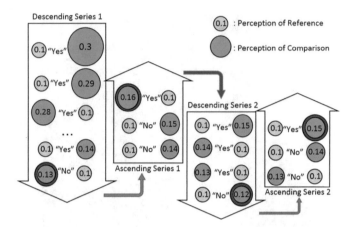

**Fig. 4.** Process for measuring the just-noticeable difference. To measure the JND of haptic sensitivity for a reference parameter setting $K_{en} = 0.1$ N/mm. Descending and ascending series are alternated. Perceptions are represented by the circle size. "Yes" or "No" between two circles are the subject's answer. Circles with red rings are the recorded $K_{en}$, at which the subject's answer transferred.

continued until the subject's answer transfers from "Yes" to "No". For example, when $K_{en} = 0.13$ for a subject, a "no" answer means that the subject could not discriminate the difference between the reference and comparison trials.

Next, we increased the comparison $K_{en}$ by every 0.01 to separate it from the reference $K_{en}$. This was the ascending series, which continued until the subject's answer transferred from "No" to "Yes", for example, in Fig. 4, when $K_{en} = 0.16$.

The ascending and descending series were repeated twice respectively, and we recorded the four $K_{en}$ in comparison trials when the subject's answer transferred between "Yes" and "No" (for example, $K_{en} = 0.13, 0.16, 0.12, 0.15$ in Fig. 4). The mean values of the four recorded $K_{en}$ (0.14 in this example) are the upper threshold parameter setting that makes the strengthening effect on haptic sensitivity of $K_{en} = 0.1$ just noticeable. Dynamic motion profiles with the reference and upper-threshold parameter settings were also recorded, from which the corresponding index values can be calculated.

### 3.3.3    Experiment Results

For ten reference parameter settings, the average LT and UT $K_{en}$ and their corresponding index values that measured from all subjects are listed in Table 1. From columns 4 and 7 of Table 1, it can be observed that when $B_m$ is increased from 0.001 to 0.004 Ns/mm, for the same $K_{en}$ level in the reference trials, subjects need more increments or decrements of the $K_{en}$ value in the comparative trials to perceive the difference. In column 6 and 9 of Table 1, $R_C$ is the change rate of index values between the reference sensation and their JND, in UT measurement, $R_C = C_{UT}/C_{ref}$; and in LT measurement, $R_C = C_{ref}/C_{LT}$.

**Table 1.** UT and LT $K_{en}$ and the corresponding index value to ten references

| No. | REFERENCE PARAMETER SETTING | MEAN OF REFERENCE C VALUE: $C_{REF}$ | MEAN OF KEN IN UT (STANDARD DEVIATION) | MEAN OF UT C VALUE: $C_{UT}$ | $R_C = C_{UT}/C_{REF}$ | MEAN OF KEN IN LT (STANDARD DEVIATION) | MEAN OF C VALUE: $C_{LT}$ | $R_C = C_{REF}/C_{LT}$ |
|---|---|---|---|---|---|---|---|---|
| 1 | $B_m = 0.001, K_{en} = 0.1$ | 0.149 | 0.148 (0.016) | 0.212 | 1.423 | 0.073 (0.007) | 0.087 | 1.712 |
| 2 | $B_m = 0.001, K_{en} = 0.2$ | 0.243 | 0.289 (0.030) | 0.340 | 1.399 | 0.144 (0.010) | 0.165 | 1.447 |
| 3 | $B_m = 0.001, K_{en} = 0.3$ | 0.365 | 0.438 (0.044) | 0.456 | 1.249 | 0.218 (0.015) | 0.253 | 1.445 |
| 4 | $B_m = 0.001, K_{en} = 0.4$ | 0.449 | 0.576 (0.071) | 0.530 | 1.180 | 0.287 (0.032) | 0.300 | 1.496 |
| 5 | $B_m = 0.001, K_{en} = 0.5$ | 0.501 | 0.702 (0.085) | 0.610 | 1.220 | 0.347 (0.042) | 0.361 | 1.385 |
| 6 | $B_m = 0.004, K_{en} = 0.1$ | 0.119 | 0.165 (0.030) | 0.185 | 1.554 | 0.061 (0.009) | 0.073 | 1.630 |
| 7 | $B_m = 0.004, K_{en} = 0.2$ | 0.202 | 0.318 (0.035) | 0.314 | 1.554 | 0.131 (0.007) | 0.138 | 1.463 |
| 8 | $B_m = 0.004, K_{en} = 0.3$ | 0.305 | 0.468 (0.048) | 0.445 | 1.459 | 0.186 (0.025) | 0.203 | 1.525 |
| 9 | $B_m = 0.004, K_{en} = 0.4$ | 0.384 | 0.597 (0.056) | 0.480 | 1.250 | 0.252 (0.041) | 0.263 | 1.460 |
| 10 | $B_m = 0.004, K_{en} = 0.5$ | 0.460 | 0.714 (0.069) | 0.552 | 1.196 | 0.328 (0.047) | 0.341 | 1.353 |

When the reference sensation ($C_{ref}$) changes from weak to strong, $R_C$ that needed for the change in haptic sensitivity to become noticeable is decreasing. We consider this situation to be caused by the difference in the receptor; when the stimuli are weak, the sensation is the tactile sensation; as it increases, kinesthetic sensation will be involved, which sensitizing the discrimination ability.

In extreme cases, compared to no sensation ($C = 0$), when a minute sensation can be just perceived, $R_C$ is infinite. When the sensation is very strong ($C \rightarrow 1$), the discrimination of haptic sensitivity is close to that of the stiffness or force. [15, 16] revealed that the discrimination threshold of stiffness or force will be close to about 10–20% higher than the reference stimulus. Therefore, in this research, we choose the intermediate value and believe that $R_C$ will finally become 1.15.

Measuring the LT to a reference can be considered as UT measuring if the contact sensation under LT parameter setting is regarded as the reference. We plot all the denominators in calculating $R_C$ as the lateral axis (including the extreme cases in which $C_{ref} = 0$ and 1), and the $R_C$ value as the vertical axis in Fig. 5, from which we can obtain the necessary change of index value to strengthen the subject's haptic sensitivity.

According the tendency in Fig. 5, the relationship between the $R_C$ and reference C ($C_{ref}$) values can be considered as a logarithmic function:

$$R_C = \log_{0.001}^{C_{ref}} + 1.15 \quad C_{ref} \in [0, 1] \tag{4}$$

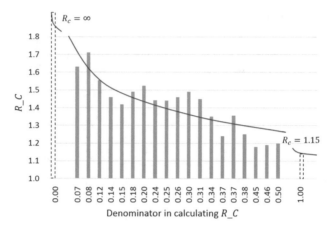

**Fig. 5.** Reference C values and necessary change rate $R_C$ from a reference C value to make the difference of two sensation become noticeable.

## 4  Validity of the Haptic Sensitivity Index

If the proposed index is valid, despite the difference in parameters setting of trials, as well as the relationship of their corresponding index values is known, the operator's haptic performance should be in accordance with the prediction that based on the index values.

According to this, in this section, another psychophysics experiment was conducted. Here, we appointed some pairs of parameter settings with designated C value relationship between the pair elements, then showed the sensation under appointed parameter setting pairs to the subjects and checked if their haptic performances were in accordance with the prediction that based on the designated C value relationship.

### 4.1  Relationship Between Parameters and C Value

For better deciding parameters with designated index values, we plotted all the parameter settings (Columns 2, 4 and 7) and their corresponding C values (Columns 3, 5 and 8) in Table 1 as Fig. 6. The curves were fitted from the plotted points, which represents the index when $K_{en}$ changes from 0 to 1 continuously under two $B_m$ levels.

### 4.2  Experimental Procedure

The experimental apparatus and subject motion were the same as described in Sect. 3.2. We enrolled 15 subjects, all males from 22 to 35 years old.

The experiment was designed based on the method of constant stimuli. Two kinds of parameter settings with designated index value relationship were set as one pair, sensations under this two parameter settings (pair elements) were shown to the subject as two trials. Every time after showing the one pair, the subject was asked in which trial

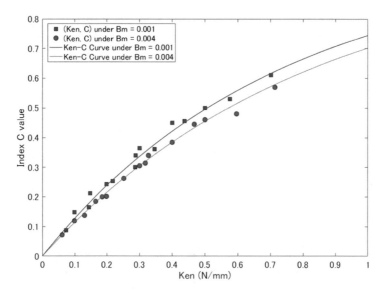

**Fig. 6.** Relationship between $K_{en}$ and C values under two $B_m$ levels.

it was easier to detect the virtual wall. Subjects could answer "Former", "Latter", or "Same".

The designated relationships of index value between pair elements were divided into two types: 1. parameter elements with the same index value, 2. parameter elements with just noticeably different index values. We appointed five parameter pairs for each type, in which the pair elements were labeled $A_1$-$B_1$ to $A_5$-$B_5$.

For each subject, we repeated each parameter setting pair 20 times. Therefore, for each parameter setting pair, we can obtained 300 answers from all 15 subjects. To prevent the subject from deducing the answers from the ongoing statistics, the two types of parameter pairs were mixed and shown in a random sequence. After obtaining all 300 answers, the proportion of undistinguishable answer was counted, which included the "same" answer and an offset between answers waving on two pair elements.

### 4.3   Experimental Results

In the experiment of Sect. 3, each trial set element was provided under the same $B_m$ level. In experiments of this section, the $B_m$ levels differed between the pair element ($B_m$ was 0.001 in A1-A5 and 0.004 in B1-B5).

#### 4.3.1   Pairs of Parameter Sets with the Same Index Value

Figure 7 shows the parameter pairs of which index C value pairs is designated as the same; the five appointed index C values of pair elements are shown in column 3 and 5 of the table in Fig. 7. The corresponding $K_{en}$ parameters in $A_1$-$B_1 \sim A_5$-$B_5$ are listed in column 2 and 4 of the table, which are calculated back from the red and blue curves.

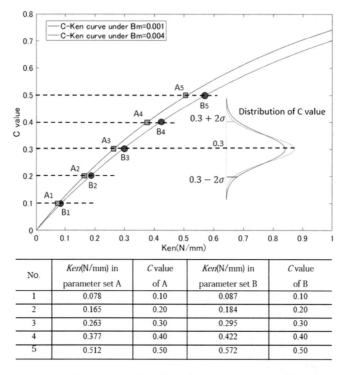

**Fig. 7.** Five parameter pairs of type 1, for which the contact sensation is expected to be same. $B_m$ in $A_1$ to $A_5$ is 0.001, but 0.004 in $B_1$ to $B_5$.

| No. | Ken(N/mm) in parameter set A | C value of A | Ken(N/mm) in parameter set B | C value of B |
|-----|------|------|------|------|
| 1 | 0.078 | 0.10 | 0.087 | 0.10 |
| 2 | 0.165 | 0.20 | 0.184 | 0.20 |
| 3 | 0.263 | 0.30 | 0.295 | 0.30 |
| 4 | 0.377 | 0.40 | 0.422 | 0.40 |
| 5 | 0.512 | 0.50 | 0.572 | 0.50 |

The parameter settings elements in each pair with the same index C value mean the sensation they provided to the subject was expected to be the same. For each parameter pair, the proportion of undistinguishable answers after 300 trials is shown in Table 2. From which we can see the proportion of undistinguishable answers is high and almost more than 80%, the contact sensations that provided by pair elements are the same.

**Table 2.** Proportion of undistinguishable answers under the same C value

| Parameter pair no. | Proportion of undistinguishable answers | Expected proportion of undistinguishable answers |
|-----|------|------|
| $A_1$, $B_1$ | 78% | 71.1% |
| $A_2$, $B_2$ | 82% | 81.3% |
| $A_3$, $B_3$ | 80% | 86.9% |
| $A_4$, $B_4$ | 83% | 87.2% |
| $A_5$, $B_5$ | 84% | 87.5% |

The third row of Table 2 is the expected proportion of undistinguishable answers. However, it is not 100% even the index values of pair elements are the same. This is because of the statistical characteristics of the index value for different operators. In Fig. 7, points of $A_{1\sim5}$ and $B_{1\sim5}$ on the curve are the mean values for different operators, the true index value in each trial may differ but will be restrained in a normal distribution (see the normally distribution curves example in Fig. 7). For example, if the true index value when showing $A_3$ was $0.3 - 2\sigma$ of the red distribution, whereas the true index value when showing $B_3$ was $0.3 + 2\sigma$ of the blue distribution, in this pair, the difference between the true sensations of $A_3$ and $B_3$ will be beyond the just-noticeable threshold, which allows the subject to distinguish the difference even their calculated index value are same.

Considering this situation, for any pair of parameter setting, the expected proportion of undistinguishable answers can be calculated by the following convolution integral, which calculated: as the true index of element A changes from $C_A - 2\sigma$ to $C_A + 2\sigma$, how much the part in distribution of element B's true index value is within the just noticeable difference of present element A's true index value: $(C/R_C, C \cdot R_C)$

$$p = \int_{C_A - 2\sigma}^{C_A + 2\sigma} \left\{ \int_C^{C + \Delta C} f_A(C) \, dC \cdot \int_{C/R_C}^{C \cdot R_C} f_B(C) \, dC \right\} dC \tag{5}$$

In (5), p is the possibility of undistinguishable answers, C is the present element A's true index value, $f_A(C)$ and $f_B(C)$ are the probability-density functions of the normal distributions of pair elements A and B (the standard deviation is derived from recorded experiment data); $C_A$ is the average index value of pair element A. $R_C$ is the necessary change rate of index value to make a different haptic sensitivity, which can be calculated by substituting $C_A$ as $C_{ref}$ into (4). For example, when calculating the expected proportion for $A_2$ and $B_2$, substitute the C value of $A_2$ (0.2) into (4) as $C_{ref}$, $R_C$ can be obtained as 1.33.

### 4.3.2  Pairs of Parameter Sets with just Noticeably Different Index Values

Figure 8 shows the parameter pairs of which C value pairs that are appointed as reaching the JND, means the change rate of index C between the pair elements is $R_C$. The five appointed index C value pairs are shown in the columns 3 and 5 of the table in Fig. 8. The $K_{en}$ parameter in $A_1$-$B_1$ to $A_5$-$B_5$ are listed in column 2 and 4 of the table.

Considering the statistical characteristics of true index value from different operators, the expected proportions calculated by (5) are about 47% (50% in psychophysics theory without considering statistical characteristics).

The proportions of undistinguishable answers from the experiment are shown in Table 3. From Table 3, when the difference between the C values of the two parameter settings is $R_C$, the proportion of undistinguishable answers is near the expected value. Therefore, the difference in haptic sensitivity for detecting the virtual wall between pair elements is just noticeable.

From the experiments result in Sect. 4. We can say that, regardless of how parameters are set to produce the contact sensation, as long as the index value is determined, the relationship between contact sensations under any parameter settings can be estimated, meaning that the numerical value of C can correctly represent an operator's sensation and haptic sensitivity.

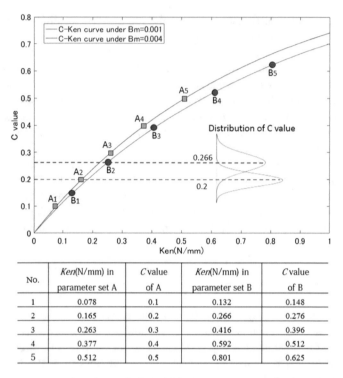

**Fig. 8.** Five parameter pairs of type 2, for which the contact sensation is expected to be just noticeable. $B_m$ in $A_1$ to $A_5$ is 0.001, but 0.004 in $B_1$ to $B_5$.

**Table 3.** Proportion of undistinguishable answers under the just noticeably different $C$ value

| Parameter pair no. | Proportion of undistinguishable answers | Expected proportion of undistinguishable answers |
|---|---|---|
| $A_1$, $B_1$ | 44% | 46.3% |
| $A_2$, $B_2$ | 40% | 47.6% |
| $A_3$, $B_3$ | 41% | 47.6% |
| $A_4$, $B_4$ | 46% | 47.6% |
| $A_5$, $B_5$ | 45% | 47.5% |

## 5   Discussion and Future Works

### 5.1   Relationship Between Haptic Sensitivity and Contact Velocity

In this research, subjects were instructed to move the master device at a comfortable velocity; thus, the contact velocity was not uniformed. There is no need to uniform the contact velocity not only because it is impossible to instruct the operator to move under a designated velocity but also because of the following reasons:

Firstly, calculation of the index value is a division operation of which the denominator is the velocity before contact, the effect of contact velocity has been calculating the index value has been neutralized. Actually, from the verbal reports by experimental participants, the contact sensation was the same, regardless of contact velocity. Secondly, as mentioned in Sect. 2, the haptic sensitivity for detecting an environment is consisted of tactile and kinesthetic sensation. [15] claims that the contact velocity wouldn't affect the tactile performance; and in kinesthetic, [10] claims that the ability to detect a change in the position of a limb is not affected by the angular velocity of the arm movement.

## 5.2  Application of the Index

First, the index can be used to estimate the operator's sensation under appointed parameter set; second, it can be used to evaluate the change of haptic sensitivity under different parameter set; third, this index can be used to guide the parameter setting for different requirements on the haptic sensitivity and different operators' preferences.

For example, while using robot-assisted surgical systems to operate on delicate organs, a surgeon may need high master damping to reduce hand shaking, but it is unclear how much the damping parameter can be increased before the surgeon's sensation and haptic sensitivity is weakened. Using the index value, we can quantify the sensation under the current parameter setting by the index value, and then use the relationship between index and damping parameter values to calculate the extent to which damping can be increased before the haptic sensitivity degrades. System parameter design becomes convenient and well-founded instead of relying upon trial and error.

## 5.3  Future Works

In this paper, we focused only upon the damping effect of the master device and the stiffness of the environment. More parameters such as stiffness and damping between the master and the slave, position, force-scale rate, and mechanical friction resistance will be enrolled in future works. With a research objective of calculating the index directly from parameters, our future task will be to derive a function to express the mathematical relationship between the multiple kinds of parameters and the index value.

# References

1. Wagner, C., Stylopoulos, N., Jackson, P., Howe, R.: The benefit of force feedback in surgery: examination of blunt dissection. Presence: Teleoperators Virtual Environ. **16**(3), 252–262 (2007)
2. Bolopoin, A., Regnier, S.: A review of haptic feedback teleoperation systems for micromanipulation and micro assembly. IEEE Trans. Autom. Sci. Eng. **10**(3), 496–502 (2013)

3. Beretta, E., Nessi, F., Ferrigno, G., De Momi, E.: Force feedback enhancement for soft tissue interaction tasks in cooperative robotic surgery. In: IEEE/RSJ International Conference on Intelligent Robots and Systems, Hamburg, Germany, 28 September–2 October 2015

4. Yamakawa, S., Abe, K., Fujimoto, H.: Conditions of force scaling methods in master–slave systems based on human perception abilities for time-variant force. In: SICE Annual Conference in Sapporo, vol. 3, pp. 2385–2388, August 2004

5. Misra, S., Okamura, A.M.: Environment parameter estimation during bilateral telemanipulation. In: Symposium on Haptic Interfaces for Virtual Environment and Teleoperator Systems, 25–26 March 2006, Alexandria, Virginia, USA (2006)

6. Son, H., Cho, J., Bhattacharjee, T., Jung, H., Lee, D.: Analytical and psychophysical comparison of bilateral teleoperators for enhanced perceptual performance. IEEE Trans. Ind. Electron. **61**(11), 6202–6212 (2014)

7. Christiansson, G.A.V., van der Linde, R.Q., van der Helm, F.C.T.: The influence of teleoperator stiffness and damping on object discrimination. IEEE Trans. Robot. **24**(5), 1252–1256 (2008)

8. Dahiya, R.S., Metta, G., Valle, M., Sandini, G.: Tactile sensing-from humans to humanoids. IEEE Trans. Robot. **26**(1), 1–20 (2010)

9. Jones, L.A.: Kinesthetic sensing. In: Human and Machine Haptics. MIT Press (2000)

10. Wei, L., Zhou, H., Nahavandi, S., Wang, D.: Toward a future with human hands-like haptics. IEEE Syst. Man Cybern. Mag. **2**(1), 14–25 (2016)

11. Gillespie, R., Cutkosky, M.: Stable user-specific haptic rendering of the virtual wall. In: Proceedings of the ASME International Mechanical Engineering Conference and Exhibition, vol. 58, pp. 397–406, November 1996

12. Abraira, V., Ginty, D.: The sensory neurons of touch. Neuron **79**(4), 618–639 (2013)

13. Bhardwaj, A., Chaudhur, S.: Estimation of resolvability of user response in kinesthetic perception of jump discontinuities. In: IEEE World Haptics Conference, 22–26 June 2015, Evanston, USA (2015)

14. Jones, L.A.: Application of psychophysical techniques to haptic research. IEEE Trans. Haptics **6**(3), 268–287 (2013)

15. Hirano, M., Maruyama, T., Nakahara, Y.: Relationship between the recognition of object's hardness by human finger and contact force under various contact conditions. In: Abstract of Dynamic and Design Conference, vol. 2000, p. 370 (2000)

16. Botturi, D., Vicentini, M., Righele, M., Secchi, C.: Perception-centric force scaling in bilateral teleoperation. Mechatronics **20**(7), 802–811 (2010)

# Operating a Robot by Nonverbal Voice Expressed with Acoustic Features

Shizuka Takahashi[(✉)] and Ikuo Mizuuchi

Tokyo University of Agriculture and Technology, Tokyo, Japan
shizuka070902@gmail.com

**Abstract.** This paper proposes methods for operating a robot by non-verbal voice. These methods enable operators to operate multi-degrees of freedom simultaneously and operate a robot intuitively by nonverbal voice by associating the nonverbal voice, tongue position and the coordinate of the robot's hand. The voice is defined by formants or Mel Frequency Cepstral Coefficients (MFCC). Formants and MFCC are acoustic features and they show the characteristics of the vocal tract such as the mouth and the tongue. We propose two methods. One is the method in which voice expressed with overlapped formants ranges are used to change variable about robots' operation. This method enables operators to operate multi-degrees of freedom simultaneously by nonverbal voice. The other is the method that operators tongue positions are distinguished by nonverbal voice. These tongue positions correspond to the coordinate of the robot's hand and it enables the operators to operate a robot intuitively. We found the feasibility of the methods through experiments of simple tasks. These methods can realize operating a robot intuitively in continuous values by voice and can be utilized for user-friendly system.

## 1 Introduction

There are many kinds of method of operating robots. Joysticks are mainly used to operate robots and the master-slave method which enables operators to operate robots intuitively is also used, but they are difficult for people who cannot use hands or whose hands are full. There are methods that utilize biological signals such as voice, brain waves, and myoelectricity recently because of technology improvement of signal processing. Appropriate ways of operating robots differ depending on the type of robot, situation where the operator is, type of tasks and so on. Operating robots by speech recognition is easy and people don't need to touch anything to operate. Especially, it is useful for people who have physically disabled and those who want to do some tasks at the same time. However, the output of speech recognition with natural language processing is a phrase which is shown in discrete symbol string. Therefore, it is difficult to operate a robot as if it was an operator's own limb because they cannot change joint angles with variables with speech recognition for the people.

© Springer Nature Switzerland AG 2019
M. Strand et al. (Eds.): IAS 2018, AISC 867, pp. 573–584, 2019.
https://doi.org/10.1007/978-3-030-01370-7_45

Although the verbal voice is widely used in operating robots by voice, non-verbal voice can be useful because it is not shown in discrete symbol string and it is expected to obtain continuous values as output. This paper proposes methods which realize operating multi-degrees of freedom simultaneously and operating a robot intuitively by nonverbal voice by associating the nonverbal voice, tongue positions and the coordinate of the robot's hand Formants and Mel Frequency Cepstral Coefficients (MFCC) are acoustic features and they show the characteristics of the vocal tract such as the mouth and the tongue.

It is expected that voice can be distinguished by various methods with formants or MFCC as same as or more than the number of expressions with language and utilized for operating many degrees of freedom at the same time and change values of variables about robots' operation in continuous values.

## 2   Nonverbal Voice for Operating Robots

### 2.1   Related Work

Speech recognition is important in human-robot interaction and operating robots, so there are much research about it. HARK [9] is open-source robot audition software which is developed to realize real-time processing in noisy surroundings for robot audition. Robustness is an important factor in speech recognition. Audio-visual integration [13] and speech and noise discrimination based on pitch estimation [1] are proposed to improve the robustness. In the other hand, there are also some proposals about utilizing nonverbal voice. Using voice as sound for computer-user interface has been proposed [5] and it has been used for mobile devices [10]. Nonverbal voice can be used also in cursor movement [7]. Voice pen, which has been proposed for writing [2,3], also uses nonverbal voice. They show that nonverbal voice is useful for the operation which needs using continuous values and for people with motor impairments. Computers which are easy to use for motor impairments are needed [11] and the same is true of operating robots. It is said that using nonverbal voice for the real-time control of a model car is better than using speech recognition [12]. There is a robotic arm which is controlled with vowels and pitch [4] using "vocal joystick" [6]. This research about the robotic arm is the first one applying nonverbal voice expressed as vowels in operating a robotic arm, but the robotic arm cannot be operated more than two degrees of freedom at the same time because these vowels and pitch are applied to each degree of freedom independently. The voice for the operation is defined as vowels, Used vowels are different from each country, so it might be difficult to operate a robot.

Therefore, this paper proposes methods which realize operating multi-degrees of freedom simultaneously and operate a robot intuitively by nonverbal voice without depending on operators' languages.

## 2.2   Using Characteristic of Frequency Domain for Operating a Robot

Formants and MFCC are acoustic features and are used for the proposed methods of operating a robot. We used formants and MFCC, which show the characteristics of the vocal tract such as the mouth and the tongue, because changing a tongue position or a shape of the mouth is easy and there are many patterns of the tongue positions or mouth shapes and compared changing the volume of voice or pitch of the voice. It is considered that combination of utilizing the characteristics of the vocal tract, the volume of voice and pitch of voice are also better for operating a robot, however, we used only the characteristics of the vocal tract in this paper to show the feasibility of operating a robot with formants and MFCC.

Formants are peaks of spectrum envelope. Spectrum envelope is an outline drawing of speech spectrum. Figure 1 shows an example of spectrum envelope and Fast Fourier Transform spectrum of voice. Voice is generated by vibration of vocal cords and resonance of the vocal tract. The sound source is generated by vocal cords and power of specific frequencies are emphasized through the vocal tract. The emphasized frequencies are defined as formants. Formants show the characteristics of the vocal tract such as the mouth and the tongue. There are some formants in utterance. The formant with the lowest frequency is called as $F1$, the second, $F2$, and the third, $F3$. $F1$ and $F2$ are often used to distinguish vowels. It is said that $F1$ is influenced by how the chin is opened and if the chin is opened more, $F1$ becomes higher. As for $F2$, it is influenced by the form of a tongue. If the tongue is located more forward, $F2$ becomes higher.

MFCC is also acoustic features and show the vocal tract characteristics. Fast Fourier Transform spectrum is taken to voice, the spectrum is mapped onto the mel scale which approximates the human auditory system's response, discrete cosine transform is taken to the mel log power which is taken from the logs of the powers and MFCC are obtained from the amplitude of the spectrum. MFCC is often used for speech recognition.

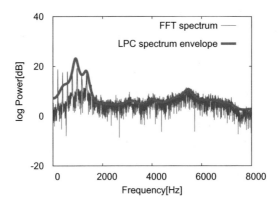

**Fig. 1.** Example of spectrum envelope and Fourier power spectrum

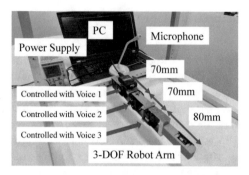

**Fig. 2.** Experimental setup

# 3   Multiplexing Operation by Nonverbal Voice

## 3.1   Three Kinds of Voice for the Operation

First, we proposed the method which an operator can operate some degrees of freedom at the same time by keeping utterance and changing the positions of operators' tongue and the form of operators' mouth. We conducted experiments which one subject who is female and in her twenties operated three-degree of freedom robotic arm with the method. We used $F1$ and $F2$ to define voice for operating the robotic arm because $F1$ and $F2$ are especially changed a lot when people change the positions of their tongue and the form of their mouth, and the peaks of $F1$ and $F2$ are sharper than those of other formants, so they are suitable to use for the operating. In order to get the ranges of $F1$ and $F2$ which can be uttered, the subject uttered various vowels, for example, 'a', 'i', 'u', 'e', 'o', for about six minutes. Figure 3 shows the result of the utterance and distribution of $F1$ and $F2$ during the utterance for six minutes. Circles in Fig. 3 show rough $F1$ and $F2$ ranges of vowels('a', 'i', 'u', 'e', 'o'). We defined the three ranges in the $F1$ and $F2$ dimension and named voice1, 2 and 3 to operate the three-degree of freedom robotic arm based on the result and $F1$ and $F2$ ranges of vowels.

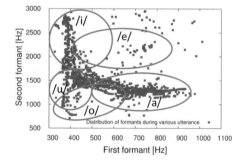

**Fig. 3.** Fast and second formants of various utterance

The three ranges are overlapped with one another. if an operator can utter the voice which is in the range of the overlapping, two or three degrees of freedom can be changed simultaneously. To sum up, we defined seven ranges in $F1$ and $F2$ dimension including the overlapped ranges. We thought operators can utter each range of voice if they start uttering vowels and change their positions of tongue and their form of mouth subtly because $F1$ becomes higher if the chin is opened more and $F2$ becomes higher if the tongue is located more forward (Fig. 4).

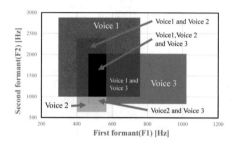

**Fig. 4.** Formants ranges of voice 1, 2 and 3

## 3.2   Possibilities of the Utterance

To confirm possibilities of uttering each of the seven kinds of voice which we defined with $F1$ and $F2$, we conducted experiments of the utterance. The seven kinds of voice consist of voice 1, 2, 3 and four overlapped ranges by voic1, 2 and 3. Table 1shows the ranges value of formants of voice 1, 2 and 3. These values would be different depends on operators, however, ranges of formants which are shown when people utter vowels are similar. Therefore, the defined voice could be uttered if other subjects try to utter.

An operator tried to keep uttering each of the voice for about 10 s and repeated it 20 times. 2 points expressed with $F1$ and $F2$ are recorded per 1 s, so

**Table 1.** Formants ranges of voice

| Voice | $F1 or F2$ | $F_{min}$ [Hz] | $F_{max}$ [Hz] |
|-------|------------|----------------|----------------|
| 1     | $F1$       | 300            | 750            |
|       | $F2$       | 1000           | 2700           |
| 2     | $F1$       | 400            | 530            |
|       | $F2$       | 600            | 2300           |
| 3     | $F1$       | 450            | 1000           |
|       | $F2$       | 830            | 1800           |

**Table 2.** Possibilities of uttering each voice

| Voice | First half [%] | Second half [%] | 20 times [%] |
|---|---|---|---|
| 1 | 42 | 44 | 43 |
| 2 | 32 | 53 | 43 |
| 3 | 18 | 43 | 30 |
| 1 and 2 | 27 | 41 | 34 |
| 1 and 3 | 49 | 53 | 51 |
| 2 and 3 | 3.8 | 8.0 | 5.9 |
| 1, 2 and 3 | 16 | 32 | 24 |

40 points expressed with $F1$ and $F2$ are recorded in one experiment. We calculated how much percentage of the points is in the ranges of the voice which we defined. Table 2 shows results of the first half, the second half and total about each of the voice. It was confirmed that it is possible to utter each of the voice, though the variation of possibilities about each of the voice is large. As for the voice which has ranges overlapped by voice 2 and 3, the percentage of the result was 5.9% and it was small. However, the percentage of the first half is 3.8% and the second half is 8.0%. The result of the second half was twice as large as that of the first half. It is considered that the second half was bigger than the first half because the operator improved the utterance during the experiments. The tendency that the percentage is higher in the second half than the first half can be seen in all experiments of uttering the seven kinds of the voice.

Therefore, there are possibilities that utterance within any formants ranges can be improved by training. It means the training may enable to operate a robot which has many joints by nonverbal voice.

## 3.3    Grasping Experiments

We conducted grasping experiments to confirm the feasibility of the multiplexing operation with the nonverbal voice. Figure 2 shows the experimental setup. This is composed of a power supply, microphone, PC which CPU has 4 core, 8 threads, and the operating frequency is 3.7 GHz and the robotic arm. This robotic arm has three RS405CB, which is radio control servo module. This system is simple and does not cost much. This is also one of the advantages of the methods which we proposed. Formants are calculated at intervals of 256 ms. If the voice which has formants in the range of voice1, 2 and 3 is detected, each joint of the robotic arm rotates 2° with 67 deg/s. If they reach maximum angle, they change the direction of rotating. The robotic arm was stopped while the operator was breathing.

A ball which has a diameter of approximately 20 cm and a can which has a diameter of approximately 7 cm are grasped by the robotic arm in the experiment. We used a ball and a can for the grasping because the form of the cylinder and the sphere are easy to grasp and each grasped object has different size. It is defined that if the ball and the can are touched by both the hand of the robotic arm where about 80 mm from the tip and a bar made of aluminum which the robotic arm is fixed by, grasping succeed. Initial positions of the ball and the can are shown in Figs. 5(a) and 6(a). Figures 5(b) and 6(b) show success in grasping. The changes of each joint angle are shown in Figs. 7 and 8.

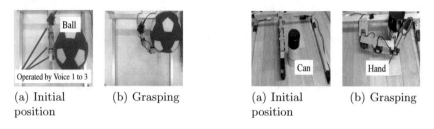

(a) Initial          (b) Grasping                 (a) Initial          (b) Grasping
position                                          position

**Fig. 5.** Grasping the ball                    **Fig. 6.** Grasping the can

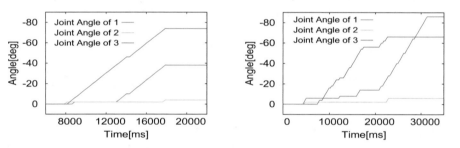

**Fig. 7.** Change of joints angles about       **Fig. 8.** Change of joints angles about
grasping the ball                               grasping the can

As for grasping the ball, the operator tried to move the joint angle of 3 first, and move the joint angle of 1 and 3. Figure 9 shows transitions of $F1$ and $F2$. Arrows in the line indicate the direction of the flow of time. Each point shows $F1$ and $F2$ every 256 ms. The line shows the operator's attempt to move the joint angle of 3 first. Many points on the line are in the range of voice 3 for the first. The utterance moved the joint angle of 3, which is shown in Fig. 7. The operator tried to move the joint angle of 1 and 3 after that. The operator could move the points of formants from the range of voice 3 to the overlapped range of voice 1 and 3.

As for grasping a can, the operator tried to move the joint angle of 3 and after that, move that of 1. Figure 10 shows the transitions of $F1$ and $F2$ about grasping the can. The blue line is the first utterance, which is for moving the joint angle of 3. It fluctuated first, but finally, reach the range of voice 3 and the operator could keep the utterance, which moved the joint angle of 3 as shown in Fig. 8.

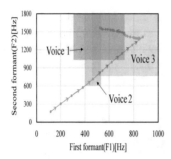
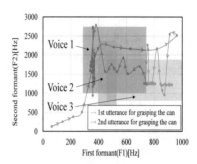

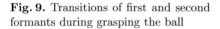

**Fig. 9.** Transitions of first and second formants during grasping the ball

**Fig. 10.** Transitions of first and second formants during grasping the can

The orange line shows the second utterance. The operator tried to move the joint angle of 1 in the second utterance. It also fluctuated first and the utterance was in overlapped ranges, but it reached the range of voice 1 finally and the utterance was kept in the range and the operator could move just joint angel of 1 as shown in Fig. 8. This operating is simple and an operator can get a hang of operation if you find the form of your mouth and the positions of your tongue which can utter the voice in the defined range of formants.

## 4 Operation with Tongue Positions Distinguished by Nonverbal Voice

It might be difficult to get how to utter nonverbal voice for the robot's operation by nonverbal voice. Therefore, we proposed a method of operating a robot with tongue positions which are distinguished by nonverbal voice. It is expected that it becomes easier to operate a robot by nonverbal voice if the tongue positions correspond to the coordinate of the robot's hand.

### 4.1 Distinguishing Tongue Positions by Nonverbal Voice

Tongue positions for the operation are distinguished by nonverbal voice. We used Mel Frequency Cepstral Coefficients (MFCC) and Deep Neural Network (DNN) to realize the method. MFCC are one of the acoustic features and they show vocal tract characteristics.

We defined 5 kinds of tongue positions. Figure 11 shows the definition. A tongue is located middle between upper teeth and bottom teeth in the middle position. A tongue is touched by upper middle teeth or bottom middle teeth in the upper or bottom position. A tongue is touched by an upper back tooth or a bottom back tooth in the upper side or bottom side position. The way to utter each voice is to utter "a" and move the tongue to each position.

These tongue positions are predicted by nonverbal voice. We used DNN for the prediction. The model architecture is shown in Fig. 12 The input is 12-dimensional MFCC. The network consisted of 2 convolutional Neural Network

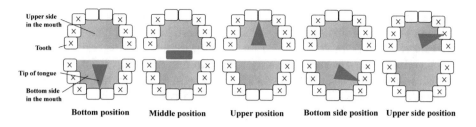

**Fig. 11.** Five kinds of tongue positions

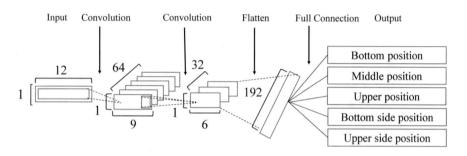

**Fig. 12.** Architecture of network for learning tongue positions

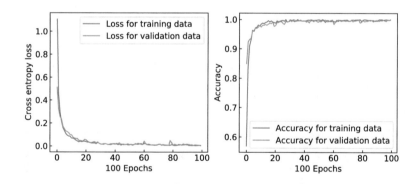

**Fig. 13.** Cross entropy loss and accuracy

(CNN) layers, a flatten layer and a full connection layer. The output is 5 kinds of tongue positions. One subject who is a female and in her twenties uttered the 5 kinds of voice and they are recorded for training the network. The voice was recorded under the condition of sampling frequency 441000 Hz. Each voice was recorded for 6 s. Fast Fourier Transform was performed with the number of data as 2048 points and the window was overlapped every 512 points. 360 samples of 12-dimensional MFCC are extracted from the recorded data per one kind of the voice. We got 1800 samples of MFCC in total. The network was trained for 100 epochs. Validation data was 40% of the 1800 samples. The cross entropy loss and accuracy of the training for 100 epochs were shown in Fig. 13. The F-measure of the learned network was 0.99. We train the network with only one subject's data in the training and the network is optimized to only the individual, however, it took only about 40 s with PC which CPU has 2 core, 4 threads, and the operating frequency is 2.7 GHz. Therefore, it is possible to train the network again with every operator in a short time.

## 4.2  Grasping Experiments

We conducted grasping experiments of operating a robotic arm by changing tongue positions distinguished by nonverbal voice. The network had high accuracy with recorded data, however, the accuracy of prediction about real-time utterance was about 50%. It due to differences between the recorded condition and real-time utterance condition, for example, the distance between mouth and mic and ambient noise. The bottom position voice and the upper position voice had high accuracy compared with another positions' voice, therefore, we used the 2 kinds of voice for operating experiments. We used Kitchen Assistant Robot [8]. We designed a system which the gripper of the robot moves up with the upper position voice and moves down with the bottom position voice with constant velocity. The robot was stopped while the operator was breathing.

We conducted grasping experiments. We used a fake apple as the grasping target. The same subject as one whose voice was used for the training the network uttered upper position voice or bottom position voice and moved the robot's hand up or down. Proximity sensors are equipped with the gripper and if the value of the proximity sensor becomes greater than a threshold, The gripper was closed and grasped the target. The subject could move the robot up or down as she expected and could grasp the target and move it up with the robot. Figure 14 shows the experiments.

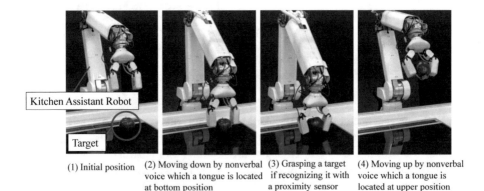

(1) Initial position    (2) Moving down by nonverbal voice which a tongue is located at bottom position    (3) Grasping a target if recognizing it with a proximity sensor    (4) Moving up by nonverbal voice which a tongue is located at upper position

**Fig. 14.** Grasping experiments with tongue positions distinguished by nonverbal voice

We could use only 2 kinds of voice in this experiments. It is expected that if each voice for training the network is recorded under the same condition as operating condition, the accuracy of the prediction about real-time utterance become greater. It enables to operate a robot with more kinds of voice.

## 5   Conclusion

We proposed two methods of operating a robot by nonverbal voice expressed with formants and MFCC in this paper.

One is the method in which formants ranges are divided into several kinds so as to overlap. An operator can move some joints of a robotic arm simultaneously with the voice which has formants within the defined ranges. We found that simple tasks such as grasping can be performed with the method.

The other is the one that tongue positions are predicted by nonverbal voice and the tongue positions correspond to the coordinate of the robot's hand. These tongue positions are corresponded to the coordinate of the robot's hand and it enables the operators to operate a robot intuitively by nonverbal voice. Experiments of grasping a target by nonverbal voice and proximity sensor were succeeded. We found that operating a robot by nonverbal voice with the proposed methods realized the intuitive operation. We also found that combination of these methods and autonomous operation with some kinds of sensors help operators more to operate a robot to do various tasks with voice intuitively.

It is needed to increase the accuracy of predicting tongue position in real-time utteranc and kinds of nonverbal voice for the operation to realize more flexible operation. One way to increase the accuracy is to use data which is recorded under the same condition as operating a robot condition when the network for the prediction is trained. Finding the better combinations of the nonverbal voice operation and some kinds of sensors' feedback operation is also the future works.

## References

1. Grondin, F., Michaud, F.: Robust speech/non-speech discrimination based on pitch estimation for mobile robots. In: 2016 IEEE International Conference on Robotics and Automation (ICRA), pp. 1650–1655, May 2016
2. Harada, S., Saponas, T.S., Landay, J.A.: Voicepen: augmenting pen input with simultaneous non-linguistic vocalization. In: the 9th Conference on Multimodal Interfaces, pp. 178–185 (2007)
3. Harada, S., Wobbrock, J.O., Landay, J.A.: Voicedraw: a hands-free voice-driven drawing application for people with motor impairments. In: The 9th Conference on Computers and Accessibility, pp. 27–34 (2007)
4. House, B., Malkin, J., Bilmes, J.: The voicebot: a voice controlled robot arm. In: The Conference on Human Factors in Computing Systems, pp. 183–192 (2009)
5. Igarashi, T., Hughes, J.F.: Voice as sound: using non-verbal voice input for interactive control. In: The 14th Symposium on User Interface Software and Technology, pp. 155–156 (2001)

6. Malkin, J., Li, X., Harada, S., Landay, J., Bilmes, J.: The vocal joystick engine v1.0. computational speech. Language **25**(3), 535–555 (2011)

7. Mihara, Y., Shibayama, E., Takahashi, S.: The migratory cursor: accurate speech-based cursor movement by moving multiple ghost cursors using non-verbal vocalizations. In: The 7th Conference on Computers and Accessibility. Assets 2005, pp. 76–83 (2005)

8. Mizuuchi, I., Fujimoto, J., Sodeyama, Y., Yamamoto, K., Okada, K., Inaba, M.: A kitchen assistant manipulation system of a variety of dishes based on shape estimation with tracing dish surfaces by sensing proximity and touch information. J. Robot. Soc. Jpn. **30**(9), 889–898 (2012)

9. Nakadai, K., Okuno, H.G., Nakajima, H., Hasegawa, Y., Tsujino, H.: An open source software system for robot audition hark and its evaluation. In: Humanoids 2008 - 8th IEEE-RAS International Conference on Humanoid Robots, pp. 561–566, December 2008

10. Sakamoto, D., Komatsu, T., Igarashi, T.: Voice augmented manipulation: using paralinguistic information to manipulate mobile devices. In: The 15th Conference on Human-Computer Interaction with Mobile Devices and Services, pp. 69–78 (2013)

11. Sears, A., Young, M.: Physical disabilities and computing technologies: An Analysis of Impairments. In: The Human-Computer Interaction Handbook, pp. 482–503 (2003)

12. Sporka, A.J., Slavík, P.: Vocal control of a radio-controlled car. In: SIGACCESS, pp. 3–8 (2008)

13. Yoshida, T., Nakadai, K., Okuno, H.G.: Automatic speech recognition improved by two-layered audio-visual integration for robot audition. In: 2009 9th IEEE-RAS International Conference on Humanoid Robots, pp. 604–609, December 2009

# Reduced Feature Set for Emotion Recognition Based on Angle and Size Information

Patrick Dunau[1(✉)], Mike Bonny[1], Marco F. Huber[1,3], and Jürgen Beyerer[2,3]

[1] USU Software AG, Rüppurer Str. 1, 76137 Karlsruhe, Germany
p.dunau@usu.de
[2] Fraunhofer Institue of Optronics, System Technologies, and Image Exploitation
(IOSB), Fraunhoferstr. 1, 76131 Karlsruhe, Germany
[3] Karlsruhe Insitute of Technology (KIT), Institute for Anthropomatics
and Robotics, Adenauerring 4, 76131 Karlsruhe, Germany
https://katana.usu.de

**Abstract.** The correct interpretation of facial emotions is important for many applications like psychology or human-machine interaction. In this paper, a novel set of features for emotion classification from images is introduced. Based on landmark points extracted from the face, angles between point-connecting lines and size information of mouth and eyes are extracted. Experiments compare the quality and reliability of the feature set to landmark-based features and facial action unit based features.

**Keywords:** Pattern recognition · Emotion recognition
Classification · Feature extraction · Neural networks

## 1 Introduction

It is well known that communication relies merely minimally on spoken words. Most of the information is delivered non-verbally, for instance by facial emotions. Thus, the accurate classification of emotions is of high importance in many applications. In the area of psychology, the ability to measure the patient's emotional state will be helpful for analysis. Furthermore, modern assistance systems like attention recognition systems in cars may observe the driver to evaluate his or her emotional state. The system needs a feature model that evaluates fast, as the facial expression changes quickly.

The human face possesses a high variability. Moreover, it is highly dynamic. To overcome these difficulties a variability minimizing representation of the facial expression is important. Facial landmark models possess the ability to concisely represent the facial expression using a limited number of points. This reduction of the feature space overcomes ethnic or shape based variations.

Kazemi et al. [1] provides a face alignment algorithm, which estimates the location of facial landmarks. This method locates the landmark positions based

© Springer Nature Switzerland AG 2019
M. Strand et al. (Eds.): IAS 2018, AISC 867, pp. 585–596, 2019.
https://doi.org/10.1007/978-3-030-01370-7_46

on a sparse subset of pixel intensities. The estimator is trained using gradient boosting for learning an ensemble of regression trees [1]. The method presented is extended to handle missing landmarks, i.e., it provides robustness against occlusion of individual landmarks. The authors also claim that the method is fast and efficient. Qu et al. [2] present a robust cascade regression algorithm to provide a method to locate facial landmarks on the face. In contrast to Kazemi et al. they propose to use RootSIFT features [2] over pixel intensities. To train the regressor model an iteratively reweighted least squares approach is employed. Furthermore, the authors pose different fitting strategies to ensure high precision of the fittings.

Jain et al. [3] use a 68 point landmark set to capture temporal shape variation for recognizing seven different emotion classes. They combine temporal features with latent-dynamic conditional random fields for the classification. For static emotion recognition shape-based features are considered. One of their findings is that happiness and surprise are best to classify and that in the static case anger and sadness are difficult to recognize. Not every feature set for emotion classification is based on facial landmarks. In his dissertation [4] Huang presents a feature set based on local binary patterns (LBP). He enhances the standard definition of local binary patterns by including temporal features for the exploitation of dynamic relations. Furthermore, he provides possibilities to recognize facial expressions in uncontrolled conditions. Lucey et al. [5] use active appearance model features together with a support vector machine to classify action unit features. Using this approach they provide a so-called baseline algorithm, which reaches promising recognition rates. Tong and Ji [6] use dynamic dependencies between action units. They feature a dynamic Bayesian network to model dynamics between action units. In their paper, they provide results that show improvements in contrast to semantic action unit models and towards AdaBoost based methods.

This paper presents a feature set based on angles and sizes. The 68 point landmark model described in Jain et al. [3] is the basis of this set. The method provided by Kazemi et al. [1] is used to extract the landmarks from the facial images. Due to shape and ethnic differences, the human face is highly variant. Thus, the landmarks need to be superimposed to increase robustness. The superimposing is done using the Generalized Procrustes Analysis (GPA) [7]. Through variance-based selection the most variant landmark points are determined. By the variance-based selection and anatomic relations, e.g., eye landmarks belong to the eye, point pairs are selected. These point pairs are used to generate lines. These lines are intersected to measure angles. The most variant angles are selected using a variance-based selection scheme. These angles are then included in the proposed feature set. Size features enhance the feature set. As the eyes and mouth regions have high importance in the definition of facial expressions, ellipses are fitted to these points to calculate size features. The angle and size features generate a facial expression description.

The organization of the paper is as follows. Section 2 provides a problem formulation. Section 3 presents the feature model creation. Section 4 contains the experimental results. Section 5 offers a discussion of the results and conclusions.

## 2   Problem Statement

Emotion classification is a pattern recognition problem. The formal process is displayed in Fig. 1. The input to this process is an image of a face carrying an emotional facial expression. The emotional facial expression is a specific facial expression that connects with an emotion. For example, a smile connects with happiness, or a grim face connects with anger.

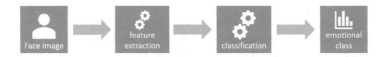

**Fig. 1.** Process flow of the emotion classification.

The next step is concerned with feature extraction, where a numerical representation capturing information about the facial emotion is determined and stored in a feature vector $\underline{x} \in \mathbb{R}^n$ with dimension $n$. In the classification step, the feature vector is mapped to one of many emotional classes $y \in \mathcal{Y}$, according to the nonlinear mapping

$$y = f(\underline{x}) \ , \tag{1}$$

where the function $f(\underline{x})$ represents the classifier. Thus, the desired output for this problem is a decision about the displayed emotion.

Ekman [8,9] provides a formal definition of prototypical emotion classes $\mathcal{Y}$. The class system for this paper has six categories: *anger, disgust, fear, happiness, sadness*, and *surprise*. Without loss of generality, these emotion classes are represented numerically by means of $\mathcal{Y} = \{0, \cdots, 5\}$.

In this paper, we focus merely on extracting a feature vector of low dimension $n$. For the classification step in Fig. 1, we utilize well-known methods from machine learning.

## 3   Feature Creation

The tension of mimic muscles is responsible for the forming of facial expressions. The mimic muscles are responsible for the way the mouth is showing or for the wrinkling of the nose. Facial landmarks provide a good measurement of the current state of the facial expression. They represent specific points on the face that provide a geometrical representation of the face and allow reliably catching the facial expression. The facial landmark detector implemented in Davis King's dlib [10] is used to extract the facial landmarks. The implementation is based on the work in Kazemi et al. [1]. Before the actual landmark extraction, the images are resized to a width of 500 pixels and converted to gray-scale. Figure 2 displays the facial landmarks fitted to a neutral, an angry, and a happy face taken from the Cohn-Kanade+ database [5, 11].

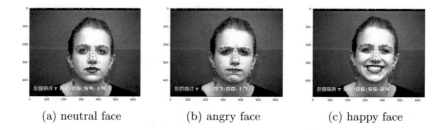

(a) neutral face                (b) angry face                (c) happy face

**Fig. 2.** Landmarks extracted from facial expressions: (a) neutral expression, (b) angry expression, and (c) happy expression. The images are used from the Cohn-Kanade+ Database  (©Jeffrey Cohn)

Figure 2 directly shows the influence of different facial expressions on the configuration of the facial landmarks. The feature points are located slightly different when compared to another facial expression. The landmark model in Fig. 2 contains 68 feature points resolving into a feature vector with $n = 136$ elements, i.e., the coordinates in $x$ and $y$ direction of each feature point. The landmark model constructs as follows: 17 points belong to the outline of the face, and 51 points belong to the inner part of the face. The number of elements in the feature vector is too large for many applications. The large number of elements contains ambiguous information and requires high computational power. Therefore, in this paper, we propose the usage of a combination of angles and sizes to reduce the number of feature points with the aim of retrieving an easy to evaluate feature set.

### 3.1  Generalized Procrustes Analysis

The landmarks are aligned with each other to enable further evaluation of the feature set. The Generalized Procrustes Analysis (GPA) superimposes the facial landmarks of the different faces and expressions. The GPA aligns the landmarks while preserving the shape. To initialize the procedure, one of the shapes to be aligned is chosen as reference shape $\mathcal{R} = \{\underline{r}_1, \dots, \underline{r}_k\}$ with landmark point $\underline{r}_i = [x_i, y_i]^{\mathrm{T}}$ with $\cdot^{\mathrm{T}}$ being the vector transpose. Then, for all remaining shapes, translation, scaling, and rotation operations are performed as follows. At first, the landmarks are translated to the origin by subtracting the mean

$$\hat{\underline{z}} = \frac{1}{k = 68} \sum_{i=1}^{k} \underline{z}_i \; . \tag{2}$$

The Root Mean Squared Distance (RMSD)

$$s = \sqrt{\frac{\sum_{i=1}^{k=68} \left\| \underline{z}_i - \hat{\underline{z}} \right\|^2}{k}} \quad , \tag{3}$$

with $\|\cdot\|$ being the $l_2$-distance, is used to scale the landmark points, such that the points have distance 1 to the origin. Then, the landmarks are rotated to match the rotation of the reference shape. The rotation angle $\theta$ is calculated by

$$\theta = \arctan \left( \frac{\sum_{i=1}^{k} (u_i y_i - v_i x_i)}{\sum_{i=1}^{k} (u_i x_i - v_i y_i)} \right) \quad , \tag{4}$$

where $\underline{z}_i = [u_i, v_i]^{\mathrm{T}}$ is the landmark with index $i$ of the shape to be aligned. Now, based on $\hat{\underline{z}}$, $s$, and $\theta$, the mean shape $\mathcal{Z} = \{\hat{\underline{z}}_i, \dots, \hat{\underline{z}}_k\}$ is calculated for the shape to be aligned. That is, the mean of every landmark over all shapes is calculated. Then, the Procrustes Distance

$$d\left(\hat{\mathcal{Z}}, \mathcal{R}\right) = \sqrt{\sum_{i=1}^{k} \left\| \hat{\underline{z}}_i - \underline{r}_i \right\|^2} \tag{5}$$

is evaluated. If the distance is above a given threshold, the mean shape is defined as the new reference shape, and the process starts over again. After the distance is below the threshold, the process is finished. The GPA guarantees that the distance is minimized with every step.

The next step calculates the covariance matrices for each facial landmark. The covariance matrix $C_i$ is computed from the distribution of each landmark $\underline{z}_i$ over all sample shapes. Using

$$\sigma_i = \sigma_{11} + \sigma_{22} \quad , \tag{6}$$

to calculate the landmark's standard deviation by summing the square root of the diagonal elements of each covariance matrix, where $i$ denotes the index of the $i$th landmark, $\sigma_{jj}$ is the standard deviation of dimension $j = 1, 2$, which refers to the dimensions $x$ and $y$, . Not all 68 landmarks contribute equally to the pattern for recognizing the emotional state. Figure 3 shows the standard deviations of the landmarks using all points. The distribution of the deviations shows that the 17 outline points dominate the variances of the 51 inner points.

## 3.2 Angle and Size Features

The GPA gives an indication about the most robust landmark points that form the basis for reducing the feature points significantly. This reduction allows for a higher classification speed. However, every feature reduction carries the risk of losing information compared to the original landmark model, which may result in worse classification performance. Thus, the reduction needs to be performed in

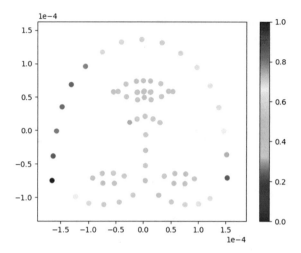

**Fig. 3.** The configuration of points shows the mean shape from the GPA. The shade of the colors indicates the standard deviation of each point.

such a way that no information gets lost. Figure 3 implies that a huge amount of information lies in the landmarks of the eyes and the mouth. Thus, we propose a feature set that extracts size information from the eyes and the mouth. In addition, angles are calculated to describe the configuration of the landmarks in a compressed form.

First, we fit ellipses to the corresponding points of the mouth and the eyes. The semi-major axis $a$ and the semi-minor axis $b$ of the ellipses are used in

$$l = \frac{b}{a} \qquad (7)$$

to calculate the size $l$ of the mouth and eyes. The ratio $l$ is a single-valued measure of the size of the ellipses. Figure 4 displays the procedure to compute the ellipses size.

The configuration of the facial feature points is different for every emotional facial expression. These configuration changes imply that the angles differ for every emotional facial expression.

Figure 3 displays the calculated standard deviations. The standard deviations are very high for the margin points of the face model. The points on the edge of the face are called margin points. The high variability stems from the superimposing of the landmarks, which contributes most to the central points. We chose the most variant landmarks of the margin as well as specific inner points. The eyebrows, as well as the eyes, are chosen, as they are highly variant in the visual inspection of facial expressions. Also, the mouth turns out to be very interesting to investigate, as the mimic muscles act strongly on the mouth. Using this selection results in 55 landmarks. In Fig. 5 the landmarks used are displayed.

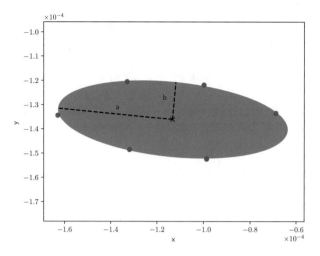

**Fig. 4.** Ellipse fitted using the six landmark points of one eye. The semi-major axis $a$ and semi-minor axis $b$ are used to calculate the axes ratio for the size features.

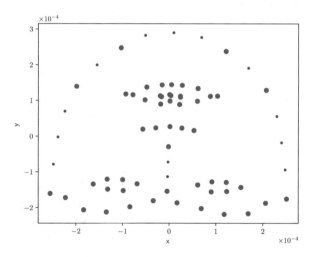

**Fig. 5.** The bold (red) circles denote the landmarks used for the angle generation. The small (blue) dots are omitted.

From these 55 landmarks pairs were chosen, which relate anatomically. These pairs were used to construct lines by calculating direction vectors with

$$\underline{w} = \underline{z} - \underline{z}' \ , \tag{8}$$

where $\underline{w}$ is a direction vector and $\underline{z}$, $\underline{z}'$ are the corresponding landmarks of the pair, i.e, the direction vector is pointing from $\underline{z}'$ to $\underline{z}$. Only the intersection of the directions is considered as the positional information is not required. The angle of intersection between two direction vectors $\underline{w}$ and $\underline{w}'$ is calculated using

$$\cos \alpha = \frac{< \underline{w}, \underline{w}' >}{\|\underline{w}\| \cdot \|\underline{w}'\|} \; , \qquad (9)$$

where $\alpha$ is the angle of intersection, and $< \cdot, \cdot >$ is the scalar product. Figure 6 presents a cropped region with some of the constructed lines.

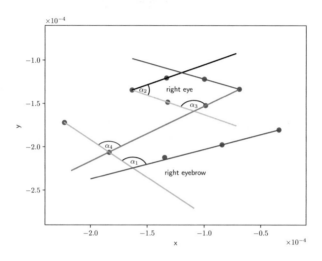

**Fig. 6.** Detailed view on the landmark points of the right eye together with the lines used for the angle computations.

The resulting angles and the size features are combined in a feature vector. The variance of each angle is calculated, and the angles with highest variance become part of the feature vector. Then the classification performance was evaluated using five-fold cross-validation. The first evaluation yielded a lower performance than the landmark feature set. Thus, further angles were included by selecting lines that connect different anatomical regions, e.g., the eye and the nose or the cheek and the eyebrow. After some iterations of this procedure, a high classification performance was reached. Finally, a feature set containing 26 angles and three size features was generated, which yielded the best results.

## 4    Experiments

In the following, the proposed feature set is denoted angle and size features (ASF). ASF is compared against landmark features in terms of classification performance and run-time. The first test evaluates the effect of reducing the raw landmarks as features on the classification metrics. This test uses the pictures with the most intense emotional expression of the Cohn-Kanade+ database. This gives a total of 443 images distributed over the six emotion classes *anger, disgust, fear, happiness, sadness,* and *surprise*. The multilayer perceptron (MLP) classifier from the python library scikit-learn [12] is used to evaluate classification metrics. This classifier is selected as it yielded the highest recognition rates compared to other classifiers tested. The MLP classifier is a neural network using hidden layers. The rectified linear unit (ReLU) activation function is used, and one hidden layer comprising 100 nodes. Both feature sets the landmark features and ASF are preprocessed using a standardization first and a principal component analysis, maintaining all components, afterward. Figure 7 shows the confusion matrices for the landmark features and ASF.

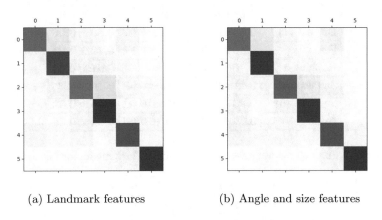

(a) Landmark features        (b) Angle and size features

**Fig. 7.** Confusion matrices showing the results of the feature tests for (a) the landmark features and (b) the proposed features.

As stated in Sect. 2, the class labels $0, \ldots, 5$ correspond to the emotions anger to surprise. The confusion matrices are almost identical, which indicates that even by reducing the feature set there is no significant loss of information.

Table 1 presents the classification metrics of the five-fold cross-validation test using the Cohn-Kanade+ database. The results show that ASF performs superior compared to the landmark features. Furthermore, the table shows that ASF outperforms the landmark features in most cases with a performance gain of up to 6% points. In the few inferior cases, the performance loss is at most merely 2% points. The main contribution of this feature set is that the size of the feature vector can be reduced to 29 compared to 136 elements when using all landmarks. The reduction impacts on the classification time. The significantly lower number of features improves the classification speed. The classification of one sample of the landmark feature set takes 8.8 ms. With the proposed ASF set the classification of one facial expression takes 4.6 ms. The results of the computation times are summarized in Table 2. The reduction in computation time for one sample results from the significant reduction in the number of feature elements.

**Table 1.** Classification metrics for the landmark features compared to the proposed angle- and size-based feature set ASF.

| Features | Class | Precision | Recall | $F_1$-Score | Accuracy |
|---|---|---|---|---|---|
| Landmarks | Anger (0) | **0.77 ± 0.11** | 0.72 ± 0.18 | **0.74 ± 0.14** | |
| | Disgust (1) | 0.81 ± 0.02 | 0.86 ± 0.14 | 0.83 ± 0.06 | |
| | Fear (2) | 0.79 ± 0.12 | 0.70 ± 0.13 | 0.73 ± 0.06 | |
| | Happiness (3) | 0.85 ± 0.04 | **0.93 ± 0.07** | 0.89 ± 0.03 | |
| | Sadness (4) | **0.83 ± 0.11** | 0.81 ± 0.13 | **0.82 ± 0.11** | |
| | Surprise (5) | 0.96 ± 0.04 | 0.93 ± 0.06 | 0.94 ± 0.02 | |
| | Average | 0.83 ± 0.06 | 0.83 ± 0.09 | 0.82 ± 0.08 | 0.84 ± 0.02 |
| ASF | Anger (0) | 0.75 ± 0.04 | **0.72 ± 0.16** | 0.72 ± 0.08 | |
| | Disgust (1) | **0.85 ± 0.12** | **0.92 ± 0.07** | **0.88 ± 0.06** | |
| | Fear (2) | **0.85 ± 0.09** | **0.76 ± 0.11** | **0.79 ± 0.04** | |
| | Happiness (3) | **0.90 ± 0.03** | 0.92 ± 0.06 | **0.91 ± 0.02** | |
| | Sadness (4) | 0.82 ± 0.08 | **0.81 ± 0.08** | 0.81 ± 0.03 | |
| | Surprise (5) | **0.98 ± 0.03** | **0.94 ± 0.04** | **0.96 ± 0.02** | |
| | Average | **0.86 ± 0.07** | **0.85 ± 0.09** | **0.84 ± 0.08** | **0.86 ± 0.02** |

**Table 2.** Average computation time of each feature model for one sample.

| Feature set | Time in ms |
|---|---|
| Landmark features | 8.8 |
| ASF model | 4.6 |

The results obtained for the given CK+ database compare to another study. Jain et al. also provide static results on the database in [3]. They only provide recognition rates without standard deviations. Table 3 lists the results presented in [3] together with the recognition results for the landmark and the ASF feature sets. The results of the baseline method refer to the facial action coding system based model described in [5].

**Table 3.** Result comparison to the static results in [3].

| Emotion | Baseline [3,5] | Shape + SVM [3] | Landmarks + MLP | ASF + MLP |
|---|---|---|---|---|
| Anger | 75.00 | 74.70 | 71.67 ± 17.95 | 71.67 ± 16.33 |
| Disgust | 94.70 | 87.13 | 86.15 ± 14.10 | 92.42 ± 6.88 |
| Fear | 65.20 | 88.77 | 70.00 ± 13.41 | 75.93 ± 10.67 |
| Happiness | 100.00 | 98.43 | 93.33 ± 6.48 | 92.22 ± 5.67 |
| Sadness | 68.00 | 63.70 | 81.33 ± 12.93 | 81.33 ± 7.77 |
| Surprise | 96.00 | 91.67 | 93.07 ± 5.66 | 94.25 ± 3.52 |
| Average | 83.15 | 84.06 | 82.59 ± 9.29 | 84.64 ± 8.81 |

The experiments show that the proposed method is capable of retrieving good results for the static emotion recognition problem. Furthermore, the features are easy to calculate. The comparison with the landmarks shows good suitability for most of the emotions.

## 5   Conclusion

In this paper a feature set extracted from a 68 point landmark model is presented. The underlying landmark model itself forms a feature set to represent emotional facial expressions. With the definition of highly variable landmarks and the composition of lines, created from pairs of landmarks, the intersection of these lines computes angles. These angles provide insights on the changes of landmark configurations. This information is given by 26 angles, together with three size ratios, being calculated from ellipses that are fitted to both eyes and to the outline of the mouth.

The experimental results show that when combined with a multilayer percepton classifier the reduced feature set reaches and often even outperforms the performance of the landmark model. Furthermore, the experiments showed a reduction of the time for classifying a single feature vector compared to the time consumed for a landmark feature vector. In the literature papers exist that use deep learning techniques for facial expression recognition yielding higher results. E.g., Lopes et al. [13] reported an overall accuracy of 96.4% on the Cohn-Kanade+ dataset using a convolutional neural network (CNN) with 7 layers and several preprocessing steps with rotation correction, face cropping, subsampling and intensity normalisation. Liu et al. presented an accuracy of 92.4% on the same dataset with a 7 layers CNN on preprocessed sequences for training in [14]. Finally, Mollahosseini et al. [15] presented 93.2% accuracy on the dataset using a sparse neural network with an incepted neural network to improve recognition rates. As the results from the deep learning based works are not compared in this paper, this comparison is part of future work.

The provided feature set has several properties that support dynamic analysis. The temporal evolution of angles and the size ratios can easily be measured. It is possible that the usage of temporal changes of angles can improve the reliability of the feature model. Consequently, the recognition rate might be improved. The angles and the knowledge about the angle creation enable the generation of a generative model. The research on the dynamic properties of the model as well as on the generative model will be part of future research.

**Acknowledgment.** The research and development project on which this report is based is being funded by the Federal Ministry of Transport and Digital Infrastructure within the mFUND research initiative.

# References

1. Kazemi, V., Sullivan, J.: One Millisecond face alignment with an ensemble of regression trees. In: Proceedings of the 2014 IEEE Conference on Computer Vision and Pattern Recognition (CVPR 2014), Columbus, OH, USA, pp. 1867–1874 (2014)
2. Qu, C., Gao, H., Monari, E., Beyerer, J., Thiran, J.-P.: Towards robust cascaded regression for face alignment in the wild. In: Proceedings of the 2015 IEEE Conference on Computer Vision and Pattern Recognition Workshops (CVPRW), Boston, MA, USA, 1–9 (2015)
3. Jain, S., Hu, C., Aggarwal, J.K.: Facial expression recognition with temporal modeling of shapes. In: Proceedings of the 2011 IEEE Conference on Computer Vision Workshops (ICCV Workshops), pp. 1642–1649 (2011)
4. Huang, X.: Methods for facial expression recognition with applications in challenging situations. Doctoral Dissertation, Acta Universitatis Ouluensis. C, Technica., Number 509 (2014)
5. Lucey, P., Cohn, J.F., Kanade, T., Saragih, J., Ambadar, Z., Matthews, I.: The extended Cohn-Kanade dataset (CK+): a complete expression dataset for action unit and emotion-specified expression. In: Proceedings of the Third International Workshop on CVPR for Human Communicative Behavior Analysis (CVPR4HB 2010), San Francisco, USA, pp. 94–101 (2010)
6. Tong, Y., Ji, Q.: Exploiting dynamic dependencies among action units for spontaneous facial action recognition. In: Emotion Recognition, pp. 47–67. Wiley (2015)
7. Gower, J.C.: Generalized procrustes analysis. Psychometrika **40**(1), 33–51 (1975)
8. Ekman, P.: An argument for basic emotions. Cogn. Emot. **6**(3/4), 169–200 (1992)
9. Ekman, P.: Basic Emotions. Handbook of Cognition and Emotion, pp. 45–60. Wiley, New York (1999)
10. King, D.E.: Dlib-ml: a machine learning toolkit. J. Mach. Learn. Res. **10**(4), 1755–1758 (2009)
11. Kanade, T., Cohn, J.F., Tian, Y.: Comprehensive database for facial expression analysis. In: Proceedings of the Fourth IEEE International Conference on Automatic Face and Gesture Recognition (FG 2000), Grenoble, France, pp. 46–53 (2000)
12. Pedregosa, F., Varoquaux, G., Gramfort, A., Michel, V., Thirion, B., Grisel, O., Blondel, M., Prettenhofer, P., Weiss, R., Dubourg, V., Vanderplas, J., Passos, A., Cournapeau, D., Brucher, M., Perrot, M., Duchesnay, E.: Scikit-learn: machine learning in Python. J. Mach. Learn. Res. **12**, 2825–2830 (2011)
13. Lopes, A.T., de Aguiar, E., De Souza, A.F., Oliveira-Santos, T.: Facial expression recognition with convolutional neural networks: coping with few data and the training sample order. Pattern Recognit. **61**, 610–628 (2017)
14. Liu, M., Li, S., Shan, S., Wang, R., and Chen, X.: Deeply learning deformable facial action parts model for dynamic expression analysis. In: Proceedings of the 12th Asian Conference on Computer Vision (ACCV 2014), Singapore (2014)
15. Mollahosseini, A., Chan, D., Mahoor, M.H.: Going deeper in facial expression recognition using deep neural networks. In: Proceedings of the 2016 IEEE Winter Conference on Applications of Computer Vision (WACV), Lake Placid, NY, USA (2016)

# Multimodal Path Planning Using Potential Field for Human–Robot Interaction

Yosuke Kawasaki[1], Ayanori Yorozu[2(✉)], and Masaki Takahashi[1]

[1] Department of System Design Engineering, Keio University,
3-14-1 Hiyoshi, Kohoku-ku, Yokohama 223-8522, Japan
y-kawasaki@keio.jp, takahashi@sd.keio.ac.jp
[2] Graduate School of Science and Technology, Keio University,
3-14-1 Hiyoshi, Kohoku-ku, Yokohama 223-8522, Japan
ayanoriyorozu@keio.jp

**Abstract.** In a human–robot interaction, a robot must move to a position where the robot can obtain precise information of people, such as positions, postures, and voice. This is because the accuracy of human recognition depends on the positional relation between the person and robot. In addition, the robot should choose what sensor data needs to be focused on during the task that involves the interaction. Therefore, we should change a path approaching the people to improve human recognition accuracy for ease of performing the task. Accordingly, we need to design a path-planning method considering sensor characteristics, human recognition accuracy, and the task contents simultaneously. Although some previous studies proposed path-planning methods considering sensor characteristics, they did not consider the task and the human recognition accuracy, which was important for practical application. Consequently, we present a path-planning method considering the multimodal information which fusion the task contents and the human recognition accuracy simultaneously.

**Keywords:** Human–robot interaction · Multimodal path planning
Potential field

## 1 Introduction

Human–robot interaction (HRI), focusing on the communication between people and robots, is a very active research field [1]. To realize HRI with the same smoothness as communication between a person and a person, it is necessary to sense a person accurately at first interaction [2]. Assuming actual operation, it is desirable to be able to perform human sensing and interaction with only the sensors mounted on robots. However, because there are the constraints of the current sensor performance, depending on the location of robots, the position of the person may not be recognized accurately. Therefore, considering the constraints of the sensor, it is necessary for robots to move to a position where robots can obtain more accurate sensor data. When robots interact with people, robots first need to detect and sense the position of the people for approaching them.

© Springer Nature Switzerland AG 2019
M. Strand et al. (Eds.): IAS 2018, AISC 867, pp. 597–609, 2019.
https://doi.org/10.1007/978-3-030-01370-7_47

(a)  Situation of handover                    (b)  Situation of voice communication

**Fig. 1.**  Situation of communication

Then, robots move to a destination for the interaction task. Furthermore, the destination changes according to the position of the people and the task content. For example, when a robot and a person perform a handover, the robot needs the posture information of the person. Because that information can be acquired from the camera image, the robot needs to move to a position where it is easy to recognize a posture of a person, as in Fig. 1(a). On the other hand, when a robot and a person conduct spoken dialogue, because the voice information of the people is necessary, the robot needs to move to a position where the voice can be easily acquired — see Fig. 1(b). To decide the above mentioned destination, the human recognition accuracy is important. In this manner, the type of the important sensor data changes depending on the human recognition accuracy and the task contents. Then, the destination and the path to the destination change based on the characteristics of the important sensor. The robot has to plan the path and destination considering not only the shortest path and obstacle avoidance but also "multimodal information" consisting of recognition accuracy, some sensor characteristics, and the task contents of the interaction.

One general path-planning method used a potential field [3]. In this method, path planning is performed by considering the gradient of the potential field expressing virtual mountains and valleys. Height of the potential evaluates the position where the robot has to move on the actual field. For example, by expressing obstacles as mountains and goals as valleys, it is possible to plan a path considering obstacle avoidance. This method can be performed with low calculation cost. In addition, Mead et al. [4] formulated a potential field called the "interaction potential" based on the recognition rate of human words (microphone) and posture (RGB-D camera). They formulated the interaction potential based on the change in the success rate of human posture and speech recognition when changing the position of the camera and the microphone relative to a person. By using interaction potential, the robot can quantitatively evaluate where the robot can sense people with high accuracy. However, in the actual human-robot interaction, because the situation such as the relative position between the robot and the person changes frequently, the robot has to change the potential field to correspond to it in real time. Furthermore, when performing HRI, it is necessary to consider task contents and recognition accuracy simultaneously. Therefore, we propose "multimodal path planning" based on multimodal information integrating such as sensor characteristics, task contents, and recognition accuracy. In proposed method, the each information is expressed using potential. Because the potentials are integrated based on "superposition principle", the robot is able to plan a path based on the potential field integrated these information. And the robot can move to the destination along the planed path to sense the necessary data for performing the interaction task.

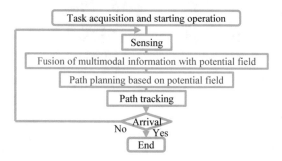

**Fig. 2.** Overview of robot system

## 2  Overview of Robot System

An overview of the proposed system is shown in Fig. 2. The robot starts operation after getting an interaction task. Then, the robot fusions the multimodal information on the potential field based on results of sensing and plans a path dynamically. Finally, it traces the planed path. By processing this flow until approaching the destination, the robot can change the potential field and the path in real time base on the sensing result.

The outline of the paper is as follows: In Sect. 3 the recognition reliability which is evaluation index of the human recognition accuracy is presented. The detail of the information integrating and path planning method is described in Sect. 4. Section 5 describes two experiments to verify the effectiveness of the proposed method. Finally, Sect. 6 presents conclusions.

## 3  Recognition Reliability

### 3.1  Human Recognition Method

In this research, as a human recognition method, we use a human joint position estimation method using an RGB-D camera. In environments where people live, there are cases where parts of the body are hidden frequently; thus, it is difficult to estimate the joint position of a person using only the depth image [6]. Therefore, we use OpenPose [7], which makes human joint recognition possible even when a part of the body is hidden (Fig. 3). However, because OpenPose estimates the position of the joint in the color image, it becomes two-dimensional coordinate information. Therefore, by using the depth image, two-dimensional joint information is converted into three-dimensional information. Then, the position of the person is estimated based on the three-dimensional joint position estimation result.

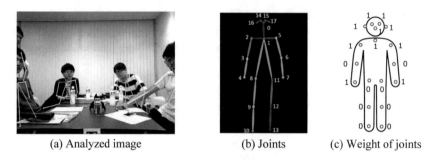

(a) Analyzed image          (b) Joints          (c) Weight of joints

**Fig. 3.** Joint position estimation using OpenPose [7, 8]

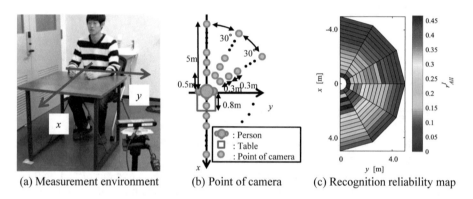

(a) Measurement environment          (b) Point of camera          (c) Recognition reliability map

**Fig. 4.** Measurement of recognition reliability

## 3.2   Recognition Reliability $r_{All}^{I}$

We define recognition reliability $r_{All}^{I}$ to evaluate quantitatively the accuracy of the estimated human position. We use $r_i$ and $r^{Depth}$ as evaluation indexes of the recognition accuracy. $r_i$ is the detection confidence for each joint that can be acquired when executing two-dimensional joint position estimation with OpenPose. In addition, the accuracy of the depth sensor decreases in proportion to the square of the distance from the recognition target [9]. $r^{Depth}$ is a formulation of its characteristics.

First, the detection confidence of human is evaluated using the joint detection confidence $r_i$. The confidence is obtained by the weighted average of each $r_i$ using the weight $W_i$ for each joint—see Fig. 3(b), (c). In the environment where people live, when it is rare that the lower half of a person can be detected clearly, we weighted the upper body. Then, it is necessary to consider the depth sensor characteristics $r^{Depth}$ of according to the distance to the estimated position of the person. Through these processes, recognition reliability $r_{All}^{I}$ is formulated by the following equation.

$$r^I_{All} = \frac{\sum\limits_{i=1}^{18} r_i \, W_i}{\sum\limits_{i=1}^{18} W_i \left(1 + C^{Depth} \, d^2\right)} \tag{1}$$

### 3.3    Experiment for Evaluating Recognition Reliability Characteristics

When sensing a person, the recognition reliability changes depending on the positional relationship between the person and the RGB-D camera. We evaluate how recognition reliability $r^I_{All}$ changes depending on the positional relationship between people and the RGB-D camera. We formulated the characteristics as recognition reliability characteristics $r^I$. As shown in Fig. 4, we conducted experiments in shooting people using RGB-D cameras placed on a circular arc. Using the three-dimensional image acquired through the experiment, we estimated human position and recognition reliability $r^I_{All}$ calculations. Figure 4(c) shows the variation of $r^I_{All}$ with respect to the shooting point. It was confirmed that $r^I_{All}$ becomes highest when shooting from the angle of the front of the person and from the position 1 to 2 m away from the person.

### 3.4    Formulation of Recognition Reliability Characteristics $r^I$

To formulate the recognition reliability characteristic $r^I$ that expresses the variation of the recognition reliability $r^I_{All}$ respected to the distance and the angle to the person, we approximate the average value of the result obtained by the experiment using the least-squares method. Because $r^I$ are a function dependent on distance and angle, we approximated the distance characteristic and the angular characteristic separately and then multiplied to form $r^I$. The result of the formulation is shown in the following equation.

$$r^I = \frac{\frac{\log C^{RGB,D}_1 d}{C^{RGB,D}_2 d} + C^{RGB,D}_3}{1 + C^{Depth} d^2} \left( C^{RGB,A}_1 \theta + C^{RGB,A}_2 \right) \tag{2}$$

where $d$ is distance from the RGB-D camera to the person, $\theta$ is angle between the RGB-D camera and the person, and $C^{RGB,D}, C^{RGB,A}$ and $C^{Depth}$ are the coefficients of each characteristic.

## 4    Multimodal Path Planning

An overview of the proposed multimodal path planning method is shown in Fig. 5. When planning paths for HRI, we generate image potential based on recognition reliability characteristics, spoken dialog potential and obstacle potential.

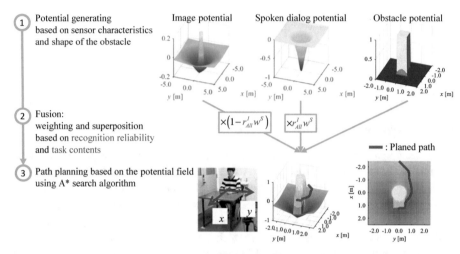

Fig. 5. Overview of multimodal path planning method

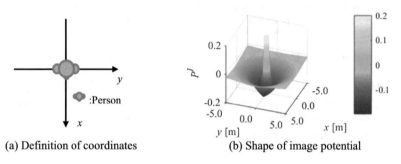

(a) Definition of coordinates          (b) Shape of image potential

Fig. 6. Image potential $P^I$

Then, after weighting each potential according to recognition reliability and task contents, they are superposed to fusion the multimodal information. Finally, we plan the path using the A * algorithm [5], setting the destination to the minimum point of the potential field and the movement cost to the gradient of the potential field. This section describes the detail of the method.

## 4.1  Design of Each Potential

**Design of Image Potential $P^I$.** We formulate the image potential based on recognition reliability characteristics $r^I$. We use the RGB-D cameras to not only recognize the posture of a person but also measure the position of a person. Therefore, it is necessary to consider the measurement of multiple people's positions. When measuring the multi-people position, it is necessary to improve the accuracy when a person has been recognized with poor accuracy compared with a person whose position can be measured accurately. Furthermore, when the measurement accuracy of the human position is low, the angle characteristic of $r^I$ should not be considered.

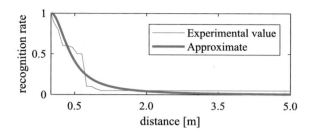

**Fig. 7.** Distance characteristics of microphone

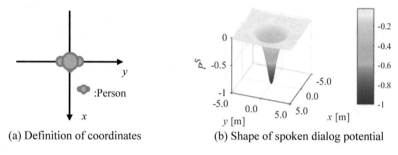

(a) Definition of coordinates          (b) Shape of spoken dialog potential

**Fig. 8.** Spoken dialog potential $P^S$

First, considering the superposition of potential, the magnitude of each potential needs to be normalized. We normalized the recognition reliability characteristics so that the minimum value becomes $-1$. Next, based on the above, the image potential is expressed by the following equation.

$$P^I = -\left(1 - r_{All}^J\right) \frac{\frac{\log C_1^{RGB,D} d}{C_2^{RGB,D} d} + C_3^{RGB.D}}{1 + C^{Depth} d^2} \left(r_{All}^{J2} C_1^{RGB,A} \theta + C_1^{RGB,A}\right) \tag{3}$$

Figure 6 shows the image potential $P^I$ assuming the situation of standing with the front of the body facing the $x$-axis direction.

**Design of Spoken Dialog Potential $P^S$.** The recognition accuracy of the microphone has been verified by experiments in previous research [10]. The results are shown in Fig. 7. Approximation is made assuming that the recognition rate decreases in proportion to the square of the distance from the person in Fig. 7. Therefore, the spoken dialog potential can be expressed by the following equation:

$$P^S = \frac{1}{1 + C^S d^2} \tag{4}$$

where, $C^S$ is the coefficient of distance characteristics, and the coefficient is determined to be trial and error. Figure 8 shows the spoken dialog potential $P^S$ assuming the situation of standing with the front of the body facing the x-axis direction.

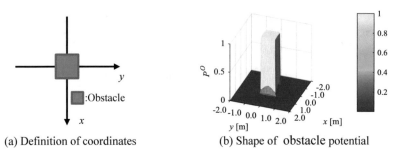

(a) Definition of coordinates            (b) Shape of  obstacle potential

**Fig. 9.**  Obstacle potential $P^O$

**Design of Obstacle Potential $P^O$.** We assume that the robot knows the information of static obstacles such as furniture before operation. We use the obstacle potential $P^O$ so that the robot will not collide with these obstacles. To prevent the robot from entering the area where obstacles are present, we designed the obstacle potential $P^O$ so that the potential of the obstacle area increases relatively. At that time, to maintain the continuity of the potential field, the obstacle potential was designed using the sigmoid function. A figure of the obstacle potential $P^O$ when the desk is an obstacle is shown in Fig. 9.

### 4.2  Weighting Each Potential

We weight each potential according to the recognition reliability $r_{All}^I$ of the people and the task contents. By doing this, we can consider task contents, human measurement accuracy, and sensor characteristics at the same time in the potential field.

When a robot performs an interaction, the robot first moves for measuring the position of the person; then, when the measurement accuracy of the person improves, the robot moves to perform the interaction. In this manner, because the importance of information changes depending on the situation, it is necessary to weight the potential according to the situation. When $r_{All}^I$ is low, human recognition is more important. On the other hand when $r_{All}^I$ is high, to perform the interaction is more important. Based on the above, we dynamically change the weight using $r_{All}^I$. Furthermore, the importance of sensor data changes depending on the task contents of the interaction. For example, when the robot performs an interaction using voice, the weight of the microphone information is high; when the robot performs interaction using the posture of a person, such as object handover, the weight of the RGB-D camera information is high.

Therefore, we weight the importance of sensor data by using the spoken dialog importance $w^S$ determined by task contents. Weighting and superposition using $r_{All}^I$ and $w^S$ are expressed by the following equation.

$$P = \left(1 - r_{All}^I w^S\right)P^I + r_{All}^I w^S P^S + P^O \tag{5}$$

### 4.3   Path Planning Considering the Potential Field

**Destination Determination Method.** We designed the potential field so that the potential of the place where the robot should go is small. Therefore, the destination of the robot is the place with the smallest potential on the potential field. When there are places with multiple minima, a position closer to the robot position is selected as the destination.

**Path Planning Method.** We need to make a path using a planner that can consider the gradient of the potential field. However, considering only the gradient of the potential field, there is a possibility of falling into a local solution. Therefore, we use the A* search algorithm, which is a global path planner that can take into consideration the distance to the destination. To consider the gradient of the potential field with the A* search algorithm, we set the gradient of the potential field as the movement cost $COST(n,m)$ of node $m$ from node $n$. The movement cost $COST(n,m)$ in consideration of the gradient of the potential field can be expressed by the following equation.

$$COST(n,m) = \alpha_A(P(m) - P(n)) \tag{6}$$

where, $P(n)$ is the potential of node $n$, and $\alpha_A$ is a coefficient.

The coefficient $\alpha_A$ is the weight of the heuristic function $h(n)$ and the movement cost $COST(n,m)$. If the coefficient $\alpha_A$ increases, it becomes a path with emphasis on movement cost $COST(n,m)$. However, if the coefficient $\alpha_A$ decreases, it becomes a path with emphasis on the moving distance. In this study, we decided coefficients $\alpha_A$ by trial and error. In addition, we used a method based on fuzzy inference for path tracking.

## 5   Experiment

To verify the effectiveness of the proposed method, a machine experiment was conducted. The method aims at moving robots to an appropriate position in consideration of sensor characteristics, task contents, and human recognition accuracy. In this experiment, we assume the environment of the common household and the environment of a café shown in Fig. 10. The experimental conditions are shown in Tables 1 and 2. We conducted the experiments in two scenarios of assuming the handover task (Experiment 1) and the dialog task (Experiment 2). At the beginning of the experiment, the robot could not see the people. Also, the robot had a two-dimensional grid map built using ROS gmapping package [11]. And the robot estimated self-pose using a particle filter to track the pose of a robot against a known map (the adaptive Monte Carlo localization approach [12]). After the robot started operation, the robot moved while updating the potential field and the path based on the result of sensing.

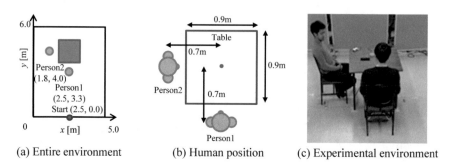

(a) Entire environment          (b) Human position          (c) Experimental environment

**Fig. 10.** Experimental environment

**Table 1.** Robot system parameters.

| OpenPose frequency | 2 Hz |
|---|---|
| Path planning frequency | 3 Hz |

**Table 2.** Experimental scenario.

| Experiment | Task | $w^S$ |
|---|---|---|
| Experiment 1 | Handover | 0.1 |
| Experiment 2 | Dialog | 0.9 |

● :Minimum 90% or less  ▬:Planned Path  ▬:Result Path  ⬆:Robot  ⬆:Person1  ⬆:Person2

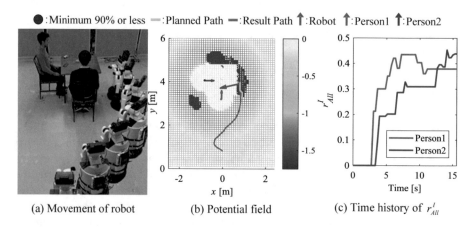

(a) Movement of robot          (b) Potential field          (c) Time history of $r^I_{All}$

**Fig. 11.** Experiment result (Experiment 1)

The result of Experiment 1 is shown in Fig. 11. Figure 11(a) and (b) show the path until the robot reaches the final goal, and Fig. 11 (c) shows the time history of each $r_{All}^I$ while the robot moves. Ultimately, the robot moved to a position on the front side of the two people. Because we imposed the task in which image information is important on the robot, the robot planned the path to acquire image information.

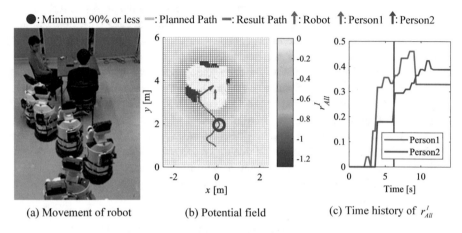

●: Minimum 90% or less  ▬: Planned Path  ▬: Result Path  ↑: Robot  ↑: Person1  ↑: Person2

(a) Movement of robot      (b) Potential field      (c) Time history of $r_{All}^I$

**Fig. 12.** Experiment result (Experiment 2)

Moreover, because $r_{All}^I$ is improved every time the robot moves, it can be confirmed that it planned not only the destination but also the path considering the accuracy of human recognition.

The result of Experiment 2 is shown in Fig. 12. Figure 12(a) and (b) show the path until the robot reaches the final goal, and Fig. 12(c) shows the time history of each $r_{All}^I$ while the robot moves. As a result, the robot moved to a position between the two people. The position is where the robot could get as close as possible to people. Because we imposed tasks in which voice information is important on the robot, the robot made a path considered the ease of acquiring voice information. In addition, when checking the path of the robot, it is found that the robot first moved on a path like the path in Experiment 1, but then the path changed to a path considered the ease of acquiring voice information (marked with a red circle in Fig. 12(b)). Since $r_{All}^I$ was low at the beginning, the robot planned the path to recognize people easily. But after $r_{All}^I$ was improved (marked with a red line in Fig. 12(c)), the robot planned the path to the position where the robot is easy to perform the interaction task.

Finally, the proposed method can be used to plan an appropriate path based on sensor characteristics, recognition status, and task contents.

## 6 Conclusion

We proposed a path-planning method using a potential field integrating multimodal information, such as sensor characteristics, task contents, and people recognition accuracy. First, to evaluate the accuracy of the result when recognizing a person using the RGB-D camera mounted on the robot, we defined the recognition reliability based on the reliability of joint estimation and the distance between the person and the robot. Using the recognition reliability, we carried out the person measurement experiment. Then, the measurement result was formulated as a recognition reliability characteristic. When planning paths for HRI, we generated image potential based on formalized recognition reliability characteristics, spoken dialog potential, and obstacle potential. Then, after weighting each potential according to task contents and recognition reliability, they were superposed to generate a potential field. Finally, we planned the path using the A* algorithm, setting the minimum point of the potential field as the destination and the gradient of the potential field as the movement cost. By updating the potential field in real time based on the sensing result and planning the path, it became possible to obtain the necessary information with high accuracy.

In future work, we will apply and verify this method for different situations, such as HRI with pedestrians.

**Acknowledgment.** This study was supported by "A Framework PRINTEPS to Develop Practical Artificial Intelligence" of the Core Research for Evolutional Science and Technology (CREST) of the Japan Science and Technology Agency (JST) under Grant Number JPMJCR14E3.

## References

1. Kanda, T.: Enabling harmonized human-robot interaction in a public space. In: Human-Harmonized Information Technology, vol. 2, pp. 115–137. Springer, Tokyo (2017)
2. Satake, S., Kanda, T., Glas, D., Imai, M.: How to approach humans?-strategies for social robots to initiate interaction. In: 4th ACM/IEEE International Conference on Human-Robot Interaction (HRI), pp. 109–116 (2009)
3. Rimon, E., Koditschek, D.E.: Exact robot navigation using artificial potential functions. IEEE Trans. Robot. Autom. **8**(5), 501–518 (1992)
4. Mead, R., Matarić, M.J.: Autonomous human–robot proxemics: socially aware navigation based on interaction potential. Auton. Robot. **41**, 1189–1201 (2017)
5. Hart, P.E., Nilsson, N.J., Raphael, B.: A formal basis for the heuristic determination of minimal cost paths. IEEE Trans. Syst. Sci. Cybern. **4**(2), 100–107 (1968)
6. Li, B., Jin, H., Zhang, Q., Xia, W., Li, H.: Indoor human detection using RGB-D images. In: IEEE International Conference on Information and Automation, pp. 1354–1360 (2016)
7. Cao, Z., Simon, T., Wei, S.E., Sheikh, Y.: Realtime multi-person 2D pose estimation using part affinity fields. In: Computer Vision and Pattern Recognition. arXiv:1611.08050 (2017)

8. CMU-Perceptual-Computing-Lab, "OpenPose". https://github.com/CMU-Perceptual-Computing-Lab/openpose. Accessed Feb 2018
9. Khoshelham, K., Elberink, S.O.: Accuracy and resolution of Kinect depth data for indoor mapping applications. In: Sensors, vol. 12, pp. 1437–1454 (2012)
10. Ishiguro, H., Miyashita, T., Kanda, T.: Science of knowledge Communication robot. Ohmsha, (Japanese) (2005)
11. gmapping, http://www.ros.org/gmapping. Accessed May 2018
12. AMCL. http://www.ros.org/amcl. Accessed May 2018

# 2D and 3D Computer Vision for Robotics

# Deep Learning Waterline Detection
# for Low-Cost Autonomous Boats

Lorenzo Steccanella, Domenico Bloisi, Jason Blum, and Alessandro Farinelli[✉]

Department of Computer Science, University of Verona,
Strada le Grazie 15, 37134 Verona, Italy
alessandro.farinelli@univr.it

**Abstract.** Waterline detection in images captured from a moving camera mounted on an autonomous boat is a complex task, due the presence of reflections, illumination changes, camera jitter, and waves. The pose of the boat and the presence of obstacles in front of it can be inferred by extracting the waterline. In this work, we present a supervised method for waterline detection, which can be used for low-cost autonomous boats. The method is based on a Fully Convolutional Neural Network for obtaining a pixel-wise image segmentation. Experiments have been carried out on a publicly available data set of images and videos, containing data coming from a challenging scenario where multiple floating obstacles are present (buoys, sailing and motor boats). Quantitative results show the effectiveness of the proposed approach, with 0.97 accuracy at a speed of 9 fps.

**Keywords:** Robotic boats · Autonomous navigation · Deep learning
Robot vision

## 1 Introduction

The water quality monitoring process costs every year more than 1 billion EUR at the European Union level. In particular, investigation looking for pollution in large rivers or lakes supplying drinking water are in the range of 150,000 to 400,000 EUR. The current approach, based on field sample and laboratory analysis, is unable to assess temporal and spatial variation in the contaminants of concern. This means that, in case of an accident, there is the risk of taking inadequate and late decisions for mitigating the pollution impacts.

The use of autonomous surface vehicles (ASVs) for persistent large-scale monitoring of aquatic environments is a valid and efficient alternative to the traditional manual sampling approach [1]. ASVs are capable of undertaking long-endurance missions and carrying multiple sensors (e.g., for measuring electrical conductivity and dissolved oxygen) to obtain water quality indicators [2].

© Springer Nature Switzerland AG 2019
M. Strand et al. (Eds.): IAS 2018, AISC 867, pp. 613–625, 2019.
https://doi.org/10.1007/978-3-030-01370-7_48

**Fig. 1.** IntCatch2020 project uses Platypus Lutra boats, about 1 m long and 0.5 m wide. A camera has been mounted on the front of the ASV.

There exist commercial ASVs specifically developed for water quality monitoring. An example is the Lutra mono hull boat produced by Platypus[1], which can mount a submerged propeller (see Fig. 1) or an air fan for propulsion. Lutra boats are used in the EU-funded project IntCatch2020[2], which will develop efficient and user-friendly monitoring strategies for facilitating sustainable water quality management by community groups and non-governmental organizations (NGOs).

To achieve true autonomous navigation, an ASV must sense its environment and localize itself within that environment. We seek to develop vision-based sensing as a first step, focusing on the domain of small, low-cost ASVs. The low-cost design goals of the IntCatch boat preclude the use of sensors commonly utilized for these tasks, such as Lidar. Avoiding water-bourne obstacles can begin by segmenting an image into water and non-water pixels. From there, we can attempt to derive a relationship between the location of non-water pixels in the image and distance from the camera to those objects. Such a relationship would be heavily dependent on the pitch and roll angles of the ASV, and the low-cost gyroscope and accelerometer available on the IntCatch boat are not very precise. To address this, a more precise measurement of the pitch and roll angles of the ASV could be determined by tracking the horizon line. With limited onboard processing power, we seek methods that can address both the pixel-wise segmentation of the image to first identify obstacles, and track the horizon line to increase the precision of pitch and roll estimates.

The dynamic nature of water is capable of producing mirror-accurate reflections of the environment or a turblent, erratically textured surface. Recent advancements in deep learning methods for computer vision, particularly Convolutional Neural Networks (CNNs) [3], show promise for segmenting images captured by an ASV into obstacles and the water that surrounds them. While CNNs have been previously introduced into other mobile robotics domains (e.g., indoor wheeled and quadrotor), and ASV obstacle avoidance has been attempted

---

[1] senseplatypus.com.

[2] www.intcatch.eu.

with classical computer vision methods such as optical flow, the two have yet to be combined.

In this paper, we present a pixel-wise deep learning method for segmenting images captured by a camera mounted on a small ASV and extracting the horizon line, which we refer to as the *waterline* in this context. Another important contribution is the creation of a publicly available data set of images and videos, called IntCatch Vision Data Set[3]. This data set contains annotated data that can be used for developing supervised approaches for object detection.

The remainder of the paper is structured as follows. Related work is discussed in Sect. 2. The proposed method is presented in Sect. 3. Experimental evaluation is shown in Sect. 4. Finally, conclusions are drawn in Sect. 5.

## 2  Related Work

Monocular vision-based obstacle detection for low-cost Autonomous Surface Vehicles (ASVs) has previously received some interest. El-Gaaly et al. [4] utilize sparse optical flow, reflection rejection, and an occupancy grid overlaid on the image. This process produced a significant number of false positives when the water surface was disturbed. Sadhu et al. [5] used grayscale histograms of pixel neighborhoods as a descriptor of texture and saliency to detect logs floating on the surface of water. However, this method ignores the shoreline.

The inherent complexity of the water scenario makes unsupervised approaches, like the above ones, effective only in a limited set of situations. In this work, we apply a supervised approach, to obtain a mapping between image pixels and the two classes *water/not-water*. In particular, we use Convolutional Neural Networks (CNNs), a type of deep, feed-forward Artificial Neural Networks (ANNs). Since ANNs are able to approximate any continuous functional mapping, they can be employed when the form of the required function is unknown [6].

CNNs have shown impressive performance on image classification [7] and object detection [8]. Recognition approaches use machine learning methods based on pre-trained models to address the various categories of objects present in general scenes. Segnet [9] is an example of deep fully convolutional neural network architecture for semantic *pixel-wise* segmentation. Its segmentation engine relies on an encoding-decoding scheme and it can be employed for scene understanding applications.

Multiple mobile robotics domains have started to incorporate CNNs into obstacle detection and navigation. Giusti et al. [10] detect forest paths for a quadrotor to navigate. Chakravarty et al. [11] trained a CNN to produce depth maps from a single image in indoor enviroments. To the best of our knowledge, CNNs have yet to be implemented for ASV applications.

U-net [12] is a encoder-decoder type of network for pixel-wise predictions. In U-net, the receptive fields after convolution are concatenated with the receptive

---

[3] goo.gl/Kxt6HP.

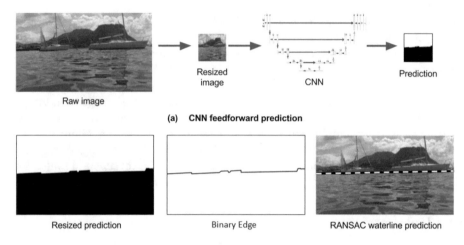

(a)   **CNN feedforward prediction**

Resized prediction          Binary Edge          RANSAC waterline prediction

(b)   **Waterline overlay on class mask edge pixels**

**Fig. 2.** Waterline detection algorithm, idealized example. (a) Raw captured image is fed to trained neural network, producing the class mask. (b) Pixels on the border between classes are isolated by detecting the edges in the mask, and the waterline prediction is created with linear regression over the edge pixels.

fields in a up-convolution process, allowing the network to use original features in addition to features after transpose convolution. Every step in the expansive path consists of *(i)* an upsampling of the feature map followed by a $2 \times 2$ transpose convolution that halves the number of feature channels, *(ii)* a concatenation with the correspondingly cropped feature map from the contracting path, and *(iii)* two $3 \times 3$ convolutions, each followed by a ReLU. This results in overall better performance than a network that has access to only features after up-convolution. U-net is the starting point for building our CNN architecture.

## 3   Methods

We use a Fully Convolutional Neural Network [13] to segment images captured by an ASV, classifying pixels as water or non-water. After this, we process the segmentation mask to detect the waterline. Figure 2 shows the overview of our approach.

### 3.1   Network Architecture

A U-net based architecture is used for the segmentation process [12]. The decision of choosing the U-net architecture has been taken for three main reasons:

1. U-net does not have any dense layer. This means that there is no restriction on the size of the input image.

2. The training stage can be carried out even with a limited amount of training data.
3. The receptive fields in encoding and decoding are concatenated. This allows the network to consider the feature after upconvolution together with the original ones.

The architecture of the net is shown in Fig. 3. The input image is downsampled to obtain a $160 \times 160$ resized image. The encoding stage is needed to create a 512 feature vector, while the decoding stage is needed to obtain the predicted mask at $160 \times 160$ pixels. The encoding stage is made of ten $3 \times 3$ convolutional layers, and by four $2 \times 2$ max pooling operations with stride 2. In particular, there is a repeated application of two unpadded convolutions, each followed by a rectified linear unit (ReLU) and a max pooling operation. The expansive path (see the right side of Fig. 3) is made of eight $3 \times 3$ convolutional layers and by four $2 \times 2$ transpose layers. There is a repeated application of two unpadded convolutions, each followed by a ReLU and a transpose operation.

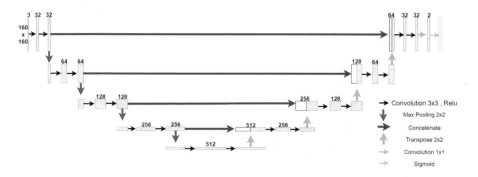

**Fig. 3.** Architecture for Full $160 \times 160$ network.

**Increasing the Speed of the Net.** Our eventual goal is autonomous obstacle avoidance for a low-cost ASV. When avoiding collisions while moving at full speed, processing time is at a premium and processing speed is critical. This motivates the pursuit of scalable methods that can be deployed with limited processing power. With the above described network architecture (denoted as *Full $160 \times 160$*), we are able to obtain a computational speed of about 4.5 fps (see next Sect. 4). We created three additional networks to explore the trade-off between performance and processing speed: Full $80 \times 80$, Half-Conv $160 \times 160$, and Half-Conv $80 \times 80$.

Full $80 \times 80$ is a network with the same architecture of Full $160 \times 160$ with an $80 \times 80$ reduced image in input. Half-Conv $160 \times 160$ and Half-Conv $80 \times 80$ are two reduced networks, meaning that they present an architecture where the convolution filter size is reduced by a factor of two. By reducing the input and convolutional filter size, it is possible to obtain a faster computation in terms of frames per second. Quantitative experimental results are shown in Sect. 4.

## 3.2   Data Set

To train the net we have used 191 labeled images coming from the IntCatch Vision Data Set[4], which has been created to store visual and sensor data collected during the IntCatch2020 project. At the moment, the data set contains 22 low and high resolution videos captured at different sites. The IntCatch Vision Data Set will be extended and updated until the end of the project, namely January 2020.

In this work, we use 8 videos coming from Lake Garda in Italy. Figure 4 shows three frames captured at different times of the day.

**Fig. 4.** Three frames captured at Lake Garda in Italy, which is one of the sites of the IntCatch2020 project.

Training has been performed on 191 labeled images and validated for early stopping on 40 labeled images taken from the first 6 videos namely:

– sequence lakegarda-may-9-prop-1
– sequence lakegarda-may-9-prop-2
– sequence lakegarda-may-9-prop-3
– sequence lakegarda-may-9-prop-4
– sequence lakegarda-may-9-prop-5
– sequence lakegarda-may-9-prop-6

While the experimental results have been obtained by considering 80 labeled images taken from:

– sequence lakegarda-may-9-prop-7
– sequence lakegarda-may-9-prop-8

**Annotation.** In order to perform supervised training of our network trough gradient descent and to evaluate the results, we annotated a total of 311 images by hand. The annotation has been performed with a custom tool created for the task. The tool takes advantage of super-pixel segmentation to give hints to the user that has to select the segments belonging to the water class. To perform super-pixel segmentation we used the SLIC algorithm [14]. This algorithm performs K-means in the 5d space of color information and image location.

---

[4] goo.gl/Kxt6HP.

The mask created at this stage has been further redefined by a brush drag and drop with the mouse.

**Data Augmentation.** Working with a data set of limited size presents a problem related to overfitting: Models trained with a small amount of data can have a limited generalization capacity. Data augmentation has become a usual practice to handle training on small data sets [15]. In this work, we have performed augmentation on the Lake Garda data to create a larger training data set. Figure 5 shows some results of the data augmentation process.

**Fig. 5.** Data augmentation.

Annotated ground truth masks were flipped horizontally, transformed through a random rotation within $-20$ to $20°$, sheared, zoomed, improved with a random brightness filter and patches from the full size original masks have been extracted producing 7 times the number of the original training samples. The dynamics of water frequently cause a small ASV to rotate (roll) significantly. While natural rolling motion is captured in the raw video footage, this effect is critical for performance, so we augment the data set with rotations as suggested in [12]. The masks obtained by flipping and rotating have been scaled down to $160 \times 160$ or $80 \times 80$ pixels, according to the input size of the CNN.

### 3.3   Training

The training step has been performed taking advantage of mini-batch gradient descent. In fact, even with our limited size data set, training over all the images (i.e., Vanilla Gradient Descent) was impossible on a commercial laptop.

Mini-batch gradient descent performs an update for every mini-batch of $n$ training examples:

$$\theta = \theta - \alpha \bigtriangledown J(\theta; x^{(z:z+bs)}; y^{(z:z+bs)}) \tag{1}$$

where $\theta$ are the weights, $\alpha$ is the learning rate, $bs$ is the mini-batch size and $J$ the cost function, computed as:

$$J(\theta, b, x^{(z:z+bs)}, y^{(z:z+bs)}) = \frac{1}{bs} \sum_{z=0}^{bs} J(\theta, b, x^{(z)}, y^{(z)}) \tag{2}$$

We use an Adam optimizer [16] with a *bs* of 30 images. Adam Optimizer performs gradient descent with momentum, involving a weighted average. The hyper parameters chosen for training our net are: a learning rate of 0.001, a $B_1$ value of 0.9 and a $B_2$ value of 0.999. In order to prevent overfitting, the nets have been trained over 20 epochs performing early stopping by monitoring the loss over the validation set [17].

As a difference with respect to the U-net formulation given in [12], where a cross entropy loss function is used, we employ the Dice Similarity Coefficient (DSC) as loss function, which is defined as follows [18].

$$J = DSC = \frac{2 \cdot TP}{2 \cdot TP + FP + FN} \tag{3}$$

where $TP$, $TN$, $FP$, and $FN$ are the number of pixel-wise true positives, true negatives, false positives, and false negatives respectively. Using the DSC loss function allows to mitigate the class imbalance problem, i.e., pixel belonging to water class are more than other pixels.

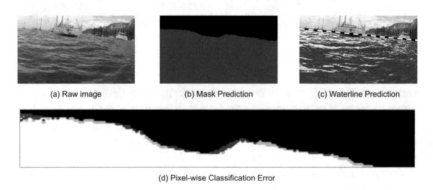

(a) Raw image        (b) Mask Prediction        (c) Waterline Prediction

(d) Pixel-wise Classification Error

**Fig. 6.** Example classification and waterline result. In (d), black is true negative, white is true positive, green is false positive, and blue is false negative. The image is zoomed to the region around the waterline.

### 3.4    Waterline Detection Algorithm

Figure 2b shows an idealized example of the waterline detection algorithm. A raw test image is resized and fed through the CNN, producing a segmentation mask. The segmentation ideally follows the contour of the boundary between the water and any obstacles and the horizon line. To effectively extract the horizon line, which we refer to as the waterline in this context, we use post-processing methods that reduce the influence of sharp changes in the contour due to obstacles.

First a median filter is applied to the segmentation mask, reducing narrow protrusions (such as sailboat masts). Binary edge detection is then used to isolate the pixels on the contour between the two classes. The probabilistic Hough transform with line primitive is applied to filter out pixels that may be part

**Table 1.** Segmentation Results

| Network | Precision | Recall | Accuracy | F₁ Score | Fps |
|---|---|---|---|---|---|
| Half-Conv $80 \times 80$ | 0.932 | 0.982 | 0.954 | 0.956 | 16 |
| Half-Conv $160 \times 160$ | 0.964 | 0.991 | 0.977 | 0.978 | 9 |
| Full $80 \times 80$ | 0.972 | 0.987 | 0.979 | 0.979 | 10 |
| Full $160 \times 160$ | **0.984** | **0.993** | **0.988** | **0.988** | **4.5** |

shorter, closed contours that appear below the horizon line (a small buoy below the horizon line, for example). Only the pixels that compose a long edge are retained. Finally, linear regression is performed, utilizing RANSAC [19] to further reduce the influence of sharp changes in the contour. Figure 6 shows an example result of the waterline detection algorithm.

## 4    Experimental Results

Lake Garda represents a challenging scenario for a low-cost ASV due to (1) the presence of large waves (compared to the dimension of the ASV) and (2) a high number of floating obstacles (boats and buoys) on the water surface.

The four networks that we developed have been written in Python, using functions included in the libraries OpenCV, TensorFlow, and Keras. The complete source code is available as part of the IntCatch AI library downloadable at goo.gl/KBSoQD.

**Quantitative Evaluation.** All the experiments have been carried out using a notebook with the following specifications: Intel Core i5-6300HQ CPU, 8 GB RAM, and a NVIDIA GeForce GTX 960M w/2 GB GDDR5 GPU. Training time varies according to the complexity of the network and number of features extracted, it takes from a minimum of 4 h to a maximum of 6 h on our notebook.

We evaluate the performance of the water pixel-wise classification with multiple metrics: Precision ($P$), Recall ($R$), Accuracy ($A$), and F1-score ($F_1$) [20]. Using $TP$, $TN$, $FP$, and $FN$ as the number of pixel-wise true positives, true negatives, false positives, and false negatives respectively, they are defined as

$$P = \frac{TP}{TP + FP} \tag{4}$$

$$R = \frac{TP}{TP + FN} \tag{5}$$

$$A = \frac{TP + TN}{TP + FP + TN + FN} \tag{6}$$

$$F_1 = 2\frac{P \times R}{P + R} \tag{7}$$

Table 1 shows the segmentation results for the four different network configurations tested. The video sequence used for computing the quality metrics for water segmentation and waterline extraction can be downloaded at goo.gl/FHgwkV.

**Waterline Error and Speedup.** A pixel-wise evaluation of classification error does not translate directly into an error in the waterline. To evaluate this, we calculated the maximum vertical distance in pixels between the ground truth waterline (a RANSAC fit over the ground truth mask) and the predicted waterline for each original size test image. The median of this distance for the entire test set is plotted against the number of frames processed through the algorithm per second, shown in Fig. 7.

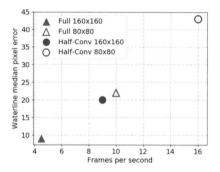

**Fig. 7.** Waterline median pixel error vs. frames per second.

**Fig. 8.** Two examples demonstrating the limitations of a water line. (a) The contour of a boat begins to appear and is classified correctly. RANSAC line sticks to dominant horizon line. (b) Waterline construct breaks down completely, motivating the use of a water contour.

**Discussion.** We used the median pixel error between the ground truth waterline and the predicted waterline to provide a more geometric measure of error. This error scales in a roughly linear fashion with the frames per second processed by the algorithm (see Fig. 7). This demonstrates the scalability of the method.

While this does not translate directly into a measure of obstacle avoidance performance, we achieve processing speeds that would be acceptable for a small, low-speed ASV. When the boundary between water and non-water pixels is dominated by the horizon line, assuming the boundary between water and obstacles is linear is reasonable. But as the distance to obstacles decreases, this assumption begins to break down. Figure 8 shows two examples of this. At a certain point, a water boundary "contour" becomes more useful. Deep learning methods, especially CNNs, could be used to learn to identify this contour directly from raw pixel data, circumventing the need for edge detection or other similar post-processing.

## 5    Conclusions

In this paper, we have shown the use of a deep learning based method for waterline detection on a low-cost ASV. Images captured by the ASV were segmented pixel-wise into water and not-water classes using a CNN. A line was fit to the edges in this binary class mask to create the waterline prediction. This can be used to infer the pose of the ASV and detect obstacles in front of the boat. The waterline extraction method has been tested on different sequences captured at Lake Garda in Italy. The chosen application scenario is challenging, due to the presence of large waves (compared to the dimension of our ASV) and a number of floating objects (buoys, sailing and motor boats).

We have demonstrated the effectiveness of the method with two different network architectures and two input layer sizes. The pixel-wise classification achieves good performance in all four cases, with accuracy and $F_1$ score ranging from 0.954 to 0.988. Reducing the input size and filter size of the CNN results in significant speedup without a significant reduction in the segmentation accuracy.

**Future Directions.** This work represents a first proof of the feasibility of a deep learning approach to horizon line detection and water segmentation on images coming from a small ASV. It paves the way for investigating the use of different types of network architectures. A Recurrent Neural Network over the latent space on our network can be used to track the feature over time, and this could provide improvements in accuracy and smoothness of transitions between frame detection. The network itself could provide directly the horizon line attaching dense layers over the encoded latent space, and having a multiple output network.

The water detection scheme proposed in this paper will be used as starting point to develop an obstacle avoidance module on an embedded board mounted on the IntCatch boat. Moreover, we intend to expand the training input and experimental results to other operational environments within the IntCatch2020 project.

**Acknowledgment.** This work is partially funded by the European Union's Horizon 2020 research and innovation programme under grant agreement No 689341+.

# References

1. Dunbabin, M., Grinham, A.: Quantifying spatiotemporal greenhouse gas emissions using autonomous surface vehicles. J. Field Robot. **34**(1), 151–169 (2017)
2. Codiga, D.L.: A marine autonomous surface craft for long-duration, spatially explicit, multidisciplinary water column sampling in coastal and estuarine systems. J. Atmos. Oceanic Technol. **32**(3), 627–641 (2015)
3. Lecun, Y., Bottou, L., Bengio, Y., Haffner, P.: Gradient-based learning applied to document recognition. In: Proceedings of the IEEE (1998)
4. El-Gaaly, T., Tomaszewski, C., Valada, A., Velagapudi, P., Kannan, B., Scerri, P.: Visual obstacle avoidance for autonomous watercraft using smartphones. In: Autonomous Robots and Multirobot Systems Workshop (2013)
5. Sadhu, T., Albu, A.B., Hoeberechts, M., Wisernig, E., Wyvill, B.: Obstacle detection for image-guided surface water navigation. In: 2016 13th Conference on Computer and Robot Vision (2016)
6. Castellini, A., Manca, V.: Learning regulation functions of metabolic systems by artificial neural networks. In: Proceedings of the 11th Annual Genetic and Evolutionary Computation Conference, pp. 193–200 (2009)
7. Krizhevsky, A., Sutskever, I., Hinton, G.E.: ImageNet classification with deep convolutional neural networks. In: Advances in Neural Information Processing Systems, vol. 25, pp. 1097–1105 (2012)
8. Erhan, D., Szegedy, C., Toshev, A., Anguelov, D.: Scalable object detection using deep neural networks. In: CVPR, pp. 2155–2162 (2014)
9. Badrinarayanan, V., Kendall, A., Cipolla, R.: Segnet: a deep convolutional encoder-decoder architecture for image segmentation. IEEE Trans. Pattern Anal. Mach. Intell. **39**(12), 2481–2495 (2017)
10. Giusti, A., Guzzi, J., Cirean, D.C., He, F.L., Rodrguez, J.P., Fontana, F., Faessler, M., Forster, C., Schmidhuber, J., Caro, G.D., Scaramuzza, D., Gambardella, L.M.: A machine learning approach to visual perception of forest trails for mobile robots. IEEE Robot. Autom. Lett. **1**(2), 661–667 (2016)
11. Chakravarty, P., Kelchtermans, K., Roussel, T., Wellens, S., Tuytelaars, T., Eycken, L.V.: CNN-based single image obstacle avoidance on a quadrotor. In: 2017 IEEE International Conference on Robotics and Automation (2017)
12. Ronneberger, O., Fischer, P., Brox, T.: U-net: convolutional networks for biomedical image segmentation. In: MICCAI, pp. 234–241 (2015)
13. Long, J., Shelhamer, E., Darrell, T.: Fully convolutional networks for semantic segmentation. In: CVPR, pp. 3431–3440 (2015)
14. Achanta, R., Shaji, A., Smith, K., Lucchi, A., Fua, P., Süsstrunk, S.: Slic superpixels compared to state-of-the-art superpixel methods. IEEE Trans. Pattern Anal. Mach. Intell. **34**(11), 2274–2282 (2012)
15. Perez, L., Wang, J.: The effectiveness of data augmentation in image classification using deep learning, arXiv preprint arXiv:1712.04621 (2017)
16. Kingma, D.P., Ba, J.: Adam: A method for stochastic optimization, arXiv preprint arXiv:1412.6980 (2014)
17. Caruana, R., Lawrence, S., Giles, C.L.: Overfitting in neural nets: Backpropagation, conjugate gradient, and early stopping. In: Advances in Neural Information Processing Systems, pp. 402–408 (2001)
18. Kline, T.L., Korfiatis, P., Edwards, M.E., Blais, J.D., Czerwiec, F.S., Harris, P.C., King, B.F., Torres, V.E., Erickson, B.J.: Performance of an artificial multi-observer deep neural network for fully automated segmentation of polycystic kidneys. J. Digital Imaging **30**(4), 442–448 (2017)

19. Fischler, M.A., Bolles, R.C.: Random sample consensus: a paradigm for model fitting with applications to image analysis and automated cartography. Commun. ACM **24**(6), 381–395 (1981)
20. Pennisi, A., Bloisi, D.D., Nardi, D., Giampetruzzi, A.R., Mondino, C., Facchiano, A.: Skin lesion image segmentation using delaunay triangulation for melanoma detection. Comput. Med. Imaging Graph. **52**, 89–103 (2016)

# Context-Aware Recognition of Drivable Terrain with Automated Parameters Estimation

Jan Wietrzykowski and Piotr Skrzypczyński[✉]

Institute of Control, Robotics, and Information Engineering,
Poznań University of Technology, ul. Piotrowo 3A, 60-965 Poznań, Poland
{jan.wietrzykowski,piotr.skrzypczynski}@put.poznan.pl

**Abstract.** This paper deals with the terrain classification problem for autonomous service robots in semi-structured outdoor environments. The aim is to recognize the drivable terrain in front of a robot that navigates on roads of different surfaces, avoiding areas that are considered non-drivable. Since the system should be robust to such factors as changing lighting conditions, mud and fallen leaves, we employ multi-sensor perception with a monocular camera and a 2D laser scanner. The labeling of the terrain obtained from a Random Trees classifier is refined by context-aware inference using the Conditional Random Field. We demonstrate that automatic learning of the parameters for Conditional Random Fields improves results in comparison to similar approaches without the context-aware inference or with parameters set by hand.

**Keywords:** Vision · Terrain classification
Conditional Random Fields

## 1 Introduction

Autonomous outdoor navigation is a challenging task for affordable service robots that are equipped with a limited number of simple sensors. The robot has to recognize the surrounding terrain, associating the particular area to one of the predefined classes [13] in order to avoid obstacles and to regard the traversability. The difficulty of recognizing drivable terrain depends on the variation of cases considered. The area that can be considered as the "road" and the non-drivable areas are often differentiated not only by their perceptual properties but also by the local neighborhood context, e.g. a narrow strip of grass between concrete panels on a pavement should not be regarded as a lawn area. Therefore, we avoid designing specific procedures that recognize those different cases upon their geometric shapes or other features. Instead, we apply the Random Trees classification method that works with a multi-sensor, yet affordable, perception system. Then, we let a human expert to define the drivable and non-drivable terrain during labelling the data sequences used to train the classifier.

© Springer Nature Switzerland AG 2019
M. Strand et al. (Eds.): IAS 2018, AISC 867, pp. 626–638, 2019.
https://doi.org/10.1007/978-3-030-01370-7_49

We deal with the problem of terrain classification from multisensory but sparse data obtained in-motion from a passive camera and an active 2D laser scanner in an outdoor, semi-structured environment. Since the perception system of the robot and the integration of the terrain classification module with the motion planning algorithm have been described in our recent paper [17], here we focus on the classification module itself and the way it exploits the local context of adjacency between the observed terrain patches. An important aspect of this approach is the automatic learning of parameters for the inference system from the provided examples.

The remainder of this work is organized as follows: in Sect. 2 we refer to the most relevant publications on terrain classification, Sect. 3 provides an overview of our classification system architecture, while Sect. 4 details the probabilistic inference mechanism. Section 5 describes the parameter learning algorithm. Results are presented in Sect. 6, and Sect. 7 concludes the paper.

## 2   Related Work

Although vision-based terrain classification was researched extensively for wheeled robots [4,6] and autonomous cars [5], some systems employ RGB-D cameras or complement the passive vision by laser scanner data [9,18]. Laser data not only allow the robot to easily infer the profile of the observed terrain but make available the intensity of the reflected light that helps distinguish particular terrain classes [20]. The laser data can be also applied to supervise learning of the camera image classifier [5]. Whereas monocular vision systems benefit much from additional depth data obtained with an active sensor, also the passive stereo depth maps can be fused with the color information for terrain classification [7].

Different types of features and classifiers are applied in the terrain classification systems known from the literature. For instance, features can have a form of a histogram of RGB values or a local binary pattern (LBP) histogram [12]. Laible *et al.* [9] classify terrain using local ternary pattern (LTP) features from RGB images fed to the Random Trees classifier [3]. Random Trees are used also in our research [17] but the most popular solution in terrain classification is the Support Vector Machine [2], which we have tested also in our system [18]. Other prominent approaches known from the literature are regression-based classifiers [6] and neural networks [13].

A terrain classification system can be enhanced by performing probabilistic inference after the main classification stage [8]. The inference step improves the classification efficiency capturing the neighborhood context – the local dependencies between terrain patches. In particular, Conditional Random Field can be used to describe the probabilistic dependencies between neighboring segments [9]. This approach is used also in our terrain classification system [17] and in [10]. However, Laible and Zell [10] combine the information gathered by the robot from different views after the inference step, reducing the computational demand but disposing also of part of the information that could be taken into consideration during inference. To infer the most likely labeling a Gibbs sampler with simulated annealing is used in [10]. The Gibbs sampler is easy to implement but does not have any obvious termination criteria and is slower comparing to the loopy belief propagation method applied in our system. These issues complicate the parameter estimation procedure, which we were able to automatize in our approach.

## 3   Terrain Classification System

### 3.1   System Structure

The classification system was designed for a wheeled, skid-steering TAPAS robot intended for delivery of goods and patrol tasks in urban environments [19]. The robot is equipped with a standard GPS module, Attitude and Heading Reference System (AHRS) that provides proprioceptive measurements and the magnetic heading, encoder-based odometry, and the exterocpetive sensors consisting of an off-the-shelf webcam and a 2D laser scanner Hokuyo UTM-30LX. The robot employs OpenStreetMaps for global path planning and uses the well-known Vector Field Histogram (VFH) algorithm to compute the local steering direction. The computed direction takes into account the terrain traversability labels assigned to particular areas by the classification system.

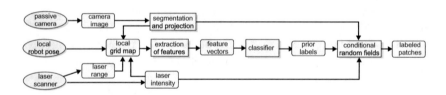

**Fig. 1.** Block scheme of the terrain classification system. Rectangles denote processing blocks, while rounded rectangles stand for data structures

The process of classification consists of several steps (Fig. 1). At first, the data from the camera image and the laser scans are segmented to obtain a data structure that is compatible with the local grid map of the terrain used by the VFH algorithm to determine the steering direction. The main classification process uses the information about the color of the image pixels and intensity values from the laser scanner. The main step classifies every patch of the terrain separately utilizing only information stored within a given cell of the grid map. Then, data representing the color and laser intensity of the cells, and the results of classification for these cells (called prior labels) are acquired from several consecutive scene views in order to ensure that a wider spatial context can be exploited in the probabilistic inference stage, which captures not only properties of the terrain patches but also their neighborhood connectivity.

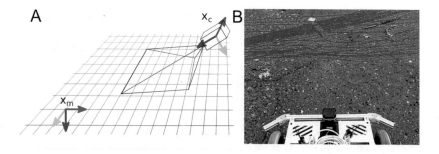

**Fig. 2.** Camera image reprojection problem (A), and laser scanner measurements acquired during rotation of the robot and reprojected onto the camera image (B)

## 3.2 Local Terrain Representation

The environment with the observed motion restrictions is represented as a robot-centric grid map that moves together with the robot [17]. Whereas the obstacles protruding from the ground are represented as elevation values obtained from the moving laser profile [1] and associated with particular cells of the grid, the labels for small patches of the terrain determine traversability. To make this representation compatible with the elevation map, images from the on-board camera are segmented accordingly to the layout of the cells in the grid. All image pixels that are observed within the same cell are treated as belonging to a single patch. The pixels are reprojected from the image frame into the 3D frame of reference of the map (Fig. 2A).

To overcome the problem of projecting points from a monocular camera image into 3D space it is assumed that the terrain the robot is moving on is locally flat and that the ground plane is parallel to the plane defined by the $x$ and $y$ axes of the robot coordinate system, which in turn is the reference system for the moving map. Then, the problem of reprojecting the pixels is written as:

$$
s \begin{bmatrix} r \\ c \\ 1 \end{bmatrix} = \begin{bmatrix} f_x & 0 & c_x & 0 \\ 0 & f_y & c_y & 0 \\ 0 & 0 & 1 & 0 \end{bmatrix} \cdot \mathbf{A}_m^c \cdot \begin{bmatrix} x_p \\ y_p \\ 0 \\ 1 \end{bmatrix},
\tag{1}
$$

where $r$ and $c$ are pixel coordinates in the image, $f_x$, $f_y$, $c_x$ and $c_y$ are intrinsic camera parameters, $\mathbf{A}_m^c$ is the homogeneous transformation matrix that transforms a vector from the map frame $X_m$ to the camera frame $X_c$, while $x_p$, $y_p$ are pixel coordinates in the map frame. Also, the laser scanner reflected intensity data are projected onto the grid map. This operation is straightforward, as the measured distance is available for each laser beam and is used to find the 3D coordinates of the data point. However, the intensity data are sparse and are available only for a fraction of the camera image pixels (Fig. 2B). The information about image pixels and laser scans is treated as valid for a limited time, due to the environment dynamics and accumulation of the local position error, as the robot is localized mainly using odometry and AHRS [17]. Hence, the robot moves the grid map position before each classification cycle and re-initializes it according to the new coordinates.

### 3.3   Extraction of Features

For each terrain patch (map cell), a set of values that characterize this patch is computed. The set is called a feature vector and it should contain values that enable to distinguish between terrain classes and should generalize description of patches of the same terrain class. The choice of a proper feature vector is crucial for successful classification. Our choice of features is motivated practically – as the acquired images are often blurred or contain rolling shutter artifacts, we avoid features that describe the observed texture and deteriorate easily in the presence of image artifacts. Instead, we use color and laser intensity features:

1. two-dimensional histogram ($4 \times 4$ bins) of the H-S components in the HSV (Hue-Saturation-Value) color space converted to a 16-dimensional vector,
2. 8 bin histogram of the V channel in the HSV space,
3. mean and covariance matrix for pixels in the HSV space converted to a single 12-dimensional vector,
4. mean and covariance matrix for values of the intensity and distance from laser scanner converted to a single 6-dimensional vector,
5. 25 bin histogram of intensity values from the laser scanner.

These features are computed for every terrain patch and concatenated to form a single vector, then fed to the classification process.

### 3.4   Classification

Designing our system we tested two classifiers: the Support Vector Machine (SVM) and the Random Trees (RT). SVM is a widely known classifier, used in many state-of-the-art solutions to terrain recognition. In our tests SVM was used with a Radial Basis Function (RBF) kernel. The RBF parameters were obtained in a cross-validation procedure. RT is a classification algorithm based on a collection of decision trees. Given the input features, classification is performed separately on every tree. The result is a class label that was chosen by the majority of trees. Whereas RT is easy to implement and fast, it allows the classification system to rank the importance of the used features [3] and provides a probability of the class prediction. In the presented system, the RT classifier is chosen to be simple and fast, hence there are 25 trees of maximal depth equal to 2.

Due to its good performance in the initial tests, the RT was selected as the default classification algorithm in our system, while SVM was used for comparison. Only results from the RT classifier were used with the probabilistic inference algorithm. The probabilities of the given patch being labeled as a certain terrain type were treated as prior labels for the inference process.

## 4  Probabilistic Inference

The probabilistic inference stage for the terrain classification problem was modeled using the Conditional Random Fields (CRF). The CRF are a variant of Probabilistic Graphical Models, which are a way to describe a probability distribution $p(\mathbf{y})$ using a factor graph. In CRF some variables $\mathbf{x}$ are always observed. This property makes describing dependence between those variables unnecessary. The model is based on an assumption that the distribution can be factorized as a product of some factors:

$$p(\mathbf{y}|\mathbf{x}) = \frac{1}{Z(\mathbf{x})} \prod_{a \in F} \Psi_a(\mathbf{y}_a, \mathbf{x}), \tag{2}$$

where $Z(\mathbf{x})$ is a normalization factor, $F$ is a set of factors, $\Psi_a(\mathbf{y}_a, \mathbf{x})$ is a value of the factor $a$ and $\mathbf{y}_a$ is a subset of random variables that the factor $a$ depends on. The values of factors are defined by an exponential family:

$$\Psi_a(\mathbf{y}_a, \mathbf{x}) = \exp\left\{ \sum_{k \in \mathcal{K}_a} \theta_k f_k(\mathbf{y}_a, \mathbf{x}) \right\}, \tag{3}$$

where $f_k(\mathbf{y}_a, \mathbf{x})$ is a function that delivers sufficient statistics [16], $\mathcal{K}_a$ is a set of features used by the factor $a$, and $\theta_k$ is a design parameter (Fig. 3).

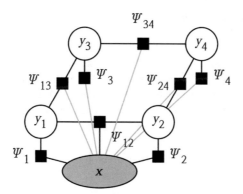

**Fig. 3.** A snippet of a factor graph (some links were grayed out to make image legible). Every round node represents a random variable, every square node represents a factor and every edge represents dependence between a factor and a variable. Gray variables are observed ones

In our solution, every terrain patch has a random variable that represents the class label $(y_i)$ assigned to it. Every patch has variables representing observations (visual features) and prior labels. Factors (in the sense of factor graph) are divided into two groups: per-node factors that drive the random variable towards the known classification results, and pairwise factors, that force neighboring patches to agree with each other as to the class label. The specific formula for a joint probability distribution represented as a graph is inspired by [9]:

$$p(\mathbf{y}|\mathbf{x}) = \frac{1}{Z(\mathbf{x})} \exp \left\{ \sum_{i \in \mathcal{P}} \sum_{l \in \mathcal{L}} \theta_N g(y_i, \mathbf{x}_{Ni}) + \sum_{(i,j) \in \mathcal{N}} \sum_f \theta_{Ef} h(y_i, y_j, x_{Eif}, x_{Ejf}) \right\}, \tag{4}$$

$$g(y_i, \mathbf{x}_{Ni}) = -\log(p(y_i|\mathbf{x}_{Ni})), \quad h(y_i, y_j, x_{Eif}, x_{Ejf}) = \mathbf{1}_{\{y_i \neq y_j\}} \exp\left(-\beta(x_{Eif} - x_{Ejf})^2\right),$$

where $\mathcal{P}$ is a set of patches, $\mathcal{L}$ is a set of labels, $\mathcal{N}$ is a set of edges connecting patches, $\mathbf{1}_{\{c\}}$ is an indicator function that equals to 1 when condition $c$ is met and 0 otherwise, $\theta_N$ is a parameter for the node factor, $\theta_{Ef}$ is a parameter for pairwise factors depending on a feature $f$, $\mathbf{x}_{Ni}$ are features for the node $i$ that were used at the classification stage, $p(y_i|\mathbf{x}_{Ni})$ is a priori label for a node $i$ that comes from the classifier, $x_{Ejf}$ is a value of a feature $f$ for a node $j$ and $\beta$ is a parameter that controls how fast pairwise factors weaken with increasing discrepancy between features. In the inference process, features from both the passive camera and laser scanner are used, if both are available. This is in opposition to [9], where only image-based features were used.

Given the model, it is straightforward to calculate a joint probability from (4). Although, in terrain classification, the aim is to compute a labeling that maximizes this joint probability. Such labeling is called the Maximum a Posteriori (MAP) assignment $\mathbf{y}^* = \arg\max_{\mathbf{y}} p(\mathbf{y}|\mathbf{x})$, where $\arg\max_{\mathbf{y}}$ is an argument of a maximum value from all combinations of variables $\mathbf{y}$. Unfortunately, for general graphs, this problem is intractable, and for larger graphs, it can't be solved exactly in a reasonable time. Therefore, an algorithm based on passing messages between factors and random variables, called loopy belief propagation was applied. The messages propagate information in a graph and enable making decisions about a labeling harnessing the whole context. It turns out, that under some conditions, the problem of passing messages can be viewed as a variational problem [16]. A message from factor $a$ to a random variable $i$ and a message from a random variable to a factor are computed as follows:

$$m_{ai}(y_i) = \max_{\mathbf{y}_a \setminus y_i} \Psi_a(\mathbf{y}_a, \mathbf{x}) \prod_{j \in \mathcal{N}_a \setminus i} m_{ja}(y_j), \quad m_{ia}(y_i) = \prod_{b \in \mathcal{N}_i \setminus a} m_{bi}(y_i), \tag{5}$$

where $\max_{\mathbf{y}_a \setminus y_i}$ denotes a maximal value from all combinations of variables $\mathbf{y}_a$ excluding variable $y_i$, $\mathcal{N}_a$ denotes all neighboring random variables of a factor $a$ and $\mathcal{N}_i$ denotes all neighboring factors of a random variable $i$. We use the Tree Reparameterization (TRP) algorithm to ensure a proper scheduling of the message passing process, which improves convergence speed [15].

## 5    Estimation of Parameters

The parameters used to control the inference process using CRF were obtained automatically in a parameter estimation procedure that maximizes a log-likelihood function according to a database $\mathcal{D}$ of manually labeled examples of terrain observations. The function is a measure of how likely a set of parameters $\boldsymbol{\theta}$ parametrizes the distribution of the dataset. The procedure follows an assumption that examples are independent of each other and identically distributed. Therefore, the likelihood becomes a product of probabilities for all examples:

$$L(\boldsymbol{\theta} : \mathcal{D}) = p(\mathcal{D}|\boldsymbol{\theta}) = \prod_{m \in \mathcal{D}} p(\mathbf{y}_m|\mathbf{x}_m, \boldsymbol{\theta}). \tag{6}$$

A logarithm of this function has the form:

$$l(\boldsymbol{\theta} : \mathcal{D}) = \sum_{m \in \mathcal{D}} \left\{ \sum_{i \in \mathcal{P}^m} \sum_{l \in \mathcal{L}} \theta_N g(y_i^m, \mathbf{x}_{Ni}^m) \right.$$

$$\left. + \sum_{(i,j) \in \mathcal{N}^m} \sum_f \theta_{Ef} h(y_i^m, y_j^m, x_{Eif}^m, x_{Ejf}^m) \quad - \log Z(\mathbf{x}^m, \boldsymbol{\theta}) \right\}. \tag{7}$$

As the problem described by (7) is concave, the parameters were found during a gradient optimization using the L-BFGS algorithm [11]. We chose the L-BFGS as it has proven its usability in fitting CRF models [14] and needs no parameters tuning. The only parameter that had to be set-up manually is $\beta$ in (4) and the value of $\beta = 2$ was chosen. It is worth noting that, in opposition to deep learning approach, the learning process is quick and takes approximately only 2 min on a standard laptop computer.

## 6   Experiments and Results

Six different variants of the classification system were tested on the same data sequences:

- $rt0$ – basic variant with RT classifier and no inference,
- $rt1$ – variant with RT classifier, only image features used during inference and parameters set manually, as in [9],
- $rt2$ – variant with RT classifier, only image features used during inference, and pairwise factors parametrized with one parameter,
- $rt3$ – variant with RT classifier and only image features used during inference,
- $rt4$ – the complete system described in the paper,
- $svm$ – alternative basic variant with SVM classifier and no inference.

The $rt1$ variant used the same inference parameter values as in [9], where there were only image features (used by the single factor that compared whole vectors) and only two parameters $\theta_N = 0.5$ and $\beta = 2.0$. The next variant, referred to as $rt2$, was enriched with an additional parameter $\theta_E$, features were treated separately, and parameter values were estimated automatically. The $rt3$ variant had separate parameter values for each color component, and all four $\theta$ parameters were estimated. The $rt4$ variant used both the camera and the laser scanner data during classification and the inference stage. It also had separate parameters for all factors – prior label, separate color components, and intensity.

The training dataset contained 164 images and was collected in Pilzno (Czech Republic), while our robot took part in the Robotour contest [19] (Fig. 4A), and the testing dataset was collected several months later in a park in Poznań (Fig. 4B). The datasets were collected at daylight, as TAPAS is not suited for night operation, but the sequences contain mixed lighting conditions due to deep shadows cast by the trees. The testing dataset was collected when the ground in the park was to a great extent covered with fallen leaves, sometimes mixed with the soil (Fig. 4C). This makes the experiments challenging, despite the fact that only two terrain classes are considered. The testing dataset is divided into eight runs (*run1* to *run8*). Each run is a separate ride along a different path in the park but only the *run4* to *run8* sequences contain both RGB (camera) and laser intensity data, while only the *run7* to *run8* ensure balanced data between the drivable and non-drivable terrain seen by the robot.

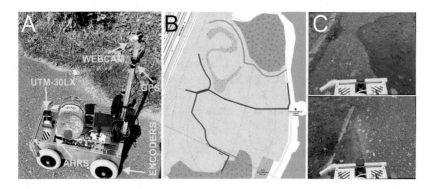

**Fig. 4.** Robot with a payload (A), OpenStreetMap presenting the test path (red) in a park (B), and example terrain views from the camera (C)

Besides the standard accuracy measure (confusion matrices), three new performance measures are proposed [17] to demonstrate the influence of the terrain classification on the performance of the wheeled robot. They all measure deviation of the motion direction devised by the VFH algorithm using the automatic terrain labeling from the direction computed by the same algorithm but using manual terrain labeling:

1. *int* - sum of direction differences (in degrees) for all classification acts, calculated as $\sum_i |d_{mi} - d_{ci}|$, where $d_{mi}$ is the direction with manual labeling at time $i$, and $d_{ci}$ is the computed labeling direction at time $i$;
2. *max* - maximal deviation (in degrees) from the manual labeling direction, calculated as $\max_i |d_{mi} - d_{ci}|$;
3. *relev* - number of relevant deviations from the manual labeling direction $|R|$, where $R = \{i : |d_{mi} - d_{ci}| \geq 30°\}$ to avoid sharp turns of the robot.

The "manual labeling direction" is the motion direction computed by the VFH algorithm on the basis of manual terrain labeling (ground truth), while the "computed labeling direction" is the direction computed when the terrain labels obtained from the classification system are used.

The accuracy was tested on balanced test data, where the number of samples was approximately equal for both terrain types. In Table 1 the accuracy of different methods is presented, whereas confusion matrices are shown in Table 2. Table 3 shows the outcome of the investigated classification system variants in the context of control commands efficiency.

**Table 1.** Accuracy results for all methods tested on balanced data

| Trajectory | run7 | | | | | | run8 | | | | | |
|---|---|---|---|---|---|---|---|---|---|---|---|---|
| System | $rt4$ | $rt3$ | $rt2$ | $rt1$ | $rt0$ | $svm$ | $rt4$ | $rt3$ | $rt2$ | $rt1$ | $rt0$ | $svm$ |
| Accuracy | 98 | 98 | 98 | 98 | 98 | 99 | 93 | 94 | 94 | 94 | 92 | 87 |

**Table 2.** Confusion matrices for three runs

| Trajectory | run4 | | | | | | run5 | | | | | | run6 | | | | | |
|---|---|---|---|---|---|---|---|---|---|---|---|---|---|---|---|---|---|---|
| System | $rt4$ | | $rt0$ | | $svm$ | | $rt4$ | | $rt0$ | | $svm$ | | $rt4$ | | $rt0$ | | $svm$ | |
| Confusion | 86 | 14 | 89 | 11 | 88 | 12 | 85 | 15 | 89 | 11 | 86 | 14 | 89 | 11 | 99 | 1 | 86 | 14 |
| Matrix | 2 | 98 | 3 | 97 | 3 | 97 | 2 | 98 | 2 | 98 | 2 | 98 | 1 | 99 | 3 | 97 | 3 | 97 |

Another test was conducted on data collected in an environment with many fallen leaves (Table 4). In this test, the laser intensity was not used to show the ability of the probabilistic inference stage in the classification system to compensate for the external disturbances in the appearance of the images.

Although the fast RT classifier provides good results in terms of the statistics (Table 2), it often generates misclassified isolated patches, which result in wrong control commands (Table 3). The CRF allows us to get rid with such isolated patches, by exploiting the connectivity information, and improves the performance in terms of the control commands. The experimental results given in Tables 3 and 4 suggest that the parametrization of the inference system doesn't play an important role – the results are very similar for the pairwise factors parametrized with one parameter, and for the independent parameters.

**Table 3.** Robot control performance measures for all methods

| Trajectory | run4 | | | | | | run5 | | | | | | run6 | | | | | |
|---|---|---|---|---|---|---|---|---|---|---|---|---|---|---|---|---|---|---|
| System | $rt4$ | $rt3$ | $rt2$ | $rt1$ | $rt0$ | $svm$ | $rt4$ | $rt3$ | $rt2$ | $rt1$ | $rt0$ | $svm$ | $rt4$ | $rt3$ | $rt2$ | $rt1$ | $rt0$ | $svm$ |
| $int$ | 330 | 400 | 400 | 640 | 430 | 1680 | 230 | 220 | 200 | 330 | 170 | 610 | 210 | 280 | 280 | 360 | 520 | 1250 |
| $max$ | 40 | 40 | 40 | 70 | 30 | 80 | 30 | 30 | 30 | 40 | 30 | 60 | 40 | 40 | 40 | 50 | 90 | 90 |
| $relev$ | 2 | 2 | 2 | 4 | 2 | 25 | 1 | 1 | 1 | 3 | 1 | 8 | 2 | 3 | 3 | 5 | 6 | 16 |

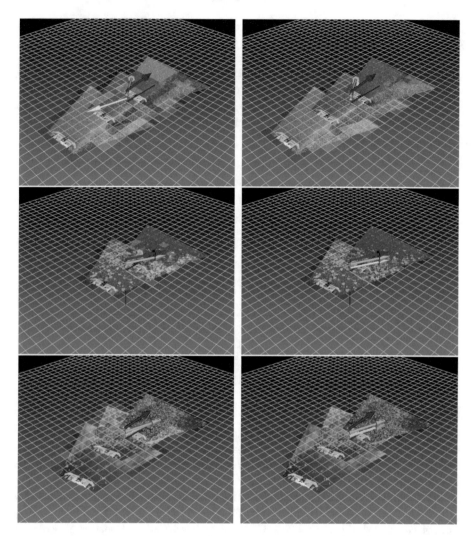

**Fig. 5.** Example VFH steering directions computed after SVM classification (left column), and RT+CRF classification (right column). Magenta arrows point goal direction, green arrows point manual labeling direction, and yellow arrows point computed labeling direction. Non-drivable areas are shown in green, while drivable areas are red

Example visualizations of the computed terrain labeling and steering directions are presented in Fig. 5. The first row shows a system behavior in the presence of lighting variance, which confuses the SVM classifier. In the second and third situation, fallen leaves made the classification problem difficult, but the RT classifier with probabilistic inference labeled the terrain correctly. Notice, that in the third situation it is hard, even for a human observer, to determine the exact border of the drivable area. It is worth noting that despite the added inference process, the overall computation time is sufficient for real-time operation, even

**Table 4.** Experiment results in severe weather conditions without laser data

| Trajectory | run1 | | | run2 | | | run3 | | |
|---|---|---|---|---|---|---|---|---|---|
| System | $rt3$ | $rt2$ | $rt1$ | $rt3$ | $rt2$ | $rt1$ | $rt3$ | $rt2$ | $rt1$ |
| $int$ | 340 | 320 | 390 | 640 | 630 | 810 | 630 | 640 | 950 |
| $max$ | 30 | 30 | 60 | 40 | 40 | 40 | 70 | 70 | 60 |
| $relev$ | 2 | 2 | 2 | 4 | 4 | 8 | 7 | 6 | 12 |

on the on-board Intel Atom D525 processor. The classification and inference time was 0.66 s on the Intel Atom, and 0.09 s on the Intel Core i5 3230M, which fits within the time needed by the robot to traverse the classified area (approx. 2 s).

# 7   Conclusions

The topology and connectivity of the terrain patches plays at least as important role as the terrain features themselves in recognition of the drivable roads neighboring to various non-drivable areas. Hence, we use the CRF in order to find dependencies between visual features and the simple RT classification results. The results show that automatic parameter estimation improves the results with respect to the parameters set by hand. Moreover, choosing the training data and the evaluation data that have been collected at different times in the year and at different locations, we have shown that our method is general enough to be taught on a representative sequence and then used in real-life scenarios.

# References

1. Belter, D., Łabęcki, P., Skrzypczyński, P.: Estimating terrain elevation maps from sparse and uncertain multi-sensor data. In: Proceedings of IEEE International Conference on Robotics and Biomimetics, Guangzhou, pp. 715–722 (2012)
2. Boser, B.E., Guyon, I.M., Vapnik, V.N.: A training algorithm for optimal margin classifiers. In: Fifth Annual Workshop on Computational Learning Theory, COLT 1992, pp. 144–152. ACM, New York (1992)
3. Breiman, L.: Random Forests. Mach. Learn. **45**(1), 5–32 (2001)
4. Chetan, J., Krishna, M., Jawahar, C.: Fast and spatially-smooth terrain classification using monocular camera. In: International Conference on Pattern Recognition, Istanbul, pp. 4060–4063 (2010)
5. Dahlkamp, H., Kaehler, A., Stavens, D., Thrun, S., Bradski, G.: Self-supervised monocular road detection in desert terrain. In: Robotics: Science and Systems, Philadelphia (2006)
6. Hadsell, R., Sermanet, P., Ben, J., Erkan, A., Scoffier, M., Kavukcuoglu, K., Muller, U., LeCun, Y.: Learning long-range vision for autonomous off-road driving. J. Field Robot. **26**(2), 120–144 (2009)
7. Happold, M., Ollis, M., Johnson, N.: Enhancing supervised terrain classification with predictive unsupervised learning. In: Robotics: Science and Systems, Philadelphia (2006)

8. Häselich, M., Arends, M., Lang, D., Paulus, D.: Terrain classification with Markov random fields on fused camera and 3D laser range data. In: European Conference on Mobile Robots, Örebro, pp. 153–158 (2011)
9. Laible, S., Khan, Y.N., Zell, A.: Terrain Classification with conditional random fields on fused 3D LIDAR and camera data. In: European Conference on Mobile Robots (ECMR), Barcelona, pp. 172–177 (2013)
10. Laible, S., Zell, A.: Building local terrain maps using spatio-temporal classification for semantic robot localization. In: International Conference on Intelligent Robots and Systems, Chicago, pp. 4591–4597 (2014)
11. Nocedal, J.: Updating quasi-newton matrices with limited storage. Math. Comput. **35**(151), 773–782 (1980)
12. Ojala, T., Pietikainen, M., Harwood, D.: Performance evaluation of texture measures with classification based on Kullback discrimination of distributions. In: International Conference on Pattern Recognition, Jerusalem, pp. 582–585 (1994)
13. Ojeda, L., Borenstein, J., Witus, G., Karlsen, R.: Terrain characterization and classification with a mobile robot. J. Field Robot. **23**(2), 103–122 (2006)
14. Sutton, C., McCallum, A.: An introduction to conditional random fields. Found. Trends Mach. Learn. **4**(4), 267–373 (2011)
15. Wainwright, M.J., Jaakkola, T.S., Willsky, A.S.: Tree-based reparameterization framework for analysis of sum-product and related algorithms. IEEE Trans. Inf. Theory **49**(5), 1120–1146 (2006)
16. Wainwright, M.J., Jordan, M.I.: Graphical models, exponential families, and variational inference. Found. Trends Mach. Learn. **1**(1–2), 1–305 (2008)
17. Wietrzykowski, J., Skrzypczyński, P.: Terrain classification for autonomous navigation in public urban areas. In: Silva, M. (ed.) Human-Centric Robotics, pp. 319–326. World-Scientific (2017)
18. Wietrzykowski, J., Belter, D.: Boosting support vector machines for RGB-D based terrain classification. J. Autom. Mob. Robot. Intell. Syst. **8**(3), 28–34 (2014)
19. Wietrzykowski, J., Nowicki, M., Bondyra, A.: Exploring openstreetmap publicly available information for autonomous robot navigation. In: Szewczyk, R., et al. (eds.) Progress in Automation, Robotics and Measuring Techniques: Volume 2 Robotics, pp. 309–318. Springer, Cham (2015)
20. Wurm, K.M., Stachniss, C., Kümmerle, R., Burgard, W.: Improving robot navigation in structured outdoor environments by identifying vegetation from laser data. In: International Conference on Intelligent Robots and Systems, St. Louis, pp. 1217–1222 (2009)

# Crop Edge Detection Based on Stereo Vision

Johannes Kneip, Patrick Fleischmann$^{(\boxtimes)}$, and Karsten Berns

Robotics Research Lab, Department of Computer Science,
Technische Universität Kaiserslautern, 67663 Kaiserslautern, Germany
{j_kneip11,fleischmann,berns}@cs.uni-kl.de

**Abstract.** This paper proposes a robust detection method of uncut crop edges which is used for automated guidance of a combine harvester. The utilized stereo vision system allows for real-time depth perception of the environment. A three-dimensional elevation map of the terrain is constructed by the point cloud acquired in this way. The heights of crop and harvested areas are estimated using Expectation Maximization and segmented using of the clustering results. In a row-wise processing step, each scan line of heights is cross-correlated with a model function to compute possible candidate points located at the very crop edge. Using robust linear regression, a linear crop edge model is calculated, modeling the spatial distribution of the candidate points. An overall crop edge model is updated via exponentially weighted moving average.

**Keywords:** Agricultural automation
Computer vision for automation · Visual-based navigation

## 1  Introduction

Research on automation of agricultural vehicles has accelerated recently, which of course is partially also a consequence of the current era of technology as computer costs decline while computation capabilities increase. In harvest processes, the operator has to supervise the field operation while steering, which requires a high level of concentration and is quite demanding over time. The market came up, with assistance systems greatly supporting the farmers. Most commonly used sensor technologies are laser scanners, the Global Positioning System (GPS), inertial positioning systems and machine vision. While guidance systems based on the GPS ensure a certain precision, the costs are rather high and tracking errors might occur from occluded satellites. The achieved accuracy usually is sufficient and within an error of less than 10 cm. Researchers from the Carnegie Mellon University used a New Holland 2550 Speedrower, the Demeter system [9], equipped with GPS and a camera for automated navigation in harvesting tasks in sorghum and alfalfa fields. [4] introduced an image processing approach based on unsupervised texture analysis by Markov segmentation with subsequent model function correlation in order to detect the crop edge in grass

© Springer Nature Switzerland AG 2019
M. Strand et al. (Eds.): IAS 2018, AISC 867, pp. 639–651, 2019.
https://doi.org/10.1007/978-3-030-01370-7_50

fields. In contrast, [2] uses a 1D scanning laser rangefinder in a correlation-based approach for automated guidance of agricultural vehicles such as a windrow harvester and combine harvester. [3] conducted tests with a laser scanner for crop row localization in a row-planted soybean field with three-dimensional field information obtained using a pan-tilt unit rotating the laser scanner. A rather inexpensive solution to this could be by utilizing a stereo vision system for local navigation and environment perception. Stereo vision systems provide 3D information from which height and volume can be estimated. Furthermore, as for local guidance, no a priori map of the area is needed and an additional safety-relevant obstacle detection system may be easily integrated. [10] successfully developed a system based on stereo vision to detect the crop edge in corn fields by using the height information. [1] considered the different mounting positions of cameras for perception during combine guidance, as each position allows for different approaches to capture crop edge characteristics. They concluded that the best position is above the cabin which, for example, is less influenced by vibrations generated from the operating machine. Unlike monocular vision, stereo can to some extent also account for shadows as the texture changes in both images simultaneously and therefore doesn't affect the stereo matching.

These properties of stereo vision, the benefits of positioning the camera on top of the cabin, together with the knowledge gained through previous investigations and the need for robust crop edge detection algorithms, motivated this research. In this paper, we present our stereo vision system setup followed by a detailed description of the crop edge detection approach. The proposed system consists of three decisive modules: *Stereo Vision System* (see Sect. 2), *Elevation Map* (see Sect. 4) and *Robust Linear Regression* (see Sect. 6) as presented in the schematic overview (see Fig. 1).

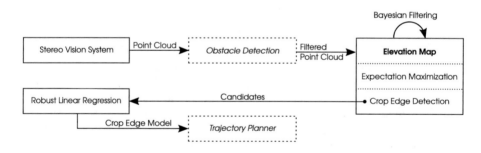

**Fig. 1.** Schematic overview of the proposed system.

## 2   Stereo Vision System

The stereo system used on our test platform (see Sect. 7) consists of two stereo cameras. Both are mounted to the cabin [1] — one in each frontal outer corner — and provide a broad overview of the harvesting area in front of the combine (see Fig. 2). To reduce the load of the CPU and overcome the difficulty to achieve real-time performance, the disparity map is generated by the *libSGM*[1] library, a CUDA implementation of the Semi-Global Matching (SGM) algorithm [7]. In comparison to the CPU-implementation, a considerably increased frame rate from 10 Hz to 40 Hz with images of size (width $\times$ height) $= (640 \times 480)$ at 64 disparity levels is achieved. We extended the library to allow for a disparity refinement step using quadratic interpolation of the calculated matching costs in the disparity space image (DSI). Using the camera's intrinsic and extrinsic parameters two point clouds (left and right stereo system) are obtained which are projected to the Robot Coordinate System (RCS) $(x^{RCS}, y^{RCS}, z^{RCS}$, see Fig. 2) using the known position of the cameras on the vehicle.

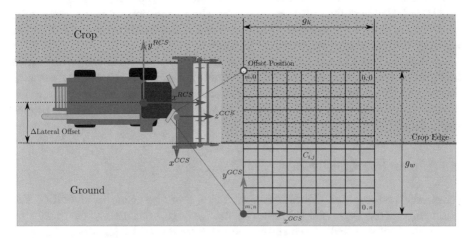

**Fig. 2.** Sketch of a harvesting scenario in bird's-eye view with the used elevation grid map. The axes of the Robot- (RCS), Camera- (CCS), and the Grid-Coordinate-System (GCS) are color-coded in red $(x)$, green $(y)$ and blue $(z)$. The two slightly outwards rotated and downwards tilted stereo cameras (green) are mounted on the roof of the cabin.

## 3   Height Segmentation

### 3.1   Motivation

Depending on the season, each crop type near maturity stipulates a different plant height. At this time, wheat has an approximate height of 0.5–1 m, rapeseed of 0.3–1.5 m and grass ranges between 0.1 m and 1 m. A generally valid height segmentation is desired to handle different types of crop, thus the segmentation

---

[1] https://github.com/fixstars/libSGM.

algorithm has to be adaptive and to some extent universal/unsupervised. The problem can be formulated as follows:

$$f(h) = \begin{cases} crop, & \text{if } h \geq T \\ ground, & \text{otherwise} \end{cases} ,$$
(1)

where the function $f$ evaluates a height value $h$ and assigns it to a class according to the threshold $T$. Reviewing the height distribution of multiple point clouds of harvesting scenarios revealed a bi-modal distribution. This property makes it possible to use automatic clustering techniques like Expectation Maximization (EM) with two Gaussian Models to come up with a threshold $T$.

A *paired t-test* with samples of each class was used to evaluate the class separation. The difference between the two classes *crop* and *ground* found to be statistically significant ($\alpha = 0.001$) as the evaluation of height estimates of several wheat and rapeseed fields revealed. These results indicate that an accurate and reliable segmentation is possible.

A drawback of this method is of course that if no observable height difference is found this method fails. Such a case could, for example, be cutting grass in an early growth state where the height difference would only range within a few centimeters. In such a case the classes variance will overlap and make a proper distinction challenging.

## 4    Elevation Grid Map

Given the stereo point cloud $P = \{p_1, \ldots, p_k\}$ of the scene in the RCS, the points are sorted into a 2D grid map of size $(g_h \times g_w)$ consisting of cells $C_{i,j}$ as shown in Fig. 3a. A cell's fixed resolution is parametrized by $(c_h \times c_w)$. The grid map frame is positioned in the horizontal $(x^{RCS}, y^{RCS})$ plane with a predefined offset position $O \in \mathbb{R}^2$ to the grid coordinate system origin. The offset $O_x$ allows to position the grid map in front of combine header and thus skip invalid and unusable measurements in this area. The offset $O_y$ is used to limit the area in which the crop edge can occur, independent of the header width. As the output of the stereo matching algorithm depends on the camera's baseline only disparities up to a specific range are useful. For this reason, the grid is limited in x-direction to be able to cut off the point cloud at a predefined depth $x_{\max}$. As the crop edge only appears in a short section of the point cloud, likewise the y-direction is limited to a certain length $y_{\max}$.

In each cell, a height representative $h_{i,j}^{\text{median}}$ of the 3D points that fall into this specific cell is calculated as the median of the points' height. The median has been chosen in preference to the mean since its a more robust measure of the central tendency. As the cells are independent of each other, this step is parallelized. This statistical measure is then stored in a circular buffer of capacity three so that at most the last three height representatives are kept. As agricultural vehicles in harvesting scenarios usually drive at low speed this is a feasible approach. If at some point in time it is not possible to compute a height

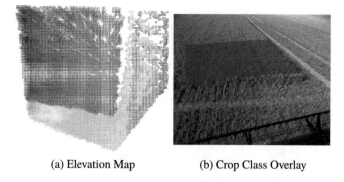

(a) Elevation Map          (b) Crop Class Overlay

**Fig. 3.** (a) Bird's-eye view of a elevation map with heights color-coded from low (blue) to high (red). (b) Based on (a) the portion of the point cloud with *crop*-membership visualized as an overlay on the stereo camera input image. Due to the short baseline of the stereo camera the predefined depth cut off is set to $x_{\max} = 12.5\,\mathrm{m}$.

representative and a cell's buffer consists of more than two representatives, one is dropped to avoid lag and distortions of the map and the real world. Empty cells (holes) in the grid, mostly caused by occlusions or by the stereo vision system itself as the depth output is only valid up to a certain range, are efficiently closed using the mean height of all heights stored in the buffer. As a result, the grid map models the current elevation of the real world in front of the combine.

## 4.1  Expectation Maximization with Gaussian Mixture Model

Based on our observations, heights of the elevation map can be modeled as a mixture of $k$ Gaussians, where $k = 2$. Here, it has to be mentioned, that large obstacles like trees, power poles and other agricultural machines are removed by a preceding filter (see [6]). An algorithm to estimate the parameters of a probability density function of a Gaussian Mixture Model is the Expectation Maximization (EM) algorithm [5]. The assumption is that the set of computed height representatives $\mathcal{H}$ of all cells holding a representative are generated from a set of Gaussians of unknown parameters. EM is utilized to estimate these parameters to determine the mixture model a single height representative $h_i \in \mathcal{H}$ originates from, by maximizing a likelihood function of the variables in the model which are unobserved. These parameters $\theta = (\pi_1, \pi_2, \mu_1, \mu_2, \sum_1, \sum_2)$ are the weight $\pi_i$ of the mixture, the mean $\mu_i$ and the covariance $\sum_i$ which are estimated iteratively.

The results of EM, apart from the estimated parameters $\theta$, are for each cell with $h_i$ a most probable explicitly assigned Gaussian component label and log-likelihood estimate $\log p(h_i \mid \theta)$. To accelerate the parameter estimation process, prior parameters $\theta$ of the last frame are used to provide an initial estimate for the subsequent calculation. This approach is especially permitted as the environment between two point clouds only changes insignificantly as the combine drives at low speed.

Considering the original problem of segmenting the harvest area into two regions, the resulting $\mu_1$ and $\mu_2$ can be interpreted as approximates of *crop* and *ground* heights, see Fig. 4. Also taking a look at the variance of each component reveals how scattered the height measurements are.

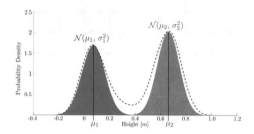

**Fig. 4.** Example of a probability distribution (black dotted line) of the heights in the elevation map estimated by a Gaussian for each class of *ground* height $\mu_1$ (blue) and *crop* height $\mu_2$ (red). In this scenario, stubbles left from wheat harvest caused a slight shift of the *ground* height from zero to $\mu_1 = 0.082$.

## 4.2   Bayesian Filtering

Similar to occupancy grid maps, each cell in the elevation map stores a binary crop value, which specifies whether it contains *crop* (1) or can be considered *ground* (0). The EM log-likelihood estimates are used in a Bayes filter to continuously update each cell's belief in currently holding points belonging to the *crop* class. We use the sigmoid function (2) to obtain a probabilistic crop map with scaled probabilities in a range between $[0, 1]$

$$p\left(C_{i,j} \mid h_i^{(1:t)}\right) = 1 - \frac{1}{1 + e^{\log p\left(h_i^{(1:t)}\mid\theta\right)}} \, , \tag{2}$$

where $h_i^{(1:t)}$ is the set of all height measurements up to time $t$. Therefore, the segmentation result label $L_{i,j}(b)$ for a cell $C_{i,j}$ can be determined based on a specified threshold level $b$ of belief in crop (3)

$$L_{i,j}(b) = \begin{cases} 1 \, , & \text{if } p\left(C_{i,j} \mid h_i^{(1:t)}\right) \geq b \\ 0 \, , & \text{otherwise} \end{cases} \, , \tag{3}$$

which was set to $b = 0.8$ in our experiments.

## 5   Crop Edge Detection

The crop edge is the transition point of standing crop and ground. Therefore, we assume that the *crop* and *ground* classes can be modeled by a model function $g$

in the $(y^{GCS}, z^{GCS})$ plane. Such a function follows the profile shape of a perfect crop edge, which is why we have used a unit step function:

$$g(x) = \begin{cases} crop \Leftrightarrow 1, & x < 0 \\ ground \Leftrightarrow 0, & x \geq 0 \end{cases} \tag{4}$$

The exact point of transition (edge) between 0 and 1 (at $x = 0$) is known a priori, so that we defined the crop edge to live on a 2D linear subspace of the elevation map at ground level height.

The proposed grid setup allows for a parallelized row-wise processing to detect the crop edge in each row, from now on referred to as scan line, separately. A scan line $\boldsymbol{S}_m$ is a $(n+1)$-dimensional vector consisting of the height segmentation results $L_{m,j}(b)$ of cells in grid row $m$. Scan lines only containing *crop* or *ground* are excluded from further processing steps since no edge is present and are marked as invalid ($\boldsymbol{S}_m = $ invalid). Now, each scan line is correlated with $g$ to locate a possible 2D location of the crop edge, in future called *Crop Edge Candidate* (CEC).

Following the work of [2–4] the correlation coefficient at some given delay $d$ is calculated using (5)

$$r(d) = \frac{\sum_{i=0}^{n} (g(i - d) - \bar{g})(\boldsymbol{S}_{m;i} - \bar{S}_m)}{\sqrt{\sum_{i=0}^{n} (g(i) - \bar{g})^2} \sqrt{\sum_{i=0}^{n} (\boldsymbol{S}_{m;i} - \bar{S}_m)^2}}, \tag{5}$$

where $\bar{g}$ and $\bar{S}_m$ are the respective means. A series of correlation coefficients is generated by calculating all possible delays $d \in [0, \dots, n]$. All sample points outside the range of the model function are ignored and their computation is skipped. The argument at which (5) reaches its maximum $r_{\max}$ corresponds to a cell position which possibly contains a crop edge point, see Fig. 5. As the above approach is performed for each scan line, a set $E$ of cells evolves (6)

$$E = \left\{ C_{i,j} \mid i = 0, \dots, m; \; j = \arg \max_{k \in [0,n]} r(k); \right\}. \tag{6}$$

Up to this point, the computed crop edge location is without regard to the cell resolution inside the grid. Therefore, additionally half of a cell's width needs to be subtracted for a *crop-ground*-edge and added for a *ground-crop*-edge. The set of real world crop edge points is given by (7)

$$\mathcal{P} = \left\{ (x, y) \mid x = E_{\text{center};x}, y = E_{\text{center};y} \pm \frac{c_w}{2} \right\}, \tag{7}$$

where $E_{\text{center}}$ is a cell's center of a cell of set $E$.

## 6    Linear Crop Edge Model

### 6.1    Motivation

In the previous step, the set $\mathcal{P}$ with 2D points in the $(x^{GCS}, y^{GCS})$ plane located at the very crop edge position are identified. We made the initial assumption

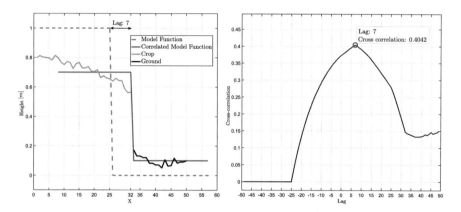

**Fig. 5.** Exemplary computation of the crop edge position in a segmented height scan line $S_0$ via cross correlation. Cell heights of *crop* (orange) and *ground* (black) are correlated with the function model $g$ (dotted blue) which has its known edge position at $x = 25$. The maximum correlation $r_{\max} = 0.4042$ is at a lag of 7, hence cell $C_{0,32}$ contains a CEC.

of the crop edge most likely being of linear nature (see [1,3,4]) and use robust linear regression to model their relationship of spatial distribution. Partially flattened crop is not uncommon and can be caused for example by bad weather. With respect to the elevation map the result can be that an entire scan line only contains *ground*-samples, in which case no CEC would be found, or the CEC may be strongly shifted. Ordinary least-squares (OLS) estimators are too sensitive to such outliers, hence the estimated model would be unstable (low distributional robustness).

## 6.2 Model Fitting

To reliably estimate a model and suppress outliers we use the *Huber* M-Estimator [8]. The set $\mathcal{P}$ is directly used for model fitting with all points $d_i \in \mathcal{P}, d_i = (x_i, y_i)$. Based on the predictor values $x_i$ and the measured responses in each grid row, namely the crop edge positions $y_i$, a predicted response $\hat{y}_i$ of the model is obtained. The robust linear regression model is then given by (8)

$$\hat{y}_i = b + mx_i + \epsilon_i \,, \tag{8}$$

where parameter $b$ is the line intercept with the $y^{GCS}$-axis, $m$ the slope of the line and the disturbance term $\varepsilon \sim \mathcal{N}\left(0, \sigma^2\right)$. The M-estimation is solved by using the iteratively re-weighted least squares (IRLS) algorithm. To to compensate the impact of a single estimated model on the combine guidance trajectory and maintain to some extent the knowledge of the crop edge gained in previous models, the exponentially weighted moving average (EWMA) (9) is used

$$\hat{Y}^t = \beta \hat{y}^t + (1 - \beta)\hat{y}^{t-1}, \tag{9}$$

where $\beta$ is the degree of weighting decrease.

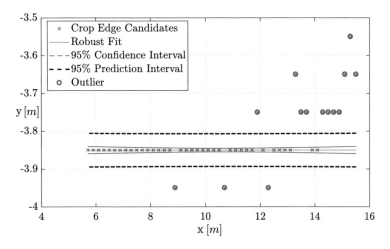

**Fig. 6.** An exemplary robust linear crop edge model with a length of 10 m. Out of the 50 CECs (plotted in RCS) found, the subset of 36 CECs (72%) contribute to the new crop edge model.

## 6.3    Model Crop Edge Filtering

The smoothed model $\hat{Y}^t$ allows for the prediction of the CEC in each grid row at time $t+1$. One could filter each new CEC based on the model and therefore use the 95% prediction interval (PI) which gives an estimate of where one can expect the next crop edge candidate estimates in 95% of the cases, see Fig. 6. But our experiments showed that this assumption forces the model to be too strict and does not properly account for unpredictable crop edge occurrences (see Sect. 7). The scan line of grid row $m$ is therefore only marked as an outlier ($S_m$ = invalid) if its corresponding CEC is not included in the interval (10)

$$[\hat{y}_m \pm \gamma] \ , \tag{10}$$

where $\gamma$ is a predefined parameter; in our experiments set to $\gamma = 5 \cdot c_w$.

## 6.4    Model Confidence

In order to assess the quality of the currently detected crop edge, we introduced a confidence value $v_{\text{confidence}}$ (11)

$$v_{\text{confidence}} = w_0 v_{\text{model\_compliance}} + w_1 v_{\text{correlation}} + w_2 v_{\text{valid}} \tag{11}$$

which is a weighted linear combination of three separate confidence values (12)–(14) within the range $[0, 1]$. The three associated parameters $w_0, w_1, w_2$ allow to adjust the confidence preferences. In our experiments, the weights were set to $w_0 = 0.4, w_1 = 0.4, w_2 = 0.2$. The first confidence value $v_{\text{model\_compliance}}$ (12)

gives an indication of how well the newly found CECs correspond with the latest established crop edge model

$$v_{\text{model\_compliance}} = 1 - \frac{\sum_{i=0}^{s-1} \left| \hat{Y}_{i,y}^{t-1} - \mathcal{P}_{i,y} \right|}{\gamma s}, \tag{12}$$

where $s = |\{x \in \{0, \ldots, m\} \,|\, S_x = \texttt{valid}\}|$ is the number of valid scan lines and $\mathcal{P}_{i,y}$ the $y$-value of a respective crop edge point in a scan line $i$. The second value $v_{\text{correlation}}$ (13) is an overall measure of how well the model function was correlated in each scan line.

$$v_{\text{correlation}} = \frac{\sum_{i=0}^{s-1} r_{\max,i}}{s} \tag{13}$$

The last indicator $v_{\text{valid}}$ (14) is a simple ratio of the number of valid scan lines and the total number of scan lines

$$v_{\text{valid}} = \frac{s}{m+1} \tag{14}$$

The combined confidence value could be used as feedback for the combine operator or further influence the subsequent guidance system.

## 7   Experiments and Results

A *John Deere Combine* equipped with an *Intel Core i7-4790K* CPU was used as a test platform. With the proposed system implemented in the C++ Real-time Robotic Framework FINROC, multiple dynamic tests have been conducted. In preparatory tests, logged data of actual harvesting in wheat and rapeseed fields has been used. For the conclusive experiments, the combine harvester was driven in manual operation mode, with the operator instructed to accurately follow the crop edge near the header divider. In order to evaluate the performance and quality of our system, the crop edge models together with the point cloud were logged. Later on, they have been first visually and then geometrically examined by comparing them in three-dimensional space.

During our tests, the elevation map was parameterized as follows: $g_h = 50$, $g_w = 75$, $c_h = 0.25$ m and $c_w = 0.1$ m. The cell width needed to be chosen with care because the disparities at the very crop edge at plant height is blurred due to stalks and ears sticking out. The chosen width was sufficient enough to capture the crop edge properties and still allowed for precise detection of the edge location. Figure 7 shows the results of a 452.5 m run of harvesting wheat at an average speed of 7 km/h with a mounted header of 8 m in width. The RMSE of the heading angle amounts to $0.9°$ and for the lateral offset 0.09 m in comparison to track driven by the DGPS-based guidance system (with RTK) used to record the data for evaluation. The average run time of the entire system adds up to 55 ms with 25 ms for disparity computation and 30 ms for the crop edge detection.

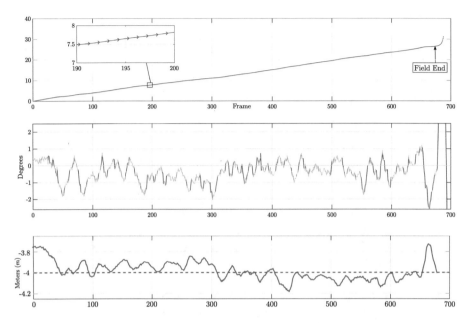

**Fig. 7.** Experiment results for the scenario depicted in 3b. *Top*: Trajectory generated. *Middle*: Crop edge model heading angle in degrees with reference with a RTK-DGPS-guidance system driving along the crop edge (color coded by gradient). *Bottom*: Lateral offset (red) of kinematic center (RCS) and crop edge model located at the header divider.

(a)                                        (b)

**Fig. 8.** (a) Scenario with a tramline present displayed with crop edge overlay. (b) 3D crop edge model (red) and point cloud of the scenario shown in (a)

The system was also capable of correctly detecting the actual crop edge in cases of irregularities where tramlines, partly flattened crop and small bushes occurred. Such a "difficult" scenario is shown in Fig. 8b where a calculated 3D

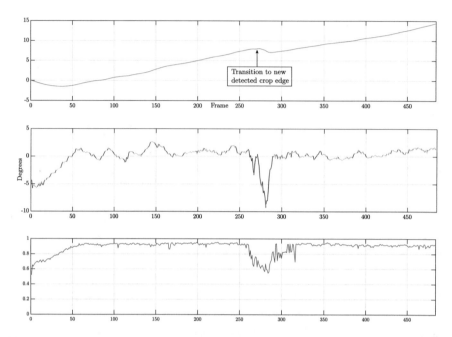

**Fig. 9.** Experiment results of the dataset shown in Fig. 8: *Top*: Trajectory generated with transition to new crop edge. *Middle*: Crop edge model heading angle in degrees with reference to GPS guidance system driving along the crop edge (color coded by gradient). *Bottom*: Confidence values.

crop edge model in a scenario with a tramline being present is depicted. The quantitative results of the experiment related to the tramline dataset are given in Fig. 9. A peak between frame 250 and 300 is noticeable as the system performed a transition to a new outer detected crop edge which occurred as a consequence of a tramline separated a stripe of wheat. In this case, also the confidence $v_{\text{combined}}$ shortly decreased. An independent decrease of $v_{\text{valid}}$ was an indication for the *end-of-crop*.

## 8    Conclusions

In this paper, we have presented a robust crop edge detection approach to allow for automated guidance of combine harvesters equipped with stereo camera(s) as perception sensor. With the help of an elevation map, the 3D point cloud is partitioned into scan lines allowing for parallelized crop edge detection. A fast height segmentation was achieved by frame-wise iterative Expectation Maximization. The detected linear crop edge was modeled via robust regression and additionally filtered. The proposed system was capable to accurately detect the crop edge

in wheat and rapeseed fields. In the future, we plan to fuse our height based approach with a texture based crop edge detection to improve the guidance precision
– for example in cases of flattened crop or grass fields when the height difference
between ground and crop is very small.

# References

1. Benson, E., Reid, J., Zhang, Q.: Machine vision-based guidance system for agricultural grain harvesters using cut-edge detection. Biosyst. Eng. **86**(4), 389–398
   (2003)
2. Chateau, T., Debain, C., Collange, F., Trassoudaine, L., Alizon, J.: Automatic
   guidance of agricultural vehicles using a laser sensor. Comput. Electron. Agric.
   **28**(3), 243–257 (2000)
3. Choi, J., Yin, X., Yang, L., Noguchi, N.: Development of a laser scanner-based
   navigation system for a combine harvester. Eng. Agric. Environ. Food **7**(1), 7–13
   (2014)
4. Debain, C., Chateau, T., Berducat, M., Martinet, P., Bonton, P.: A guidance-
   assistance system for agricultural vehicles. Comput. Electron. Agric. **25**(1–2), 29–
   51 (2000)
5. Dempster, A.P., Laird, N.M., Rubin, D.B.: Maximum likelihood from incomplete
   data via the em algorithm. J. Roy. Stat. Soc.: Ser. B (Methodol.) **39**(1), 1–38
   (1977)
6. Fleischmann, P., Berns, K.: A stereo vision based obstacle detection system for
   agricultural applications. In: Wettergreen, D.S., Barfoot, T.D. (eds.) Field and Service Robotics: Results of the 10th International Conference, pp. 217–231. Springer
   International Publishing, Cham (2016)
7. Hirschmüller, H.: Stereo processing by semiglobal matching and mutual information. IEEE Trans. Pattern Anal. Mach. Intell. **30**(2), 328–341 (2008)
8. Huber, P.J.: Robust Statistics. Wiley Series in Probability and Statistics. Wiley,
   Hoboken (2005)
9. Pilarski, T., Happold, M., Pangels, H., Ollis, M., Fitzpatrick, K., Stentz, A.: The
   demeter system for automated harvesting. Auton. Robots **13**(1), 9–20 (2002)
10. Rovira-Más, F., Han, S., Wei, J., Reid, J.: Autonomous guidance of a corn harvester
    using stereo vision. Agric. Eng. Int. CIGR J. (2007)

# Extracting Structure of Buildings Using Layout Reconstruction

Matteo Luperto$^{(\boxtimes)}$ and Francesco Amigoni

Politecnico di Milano, 20133 Milan, Italy
{matteo.luperto,francesco.amigoni}@polimi.it

**Abstract.** Metric maps, like occupancy grids, are the most common way
to represent indoor environments in mobile robotics. Although accurate
for navigation and localization, metric maps contain little knowledge
about the structure of the buildings they represent. However, if explicitly identified and represented, this knowledge can be exploited in several
tasks, such as semantic mapping, place categorization, path planning,
human robot communication, and task allocation. The *layout* of a building is an abstract geometrical representation that models walls as line
segments and rooms as polygons. In this paper, we propose a method to
reconstruct two-dimensional layouts of buildings starting from the corresponding metric maps. In this way, our method is able to find regularities
within a building, abstracting from the possibly noisy information of the
metric map. Experimental results show that our approach performs effectively and robustly on different types of input metric maps, characterized
by noise, clutter, and partial data.

## 1 Introduction

Understanding the environments in which they operate is an important capability for autonomous mobile robots. Buildings are strongly structured environments that are often organized in regular patterns. Metric maps, like grid maps
[1], which are the usual environment representation employed in mobile robotics,
do not explicitly contain any knowledge about the building structure, but just
represent the space occupation for navigation and localization purposes. However, the identification of the building structure is useful for several tasks, such
as semantic mapping, place categorization, path planning, human robot communication, and task allocation [2]. For instance, knowledge about the building
structure can be useful to efficiently spread the robots to incrementally build
the map of an initially unknown environment, in the context of coordinated
multirobot exploration [3].

The *layout* of a building is a geometrical representation of its walls and
rooms. Each room is represented either by a polygon (in 2D) or by a box model
or a set of planes (in 3D). Walls are accordingly represented as line segments or
planes. A layout of a building thus represents rooms that compose the building
disregarding information about furniture and noisy and missing data, which

© Springer Nature Switzerland AG 2019
M. Strand et al. (Eds.): IAS 2018, AISC 867, pp. 652–667, 2019.
https://doi.org/10.1007/978-3-030-01370-7_51

are often affecting metric maps. It thus provides a "clean" and stable knowledge about the structure of the building. *Layout reconstruction* is the task of retrieving the layout from a metric representation of the building and can be performed at room or at floor level, starting from 2D grid maps [4] or from 3D point clouds [5,6].

In this paper, we propose a method that reconstructs the layouts of buildings starting from 2D metric maps. Most of the methods for reconstructing the building layout start from scans that are 3D point clouds, often assuming perfect alignment between them and precise knowledge of their poses [6]. As a consequence, many of these methods are not easily applicable to maps typically acquired by mobile robots. Our proposed method considers a different type of input, namely 2D metric maps obtained by autonomous mobile robots using 2D laser range scanners and SLAM algorithms [1]. Such maps can present several inaccuracies that our method is able to address, such as partial, missing, unaligned, and noisy data.

The proposed method identifies the *representative lines* along which the walls of a building are aligned and uses these lines to segment the area in smaller parts, called *faces*, which are finally clustered in rooms. The representative lines allow to find regularities between parts of the same building; for instance, two rooms placed at the opposite sides of the building can have aligned walls or a long corridor can connect rooms with the same shape and sharing the same wall.

Our approach is experimentally validated in an extensive range of settings. In particular, to show a possible use of our method for layout reconstruction, we consider a setting in which a robot is exploring an initially unknown building. Starting from the current partial grid map, our method is able to distinguish between fully and partially explored rooms and to provide a rough estimate of the geometrical shape and of the size of the latter ones.

## 2   Related Work

Automatic analysis of representations of floors of buildings in order to extract structural information is an interdisciplinary topic addressed in different research fields, such as robotics, architecture, computer vision, and image analysis. While an exhaustive survey is out of the scope of this paper, here we cover a significant sample of methods, focusing on those developed for mobile robots.

*Room segmentation* methods typically start from 2D metric maps obtained from laser range scanners and separate them into parts, each one corresponding to a different room. Recently, the authors of [2] presented a survey of room segmentation techniques for mobile robots. They identify four main families of approaches, namely Voronoi-based partitioning [7], graph partitioning [8], feature-based segmentation [9–11], and morphological segmentation [12]. Representative methods of the four families are compared in [2] with no method clearly outperforming the others, although Voronoi-based segmentation techniques appear to be the most accurate.

A system for room segmentation that shares some similarities with our layout reconstruction approach can be found in [13]. Similarly to our method, Canny edge transform and Hough line transform are used to extract lines from a grid map. The main difference from our method is that, while we extract a small set of representative lines that are used for identifying walls, in [13] a larger set of lines are used for obtaining a scalable grid representation of the environment, similar to an octo-map and called *A-grid*, which is eventually used to perform segmentation. Moreover, differently from our method, that of [13] does not perform layout reconstruction.

In general, our approach differs from room segmentation methods, which typically partition the metric maps in rooms, because it extracts from the metric map a more abstract representation in which rooms are modeled using geometrical primitives. In our layout representation, a room is a polygon, while, in room segmentation, a room is a set of cells of the grid map. One could say that layout reconstruction provides a qualitative representation of the structure of a building that does not have to be metric exact. Although it is possible to recover the polygon representing a room from the set of cells associated to that room [6] (i.e., as a post-processing of room segmentation), this is largely based on information local to the rooms, hardly capturing global regular features, like alignment of rooms along a corridor, which our approach is able to consider. Most of the methods that perform layout reconstruction require aligned 3D point clouds and usually exploit the knowledge of the poses from where the scans are taken. Our method is inspired by such approaches, but it is adapted to be used on 2D maps obtained by a mobile robot. More precisely, we extend and adapt the method presented in [6], as explained extensively in Sect. 3.

Authors of [5] segment and reconstruct rooms of an indoor environment from a 3D point cloud perfectly aligned to a reference system. Points are projected on the 2D plan, in order to find walls as sets of points whose projections are close to each other. The entire floor is segmented in areas using these projections of walls. Areas that are adjacent but not separated by a "peak-gap-peak" pattern in the projected points distribution are merged in the final segmentation.

In [14], wall lines are retrieved from an analysis of the distribution of points of a 3D point cloud projected in 2D along the $z$ axis. The structure is then decomposed into its horizontal and vertical parts, which are used to extract a volumetric model trough an energy minimization process solved by a Graph-Cut method.

Two methods which present similarities with our approach but are based on aligned 3D point clouds are [15,16]. In [15], 3D point cloud scans (and their poses) are used to recover the 3D model of a building. Walls are detected by projecting on a plane the wall surfaces perceived in different rooms. Similarly to our approach, walls are used for dividing the space. In particular, wall centerlines are used to reconstruct a planar graph that is later segmented into different rooms by an energy minimization approach. A similar, but improved, method is presented in [16]. It projects a 3D point cloud on a 2D plane to retrieve walls. The 2D map is segmented into rooms by identifying door/openings along walls and by selecting a set of viewpoints from the Voronoi graph built on the 2D projection. Finally, rooms are found by solving an energy minimization problem on a cell planar graph obtained from the lines of the wall segments. Viewpoints are used to initialize the graph potentials following the intuition that points that can be seen from the same viewpoint are more likely to be part of the same room.

Similarly to our work, [4] proposes a method that segments a 2D metric map of an indoor environment built by a robot while reconstructing its layout. It uses a framework based on Markov Logic Networks and data-driven Markov Chain Monte Carlo (MCMC) sampling. Using MCMC, the system samples many possible semantic worlds (layouts of the environment) and selects the one that best fits the sensor data. Differently from our work, which identifies the building structure by analyzing the entire metric map, each transition from a state to another state in the MCMC is based on local edit operations on the layout of a single room, ignoring other parts of the building.

## 3   Our Method

The method we propose in this paper starts from a 2D metric map, identifies the walls in it, and uses them for finding rooms and for reconstructing the layout of the environment. The method is composed of a number of steps executed sequentially, sketched in Algorithm 1 and detailed in the following with the help of a running example (Fig. 1). Some of the steps are inspired by the approach of [6]. More precisely, we use the method developed in [6] for dividing the metric map in smaller parts (which are used to identify rooms) using a set of lines. However, as already discussed, [6] reconstructs the 3D layout of a building from point clouds precisely aligned and registered in the same coordinate system, knowing the number and the poses of scans. Differently from [6], our method is intended to be used on 2D metric maps obtained from data acquired by laser range scanners and processed by SLAM algorithms, which can present misalignments and artifacts. Moreover, our method can be used effectively on different types of metric maps obtained from different sources, such as incomplete metric maps, blueprints, and evacuation maps, as described in Sect. 4.

The starting point is a metric map $M$ representing an indoor environment, typically a floor of a building. Without loss of generality, we assume that $M$ is a 2D grid map, as the one in Fig. 1a, namely a two-dimensional matrix of cells (pixels), each one representing the probability that the corresponding area is occupied by an obstacle.

The first operation on $M$ is the detection of significative edges using the Canny edge detection algorithm [17], which partitions the cells of $M$ into free cells and obstacle cells. The binary metric map resulting from the application of the Canny edge detection algorithm is called $M'$ and an example is shown in Fig. 1b.

The set of edges (obstacle cells) is then processed by a probabilistic Hough line transform algorithm [18] to detect the line segments $S$ that approximate the edges in $M'$. Figure 1c shows such line segments in green. Then, the contour of the map is obtained, using the contour detection algorithm of [19], after the application of a threshold $q$ that divides cells of the original map $M$ in free or occupied. Figure 1d shows in yellow the area inside the map border.

**Input:** a grid map $M$
**Output:** the reconstructed layout $\mathcal{L}$
/* Compute features from the map                                              */
$M' \leftarrow CannyEdgeDetection(M)$
$S \leftarrow pHoughLineTransform(M')$

/* Obtain contour of the map                                                  */
$M'' \leftarrow thresholdMap(M, q)$
$Inner \leftarrow computeMapContour(M'')$

/* Create clusters of collinear segments                                      */
$\mathcal{C} \leftarrow meanShiftClustering(S)$
$\mathcal{C}' \leftarrow spatialClustering(\mathcal{C}, line\_distance\_threshold)$

/* Find faces from representative lines                                        */
$lines \leftarrow getsRepresentativeLines(\mathcal{C}')$
$F \leftarrow findFaces(lines)$

/* Compute spatial affinity between faces                                      */
$L \leftarrow computeAffinityMatrix(F)$
$D \leftarrow diag(\sum_{j=1}^{n} L_{i,j})$
$A \leftarrow D^{-1}L$

/* Remove external faces outside border                                        */
**for** $f \in F$ **do**
$\quad$ **if** $area(f \cap Inner) < \delta$ **then**
$\quad\quad$ | $F \leftarrow F \setminus f$
$\quad$ **end**
**end**

/* Cluster together faces into rooms                                           */
$\mathcal{L} \leftarrow DBSCAN(F, A, \epsilon, minPoints)$
**return** $\mathcal{L}$

**Algorithm 1.** Our method for layout reconstruction.

The line segments $S$ are then clustered together according to the angular coefficients of their supporting lines using the mean shift clustering algorithm [20]. At the end of the angular clustering, we obtain a set of clusters $\mathcal{C} = \{C_1, C_2, \ldots\}$ such that each $C_j \subseteq S$ and $C_1 \cup C_2 \cup \ldots = S$. Each cluster $C_j$ represents the set of line segments with similar angular coefficient $\alpha_j$, namely with similar direction, independent of spatial proximity.

Next, the line segments belonging to the same angular cluster $C_j$ are further clustered according to their spatial separation. Consider two line segments $s$ and $s'$ belonging to an angular $C_j$ with angular coefficient $\alpha_j$ and call $l$ and $l'$ the lines passing through the middle points of $s$ and $s'$ and with angular coefficient $\alpha_j$. If the distance between the parallel lines $l$ and $l'$ is less than a threshold (set, after some initial tests, to 90 cm in our experiments; note that 90 cm is approximately the width of a doorway), then $s$ and $s'$ are put in the same spatial cluster. At the end of this step, we have a set of clusters $\mathcal{C}' = \{C_{1,1}, C_{1,2}, \ldots, C_{2,1}, C_{2,2}, \ldots\}$, such that $C_1 = C_{1,1} \cup C_{1,2} \cup \ldots$ and $C_2 = C_{2,1} \cup C_{2,2} \cup \ldots$, and so on. Figure 1e shows the results of the spatial clustering where the line segments belonging to the same cluster in $\mathcal{C}'$ are displayed with the same color.

At this point, for each cluster $C_{j,k}$, a *representative line* $l_{j,k}$ that represents all the line segments in $C_{j,k}$ is determined. This representative line is computed

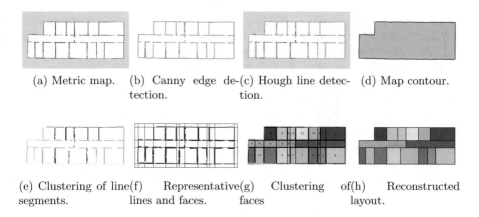

(a) Metric map.     (b) Canny edge de-(c) Hough line detec-(d) Map contour.
                    tection.               tion.

(e) Clustering of line(f)     Representative(g)     Clustering     of(h)     Reconstructed
segments.               lines and faces.       faces               layout.

Fig. 1. An example run of our method.

as the line with angular coefficient $\alpha_j$ associated to $C_j$ and that passes through the median of the set of middle points of the line segments in $C_{j,k}$. Each representative line, in red in Fig. 1f, indicates the direction of a wall within the building. The intersections between all lines divide the area of map $M'$ into different areas, called *faces*, as shown in Fig. 1f. We call $F$ the set of faces.

Rooms are determined by grouping faces together. Adjacent faces that are separated by an edge corresponding to a wall should belong to different rooms, while adjacent faces that are separated by an edge not corresponding to any wall should be grouped together in the same room. More precisely, for each pair of faces $f$ and $f'$ that share a common edge, we compute a weight $w(f, f')$ as follows. Given an edge $e_{f,f'}$ shared by two faces $f$ and $f'$ (and belonging to the representative line $l_{j,k}$ of a spatial cluster $C_{j,k}$), its weight is calculated (as described in [6]) as $w_{f,f'} = cov(e_{f,f'})/len(e_{f,f'})$, where $len(e_{f,f'})$ is the length of $e_{f,f'}$ and $cov(e_{f,f'})$ is the length of the projections, on $e_{f,f'}$, of the line segments in $C_{j,k}$. The larger the weight $w_{f,f'}$, the stronger the hypothesis that there is a wall (obstacle) along $e_{f,f'}$ (namely, between faces $f$ and $f'$). If an edge is completely covered by projections of line segments in $C_{j,k}$, then its weight is 1. Following the definition of [6], weighted edges are used to compute an affinity measure $L$ between all pairs of faces. $L$ is similar to a Laplacian, and its entries $L_{f,f'}$ are defined as:

$$L_{f,f'} = \begin{cases} e^{-w(f,f')/\sigma} & \text{if } f \neq f' \text{ and } f \text{ and } f' \text{ are adjacent} \\ 1 & \text{if } f = f' \\ 0 & \text{otherwise} \end{cases}$$

where $\sigma$ is a regularization factor. From the matrix $L$, a local affinity matrix $A$ is defined as $A = D^{-1}L$, with $D = diag(\sum_{j=1}^{n} L_{i,j})$, where $n$ is the number of faces in $F$ and $i$ is the row of the matrix. Each element $A_{f,f'}$ indicates an affinity value considering the local connectivity between faces $f$ and $f'$. The matrix $A$ is used as input for DBSCAN [21], which clusters faces. DBSCAN groups together faces that are close to each other in a dense portion of the feature space represented by matrix $A$. (In our experiments, we set the two DBSCAN parameters to $\epsilon = 0.85$ and $minPoints = 1$.) Note that DBSCAN detects rooms without additional information, while the method of [6] relies on the availability of scan poses. In this sense, our approach can more flexibly adapt to different sources for input metric maps, as shown in the next section.

Before applying DBSCAN, some faces are discarded. Specifically, discarded faces are those called *external* and such that the area of their intersection with the inner area of $M$ (obtained from the contour) is smaller than a threshold $\delta$. Remaining faces are called *partial* if they are adjacent to an external face via an edge whose weight is less than a threshold (0.2 in our experiments) and *internal* otherwise. Partial faces cover the area of rooms that are not fully known according to the data collected in $M$, as for example during the exploration of a building. Partial faces contain *frontiers*, namely boundaries between known and unknown parts of the environment. Frontiers are present due to the limited range of sensors [22]. Examples of partial faces can be seen, in gray, in Fig. 5.

Then, DBSCAN is applied to internal faces to obtain a set of clusters $R = \{F_1, F_2, \ldots\}$ of $F$. Each cluster $F_i$ corresponds to a room $r_i$, which is represented as a polygon obtained by merging together all the faces belonging to $F_i$. Figure 1g shows the results of DBSCAN for our example. The number $i$ in each face indicates its cluster $F_i$. Different clusters have also different colors.

The final reconstructed layout $\mathcal{L} = \{r_1, r_2, \ldots\}$ is finally displayed in Fig. 1h. Note that it is a "clean" and abstract representation of the original grid map of Fig. 1a, which nevertheless retains the main structural features.

A post-processing refinement could be used to remove unconnected rooms in maps obtained from blueprints, by checking if each room is connected to the Voronoi graph obtained from $M$ following the approaches presented in [2].

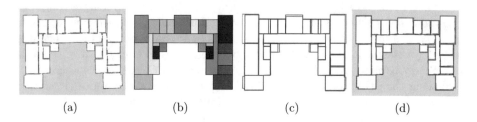

|       |       |       |       |
|-------|-------|-------|-------|
| (a)   | (b)   | (c)   | (d)   |

**Fig. 2.** An example of a layout reconstruction. (a) Grid map. (b) Reconstructed layout. (c) Reconstructed layout (black) and ground truth (red). (d) Grid map and ground truth (red).

## 4    Experimental Results

In this section, we evaluate our method for reconstructing the layouts of buildings from 2D metric maps. We note that a comparison of our approach against similar methods such as [6,15,16] requires radically different input data (e.g., aligned 3D point scans vs. 2D grid maps) and is not reported here. Reconstructing the layout of a building with our approach takes up to few seconds on a commercial laptop. Further results, beyond those presented in the following, are shown in the video at https://youtu.be/ENcbDsXknZE.

## 4.1   Results with Grid Maps from Stage

A reconstructed layout is expected to present two main characteristics: (1) all the rooms of the real building (and only them) should be in the layout; (2) the shape of each reconstructed room should match that of its real counterpart. Evaluation is performed both visually and quantitatively, comparing the reconstructed layout $\mathcal{L}$ and a ground truth layout $Gt$. Following the approach of [2], we introduce two matching functions between rooms in $\mathcal{L}$ and in $Gt$, namely *forward coverage FC* and *backward coverage BC*. $FC$ represents how well the reconstructed layout $\mathcal{L}$ is described by the ground truth layout $Gt$, while $BC$ represents how well the ground truth layout $Gt$ is described by the reconstructed layout $\mathcal{L}$. Specifically:

$$FC : r \in \mathcal{L} \mapsto r' \in Gt \qquad\qquad BC : r' \in Gt \mapsto r \in \mathcal{L}$$

For each room $r \in \mathcal{L}$, $FC$ finds the room $r' \in Gt$ that maximally overlaps $r$; conversely, for each room $r' \in Gt$, $BC$ finds the room $r \in \mathcal{L}$ with the maximum overlap with $r'$. Calling *area*() a function that computes a polygon area, the overlap between a room $r \in \mathcal{L}$ and a room $r' \in Gt$ is defined as $area(r \cap r')$. Matching functions $FC$ and $BC$ are used for computing two measures of accuracy called *forward accuracy* $A_{FC}$ and *backward accuracy* $A_{BC}$:

$$A_{FC} = \frac{\sum_{r \in \mathcal{L}} area(r \cap FC(r))}{\sum_{r \in \mathcal{L}} area(r)} \qquad\qquad A_{BC} = \frac{\sum_{r' \in Gt} area(BC(r') \cap r')}{\sum_{r' \in Gt} area(r')}$$

The numbers of rooms in $\mathcal{L}$ and $Gt$ can be different, due to over- or under-segmentation. Over-segmentation results in high $A_{FC}$ and low $A_{BC}$, while under-segmentation results in high $A_{BC}$ and low $A_{FC}$.

We consider 20 grid maps obtained running the ROS implementation[1] of the GMapping algorithm [23] on data collected by a robot equipped with a laser range scanner and moving autonomously in 20 school buildings simulated in Stage[2]. It is important to point out that the evaluation is performed by comparing the layouts reconstructed from the grid maps built by the robot and the *actual* layouts of the simulated buildings fed to Stage. This allows us to evaluate if our layout reconstruction approach is able to cope with noise and errors introduced in the mapping process as a result of noisy readings from the sensor, errors in localization, or errors due to inaccurate odometry readings. In this sense, we consider simulated but realistic metric maps obtained by robots.

---

[1] http://wiki.ros.org/gmapping.
[2] http://wiki.ros.org/stage.

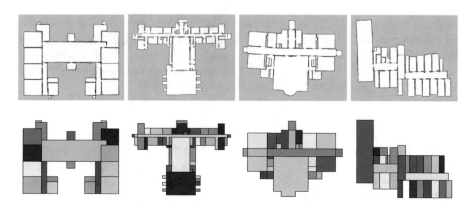

**Fig. 3.** Examples of layout reconstructions.

Layout reconstruction accuracies are the following: $A_{BC} = 87.8\% \pm 4.5\%$ and $A_{FC} = 87.6\% \pm 5.7\%$. On the set of 20 worlds mapped by the simulated robot, our method is able to reconstruct successfully and with good accuracy the layout of the original buildings. For example, Fig. 2 presents a grid map obtained by the robot (Fig. 2a) and the layout reconstructed using our method (Fig. 2b). For this particular example, we have $A_{FC} = 90.1\%$ and $A_{BC} = 93.3\%$. In Fig. 2c, the reconstructed layout (in black) and the ground truth layout (in red) are superimposed. Although the layout is generally accurate, there all small differences in the geometry of rooms, due to approximations introduced by our method, which result in a slight performance degradation. In Fig. 2d the grid map and the ground truth building layout (in red) are superimposed. While the grid map provides a good representation of the environment, some inaccuracies, such as irregular gaps between walls, are present. Our method tries to filter out those inaccuracies when reconstructing the layout.

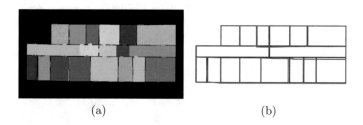

(a)                              (b)

**Fig. 4.** Segmentation by a Voronoi-based approach, from [2] (a) and comparison between our reconstructed layout (black) and the ground truth layout (red) (b) for the metric map of Fig. 1a.

Figure 3 shows four other examples of reconstructed layouts from grid maps. In all the four cases, our method is able to correctly reconstruct the layout of the environment, coping well with alignment and rotation errors and with gaps between rooms in the metric maps (e.g., see the third grid map). Few inaccuracies are introduced, like long corridors split into smaller units, thus producing over-segmentation. Another error is made sometimes when there is a small gap within the building (e.g., due to large wall or a pillar), which is misclassified as internal face and subsequently added to an adjacent room or considered as a small independent room. However, in general, our method can successfully retrieve the layouts of large-scale buildings, as in the second example of Fig. 3, where the rooms connected to the top corridor (light blue) are the classrooms of a high school.

**Table 1.** Layout reconstruction results on datasets from [2].

| No furniture | | Furniture | |
|---|---|---|---|
| Recall | 90.0% ± 4.1% | Recall | 89.5% ± 6.2% |
| Precision | 90.1% ± 6.3% | Precision | 94.0% ± 2.2% |

Since our method does not assume a Manhattan world, it can be potentially used in non-Manhattan environments, such as those with round or diagonal walls. However, since walls are approximated by straight lines, round walls are approximated by polylines (see, again, the third example of Fig. 3). Additional results are reported in the video.

Maps that present (relatively) small misalignments are adjusted by our method, as shown by the last two examples of Fig. 3. A trade-off exists between the accuracy of alignment that can be reached by our method and its ability to approximate round walls with polylines. This trade-off can be set by modifying the parameters of the line segment clustering step (like the 90 cm threshold), where collinear walls are clustered together and representative lines are identified. In all the examples presented in this paper we preferred strong alignment over good approximation of round walls.

## 4.2  Results with Publicly Available Datasets

Here, we present the results of the evaluation of our method on the two datasets (each composed of 20 metric maps) used in [2] for comparing different methods for room segmentation. The datasets contain metric maps with and without furniture, respectively, and, according to [2], are evaluated employing precision and recall metrics, which are defined similarly to $A_{FC}$ and $A_{BC}$:

$$Precision = \frac{\sum_{r \in \mathcal{L}} area(r \cap FC(r))/area(r)}{|\mathcal{L}|}$$

$$Recall = \frac{\sum_{r' \in Gt} area(BC(r') \cap r')/area(r')}{|Gt|}$$

When the dataset with furniture is used, our method filters out some of the clutter introduced by furniture by automatically clustering and removing isolated obstacle cells using DBSCAN directly on the original metric map $M$ (this is a very simple way to handle furniture, which can be definitely improved). Results, reported in Table 1, are good and confirm those obtained with our dataset.

A comparison with the results obtained by room segmentation methods of [2] is unfair, because these methods directly partition the cells of the metric maps, while our method uses a more abstract representation based on representative lines and faces, thus introducing some approximations. An example of the approximations introduced by our layout reconstruction with respect to room segmentation is displayed in Fig. 4. Figure 4a shows the Voronoi-based segmentation of the metric map of Fig. 1a, as reported in [2]. It can be observed that Voronoi-based methods tend to produce over-segmentation. In Fig. 4b, we compare our reconstructed layout (in black) with the real structure of the building (in red). Despite being able to correctly capture the building layout, for this map we have $A_{FC} = 93.6$ and $A_{BC} = 94.6$, due to some visible approximations. That said, comparing Table 1 with Table II of [2], we can say that our method produces results comparable (within 1 sigma) with those of the methods surveyed in [2], further confirming that the reconstructed layout actually captures the structure of the building. Moreover, the availability of the layout enables a number of possible tasks, as suggested in the following. Additional results obtained on the datasets of [2] are reported in the video.

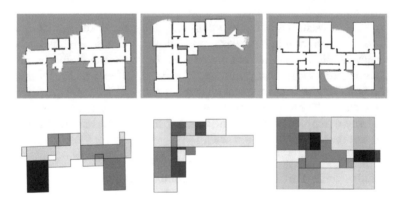

**Fig. 5.** Examples of layout reconstructions from partial grid maps (partially explored rooms are in gray).

### 4.3   Results with Partial Grid Maps

Figure 5 shows three examples of layout reconstruction originally performed by our approach starting from partial grid maps. Partially explored rooms are identified correctly and marked in gray. The faces composing a partially explored room are added to the reconstructed layout, providing a guess of the unknown room shape. If a partially explored room is not "closed" by any line segment found in other rooms, then the room is arbitrarily completed with a wall parallel to the closest line segment.

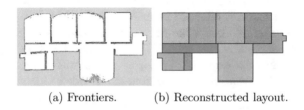

(a) Frontiers.          (b) Reconstructed layout.

**Fig. 6.** Frontiers (highlighted in Fig. 6a) in almost fully explored rooms and frontiers that lead to unexplored parts of the building (in red in Fig. 6b) can be identified.

These results suggest that our method can be used for reconstructing the shape of partially seen rooms, on the basis of the layout reconstructed from of the rest of the building. Such knowledge can be potentially used for speeding-up exploration, as it can be seen from the example of Fig. 6. By reconstructing the shape of partial rooms, the three frontiers on the left are correctly identified as being inside almost fully explored rooms, while the frontier that appears to be more promising to continue the exploration is that on the right, which is in a room that is not closed by any representative line (highlighted in red). The entire layout of the building can be seen for reference in the first example of Fig. 3 (which is rotated by 90° with respect to Fig. 6). This example shows a possible use of the outcomes of our method. A complete exploration method that exploits layout reconstruction is currently being developed and its description is out of the scope of this paper.

### 4.4 Results with Blueprints and Evacuation Maps

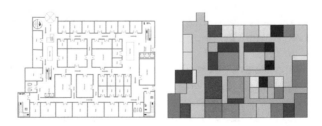

**Fig. 7.** Reconstructed layout from a blueprint of a highly structured building.

Our method can be used also for reconstructing the layout of a building from its blueprint or evacuation map. Blueprints and evacuation maps are usually images in which the walls of buildings are represented together with symbols and words that explain the meaning and the locations of some features (e.g., the functions of the rooms and the presence of fire extinguishers). We pre-process these maps as follows. Words (like the name of the building or the functions of rooms) are recognized and filtered out using a standard OCR method (Tesseract). Doors and other symbols (like fire extinguishers) are easily retrieved and eliminated, since they are represented with standard symbols, by using template-matching algorithms. An example of a layout reconstructed from a blueprint is shown in Fig. 7. Other examples are reported in the video. Quantitative results of layout reconstruction starting from partial grid maps and from blueprints (and evacuation maps) are similar to those obtained with datasets from [2] and are not reported here.

The reconstructed layout of a building obtained from a blueprint can be a useful source of knowledge, especially when the environment in which robots operate is initially unknown, as in a situation where a team of robots perform search and rescue operations and an evacuation map of the environment can be seen on a wall. One of the possible uses of such knowledge is localization, as done in [24], where a robot can localize itself using a floor plan instead of a metric map. A reconstructed layout may introduce some approximations compared to the real shape of the environment, as explained previously. However, [25,26] show how partially inaccurate hand-drawn sketch maps can be effectively used by robots for localization. In principle, using similar approaches, a reconstructed layout of a building can be used for localization even if it is slightly inaccurate.

### 4.5   Results with a Grid Map Acquired by a Real Robot

Figure 8 shows the application of our method to a grid map obtained from the Radish repository [27]. We used the same parameter values set for the maps of the previous datasets (demonstrating the robustness of the approach). Clutter is initially and partially removed by filtering out isolated obstacle cells retrieved and clustered using DBSCAN (as in Sect. 4.2). The figure clearly shows that our method is able to reliably reconstruct the layout of the environment starting from noisy and cluttered maps. Note that we reconstruct the layout of partial rooms, like the one at the bottom left of the map. Moreover, we highlight with bold lines the reconstructed walls, namely the edges between faces whose weight $w(f, f')$ is close to 1.

By visual comparison of our result to those of [4] obtained on the same map of Fig. 8, it can be noticed that, while the method of [4] effectively reconstructs the layout of all fully explored rooms (see Fig. 7 of [4] for reference), our method is able to estimate the layout of both fully explored and partially explored rooms, thus reconstructing the layout of the entire floor, and that representative lines are used to align together rooms that share one or more walls. Open doors in the map, which result in diagonal line segments close to the doorway, are aligned by our method to the closest representative lines. While the identification of walls

**Fig. 8.** Reconstructed layout for a metric map from [27]. (Walls are in bold.)

and representative lines can be performed effectively in both real and simulated environments (without changing values of parameters), the clustering of faces into rooms may result in under-segmentation when used on real-world maps, because partially-explored faces are added to the closest room (as it happens to the corridor at the bottom of Fig. 8).

## 5    Conclusions

In this paper, we have presented a method for reconstructing the layout of a building given its 2D metric map obtained from laser range scanners onboard of autonomous mobile robots. Experimental results show that our method performs well and is able to cope with complete and partial metric maps of large-scale buildings and also with blueprints and evacuation maps provided as input.

Future work includes the combination of our method with room segmentation approaches to investigate if better performance can be obtained by looking both at the global structural features and at the local features of the buildings. Moreover, we would like to more precisely assess the performance of our method for very large buildings and in presence of clutter and noise in metric maps. Our method can be also extended to input maps that can be transformed in 2D maps composed of line segments (e.g., 3D point clouds). The use of sketch maps [25,26] as input is also a promising further development. Finally, we plan to use the reconstructed layouts for semantic classification of rooms and for predicting the shape of partially explored rooms to improve performance of exploring robots.

## References

1. Thrun, S., Burgard, W., Fox, D.: Probabilistic Robotics. The MIT Press, Cambridge (2005)
2. Bormann, R., Jordan, F., Li, W., Hampp, J., Hägele, M.: Room segmentation: Survey, implementation, and analysis. In: Proceedings of ICRA, pp. 1019–1026 (2016)
3. Quattrini Li, A., Cipolleschi, R., Giusto, M., Amigoni, F.: A semantically-informed multirobot system for exploration of relevant areas in search and rescue settings. Auton. Robot. **40**(4), 581–597 (2016)
4. Liu, Z., von Wichert, G.: A generalizable knowledge framework for semantic indoor mapping based on Markov logic networks and data driven MCMC. Futur. Gener. Comput. Syst. **36**, 42–56 (2014)

5. Armeni, I., Sener, O., Zamir, A., Jiang, H., Brilakis, I., Fischer, M., Savarese, S.: 3D semantic parsing of large-scale indoor spaces. In: Proceedings of CVPR, pp. 1534–1543 (2016)
6. Mura, C., Mattausch, O., Villanueva, A.J., Gobbetti, E., Pajarola, R.: Automatic room detection and reconstruction in cluttered indoor environments with complex room layouts. Comput. Graph. **44**, 20–32 (2014)
7. Thrun, S.: Learning metric-topological maps for indoor mobile robot navigation. Artif. Intell. **99**(1), 21–71 (1998)
8. Brunskill, E., Kollar, T., Roy, N.: Topological mapping using spectral clustering and classification. In: Proceedings of IROS, pp. 3491–3496 (2007)
9. Mozos, O.: Semantic Labeling of Places with Mobile Robots. Springer Tracts in Advanced Robotics, vol. 61. Springer (2010)
10. Friedman, S., Pasula, H., Fox, D.: Voronoi random fields: Extracting the topological structure of indoor environments via place labeling. In: Proceedings of IJCAI, pp. 2109–2114 (2007)
11. Sjoo, K.: Semantic map segmentation using function-based energy maximization. In: Proceedings of ICRA, pp. 4066–4073 (2012)
12. Buschka, P., Saffiotti, A.: A virtual sensor for room detection. In: Proceedings of IROS, pp. 637–642 (2002)
13. Capobianco, R., Gemignani, G., Bloisi, D., Nardi, D., Iocchi, L.: Automatic extraction of structural representations of environments. In: Proceedings of IAS-13, pp. 721–733 (2014)
14. Oesau, S., Lafarge, F., Alliez, P.: Indoor scene reconstruction using feature sensitive primitive extraction and graph-cut. ISPRS J. Photogramm. **90**, 68–82 (2014)
15. Ochmann, S., Vock, R., Wessel, R., Klein, R.: Automatic reconstruction of parametric building models from indoor point clouds. Comput. Graph. **54**, 94–103 (2016)
16. Ambruş, R., Claici, S., Wendt, A.: Automatic room segmentation from unstructured 3-D data of indoor environments. IEEE Robot. Autom. Lett. **2**(2), 749–756 (2017)
17. Canny, J.: A computational approach to edge detection. IEEE Trans. Pattern Anal. Mach. Intell. **8**(6), 679–698 (1986)
18. Kiryati, N., Eldar, Y., Bruckstein, A.M.: A probabilistic hough transform. Pattern Recogn. **24**(4), 303–316 (1991)
19. Suzuki, S., Abe, K.: Topological structural analysis of digitized binary images by border following. Comput. Vision Graph. **30**(1), 32–46 (1985)
20. Comaniciu, D., Meer, P.: Mean shift: a robust approach toward feature space analysis. IEEE Trans. Pattern Anal. Mach. Intell. **24**(5), 603–619 (2002)
21. Ester, M., Kriegel, H.-P., Sander, J., Xu, X., et al.: A density-based algorithm for discovering clusters in large spatial databases with noise. In: Proceedings of KDD, pp. 226–231 (1996)
22. Yamauchi, B.: A frontier-based approach for autonomous exploration. In: Proceedings of CIRA, pp. 146–151 (1997)
23. Grisetti, G., Stachniss, C., Burgard, W.: Improved techniques for grid mapping with Rao-Blackwellized particle filters. IEEE Trans. Robot. **23**, 34–46 (2007)
24. Winterhalter, W., Fleckenstein, F., Steder, B., Spinello, L., Burgard, W.: Accurate indoor localization for RGB-D smartphones and tablets given 2D floor plans. In: Proceedings of IROS, pp. 3138–3143 (2015)
25. Behzadian, B., Agarwal, P., Burgard, W., Tipaldi, G.D.: Monte Carlo localization in hand-drawn maps. In: Proceedings of IROS, pp. 4291–4296 (2015)

26. Boniardi, F., Behzadian, B., Burgard, W., Tipaldi, G.D.: Robot navigation in hand-drawn sketched maps. In: Proceedings of ECMR, pp. 1–6 (2015)
27. Howard, A., Roy, N.: The robotics data set repository (Radish) (2003). http:// radish.sourceforge.net/

# Unknown Object Detection by Punching: An Impacting-Based Approach to Picking Novel Objects

Yusuke Maeda[1]([✉]), Hideki Tsuruga[2], Hiroyuki Honda[2], and Shota Hirono[2]

[1] Faculty of Engineering, Yokohama National University, 79-5 Tokiwadai,
Hodogaya-ku, Yokohama 240-8501, Japan
maeda@ynu.ac.jp
http://www.iir.me.ynu.ac.jp/
[2] Graduate School of Engineering, Yokohama National University, Yokohama, Japan

**Abstract.** In this paper, a method for unknown object detection based on impacting and keypoint tracking is presented. In this method, a robot perturbs object positions by punching the floor on which the objects are placed, to detect each of the objects individually from camera images before and after the punching. The detection method utilizes consistent movements of the keypoints of each object according to its rigid-body motion. After the detection, a grasp of each of the detected objects is planned based on extracting its two parallel edges. The proposed method is successfully applied to picking up of mahjong tiles by an industrial manipulator.

**Keywords:** Interactive perception · Segmentation · Picking

## 1 Introduction

Traditionally robots are used in structured environments to perform simple and repetitive operations. However, to perform complex and elaborate tasks like humans, robots have to understand non-structured environments. A straightforward solution to understand cluttered and non-structured scenes is to enhance the sensing ability of robots, but it has a limitation in nature: for example, it is impossible to distinguish between movable objects and immobile projections only from their appearance. Thus active sensing or interactive perception [3] to understand scenes through sensing and manipulation is important.

Metta and Fitzpatrick showed that simple operations of a robot such as poking and prodding bring better visual information to aid object recognition [12]. Chang et al. proposed adaptive pushing to singulate objects from a pile, which helps to improve picking success rates [4]. Hermans et al. proposed guided pushing to singulate objects on a table [9]. Gupta et al. presented spreading of piled objects to singulate them and aid object recognition for robotic sorting [6].

© Springer Nature Switzerland AG 2019
M. Strand et al. (Eds.): IAS 2018, AISC 867, pp. 668–678, 2019.
https://doi.org/10.1007/978-3-030-01370-7_52

Katz et al. used pushing to singulate piled objects to clear them [10]. Schiebener et al. studied object segmentation using manipulation by a humanoid robot [13]. In this method, unknown rigid objects in a cluttered environment are pushed by the robot and detected through analysis of motion of color-annotated point cloud obtained by stereo vision.

The previous studies demonstrated the usefulness of interactive perception by perturbing environments through robot operations. However, in these methods, perturbed region in a scene is limited because they use robot operations through direct contacts with objects, like pushing. Moreover, occlusion by robot bodies during operations can be a nasty problem for object detection. In this paper, we introduce a new approach to robotic interactive perception so that a robot can understand a broader region in a scene: impacting. By punching the floor on which the objects are located, objects in a large area can be perturbed without direct contacts between the robot and the objects (Fig. 1). Even novel objects can be detected through keypoint tracking between scene images before and after impacting (Fig. 2). By selecting appropriate punching points, occlusion can be avoided.

In this paper, for simplicity, we use a monocular camera to detect unknown objects on a floor. Some experiments on picking of them are shown. This paper is organized as follows: this section is introduction; our proposed method is overviewed in Sect. 2; its details are described in Sect. 3; some experimental results are shown in Sect. 4; Sect. 5 summarizes this paper.

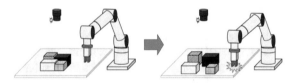

**Fig. 1.** Impacting for detection of unknown objects

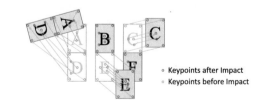

**Fig. 2.** Keypoint tracking for object segmentation

## 2    Overview of Impacting-Based Object Detection

Our impacting-based object detection can be summarized as follows:

1. A scene image is obtained from a camera. Then a robot makes an impact to the scene to perturb the object positions: for example, punching a table on which objects are located. After the impacting, a new scene image is obtained from the camera.
2. The images before and after impacting are analyzed. First, feature points in both of the images are extracted. Here we use SIFT keypoints [11], which are invariant to orientation changes. We also use Harris corners [8] to detect object corners mainly for textureless objects. Then we find one-to-one correspondences of the keypoints between the images before and after impacting, which stand for the movements of the keypoints by impacting.
3. The keypoints are grouped so that each group corresponds to an object. A RANSAC-based method is used to estimate the homogeneous transformation matrix that stands for the rigid-body motion of each object by impacting. We adopt $\alpha$-shape [5] to segment each object, which may be concave.
4. The robot picks up one of the detected objects. A grasp for the object is synthesized based on the segmentation result.

## 3    Details of Object Detection

### 3.1    Keypoint Detection

In this study, SIFT keypoints and Harris corners are detected for novel object detection. The former is used for textured objects and the latter is mainly for textureless objects. 128-dimensional SIFT descriptors are calculated not only for the SIFT keypoints but also for the Harris corners for keypoint tracking.

### 3.2    Keypoint Matching

We find keypoint correspondence between images before and after impacting. Euclidean distance between SIFT descriptors is adopted as the index for the correspondence. Additionally the positional difference between two keypoints is considered in finding the correspondence. Because keypoint movements by impacting should be small, we ignore keypoint pairs with large positional distance to omit incorrect correspondence even when there are two or more objects with identical textures. Additionally we ignore keypoint pairs with too small positional distance to omit background keypoints.

For each of the keypoints detected in the image before impacting, we calculate the Euclidean SIFT distances and the positional distances with all the keypoints detected in the image after impacting. The keypoint that has the minimum SIFT distance and the positional distance larger than a threshold $K_{\mathrm{bg}}$ and smaller than a threshold $K_{\mathrm{dist}}$ is adopted as the correspondent, if any.

### 3.3   Keypoint Pair Filtering

The keypoint pairs found in the matching above are filtered so that the keypoint correspondences are one-to-one. If a keypoint in the image after impacting has multiple correspondents in the image before impacting, we keep only the pair with the minimum positional distance and remove the others (Fig. 3).

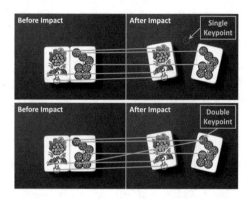

**Fig. 3.** Keypoint pair filtering

### 3.4   Object Tracking

We perform object tracking using RANSAC (Random Sampling Consensus) based on the result of keypoint tracking above as follows:

1. Two keypoint pairs are selected at random.
2. If the positional distance between the keypoints of the pairs in the image before impacting $d_{\mathrm{bef}}$ is almost equal to that in the image after impacting $d_{\mathrm{aft}}$, the two pairs can be a part of the rigid-body motion of an object. Concretely, if $\|d_{\mathrm{bef}} - d_{\mathrm{aft}}\| < K_{\mathrm{len}}$, where $K_{\mathrm{len}}$ is a threshold, we calculate the 2D homogeneous transformation matrix $\boldsymbol{H}$ that maps the keypoints of the two pairs in the image before impacting on those in the image after impacting.
3. If $\boldsymbol{H}$ is almost equal to a previously calculated one $\boldsymbol{H}'$, $\boldsymbol{H}'$ gets one vote. Concretely, if the difference of translation between $\boldsymbol{H}$ and $\boldsymbol{H}'$ is smaller than a threshold $K_{\mathrm{trans}}$ and the difference of rotation between $\boldsymbol{H}$ and $\boldsymbol{H}'$ is smaller than a threshold $K_{\mathrm{rot}}$, we regard $\boldsymbol{H}$ and $\boldsymbol{H}'$ as identical. Otherwise, $\boldsymbol{H}$ gets one vote.
4. Repeat the above $N_{\mathrm{vote}}$ times.

The homogeneous transformation matrices that get more than $K_{\mathrm{vote}}$ votes are considered likely to correspond to the rigid-body motions of objects, where $K_{\mathrm{vote}}$ is a threshold.

Then we group keypoint pairs for the elected homogeneous transformation matrices for object tracking. For each of the elected homogeneous transformation matrices, $\boldsymbol{H}^*$, we find all the keypoint pairs that match $\boldsymbol{H}^*$. If a keypoint pair $(\boldsymbol{p}_{\mathrm{bef}}, \boldsymbol{p}_{\mathrm{aft}})$ satisfies

$$\|\boldsymbol{H}^*\boldsymbol{p}_{\text{bef}} - \boldsymbol{p}_{\text{aft}}\| < K_{\text{diff}}, \tag{1}$$

where $K_{\text{diff}}$ is a threshold, we include it in the group of $\boldsymbol{H}^*$. All the grouped keypoint pairs stand for a tracked object.

However, $\boldsymbol{H}^*$ could have a certain error because it is calculated only from two keypoint pairs in the beginning. Thus we recalculate the homogeneous transformation matrix $\boldsymbol{H}^{**}$ from all the keypoint pairs that match $\boldsymbol{H}^*$ with the method by Arun et al. [2]. Then we group all the keypoint pairs that match $\boldsymbol{H}^{**}$ such that

$$\|\boldsymbol{H}^{**}\boldsymbol{p}_{\text{bef}} - \boldsymbol{p}_{\text{aft}}\| < K'_{\text{diff}}, \tag{2}$$

where $K'_{\text{diff}}(< K_{\text{diff}})$ is a threshold. All the keypoint pairs that match $\boldsymbol{H}^{**}$ give us an improved result of object tracking.

### 3.5   Object Segmentation

We segment a tracked object for each of the keypoint pair groups in the above. For the keypoints in the image after impacting in a group of keypoint pairs, we calculate $\alpha$-shape [5], which can represent not only convex but also concave shapes. The minimum value of $\alpha$ such that all the keypoints in the group are included in the $\alpha$-shape is adopted. The $\alpha$-shape approximates the shape of the tracked object.

### 3.6   Picking

Here we select the $\alpha$-shape with the minimum area as the picking target. The minimum-area criterion is adopted to avoid selecting slightly moved background.

Then we consider how to grasp the $\alpha$-shape with a parallel gripper for picking. Harada et al. proposed a method to synthesize parallel gripper grasps by finding near-parallel surfaces of 3D objects [7]. In this study we simplify their method for its application to 2D objects (Fig. 4).

The grasp synthesis can be summarized as follows:

1. Consider three adjacent points in the $\alpha$-shape. If the three points are nearly in line, that is, the angle made by the three points is larger than a threshold $\theta_{\text{line}}$, the midpoint of the three is removed. Repeat this process to obtain a simplified $\alpha$-shape.
2. Find clusters of adjacent line segments that are located within two parallel lines with distance $h_{\text{max}}$. Each of the found clusters is considered as a graspable region and approximated as a line segment.
3. Consider one of the obtained clusters in the descending order of the length of its approximated line segment. Then find its opposing cluster, if any. If there are multiple opposing clusters, the cluster closest to parallel is selected.
4. Synthesize a parallel grasp for the opposing clusters. The gripper is rotated to align with the two approximated line segments and translated to the center of the two approximated line segments.

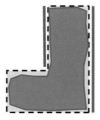

**Fig. 4.** Finding near-parallel edges for grasping

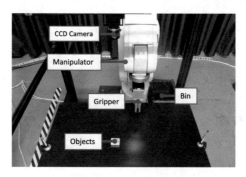

**Fig. 5.** Experimental setup

## 4    Experiments

### 4.1    Experimental Setup

The experimental setup is shown in Fig. 5. A 6-axis manipulator (RV-1A by Mitsubishi Electric) equipped with a parallel gripper (ESG1-SS-2815 by Taiyo) is used. A spring mechanism is installed between the manipulator and the gripper to absorb shock in impacting. A grayscale camera (Flea2 FL2G-13S2M by Point Grey Research, 1296 × 964 pixels) is placed above the manipulator. A Linux PC with Intel Core i7-3770K (3.50 GHz) is used to control the manipulator and the camera. OpenCV [1] is used for image processing implementation.

Mahjong tiles are used as objects to be picked up because they have a wide variety in texture, from textureless to highly textured. The size of each of the mahjong tiles is 19.4 [mm] (width) × 26.5 [mm] (depth) × 16.1 [mm] (height). Its weight is about 14 [gf].

### 4.2    Assumptions

We set an impacting point on the floor, to be punched by moving down the gripper of the manipulator vertically. The floor is sloped about five degrees to make object movement by impacting easier. The position of the floor is known.

All the objects are placed on the floor. Thus picking is possible if the objects are localized on the floor with the camera calibrated in advance.

Parameters used in the experiments are shown in Table 1.

**Table 1.** Method parameters

| $K_{\text{dist}}$ | 20 [pixel] |
|---|---|
| $K_{\text{bg}}$ | 2 [pixel] |
| $K_{\text{len}}$ | 2 [pixel] |
| $K_{\text{trans}}$ | 4 [pixel] |
| $K_{\text{rot}}$ | 10 [deg] |
| $K_{\text{diff}}$ | 2 [pixel] |
| $K'_{\text{diff}}$ | 1 [pixel] |
| $N_{\text{vote}}$ | 1000 |
| $K_{\text{vote}}$ | 100 |
| $\theta_{\text{line}}$ | 170 [deg] |
| $h_{\text{max}}$ | 3 [mm] |

**Table 2.** Result of detection experiments

| Successful | 14/18 |
|---|---|
| Unsuccessful | 4/18 |
| Success Rate | 78% |

### 4.3   Experiments of Object Detection

Here we show the experimental results of object detection. Our proposed method is applied to eighteen sample image pairs before and after impacting.

Results are shown in Table 2 and Fig. 6. The success/failure of detection shown in Table 2 is determined subjectively according to the difference between the outer shape of the target object and its $\alpha$-shape segmentation in terms of picking feasibility.

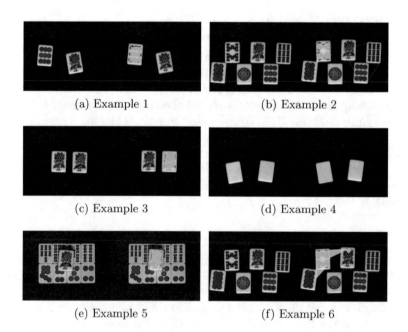

(a) Example 1          (b) Example 2

(c) Example 3          (d) Example 4

(e) Example 5          (f) Example 6

**Fig. 6.** Experiments of object detection

Figure 6 shows some examples of the segmentation results. The left and right are images before and after impacting, respectively. Segmented objects are specified by green regions. Figure 6a includes two differently textured objects; Fig. 6b and f include six differently textured objects; Fig. 6c includes two identically textured objects; Fig. 6d includes two textureless objects; Fig. 6e includes one textured object placed on the photo of mahjong tiles (one genuine object on background fake objects). Figure 6a–e show successful object segmentation.

On the other hand, Fig. 6f shows unsuccessful object segmentation, which includes not only the upper-left target object but also outliers on the neighbor objects. The outliers are keypoints that moved consistently with the movement of the target object and it is difficult to omit them when using only one image after impacting. Thus to solve this problem, calculating intersection of segmented regions using multiple frames after impacting would be useful.

### 4.4 Picking Experiments

We performed picking experiments with our proposed method. Success of picking is defined as transportation of the picked object to a goal bin. Experiments are terminated when no object movements by impacting are found.

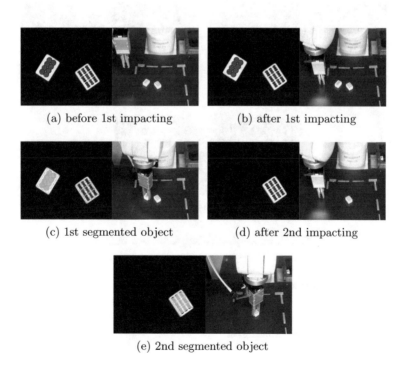

(a) before 1st impacting          (b) after 1st impacting

(c) 1st segmented object          (d) after 2nd impacting

(e) 2nd segmented object

**Fig. 7.** Picking differently textured objects

**Differently Textured Objects.** Here we show a result of picking two differently textured objects. Some scenes in this experiment are shown in Fig. 7. The left halves of the subfigures are photos captured by the camera for image processing. Segmented objects are specified by interposed green regions. The two objects were successfully picked up after impacting twice in total.

**L-Shaped Object.** Here we show a result of picking objects including an L-shaped one, which is composed of three connected mahjong tiles. Some scenes in this experiment are shown in Fig. 8. Two faces of the L-shaped object were grasped and its picking was successful.

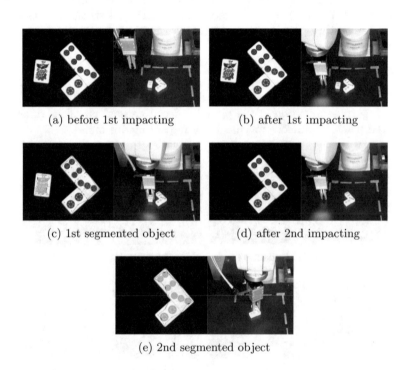

Fig. 8. Picking including L-shaped object

**Picking with a Textured Background.** Here we show a result of picking an object placed on a textured background. Some scenes in this experiment are shown in Fig. 9. In this case, the object placed on eight fake objects was found correctly and picking was successful.

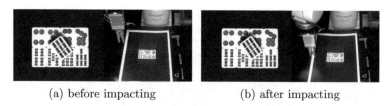

(a) before impacting          (b) after impacting

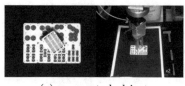

(c) segmented object

**Fig. 9.** Picking with a textured background

## 5   Conclusion

This paper presented a method to detect novel objects by impacting. In contrast to most of previous studies on interactive perception through pushing, the method can detect objects in wider region and suffers less from occlusion because impacting does not need direct contacts with the objects. We showed that it was possible to detect novel objects including textured and textureless mahjong tiles in experiments. Moreover, by grasp synthesis based on object segmentation, we showed robotic picking of the detected novel objects in experiments.

The success rate of object detection should be improved in future work. Using three or more frames in keypoint tracking would be useful. Grasp synthesis should also be improved for more reliable picking.

Currently, our method is applicable only to 2D objects. Its extension to 3D objects should also be addressed. It would be possible through 2D keypoint tracking with stereo vision, or 3D keypoint tracking with depth sensing.

## References

1. OpenCV. http://opencv.org/
2. Arun, K.S., Huang, T.S., Blostein, S.D.: Least-squares fitting of two 3-D point sets. IEEE Trans. Pattern Anal. Mach. Intell. **9**(5), 698–700 (1987)
3. Bohg, J., Hausman, K., Sankaran, B., Brock, O., Kragic, D., Schaal, S., Sukhatme, G.S.: Interactive perception: leveraging action in perception and perception in action. IEEE Trans. Robotic. **33**(6), 1273–1291 (2017)
4. Chang, L., Smith, J.R., Fox, D.: Interactive singulation of objects from a pile. In: Proceedings of 2012 IEEE International Conference on Robotics and Automation, pp. 3875–3882 (2012)
5. Edelsbrunner, H., Kirkpatrick, D., Seidel, R.: On the shape of a set of points in the plane. IEEE Trans. Inf. Theor. **29**(4), 551–559 (1983)

6. Gupta, M., Müller, J., Sukhatme, G.S.: Using manipulation primitives for object sorting in cluttered environments. IEEE Trans. Autom. Sci. Eng. **12**(2), 608–614 (2015)
7. Harada, K., Tsuji, T., Nagata, K., Yamanobe, N., Maruyama, K., Nakamura, A., Kawai, Y.: Grasp planning for parallel grippers with flexibility on its grasping surface. In: Proceedings of IEEE International Conference on Robotics and Biomimetics, pp. 1540–1546 (2011)
8. Harris, C., Stephens, M.: A combined corner and edge detector. In: Proceedings of Fourth Alvey Vision Conference, pp. 147–151 (1988)
9. Hermans, T., Rehg, J.M., Bobick, A.: Guided pushing for object singulation. In: Proceedings of 2012 IEEE/RSJ International Conference on Intelligent Robots and Systems, pp. 4783–4790 (2012)
10. Katz, D., Venkatraman, A., Kazemi, M., Bagnell, J.A., Stentz, A.: Perceiving, learning, and exploiting object affordances for autonomous pile manipulation. Auton. Robots **37**(4), 369–382 (2014)
11. Lowe, D.G.: Object recognition from local scale invariant features. In: Proceedings of 1999 IEEE International Conference on Computer Vision, pp. 1150–1157 (1999)
12. Metta, G., Fitzpatrick, P.: Better vision through manipulation. Adapt. Behav. **11**(2), 109–128 (2003)
13. Schiebener, D., Ude, A., Asfour, T.: Physical interaction for segmentation of unknown textured and non-textured rigid objects. In: Proceedings of 2014 IEEE International Conference on Robotics and Automation, pp. 4959–4966 (2014)

# Efficient Semantic Segmentation
# for Visual Bird's-Eye View Interpretation

Timo Sämann$^{(\boxtimes)}$, Karl Amende$^{(\boxtimes)}$, Stefan Milz$^{(\boxtimes)}$, Christian Witt$^{(\boxtimes)}$,
Martin Simon$^{(\boxtimes)}$, and Johannes Petzold$^{(\boxtimes)}$

Valeo Comfort and Driving Assistance, Site Kronach (Germany),
Hummendorfer Str. 72, 96317 Kronach, Germany
{timo.saemann,karl.amende,stefan.milz,christian.witt,
martin.simon,johannes.petzold}@valeo.com

**Abstract.** The ability to perform semantic segmentation in real-time capable applications with limited hardware is of great importance. One such application is the interpretation of the visual bird's-eye view, which requires the semantic segmentation of the four omnidirectional camera images. In this paper, we present an efficient semantic segmentation that sets new standards in terms of runtime and hardware requirements. Our two main contributions are the decrease of the runtime by parallelizing the ArgMax layer and the reduction of hardware requirements by applying the channel pruning method to the ENet model.

**Keywords:** Efficient semantic segmentation · Channel pruning
Embedded systems · Bird's-eye view generation

## 1 Introduction

The understanding of scenes plays a key role in the technical realization of self-driving vehicles, home-automation devices and augmented reality wearables. A prerequisite for understanding scenes based on cameras is the semantic segmentation. The aim of semantic segmentation is to classify every pixel of an image into meaningful classes. This task is typically realized with Deep Neural Networks (DNNs). The generation of a top view of a vehicle by using four omnidirectional cameras provides a 360° surrounding bird's-eye view. The perception of the fully surrounding environment is important in many traffic situations for automated driving, e.g. autonomous parking. As shown in Fig. 1 the interpretation and understanding of such a surround view could be done by DNN based semantic segmentation.

To enable the operation of DNNs on low power devices such as embedded systems in real-time, they need to be implemented efficiently. [1] represents a Deep Neural Network architecture (ENet) for real-time semantic segmentation. The ENet is listed on the Cityscapes benchmark as the fastest model,

© Springer Nature Switzerland AG 2019
M. Strand et al. (Eds.): IAS 2018, AISC 867, pp. 679–688, 2019.
https://doi.org/10.1007/978-3-030-01370-7_53

while provides a respectable quality that is sufficient for many application [2]. However, we were able to show that the computational and memory requirements are too high for generating the semantically segmented top view image on the NVIDIA Jetson TX2 board in real-time.

In terms of runtime, the ArgMax layer represents a bottleneck on models for semantic segmentation. It determines the indices of the maximum values along the depth axis for the output feature maps. In most publications, this calculation is excluded from the runtime measurement [1,3,4], since this calculation is very time-consuming. In the case of a real-time application, this calculation is relevant and cannot be ignored. By parallelizing the ArgMax calculation on the GPU, the runtime of this layer can be drastically reduced on the NVIDIA TX2 board compared to common CPU implementations.

Since we have to calculate the semantic segmentation four times to generate the top view image and the available resources on embedded systems are scarce, we reduced the number of parameters and thus the required GPU memory of the ENet by a variant of the channel pruning method [5]. The idea of this method is to prune channels[1] of convolutional layer by a LASSO regression based channel selection followed by a fine-tuning step for recover the weights. We extend the idea of channel pruning for image classification to the task of semantic segmentation and prune the ResNet based network ENet. For our experiments, we used the ENet implementation of Caffe which is publicly available on GitHub[2].

## 2   Related Work

Efficiency is one of the key research areas for automated vehicles. Since the real-time capability is needed it has become a mandatory requirement for DNN applications. ENet is one of the most efficient Deep Neural Networks for semantic segmentation [1]. The ENet consists of an encoder-decoder structure. Unlike SegNet [4], which uses a symmetric encoder-decoder structure, the ENet uses a larger encoder and a smaller decoder, which reduces the computational effort. Additionally, ENet places great importance on early reduction of input information. Calculation operations on input images with a lower resolution are less complex and require less time. Furthermore, it uses asymmetric convolution presented in [6]. An $n \times n$ filter is divided into an $n \times 1$ and $1 \times n$ filter. Both filters applied one after another which results in the same output as a $n \times n$ filter once applied, with the advantage of lower computational effort.

A massive amount of work on DNN acceleration has been done in the following three fields [5]: 1. Optimized implementation [7], 2. Quantization [8] and 3. Structured simplification [9]. However, the choice of the right method strongly depends on the application task and the basic architecture of the used DNN.

---

[1] The term *channels* is synonymous with *feature maps*.
[2] https://github.com/TimoSaemann/ENet.

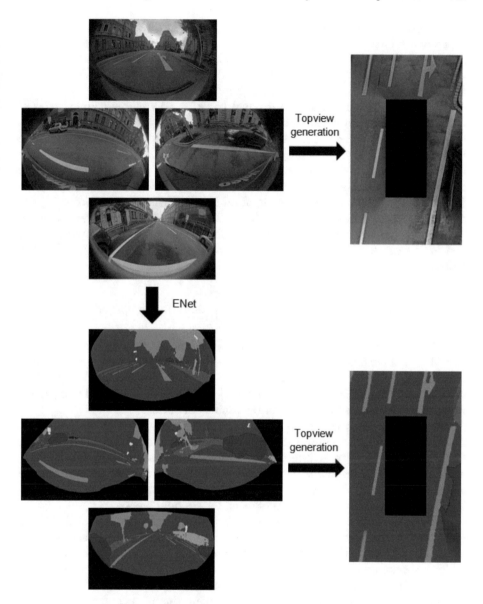

**Fig. 1.** Illustration of top view generation on raw fisheye camera and semantically segmented images by projecting the images on a plane parallel to the ground using the camera model.

An optimized implemented method speeds up convolutions by special convolution operations or approximations. Similar to this, quantization tries to approximate large floating point multiplications by less complex reduced floatings points or single bit operations [8]. There are methods for sparse connection [10] or tensor factorization [9], which decompose weights into subsets.

A famous method to improve residual block based architectures in terms of efficiency and memory consumption is channel pruning [5]. The basic idea of this method is to reduce the number of channels that serve a convolutional layer as an input while maintaining the output of the layer. This means that only channels are removed, which have a minor impact on the output. Those channels can be found by performing a LASSO regression. Formally, suppose we apply the filter $W$ with $n \times c \times k_h \times k_w$ to a sample of an input $X$ with $N \times c \times k_h \times k_w$, the output $Y$ results with an output size $N \times n$. The letter $c$ represents the number of channels, $n$ the number of output feature maps, $N$ the number of input samples and $k_h$, $k_w$ are the filter size. The channel pruning method can be described as follows:

$$\arg\min_{\beta,W} \frac{1}{2N} \left\| Y - \sum_{i=1}^{c} \beta_i X_i W_i^T \right\|_F^2 \text{ subject to } \|\beta\|_0 \le c'$$

$\|\cdot\|_F$ designates the Frobenius norm. $X_i$ and $W_i$ designates the input data and filter of a channel $c$ with $i = 1, \ldots, c$, respectively. $\beta$ is a vector of size $c$ and takes either 0 or 1 for each element. If $\beta_i = 0$, the channel with the corresponding index $i$ gets removed from the feature map. $\|\beta\|_0$ is less than or equal to $c'$, which represents the maximum number of remaining channels.

Channel pruning could be separated into training based methods and inference-time based methods [5], whereas the latter is extremely challenging especially for very deep architectures, e.g. residual networks like ENet. [5] mentions a bottom-up technique, where first channel pruning is applied to a single convolutional layer. Afterwards, the method is stretched to the whole model. The results mentioned in [5] are promising for residual blocks.

The top or birds-eye view generation is a common state of the art feature in almost every 360° surround view application for advanced driver assistance systems [11]. The basic idea is a texture mapping of four omnidirectional cameras, which are mounted with different viewing directions into a top view plane (see Fig. 1). The aim of this feature is a better environmental perception. Therefore, [12] fundamentally studied the semantic segmentation of such a top-view. This is helpful for freespace and road marking detection within automated driving applications. An efficiency investigation for such an application is missing. The aim of this work is to provide an efficient solution to enable semantic top view interpretation for automated driving.

## 3    Methods

In this section, we first propose an efficient ArgMax implementation to accelerate the forward pass of the ENet. Then we describe how we prune the ENet architecture using the channel pruning method.

### 3.1    ArgMax Implementation

The ArgMax layer is the last layer in a model for semantic segmentation. For every pixel, the index of the maximum value along the depth axis is determined (see Fig. 2). The resulting index corresponds to the class that the pixel will be assigned to. Due to the serial implementation on the CPU, which is used by Deep Learning Frameworks such as Caffe [13], the layer becomes a bottleneck especially for embedded systems such as the NVIDIA TX2 board. In order to get a high frame rate on embedded hardware, we need to implement the ArgMax layer on the GPU.

The ArgMax calculation for a pixel requires the values along the depth axis, as shown in Fig. 2. In theory, it is possible to calculate the ArgMax for all pixels simultaneously without conflicts. We used this observation and implemented a GPU version for the ArgMax layer with CUDA in Caffe. We implemented our custom ArgMax kernel which calculates the ArgMax for a given pixel along the depth axis. Every CUDA thread computes the ArgMax for exactly one pixel. Since there are no dependencies or conflicts between the pixels during the ArgMax calculation, we can use the maximum number of threads and have to read each value from the input exactly once. In this way, we achieve a high degree of parallelism. The results and speed of our implementation are presented in the Sect. 5, Table 1.

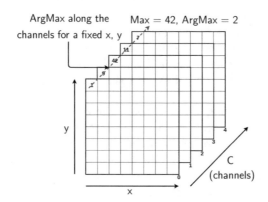

**Fig. 2.** Example of the ArgMax operation for one pixel.

## 3.2   ENet's Channel Pruning

ENet's Channel Pruning consists mainly of two steps. The first step is to select the channels that can be pruned and the second step is the fine-tuning step where the weight parameters are recovered in a fine-tuning.

The ENet architecture is based on many consecutive residual blocks. An example residual block is shown in Fig. 3. Please note that the batch normalization and dropout layer have been merged into the convolutional filters. In each residual block, every second and third convolutional (conv) layer was pruned.

In the middle layer ($3 \times 3$ conv), the number of channels was reduced by a certain ratio. The depth of the filter of the following convolutional layer ($1 \times 1$ conv) has been adapted accordingly. The selection of the channels and filter depths to be pruned was done using the LASSO regression as described in Sect. 2.

The first layer of the residual block was not pruned because for the practical implementation a so-called *feature map sampling* layer has to be applied before the first convolutional layer.

[5] claims that the runtime of the *feature map sampling* layer is negligible, but we found that the runtime is longer than the saved runtime due to the lower computational effort resulting from the smaller number of channels.

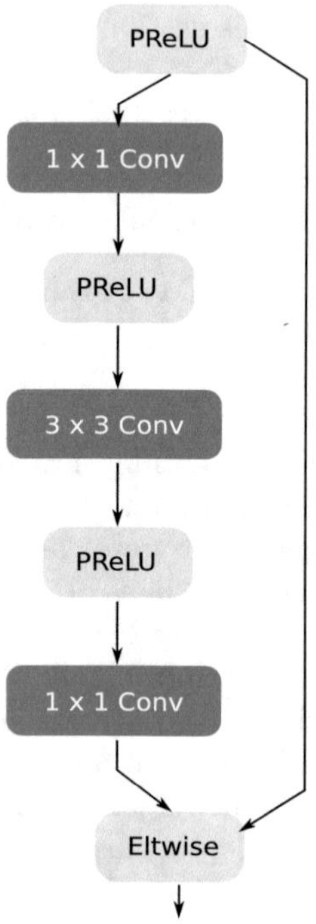

The number of channels to be pruned depends on the channel factor. This value is divided by the number of existing channels and thus determines the ratio of the pruned channels. The

**Fig. 3.** Example residual block as it occurs several times in the ENet architecture.

selection of the appropriate channel factor value is crucial for the success of the pruning. Therefore, the choice of this value is discussed in more detail in the next Sect. 4. In the last residual block, no pruning was performed, because the number of channels is only 4 and a reduction would affect the quality very negatively.

In the fine-tuning step, we used again the customize training data set. In contrast to the previous training from the scratch, which passed through 150 epochs, the fine-tuning step can be limited to a few epochs (3 to 5) in order to achieve network convergence. Since the pruned ENet requires less GPU memory for training, the batch size can be increased from 6 to 11 per GPU. The learning rate was set to $10^{-8}$, which is ten times higher than the learning rate after 150 epochs of training from the scratch. The remaining training parameters have been adopted from [1].

## 4    Channel Factor Selection

In our experiments, we tested various channel factors which we use to reduce the feature maps. We followed the approach of [5] and chose a high channel factor of 1.5 for the shallower residual blocks and a lower factor of 1.25 for the deeper residual blocks. We found out that the quality of the model decreased sharply and reversed the ratio of the channel factor. Now the quality of the model was only slightly reduced. This leads us to the assumption that the number of channels in the shallower residual blocks should not or only slightly be reduced. The reason for this might be, that the number of feature maps in the shallower residual blocks is lower than in the deeper ones. Therefore, we reduced the channel factor of the shallower residual blocks from 1.25 to our final value of 1.1. Since the number of channels in the shallower residual blocks is quite small (16), the benefit of a larger reduction is low. As a result, we find these values as a better compromise between saving computational effort and losing quality.

## 5    Results

By parallelizing the ArgMax calculation on the GPU, the runtime of this layer can be drastically reduced on the NVIDIA TX2 board compared to common CPU implementations. For an input image with a resolution of 640 px × 400 px, the runtime can be reduced from 92 ms to 0.05 ms. A comparison of the performance for the respective CPU and GPU implementation of the ArgMax layer can be found in Table 1. For larger image resolutions, the factor is even greater, since the parallelization can be better utilized.

The channel pruning method applied to the ENet allowed us to reduce the required GFLOPs from 1.87 to 1.34 for an input size of 640 px × 400 px as shown in Table 2. Furthermore, we were able to reduce the number of parameters from 363 k to 255 k. Accordingly, the required memory of the parameters decreases from 1.49 MB to 1.06 MB (FP32). Due to the lower number of FLOPs, the inference time could be increased on the CPU by 17.7%. After all, a runtime improvement from 11.06 fps to 11.53 fps could be achieved on the GPU. All computational measurements were done on the NVIDIA TX2 board with CUDA 9.0 and cuDNN 7.0.

**Table 1.** Comparison of performance using CPU and GPU implementation of ENet for an input image size of 640 px × 400 px.

| Network | ArgMax layer | Performance |
|---------|--------------|-------------|
| CPU | CPU | 0.23 fps |
| GPU | CPU | 5.46 fps |
| GPU | GPU | 11.06 fps |

**Table 2.** Comparison of hardware requirements and performance of ENet (including ArgMax calculation) for an input image size of 640 px × 400 px.

| Model | GFLOPs | Parameter | Model size (FP32) | Performance CPU | Performance GPU |
|-------|--------|-----------|-------------------|-----------------|-----------------|
| ENet | 1.87 | 363 k | 1.49 MB | 0.23 fps | 11.06 fps |
| ENet pruned | 1.34 | 255 k | 1.06 MB | 0.28 fps | 11.53 fps |

To compare quality results we used mean intersection over union (mIoU) and global accuracy. The mIoU of our 20 classes dropped from 53.8% to 51.4% of our custom fisheye test dataset. The IoU values for each class are shown in Table 3. Interestingly, the IoU decreases especially for classes with a small pixel density (e.g. pole). For classes with a high pixel density, the value remains the same or even increases (e.g. road), which explains the slightly increased global accuracy, which has increased from 94.04% to 94.12%.

**Table 3.** Representation of the Intersection over Union (IoU) values per class for comparison of ENet before and after pruning.

| Classes | IoU ENet | IoU ENet pruned |
|---------|----------|-----------------|
| Road | 95.4% | 95.6% |
| Sidewalk | 75.1% | 75.2% |
| Building | 87.3% | 87.2% |
| Wall | 65.3% | 63.8% |
| Fence | 45.1% | 42.7% |
| Pole | 30.7% | 23.4% |
| Traffic light | 41.5% | 39.1% |
| Traffic sign | 27.1% | 24.9% |
| Vegetation | 80.1% | 80.2% |
| Terrain | 24.6% | 25.0% |
| Sky | 95.9% | 96.1% |
| Person | 42.2% | 40.3% |
| Rider | 13.6% | 07.3% |
| Car | 83.1% | 82.2% |
| Truck | 40.3% | 35.4% |
| Bus | 51.3% | 47.3% |
| Motorcycle | 15.0% | 07.5% |
| Bicycle | 47.9% | 43.0% |
| Road markings | 61.2% | 61.2% |
| **Mean IoU** | **53.8%** | **51.4%** |

# 6  Conclusion

We have proposed a parallelized ArgMax Layer implementation that dramatically improves runtime for semantic segmentation models. For an input image size of 640 px × 400 px, the runtime of the ArgMax layer on the NVIDIA TX2 board could be reduced by a factor of 1840.

In addition, the hardware requirements for the ENet could be significantly reduced by channel pruning. The number of required GFLOPs could be reduced by about 30%, which allows a theoretical speed up of 1.4.

Despite this significant reduction in the hardware requirements of the already efficient ENet model, the IoU value has fallen slightly only for small classes (e.g. pole). These results are essential for embedded systems to use semantic segmentation in a real-time capable application such as the generation of the semantically segmented birds-eye view. In the future, we plan to extend the channel pruning method to additional layers of the ENet to further reduce hardware requirements. Furthermore, we plan to work on a more comprehensive fine-tuning step to maintain the quality of the ENet at a similar level as before pruning.

**Acknowledgments.** We would like to thank Senthil Yogamani and our colleagues at Valeo Vision Systems in Ireland for collaboration on our dataset using automotive fisheye cameras. We would like to thank Valeo, especially Jörg Schrepfer, for the opportunity doing fundamental research.

# References

1. Paszke, A., Chaurasia, A., Kim, S., Culurciello, E.: ENet: a deep neural network architecture for real-time semantic segmentation. arXiv preprint arXiv:1606.02147 (2016)
2. Cordts, M., Omran, M., Ramos, S., Rehfeld, T., Enzweiler, M., Benenson, R., Franke, U., Roth, S., Schiele, B.: The cityscapes dataset for semantic urban scene understanding, pp. 3213–3223 (2016)
3. Long, J., Shelhamer, E., Darrell, T.: Fully convolutional networks for semantic segmentation. In: Proceedings of the IEEE Conference on Computer Vision and Pattern Recognition, pp. 3431–3440 (2015)
4. Badrinarayanan, V., Kendall, A., Cipolla, R.: SegNet: a deep convolutional encoder-decoder architecture for image segmentation. IEEE Trans. Pattern Anal. Mach. Intell. **39**(12), 2481–2495 (2017)
5. He, Y., Zhang, X., Sun, J.: Channel pruning for accelerating very deep neural networks. In: International Conference on Computer Vision (ICCV), vol. 2, p. 6 (2017)
6. Szegedy, C., Vanhoucke, V., Ioffe, S., Shlens, J., Wojna, Z.: Rethinking the inception architecture for computer vision. In: Proceedings of the IEEE Conference on Computer Vision and Pattern Recognition, pp. 2818–2826 (2016)
7. Bagherinezhad, H., Rastegari, M., Farhadi, A.: LCNN: lookup-based convolutional neural network (2016)
8. Rastegari, M., Ordonez, V., Redmon, J., Farhadi, A.: XNOR-Net: ImageNet classification using binary convolutional neural networks. CoRR abs/1603.05279 (2016)

9. Jaderberg, M., Vedaldi, A., Zisserman, A.: Speeding up convolutional neural networks with low rank expansions. CoRR abs/1405.3866 (2014)
10. Han, S., Pool, J., Tran, J., Dally, W.J.: Learning both weights and connections for efficient neural networks. CoRR abs/1506.02626 (2015)
11. Zhang, B., Appia, V.V., Pekkucuksen, I., Liu, Y., Batur, A.U., Shastry, P., Liu, S., Sivasankaran, S., Chitnis, K.: A surround view camera solution for embedded systems. In: 2014 IEEE Conference on Computer Vision and Pattern Recognition Workshops, pp. 676–681 (2014)
12. Deng, L., Yang, M., Li, H., Li, T., Hu, B., Wang, C.: Restricted deformable convolution based road scene semantic segmentation using surround view cameras (2018)
13. Jia, Y., Shelhamer, E., Donahue, J., Karayev, S., Long, J., Girshick, R., Guadarrama, S., Darrell, T.: Caffe: convolutional architecture for fast feature embedding. arXiv preprint arXiv:1408.5093 (2014)

# Concept Study for Vehicle Self-Localization Using Neural Networks for Detection of Pole-Like Landmarks

Achim Kampker, Jonas Hatzenbuehler$^{(\boxtimes)}$, Lars Klein, Mohsen Sefati, Kai D. Kreiskoether, and Denny Gert

PEM RWTH Aachen, 52074 Aachen, NRW, Germany
a.kampker@pem.rwth-aachen.de, jonas.hatzenbuehler@gmx.de
https://www.pem.rwth-aachen.de/

**Abstract.** This paper discusses and showcases a software framework for the self-localization of autonomous vehicles in an urban environment. The general concept of this framework is based on the semantic detection and observation of objects in the surrounding environment. For the object detection three different perception approaches are compared; LiDAR based, stereo camera based and mono camera based using a neural net. The investigated objects all share the same geometrical shape; they are vertical with a high aspect ratio. To compute the pose of the vehicle an Adaptive Monte-Carlo Algorithm has been implemented. Hence it is necessary to create a high-precision digital map this is done with a dense map, the detected objects and the LiDAR point cloud. Comparison with an earlier paper have shown that this approach keeps the global positioning accuracy around 0.50 m and leads to more robust results in highly dynamic scenarios where a small amount of objects can be detected.

**Keywords:** Object detection · Mapping · Localization
Neural networks

## 1 Introduction

Self-localization in a dynamic and urban environment remains one of the main challenges in autonomous driving. Especially in urban environment there are many landmarks which can be used for the localization of a vehicle. Pole like landmarks are especially interesting if there are no typical road markings available, this is in housing areas the case where many autonomous driving vehicles will be driving. This paper proposes a localization framework which gives on possible solution for this research issue, it extends the previous work [33] with a new perception approach using a neural net. Current solutions for global and local localization methods in research vehicles as well in commercially available products make use of Global Navigation Satellite Systems (GNSS) for most of

© Springer Nature Switzerland AG 2019
M. Strand et al. (Eds.): IAS 2018, AISC 867, pp. 689–705, 2019.
https://doi.org/10.1007/978-3-030-01370-7_54

the localization tasks. This is a liable solution for situations where there is no obstruction of sight, no house canyons or other sources for multi-pathing or interferences. Once these conditions are not met the accuracy and reliability of the self-localization will drop. This paper mainly deals with overcoming these shortcomings and build upon a landmark based approach [2, 21]. One of the first works in this field is described in [11, Chap. 7]. The main concept behind this approach is the detection of static objects, so called landmarks, mapping these objects on a high-precision representation of the environment and finally use this map for the localization. In the scope of this work the landmarks which are detected are shaped vertically with a high aspect ratio, this could be lanterns, tree trunks, traffic signs, traffic lights or small street poles. Therefore, the proposed method is designed to work well in urban areas, forest-like areas and special purpose areas. In Fig. 1 such an area is displayed.

**Fig. 1.** Illustration of the landmark detection

The main motivation behind this paper is to improve the robustness of an existing approach detecting landmarks based on pure LiDAR point clouds and stereo image analysis as described in Sect. 5.2. With this new approach we show that object detection based on Convolutional neural Net (CNN) results in a more robust localization and hence can outperform current landmark based localization algorithms. For the test, training and evaluation of the CNN different data sets were used. We utilized the KITTI Vision Benchmark Suite [14], IJRR (Ford) [24] and CityScapes [9]. All of which have data for challenging urban scenarios. Different camera models and light settings are implicitly proofing the robustness of the proposed framework. For the object detection the single input camera pictures where used, the depth map was created using available stereo images. The self-localization approach is developed based on an Adaptive Monte Carlo Localization (AMCL) technique and its underlying digital map is created by using the presented landmark detection method. The results of the localization have been validated using an OXTS RT 3003 RTK-GPS unit which was available in the open source databases. The paper is divided in the following sections:

After discussing related works in Sect. 2, an overview about the presented approach is given in Sect. 3. The implementation of this approach is discussed in Sect. 4 and results are illustrated in Sect. 5. A conclusion in the last chapter end this paper.

## 2   Related Work

For the self-localization of vehicles many different approaches have been investigated in the last years. Most of them utilized GNSS and odometry information to precisely determine the pose of the vehicle with minimal drift and good results in various environments. However, the latest developments are trying to remove the importance of GNSS information since it is not a reliable information source in dense urban areas [25,38]. Hence the importance of other sensor systems like LiDAR, radar and camera systems has steadily increased. Many vehicles are equipped with advanced camera systems nowadays which are used for object detection and tracking. This trend has led to the development of three different localization categories namely, point wise matching e.g. [17,34], dense map matching [16,26,27] and landmark based localization [21,39]. These algorithms differ in the way sensor data is grouped and interpreted. Point wise matching algorithms are highly computationally expensive and extract gradient based features, which makes them applicable in many different use cases especially in symmetric environments (e.g. indoor robotics and logistics). The dense based methods have mainly been investigated in [7] and improve robustness and minimize storage space for dense point clouds. These algorithms transfer the point cloud in a representation in the frequency space [26] and post-processing is done using mathematical manipulation of this representation. These approaches have shown to be more robust to sensor errors and vehicle speed, e.g. urban areas. However, these algorithms tend to be very sensitive to dynamic objects hence an extensive preprocessing to filter the dynamic objects is necessary for a stable overall performance. The third category is using the most abstract representation of the sensor input data. By utilizing semantic information, the objects can be categorized and stored accordingly. This representation is reducing the amount of data required for self-localization, hence it is computationally efficient. With this work we want to extend these approaches and show the effect of robust object detection on landmark based localization algorithms.

In this paper we will focus on the object detection using only mono camera pictures. In Sect. 4.3 the extraction of depth information for the localization framework is described in greater detail. In order to perform image based object detection one can decide for classical approaches from the image processing community e.g. [8]. Another approach heavily used for facial detection, pattern detection and data analysis is the training of a neural net. With the great advances in the last years [35] complex and high performance neural nets are available for a wider use. A good overview over the existing CNN libraries and resulting performance values can be seen in [19]. State of the art is the SDD approach from [22] which has shown excellent results in many different applications. Another network which has shown good results is YOLO. In this paper we have decided to use the this framework [28]. That is mainly due to the implementation of the localization framework and the strong customizability of the YOLO structure. With the classification of images neural networks currently reach human accuracy [12]. With object detection one also has to localize these object on an image. Several algorithms like RCNN, Fast-RCNN and Faster-RCNN have been established. In the next paragraphs a short overview over these algorithms is given. The modular algorithm RCNN [4] builds upon this necessity. First, regions of interest (ROI) are extracted from an image using a region proposal step. In the next step a neural net is used for feature extraction. From the input picture several section proposals have to be made in order for the classifier to identify actual objects in each proposal. Typically, a regression step has to be done to improve the section proposal step. In RCNN this is realized through a non-max suppression approach. The region proposal step is the essential part of the RCNN. Usually many regions are being proposed to improve the quality of the prediction. ROIs which do not fit can be neglected, hence the network can find specific objects better. The problem with many region proposals is the computation time, RCNN takes up to several minutes for one object detection step. An optimized approach can be seen in the Fast-RCNN algorithm [15]. In this approach an activation map is being created based on extracted features from the original image. The ROIs are then matched on this activation map not on the original image which means that the computation time for the convolutional layer only has to be done once instead of multiple times for each ROIs [15]. For the classification and regression steps Fast-RCNN uses neural networks instead of SVM (Support Vector Machines) and linear regression methods. The cost function can evaluate the classification and regression steps simultaneously, hence it is designed for simultaneous optimized feature extraction.

Faster-RCNN [29] is another step towards reusing the CNN for multiple tasks. Here the same network is used for the detection of ROIs as well as for their classification. Through this step features have to be extracted once form the input image, which saves further computation time [29]. Another relevant neural network architecture in the context of object detection is SSD [22]. With this an image can be analyzed within one single forward pass step and the algorithm outputs a list of matched objects. This is improving the performance significantly [19]. The cost function for this framework consists of three aspects: finding objects in an input image, classifying these objects and fitting a bounding box around the classified objects. This concept is also realized in the YOLO – You Only Look Once [28] architecture which is utilized in this paper. Similar to SSD, YOLO is using a monolithic approach for object detection with a neural network. The cost function is also modularly composed and consists of multiple terms. Recently a new version of YOLO has been published [28] where the authors show within several tests that YOLOv2 outperforms YOLO with respect to precision and computation performance. This is the reason the YOLOv2 architecture has been used in the scope of this work. In Sect. 4.3 the YOLO architecture as used in this work is described in greater detail, also the cost function is documented in detail.

## 3    Overview of the Approach

In this section an overview of the developed localization framework is given. Figure 2 shows the conceptual structure of the new approach.

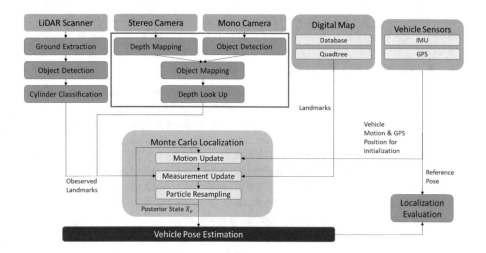

**Fig. 2.** New localization framework

In the old approach the main shape and therefore category of detected landmarks was limited to pole-like cylindrically shaped objects. This was due to the implemented object detection algorithms which are specialized on circle fit on 2D point clouds as well as on fitting a cylindrical volume in a 3D point cloud. With the use of CNN more complex landmarks can be used for the newly proposed localization framework. These characteristics were the main motivation for the extension described in this work. As can be seen in Fig. 2 the new localization framework has mainly changed the way the stereo camera images are analyzed. Here the proposed machine learning algorithm was integrated. With this addition the user does not limit itself to simple structures like trees and lanterns but one can also utilize variations of these objects as they appear in the real world, e.g benches and houses. Furthermore, the localization framework is now easily extendable for the integration of these other landmarks. Finally, the integration of machine learning algorithms in this framework increases the number of landmarks detected and therefore stored in the SQL database, which will then improve the robustness and accuracy of the estimated ego pose. For the training of the network 1000 pictures of the KITTI database were manually labeled, here we made sure that test and training data are from different sequences to prevent over-fitting. To tests the training quality of the neural net tests on multiple databases were done to minimize the effect of different perspectives, lightning effect or color shifts on the image itself.

# 4    Implementation

This chapter describes the implementation of the proposed self-localization framework. Hereby, the different perception concepts are explained while the focus lies on the perception via mono camera. Here the creation of the dense map as well as the use of the CNN framework is discussed in detail. The chapter terminates with the description of the mapping and localization implementation as shown in Fig. 2.

## 4.1    Perception via LiDAR

The detection of abstract landmarks using LiDAR sensors is limited to the detection of specific shapes and volumes in the sensor output. In most cases the sensor output is represented as a 3D point cloud, this is also the case in this paper. This limitation restricts this perception method to detect spatially large and clearly visible landmarks like tree stems and lanterns. Landmarks like traffic signs, delineators, reflector posts and other smaller objects cannot be detected reliably with this method. The object detection pipeline as implemented in this work consists of four main steps. First, the ground plane is removed. This can be done with the algorithm proposed in [5,13], which is a sloped-based channel classification based on polar grids. With this preprocessed point cloud, a horizontal 2D grid map is created which projects the 3D point cloud on a 2D layer. In this grid map the height of detections and occupancy of each cell is stored. In the third step the circular shapes are detected on this 2D grid map. These detections are considered as possible landmark candidates and are evaluated through multiple filter steps. By computing the aspect ratio of detections one exclude thicker and short objects as false positives, also comparing the detection with a cylindrical template shape will eliminate false positives. For better results one can create multiple 2D grids for different heights in the original 3D point cloud and do a final merging processing step by averaging the different layers to end up with the cylindrical approximation of the landmark.

## 4.2    Perception via Stereo Camera

According to [10,31] using trees as semantic landmarks for self-localization can also be done using stereo cameras, especially in forest areas this approach is beneficial. However, for different landmark categories there are different algorithms in use. For bigger landmarks, e.g. for trees, a flood fill approach can lead to good results since the area of the landmark is easily distinguishable from the background and noisy depth maps. In Fig. 3 the different steps from taking a stereo image to receiving a list of landmarks with distance information are visualized. These steps make this approach expensive to implement and its lacking adjust ability related to different landmarks and different environments. Using stereo camera images to create a disparity map for subsequently estimating the position of this traffic sign landmark via stereo triangulation [3] will output a set of landmarks with information about their. This list can then be used to for the mapping step (see Sect. 4.3).

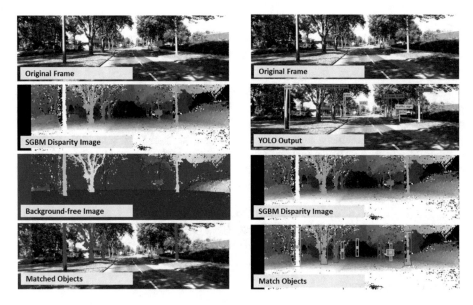

**Fig. 3.** Stereo image frame [33]          **Fig. 4.** Depth image look up

### 4.3   Perception via Mono Camera

As described in Sect. 3 the landmark detection based on mono cameras is
the main contribution to this paper. Here the different algorithms have been
merged to form one flexible, robust and extendable software framework for self-
localization based on predefined landmarks. As was shown in Sect. 4.2 the position of a landmark relative to the vehicle can be determined by detecting ROIs
on a picture and mapping these on a depth map. This approach can be extended
by integrating machine learning concepts especially in the object detection step.

To integrate YOLO in the self-localization framework and analyze the via-
bility of this approach multiple steps were defined. First the different databases
for labeling, training and testing were defined. Here CityScapes [9], IJRR (Ford)
[24] and KITTI database [14] have been used. While the entire pipeline is tested
on the KITTI database. With CityScapes and IJRR (Ford), the object detection
is verified.

**CNN/YOLO** [4,23]. The basic concept behind YOLO is that of a fully convolutional network, it does not consist of fully-connected layers. The networks
output is the activation map from the last convolution. For the network to work,
so called anchor boxes have to be defined. An anchor box is a rectangle with
fixed width and height. Each coordinate of the activation map consists of a vector of data. These vectors carry the following information for each predefined
anchor  box:

Objectiveness: At this coordinate and in this anchor box there exists an Object

Coordinates: Translation and scaling of the anchor box so that it frames the detected object precisely. Hence the detection is basically a correction value which describes the optimal anchor box transformation.

Class: Type/Label of detected object.

Each component in every anchor box has its own filter, meaning that each anchor box can be used to learn specific shaped objects. E.g. anchor boxes with a small aspect ratio can be used to detect flat and wide objects like benches, while anchor boxes with a high aspect ratio can be used to detect high and thin objects like lanterns. This architecture allows for a flexible and efficient object detection. The network is trained end-to-end with a modular loss function. For each prediction type (class, objectiveness and coordinates) one cost function

$$
\begin{aligned}
C_{coords} &\equiv Coordinates \\
C_{obj} &\equiv ROI_{positve} \\
C_{no\ Obj.} &\equiv ROI_{negative} \\
C_{class} &\equiv Classifikation
\end{aligned}
\tag{1}
$$

and weighting factor

$$
\begin{aligned}
\lambda_{coords} &= 2 \\
\lambda_{obj} &= 2 \\
\lambda_{no\ Obj.} &= 5 \\
\lambda_{class} &= 0.5
\end{aligned}
\tag{2}
$$

exists to compute the entire cost function as

$$
C_{total} = \lambda_{coords} \cdot C_{coords} + \lambda_{class} \cdot C_{class} + \lambda_{obj} \cdot C_{obj} + \lambda_{no.\ Obj.} \cdot C_{no.\ Obj.} \tag{3}
$$

The loss function combines all predictions for each cell and each anchor box. For the training the output value is compared with the mean-squared-error of the correct value. For the classification a softmax function is used.

For the training back propagation is used. Once the training is successfully terminated the anchor box predictions can be exported as classified objects – landmarks. With the definition of geometric shape of the anchor boxes and different object classes this approach is very flexible for more specific semantic object detection and hence can be used in a greater variety of scenarios.

**Dense Map.** For the correct localization in a horizontal 2D plane information about the distance of the detected landmarks is required. In a previous section the creation of a dense map is explained. For the perception with mono camera the dense map as a used as a look up table for the ROI of each object. The brightness level corresponds to a distance value. For the entire ROI the color distribution is computed and the mean value taken as the actual object distance. Fig. 4 shows one example distance computation. The dense map was created using the SGBM disparity image algorithm as described in [18].

*Mapping.* The described self-localization approach is based on a high precision offline map of landmarks. Therefore, a map generating car ride has to be performed in order to use this proposed framework. For the map generation the same sensor and software concept as when driving autonomously can be used. Since a fast read access to the offline map is crucial for the performance of the framework, the detected landmarks are stored with its characteristic values in an SQL database. In this work the characteristic values are limited to the 2D position of each landmark. To improve access time a limited subset of 2D positions is cached in a quad tree structure [32].

For all the described perception methods, the SQL database communication looks similar and can be used interchangeable.

*Localization.* The main task and overall goal of the framework in this work is the localization of an automated vehicle in a complex and dynamic environment. The global position of a vehicle can be described by a global pose $(x, y, \theta)$ at a specific time $t$. Many different algorithms [6,20,30] have been designed to purely get this time variant pose. A main challenge for the pose approximation is the inherent measurement inaccuracies which are present at every step of the framework (sensor noise of LiDAR, camera, odometry and GNSS sensors) computation errors due to algorithm design as well as numerical errors due to the computational hardware capabilities. A simple way to account for this inaccuracies during the pose estimation is the concept of Bayesian filtering [1]. In this work a more sophisticated approach the Adaptive or Augmented MCL has been implemented. In contrary to the Bayesian filter the AMCL employs a particle filter, which maintains a pose estimate by a set of weighted hypotheses (particles), that approximate the posterior probability distribution of a robot's belief. The adaptive part is an extension of the standard MCL which incorporates continuously adding new pose hypotheses for each re-sampling step and as such, is less prone to global localization failures [37, p. 206].

For the successful localization the initial pose has to be known to the vehicle, a digital map has to be present and the initial weights for each particle have to be evenly distributed over the entire sample. In a re-sampling step the particles weights are updated based on environment perceptions, and after that redistributed according to their newly assigned weights. This procedure is widely studied and applied in many prominent literature [1,37]. As presented in [36], the likelihood of each pose is calculated by building the product of likelihoods for all observed landmarks. For successfully associated landmarks this probability is proportional to the distance between observed and corresponding mapped landmark, while for unassociated landmark detections the likelihood is depending on the detection probability of utilized sensors.

## 5   Results

This section is a comprehensive overview of the different tests which have been done for the evaluation of the new framework. All test results were done on the same KITTI scene to allow comparability. However, the individual perception models were tested on multiple scenes to show general applicability. All of these test was performed to investigate the robustness of the YOLO framework and its trained weights on different lightning and camera settings. To test this, the trained network weights were applied on data sets form CityScapes and the open source data set IJRR (Ford) [24]. All of these images were fed in the YOLO framework with their original aspect ratio and without any preprocessing. The second paragraph then compares the old results discussed in [33] with the new approach and gives a better understanding of the potential of YOLO in the context of mapping and localization.

## 5.1   Old Approach

In Fig. 5 the histogram of the old approach is shown. Most significant is the divergence between these two approaches where the landmarks detected by the Laser Scanner are around 30 cm and the landmarks diameter of the Stereo Camera approach is averaging 40 cm. Compared to the new approach the deviation around the mean value is high and does not represent the real world values. The absolute localization error of the old approach can be seen in the Fig. 6.

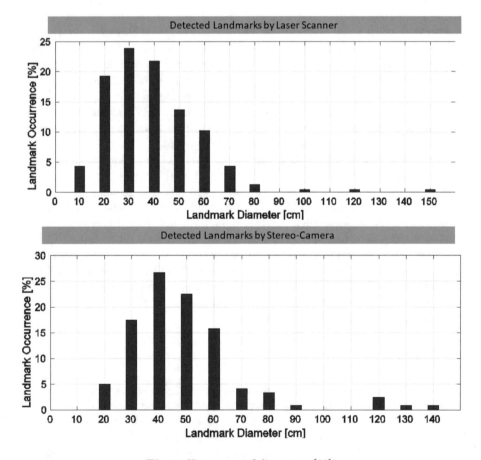

**Fig. 5.** Histogram of diameters [33]

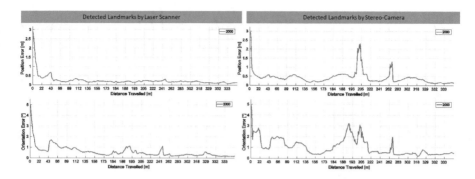

**Fig. 6.** Comparison LiDAR and stereo camera approach [33]

## 5.2   New Approach

The new approach as described in this paper was tested in a two stage set up. The system which was used for this setup was a 2017 Intel i7-7700K CPU, 16 GB DDR5 RAM and a Nvidia GeForce GTX 1070Ti GPU. First the performance of the YOLO object detection framework was tested and evaluated against different open source available data sets. Second the integration of the YOLO based perception approach in the localization framework was tested. As can be seen in Fig. 7 a similar distribution of landmarks can be seen, however the occurrence is far more concentrated around the 10–30 cm diameter, which is correct if we look at the underlying data set. Since we only detected pole-like landmarks in this case. For the first test, the trained YOLO weights and configuration files were used to detect objects on multiple open source data sets.

First, the trained network was applied on pictures from IJRR (Ford). The images were recorded with a Point Grey Ladybug3 omnidirectional camera system. The data was collected on the Ford Research campus and downtown Dearborn, Michigan during November/December 2009. In Fig. 8 you can see a picture which stays representative for our overall findings on this particular data set. You can see that the network is able to detect objects and also label the object correctly, however compared to the other data sets the absolute number of detected objects is very little. Typically, there was approx. 1–2 objects detected per image. For the localization framework this is not sufficient. Tests from the other approaches have shown that at least 5 landmarks are necessary to get a good accuracy values. The little amount of detected landmarks can be explained by the way the network was trained and the difference in the input data to the training data. The fish-eye effect of the Ladybug3 camera is very present and hence the objects are highly transformed and therefore very hard for the network to identify. Lastly the image ratio of 616 × 1616 is a challenge for the YOLO framework. YOLO is subdividing the input images in same sized sections and computes the outputs based on these sections. An overlaying and merging of each section allows for correct object detection but an input image which

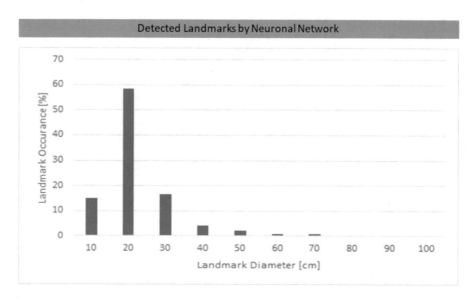

**Fig. 7.** Histogram of diameters neural network approach

is more vertical than horizontal minimizes the positive detection rate tremendously. Additionally, it can be stated that there was no false positive detection on the entire tested data set.

The second database the YOLO framework was tested on has shown the general functionality of such a framework for the robust and reliable detection of pole like objects. The data set was recorded with an automotive-grade 22 cm baseline stereo camera using 1/3 in CMOS 2 MP sensors (OnSemi AR0331) with rolling shutters at a frame-rate of 17 Hz. The sensors were mounted behind the windshield and yield high dynamic-range (HDR) images with 16-bit linear color depth. The resolution of these images is 2048 × 1024. Here the network detected in average 2–3 objects correctly (Fig. 9). No false positives were detected in the tested data set. To explain the better results, the same reasoning as in the above paragraph can be applied. Especially the color scheme of these images is different to the KITTI data set images which is a challenge for the neural net. The resolution and aspect ratio of the CityScapes images allow for better object detection; however, the achieved results are not sufficient enough for a self-localization purely based on the mono camera perception. It is important to state that no pre-labeled image annotations from the CityScapes database were used for the creation of these results.

The last data set which was tested to evaluate the functionality of the YOLO framework against is KITTI (Fig. 10). The hardware used is 2 Point Grey Flea 2 (FL2-14S3C-C) cameras which are able to output color and gray scale images. The resolution of these images is 1392 × 512. One example evaluation frame of the detected objects can be seen in Fig. 10. Approximately 10 landmarks were detected on every image throughout the entire evaluation data set. With this

**Fig. 8.** YOLO on IJRR (Ford)

amount of detected objects, a robust localization is possible. The accuracy and quality of object detection purely based on input images is highly dependent on the quality and variety of the given training data. The computation time for processing one single image is 0.13–0.14 s. In general, it can also be stated the detected landmarks are not depended on the diameter of the actual object. This leads to a more flexible and broader application of the perception algorithm. The second test stage was the successful integration of the neural network into the self-localization framework. To do this the SQL database was utilized. The detected landmarks were written in the same SQL database as the landmarks detected from the old approach. The entire detection, storing and accessing process took ∼0.02 s. With this the self-localization pipeline could be extended with the third perception approach as described above. We here showed the successful training and integration of a neural network into a localization framework. It has

**Fig. 9.** YOLO on CityScapes

**Fig. 10.** YOLO in KITTI dataset

shown promising results on pole-like landmarks and leads to the conclusion that with further studies this approach can be lead to a robust standalone localization framework for dynamic urban environments.

## 6    Conclusion

In this paper a self-localization framework for autonomous driving vehicles was presented. The framework is using a highly detailed offline map which is built using a SQL Database. This database stores landmarks which are computed with a novel approach using either LiDAR sensor data, stereo camera images or mono camera images. In this work we focused on cylindrical, pole-like landmarks e.g. tree trunks, lanterns, traffic lights and traffic signs. The approach as described is applicable all over the world and is expected to function in many different scenarios. For future work it is very promising to extend the landmark detection to more complex objects like entire buildings, benches, gates or house facades. It could be shown that these approaches lead to sufficient accuracy and the novel approach of utilizing machine learning algorithms for the object detection has shown the assumed effect of detecting more reliably and a higher number of landmarks. The evaluations in this paper were done in two steps. Once the KITTI data was utilized for the comparison of the different perception approaches, second the machine learning based perception was validated and tested on two other data sets, namely the IJRR (Ford) and CityScapes. For further studies it would be interesting to investigate different machine learning concepts and improve the accuracy of the bounding boxes or the matching algorithm on the depth map. Therefore, one approach could be semantic segmentation algorithms which are currently used in free space detection applications. These approaches could minimize the bottleneck of this approach, the depth mapping, and improve the accuracy even further. This approach has shown to be working in different scenarios and with different hardware setup, hence it is a valid addition to current existing solutions and can be used as a second complimenting strategy.

# References

1. Arulampalam, M.S., Maskell, S., Gordon, N., Clapp, T.: A tutorial on particle filters for online nonlinear/non-gaussian Bayesian tracking. IEEE Trans. Sig. Process. **50**(2), 174–188 (2002)
2. Bais, A., Sablatnig, R., Gu, J.: Single landmark based self-localization of mobile robots. In: The 3rd Canadian Conference on Computer and Robot Vision, p. 67. IEEE, Piscataway (2006)
3. Balali, V., Golparvar-Fard, M.: Recognition and 3D localization of traffic signs via image-based point cloud models. In: Ponticelli, S., O'Brien, W.J. (eds.) Computing in Civil Engineering 2015, pp. 206–214. American Society of Civil Engineers, Reston (2015)
4. Bappy, J.H., Roy-Chowdhury, A.K.: CNN based region proposals for efficient object detection. In: 2016 IEEE International Conference on Image Processing, pp. 3658–3662. IEEE, Piscataway (2016)
5. Bazin, J., Laffont, P., Kweon, I., Demonceaux, C., Vasseur, P.: An original approach for automatic plane extraction by omnidirectional vision. In: The IEEE/RSJ International Conference on Intelligent Robots and Systems, pp. 752–758. IEEE, Piscataway (2010)
6. Bernay-Angeletti, C., Chabot, F., Aynaud, C., Aufrere, R., Chapuis, R.: A top-down perception approach for vehicle pose estimation. In: 2015 IEEE International Conference on Robotics and Biomimetics, pp. 2240–2245. IEEE, Piscataway (2015)
7. Biber, P., Strasser, W.: The normal distributions transform: a new approach to laser scan matching. In: 2003 IEEE/RSJ International Conference on Intelligent Robots and Systems, pp. 2743–2748. IEEE, Piscataway (2003)
8. Bora, D.J., Gupta, A.K., Khan, F.A.: Comparing the performance of L*A*B* and HSV color spaces with respect to color image segmentation, 04 June 2015. http://arxiv.org/pdf/1506.01472
9. Cordts, M., Omran, M., Ramos, S., Rehfeld, T., Enzweiler, M., Benenson, R., Franke, U., Roth, S., Schiele, B.: The cityscapes dataset for semantic urban scene understanding, 07 April 2016. http://arxiv.org/pdf/1604.01685

10. Dailey, M.N., Parnichkun, M.: Landmark-based simultaneous localization and mapping with stereo vision (2005)
11. Denzler, J., Notni, G., Süße, H.: Pattern Recognition, vol. 5748. Springer, Heidelberg (2009)
12. Everingham, M., van Gool, L., Williams, C.K.I., Winn, J., Zisserman, A.: The pascal visual object classes (VOC) challenge. Int. J. Comput. Vis. **88**(2), 303–338 (2010)
13. Feng, C., Taguchi, Y., Kamat, V.R.: Fast plane extraction in organized point clouds using agglomerative hierarchical clustering. In: 2014 IEEE International Conference on Robotics and Automation (ICRA), pp. 6218–6225. IEEE, Piscataway (2014)
14. Geiger, A., Lenz, P., Stiller, C., Urtasun, R.: Vision meets robotics: the kitti dataset. Int. J. Robot. Res. **32**(11), 1231–1237 (2013)
15. Girshick, R.: Fast R-CNN (2015). http://arxiv.org/pdf/1504.08083
16. Green, W.R., Grobler, H.: Normal distribution transform graph-based point cloud segmentation. In: Proceedings of the 2015 Pattern Recognition Association of South Africa and Robotics and Mechatronics International Conference (PRASA-RobMech), pp. 54–59. IEEE, Piscataway (2015)
17. He, Z., Wang, Y., Yu, H.: Feature-to-feature based laser scan matching in polar coordinates with application to pallet recognition. Procedia Eng. **15**, 4800–4804 (2011)
18. Hirschmüller, H.: Stereo processing by semiglobal matching and mutual information. IEEE Trans. Patt. Anal. Mach. Intell. **30**(2), 328–341 (2008)
19. Huang, J., Rathod, V., Sun, C., Zhu, M., Korattikara, A., Fathi, A., Fischer, I., Wojna, Z., Song, Y., Guadarrama, S., Murphy, K.: Speed/accuracy trade-offs for modern convolutional object detectors (2016). http://arxiv.org/pdf/1611.10012
20. Jang, C., Kim, Y.K.: A feasibility study of vehicle pose estimation using road sign information. In: ICCAS 2016, pp. 397–401. IEEE, Piscataway (2016)
21. Jayatilleke, L., Zhang, N.: Landmark-based localization for unmanned aerial vehicles. In: 2013 IEEE International Systems Conference (SysCon 2013), pp. 448–451. IEEE, Piscataway (2013)
22. Liu, W., Anguelov, D., Erhan, D., Szegedy, C., Reed, S., Fu, C.Y., Berg, A.C.: SSD: single shot multibox detector (2015). http://arxiv.org/pdf/1512.02325
23. Macukow, B.: Neural networks - state of art, brief history, basic models and architecture. In: Saeed, K., Homenda, W. (eds.) Computer Information Systems and Industrial Management. Lecture Notes in Computer Science, pp. 3–14. Springer, Cham (2016)
24. Pandey, G., McBride, J.R., Eustice, R.M.: Ford campus vision and lidar data set. Int. J. Robot. Res. **30**(13), 1543–1552 (2011)
25. Rademakers, E., de Bakker, P., Tiberius, C., Janssen, K., Kleihorst, R., Ghouti, N.E.: Obtaining real-time sub-meter accuracy using a low cost GNSS device. In: 2016 European Navigation Conference (ENC), pp. 1–8. IEEE, Piscataway (2016)
26. Rapp, M., Barjenbruch, M., Hahn, M., Dickmann, J., Dietmayer, K.: Clustering improved grid map registration using the normal distribution transform. In: 2015 IEEE Intelligent Vehicles Symposium (IV), pp. 249–254. IEEE, Piscataway (2015)
27. Reddy, B.S., Chatterji, B.N.: An FFT-based technique for translation, rotation, and scale-invariant image registration. IEEE Trans. Image Process. **5**(8), 1266–1271 (1996)
28. Redmon, J., Farhadi, A.: YOLO9000: better, faster, stronger (2016). http://arxiv.org/pdf/1612.08242

29. Ren, S., He, K., Girshick, R., Sun, J.: Faster R-CNN: towards real-time object detection with region proposal networks. In: Cortes, C., Lawrence, N.D., Lee, D.D., Sugiyama, M., Garnett, R. (eds.) Advances in Neural Information Processing Systems 28, pp. 91–99. Curran Associates, Inc (2015). http://papers.nips.cc/paper/5638-faster-r-cnn-towards-real-time-object-detection-with-region-proposal-networks.pdf

30. Rohde, J., Jatzkowski, I., Mielenz, H., Zöllner, J.M.: Vehicle pose estimation in cluttered urban environments using multilayer adaptive monte carlo localization. In: 2016 19th International Conference on Information Fusion (FUSION), pp. 1774–1779 (2016)

31. Rossmann, J., Sondermann, B., Emde, M.: Virtual testbeds for planetary exploration: the self-localization aspect. 11th Symposium on Advanced Space Technologies in Robotics and Automation. ASTRA, pp. 1–8. ESA/ESTEC, Noordwijk (2011)

32. Schindler, A.: Vehicle self-localization with high-precision digital maps. In: 2013 IEEE Intelligent Vehicles Symposium workshops (IV workshops), pp. 134–139. IEEE, Piscataway (2013)

33. Sefati, M., Daum, M., Sondermann, B., Kreiskother, K.D., Kampker, A.: Improving vehicle localization using semantic and pole-like landmarks. In: 28th IEEE Intelligent Vehicles Symposium, pp. 13–19. IEEE, Piscataway (2017)

34. Shu, L., Xu, H., Huang, M.: High-speed and accurate laser scan matching using classified features. In: Ben-Tzvi, P. (ed.) 2013 IEEE International Symposium on Robotic and Sensors Environments (ROSE), pp. 61–66. IEEE, Piscataway (2013)

35. Sindagi, V.A., Patel, V.M.: A survey of recent advances in CNN-based single image crowd counting and density estimation. Patt. Recogn. Lett. **107**, 3–6 (2018). https://doi.org/10.1016/j.patrec.2017.07.007

36. Spangenberg, R., Goehring, D., Rojas, R.: Pole-based localization for autonomous vehicles in urban scenarios. In: IROS 2016, pp. 2161–2166. IEEE, Piscataway (2016)

37. Thrun, S., Burgard, W., Fox, D.: Probabilistic Robotics. Intelligent Robotics and Autonomous Agents. MIT Press, Cambridge (2010). Mass. [u.a.], [nachdr.] edn

38. Xie, P., Petovello, M.G.: Measuring GNSS multipath distributions in urban canyon environments. IEEE Trans. Instrum. Measur. **64**(2), 366–377 (2015)

39. Zhang, H., Zhang, L., Dai, J.: Landmark-based localization for indoor mobile robots with stereo vision. In: 2012 Second International Conference on Intelligent System Design and Engineering Application (ISDEA), pp. 700–702. IEEE, Piscataway (2012)

# Cross Domain Image Transformation Using Effective Latent Space Association

Naeem Ul Islam and Sukhan Lee[✉]

Intelligent Systems Research Institute, Sungkyunkwan University, P.O. Box: 15000,
Seoul 440-746, Republic of Korea
{naeem,lsh1}@skku.edu

**Abstract.** Cross-domain image to image translation task aims at learning the joint distribution of images from marginal distributions in their respective domains. However, estimation of joint distribution from marginal distribution is a challenging problem as there is no, one to one correspondence. To address this problem, we propose a general approach based on variational autoencoders along with latent space association network (VAE-LSAN). The variational autoencoders learn the marginal distribution of the images in the individual domain whereas, the association network provides the correspondence between the marginal distributions of the cross domains. Our architecture effectively performs mapping of images in individual as well as cross domains. Experimental results show state of the art performance of our framework on different datasets in term of synthesizing images within its respective domain as well as cross domains.

**Keywords:** Generative models · Variational autoencoder
Association network

## 1 Introduction

Different domains around us are interconnected and their relationship with each other can be categorized considering different aspects. Such cross-domain relations are natural to humans in which we recognize these relationships. A scene can be rendered as an RGB image, an edge map, a gradient field, etc, in the same way as a concept can be represented in different languages. We define the image to image translation task as a problem of translating representation of one possible scene in one domain to a corresponding scene in another domain, given a sufficient training data. Image to image translation may be computer vision related task where the mapping is from many to one or it may be one to many which are related to computer graphics. Despite the same nature of these tasks, they have been tackled separately by Efros et al. (2001); Hertzmann et al. (2001); Fergus et al. (2006); Buades et al. (2005); Chen et al. (2009); Shih et al. (2013); Laffont et al. (2014); Long et al. (2015); Eigen and Fergus (2015); Xie and Tu (2015) and Zhang et al. (2016). However, in our approach, we tackled these problems in a unified framework.

© Springer Nature Switzerland AG 2019
M. Strand et al. (Eds.): IAS 2018, AISC 867, pp. 706–716, 2019.
https://doi.org/10.1007/978-3-030-01370-7_55

A significant amount of research has been done in this area from the perspective of regression, with convolutional neural networks (CNNs) as the basic platform for a wide variety of image prediction problems. Secondly, the availability of a large amount of paired data for the image to image translation task makes Convolutional Neural Networks (CNNs) well-suited for this particular task. CNN-based regression methods have been able to successfully translate images from one domain to its corresponding domain, surpassing the performance of non-CNN state of the art models without expert knowledge such as the work of Cheng et al. (2015) and Iizuka et al. (2016). Regression methods using standard L1 or L2 losses, however, cannot handle generalization problems and produce very blurry results as shown by Pathak et al. (2016) and Zhang et al. (2016).

Variational autoencoder (VAE) by Kingma et al. (2013) provides an additional way to handle generalization problem while producing realistic images. VAE learns a loss function which is based on feature matching, which makes the network being able to reconstruct the input distribution realistically along with the loss function based on KL divergence between the zero mean and unit variance normal distribution and the encoder output distribution. This loss term ensure generalization hence, makes the network more robust to the noise in the input distribution.

VAE has been studied extensively for the image to image translation task in their respective domains. However, none of the previous approaches used VAE for cross-domain image to image translation tasks, so, it has remained unclear how effective variational autoencoder can be as a general-purpose solution for image to image translation tasks.

In order to discover the relationship between two visual domains, we introduce a model based on association network along with two variational autoencoder networks. The variational autoencoders individually learn the transformations in their individual domains, whereas, the association network discovers cross-domain transformations. The association network maps the latent space of images in one domain to the corresponding latent space of images in another domain. Through visualization results from various experiments of image translation tasks, we verify the effectiveness of the proposed network.

## 2   Related Work

A significant amount of work has been done in term of generative models where the goal is to make the model able to generate realistic images. Several deep generative models have been recently proposed such as VAEs by Kingma et al. (2013) and Rezende et al. (2014), Generative Adversarial Network (GAN) by Goodfellow et al. (2014), moment matching networks by Li et al. (2015), Pixel-CNN by van den Oord et al. (2016), and Plug&Play Generative Networks by Nguyen et al. (2016). As the recent work is mostly based on the combination of GAN and VAE, so, we first review VAE and GAN based works and then discuss related image translation works.

GAN is based on adversarial learning, where the discriminator tries to discriminate between real and synthetic images generated by the generator. The generator, on the other hand, tries to fool discriminator by generating realistic images. The training continues until the discriminator fails to discriminate between real and synthetic images. Several variants of GAN have been proposed such as LapGAN by Denton et al. (2015), DCGAN by Radford et al. (2015) and WGAN by Arjovsky et al. (2017).

VAEs optimize a variational lower bound of the input distribution likelihood function. Better estimation of input distribution by the model depends upon the estimation of variational lower bound as shown by Maaløe et al. (2016) and Kingma et al. (2016). To improve image generation quality of VAE, a VAE-GAN network has been proposed by Larsen et al. (2015). Yan et al. (2016) proposed VAE based architectures for translating face image attributes.

Image Translation has a wide range of applications. One may translate an image from a modality which is difficult to understand to a corresponding modality where it becomes easy to understand and visualize. Similarly, for training classifier in the target domain where less labeled data is available one can create more training data from labeled images by translating from corresponding domain to the target domain.

Most of the approaches for mapping images from one domain to its corresponding domain is based on the conditional generative model, such as the work of Ledig et al. (2016) and Isola et al. (2016), where exact correspondence for training the network in two domains are required. Recently, Taigman et al. (2016), proposed domain transformation network (DTN) for low-resolution digits and face translation tasks, where they achieved promising results. Shrivastava et al. (2016) proposed a conditional generative adversarial network-based approach for translating rendering images to a real image for gaze estimation. The objective here was to minimize the cost function based on the $L_1$ distance between rendering images and real images. As translating from rendering images to real images is a simple task, where minimizing $L_1$ distance produces the realistic translation. On the other hand, when it comes to natural images it becomes a challenging task and minimizing $L_1$ distance will not work alone, both in term of realistic translation and generalization.

However, unlike these prior works which use the combination of variational autoencoder and GAN for the image to image translation tasks, our approach is solely based on variational autoencoder, which has not been analyzed explicitly for this application yet. Secondly, our method also differs from the previous works in architecture style in term of association between different domains.

## 3    Model

**Network Architecture:** We propose an image translation network for image to image translation task. The framework, which is illustrated in Fig. 1, is motivated by recent advances in deep generative models called variational autoencoders (VAEs) by Kingma et al. (2013); Rezende et al. (2014) and Larsen et al. (2015).

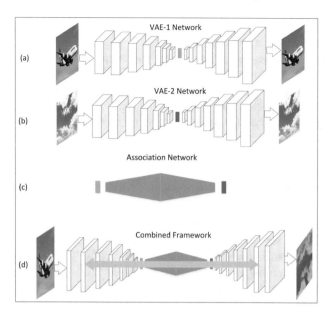

**Fig. 1.** $(a), (b)$ shows variational autoencoder in their respective domains, $(c)$ shows association network, whereas $(d)$ shows the combined cross domain transformation framework

Our image translation model is based on two variational auto encoder networks and one association network. The VAEs relate translated images with input images in its respective domains, whereas the association network relates the latent spaces of the cross domains.

**VAEs:** The first VAE takes input an RGB image $x_1$ from $X_1$ domain and maps it to the mean vector $\mu(x_1)$ and a variance vector, $\sigma^2(x_1)$, where the distribution of the latent space $z_1$ is given by $q_1(z_1|x_1) \equiv N(z_1|\mu(x_1), diag(\sigma^2(x_1)))$. Following the design of Kingma et al. (2013), the KL Divergence between the encoder distribution, $q_1(z_1|x_1)$ and the prior distribution $p(z)$, which is zero mean and unit variance is minimized. The latent space is then sampled from this distribution which is decoded to the input reconstructed image by the decoder of VAE-1, $x_1'=D_1(z_1|x_1)$). Similarly, the second VAE takes input image $x_2$ from $X_2$ domain and maps it to the mean vector $\mu(x_2)$ and a variance vector, $\sigma^2(x_2)$ where the distribution of the latent code $z_2$ is given by $q_2(z_2|x_2) \equiv N(z_2|\mu(x_2), diag(\sigma^2(x_2)))$. The latent space $z_2$ is then sampled from this distribution, which is decoded to the input reconstructed image by the decoder of VAE-2, $x_2' = D_2(z_2 \sim q_2(z_2|x_2))$.

$$L_{VAE_s} = D_{KL(q_1(z_1|x_1)||p(z))} + ||x - D_1(z_1|x_1)|| + D_{KL(q_2(z_2|x_2)||p(z))} + ||x - D_2(z_2|x_2)|| \tag{1}$$

The first and third terms on the right-hand side of Eq. (1) shows KL Divergence between the prior distributions and the encoder output of $VAE_s$, whereas the second and fourth terms represent the $L_2$ loss between the input and reconstructed images in their respective domains.

Both variational autoencoders having the same number of parameters and the same number of layers. We used seven encoder layers and seven decoder layers where the input to VAEs are $[128 \times 128]$ RGB images. In the initial layer, we used 64 number of filters with size $[4 \times 4]$ and stride 2. The number of filters are doubled in the preceding layers up to $4th$ layer and then in the rest of the layers, we keep them same. In each layer, convolution is followed by batch normalization proposed by Ioffe et al. (2015) and activation layer. We used Leaky Relu activation function proposed by Maas et al. (2013) activation function in each of these layers except the last layer, where we used sigmoid activation function.

**Association Network:** In order to a build cross-domain relationship between the two VAEs, we use an association network. The association network makes an association between the latent space of both VAEs. The association network is composed of six fully connected layers, where the input to the network is the latent space from one of the VAEs from one domain and the output of the association network is the translated version of latent space of second VAE in the next domain.

$$L_{LSAN} = q_1(z_1|x_1) - LSAN(q_2(z_2|x_2) + q_2(z_2|x_2) - LSAN(q_1(z_1|x_1) \quad (2)$$

Where $q_1(z_1|x_1)$ is the encoder output of VAE-1, $q_2(z_2|x_2)$ is the encoder output of VAE-2 and LSAN is the association between two latent spaces. This loss function is minimized using stochastic gradient method. The combined loss of VAE-LSAN is:

$$L_{VAE-LSAN} = L_{VAE_s} + L_{LSAN} \quad (3)$$

## 4   Training Details

We have three phases of training our framework. In the first phase, we trained VAEs of each domain separately, where each VAE relates the reconstructed images and the input images in the respective domains. Once the VAEs are trained on the images of their respective domains, the latent space of the trained VAE represents the relation between the input images and reconstructed images of the respective domains. The second phase is to make correspondence between the latent spaces of two domains, which is related to the image to image translation task. To make an association between the latent spaces we train the association network. The input to the association network is the latent space in one domain and the output is the latent space in the corresponding domain. Once the transformation of latent space from one domain to the next domain is done, the next phase is to make the VAEs being able to produce the cross-domain images. In our framework, we do not use separate VAE for producing the cross-domain images but instead, we use sharing weight concept to train the decoder part of the VAEs based on the transformed latent from the association network. The minibatch size during training our framework was set to 32.

Learning rate also plays an important role in the successful training of the network. Too high learning rate will cause the network to overshoot from the desired optima and too small learning rate can keep the network around local minima. In our experiments we set the learning rate 0.0015 for both variational autoencoder in the first phase, 0.0001 for training association network in the second phase and 0.0001 for the final learning of decoder part of both variational autoencoders. Similarly, we used ADAM optimizer Kingma and Ba (2014) for optimization the parameters of the network with beta = 0.5.

## 5   Experiments

To evaluate the performance of our framework whether it can effectively learn the cross-domain relation, we show the results of several, image to image translation tasks of our network. As the purpose of our work is to translate the image from one domain to other domain which has some natural correspondence, so, first we build the datasets of different corresponding pairs such as airplane-sky, car-wheel, and flower-honey bee. We selected aeroplane, car, wheel, and honey bee from ImageNet dataset developed by Deng et al. (2009). The ImageNet dataset is composed of 1000 different classes of RGB and gray scale images with around 1300 images per class. For our experiments, we used only RGB images and removed gray scale images. We also removed unrealistic images from this dataset, so we left with around 1200 images per category. From 1200 images we used 1100 as training samples and remaining 100 as testing samples. The sky and flower classes are not available in the ImageNet dataset. For flower images we used oxford flower dataset made by Nilsback et al. (2008) which contains 17 different categories flowers with 80 images per category. As from ImageNet dataset, we have 1100 training samples per category so we selected the same number of flowers samples from this dataset. In terms of sky images, we used skyfinder dataset of Mihail et al. (2016). The Skyfinder dataset consists of about 90,000 labeled outdoor images which are captured under a wide range of illumination and weather conditions. However, we only selected 1100 images for training and 100 images for testing. Apart from the above datasets, we also used Map dataset from Isola et al. (2016) and EMNIST dataset developed by Cohen et al. (2017) for learning the correspondence between map and aerial views as well as small and capital letters respectively. The resolution of the RGB images we used in our framework is [128 128,3]. After training, we applied our learned framework to the image to image translation task for the above datasets. The results are shown in Figs. 2, 3 and 4. As our framework is bidirectional, it performs realistic image to image translation in both directions.

### 5.1   Results

Results of our frame work on different tasks are shown in Figs. 2, 3 and 4.

**Airplane to sky translation:** For airplane and sky correspondence we picked airplane images from the ImageNet dataset. The ImageNet dataset does not contain sky images so we picked sky images from skyfinder dataset of Mihail et al. (2016).

We scale down these images to [128 128 3] resolutions. Figure 2a shows the results of translation from airplane to sky whereas from the sky to airplane transformation is shown in Fig. 2d. Our experimental results show that it produces realistic translated images.

**Flower to honey bee translation:** From the ImageNet dataset we selected RGB images of the honey bee, whereas the corresponding flower images, we picked from Oxford Flower dataset developed by Nilsback et al. (2008). The results of cross and respective domains translation from flower to honey bee and flower to flower are shown in Fig. 2b, whereas Fig. 2e shows the results of transformation in the opposite direction.

**Car to Wheel Translation:** We also picked corresponding RGB images of cars and wheels from the ImageNet dataset. The number of training and testing samples for this correspondence is same as that for the airplane and sky images. Figure 2c shows car to car translation within its respective domain and car to wheel translation in the cross domain. Similarly, Fig. 2f shows cross domain translation from wheel to car and respective domain translation from wheel to wheel.

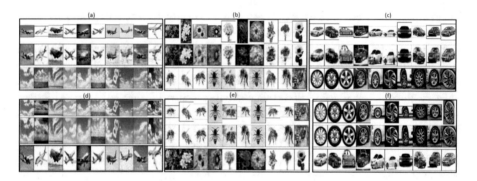

**Fig. 2.** First row of (a), (b), (c), (d), (e) and (f) shows Ground truth, second row of (a), (b) and (c) shows VAE1 output whereas second row of (d), (e) and (f) shows VAE2 output in their respective domains, last rows shows cross domain transformations

**Arial to Map:** For correspondence between Arial images and the maps, we selected the map dataset from Isola et al. (2016). This dataset contains 2194 corresponding images for training and testing. Our framework effectively learns the correspondence between these images. This correspondence is clear from Figs. 3a and 3b.

**Capital to Small letters:** Transformation from the capital to small letters as well as from small to capital letters, we used EMNIST dataset of Cohen et al. (2017). Our framework effectively transformed the data from one domain to the next domain as shown in Fig. 4.

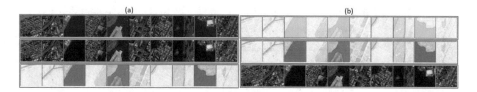

**Fig. 3.** First row of (a), (b) shows Ground truth, second row shows transformation in their individual domains, whereas last rows shows cross domain transformations

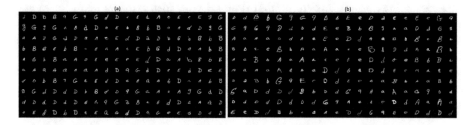

**Fig. 4.** Odd columns of (a) shows ground truth input samples of small letters whereas the Even column shows the transformed capital letters, similarly Odd columns of (b) shows ground truth input samples of capital letters whereas the Even column shows the transformed small letters

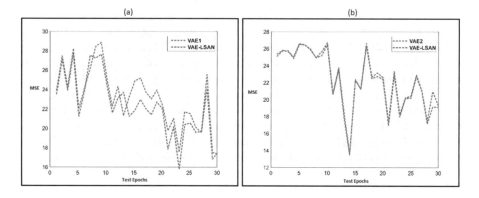

**Fig. 5.** (a) MSE of VAE-1 and VAE-LSAN, (b) MSE of VAE-2 and VAE-LSAN

We quantify these observations in term of image synthesis accuracy (Table 1) using ImageNet dataset correspondence. The association based on the cross-domain transformation shows higher performance as compared to the others.

In some of our experiments, during the first phase of training, we found out that sometimes VAE-1 converges whereas VAE-2 diverge from global optima. Also after individual domain transformation, the cross-domain transformation is more challenging in term of combining the marginal distribution of the individual domain to joint distribution of the cross-domains. The cross-domain transformation is sensitive to the hyper-parameters setting as the parameters of the network are

**Table 1.** Image synthesis accuracy on ImageNet dataset correspondence

|  | Per-pixel accuracy |
|---|---|
| AE | 0.430 |
| VAE | 0.454 |
| VAE (Skip) | 0.541 |
| VAE-LSAN(Skip) **Ours** | 0.563 |

already at its optimal for individual domain transformation, so careful hyper-parameters tuning play an important role in the successful learning of the correspondence. The qualitative performance of both VAEs and VAE-LSAN networks are shown in Fig. 5a and b, in terms of mean square error of the bidirectional reconstructed images from the cross-domain and the respective domain for different test epochs. From our qualitative and quantitative results, it is clear that our framework performs cross and respective domain translation effectively.

## 6   Conclusion

In this paper, we proposed VAE-LSAN framework for many image-to-image translation tasks, especially those involving highly structured graphical outputs. We made the association between the latent space of two domains which has natural correspondence using association network. From our experimental analysis, it shows that our framework produces realistic transformation from one domain to its corresponding another domain. As our future work, we will extend this relationship from image to image translation to 3D-2D translation along with pose information. Furthermore, as the variational autoencoder learns the data distribution with no specific boundaries, which causes VAE to produce blurry results. Generative adversarial network, on the other hand, having fine boundaries but it is very hard to train it because it suffers from mode collapse problem. Our approach is to limit the boundaries of generated data distribution by using generative adversarial concept while avoiding collapsing problem.

**Acknowledgments.** Sukhan Lee proposed the concept of latent space association for image to image translation while Naeem Ul Islam implements the concept and carries out experimentation. This research was supported, in part, by the "3D Recognition Project" of Korea Evaluation Institute of Industrial Technology (KEIT) (10060160) and, in part, by the "Robot Industry Fusion Core Technology Development Project" of KEIT (10048320), sponsored by the Korea Ministry of Trade, Industry and Energy (MOTIE).

## References

Arjovsky, M., Chintala, S., Bottou, L.: Wasserstein gan. arXiv preprint arXiv:1701.07875 (2017)

Buades, A., Coll, B., Morel, J.M.: A non-local algorithm for image denoising. In: IEEE Computer Society Conference on Computer Vision and Pattern Recognition. CVPR 2005. vol. 2, pp. 60–65. IEEE (2005)

Chen, T., Cheng, M.M., Tan, P., Shamir, A., Hu, S.M.: Sketch2photo: internet image montage. ACM Trans. Graphics (TOG) **28**(5), 124 (2009)

Cheng, Z., Yang, Q., Sheng, B.: Deep colorization. In: Proceedings of the IEEE International Conference on Computer Vision, pp. 415–423 (2015)

Cohen, G., Afshar, S., Tapson, J., van Schaik, A.: Emnist: an extension of mnist to handwritten letters. arXiv preprint arXiv:1702.05373 (2017)

Deng, J., Dong, W., Socher, R., Li, L.J., Li, K., Fei-Fei, L.: ImageNet: a large-scale hierarchical image database. In: CVPR09 (2009)

Denton, E.L., Chintala, S., Fergus, R., et al.: Deep generative image models using a laplacian pyramid of adversarial networks. In: Advances in Neural Information Processing Systems, pp. 1486–1494 (2015)

Efros, A.A., Freeman, W.T.: Image quilting for texture synthesis and transfer. In: Proceedings of the 28th Annual Conference on Computer Graphics and Interactive Techniques, pp. 341–346. ACM (2001)

Eigen, D., Fergus, R.: Predicting depth, surface normals and semantic labels with a common multi-scale convolutional architecture. In: Proceedings of the IEEE International Conference on Computer Vision, pp. 2650–2658 (2015)

Fergus, R., Singh, B., Hertzmann, A., Roweis, S.T., Freeman, W.T.: Removing camera shake from a single photograph. In: ACM Transactions on Graphics (TOG). vol. 25, pp. 787–794. ACM (2006)

Goodfellow, I., Pouget-Abadie, J., Mirza, M., Xu, B., Warde-Farley, D., Ozair, S., Courville, A., Bengio, Y.: Generative adversarial nets. In: Advances in Neural Information Processing Systems, pp. 2672–2680 (2014)

Hertzmann, A., Jacobs, C.E., Oliver, N., Curless, B., Salesin, D.H.: Image analogies. In: Proceedings of the 28th Annual Conference on Computer Graphics and Interactive Techniques, pp. 327–340. ACM (2001)

Iizuka, S., Simo-Serra, E., Ishikawa, H.: Let there be color!: joint end-to-end learning of global and local image priors for automatic image colorization with simultaneous classification. ACM Trans. Graphics (TOG) **35**(4), 110 (2016)

Ioe, S., Szegedy, C.: Batch normalization: accelerating deep network training by reducing internal covariate shift. In: International Conference on Machine Learning, pp. 448–456 (2015)

Isola, P., Zhu, J.Y., Zhou, T., Efros, A.A.: Image-to-image translation with conditional adversarial networks. arXiv preprint arXiv:1611.07004 (2016)

Kingma, D., Ba, J.: Adam: a method for stochastic optimization. arXiv preprint arXiv:1412.6980 (2014)

Kingma, D.P., Salimans, T., Welling, M.: Improving variational inference with inverse autoregressive flow. arXiv preprint arXiv:1606.04934 (2016)

Kingma, D.P., Welling, M.: Auto-encoding variational bayes. arXiv preprint arXiv:1312.6114 (2013)

Laffont, P.Y., Ren, Z., Tao, X., Qian, C., Hays, J.: Transient attributes for high-level understanding and editing of outdoor scenes. ACM Trans. Graphics (TOG) **33**(4), 149 (2014)

Larsen, A.B.L., Sønderby, S.K., Larochelle, H., Winther, O.: Autoencoding beyond pixels using a learned similarity metric. arXiv preprint arXiv:1512.09300 (2015)

Ledig, C., Theis, L., Huszár, F., Caballero, J., Cunningham, A., Acosta, A., Aitken, A., Tejani, A., Totz, J., Wang, Z., et al.: Photo-realistic single image super-resolution using a generative adversarial network. arXiv preprint arXiv:1609.04802 (2016)

Li, Y., Swersky, K., Zemel, R.: Generative moment matching networks. In: Proceedings of the 32nd International Conference on Machine Learning (ICML 2015), pp. 1718–1727 (2015)

Long, J., Shelhamer, E., Darrell, T.: Fully convolutional networks for semantic segmentation. In: Proceedings of the IEEE Conference on Computer Vision and Pattern Recognition, pp. 3431–3440 (2015)

Maaløe, L., Sønderby, C.K., Sønderby, S.K., Winther, O.: Auxiliary deep generative models. arXiv preprint arXiv:1602.05473 (2016)

Maas, A.L., Hannun, A.Y., Ng, A.Y.: Rectifier nonlinearities improve neural network acoustic models. In: ICML Workshop on Deep Learning for Audio, Speech, and Language Processing (2013)

Mihail, R.P., Workman, S., Bessinger, Z., Jacobs, N.: Sky segmentation in the wild: an empirical study. In: IEEE Winter Conference on Applications of Computer Vision (WACV), pp. 1–6 (2016)

Nguyen, A., Yosinski, J., Bengio, Y., Dosovitskiy, A., Clune, J.: Plug & play generative networks: conditional iterative generation of images in latent space. arXiv preprint arXiv:1612.00005 (2016)

Nilsback, M.E., Zisserman, A.: Automated flower classification over a large number of classes. In: Proceedings of the Indian Conference on Computer Vision, Graphics and Image Processing, December 2008

van den Oord, A., Kalchbrenner, N., Espeholt, L., Vinyals, O., Graves, A., et al.: Conditional image generation with pixelcnn decoders. In: Advances in Neural Information Processing Systems, pp. 4790–4798 (2016)

Pathak, D., Krahenbuhl, P., Donahue, J., Darrell, T., Efros, A.A.: Context encoders: feature learning by inpainting. In Proceedings of the IEEE Conference on Computer Vision and Pattern Recognition, pp. 2536–2544 (2016)

Radford, A., Metz, L., Chintala, S.: Unsupervised representation learning with deep convolutional generative adversarial networks. arXiv preprint arXiv:1511.06434 (2015)

Rezende, D.J., Mohamed, S., Wierstra, D.: Stochastic backpropagation and variational inference in deep latent gaussian models. In: International Conference on Machine Learning (2014)

Shih, Y., Paris, S., Durand, F., Freeman, W.T.: Data-driven hallucination of different times of day from a single outdoor photo. ACM Trans. Graphics (TOG) 32(6), 200 (2013)

Shrivastava, A., Pfister, T., Tuzel, O., Susskind, J., Wang, W., Webb, R.: Learning from simulated and unsupervised images through adversarial training. arXiv preprint arXiv:1612.07828 (2016)

Taigman, Y., Polyak, A., Wolf, L.: Unsupervised cross-domain image generation. arXiv preprint arXiv:1611.02200 (2016)

Xie, S., Tu, Z.: Holistically-nested edge detection. In: Proceedings of the IEEE International Conference on Computer Vision, pp. 1395–1403 (2015)

Yan, X., Yang, J., Sohn, K., Lee, H.: Attribute2image: cimage generation from visual attributes. In: European Conference on Computer Vision, pp. 776–791. Springer (2016)

Zhang, R., Isola, P., Efros, A.A.: Colorful image colorization. In: European Conference on Computer Vision, pp. 649–666. Springer (2016)

# Global Registration of Point Clouds for Mapping

Carlos Sánchez[(⊠)], Simone Ceriani, Pierluigi Taddei, Erik Wolfart,
and Vítor Sequeira

European Commission, Joint Research Centre (JRC), Via Enrico Fermi, 2749 Ispra,
VA, Italy
{carlos.sanchez,simone.ceriani,pierluigi.taddei,erik.wolfart,
vitor.sequeira}@ec.europa.eu

**Abstract.** We present a robust Global Registration technique focused
on environment survey applications using laser range-finders. Our approach works under the assumption that places can be recognized by analyzing the projection of the observed points along the gravity direction.
Candidate 3D matches are estimated by aligning the 2D projective representations of the acquired scans, and benefiting from the corresponding
dimensional reduction. Each single candidate match is then validated
exploiting the implicit empty space information associated to scans. The
global reconstruction problem is modeled as a directed graph, where scan
poses (nodes) are connected through matches (edges). This is exploited
to compute local matches (instead of global ones) between pairs of scans
that are in the same reference frame. As a consequence, both performance
and recall ratio increase w.r.t. using only global matches. Additionally,
the graph structure allows formulating a sparse global optimization problem that optimizes scan poses, considering simultaneously all accepted
matches. Our approach is being used in production systems and has been
successfully evaluated on several real datasets.

**Keywords:** Global registration · Loop detection · Place recognition
SLAM

## 1  Introduction

Retrieving relative poses between pairs of 3D point clouds, when no prior initialization exists, is a key component of mapping applications. This is the typical
situation when no global reference positioning systems are available during the
survey of an environment. In some cases, relative odometry systems can be used
to track the sensor pose as it moves (i.e. SLAM approaches). However, since the
pose estimation process is incremental, it accumulates drift. In this case, Global
Registration mitigates this effect by exploiting loop closures (revisited places).

Given a pair of maps, the Global Matcher has to (1) determine whether both
are observing the same place and, in this case, (2) it has to provide the relative

© Springer Nature Switzerland AG 2019
M. Strand et al. (Eds.): IAS 2018, AISC 867, pp. 717–729, 2019.
https://doi.org/10.1007/978-3-030-01370-7_56

transformation between the maps poses. Its main goal is to retrieve as many correct associations between maps as possible (high precision rate): missing some matches is not as critical as accepting wrong ones, since the alignment problem can be solved with just the minimum number of connections associating all the maps between themselves. In this sense, low recall may not affect the final result as badly as low precision, simplifying outlier rejection and robustification tasks.

## 1.1   Related Work

Global matching for mapping applications is strongly related to Place Recognition approaches in the literature. However, most of the state-of-the-art works focus on image-based techniques due to the benefits of working with cameras: they provide large amounts of data without distance limitations and a (roughly) uniform angular sampling. Even though pixels do not contain range information, their relative intensities can be assumed to be invariant to viewpoint and/or distance. Relatively small sets of pixel values can make up very descriptive and unique features exploited for place recognition purposes.

Image-based techniques use local descriptors like SIFT [12], SURF [1] or BRIEF [3] to characterize features. Then, two main methodologies are generally used to compare images: voting schemes (i.e. [9,13,21]) or the Bag-of-Words (BoW) model (i.e. [5,6,8]).

Adding 3D geometric information to the places described by features improves the robustness of the matching. Unlike appearance-based models and image features (which work in the visual domain), 3D data makes spatial verification straightforward. Some representative examples can be found in [5,16,17].

However, using 3D data alone makes place recognition extremely challenging: point clouds retrieved by 3D sensors make keypoint detection computationally more complex due to the higher dimensionallity. Some of the 3D keypoint detectors proposed are adapted versions of 2D ones (i.e. [18,23]), others have been explicitly developed for 3D data but are working over range images, like [25] and some others are native 3D detectors like [11,28]. For those keypoints, a variety of 3D feature descriptors has been proposed, like [19] or [27]. Once keypoints and features are available, *local* techniques match them together, retrieving rigid transformations associated with a *score*, which is calculated using voting schemes (i.e. [2]) or Bag of Words models (i.e. [4,24]).

Alternatively, *global* techniques like [14,15,26] use compact histograms to fully characterize a place as the sensor saw it. Similarity between non-consecutive observations is measured using a distance function in the descriptor space.

## 1.2   Overview

Considering the complexity of using 3D data alone to retrieve the relative poses between point clouds, we initially cast the problem as a 2D matching one. To do so, points observed by the sensor are projected into a plane tangent to the environment's manifold. For ground motion, this corresponds to the $X-Y$ plane, where the gravity direction, $\mathbf{g}$, is aligned with the $-Z$ axis. The direction of $\boldsymbol{g}$

can be estimated either assuming that the sensor was placed horizontally during the acquisition (laser scanners are typically mounted in well balanced tripods) or exploiting the information provided by IMU's, which are normally included in commercial rangefinders.

This assumption implies that only the ground distribution and projective shape of elements (i.e. walls, furniture, trees, etc.) can be exploited for matching purposes. However, the loss of information provides two great advantages: first, sensor observations can be projected into 2D bitmaps, where standard computer vision feature detectors can be applied. Second, global matching considers only 2D rigid transformations, reducing the complexity of the problem from six degrees of freedom to only three.

Once 2D matches are computed, the two additional degrees of freedom associated to rotations are provided by the projection direction, $\mathbf{g}$, and the third degree of freedom of translations over the $Z$ axis is estimated by a constrained 3D local registration initialized with the estimated 2D rigid alignment.

Given a potential 3D match, and assuming a static environment, a validation step is performed to determine if it makes geometrical sense: regions of space observed empty from one map are also supposed to be empty from the other (if observed). If so, the match is accepted, placing the two maps in the same reference frame. If not, we continue with the next best global match until a termination criteria is satisfied (e.g. a correct match is found, no more potential matches can be computed, both maps are in the same reference frame by matches with other maps or a maximum number of iterations is reached).

Finally, for all maps in the same reference frame, missing matches between pairs are retrieved with a local initialization (relative poses between themselves), and accepted if the same validation step is satisfied. With all the confirmed matches, a global registration is performed in order to get a drift-free reconstruction of the environment.

Major contributions of this paper are: (1) a global registration technique for 3D point clouds that benefits from the 2D reprojection over the gravity direction, (2) a robust match validator that ensures high precision rates and (3) a local matching strategy that, given a set of maps in the same reference frame, increases recall rates and feeds a global optimizer.

## 2   Approach

Our technique proceeds in four steps detailed, respectively, in the next subsections: (1) for all maps that are not in the same reference frame, the global matcher retrieves candidate alignments, (2) for the ones in the same reference frame, but not matched, the local matcher retrieves candidate alignments, (3) the validator accepts or rejects the potential matches coming from both, the local and the global matcher. Steps 1, 2, and 3 are repeated until a termination criteria is satisfied (e.g. all maps are connected with the rest, a maximum number of iterations is reached or all candidate alignments have been evaluated). Finally, (4) the global optimizer solves the full problem, considering all accepted matches.

Notice that, initially, each map is expressed in its own local reference frame. Consequently, first pairwise matches will be all computed by the global matcher. Only when maps start getting connected between themselves through global matches, the local matcher will start producing potential alignments. This way, if the first iteration manages to put all maps in the same reference frame, it will take another iteration for the local matcher to evaluate the pairwise associations that were discarded by the global matcher.

## 2.1  Global Matcher

Given a set of maps, $\Pi = \{\pi_1, \pi_2 \dots \pi_n\}$ and an relative estimate of the gravity direction $G = \{\mathbf{g}_1, \mathbf{g}_2 \dots \mathbf{g}_n\}$ the global matcher retrieves the set of 3D rigid transformations $\Gamma = \{\boldsymbol{\Gamma}_1, \boldsymbol{\Gamma}_2 \dots \boldsymbol{\Gamma}_n\}$ such that each relative pose $\boldsymbol{\Gamma}^i_j = \boldsymbol{\Gamma}_i^{-1}\boldsymbol{\Gamma}_j$, aligns map $\pi_j$ w.r.t. $\pi_i$.

To perform the global registration in the 2D domain, points from a single cloud, $\mathbf{p}_k \in \pi_i$ are projected along its associated gravity direction, $\hat{\boldsymbol{g}}_i$ as:

$$\begin{bmatrix} u_k \\ v_k \\ z_k \\ 1 \end{bmatrix} = \left\lfloor \mathbf{P}\,\mathbf{R}_{\hat{g}_i}\boldsymbol{p}_k \right\rfloor \qquad \mathbf{P} = \begin{bmatrix} \frac{1}{\text{res}} & 0 & 0 & r_{\max} \\ 0 & \frac{1}{\text{res}} & 0 & r_{\max} \\ 0 & 0 & 1 & 0 \\ 0 & 0 & 0 & 1 \end{bmatrix}$$

where, $[u_k, v_k]^T$ are the resulting 2D coordinates of point $\mathbf{p}_k$ expressed in homogeneous coordinates, $\mathbf{R}_{\hat{g}_i}$ is the rotation matrix that aligns the gravity vector to the $-Z$ axis of the map, $r_{\max}$ is the maximum distance to be represented in the image and res sets the scale factor (meters per pixel).

The intensity of each pixel is computed as the weighted sum of all points that fall inside it, being the weight of a single point, $w_k$, computed as follows:

$$w_k = \max(0, w_k^1 \cdot w_k^2), \quad w_k^1 = \text{angle}_{\max} - \text{asin}\left(|\bar{\mathbf{n}}_{kz}|\right), \quad w_k^2 = \|\boldsymbol{p}_k - \boldsymbol{c}_k\|^2$$

where $w_k^1$ evaluates negative if the angle between the normal and the gravity vector is smaller than $\text{angle}_{\max}$ (typically, horizontal surfaces which provide no meaningful information when projected over the $X - Y$ plane). $w_k^2$ compensates the quadratic decay of points density w.r.t. distance, where $\boldsymbol{c}_k$ is the position of the sensor when $\boldsymbol{p}_k$ was observed. Finally, intensities in the bitmap are normalized (Fig. 1(b)).

Given the 2D projective image of a point cloud, keypoints are detected using the well known Shi-Tomasi corner detector [22], where the minimum eigenvalue associated to each pixel is used to score it (the larger the better) followed by a local non maxima suppression. Results are shown as red dots in Fig. 1(b).

We generate a simple and fast to compute descriptor for each corner which is invariant to rotations. It consists of a histogram, where each bin, $b_i, 0 \le i < n$, stores the normalized sum of all pixel intensities that fall in the distance range defined by $\left[i\frac{l_{\max}}{n}, (i+1)\frac{l_{\max}}{n}\right)$, being $l_{\max}$ the size of the local neighborhood used to characterize a corner.

Potential correspondences between corners of different maps are selected by performing a radius search in the descriptor space and the largest subset of corner associations that is geometrically consistent is selected using a strategy based on [20]. The score associated to the match corresponds to the number of corners that were found compatible (red dots in Fig. 1(c)).

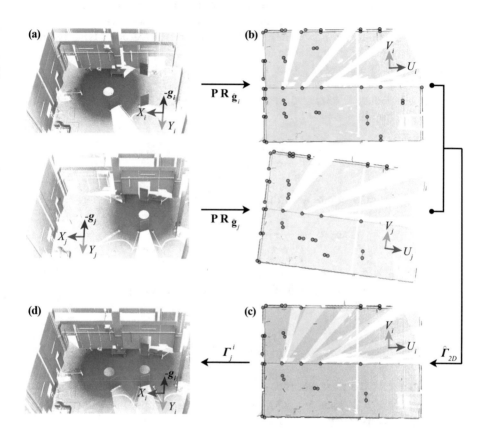

**Fig. 1.** Global registration. (a) Input point clouds and (b) their corresponding projections. Notice how the $Z$ axis is shared between both clouds whilst $X$ and $Y$ are arbitrarily defined. Red points in the 2D projections represent keypoints detected in each single map. (c) 2D match retrieved, in $\pi_i$ reference frame. Red dots represent matched keypoints. (d) 3D transformation retrieved from the 2D match, where $\pi_i$ is in its local reference frame, and $\pi_j$ has been pre-multiplied by $\boldsymbol{\Gamma}_j^i$.

Given a 2D match between two maps, $\pi_i$ and $\pi_j$, its associated 3D transformation is recovered applying the following three steps:

1. We compute a full 3D transformation, $\hat{\boldsymbol{\Gamma}}_1$, that aligns both maps in all degrees of freedom but $Z$ translations, since this information was lost during the projection. To do so, gravity vectors estimated for each map, $\hat{\boldsymbol{g}}_i$ and $\hat{\boldsymbol{g}}_j$, are

combined with the 2D transformation that comes from the 2D match, $\hat{\boldsymbol{\Gamma}}_{2D}$. $\hat{\boldsymbol{\Gamma}}_1$ is calculated as follows:

$$\hat{\boldsymbol{\Gamma}}_1 = \left(\mathbf{PR}_{\hat{g}_i}\right)^{-1} \hat{\boldsymbol{\Gamma}}_{2D} \, \mathbf{PR}_{\hat{g}_j}$$

2. A second transformation, $\hat{\boldsymbol{\Gamma}}_2$, is estimated starting from $\hat{\boldsymbol{\Gamma}}_1$, attempting to recover the displacement along the gravity direction. This is done by transforming $\pi_j$ points and normals to the $\pi_i$ reference frame. Then, for both maps, only points that provide meaningful information to estimate the vertical translation are selected (e.g. those whose associated normal vertical component is dominant). For each selected point, $\mathbf{p}_k \in \pi_j$, its nearest neighbour in the other map is retrieved, $\mathbf{nn}_k \in \pi_i$. The magnitude of the vertical translation is computed as:

$$\Delta_Z = \frac{\sum_k w_k \left[\left(\hat{\boldsymbol{\Gamma}}_1 \mathbf{p}_k - \mathbf{nn}_k\right) \cdot \hat{g}_i\right]}{\sum_k w_k}$$

where $w_k$ weights points according to their distance to the sensor:

$$w_k = \frac{1}{1 + \|\mathbf{p}_k - \mathbf{c}_k\|^2 + \|\mathbf{nn}_k - \mathbf{c}_{\mathbf{nn}_k}\|^2}$$

where $\mathbf{c}_k$ and $\mathbf{c}_{\mathbf{nn}_k}$ are the positions of the sensor when $\mathbf{p}_k$ and $\mathbf{nn}_k$ were observed, respectively. The final value of $\hat{\boldsymbol{\Gamma}}_2 = \begin{bmatrix} \mathbf{R}_2 \ \mathbf{t}_2 \end{bmatrix}$, given $\hat{\boldsymbol{\Gamma}}_1 = \begin{bmatrix} \mathbf{R}_1 \ \mathbf{t}_1 \end{bmatrix}$ is then computed as:

$$\mathbf{R}_2 = \mathbf{R}_1$$
$$\mathbf{t}_2 = \mathbf{t}_1 - (\hat{g}_i \Delta_Z)$$

3. The final relative pose between both maps, $\boldsymbol{\Gamma}_j^i$ is retrieved by performing a point-plane ICP local registration over the two maps, initialized with $\hat{\boldsymbol{\Gamma}}_2$ (Fig. 1, bottom-left).

## 2.2   Local Matcher

Considering the transitive properties of 3D rigid transformations, two maps can be placed in the same reference frame even if no global match has been explicitly detected between them. For example, if $\pi_i$ is matched with $\pi_j$ and $\pi_j$ is matched with $\pi_k$, but there is no global match between $\pi_i$ and $\pi_k$, the relative pose between these maps can be expressed as $\boldsymbol{\Gamma}_k^i = \boldsymbol{\Gamma}_j^i \boldsymbol{\Gamma}_k^j$.

To benefit from this property, we model the global alignment problem as a directed graph, where nodes correspond to map poses and edges to matches, connecting the two matched maps. This representation allows knowing if two maps are in the same reference frame (e.g. it exists a path in the graph that connects them) and, thus, deciding which matcher to use to align them.

When two maps are in the same reference frame, and no match between them exists, instead of attempting to compute a global match, the local matcher runs a point-plane ICP initialized with their deduced relative pose. By doing so, two major improvements are achieved: (1) computation time is reduced, since there is no need for 2D matching and 3D transformation computation. (2) recall rate increases considerably, since we directly rely on 3D points matching, and it cannot be ensured that common 2D keypoints will be found in both maps.

## 2.3   Match Validation

Given a potential match, $\boldsymbol{\Gamma}_j^i$, its correctness is evaluated exploiting the implicit empty space observed by the sensor in a polar representation. To do so, given a 3D point in local coordinates, $\mathbf{p} \in \pi_i$, its corresponding position in a polar range map, $\mathbf{p}_\alpha = [\mathbf{u}_p \mathbf{v}_p]^T$, and its associated depth value, $d_p$, are computed as:

$$\mathbf{u}_p = \left( \frac{\mathrm{w}}{2\pi} \left( \mathrm{atan2}\left(\mathbf{p}_y, \mathbf{p}_x\right) + \pi \right) \right) \bmod \mathrm{w}$$

$$\mathbf{v}_p = \left( \frac{\mathrm{h}}{\pi} \left( \mathrm{asin}\left( \frac{\mathbf{p}_z}{\|\mathbf{p}_x + \mathbf{p}_y\|} \right) + \frac{\pi}{2} \right) \right)$$

$$d_p = \|\mathbf{p}\|$$

where w and h are the polar range map width and height, respectively.

For all points in $\pi_i$, their polar coordinates and depth values are computed and, when two of them fall in the same position, the one with smaller depth gets stored. It can then be stated that, given a direction and its associated $d_p$ value, all points laying on the ray and closer than $d_p$ to the origin are empty.

To validate a match, points $Q^j = \{\mathbf{q}_k\} \in \pi_j$, are moved into $\pi_i$ reference frame using the associated transformation, $Q^i = \boldsymbol{\Gamma}_j^i Q^j$. Then, for each single point, $\mathbf{q}_k^i$ its polar coordinates in $\pi_i$ range map are computed and the associated $d_q^i$ value is compared w.r.t. the stored one, $d_p$. If no $d_p$ value is available (the scanner did not see any point in this coordinate), $\mathbf{q}_k^i$ is ignored. Otherwise, it is considered to be overlapping with $\pi_i$. $\mathbf{q}_k$ is considered to be correct if $d_q^i - d_p > -\theta_v$ and incorrect otherwise, where $\theta_v$ is the tolerance to accept close-enough points (typically in the order of magnitude of the sensor's accuracy).

This process is, then, repeated inversely (projecting points of $\pi_i$, into $\pi_j$'s range map) and the final confidence of the match is expressed as the ratio between correct points w.r.t. the number of overlapping points. The match is accepted if this confidence is greater than a given threshold (typically 95%).

## 2.4   Global Optimization

The pose graph problem formulation can be optimized using a graph solver (e.g. g$^2$o [10]), benefiting from the intrinsic sparse nature of the graph (typically a single map does not match all the others). Formally, in our graph, $\mathcal{G} = (V, E)$, nodes, $V$, are defined as the $\mathbb{SE}3$ poses of maps with small updates in $\mathfrak{se}3$:

$\tilde{\boldsymbol{\Gamma}}_i = \exp\left(\boldsymbol{\gamma}_i^{\wedge}\right)\boldsymbol{\Gamma}_i$. Edges, $E$, are binary relations that express point-plane errors between a pair of nodes as:

$$
\mathbf{e}_i = \begin{bmatrix} \left(\tilde{\boldsymbol{\Gamma}}_i\mathbf{p}_{i_k} - \tilde{\boldsymbol{\Gamma}}_j\mathbf{p}_{j_k}\right)\tilde{\mathbf{R}}_i\bar{\mathbf{n}}_{i_k} \\ \left(\tilde{\boldsymbol{\Gamma}}_i\mathbf{p}_{i_k} - \tilde{\boldsymbol{\Gamma}}_j\mathbf{p}_{j_k}\right)\tilde{\mathbf{R}}_j\bar{\mathbf{n}}_{j_k} \end{bmatrix}
$$

where $\tilde{\boldsymbol{\Gamma}}_i$ and $\tilde{\boldsymbol{\Gamma}}_j$ are the estimated poses of maps $\pi_i$ and $\pi_j$, respectively, $\langle\mathbf{p}_{i_k}, \bar{\mathbf{n}}_{i_k}\rangle$ and $\langle\mathbf{p}_{j_k}, \bar{\mathbf{n}}_{j_k}\rangle$ are compatible points and normals, associated between maps $\pi_i$ and $\pi_j$, respectively, and $\tilde{\mathbf{R}}_i$ and $\tilde{\mathbf{R}}_j$ are the rotational parts of $\tilde{\boldsymbol{\Gamma}}_i$ and $\tilde{\boldsymbol{\Gamma}}_j$, respectively.

## 3   Results

To validate the proposed technique, various environments were acquired and four are reported here (shown in Fig. 2). The first three correspond to the evaluation areas of the Microsoft Indoor Localization Competition (IPSN) of 2015, 2016 and 2017. IPSN-2015, is a large exhibition area with different rooms, where walls completely occlude some maps from the others (19 maps). IPSN-2016, is a two-level exhibition area where floors do not overlap (25 maps). IPSN-2017, is also a two-level exhibition area where different floors overlap and where the lower level is not fully visible from the upper one (20 maps). ISF, is another exhibition area with the particularity of being highly symmetric: only furniture can help to distinguish the side of the building the sensor is in (21 maps).

The algorithm was executed on a Windows 7 computer with an Intel Xeon ES-2650 CPU at 2.00 GHz with 32 GB of RAM.

For each building a ground truth point cloud was available, where maps had been manually registered and validated. Our test consisted on running the full pipeline and comparing the final relative poses between maps w.r.t. the ground truth ones. In each execution, precision and recall rates were computed for the global matching technique at the first iteration (selecting only the best match between a pair of maps and accepting it if the validation test was passed) and at the 10th iteration (if best match was not accepted, retrieving the next best ones and accepting them only if the validation test was passed). For this analysis, we considered *true positives* the accepted matches where the relative poses between maps were, at most, $10\,\text{cm}/1°$ wrong w.r.t. the ground truth ones and *false positives* the others. *False negatives* are the non identified matches between pairs that, in the ground truth, were overlapping 50% or more.

The overlap between two maps, $\pi_i$ and $\pi_j$, in their ground truth poses, is the percentage of points in $\pi_i$ whose nearest neighbor in $\pi_j$ is closer than $50\,\text{cm}$.

Figure 3 shows the achieved results w.r.t. the validator threshold (minimum ratio of correct reprojected points between maps to accept the match). Notice how, for the first three datasets, 100% precision is achieved with relatively low threshold values. The fourth one, however, requires 95% threshold to properly disambiguate the extreme symmetries of the environment.

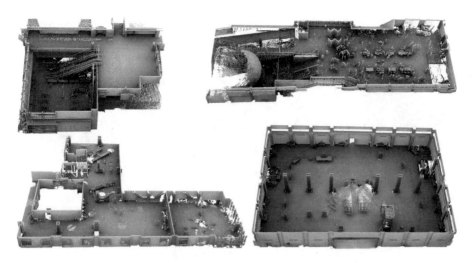

**Fig. 2.** Ground truth point clouds of the buildings used for evaluation. (top-left) IPSN-2017. (top-right) IPSN-2016. (bottom-left) IPSN-2015. (bottom-right) ISF.

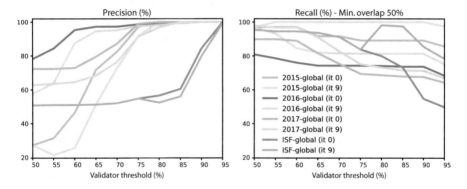

**Fig. 3.** Precision and recall for the evaluation datasets with 1 and 10 global matching iterations w.r.t. the validator threshold.

The system was also evaluated when local match detection was enabled. Figure 4 shows the matches selected using the global matcher with 10 iterations w.r.t. the ones selected using the local matcher together with the global matcher. Results shown in Table 1 make evident the benefits of enabling this feature: execution times get considerably reduced (around one order of magnitude), whilst recall rates and selected matches count are significantly boosted. In all the experiments, precision remained at 100% when using a 95% validator threshold. Interestingly, increasing the number of iterations for the global matcher increases execution time considerably, whilst the effect on recall and precision is less significant. Given this observation, once the graph is fully connected, it makes more sense to switch to the local matcher to enrich the final graph with missing edges.

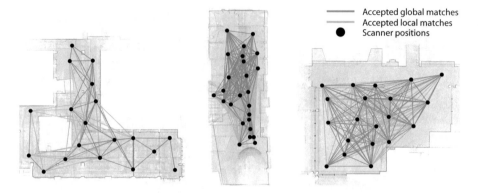

**Fig. 4.** Matches retrieved using the local matcher in the Microsoft Indoor Localization Competition datasets (2015, 2016 and 2017, from left to right, respectively).

**Table 1.** Number of matches selected, recall and execution time, setting the validator threshold to 95% and considering that two maps should be matched if their overlapping is, at least, 50%. Notice that the number of selected matches considers also the ones with a lower overlapping (not taken into account for computing recall values).

| Building | Strategy | Iterations | Selected (#) | Recall (%) | Time (s) |
|---|---|---|---|---|---|
| IPSN-2015 | Global | 1 | 30 | 91.67 | 5.902 |
| IPSN-2015 | Global | 10 | 40 | 97.56 | 44.357 |
| IPSN-2015 | Global + Local | 2 | 69 | 98.57 | 4.901 |
| IPSN-2016 | Global | 1 | 124 | 69.94 | 13.247 |
| IPSN-2016 | Global | 10 | 140 | 75.27 | 44.357 |
| IPSN-2016 | Global + Local | 2 | 262 | 95.27 | 13.789 |
| IPSN-2017 | Global | 1 | 81 | 65.08 | 8.588 |
| IPSN-2017 | Global | 10 | 85 | 66.93 | 25.743 |
| IPSN-2017 | Global + Local | 2 | 180 | 97.3 | 9.863 |
| Exhibition | Global | 1 | 104 | 59.05 | 17.849 |
| Exhibition | Global | 10 | 164 | 80.38 | 60.836 |
| Exhibition | Global + Local | 2 | 200 | 95.24 | 13.105 |

Finally, to prove the potential of our global matcher in SLAM approaches, and also in outdoor environments, we processed the tracks of the Kitti dataset [7]. Using the ground truth trajectory provided, we fused the point clouds retrieved by the sensor into maps that covered segments of 10 m each. Figure 5 shows the overlaps between these maps (left) w.r.t. the score associated to the 2D matches retrieved by our technique (right). Notice how loop closures correspond to high overlap areas, that clearly map into high score values in our matcher.

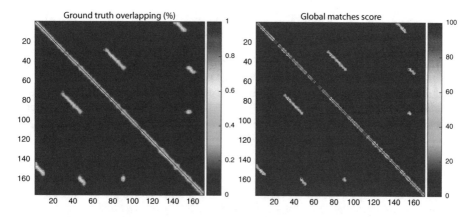

**Fig. 5.** Sample Kitti dataset track (track #5). (left) Overlap between *maps* along the car trajectory. (right) Associated 2D match score retrieved by our technique.

## 4    Conclusions and Future Works

We have presented a point cloud global registration technique for mapping applications. Taking advantage of a fast 2D matcher and a robust match validator, we have shown that the technique allows solving, in a fully automatic way, complex environments. Its potential for SLAM approaches has also been illustrated, where loop closures can be identified by the 2D match scores.

Future works include two main interventions in the technique: (1) defining a match validator that does not consider only pairwise local matches, but the global graph including all maps and (2) extending the technique to SLAM approaches in an effective way. The first goal aims to benefit from the information stored in the global reconstruction graph, where the correctness of a potential match could be dissambiguated by exploiting the already selected ones. The second goal aims to benefit from the temporal information associated to a continuous acquisition. This way, by modeling the drift associated to the incremental pose estimation of SLAM approaches, it could be decided which scans are worth matching within, and which matches make sense w.r.t. the overall trajectory of the sensor.

## References

1. Bay, H., Ess, A., Tuytelaars, T., Gool, L.V.: Speeded-up robust features (surf). Comput. Vis. Image Underst. **110**(3), 346–359 (2008)
2. Bosse, M., Zlot, R.: Place recognition using keypoint voting in large 3d lidar datasets. In: ICRA (2013)
3. Calonder, M., Lepetit, V., Ozuysal, M., Trzcinski, T., Strecha, C., Fua, P.: Brief: computing a local binary descriptor very fast. IEEE Trans. Pattern Anal. Mach. Intell. **34**(7), 1281–1298 (2012)

4. Collier, J., Se, S., Kotamraju, V., Jasiobedzki, P.: Real-time lidar-based place recognition using distinctive shape descriptors. In: SPIE Unmanned Systems Technology (2012)
5. Cummins, M., Newman, P.: Appearance-only slam at large scale with fab-map 2.0. Int. J. Robot. Res. **30**(9), 1100–1123 (2011)
6. Filliat, D.: A visual bag of words method for interactive qualitative localization and mapping. In: Proceedings 2007 IEEE International Conference on Robotics and Automation, pp. 3921–3926, April 2007
7. Geiger, A., Lenz, P., Urtasun, R.: Are we ready for autonomous driving? the kitti vision benchmark suite. In: CVPR (2012)
8. Ho, K.L., Newman, P.: Detecting loop closure with scene sequences. Int. J. Comput. Vis. **74**(3), 261–286 (2007)
9. Košecká, J., Li, F., Yang, X.: Global localization and relative positioning based on scale-invariant keypoints. Robot. Auton. Syst. **52**(1), 27–38 (2005)
10. Kümmerle, R., Grisetti, G., Strasdat, H., Konolige, K., Burgard, W.: G2o: a general framework for graph optimization. In: ICRA (2011)
11. Lee, C.H., Varshney, A., Jacobs, D.W.: Mesh saliency. ACM Trans. Graph. **24**(3), 659–666 (2005)
12. Lowe, D.G.: Object recognition from local scale-invariant features. In: ICCV (1999)
13. Lynen, S., Bosse, M., Furgale, P., Siegwart, R.: Placeless place-recognition. In: 3DV (2014)
14. Magnusson, M., Andreasson, H., Nüchter, A., Lilienthal, A.J.: Automatic appearance-based loop detection from three-dimensional laser data using the normal distributions transform. J. Field Robot. **26**(11–12), 892–914 (2009)
15. Muhammad, N., Lacroix, S.: Loop closure detection using small-sized signatures from 3d lidar data. In: 2011 IEEE International Symposium on Safety, Security, and Rescue Robotics, pp. 333–338, November 2011
16. Newman, P., Sibley, G., Smith, M., Cummins, M., Harrison, A., Mei, C., Posner, I., Shade, R., Schroeter, D., Murphy, L., Churchill, W., Cole, D., Reid, I.: Navigating, recognizing and describing urban spaces with vision and lasers. Int. J. Rob. Res. **28**(11–12), 1406–1433 (2009)
17. Paul, R., Newman, P.: Fab-map 3d: Topological mapping with spatial and visual appearance. In: ICRA (2010)
18. Rusu, R.B., Cousins, S.: 3d is here: Point cloud library (PCL). In: ICRA (2011)
19. Rusu, R.B., Blodow, N., Beetz, M.: Fast point feature histograms (FPFH) for 3D registration. In: ICRA (2009)
20. Sánchez-Belenguer, C., Vendrell-Vidal, E.: An efficient technique to recompose archaeological artifacts from fragments. In: VSMM (2014)
21. Schindler, G., Brown, M., Szeliski, R.: City-scale location recognition. In: CVPR, pp. 1–7 (2007)
22. Shi, J., Tomasi, C.: Good features to track. In: CVPR (1994)
23. Sipiran, I., Bustos, B.: Harris 3D: a robust extension of the harris operator for interest point detection on 3D meshes. Vis. Comput. **27**(11), 963–976 (2011)
24. Steder, B., Ruhnke, M., Grzonka, S., Burgard, W.: Place recognition in 3D scans using a combination of bag of words and point feature based relative pose estimation. In: IROS (2011)
25. Steder, B., Rusu, R.B., Konolige, K., Burgard, W.: Point feature extraction on 3D range scans taking into account object boundaries. In: ICRA (2011)
26. Taddei, P., Sánchez, C., Rodríguez, A.L., Ceriani, S., Sequeira, V.: Detecting ambiguity in localization problems using depth sensors. In: 3DV (2014)

27. Tombari, F., Salti, S., Di Stefano, L.: Unique signatures of histograms for local surface description. In: ECCV (2010)
28. Yao, J., Ruggeri, M., Taddei, P., Sequeira, V.: Robust surface registration using n-points approximate congruent sets. EURASIP J. Adv. Signal Process. (2011)

# Cluster ICP: Towards Sparse
# to Dense Registration

Mohamed Lamine Tazir[1(✉)], Tawsif Gokhool[2], Paul Checchin[1],
Laurent Malaterre[1], and Laurent Trassoudaine[1]

[1] Université Clermont Auvergne, CNRS, SIGMA Clermont, Institut Pascal,
63000 Clermont-Ferrand, France
tazir.med@gmail.com
{paul.checchin,laurent.malaterre,laurent.trassoudaine}@uca.fr
[2] Université de Picardie Jules Vernes, Amiens, France
tawsif.gokhool@u-picardie.fr

**Abstract.** Normal segmentation of geometric range data has been a
common practice integrated in the building blocks of point cloud registration. Most well-known point to plane and plane to plane state-of-
the-art registration techniques make use of normal features to ensure a
better alignment. However, the latter is influenced by noise, pattern scanning and difference in densities. Consequently, the resulting normals in
both a source point cloud and a target point cloud will not be perfectly
adapted, thereby influencing the alignment process, due to weak inter
surface correspondences. In this paper, a novel approach is introduced,
exploiting normals differently, by clustering points of the same surface
into one topological pattern and replacing all the points held by this
model by one representative point. These particular points are then used
for the association step of registration instead of directly injecting all the
points with their extracted normals. In our work, normals are only used
to distinguish different local surfaces and are ignored for later stages of
point cloud alignment. This approach enables us to overcome two major
shortcomings; the problem of correspondences in different point cloud
densities, noise inherent in sensors leading to noisy normals. In so doing,
improvement on the convergence domain between two reference frames
tethered to two dissimilar depth sensors is considerably improved leading
to robust localization. Moreover, our approach increases the precision as
well as the computation time of the alignment since matching is performed on a reduced set of points. Finally, these claims are backed up
by experimental proofs on real data to demonstrate the robustness and
the efficiency of the proposed approach.

**Keywords:** Registration · Dense to sparse · Selection · Clustering
Matching

## 1 Introduction

In a generic representation of the environment, point clouds can be viewed as a
collection of 3D point entities bearing a color or intensity information depending

© Springer Nature Switzerland AG 2019
M. Strand et al. (Eds.): IAS 2018, AISC 867, pp. 730–747, 2019.
https://doi.org/10.1007/978-3-030-01370-7_57

on the acquisition sensor (LiDAR, RGBD or time of flight cameras). To obtain a more meaningful information about the semantic structure of the environment, it is more instructive to rather consider a collective set of points representing the same surface. From there, several surface indices can be extracted such as their normals, curvatures and region bounds for example. Moreover, due to limitation of the field of view of 3D sensors, coupled with the complex geometry of the scanned surrounding, registration methods are required to be more robust in order to deal with data taken from large viewpoints as well as different sensor resolution.

Extraction of surface normals is a double-edged tool, which guarantees good results if accurately exploited, but can also lead to divergence of the alignment process if badly used. Since normal features are based essentially on the estimation of neighbouring points attributed to the same surface in general, they are however subjected to sensor noise, resolution and scanning patterns. Consequently, this reverberates on the alignment process due to weak inter surface correspondences between the source and the target point cloud. Furthermore, with the advancement of 3D sensors in the market, the problematic of sparse to dense registration has emerged out [1]. In this trend, software packages such as PCL [25] has made the identification and the treatment of the above mentioned problem more accessible. Eventually, an elegant solution provided by a successful sparse to dense point cloud alignment, results in interesting robotics applications such as the case of a monocular camera localization in a 3D model [7,33] or augmenting the environment with more consistent data as in [17].

In order to support the claims stated above, an illustration of sparse to dense registration is given in Fig. 1. A dense point cloud is obtained from a 3D LiDAR Leica P20[1] scanner, whilst the sparser one is extracted from an HDL-32E Velodyne[2]. Figures 1(g), (h), (i), (j) are samples of various places in a scene. The 3D points of the source and target clouds are represented in blue and green respectively, whilst their normals are in white and red. Because of the large discrepancies in density between the two point clouds, registration methods based on the classical point-to-point ICP metrics fail to provide a good pose estimate. The difficulty lies in the fact that there are no direct correspondences between the source and the target point clouds. Moreover, these figures also depict the dissimilarity between normals pertaining to the same surface, which theoretically should have the same orientations. This change is due mainly to the presence of wide amount of noise, pattern scanning, distortion and varying resolutions. This is the major problem of the methods that use geometric features according to [13,28].

In this paper, a novel registration method is introduced, exploiting normals differently. It does not seek for each point its nearest neighbor sharing the same normal, nor introducing normals in the error function or in the minimization process. To remediate for the disturbances in the registration framework, hence inaccuracies in the final result, a voxelization is performed on both clouds to

---

[1] Leica P20: http://leica-geosystems.com/.
[2] Velodyne LiDAR: http://velodynelidar.com/.

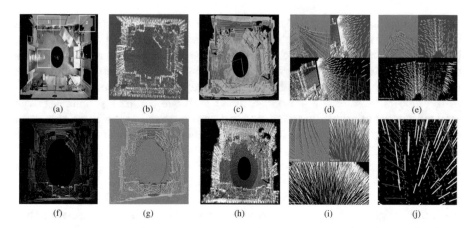

**Fig. 1.** Dense to sparse registration: (a) point cloud obtained from the Leica P20 LiDAR with 88 556 380 points; (f) point cloud obtained from an HDL-32E Velodyne with 69 984 points; (b) and (g) are their corresponding point clouds with normal vectors; (c) registration result of point cloud (a) and (f) using our proposed method; (h) normal vectors corresponding to (c); (d), (e) and (i) are exploded views of places indicated in (a); (j) is a close up view of (i).

maintain the topological details of the scene. Then for each voxel, a normal-based classification of points is done. Thereafter, only one point of each local surface is maintained for the association step. This process results into points which offers better compatibility in terms of surface representation between the source and the target point clouds. This approach is not about points sampling, but rather an improved selection of points is achieved for the matching stage. Thereon, the process evolves in a classical ICP like framework for pose estimation.

The contributions of this paper are threefold. The first main contribution is to perform an efficient voxelization method operating on top of a clustering technique for electing one representative point for each local surface. This process aims to establish two sets of points which are most likely to be matched, hence providing good correspondences. It consists of three stages: voxelization, clustering and matching. As a result, it:

1. reduces the amount of data to be processed during the matching phase,
2. improves the matching robustness by allowing only the association of compatible points,
3. avoids wrong associations that decrease alignment accuracy,
4. does not use unnecessary points that do not provide further information more than the ones used,
5. improves convergence and accuracy simultaneously.

The second contribution is that the proposed method is totally independent of the density (number of points, scanning resolution) of the two clouds, scanning patterns (nature of sensors). It takes as input point clouds of different resolution, gathered by different sensors, or with the same sensor. It is also based only on the geometric characteristics of the points, which makes it independent of weather and illumination conditions. Thirdly, normals are computed once before starting the process and are used only to distinguish the different local surfaces. They are not used in the alignment process.

The rest of this paper is organized as follows: in Sect. 2, an overview of the state of the art of registration methods is given. Section 3 details the proposed method. This is followed by experiments and a comparison with the state-of-the art methods in order to evaluate the proposed approach. Finally,conclusions and suggestions of future works are discussed in Sect. 5.

## 2 Related Work

Registration algorithms assemble two representations of an environment in a single reference frame. The problem of registration has been dealt with extensively in several studies over the last 25 years. This started with geometric approaches leading to the appearance of the Iterative Closest Point (ICP) algorithm [4,8]. ICP is used to calculate the optimal transformation fitting two point clouds by a two-step process: matching of points and minimizing a metric describing the misalignment [19]. These two-steps iterate to minimize the matching error and thus improve alignment. In the literature, two main groups of registration methods are identified:

- feature-based methods (approaches based on features extraction);
- dense methods (approaches exploit all the points in the cloud).

### 2.1 Feature-Based Approaches

Feature-based methods are generally used in outdoor environments [17]. They are based on the use of features, which may be points that are easily identified by their apparent character (position, local information contents, mathematical definition, etc.) with respect to the other points. A good feature requires stability and distinctiveness [28]. In other words, detected features should be consistent in all the frames. They should be robust to noise and invariant to rotation, perspective distortion and changes of scale [9,11,28].

Features extraction from point cloud representation is well documented literature. One can find the 3D Scale Invariant Feature Transform (3DSIFT), which is an extension of the 2D version proposed by Lowe in 1999 [15]. The 3D version was adapted by the PCL [25] community using the curvature of points instead of the intensity of pixels [12]. The method uses a pyramidal approach to reach the scale invariance characteristic of features. To achieve invariance against rotation, it assigns orientations to keypoints. This adds to an incomplete list of features such as FPFH [23], VFH (Viewpoint Feature Histogram) [24], CVFH (clustered viewpoint feature histogram) [2] to name a few.

However, feature extraction techniques are often cumbersome to determine and pose a problem to real-time applications [9], making them unsuitable for applications that require efficiency. Furthermore, the necessity of very dense clouds are required in order to obtain good features, which compromises with the use of sparse clouds [1,28,31,34]. More importantly, these methods are environment specific [6], which may result in the rejection of good data [20].

## 2.2   Dense Approaches

Dense approaches make use of all the points from both clouds, and require an initial guess (transformation) between the two clouds, which makes them sensitive to wrong initialization [9,27,35]. Despite the use of all the points, these methods are generally faster than feature-based approaches [27].

Dense techniques are however well adapted to a classical ICP framework. As pointed out by Pomerleau [21], its easy implementation and simplicity, are both its strength and its weakness. This led to the emergence of many variants of the original solution, adapted in many ways, throughout the years. At the very outset, Chen [8], improved the standard ICP by using point-to-plane metric instead of the Euclidean distance error. This approach takes advantage of surface normal information to reject wrong pairing. However, this approach fails when dealing with clouds of different densities, since normals computation are affected by the change in resolution, presence of noise and distortion [10,13].

The Normal Distributions Transform (3D-NDT) [18] discretizes the 3D points with their normals in cells, where each one is modeled by a matrix, representing the probability of occupation of its points (linear, planar and spherical). Then, a non-linear optimization is performed to calculate the transformation between the two clouds. However, according to [10], the NDT is not suitable for systems with low computing power capability.

An efficient approach for dense 3D data registration was presented in [26]. This probabilistic version of ICP called Generalized ICP (GICP) is based on a Maximum Likelihood Estimation (MLE) probabilistic model. It exploits local planar patches in both point clouds, which leads to plane-to-plane concept. Since this algorithm is point-to-plane variant of ICP, it has similar drawbacks, especially those related to normals computation. For instance in [13], it is shown that the non-uniform point densities cause inaccurate estimates, which degrade the performance of the algorithm. Moreover, Agamennoni [1] affirmed that the GICP does not work well in outdoor unstructured environment.

Serafin [27] extended the GICP algorithm by using the normals in the error function and in the selection of correspondences, which according to the authors, increases the robustness of the registration.

Our approach, called CICP for Cluster Iterative Closest Point, uses an (NDT and NICP)-like representation, however, it is different from the NDT in the way it uses the points of each voxel to determine local surfaces and get one representative point from each local surface to the matching process. In contrast, NDT computes a Gaussian distribution in points of each voxel using the vicinity of each point. Whereas, NICP uses an image projection of the voxel grid representation to compute statistics, and considers each point with the local features of the surrounding surface. These features, namely normal and curvature, are calculated for each point from its neighboring points, with a computational complexity of $\mathcal{O}(K \times N)$, where $K$ is the number of the neighboring points used to compute each normal and $N$ the total number of points. Additionally, these features are used later in the process of point matching between the two clouds, as opposed to our method, that does not use normals in the matching process. Because of the difference in density, pattern scanning, and presence of noise, will lead to noisy normals and, hence, inaccurate results.

## 3   Proposed Method

In this paper, a novel registration method exploiting normals is introduced. We adopt the Rusinkiewicz [22] decomposition and propose a new selection strategy, which aims to improve the pairing process. Figure 2 illustrates the pipeline of the proposed method.

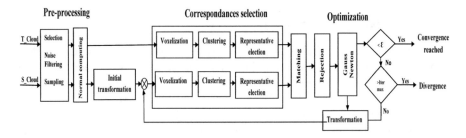

**Fig. 2.** Overview of the CICP pipeline. Given two point clouds, CICP starts by computing the surface normals of the two clouds. It looks for points sharing the same local properties, and then elects one representative point from each local cluster. This election process is based on 3D position of points and their normals. It consists of three sub-tasks: (1) Voxelization: a set of 3D cubic regions (voxels) is generated where all voxel points have very close spatial positions. (2) Clustering: classify all points of each voxel according to their normals. (3) Matching: once this grouping step is completed, the last task consists in selecting one point from each cluster (local surface) in each voxel. Representative points serve as candidates for correspondence process. As a result, few points are used in the matching process, but which are most likely to be associated, thereby, improving on convergence and accuracy simultaneously.

### 3.1   Surface Normal Segmentation

CICP starts with the estimation of normals of the source and target point clouds using Principal Component Analysis (PCA) [14] follows:

$$C = \frac{1}{k}\Sigma_{i=1}^{k}(p_i - \bar{p})^T(p_i - \bar{p}), \tag{1}$$

where, $C$ is the covariance matrix of the nearest neighbors (NNs), $k$ is the number of considered nearest neighbours $p_i$, and $\bar{p}$ is its corresponding centroid along the tangent plane.

Thereafter, the target cloud is subdivided into small voxels. Points belonging to each voxel are subjected to a classification process based on their normals, giving rise to different groups of points, according to the geometric variation of each voxel. Each group of points represents a local surface since they share the same normal vector. A single point is chosen from a local surface extract to be used for the matching process. The closest point to the centroid of each local surface is elected a winner.

Similarly, the source point cloud is first transformed into the reference frame of the target cloud using a pose estimate before undergoing a similar process; voxelization, normals-based classification, designation of point's representatives. At the end of these steps, the method results into two improved sets of points from the corresponding clouds. Each set contains the most probable points to match with the points of the second set (this is more particularly in the overlapping area of the two clouds, as it reflects the same geometry seen from two different viewpoints).

## 3.2   CICP Matching Pipeline

The main contribution of this paper is the proposal of a new selection strategy. As mentioned above, instead of matching point-to-point as the classical ICP variants, points pass through an election process, which gives rise to one representative point for each small region. These representatives appear as the most likely points to be matched between each other. These good matches ultimately result in an accurate motion between the two clouds (shown in the results section). This election process is based on 3D position of points and their normals. It consists of three sub-tasks: (1) voxelization, (2) clustering and (3) representative election. The first task performs a spatial grouping which attempts to preserve the topological information based on the 3D position of the points. A set of 3D cubic regions (voxels) is generated where all points within the voxel have very close spatial positions. The second task bundles all points of each voxel based on their normals. Once this grouping is done, we perform the last task, which selects one point for each cluster (local surface) in the voxel for the matching process. Algorithm 1 depicts the workflow of CICP.

**Voxelization.** It is applied in order to maintain the topological details of the scrutinized surface. As normal computation depends on the number of neighbouring points and as the resolution of points of the two clouds is different, voxelization with the same voxel size aims to generate equivalent local regions in the two clouds. A common criterion of comparison now becomes feasible. Therefore, the voxel size parameter is of paramount importance for our technique and it should be chosen carefully in order to keep the fundamental characteristics of both point clouds; be it dense or sparse with topological details. A voxel grid with cell size d is generated, where the following set of rules are verified:

**Definition 1 (Sparse Cloud).** *A sparse cloud is a cloud* $C = (V, P)$ *in which:* $|P| = \mathcal{O}(|V|)$.

**Definition 2 (Dense Cloud).** *A dense cloud is a cloud* $C = (V, P)$ *in which:* $|P| = \mathcal{O}(k * |V|)$, *with* $k > 2$.

whereby,

$V$: set of voxels, $P$: set of points, $\mathcal{O}$: proportionality operator.

Definitions 1 and 2 are proposed to frame the notions of sparsity and density of point clouds. The voxel size is set according to the number of points in the sparse cloud, so that each voxel contains at least one point. This choice ensures a significant difference in density between the two clouds. A dense cloud, in our case, contains at least twice as many points as the sparse cloud. Otherwise, they are considered as equivalent.

At the beginning, the procedure applies a bounding box to the entire sparse cloud by finding the minimum and maximum positions of points along the three axes $X$, $Y$ and $Z$. The number of voxels for this bounding box is determined by the number of points and the voxel size is deduced. The same procedure is applied to the dense cloud.

1. Voxel assignment: each voxel is identified by a unique linear index. If $i$, $j$, $k$ represent the voxel indices in the $X$, $Y$, $Z$ dimensions, respectively, $numDivX$, $numDivY$ are the number of voxels along $X$ and $Y$ axes, the formula to encode the linear index is [29]:

$$idx = i + j \times numDivX + k \times numDivX \times numDivY \qquad (2)$$

According to (2), we assign an index $idx$ to each point. This relationship allows direct access to the desired voxel, thereby avoiding a linear search as in [32].

2. Voxel suppression: as the shape of the point cloud is arbitrary, the step of delimiting points by a bounding box creates many empty voxels which are later pruned out. Eventually, voxelization helps to filter noise from voxels where there is insufficient occupational evidence. An illustration of the described approach is given in Fig. 3.

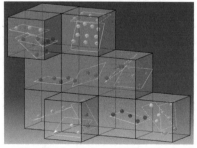
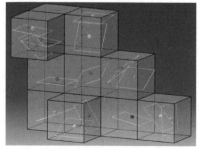

(a) Voxelized and clusterized point cloud     (b) Electing one point from each cluster for the matching phase

**Fig. 3.** Voxelized/normal-based clustering for matching process.

**Clustering.** The process of electing one point from each local surface makes them good candidates for point correspondence searching, thereby rejecting wrong matches impacting alignment accuracy. At first, all the "voxelized" points are taken and a classification method is applied to identify points belonging to the same surface. In our work, k-means clustering [3] is used as the classification technique based on the normal of each point. The appropriate number of clusters (local surfaces) within each voxel is determined using the Elbow method [29,30]. An illustration of the described approach is given in Fig. 3.

Grouping the point clouds using their normal aims at:

- improving the robustness of the matching step by only allowing the pairing of compatible points,
- reducing the amount of data to be processed during the matching stage.

**Matching.** The clustering process generates a reduced, but different number of points in both clouds. These two sets of points are used for matching. To boost up the matching process, an off shelf PCL [25] implementation of the $k$-d trees is used, whereby matching is achieved using $L_2$ norm. Outliers between the dense and sparse sets are handled using a suitable threshold.

### 3.3   Optimization Framework

In the case of a point-to-point metric, the error function to be minimized is given by:

$$E\left(x\right) = \sum_{i=1}^{N} \|T\left(\tilde{x}\right) p_i - q_i\|^2 \tag{3}$$

The localization problem of a sparse to a dense point cloud (or vice-versa) resolves to estimating the relative transformation $T(\tilde{x})$ between point clouds $\{p, q\} : \forall \{p_i, q_i\} \in \mathbb{R}^3$. The principle of rigid body motion is applied where the transformation of a point tethered to a coordinate frame represent the whole compact body motion. For any point pair lying on the body, metric properties such as distances and orientation are preserved. This kind of body motion, discussed subsequently forms part of the special euclidean group $\mathbb{SE}(3)$.

Inter-frame incremental displacement is further defined as an element of the Lie groups applied on the smooth differential manifold of $\mathbb{SE}(3)$ [5], also known as the group of direct affine isometries. Motion is parametrized as a twist (a velocity screw motion around an axis in space), denoted as $\mathbf{x} = \{[\boldsymbol{\omega}, \boldsymbol{v}] | v \in \mathbb{R}^3, \hat{\omega} \in so(3)\} \in se(3): \boldsymbol{\omega} = [\omega_x \ \omega_y \ \omega_z], \boldsymbol{v} = [v_x \ v_y \ v_z]$, with $so(3) = \{\hat{\omega} \in \mathbb{R}^{3 \times 3} | \hat{\omega} = -\hat{\omega}^\top$, where $\omega$ and $v$ are the angular and linear velocities respectively. The reconstruction of a group action $\hat{\mathbf{T}} \in \mathbb{SE}(3)$ from the twist consists of applying the exponential map using Rodriguez formula [16].

Equation (3) is solved iteratively in a Gauss Newton fashion, where at each iteration, a new error $E$ and a new Jacobian matrix $J(0)$ are computed in order to obtain the update $x$ by:

$$x = -\left(J(0)^T J(0)\right)^{-1} J(0)^T e(x) \tag{4}$$

and the rigid transformation is updated as follows:

$$\hat{T} \longleftarrow \hat{T}T(x) \tag{5}$$

Minimization is stopped when the error: $\| e \|^2 < \alpha$ occurs, or when the calculated increment becomes too small: $\|x\|^2 < \varepsilon$, where $\alpha$ and $\varepsilon$ are predefined stop criteria.

## 4  Results

Our CICP approach is implemented in C++ without code optimization and our algorithm is thoroughly evaluated by conducting multiple experiments. The computational efficiency of the algorithm is beyond the scope of this paper. We rather focus on the methodology. The proposed method does not require any knowledge about the external orientation of the sensors at the time of acquisition, their position and orientation are estimated by the algorithm, the only requirement is that the two clouds share a tolerable overlap.

The experimental is set up as shown in Fig. 4. The centre of the two sensors; Velodyne HDL32 and that of the Leica P20 are perfectly superimposed with the help of the STANLEY Cubix cross line laser. The velodyne is then physically displaced and rotated by known translations and rotations from the graduated set up in order to perturb the 6 degrees of freedom (dof) transformation. Data acquisition is then performed under different scenarios in order to exert our

---

**Algorithm 1.** CICP Algorithm.

---

**Input**: targetCloud, sourceCloud ; voxelSize, $\hat{T}$
**Output**: Optimal $T$
1  **Intialize:** NormalXYZ T_normals, S_normals; PointXYZ T_match, S_match
2  **begin**
3      T_normals = normalComputing (targetCloud)
4      S_normals = normalComputing (sourceCloud)
5      T_match = normalClustering (targetCloud, T_normals, voxelSize)
6      **while** (*iteration* < *iter*_max$\|\|x\| > \varepsilon$) **do**
7          sourceCloud = transform (sourceCloud, S_normals, $\hat{T}$)
8          S_match = normalClustering (sourceCloud, S_normals, voxelSize)
9          EstablishCorrespondences (T_match, S_match)
10         distanceRejection (distThreshold)
11         compute the Jacobian **J**
12         compute the error vector $e(x)$ (3)
13         compute the increment $x$ (4)
14         update the pose **T** (5)
15         iteration $\leftarrow$ iteration + 1
16     **end**
17     **return** $T$
18 **end**

---

**Fig. 4.** Experimental set up for data collection from Velodyne HDL32 (left) and Leica P20 (right) sensors.

CICP algorithm. Table 1 below summarizes the various experiments performed in a controlled environment. For each experiment, CICP is initialized at Identity, i.e. $x = [0, 0, 0, 0, 0, 0]$.

### 4.1  Dense-Sparse Registration with CICP

Two clouds are acquired with different sensors; the denser cloud produced by a 3D LiDAR Leica P20 laser scanner and the sparser cloud with an HDL-32E Velodyne LiDAR sensor. A Leica P20 generates very detailed and dense point clouds. Depending on the resolution chosen during the scanning process, these clouds can exceed 100 millions of points for a single scan. All computations are performed on a laptop with the following specifications; Intel Core i74800MQ processor, 2.7 GHz, and 32 GB of RAM. For reasons of computational resources, we perform a sampling process using [29] in order to reduce the number of points to the order of few millions without losing useful information. Figure 5(a), (d) illustrate the output of the sampling process with 986 344 and 2 732 783 points for the office and PAVIN[3] scenes, respectively.On the other hand, the HDL-32E Velodyne sensor generates sparse point clouds that do not exceed 70 000 points. This represents a ratio of 14 times between the two clouds from the first environment and a ratio of 40 times, for clouds of the second environment.

Figure 5 shows the registration process of such point clouds using the CICP method. On the left, the green cloud is from a LiDAR Leica P20 and the blue cloud is from an HDL32-E Velodyne. The corresponding results are shown on the right.

In order to verify the convergence of the optimization, a comparison between the two clouds at the start and the end of the registration process is recorded together with the convergence profile obtained from the evolution of the RMSE error as a function of the number of iterations to convergence (see Fig. 6). An exit loop condition is imposed on the translation ($10^{-3}$) and rotation rates ($10^{-4}$).

---

[3] PAVIN:       http://www.institutpascal.uca.fr/index.php/en/the-institut-pascal/ equipments.

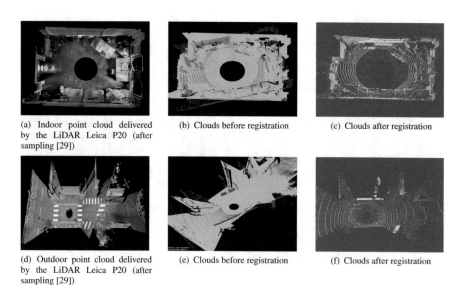

(a) Indoor point cloud delivered by the LiDAR Leica P20 (after sampling [29])

(b) Clouds before registration

(c) Clouds after registration

(d) Outdoor point cloud delivered by the LiDAR Leica P20 (after sampling [29])

(e) Clouds before registration

(f) Clouds after registration

**Fig. 5.** Registration results with CICP algorithm.

**Table 1.** CICP registration applied to mainly two compiled data sets; OFFICE and PAVIN. The resolutions of corresponding dense and sparse cloud are given along with the initial physical measured transformation from our set up given by the first row of each experiment, whilst the second row depicts the results output by our algorithm. Convergence is evaluated from the RMSE and the number of iterations required for full registration.

| Expt | Environment | # dense cloud | # sparse cloud | $t_x$ (mm) | $t_y$ (mm) | $t_z$ (mm) | $\theta_x$ (°) | $\theta_y$ (°) | $\theta_z$ (°) | $RMSE$ (m) | # iter. |
|---|---|---|---|---|---|---|---|---|---|---|---|
| 1 | Office 1 | 411 924 | 69 952 | 150.0 | 170.0 | 35.0 | 5.0 | 0.0 | 0.0 | - | - |
| | | | | 155.0 | 172.9 | 34.8 | 4.7 | 0.4 | 0.1 | 0.0198 | 40 |
| 2 | PAVIN 1 | 665 260 | 67 488 | 0.0 | 30.0 | 200.0 | 5.0 | 5.0 | 0.0 | - | - |
| | | | | 0.2 | 29.8 | 191.9 | 4.8 | 4.8 | 0.1 | 0.0188 | 36 |
| 3 | Office 2 | 986 344 | 69 984 | 350.0 | 350.0 | 0.0 | 0.0 | 0.0 | 0.0 | - | - |
| | | | | 357.3 | 342.8 | 0.3 | 0.3 | 0.1 | 0.8 | 0.0199 | 39 |
| 4 | PAVIN 2 | 1 364 245 | 67 768 | 65.0 | 45.0 | 200.0 | 0.0 | 7.0 | 5.0 | - | - |
| | | | | 64.1 | 45.7 | 200.9 | 0.3 | 6.9 | 4.9 | 0.0184 | 34 |
| 5 | Office 3 | 2 550 564 | 69 728 | 20.0 | 110.0 | 70.0 | 10.0 | 5.0 | 0.0 | - | - |
| | | | | 21.7 | 111.1 | 66.9 | 10.1 | 4.2 | 0.1 | 0.0191 | 39 |
| 6 | PAVIN 3 | 3 218 879 | 67 936 | 0.0 | 0.0 | 0.0 | 10.0 | 10.0 | 0.0 | - | - |
| | | | | 0.5 | 0.1 | 0.3 | 9.4 | 9.8 | 0.2 | 0.0184 | 55 |
| 7 | Office 4 | 4 490 859 | 69 996 | 30.0 | 470.0 | 300.0 | 20.0 | 0.0 | 0.0 | - | - |
| | | | | 27.8 | 470.3 | 307.3 | 19.6 | 0.1 | 0.1 | 0.0190 | 57 |
| 8 | PAVIN 4 | 5 025 457 | 67 904 | 50.0 | 50.0 | 50.0 | 5.0 | 5.0 | 5.0 | - | - |
| | | | | 52.1 | 48.2 | 53.3 | 5.3 | 7.5 | 5.3 | 0.0152 | 56 |
| 9 | Office 5 | 7 076 192 | 69 760 | 50.0 | 50.0 | 50.0 | 5.0 | 5.0 | 5.0 | - | - |
| | | | | 55.2 | 50.6 | 53.9 | 5.5 | 4.2 | 4.6 | 0.0190 | 35 |
| 10 | PAVIN 5 | 19 615 433 | 67 488 | 0.0 | 50.0 | 200.0 | 10.0 | 0.0 | 5.0 | - | - |
| | | | | 0.1 | 49.8 | 200.3 | 9.7 | 0.1 | 5.2 | 0.0169 | 48 |

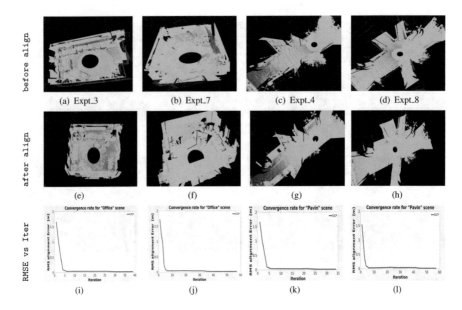

**Fig. 6.** CICP results applied to different data sets from various environments.

We run the algorithm for several indoor and outdoor scenes, with different viewpoints, as depicted in Table 1 and shown in Fig. 6. Each experiment is performed more than 20 times. For instance, for the experiment 1 (Expt_1), the displacement between the two clouds is $[150, 170, 35, 5, 0, 0]$, where the first three values correspond to the translation in millimeters and the last three to the rotation in degrees. As for the fourth experiment, the displacement is $[65, 45, 200, 0, 7, 5]$, which took 34 iterations for the algorithm to converge. From this table, three elements are identified that influence the registration results; inter-frame displacement, difference in density between the two clouds, as well as the nature of environment (indoor or outdoor). Overall, we would like to highlight the fact that displacement between two viewing angles are quite consequent keeping in mind that dense techniques generally require an inter frame displacement since the cost function is linearized around $x = 0$.

Continuing with our discussion on the influence of initial displacement, and let us take the case of experiments Expt_3 and Expt_6, which represent a pure translation and a pure rotation, respectively. These two experiments as a sample of several experiments that we carry out, show that generally, the pure rotation requires more energy to reach the convergence with respect to the case of pure translation. Regarding the influence of density, a quick look shows that the denser the clouds becomes, the more the RMSE decreases, leading to better registration. Finally, we observe that CICP performance depends on the scene. In fact, the impact of scene affects the performance of registration as observed by the difference of the number of iterations required to reach the convergence domain between PAVIN's and office data sets. It is clear that the indoor environment performs better registration than the outdoor scene. This is possibly caused by the richness in planar regions of the former. It should not be overlooked that the outdoor environment contains a large amount of noise and outliers. This can be seen on Expt_8 and Expt_9, in which the initial displacement is the same in both experiments. However, the alignment for the office dataset requires 34 iterations to converge instead of 56 for the PAVIN data set.

For the sake of illustration, we take four experiments arbitrarily (Expt_3, Expt_4, Expt_7, Expt_8), and showe their state before and after registration with their convergence profile in Fig. 6. A closer look to the RMSE curves in the third row of this figure reveals that the residues which are far away are successfully minimized. However, the convergence begins very quickly and then stabilizes for a while before it reaches its minimum. This is mainly due to the fact that there is not a perfect point-to-point equivalence in the two pairing sets. This is quite logical and it can be explained by the large difference in density between the two clouds, the noise, and the clustering defects on the two clouds.

## 4.2   Comparison with Existing Methods

In order to compare our method with the existing state-of-the-art methods, we use implemented routines of PCL [25] library for the NDT algorithm, GICP, point-to-plane ICP and simple ICP for dense methods. For the case of feature-based methods (methods based on features extraction) we also use PCL implementations of SIFT3D and FPFH to extract characteristic points from the two clouds, and use simple ICP to perform matching. The performance of each method is evaluated using three metrics: the accuracy, the relative translational error and the relative rotational error. The former describes the evolution of the root-mean-square point-to-point distance; this can be expressed mathematically as:

$$RMSE = \sqrt{\frac{1}{n}\Sigma_{i=1}^{n} \parallel E_i \parallel^2} \tag{6}$$

where $n$ is the number of points and $E_i$ is the distance error between the source points and its correspondent in the target cloud in each iteration. This can be expressed as follows:

$$E_i = \Sigma_{i=0}^{m} p_i - q_i \tag{7}$$

where $m$ is the total number of points in the sparse cloud. $p_i$ and $q_i$ which represent two points of the source and target cloud, respectively.

The second metric is the Relative Translational Error (RTE), which measures the translation gap between the ground truth $(t_{GT})$ and the estimated $(t_E)$ translation vectors.

$$RTE = \|t_{GT} - t_E\|_2 \tag{8}$$

The Relative Rotational Error (RRE) is the sum of the absolute differences of the three Euler angles, calculated from the two rotation matrices $R_{GT}$ (ground truth rotation matrix) and $R_E$ (estimated rotation matrix). RRE is calculated by the Eq. 9 as:

$$RRE = |Roll\left(R_{GT}^{-1}R_E\right)| + |Pitch\left(R_{GT}^{-1}R_E\right)| + |Yaw\left(R_{GT}^{-1}R_E\right)| \tag{9}$$

**Table 2.** Comparison with the state-of-the-art methods.

| | | Dense | | | | | Feature-based | |
|---|---|---|---|---|---|---|---|---|
| | | ICP | pt2pl ICP | NDT | GICP | CICP | SIFT 3D+ICP | FPFH +ICP |
| Office | RMSE (m) | 0.0602 | 0.0620 | 0.0636 | 0.0636 | **0.0299** | 0.0516 | Failed |
| $[m, m, m, °, °, °]$ | RTE (m) | 0.2811 | 0.2482 | 0.2019 | 0.2178 | **0.0169** | 0.0191 | Failed |
| $[0, 0.5, 0.5, 20, 0, 10]$ | RRE (°) | 1.8476 | 0.5243 | 1.0517 | 0.0978 | **0.0144** | 0.7660 | Failed |
| PAVIN | RMSE (m) | 0.0836 | 0.0804 | 0.0824 | 0.0860 | **0.0347** | 0.0678 | Failed |
| $[m, m, m, °, °, °]$ | RTE (m) | 0.0365 | 0.0253 | 0.0315 | 0.0642 | **0.0092** | 0.021 | Failed |
| $[0, 0.5, 0.3, 0, 0, 10]$ | RRE (°) | 0.1919 | 0.1747 | 0.2048 | 0.3211 | **0.1198** | 0.1989 | Failed |

Table 2 presents the results gathered in processing two indoor and outdoor scenes with the state-of-the-art methods. Bold values show the best result. Quantitatively, the RMSE value of the indoor scene reaches 6 cm in the case of point-to-point, 6.2 cm point-to-plane ICP, 6.3 cm for the NDT and the GICP, more than 5 cm for SIFT3D and less than 3 cm for the proposed method. The maximum number of iterations for each test is fixed at 500 beyond which the algorithm is considered as not having converged if it reaches that ceiling, as is the case of the FPFH method.

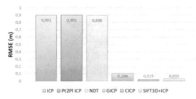

(a) Comparison of RMSE results of the different registration methods

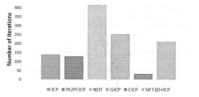

(b) Comparison of number of iterations achieved at convergence of different registration methods

**Fig. 7.** Convergence comparison between different registration methods.

Figure 7 shows comparison of convergences between different registration methods. Again CICP outperforms the state of the art. In addition to that, it is shown that CICP is robust against scene variation.

### 4.3    Dense-to-Dense Data

Figure 8 shows the state of the two dense clouds before and after the registration. Despite the large number of points, the final result is correctly aligned. Here is another benefit of our approach, the fact of not considering the entire set of points for matching, but only a collected set of points from each local surface, which improves the convergence speed.

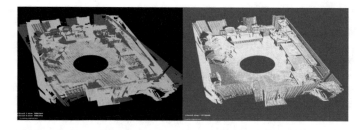

**Fig. 8.** Registration of two dense clouds of indoor scene captured by Leica P20 sensor.

## 5   Conclusion

In this paper, a novel selection technique is introduced on the forefront of an ICP framework. Throughout the experimental section, the performance of the algorithm has been demonstrated where otherwise classical state-of-the art techniques fails or performs poorly. The main highlight of this work is the way 3D surfaces are segmented in a point cloud representation using voxelization and clustering approaches. The advantages are multifold; both sparse and dense point clouds are subsampled by maintaining the geometry of the surface. Moreover, better point estimates obtained are used for later stages of matching and registration. In so doing, the convergence domain of the cost function is greatly improved leading to faster convergence of the algorithm where again, classical techniques fails. Additionally, our method of normal based segmentation not only improves on the weakness of the heterogeneous problem of sparse to dense registration but also deals with sensor noise leading to noisy normal extraction. Finally, our CICP approach improves on the precision of registration and outperforms the state of the art. This work can also be viewed as a direct solution to localization problems of a mobile robot in an *a priori* map constructed with different depth sensors. Further work includes an extension of our approach to vast scale outdoor mapping and localization as well as optimizing the computational time on the CPU for real time applications.

## References

1. Agamennoni, G., Fontana, S., Siegwart, R.Y., Sorrenti, D.G.: Point clouds registration with probabilistic data association. In: IEEE/RSJ International Conference on Intelligent Robots and Systems (IROS), pp. 4092–4098 (2016)
2. Aldoma, A., Vincze, M., Blodow, N., Gossow, D., Gedikli, S., Rusu, R.B., Bradski, G.: CAD-model recognition and 6DOF pose estimation using 3D cues. In: 2011 IEEE International Conference on Computer Vision Workshops (ICCV Workshops), pp. 585–592. IEEE (2011)
3. Arthur, D., Vassilvitskii, S.: K-means++: the advantages of careful seeding. In: Proceedings of the Eighteenth Annual ACM-SIAM Symposium on Discrete Algorithms, SODA 2007, pp. 1027–1035. Society for Industrial and Applied Mathematics, Philadelphia, PA, USA (2007)

4. Besl, P.J., McKay, N.D.: A method for registration of 3-D shapes. IEEE Trans. Pattern Anal. Mach. Intell. **14**(2), 239–256 (1992)
5. Blanco, J.L.: A tutorial on se(3) transformation parameterizations and on-manifold optimization. Technical report, University of Malaga (2010)
6. Cadena, C., Carlone, L., Carrillo, H., Latif, Y., Scaramuzza, D., Neira, J., Reid, I., Leonard, J.: Past, present, and future of simultaneous localization and mapping: towards the robust-perception age. IEEE Trans. Robot. **32**(6), 1309–1332 (2016)
7. Caselitz, T., Steder, B., Ruhnke, M., Burgard, W.: Monocular camera localization in 3D LiDAR maps. In: IEEE/RSJ International Conference on Intelligent Robots and Systems (IROS), pp. 1–6 (2016)
8. Chen, Y., Medioni, G.: Object modeling by registration of multiple range images. In: IEEE International Conference on Robotics and Automation, pp. 2724–2729 (1991)
9. Costa, C.M., Sobreira, H.M., Sousa, A.J., Veiga, G.M.: Robust 3/6 DoF self-localization system with selective map update for mobile robot platforms. Robot. Auton. Syst. **76**(C), 113–140 (2016)
10. Das, A., Diu, M., Mathew, N., Scharfenberger, C., Servos, J., Wong, A., Zelek, J.S., Clausi, D.A., Waslander, S.L.: Mapping, planning, and sample detection strategies for autonomous exploration. J. Field Robot. **31**(1), 75–106 (2014)
11. Feng, Y., Schlichting, A., Brenner, C.: 3D feature point extraction from LiDAR data using a neural network. In: International Archives of the Photogrammetry. Remote Sensing and Spatial Information Sciences-ISPRS Archives 41, vol. 41, pp. 563–569 (2016)
12. Hänsch, R., Weber, T., Hellwich, O.: Comparison of 3D interest point detectors and descriptors for point cloud fusion. In: ISPRS Annals of Photogrammetry, Remote Sensing and Spatial Information Sciences, vol. II-3, pp. 57–64 (2014)
13. Holz, D., Ichim, A.E., Tombari, F., Rusu, R.B., Behnke, S.: Registration with the point cloud library PCL. IEEE Robot. Autom. Mag. **22**(4), 1–13 (2015)
14. Jolliffe, I.T.: Principal Component Analysis for Special Types of Data, pp. 199–222. Springer, New York (1986)
15. Lowe, D.G.: Object recognition from local scale-invariant features. In: Proceedings of the Seventh IEEE International Conference on Computer Vision, pp. 1150–1157 (1999)
16. Ma, Y., Soatto, S., Košecká, J., Sastry, S.S.: An Invitation to 3-D Vision. Springer, Dordrecht (2004)
17. Maddern, W., Newman, P.: Real-time probabilistic fusion of sparse 3d lidar and dense stereo. In: 2016 IEEE/RSJ International Conference on Intelligent Robots and Systems (IROS), pp. 2181–2188. IEEE (2016)
18. Magnusson, M., Lilienthal, A., Duckett, T.: Scan registration for autonomous mining vehicles using 3D-NDT. J. Field Robot. **24**(10), 803–827 (2007)
19. Marani, R., Reno, V., Nitti, M., D'Orazio, T., Stella, E.: A modified iterative closest point algorithm for 3D point cloud registration. Comput. Aided Civ. Infrastruct. Eng. **31**(7), 515–534 (2016)
20. Nieto, J., Bailey, T., Nebot, E.: Scan-SLAM: combining EKF-SLAM and scan correlation. In: Springer Tracts in Advanced Robotics, vol. 25, pp. 167–178 (2006)
21. Pomerleau, F., Colas, F., Siegwart, R.: A review of point cloud registration algorithms for mobile robotics. Found. Trends Robot. **4**(1–104), 1–104 (2015)
22. Rusinkiewicz, S., Levoy, M.: Efficient variants of the ICP algorithm. In: Proceedings of International Conference on 3-D Digital Imaging and Modeling, 3DIM, pp. 145–152 (2001)

23. Rusu, R.B., Blodow, N., Beetz, M.: Fast Point Feature Histograms (FPFH) for 3D registration. In: IEEE International Conference on Robotics and Automation, pp. 3212–3217 (2009)

24. Rusu, R.B., Bradski, G., Thibaux, R., Hsu, J.: Fast 3d recognition and pose using the viewpoint feature histogram. In: 2010 IEEE/RSJ International Conference on Intelligent Robots and Systems (IROS), pp. 2155–2162. IEEE (2010)

25. Rusu, R.B., Cousins, S.: 3D is here: point cloud library. In: IEEE International Conference on Robotics and Automation (ICRA), pp. 1–4 (2011). http://pointclouds.org/

26. Segal, A., Haehnel, D., Thrun, S.: Generalized-ICP. In: Robotics: Science and Systems (2009)

27. Serafin, J., Grisetti, G.: NICP: dense normal based point cloud registration. In: IEEE International Conference on Intelligent Robots and Systems, vol. 2015, pp. 742–749 (2015)

28. Serafin, J., Olson, E., Grisetti, G.: Fast and robust 3D feature extraction from sparse point clouds. In: IEEE/RSJ International Conference on Intelligent Robots and Systems (IROS), pp. 4105–4112 (2016)

29. Tazir, M.L., Checchin, P., Trassoudaine, L.: Color-based 3D point cloud reduction. In: the 14th International Conference on Control, Automation, Robotics and Vision, ICARCV, pp. 1–7 (2016)

30. Tibshirani, R., Walther, G., Hastie, T.: Estimating the number of clusters in a data set via the gap statistic. J. R. Stat. Soc. Ser. B (Stat. Methodol.) **63**(2), 411–423 (2001)

31. Velas, M., Spanel, M., Herout, A.: Collar line segments for fast odometry estimation from velodyne point clouds. In: IEEE International Conference on Robotics and Automation (ICRA), pp. 4486–4495 (2016)

32. Wiemann, T., Mrozinski, M., Feldschnieders, D., Lingemann, K., Hertzberg, J.: Data handling in large-scale surface reconstruction. In: 13th International Conference on Intelligent Autonomous Systems, pp. 1–12 (2014)

33. Wolcott, R.W., Eustice, R.M.: Visual localization within LIDAR maps for automated urban driving. In: 2014 IEEE/RSJ International Conference on Intelligent Robots and Systems (IROS 2014), pp. 176–183. IEEE (2014)

34. Yang, B., Dong, Z., Liang, F., Liu, Y.: Automatic registration of large-scale urban scene point clouds based on semantic feature points. ISPRS J. Photogramm. Remote. Sens. **113**, 43–58 (2016)

35. Yang, J., Li, H., Jia, Y.: Go-ICP: solving 3d registration efficiently and globally optimally. In: Proceedings of the IEEE International Conference on Computer Vision, pp. 1457–1464 (2013)

# Markerless Ad-Hoc Calibration of a Hyperspectral Camera and a 3D Laser Scanner

Felix Igelbrink[1]([⊠]), Thomas Wiemann[1], Sebastian Pütz[1],
and Joachim Hertzberg[1,2]

[1] Knowledge Based Systems Group, Osnabrück University, Wachsbleiche. 27,
49090 Osnabrück, Germany
{figelbrink,twiemann,spuetz}@uni-osnabrueck.de
[2] DFKI Robotics Innovation Center, Osnabrück Branch, Albert-Einstein-Str. 1,
49076 Osnabrück, Germany
joachim.hertzberg@dfki.de

**Abstract.** Integrating 3D data with hyperspectral images opens up novel approaches for several robotic tasks. To that end, we register hyperspectral panoramas to cylindrically projected laser scans. With our approach, the required calibration can be done on board a mobile robot without the need of external markers using Mutual Information. Qualitative results show the robustness of the presented approach, and an application example demonstrates possible future applications for hyperspectral point clouds.

**Keywords:** Mobile robot · 3D laser scanning · Hyperspectral data
Image registration

## 1  Introduction

These days terrestrial laser scanners are able to acquire billions of points in a single scan. Laser scanning is used for building 3D environment models [10] in many applications. In robotics, these models are used for many purposes, e.g., localization, mapping, manipulation, and reasoning to safely interact with the environment.

Integrating spectral information into such models is highly desirable. Hyperspectral cameras split up the spectrum of the incoming light into buckets of wavelength intervals to capture the intensity distribution. These cameras are usually built as line cameras, where one dimension of a frame refers to the spatial image component and the other to the spectral distribution.

Mapping hyperspectral information to high resolution 3D point clouds makes the grounding of material characteristics possible. Currently, hyperspectral images are mainly collected in remote sensing from airplanes, drones or satellites. This limits the level of detail to a rough estimation of the location of

© Springer Nature Switzerland AG 2019
M. Strand et al. (Eds.): IAS 2018, AISC 867, pp. 748–759, 2019.
https://doi.org/10.1007/978-3-030-01370-7_58

detected materials due to the limited resolution. The resulting 2D maps have been used for identifying forests, fields, rivers, residential areas, and other structures of interest. In robotics, such segmentation and classification could be used to complement a plethora of applications, e.g., navigability estimation, semantic mapping and localization. In particular for navigation for and exploration purposes, it is important not only to know which materials are present in a given environment, but also precisely *where* they are located.

To combine 3D spatial information from terrestrial laser scanners and hyperspectral images, the two sensors need to be calibrated with respect to each other. In this paper, we present a method to automatically calibrate a hyperspectral image of a line camera to a high-resolution point cloud without using external calibration patterns. It is fully integrated into a mobile robot, controlled by the Robot Operating System (ROS). We present and evaluate the used calibration technique and the present preliminary results, e.g., using the well-known NDVI (Normalized Difference Vegetation Index) to segment pathways in a hyperspectral point cloud for navigation. Furthermore, we extended and enhanced the state-of-the-art calibration techniques to satisfy the time constraints, which need to be observed in a robotics setting.

## 2   Related Work

Fusing point clouds and color data from RGB camera images is state of the art in the fields of robotics and photogrammetry. The most common solution for the required extrinsic calibration is to use specific calibration patterns with detectable common feature points, e.g., a checkerboard. These correspondences are used to compute the extrinsic parameters by linear transformation and Levenberg-Marquardt optimization [12].

To align point clouds and RGB or hyperspectral image data with such a model, multiple methods have been presented: In [3], a calibrated RGB camera on top of a laser scanner is used to create an RGB-colored point cloud, which is registered to the hyperspectral image using SIFT features and a piecewise linear transformation. However, this approach requires to mount an additional camera, which may not be possible on mobile robots due to space constraints.

In [1] this is solved by using manually placed high-reflectance markers in the scene that can be detected in both the hyperspectral image and the reflectance channel of the laser scanner. These markers and additional automatically found correspondence points are used for the image matching. The use-case was further simplified by capturing only images with small panorama angles. However, for in-place recalibration on 360° images, placing markers is no option. Markerless solutions [7,13] rely solely on automatically detected correspondence points using SIFT and SURF features. However, as the image domains of reflectance data and the hyperspectral imaging are different, we found that establishing stable correspondences in both images is strongly dependent on the environment and on lighting conditions. An approach for registering image data from different domains without requiring the computation of special features is Mutual

Information (MI). The MI metric is derived from the idea of Shannon entropy [6]; it is state of the art for registering of *CT* and *MRI* images.

In [9], Normalized Mutual Information (NMI) is applied to the registration of laser scans and hyperspectral images. It requires no key-point search or manual correspondences and can therefore be used in our context. Therefore, we deem this approach as a suitable starting point for our own registration problem.

## 3    Generating Hyperspectral Panoramas

We installed our hyperspectral scanning system on the mobile robot *Pluto*, based on a VolksBot XT platform. It features a *Riegl VZ400i* high resolution 3D laser scanner and a *Resonon Pika L* hyperspectral line camera, mounted on top of the rotating laser scanner. The camera records up to 297 independent spectral channels between 400 nm and 1000 nm, which are accumulated to a hyperspectral panorama image, while the scanner is rotating and recording the point cloud data.

Unfortunately, the camera does not deliver time stamps and guarantees no continuous frame rate. Hence, synchronization of the arriving frames with the angular information from the laser scanner, which delivers time-stamped data, is required to produce consistent panoramas. During a scan, all frames from the hyperspec-

**Fig. 1.** *Pluto* equipped with a *Riegl VZ400i* terrestrial laser scanner and a *Resonon Pika L* hyperspectral camera.

tral camera are buffered and associated with the current scan angle. After a scan is completed, the panorama is built up from the buffered line images. To produce a coherent aspect ratio, the resulting panorama is scaled according to the angular resolution of the laser scan. Missing frames are automatically corrected by interpolating adjacent lines, to produce a consistent hyperspectral panorama image, from which a pseudo-RGB image is computed for the registration. For later use, the full spectral data is saved separately. Figure 2 presents an example for this panorama generation in pseudo-RGB (derived from the appropriate channels).

**Fig. 2.** Pseudo-RGB section of a panorama generated by the system.

## 4    Extrinsic Calibration

The extrinsic calibration of the laser scanner and the hyperspectral panorama is done in 3 successive steps, which will be presented in the following sections.

### 4.1    Cylindric Camera Model

The first step in the extrinsic calibration is to find an appropriate projection of the 3D scan points into a panoramic image plane that can be compared with the hyperspectral panorama. Since the hyperspectral camera is rotating together with the laser scanner, the image geometry cannot be described by a pinhole model and a central perspective projection [1,2]. Central projection is only valid for the across-track direction, while an angular component needs to be considered in the along-track direction. This is achieved by the cylindric camera model proposed in [5]. It describes the projection of 3D points into a cylindrical

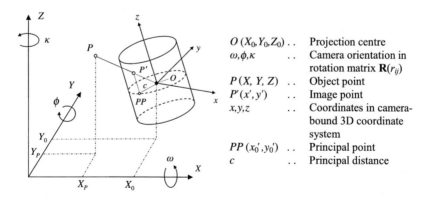

**Fig. 3.** Cylindric camera model principle [5].

image as shown in Fig. 3. For each data point $P(X, Y, Z)$ in the 3D scan, the following steps are performed to compute its pixel coordinates:

1. The world coordinates are transformed into camera coordinates:

$$p = \mathbf{R}^{-1}(P - O) \tag{1}$$

where $O$ is the camera origin in world coordinates, $\mathbf{R}(\omega, \phi, \kappa)$ is the rotation matrix, and $p(x, y, z)$ is the resulting 3D point in camera coordinates.
2. Each $p$ is projected to a cylinder surface surrounding the camera origin:

$$x'_i = x_p - c \arctan\left(\frac{-y_i}{x_i}\right) + \Delta_{x'}$$

$$y'_i = y_p - \frac{cz_i}{\sqrt{x_i^2 + y_i^2}} + \Delta_{y'} \tag{2}$$

Here $x_p$ and $y_p$ denote the $x$ and $y$ components of the principal point, $c$ the principal distance and $\Delta_{x'}$, $\Delta_{y'}$ are optional correction terms to include compensation for radial distortion of the camera lens and imperfectly aligned axes [5]. For our system, we do not include them into the optimization problem, since we are dealing with range measurements.
3. Finally, the obtained 2D coordinates are converted into the pixel coordinates of the resulting image:

$$x_i^{image} = \frac{x'_i + \min_j x'_j}{I_x} \qquad I_x = \frac{\max_i x'_i - \min_i x'_i}{R_x - 1}$$

$$\text{, where} \tag{3}$$

$$y_i^{image} = \frac{N}{2} - \frac{y'_i}{I_y} \qquad I_y = \frac{\max_i y'_i - \min_i y'_i}{(R_x \cdot \xi) - 1}$$

are the value increments for one image pixel in $x$ and $y$ directions. $R_x$ is the horizontal resolution of the projected image, and $\xi$ is the aspect ratio of the resulting image, which is computed from the projected coordinates $x'_i$, $y'_i$.

The height of the resulting point cloud panorama is usually greater than that of the hyperspectral panorama due to the limited aperture angle of the hyperspectral camera. Therefore, the regions not covered by both sensors are automatically cut from the image after the projection. A section of a panorama image generated from the point cloud using the reflectance channel as intensity values is shown in Fig. 4.

## 4.2 Mutual Information

After the projected scan image is computed, we compare it with the hyperspectral image using the Mutual Information metric. This metric measures the statistical dependence between two random variables. It is derived from the Shannon entropy [6]. It is especially well suited for multimodal images because it accounts for areas having different intensity values in both images.

(a)                                              (b)

**Fig. 4.** Comparison between a grayscale section of the hyperspectral panorama (a) and the projected point cloud (b). The black borders of nearby objects in the point cloud are caused by the transformation into the camera coordinate system.

Normalized Mutual Information (NMI) [8] is an extension of the regular Mutual Information that attempts to remove dependency of regular MI on the total amount of information contained in both images, as this dependency might cause MI to produce false global maxima where the overlap of both images is small. NMI is computed from the Shannon entropies as follows:

$$NMI\,(M,N) = \frac{H\,(N) + H\,(M)}{H\,(M,N)}, \tag{4}$$

where

$$H\,(M) = -\sum_{m\in M} p_M\,(m)\log\frac{1}{p_M\,(m)}$$

$$H\,(N) = -\sum_{n\in N} p_N\,(n)\log\frac{1}{p_N\,(n)} \tag{5}$$

$$H\,(M,N) = -\sum_{m\in M}\sum_{n\in N} p_{M,N}\,(m,n)\log\frac{1}{p_{M,N}\,(m,n)}$$

are the individual and joint entropies, respectively. $M, N$ are the discrete random variables, $p_M, p_N$ is the probability distribution over $M$ and $N$, respectively, and $p_{M,N}$ is the joint probability distribution of both variables. For image registration, the probability distributions $p_M, p_N$, and $p_{M,N}$ can be approximated by a histogram of the intensity values

$$\hat{p}\,(X = k) = \frac{1}{n}\sum_{i=1}^{n}\phi_k\,(X_i)\,, \; k \in [0,255]\,, \text{ where } \phi_k\,(x) = \begin{cases} 1 & \text{if } x = k \\ 0 & \text{otherwise} \end{cases} \tag{6}$$

Now, the objective function for the optimization problem can be defined as:

$$\hat{\Theta} = \arg\max_{\Theta} NMI\,(M,N;\Theta) \tag{7}$$

which has its global maximum at the optimal parameters of the camera model.

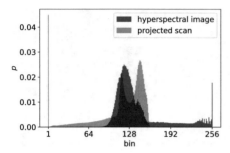
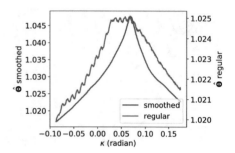

**Fig. 5.** Histograms of the hyperspectral image and the projected scan with 256 bins. The projected scan uses the reflectances mapped to [0.255].

**Fig. 6.** Example of the resulting objective functions from the regular (un-smoothed) and the smoothed histograms.

### 4.3    Smoothing the Objective Function

The basic algorithm, detailed in the previous sections, produces a nonlinear, non-convex objective function $\hat{\Theta}$ with many local maxima (see Fig. 6). Although this function can be solved on its own already, it requires expensive exhaustive optimization algorithms like, e.g., particle swarm optimization [9] or simulated annealing for robust conversion.

To enable fast optimization, the objective function $\hat{\Theta}$ has to be smoothed. Analysis shows that the non-convex behavior is mainly caused by the histogram approximation of the probability distributions, which suffers from a high mean-squared error (MSE). Smoothing $\hat{\Theta}$ aims at reducing the MSE, which can be achieved by various means. A common approach is to smooth the histograms by using a continuous interpolation method, e.g., kernel density estimation (KDE) or B-Splines, for approximating the probability distributions [4,11].

This smoothes the objective function and makes the target function differentiable, enabling to use efficient robust gradient-descent-based optimization algorithms. However, such an interpolation is expensive to compute for high-resolution images like the 360° panoramas in our use case. Additionally, the authors in [4] use multiple image pairs simultaneously to further smooth out the function. This is quite effective, but results in a significantly increased computational load, because the projection of the laser points has to be repeated for each scan in every step.

In order to maintain computational efficiency, we use a much simpler approach for our method. First, we smooth out sharp image edges using a Gaussian filter on the projected images at the borders of objects, which are produced by distant points in the laser scan with potentially very different reflectance values or with no corresponding scan points at all due to the transformation as shown in Fig. 4b. This in turn smoothes the distribution differences in the computed histograms between consecutive iterations, resulting in a smoother target function.

Additionally, we observed that the histograms from the projected reflectance images tend to be very sparse when using one bin for each possible intensity value. In Fig. 5, the projected scan image mostly uses the lower half of the bins, while the hyperspectral image covers the upper half with many noisy local maxima. These maxima have a significant effect on the entropy values. Reducing the number of bins therefore has a significant smoothing effect on the objective function, while preserving most of the contained information. In our experiments, we reduced the number of bins to 16. To smooth the function even further, we convolve the resulting histogram with a Gaussian kernel. These steps result in a very smooth objective function with a clear global maximum and significantly reduced local maxima, as shown in the blue plot in Fig. 6, while still being computationally efficient.

The smoothed objective function can be optimized using standard optimization algorithms that do not require the gradient or Hessian, as the model parameters $\theta$ are not involved in $\hat{\Theta}$ in a differentiable way, due to the histogram step. We use the Nelder-Mead algorithm with the initial simplex spread over the entire search space. This algorithm converges to the global optimum after around 100–200 evaluations of the target function.

When multiple scan/image pairs are available from the same setup, a similar technique as in [4] can be utilized. Although not necessary for convergence, using multiple scans may result in more robust calibration parameters.

## 5    Experiments

We have implemented our method in Python using the numpy[1] and scipy[2] libraries. As the projection of the point cloud into an image and the computation of the histograms are by far the most costly parts of the method, we have also implemented a GPU version of these parts using NVIDIA CUDA. This reduces the necessary computation time for one iteration significantly to about 50ms. The registration of one scan/image pair usually converges within 10s from the start which is significantly faster than the time to acquire one scan.

To evaluate our method, we acquired a set of outdoor scans using the setup depicted in Fig. 1. We kept the setup fixed, so the model parameters are assumed to be the same for all scans. We constructed the search space from the offset along the $z$-axis $Z_0$ from 0.0 m to 0.5 m, the yaw angle $\kappa$ from $-30°$ to $30°$, and the vertical component of the principle point $y_p$ from $-0.2$ to $0.2$, derived from rough estimations of the setup's alignment.

All other model parameters are irrelevant for our setup, as the camera is mounted directly on top of the laser scanner; so they are assumed to be 0 (1 for the principle distance $c$), but are still included in the implemented model to support alternative setups. This initialization limits the search space as well as the runtime and allows for in-field registration on the mobile robot platform. A global registration using no prior knowledge about the parameters is not possible

---

[1] http://www.numpy.org.

[2] https://www.scipy.org.

using our method, because our smoothing of the histograms does not remove the nonlinearity of the objective function entirely.

For our experiments, we recorded scan/image pairs at several different outdoor locations, each including both natural and human-made structures. Each scan contains roughly 11 to 14 million points and each hyperspectral panorama has a size of approx. $7500 \times 900$ pixels. The number of histogram bins was set to 16 for all scans. All images were down-sampled to a horizontal ($x$) resolution of 6000 pixels as a compromise between image quality and accuracy of the final result. The calibrations were executed on a PC using an Intel i7 4930K CPU as well as a NVIDIA 770GTX GPU for the CUDA Implementation.

In Sect. 5.1, we present qualitative results of our calibration as well as a short analysis of the obtained parameters. We evaluate the run time of our method in Sect. 5.2. Finally, we outline possible applications for our setup by detecting plants and pathways using the hyperspectral point cloud in Sect. 5.3.

### 5.1  Qualitative Results

To visualize the results of the registration, some scan sections with added hyperspectral data are visualized in Fig. 7. The colors are switched, so that the red colors indicate strong reflectance in the infrared spectral bands to highlight areas with many plants (chlorophyll) and the green and blue color channels were flipped, so that blue areas indicate regions with less plants. The white areas in the images are sky pixels from the panorama, which end up as the colors of tree twigs, because the laser scanner produces high noise in such areas with many small structures. Some mis-registration is visible in the left image on the metal structure near the scan position. This is caused by the perspective difference between the laser scanner and the hyperspectral camera, producing images with differing information, especially for close objects. So to obtain good registration results, most larger objects should be a few meters away from the scanner

**Fig. 7.** Calibration results in false colors for several different scans (best viewed in color). The circular black spot represents the scanning position.

position, as nearby objects are represented by a larger amount of pixels than far-away objects and therefore contribute more to the NMI.

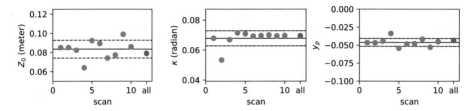

**Fig. 8.** Variation of the estimated parameter values in different scans (blue) using the same setup and all scans combined (red). The continous lines show the mean values for all scans and the dashed lines the standard deviation in both directions.

Figure 8 shows the parameter variation for all scans. The variance of the translation $Z_0$ is clearly much larger than the other parameters. This is expected [4], as the slight translation between laser scanner and camera has less influence as the points are far away from the scanning position and as the angular difference between corresponding points vanishes. Both other parameters are very stable for almost all scans. We did another registration using all 10 scans simultaneously by summing up their histograms (red points in Fig. 8). Especially the translation estimation benefits from adding more data from different scan positions.

## 5.2   Run Time Analysis

We analyzed the run time of our method and the base version without any smoothing and compared the pure CPU and GPU solutions. The results are shown in Table 1. The horizontal resolution was set to 6000 pixels for all scans. The run time for the basic version of the algorithm on the CPU for 10 scans was extrapolated from the run time of the GPU version of the same algorithm due to the very high run time.

**Table 1.** Run times of the different approaches for one calibration on a CPU and on the GPU implementation. The last column shows the resulting average pixel error of each method for 10 manually chosen correspondence points.

| Method | 1 Scan | 10 Scans | Avg. Error |
|---|---|---|---|
| CPU PSO | 6.3 h | > 40 h | ≈ 1.37 |
| GPU PSO | 312.23 s | 2379.28 s | 1.37 |
| GPU, smoothed, 256 bins | 5.62 s | 60.51 s | 794.77 |
| GPU, smoothed, 16 bins | 9.03 s | 84.88 s | 1.90 |

Obviously, the pure CPU version with particle swam optimization (PSO) is not feasible due to the high run time. Smoothing and reduction of the histogram size does not increase the error significantly while decreasing the overall run time, because Nelder-Mead optimization requires less evaluations of the target function. Note that the GPU version using all 256 bins converged faster than the solution using 16 bins, but did so on a local maximum only, resulting in a very high error. The GPU version using 16 bins converged to the global maximum reliably in about 9 s for a single scan. It is fast enough to provide a markerless in-field extrinsic calibration of the camera at high resolutions.

### 5.3   Application Example

To demonstrate the practical benefits of hyperspectral point clouds, we demonstrate an application example, where we used the well-known NDVI index for segmentation of a path way. For that, we computed the NDVI index for the point cloud taken together with the panorama presented in Fig. 2. Using a simple threshold filter, we were able to detect the existing pathway as presented in Fig. 9. Comparing the two pictures, it is obvious that such an easy segmentation would be hard to do based on a RGB image. This example is just one of the new possibilities that open up when using hyperspectral 3D data in robotic applications, which will be further explored in future work.

(a)                                    (b)

**Fig. 9.** Point cloud viewed in RGB (a) and NDVI (b). Yellow and red colors in the NDVI image indicate plants, while drivable, vegetation-free pathways are highlighted in magenta. Best viewed in color.

## 6   Conclusion

In this paper, we have presented an approach to calibrate a hyperspectral line camera against a terrestrial laser scanner on a mobile robot. For this, we used the well-known Normalized Mutual Information (NMI) approach with smoothing of the objective function to increase the robustness and reduce the run time. We have shown that the reduction of the number of bins in the histograms delivers more stable results and that the use of several scans can be beneficial for improving the quality of the estimated parameters. The combination of the these approaches allows to achieve accurate calibration on a mobile system without the need for key-point detection or artificial markers in the scene, which is extremely beneficial for real life applications, since a calibration can be done on demand in the field. If the system is equipped with a GPU, the computation can be sped up significantly using our CUDA implementation, making it possible to compute the registration ad-hoc if needed.

# References

1. Buckley, S., Kurz, T., Howell, J., Schneider, D.: Terrestrial lidar and hyperspectral data fusion products for geological outcrop analysis. Comput. Geosci. **54**, 249–258 (2013)
2. Luhmann, T., Robson, S., Kyle, S., Harley, I.: Photogram. Rec. Close Range Photogrammetry: Principles, Methods And Applications **25**, 203–204 (2010)
3. Nieto, J., Monteiro, S., Viejo, D.: 3D geological modelling using laser and hyperspectral data. In: 2010 IEEE International Geoscience and Remote Sensing Symposium (IGARSS), pp. 4568–4571. IEEE (2010)
4. Pandey, G., McBride, J., Savarese, S., Eustice, R.: Automatic extrinsic calibration of vision and lidar by maximizing mutual information. J. Field Robot. **32**(5), 696–722 (2015)
5. Schneider, D., Maas, H.: A geometric model for linear-array-based terrestrial panoramic cameras. Photogram. Rec. **21**(115), 198–210 (2006)
6. Shannon, C.: A mathematical theory of communication. ACM SIGMOBILE Mob. Comput. Commun. Rev. **5**(1), 3–55 (2001)
7. Sima, A., Buckley, S., Kurz, T., Schneider, D.: Semi-automated registration of close-range hyperspectral scans using oriented digital camera imagery and a 3D model. Photogram. Rec. **29**(145), 10–29 (2014)
8. Studholme, C., Hill, D., Hawkes, D.: An overlap invariant entropy measure of 3D medical image alignment. Pattern Recogn. **32**(1), 71–86 (1999)
9. Taylor, Z., Nieto, J.: A mutual information approach to automatic calibration of camera and Lidar in natural environments. In: Australian Conference on Robotics and Automation, pp. 3–5 (2012)
10. Wiemann, T., Annuth, H., Lingemann, K., Hertzberg, J.: An extended evaluation of open source surface reconstruction software for robotic applications. J. Intell. Robot. Syst. **77**(1), 149–170 (2015)
11. Xu, R., Chen, Y., Tang, S., Morikawa, S., Kurumi, Y.: Parzen-window based normalized mutual information for medical image registration. IEICE Trans. Inf. Syst. **91**(1), 132–144 (2008)
12. Zhang, Q., Pless, R.: Extrinsic calibration of a camera and laser range finder (improves camera calibration). In: Proceedings of the 2004 IEEE/RSJ International Conference on Intelligent Robots and Systems. (IROS 2004), vol. 3, pp. 2301–2306. IEEE (2004)
13. Zhang, X., Zhang, A., Meng, X.: Automatic fusion of hyperspectral images and laser scans using feature points. J. Sens. **1–9**, 2015 (2015)

# Predicting the Next Best View for 3D Mesh Refinement

Luca Morreale, Andrea Romanoni, and Matteo Matteucci[✉]

Politecnico di Milano, Milan, Italy
luca.morreale@mail.polimi.it, {andrea.romanoni,matteo.matteucci}@polimi.it

**Abstract.** 3D reconstruction is a core task in many applications such as robot navigation or sites inspections. Finding the best poses to capture part of the scene is one of the most challenging topic that goes under the name of Next Best View. Recently many volumetric methods have been proposed; they choose the Next Best View by reasoning into a 3D voxelized space and by finding which pose minimizes the uncertainty decoded into the voxels. Such methods are effective but they do not scale well since the underlaying representation requires a huge amount of memory. In this paper we propose a novel mesh-based approach that focuses the next best view on the worst reconstructed region of the environment. We define a photo-consistent index to evaluate the model accuracy, and an energy function over the worst regions of the mesh that takes into account the mutual parallax with respect to the previous cameras, the angle of incidence of the viewing ray to the surface and the visibility of the region. We tested our approach over a well known dataset and achieve state-of-the-art results.

## 1 Introduction

Two of the most challenging tasks for any robotic platform are the exploration and the mapping of unknown environments. When a surveying vehicle, such as a drone, needs to autonomously recover the map of the environment, it has often time and power restrictions to fulfill. A big issue is to incrementally look for the best pose to acquire a new measurement seeking a proper trade-off between the exploration of new areas and the improvement of the existing map reconstruction. This problem is the so called Next Best View (NBV).

NBV is usually addressed as an energy minimization/maximization problem. Most of the existing methods in 3D reconstruction rely on a volumetric reconstruction technique where each voxel is associated to its uncertainty and this information is used to compute the NBV. From Multi-View Stereo literature it is well-founded that volumetric reconstructions are not able to scale well and they do not obtain accurate reconstructions with large datasets [31]. Indeed on large scale problems the output of volumetric NBV algorithm has a coarse block-wise appearance.

© Springer Nature Switzerland AG 2019
M. Strand et al. (Eds.): IAS 2018, AISC 867, pp. 760–772, 2019.
https://doi.org/10.1007/978-3-030-01370-7_59

Alternative reconstruction approaches base the estimation on 3D mesh representations; these methods do not need to store and reason about uncertainty on a volume, since they reason about a 2D manifold; therefore, they result to be more scalable. Moreover the underlaying representation can coincide with the output of accurate meshing algorithms such as [31]. Mesh-based methods have been already used for exploration purposes: by navigating towards the boundary of the mesh a robot is able to explore unknown regions of the environment. However, in mapping scenarios, the focus has to be more on reconstruction accuracy, rather than exploration, therefore we cannot limit the algorithm to consider the boundary of the mesh, but a more comprehensive approach that takes into account the reconstruction accuracy, i.e., map refinement, is needed.

We propose a novel holistic approach to solve the NBV problem on 3D meshes which spots the worst regions of the mesh and looks for the best pose that improves those, specifically:

- a set of estimators to compute the accuracy of a given mesh, see Sect. 3.1,
- a novel mesh-based energy function to look for the NBV as a trade-off between exploration and refinement, see Sect. 3.2.

## 2   Related Works

The Next-Best View (NBV) has been addressed by many researcher and it is hard to define a fixed taxonomy. Since different domains have used different terms, in the following we adopt the classification proposed in the survey in [28]. We refer the reader also to more recent surveys in [4,6].

Model-based NBV algorithms assume a certain knowledge about the environment, for instance, Schmid et al. [27] rely on a digital surface model (DSM) of the scene. This assumption is restrictive and often not easy to fulfill, for this reason, in the following, we focus on non model-based algorithms that build a representation while they estimate the next best view.

One of the simplest method to estimate non-model-based NBV has been proposed in [32]. Among a fixed set of pre-computed poses, the authors choose the pose that improves three statistics about the covariance matrices of the reconstructed 3D points: the determinant, the trace and the maximum eigenvalue. The main drawback of point-based methods, is that occlusions are not considered and the search space is limited to sampled poses.

A more robust and widespread approach relies on a volumetric representation of the scene by means of a 3D lattice of voxels, usually estimated by means of OctoMap [11]; each voxel collects the information needed to define the NBV. Connelly [5] and Banta [1] classify each voxel as occupied, freespace or unknown, and they base the next best view prediction on the number of unknown voxels that the camera would perceive. Similarly, Yamauchi [33] counts the so called frontier voxels, which are the voxels between free and unknown space. An extension of the frontier-based NBV was proposed by Bircher et al. [2]: they use a receding horizon NBV scheme to build a tree which is explored and exploited efficiently.

Previous methods focus on scene exploration, neglecting the problem of refining the model estimated on which we are more interested in a mapping scenario. Vasquez *et al.* [30] plan the next view for a range sensor by relying on frontier voxels together with sensor overlapping, to also refine the model of the scene. This approach is suitable for time of flight or RGB-D cameras, but in case of RGB images the overlapping is not sufficient to evaluate if a new pose would improve the reconstructed model. Indeed, to ensure good parallax, we also need to consider the 3D position and orientation with respect to the other cameras.

While frontier-based methods are usually based on a counting metric, a different, probabilistic, approach to volumetric NBV has been proposed more recently in [12,14,21,25]. In [25] the authors collect the information about the unknown voxels, the visibility, and the occlusions in a Bayesian fashion. Kriegel *et al.* [14] combine the use of a volumetric representation of the space to plan a collision free path and a mesh representation used to compute the region that requires exploration. Isler *et al.* [12] propose a flexible framework that uses four information gain functions collected in the volumetric space. Recently Mendez *et al.* [21] propose an efficient method to compute, among a set of candidate poses, both the NBV by considering only a part of the scene, and a method to add a further camera that forms a good stereo pairs with the NBV. The method has been extended in [20] to explore the continuous space of the scene, instead of limiting to a precomputed set of images.

Volumetric methods have shown to be effective and to properly deal with occlusions differently from point based ones. However they require the boundary of the space to be known in advance and, above all, their voxel-based representation does not scale with large scenes, as underlined in the Multi-View Stereo literature [15–17,31]. Another class of NBV algorithms, named mesh-based and which directly relies on a 3D mesh reconstruction, has the advantage to directly outputs the model of the scene, in addition to scalability. Dunn and Frahm [7] define the 3D mesh reconstruction uncertainty and look for new poses which improve accuracy resolution and texture coherence. However, they require to estimate at each iteration a new 3D mesh model. With a similar approach Mauro *et al.* [18] aggregate 2D saliency, 3D points uncertainty, and point density to define the NBV, however their method relies on a point cloud representation and cannot cope with occlusions.

Differently from the literature hereafter, in this paper, we propose a mesh-based approach to estimate the photometric uncertainty of a mesh and to find the next best view which improves the worst regions. According to this metric the proposed approach is independent from the reconstruction algorithm adopted and it is able to predict the NBV such that it focuses and improves the worst part of the reconstruction. Even if the method is particularly focused on mesh refinement, when the worst regions are located at the boundary of the mesh, it also explores new regions of the scene.

## 3   Proposed Method

In this section we describe our proposed NBV algorithm which relies on a mesh representation of the environment. In the first step we estimate the 3D mesh uncertainty by computing a photo-consistent measure for each facet. Then, we select the worst facets and look for the pose which improves the accuracy of the reconstruction and, in case the selected facets lay near the boundary of the mesh, it also automatically improves the coverage. In Fig. 1 we illustrate the pipeline of the proposed method.

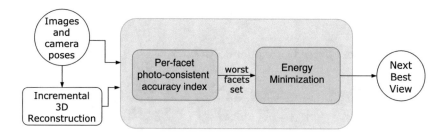

**Fig. 1.** Architecture of the proposed Next Best View System

To keep the mesh updated, we estimate new 3D points from each new pose, by means of the openMVG framework [22] and we apply the incremental reconstruction algorithm proposed in [26] and extended in [24]. The reconstruction algorithm builds a Delaunay Triangulation upon the 3D points and classifies the tetrahedra as free or occupied, the boundary between them represents the 3D mesh. As new points are estimated the triangulation and the reconstructed mesh are updated accordingly.

### 3.1   Photo-Consistent Reconstruction Index

As a first step we look for the regions of the model poorly reconstructed. Since surfaces close to the actual surface project on the images in similar patches, and, in turn, surfaces far from it likely project in patches with different appearance, we adopt photo-consistency on those patches to check for poorly reconstructed regions.

Given a triangular mesh, we consider a facet $f$ and a pair of images $I_1$ and $I_2$ where the facet $f$ is visible; $f$ projects in two triangular patches $P_1 \in I_1$ and $P_2 \in I_2$. We verify if $f$ is visible from an image $I$ by verifying that all its vertices are inside $I$ and that they are not occluded by other facets in the mesh. To check the photo-consistency of $f$ with respect to $I_1$ and $I_2$, we compare $P_1$ and $P_2$. Since their shape and dimension are arbitrary, we map them into a equilateral triangle with unitary sides; $P_1^f$ and $P_2^f$ become the mapped patches which we compare with a similarity measure $sim(P_1^f, P_2^f)$ and then we average among the whole set of images containing the same facet. This leads to the following formulation of the Photo-consistency Reconstruction Index (PRI) of facet $f$:

$$PRI(f) = \frac{1}{|\mathcal{I}^f|} \sum_{P_i^f, P_j^f \in \mathcal{I}^f} sim(P_i^f, P_j^f), \tag{1}$$

computed between each couples, where $\mathcal{I}^f$ is the pair of images where $f$ is visible.

We tested this estimator using as similarity measures the Sum of Squared Difference (SSD), i.e.,

$$sim(P_1^f, P_2^f) = \sum_{x', y'} (P_1^f(x', y') - P_2^f(x', y'))^2, \tag{2}$$

and the Normalized Cross Correlation (NCC), i.e.,

$$sim(P_1^f, P_2^f) = \frac{\sum_{x,y}(P_1^f(x, y) - \bar{P}_1^f)(I_2(x, y) - \bar{P}_1^f)}{\sqrt{\sum_{x,y}(P_1^f(x, y) - \bar{P}_1^f)^2 \sum_{x,y}(I_2(x, y) - \bar{P}_1^f)^2}}, \tag{3}$$

where $\bar{P}_1^f$ and $\bar{P}_2^f$ represent the mean values of the patches. We restricted the comparison of photo-consistency measures only to NCC and SSD since they are successfully adopted by the 3D reconstruction community, especially in multi-view stereo algorithms. Nevertheless the approach can be used with any similarity metrics.

Concerning the scalability, our approach can clearly scale very well spatially since we use only a very small part of the image, thus even using a huge number of images the memory consumption is still low. On the other hand, the time complexity is $\mathcal{O}(n^2)$, where $n$ is the number of photos, thus it does not scale well. However, this issue can be solved considering a fixed number of views making the complexity constant, or, alternatively, computing the $PRI(f)$ only for the facets modified during the reconstruction.

## 3.2   Next Best View

After we compute the per-facet accuracy, we select the $K = 10$ facets with the lower $PRI$ and we collect them in the set $\mathcal{F}_w$. In the following we look for the Next Best View that is able to increase the accuracy of these facets. To do this we propose a novel energy maximization formulation; it combines different contributions to cope with different requirements that a camera pose has to fulfill to improve both the existing reconstruction and to explore new parts of the environment. In the following we refer to $\mathcal{V}_w$ as the set of vertices belonging to the facets in $\mathcal{F}_w$, and to $\mathcal{K}$ as penalization parameter. Specifically we experimentally choose $\mathcal{K} = -10$ in order to enforce the presence of all terms of the energy function, since the violation of one of them would not result in a positive energy.

*Occlusion Term:* The first term of the energy we want to minimize is named Occlusion Term, and it promotes the poses $P$ that sees the region we have to improve, i.e., the facets in $\mathcal{F}_w$ without occlusions. We define this term:

$$O(P, v) = \begin{cases} 1 & \text{if } v \in \mathcal{V}_w \text{ is not occluded,} \\ K & \text{otherwise} \end{cases}. \tag{4}$$

This term implicitly favors exploration. Indeed the facets in $\mathcal{F}_w$ are very likely located at the boundary of the mesh, and the Occlusion term leads the new view to focus on the region around $\mathcal{F}_w$. Furthermore, this formulation does not necessarily require to use GPU for the computation while computing the percentage of overlapping does.

*Focus Term:* The Focus Term represents the idea that the region around $\mathcal{F}_w$ preferably projects to the center of the image. By favoring a projection around the center we also capture the surrounding regions and, since we have selected the worst reconstructed facets, we can fairly assume that also the nearby regions require an improvement.

To account for displacements with respect to the image center we weight the projection of a vertex $v \in \mathcal{V}_w$ with a 2D Gaussian distribution centered in the center of the image. We formalize this term of the energy function as follows:

$$\alpha = -\frac{\left(v_x^P - x_0\right)^2}{2\sigma_x^2} - \frac{\left(v_y^P - y_0\right)^2}{2\sigma_y^2}, \tag{5}$$

where $v_x^P$ and $v_y^P$ are the coordinates of $v$ projected on the camera $P$, and $(x_0, y_0)$ is the center of the image. We fix, $\sigma_x = \frac{W}{3}$ and $\sigma_y = \frac{H}{3}$ where $W$ and $H$ are the width and height of the image. To take into account also the case in which $v$ projects outside the image, we rewrite Equation (5) as follows:

$$F(P, v) = \begin{cases} \alpha & \text{if v is projected inside the image,} \\ K & \text{otherwise} \end{cases}. \tag{6}$$

*Parallax Term:* With the Parallax Term we favor poses that capture the mesh with a significantly parallax with respect to the other images. We base this term on the base-to-height (BH) constrain adopted in Aerial Photogrammetry [8] defined as $\frac{B}{H} > \delta$, where $B$ is the baseline, i.e., the distance between two poses, and $H$ represents the distance between the pose under evaluation and a point $v$, e.g., approximatively the distance between the robot and the surface, or the height of the drone in aerial surveys [8]. Finally $\delta$ is a threshold that we experimentally fixed as $\delta = 0.33$. The Parallax term becomes:

$$P(P, C) = \begin{cases} 1 & \text{if } \frac{B}{H} > \delta \\ K & \text{if } \frac{B}{H} \leq \delta \end{cases}. \tag{7}$$

We compute this term with respect to each other camera $C$.

*Incidence Term:* The last term, named the Incident Term, of our energy function encourages poses that observe the interested surface from an angle of incidence between 40° and 70°. The choice of such angles comes from the experience of the photogrammetric community as explained in [9, 19, 23]. As presented in [19], the smaller the incidence angle is the more distorted the information captured by an image. Since we deal with angular quantities, we formalize it by means of a von Mises distribution from directional statistics:

$$I(P, v) = log(\frac{e^{\kappa \cdot cos(x-\mu)}}{2\pi I_0(\kappa)}), \tag{8}$$

where $x$ represents the angle between the normal of the facet and the ray from facet barycenter to camera, $\mu$ is the angle with the highest probability (in our case $\mu = \frac{40+70}{2}$) and $\kappa$ is a measure of the concentration of the distribution, analogous to the inverse of the variance. $I_0$ is the Bessel function of order zero:

$$I_0(k) = \sum_{m=0}^{\infty} \frac{(-1)^m}{m! \, \Gamma(m+1)} \left(\frac{\kappa}{2}\right)^{2m}. \tag{9}$$

*Next Best View Energy:* We combine the previous four terms as it follows to obtain the energy function we want to minimize with the NBV algorithm:

$$NBV(P, v) = \mu_1 O(P, v) + \mu_2 F(P, v) +$$
$$\mu_3 \sum_{c \in C} P(P, c) - \mu_4 I(P, v), \tag{10}$$

where $\mu_1$, $\mu_2$, $\mu_3$ and $\mu_4$ are the weights of the different terms, that must be experimentally tuned. Then we define the energy over all vertices $v \in \mathcal{V}_w$ as:

$$E(P) = \sum_{v \in \mathcal{V}_w} NBV(P, v). \tag{11}$$

Although in the literature the energy is considered as cost, and thus the optimal pose must usually have the lowest energy, in the proposed approach terms used to compose the energy are higher for better poses. To frame the NBV as minimization we compute the negative of their sum.

## 4    Experimental Validation

To asses the quality of our method we tested it against both a custom synthetic dataset and the real dataset provided in [29]. We run both experiments over a laptop equipped with Ubuntu 16.04 LST and Intel i7-7700HQ without the use of GPU.

Our synthetic dataset contains four photo-realistic environments generated by means of PovRay [3] and depicted in Fig. 2; the big advantage of using such dataset is the possibility to capture an image from any point of view and so to

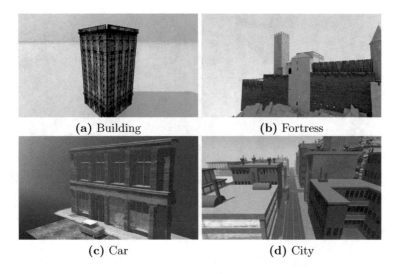

**(a)** Building          **(b)** Fortress

**(c)** Car          **(d)** City

**Fig. 2.** Images from the four synthetic datasets.

simulate a real scenario where a robot, e.g. a drone, needs to navigate and explore the environment. Moreover the ground-truth model is available and therefore the reconstruction accuracy and completeness can be evaluated.

The first step of our algorithm computes the accuracy for each facet; in Fig. 3 we illustrate the errors estimated on the reconstructed mesh with SSD and NCC as the color scale used, i.e., the hue of the HSV color space. As shown in Fig. 3(b), the NCC estimator provide a uniform underestimation of the accuracy. This estimation does not allow to distinguish clearly which areas need refinement. An appropriate estimation would show a lower accuracy around the corners and border, as in Figure Fig. 3(c). Contrary, the SSD estimator does not provide a uniform estimation but properly indicate areas, like corners, as regions which need more refinement.

**Table 1.** Similarity measure computation time (on CPU).

| Similarity measure | # facets | Per-facet | | Total | |
|---|---|---|---|---|---|
| | | NCC | SSD | NCC | SSD |
| Building | 3440 | 0,043 ms | 0,043 ms | 150 ms | 150 ms |
| Fortress | 231 | 0,012 ms | 0,012 ms | 3 ms | 3 ms |
| Car | 314 | 0,038 ms | 0,035 ms | 12 ms | 11 ms |
| City | 43 | 0,651 ms | 0,302 ms | 28 ms | 13 ms |

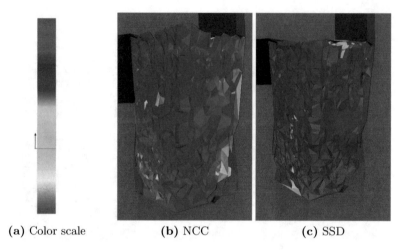

(a) Color scale          (b) NCC                 (c) SSD

**Fig. 3.** Accuracy estimates in the synthetic dataset (first iteration). In (a) we show the color scale adopted, green corresponds to the optimal accuracy whilst red to the worst one.

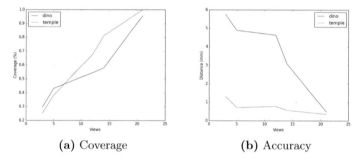

(a) Coverage             (b) Accuracy

**Fig. 4.** Reconstruction coverage (a) and accuracy (b) for the dino and temple datasets with different numbers of views.

In Table 1 we present the time required to estimate the NCC and SSD on the first reconstructed model, before estimating the first NBV. Even if we process the mesh on CPU, our method is able to rapidly estimate the accuracy.

To prove the effectiveness of our approach we tested the whole system, i.e., accuracy estimation and energy function minimization, with a real dataset. Furthermore, given the considerations on the estimators, in the following we use SSD estimator. Concerning the dataset, we use the datasets dinoRing and templeRing presented in [29] (48 images for each sequence). In this test we have a finite set of images and their relative position in the space. Our goal is, starting from an initial set of images, incrementally select views and improve the reconstruction of the scene.

We bootstrap by computing the reconstruction of the scene using three views, then we use the NBV approach previously presented to incrementally select new

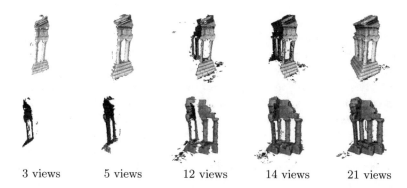

3 views          5 views          12 views          14 views          21 views

**Fig. 5.** Reconstructions of the temple with an incremental number of views.

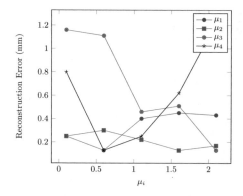

**Fig. 6.** The plot shows the reconstruction error at different values of the weights. Each curve represent a specific weight while the other are kept at their optimal value. Each weight has been tested in five different values: 0.1 0.6 1.1 1.6 2.1.

images. The system estimates which are the areas poorly reconstructed and uses the energy function to select a view, among those available, to enhance the mesh. The view selected is the one having the highest energy score.

We run multiple time the selection process to tune the parameters $\mu_1$, $\mu_2$, $\mu_3$ and $\mu_4$ presented in Equation (10). The optimal weights are those providing the best mesh reconstruction using the lowest number of images. Since no ground-truth is publicly available, we built a reference model with the reconstruction software Photoscan. For each mesh associated to a set of parameters we computed the accuracy of the 3D model generated with Photoscan and compared it with the reference model. In Fig. 4 we report the reconstruction coverage and the accuracy obtained with the optimal parameters with respect to the real ground truth model; the proposed method is able to choose the convenient poses and rapidly improves both accuracy and coverage. In Fig. 5 we illustrate the meshes we reconstructed with the increasing number of views selected by our algorithm for the temple sequence.

The best results were obtained with $\mu_1 = 0.6$, $\mu_2 = 1.6$, $\mu_3 = 2.1$ and $\mu_4 = 0.6$. To derive such values we systematically evaluated the reconstruction error, on dinoRing, changing the varying the weights and then test them with templeRing; we present the observed behavior in Fig. 6. Our approach is able to effectively select a set of images, from a larger set, that properly represent the scene and allow an accurate reconstruction of it. Table 2 shows the accuracy and completeness for the dino dataset. Using less than half of the images in the dataset we are able to achieve comparable results with the other state of the art methods, reported in [21], that use only a subset of the images (as reported in [21]), in particular with the volumetric method in [21] although using 12% less images. We employed Photoscan[1] to generate the final model for both scenes. In Table 2 we also reported the results on the temple dataset. We compare only with [13], since the other methods listed in Table 2 do not provide the results on the temple dataset. Our method is able to provide a significantly better coverage with comparable accuracy even using less images than state-of-the-art approaches.

**Table 2.** Evaluation of different approaches.

|              |            | Dino | | | | | Temple | |
|--------------|------------|------|------|---------|------|----------|---------|----------|
|              | Thresholds | [10] | [10] | [13]    | [21] | Proposed | [13]    | Proposed |
|              |            | Uniform | NBV | NBS   | NBS  |          |         |          |
| Num. frames  | -          | 41   | 41   | Unknown | 26   | 23       | Unknown | 23       |
| Error (mm)   | 80%        | 0.64 | 0.59 | 0.64    | 0.53 | 0.61     | 0.51    | 0.57     |
|              | 90%        | 1.0  | 0.88 | 0.91    | 0.74 | 1.03     | 0.7     | 0.76     |
|              | 99%        | 2.86 | 2.08 | 1.89    | 1.68 | 3.27     | 1.85    | 1.72     |
| Coverage (%) | 0.75 mm    | 79.5 | 82.9 | 72.9    | 87.3 | 85.0     | 78.9    | 84.9     |
|              | 1.25 mm    | 90.2 | 93.0 | 73.8    | 96.4 | 93.6     | 78.9    | 95.0     |
|              | 1.75 mm    | 94.3 | 96.9 | 73.9    | 98.4 | 97.2     | 78.9    | 97.5     |

# 5    Conclusions and Future Works

In this paper we proposed a mesh-based algorithm to build a 3D mesh reconstruction by incrementally selecting the views that mostly improve the mesh. Our approach estimates the reconstruction accuracy and then selects the views that enhance the worst regions maximizing a novel energy function. We have demonstrated that our approach is able to achieve coverage and accuracy comparable to the state of the art, selecting incrementally images from a predefined set. Our approach has the advantage to be independent from the reconstruction method used to build the model, as long as the output is a 3D mesh. As a future work, we plan to design a technique that, using the energy function proposed, defines the Next Best View in the 3D space instead of selecting a view from a set of precomputed poses. Moreover we would test our approach with mesh-based methods as [31].

---
[1] http://www.agisoft.com/.

# References

1. Banta, J.E., Wong, L., Dumont, C., Abidi, M.A.: A next-best-view system for autonomous 3-d object reconstruction. IEEE Trans. Syst. Man Cybern. Part A Syst. Hum. **30**(5), 589–598 (2000)
2. Bircher, A., Kamel, M., Alexis, K., Oleynikova, H., Siegwart, R.: Receding horizon "next-best-view" planner for 3d exploration. In: 2016 IEEE International Conference on Robotics and Automation (ICRA), pp. 1462–1468. IEEE (2016)
3. Buck, D.K., Collins, A.A.: POV-Ray - The Persistence of Vision Raytracer. http://www.povray.org/
4. Chen, S., Li, Y., Kwok, N.M.: Active vision in robotic systems: a survey of recent developments. Int. J. Robot. Res. **30**(11), 1343–1377 (2011)
5. Connolly, C.: The determination of next best views. In: Proceedings of the 1985 IEEE International Conference on Robotics and Automation, vol. 2, pp. 432–435. IEEE (1985)
6. Delmerico, J., Isler, S., Sabzevari, R., Scaramuzza, D.: A comparison of volumetric information gain metrics for active 3d object reconstruction. Auton. Robot. **42**, 1–12 (2017)
7. Dunn, E., Van Den Berg, J., Frahm, J.M.: Developing visual sensing strategies through next best view planning. In: IEEE/RSJ International Conference on Intelligent Robots and Systems, IROS 2009. IEEE (2009)
8. Egels, Y., Kasser, M.: Digital Photogrammetry. CRC Press, London (2003)
9. Fraser, C.S.: Network design considerations for non-topographic photogrammetry. Photogramm. Eng. Remote. Sens. **50**(8), 1115–1126 (1984)
10. Hornung, A., Zeng, B., Kobbelt, L.: Image selection for improved multi-view stereo. In: IEEE Conference on Computer Vision and Pattern Recognition, CVPR 2008, pp. 1–8. IEEE (2008)
11. Hornung, A., Wurm, K.M., Bennewitz, M., Stachniss, C., Burgard, W.: Octomap: an efficient probabilistic 3d mapping framework based on octrees. Auton. Robot. **34**(3), 189–206 (2013)
12. Isler, S., Sabzevari, R., Delmerico, J., Scaramuzza, D.: An information gain formulation for active volumetric 3d reconstruction. In: 2016 IEEE International Conference on Robotics and Automation (ICRA), pp. 3477–3484. IEEE (2016)
13. Jancosek, M., Shekhovtsov, A., Pajdla, T.: Scalable multi-view stereo. In: 2009 IEEE 12th International Conference on Computer Vision Workshops (ICCV Workshops), pp. 1526–1533. IEEE (2009)
14. Kriegel, S., Rink, C., Bodenmüller, T., Suppa, M.: Efficient next-best-scan planning for autonomous 3d surface reconstruction of unknown objects. J. R. Time Image Process. **10**(4), 611–631 (2015)
15. Labatut, P., Pons, J.P., Keriven, R.: Efficient multi-view reconstruction of large-scale scenes using interest points, delaunay triangulation and graph cuts. In: IEEE 11th International Conference on Computer Vision, ICCV 2007, pp. 1–8. IEEE (2007)
16. Li, S., Siu, S.Y., Fang, T., Quan, L.: Efficient multi-view surface refinement with adaptive resolution control. In: European Conference on Computer Vision, pp. 349–364. Springer (2016)
17. Li, S., Siu, S.Y., Fang, T., Quan, L.: Efficient multi-view surface refinement with adaptive resolution control. In: Leibe, B., Matas, J., Sebe, N., Welling, M. (eds.) Computer Vision - ECCV 2016, pp. 349–364. Springer International Publishing, Cham (2016)

18. Mauro, M., Riemenschneider, H., Signoroni, A., Leonardi, R., Van Gool, L.: A unified framework for content-aware view selection and planning through view importance. In: Proceedings of the BMVC 2014, pp. 1–11 (2014)

19. Meixner, P., Leberl, F.: Characterizing building facades from vertical aerial images. Int. Arch. Photogramm. Remote. Sens. Spat. Inf. Sci. **38**(PART 3B), 98–103 (2010)

20. Mendez, O., Hadfield, S., Pugeault, N., Bowden, R.: Taking the scenic route to 3d: optimising reconstruction from moving cameras. In: Proceedings of the IEEE Conference on Computer Vision and Pattern Recognition (2017)

21. Mendez, O., Hadfield, S., Pugeault, N., Bowden, R.: Next-best stereo: extending next best view optimisation for collaborative sensors. In: Proceedings of BMVC 2016 (2016)

22. Moulon, P., Monasse, P., Marlet, R., Others: Openmvg. an open multiple view geometry library. https://github.com/openMVG/openMVG

23. Nocerino, E., Menna, F., Remondino, F.: Accuracy of typical photogrammetric networks in cultural heritage 3d modeling projects. Int. Arch. Photogramm. Remote. Sens. Spat. Inf. Sci. **40**(5), 465 (2014)

24. Piazza, E., Romanoni, A., Matteucci, M.: Real-time CPU-based large-scale 3d mesh reconstruction. In: 2018 IEEE International Conference on Robotics and Automation (ICRA). IEEE (2018)

25. Potthast, C., Sukhatme, G.S.: A probabilistic framework for next best view estimation in a cluttered environment. J. Vis. Commun. Image Represent. **25**(1), 148–164 (2014)

26. Romanoni, A., Matteucci, M.: Incremental reconstruction of urban environments by edge-points delaunay triangulation. In: IEEE/RSJ International Conference on Intelligent Robots and Systems (IROS), pp. 4473–4479. IEEE (2015)

27. Schmid, K., Hirschmüller, H., Dömel, A., Grixa, I., Suppa, M., Hirzinger, G.: View planning for multi-view stereo 3d reconstruction using an autonomous multicopter. J. Intell. Robot. Syst. **65**(1), 309–323 (2012)

28. Scott, W.R., Roth, G., Rivest, J.F.: View planning for automated three-dimensional object reconstruction and inspection. ACM Comput. Surv. (CSUR) **35**(1), 64–96 (2003)

29. Seitz, S.M., Curless, B., Diebel, J., Scharstein, D., Szeliski, R.: A comparison and evaluation of multi-view stereo reconstruction algorithms. In: 2006 IEEE Computer Society Conference on Computer vision and pattern recognition, vol. 1, pp. 519–528. IEEE (2006)

30. Vasquez-Gomez, J.I., Sucar, L.E., Murrieta-Cid, R., Lopez-Damian, E.: Volumetric next-best-view planning for 3d object reconstruction with positioning error. Int. J. Adv. Robot. Syst. **11**, 159 (2014)

31. Vu, H.H., Labatut, P., Pons, J.P., Keriven, R.: High accuracy and visibility-consistent dense multiview stereo. IEEE Trans. Pattern Anal. Mach. Intell. **34**(5), 889–901 (2012)

32. Wenhardt, S., Deutsch, B., Angelopoulou, E., Niemann, H.: Active visual object reconstruction using d-, e-, and t-optimal next best views. In: IEEE Conference on Computer Vision and Pattern Recognition, CVPR 2007. IEEE (2007)

33. Yamauchi, B.: A frontier-based approach for autonomous exploration. In: Proceedings of the 1997 IEEE International Symposium on Computational Intelligence in Robotics and Automation, CIRA 1997, pp. 146–151. IEEE (1997)

# Robotic Applications

# Grasping Strategies for Picking Items in an Online Shopping Warehouse

Nataliya Nechyporenko, Antonio Morales, and Angel P. del Pobil$^{(\boxtimes)}$

Robotic Intelligence Lab., Universitat Jaume I, Castellón, Spain
pobil@uji.es

**Abstract.** The purpose of this study is to investigate the most effective methodologies for the grasping of items in an environment where success, robustness and time of the algorithmic computation and its implementation are key constraints. The study originates from the Amazon Robotics Challenge 2017 (ARC'17) which aims to automate the picking process in online shopping warehouses where the robot has to deal with real world problems of restricted visibility and accessibility. A two-finger and a vacuum grippers were chosen for their practicality and ubiquity in industry. The proposed solution to grasping was retrieval of a final position and orientation of the end effector using an Xbox 360 Kinect sensor information of the object. Antipodal Grasp Identification and Learning (AGILE) and Height Accumulated Features (HAF) feature based methods were chosen for implementation on the two finger gripper due to their ease of applicability, same type of input, and reportedly high success rate. A comparison of these methods was done.

## 1 Introduction

Amazon Robotics aims to automate the task of customer order placement and delivery of its products. Amazon's automated warehouses successfully remove the walking and searching for the object but automated picking still remains a difficult challenge. In order to spur advancement of these fundamental technologies, that in the end can be used at warehouse all over the world, Amazon organizes the Amazon Robotics Challenge. The challenge tasks entrants to build their own robot hardware and software that can attempt simplified versions of the general task of picking and stowing items on shelves. The challenge event consists of three tasks; the pick task, the stow task, and the final round task [1]. The robots are scored based on how many items are picked and stowed, in a fixed amount of time, from a storage system into a box/bin and vice versa. After the competition is complete, the teams share and disseminate their approach to improve future challenge results and industrial implementations.

### 1.1 Application

The items to be picked by the robot have been selected not only because of their common occurrence in warehouses and our households but also because of their

© Springer Nature Switzerland AG 2019
M. Strand et al. (Eds.): IAS 2018, AISC 867, pp. 775–785, 2019.
https://doi.org/10.1007/978-3-030-01370-7_60

varied form and composition. Figure 1 shows four out of forty known items in the ARC'17 database. In terms of grasping, the difficulty lies in item dimensions, texture, and point cloud representation. Some items, such as the bath sponge can easily slip through the fingers of the gripper and others, like the marbles, lets through vacuum air pressure. These complications call for algorithm and gripper combinations that are robust to changes in object orientation, shape and texture.

**Fig. 1.** Example items from the ARC'17 dataset [1].

The Robotic Intelligence Lab (RobinLab) located at Jaume I University (UJI) competed with a Rethink Robotics Baxter and a shelving design based on a reliable industrial solution. Bins can smoothly slide on a system of free-rotating rollers that are actuated by an external mechanism attached to the robot system. The system setup can be seen on Fig. 2.

The ARC'17 task is in line with warehouse, production line, laboratory, and household applications where a robot would have to analyze the table-like scene in front of it and then manipulate objects. The robot has to safely operate in a restricted work space, which requires precise yet easy kinematic configurations to allow for predictability and operability. There is low visibility within the environment and the robot must handle objects in a clutter of a small box. Given that the objects will be both known and unknown and the robot operates in real time, the computation has to be done quickly and efficiently. Low-cost convenient grippers with a fast acting algorithm has to be paired for the highest success of object picking.

## 1.2   Summary

The grasping algorithm receives as input from the vision team, the object that has been identified and the location of the object through its approximate point cloud. The algorithm has to output the position and orientation of one or several grasps that can be sent to the manipulation algorithm. The robot then has to move the arm such that the end effector ends up in the desired location to grasp the object. Two grippers have been mounted on the Baxter robot. The first gripper is a two-finger gripper with a limited opening width, for which it has been named the *Pincher*. The second gripper is a vacuum gripper that uses air pressure to pick up objects. The goal of the following work is to implement software in order to make the most use of the grippers in a warehouse environment.

**Fig. 2.** UJI RobinLab robot platform setup

Two algorithms for the two finger gripper have been compared and the most suitable one has been selected for further testing due to the analysis of time, robustness, and success. The chosen algorithm uses height features of point cloud in order to find a final grasping position and orientation. A vacuum gripper approach has been created based on the estimated normals of object surfaces. This algorithm proved to be fast and hence has also been implemented with the gripper on deformable objects such as a towel for which the orientation of the gripper's x-axis does not play a role. Finally, an optimized table has been devised to maximize the manipulation of an object by customizing the algorithms to each specific object.

Given the introduction for the motivation of the work, the next section, will present the most relevant research works targeted at solving grasping problems similar to those presented by ARC'17. Section 3 defines the chosen approach more in depth. The experimentation documented in Sect. 4 aims to isolate the algorithms and compare their strength and weaknesses such that the best approach to the grasping problem of ARC'17 can be chosen. Finally, Sect. 5 discusses the results.

## 2   Related Work

*Grasp synthesis* refers to the problem of finding a grasp configuration that satisfies a set of criteria relevant for the gasping task [2]. The generation, evaluation, and selection of grasps can be done in various ways but the following review will explore research on the topic of robotic grasping such that the work is applicable to object picking from a tote or a bin.

Task-based grasping has been separately studied in context of Bayesian networks for encoding the probabilistic relations among various task-relevant variables [3]. The synthesis of category and task has been performed based on 2D and 3D data from low-level features [4,5]. Instead of relying on sensor data points the proposition is to synthesize grasps based on semantics, which are stable grasps that are functionally suitable for specific object manipulation tasks that use overall object features [6,7].

One study uses shape primitives like spheres, cones and boxes to approximate object shape and used the simulation environment GraspIt but not on a real life robot application [8]. Saxena et al. proposed supervised learning with local patch-based image and depth features for grasping novel objects in cluttered environments [9]. Fischinger and Vincze develop another idea of features through height-maps where they report indicates a 92% single object grasp success rate while taking only 2–3 s of time [10, 11]. The application to Baxter shows especially successful results, which is important given that Baxter is a platform with a simple gripper and only 1 cm precision. The research done on the method shows robustness, repeatability, speed, and ease of access since the work is an open source code on Github, hence was one of the chosen methods. Features of geometry have been explored by ten Pas et al. and shows a unique combination of analytical understanding and data-driven applications and can easily be compared to the first method [2, 12], It has been inspired by the autonomous checkout robot but is supplemented with an SVM learning mechanism [13]. Another approach also uses features but rather than those of the 3D sensor, relies on supervised deep learning of 2D RGB images [14]. This method requires a large dataset and large amount of training hours.

Finally, the mechanism for a vacuum gripper in combination with vision has been previously tested in warehouse and competition environments. The results of this approach have appeared in previous ARC competitions and have been briefly described or presented with complete practical and theoretical analysis [15–17].

# 3   Proposed Approach

The following section provides theoretical knowledge for the algorithms that have been chosen for implementation. From a bird's eye perspective, the separate algorithms have their individual environments in which each one can exhibit its strength. HAF grasping takes into account the height of the objects and is used with a top grasp thereby reducing the dimensionality of the object. AGILE grasping explores the geometry of the whole object in order to find handle-like sections to exploit for grasping and thus it is often aiming at side grasps. These feature-based algorithms can easily be compared to a baseline Centroid Normals Approach (CNA), which is computationally less heavy and thus faster and more capable of grasping very flat objects like books for which a two-finger gripper with limited opening width cannot be used.

## 3.1   HAF

As the name suggests, the HAF algorithm utilizes the heights of surface points, gathered from the point cloud data, relative to their neighbors in order to learn how to grasp the objects. The authors stress three important advantages of the algorithm; *segmentation independent, integrated path planning,* and use of *known depth regions* [11, 18].

The term *height* refers to the measure of the perpendicular distance from the table plane to the points on the top surface of the object. The input point cloud is first discretized and the height grid $H$ now contains a $1 \times 1\,cm^2$ cell that saves the highest z-valued points with corresponding x and y values [18]. HAF features are defined similarly to Haar Basis functions. All height grid values of each region, $R_i$, on a height grid $H$, are summed up. The sums $r_i$ are individually weighted by $w_i$ and then summed up. The regions and weights are dependent on the HAF feature that are defined by an SVM classification. A feature value, $f_i$, is defined as the weighted sum of all regions. The $j^{th}$ HAF value $f_i$ is calculated as:

$$f_j = \sum_{i=1}^{nr\,Regions_j} w_{i,j} \cdot r_{i,j} \tag{1}$$

$$r_{i,j} = \sum_{k,l \in \mathbb{N}: H(k,l) \in R_{i,j}} H(k,l) \tag{2}$$

The paper claims to have tested 71,000 features (70,000 of which are automatically generated) and finally selected 300 to 325 with an F-score selection [11]. After implementation, the visualization on Fig. 3 shows the calculations in process.

**Fig. 3.** (left) HAF visualization on tennis ball container. View from the Kinect 2.0. (right) view from Baxter.

The long red line indicates the closing direction for a two-finger grippe. The green bars indicate identified potential grasps available. The height of the bars indicates the grasp evaluation score.

## 3.2 AGILE

AGILE grasping is an algorithm that uses a point cloud to predict the presence of geometric conditions that are indicative of good grasps on an object [12]. First, geometry is used to reduce the size of the sample space by applying the condition that for a grasp to exist the hand must be collision free and part of the object surface must be contained between two fingers and is antipodal. A pair of point contacts with friction is *antipodal* if and only if the line connecting the contact points lies inside both friction cones [19]. Then, the remaining grasps are classified, based on a feature, using machine learning for which geometry is used in order to automatically label the training set.

Grasp geometry is quantified by certain parameters. The reason that this algorithm is easy to implement is that these parameters are easy to tune depending on the dimensions of the two finger gripper. The gripper is specified by the parameters $\theta = (\theta_l, \theta_w, \theta_d, \theta_t)$ which respectively stand for gripper length, width, the distance between two fingers, and the thickness of fingers. The method, as mentioned before relies on features. Classification of hand hypothesis uses a feature descriptor of a hand hypothesis as seen on Fig. 4. In the Histogram of Gradients (HOG) feature descriptor, the distribution (histograms) of directions of gradients (oriented gradients) are used as features. Gradients (x and y derivatives) of an image are useful because the magnitude of gradients is large around edges and corners (regions of abrupt intensity changes) and edges and corners pack in a lot more information about object shape than flat regions [20].

**Fig. 4.** HOG feature representation [12]

## 4     Experimentation and Results

### 4.1     Experimentation Procedures

The following section presents the experimental setup for testing the grasping algorithms as well as the results. During the pick and stow tasks, the grasping algorithms received a single point cloud of an object to be manipulated. The algorithm for grasping had to be able to receive a point cloud of a single object and output a suitable grasp vector. During the stow task, the tote contains 20 objects in a mixed jumble. The vision pipeline, in this case, does not guarantee the segmentation of a single object and can include parts of other objects. This noisy data requires the algorithms to be robust. The grasp must also be calculated quickly and successfully. Knowing which algorithm performs best under which circumstance is key to devising a final structure that can pick up the maximum amount of objects.

### 4.2     Preliminary Object to Gripper Matching

Before applying the testing of algorithms it is important to know which gripper works well with which object. In a perfect scenario, the best grasp algorithm will output a similar result to what a human would choose.

The procedure for finding out whether the object can be grasped by the vacuum or two-finger gripper:

1. Place your hand on the wrist of the Baxter robot and approach the item with the gripper
2. Attempt to lift the gripper along with the item
3. If the item can be lifted easily then it's graspable by the given gripper
4. If the object falls down or slips then the object cannot be grasped by the given gripper

The matching procedure allows the task planning team to know which arm to use for object manipulation. If the object can be lifted using vacuum pressure on most sides of the object then the right arm will be used with suction. If the object is best grasped with the gripper then left arm will be used to manipulate the object.

### 4.3 Preliminary Implementation of Algorithms

For object manipulation the algorithm options are either AGILE or HAF. The AGILE and HAF grasping are used with the same gripper and produce similar results. In order to understand the potential of the grasping algorithms, it is first important to implement them and obtain results to see whether the algorithm is capable in dealing with the point cloud presented by the vision pipeline. The goal of this testing is to see how the algorithm behaves and fits once it is integrated into the whole system.

It is important to see the output of the final obtained approach vector as well as the time taken to perform the calculation of the given vector or vectors. Using the Open Multi-Processing library (OMP) the time, before the call of the algorithm and after the output, has been calculated using the simple commands $start = omp\_get\_wtime()$ and $end = omp\_get\_wtime()$. Then compared the differences between the algorithms. Figure 5 shows the approximate calculation time of each algorithm. With this information further steps have been taken to simplify the grasping optimization. Furthermore, Fig. 6 shows the qualitative analysis of robustness.

### 4.4 Testing Objects in Isolation

Testing objects in isolation is relevant for the ARC'17 challenge since the point cloud given from the vision pipeline is an imperfect point cloud of a single object. The evaluation was done based on 10 grasping attempts, each of which was scored on a binary 0/1 system to mark whether the object has been grasped and lifted or not. The setting of grasping is exactly as would be during the competition and in the warehouse. If the point cloud or hardware is not perfect then the algorithm has to be able to deal with this. The scenario is not fully controlled but rather as realistic as possible. The successful grasps were added up and the $\frac{success}{num.attempt} \times 100\%$ was calculated. For demonstration videos of robot grasping,

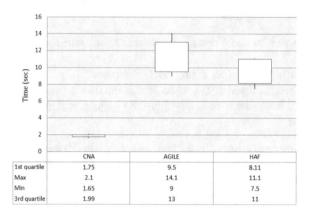

Fig. 5. Time each algorithm takes to calculate a set of grasps

| | CNA | AGILE | HAF |
|---|---|---|---|
| 1st quartile | 1.75 | 9.5 | 8.11 |
| Max | 2.1 | 14.1 | 11.1 |
| Min | 1.65 | 9 | 7.5 |
| 3rd quartile | 1.99 | 13 | 11 |

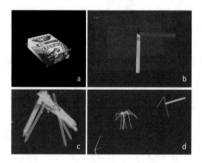

Fig. 6. Robustness comparison of AGILE and HAF

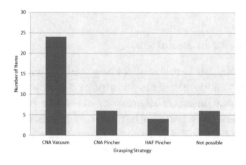

Fig. 7. Number of items per grasping strategy

please visit the referenced website [21]. The experiment ran according to the following rules:

1. All objects, originating from the ARC'17 competition, can be grasped by the gripper in their respective category.
2. The point cloud is as provided by the vision pipeline. No changes to make the object clearer than what vision sees it. The testing is done with full integration of the whole system architecture.
3. The environment is exactly as the system would predict to be in a real-world setting. A failure of gripper arm orientation or positioning is considered a failure for grasping.

Figures 7 and 8 show the number of items included in each strategy and the success per strategy. The success rate can be analyzed in terms of the success of the algorithm as well as the percent chance of being able to pick up the object if it is assigned for the pick and stow tasks.

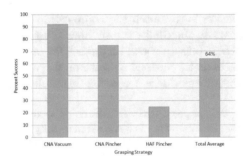

**Fig. 8.** Percent success per grasping strategy

## 5 Conclusion

The UJI RobinLab team took on the ARC'17 in order to automate the warehouse environment. One of the necessary accomplishments for the robot is to be able to grasp an object, which requires a software algorithm. The paper proposes a solution for the object grasping and using two algorithms that take a point cloud as an input outputs a position and orientation of the gripper given the hardware specifications and restrictions.

A two-finger method, has been widely researched and thus it remained to implement current work on the robot. AGILE and HAF grasping have been chosen since they both use a point cloud as an input and a grasp position and orientation as output. One contribution was the implementation of Antipodal Grasp Identification and Learning (AGILE) and Height Accumulated Features (HAF) grasping and quantitatively comparing their computation time and qualitatively comparing their robustness and success. Through simulation and arm approach, it was shown that using the AGILE feature approach was less robust and took on average 3s more than HAF grasping. HAF grasping gave fewer grasp options with only vertical grasps but showed better robustness and computational speed. Table 1 shows the strengths and weaknesses of the methods in comparison with each other. Given the results, it was decided that the AGILE feature did not add significant contribution to the task and was eliminated from the system.

**Table 1.** Comparing AGILE and HAF

| Criteria | AGILE | HAF |
|----------|-------|-----|
| Time | − | + |
| Robustness | − | + |
| Success | + | + |

The final contribution is to match the competition objects to the grasping techniques and grippers to maximize the number of grasped items. This table exposes the analysis that one universal algorithm for one universal gripper has not been found and currently various algorithms for various grippers was the best solution. Demonstration videos are available online [21].

**Acknowledgement.** This paper describes research done at UJI Robotic Intelligence Laboratory. Support for this laboratory is provided in part by Ministerio de Economía y Competitividad (DPI2015-69041-R, DPI2014-60635-R, DPI2017-89910-R) and by Universitat Jaume I.

# References

1. (2017). https://www.amazonrobotics.com/roboticschallenge
2. Bohg, J., Morales, A., Asfour, T., Kragic, D.: Data-driven grasp synthesis-a survey. IEEE Trans. Robot. **30**(2), 289–309 (2013)
3. Song, D., Ek, C.H., Huebner, K., Kragic, D.: Task-based robot grasp planning using probabilistic inference. IEEE Trans. Robot. **31**(3), 546–561 (2015)
4. Madry, M., Song, D., Kragic, D.: From object categories to grasp transfer using probabilistic reasoning (2012)
5. Huebner, K., Kragic, D.: Selection of robot pre-grasps using box-based shape approximation. In: 2008 IEEE/RSJ International Conference on Intelligent Robots and Systems, IROS, pp. 1765–1770 (2008)
6. Dang, H., Allen, P.K.: Semantic grasping: planning task-specific stable robotic grasps. Auton. Robots **37**(3), 301–316 (2014)
7. Nikandrova, E., Kyrki, V.: Category-based task specific grasping. Robot. Auton. Syst. **70**, 25–35 (2015)
8. Miller, A.T., Knoop, S., Christensen, H.I., Allen, P.K.: Automatic grasp planning using shape primitives, pp. 1–7 (2003)

9. Saxena, A., Wong, L.L.S., Ng, A.Y.: Learning grasp strategies with partial shape information (2008)
10. Fischinger, D., Vincze, M.: Learning grasps for unknown objects in cluttered scenes, pp. 609–616 (2013)
11. Weiss, A., Fischinger, D., Weiss, A., Vincze, M.: Learning grasps with topographic features (2016)
12. Platt, R.: Using geometry to detect grasps in 3D point cloud (2015)
13. Klingbeil, E., Rao, D., Carpenter, B., Ganapathi, V., Ng, A.Y., Khatib, O.: Grasping with application to an autonomous checkout robot, pp. 2837–2844 (2011)
14. Jiang, Y., Moseson, S., Saxena, A.: Efficient grasping from RGBD images: learning using a new rectangle representation (2011)
15. Correll, N., Member, S., Bekris, K.E., Berenson, D., Brock, O., Member, S., Causo, A., Hauser, K., Okada, K., Rodriguez, A., Romano, J.M., Wurman, P.R., Jan, R.O.: Lessons from the Amazon picking. Challenge 6(1), 1–14 (2007)
16. Eppner, C., Sebastian, H.: Lessons from the Amazon picking challenge: four aspects of building robotic systems (2016)
17. Yu, K.t., Fazeli, N., Chavan-dafle, N., Taylor, O., Donlon, E., Lankenau, G.D., Rodriguez, A.: A summary of Team MIT's approach to the Amazon picking challenge 2015 (2015)
18. Fischinger, D., Vincze, M.: Empty the basket - a shape based learning approach for grasping piles of unknown objects (2012)
19. de Nguyen, V.: Constructing force-closure, pp. 1368–1373 (1986)
20. Mallick, S.: Histogram of oriented gradients (2016)
21. (2017). https://vimeo.com/grasps

# Learning Based Industrial Bin-Picking Trained with Approximate Physics Simulator

Ryo Matsumura[1], Kensuke Harada[1,2(✉)], Yukiyasu Domae[2], and Weiwei Wan[1,2]

[1] Graduate School of Engineering Science, Osaka University, Suita, Japan
[2] Intelligent Systems Research Institute, National Institute of Advanced Industrial Science and Technology (AIST), Tsukuba, Japan
harada@sys.es.osaka-u.ac.jp

**Abstract.** In this research, we tackle the problem of picking an object from randomly stacked pile. Since complex physical phenomena of contact among objects and fingers makes it difficult to perform the bin-picking with high success rate, we consider introducing a learning based approach. For the purpose of collecting enough number of training data within a reasonable period of time, we introduce a physics simulator where approximation is used for collision checking. In this paper, we first formulate the learning based robotic bin-picking by using CNN (Convolutional Neural Network). We also obtain the optimum grasping posture of parallel jaw gripper by using CNN. Finally, we show that the effect of approximation introduced in collision checking is relaxed if we use exact 3D model to generate the depth image of the pile as an input to CNN.

## 1 Introduction

Randomized bin-picking refers to the problem of automatically picking an object from randomly stacked pile. If randomized bin-picking is introduced to a production process, we do not need any parts-feeding machines or human workers to once arrange the objects to be picked by a robot. Although a number of researches have been done on randomized bin-picking such as [1–10], randomized bin-picking is still difficult due to the complex physical phenomena of contact among objects and fingers. To cope with this problem, learning based approach has been researched by some researchers such as [11,12]. By using a learning based approach, it is expected that the complex physical phenomena can automatically be learned and that the robotic bin-picking can be realized with high success rate (Fig. 1).

In this paper, we research a learning based approach for robotic bin-picking. We introduce CNN (Convolutional Neural Network) to predict whether or not a robot can successfully pick an object from the pile for given depth image of the pile and grasping pose of a parallel jaw gripper. Since our CNN outputs the success rate of picking, we search for the grasping pose maximizing

© Springer Nature Switzerland AG 2019
M. Strand et al. (Eds.): IAS 2018, AISC 867, pp. 786–798, 2019.
https://doi.org/10.1007/978-3-030-01370-7_61

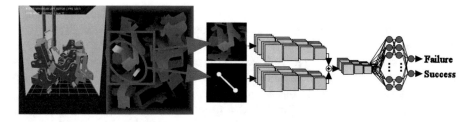

**Fig. 1.** Overview of learning based randomized bin-picking trained with physics simulator

the success rate. However, learning based bin-picking trained with CNN usually requires extremely large number of training data. To cope with this problem, this research aims to effectively collect enough number of training data by introducing a physics simulator. Here, physics simulation on randomly stacked objects with complex shape usually takes longer time than the physics simulation of simple shaped objects. For the purpose of shortening the calculation time of physics simulation used to collect the training data, we consider approximating the shape of objects. This approximation is applied for checking collision among objects and fingers. Here, although we introduce approximation in physics simulation, we do not want to reduce the accuracy of prediction made by CNN. One of the goals of our research is to give an answer to the question: *how we can relax the effect of object shape approximation on the accuracy of prediction.*

While the approximated shaped objects are used for checking collision in physics simulation, original shaped objects are used to construct the simulated depth image of the pile used as an input to CNN. To check the effect of approximation on the accuracy of prediction, we consider focusing on some cases included in the training data where a robot successfully picked an object with approximated shape while a robot may fail in picking the same object with original shape. Our finding is that *even if we approximate the object shape in collision checking, the effect of approximation can be relaxed if we use original shaped object to construct simulated depth image of the pile as an input to CNN.*

The rest of this paper is organized as follows: After introducing previous works in Sect. 2, we show the overview of our physics simulator in Sect. 3. In Sect. 4, we explain our learning based bin-picking method. In Sects. 5 and 6, we show results by using our learning based method.

## 2   Related Works

So far, research on industrial bin-picking has been mainly done on image segmentation [1,2], pose identification [3–6], and picking method [7–10].

As for the research on bin-picking method, Ghita and Whalan [3] proposed to pick the top most object of the pile. Domae et al. [7] proposed a method for determining the grasping pose of an object directly from the depth image of the pile. Some researchers such as [5,8–10] proposed methods for identifying the poses of multiple objects of the pile and picking one of them by using a grasp planning method. However, in conventional bin-picking methods, we have to carefully set up several parameters used in both visual recognition and grasp planning corresponding to each object to be picked. Since a robot usually has to pick a lot of objects to assemble a product, it is not easy for setting up parameters for all objects to be picked.

On the other hand, learning based approaches on randomized bin-picking is expected to break this barrier existing in the conventional randomized bin-picking [11,12,15,16]. Levine et al. [11] proposed an end-to-end approach by using deep neural network whose input is a 2D RGB image. However, they need extremely large number of training data which was collected 800,000 times of picking trials for two months by using 2D RGB image of the pile. Recently, there are some trials on reducing the effort to collect a number of training data by using a method so called GraspGAN [16] and cloud database [15]. On the other hand, this research aims to collect enough number of training data within reasonable time by introducing an approximate physics simulation. Our method searches for the grasping posture with maximum success rate of picking.

The learning approach has also been used for grasping a novel daily object placed on a table [18–21] and for warehouse automation [13,14]. Pas et al. [20] developed a method for learning an antipodal grasp of a novel object by using the SVM (Support Vector Machine). Lenz et al. [19] used deep learning to detect the appropriate grasping pose of an object. Zeng et al. [13] proposed a learning based picking method used for warehouse automation. However, industrial bin-picking is different from the warehouse application since the grasped object is not existing in our daily life and it is impossible to use the generalized object recognition methods.

## 3    Physics Simulator

In this section, we show an overview of the physics simulator used in this research. We use PhysX as a physics engine. The overview of the simulator is shown in Fig. 2. In the simulation world, we assume a tray where its bottom surface is horizontally flat. We also assume the gravity acceleration acting in a vertically downward direction. We use two rectangular shaped objects simulating a two-fingered parallel jaw gripper where a gripper can translate, rotate about the vertical axis and open/close the fingers.

To collect training data, we first consider dropping predefined number of objects from predefined height with randomly defined poses. Then, we consider obtaining a simulated depth image of the pile assuming that a simulated 3D depth sensor is facing vertically downward direction. We furthermore define the gripper's horizontal position and orientation about the vertical axis for the gripper to grasp the top most object. To pick an object, the gripper first moves in the vertically downward direction, closes the fingers, and then moves in the vertically upward direction. After the gripper moves up, we judge whether or not an object is successfully picked by checking the vertical position of objects. At each picking trial, we collect the following three information: (1) a depth image of the pile, (2) gripper's horizontal position and orientation about the vertical axis, and (3) success/failure of picking.

As explained in the introduction, physics simulation of randomly stacked objects with complex shape usually takes longer time than the simulation of simple shaped objects since the calculation time of physics simulation usually depends on the number of contact points included in the simulation world. For the purpose of shortening the calculation time of physics simulation used to collect training data, we consider approximating the shape of an object. This approximation is used just for checking collision among objects and fingers. In our method, the shape of an object is approximated by a set of shape primitives such as rectangular. Figure 3 shows our method for approximating an object shape. For a given polygonal model of an object, we consider applying the convex decomposition [29] where it is decomposed into a set of convex shaped polygons. Then, for each convex shaped polygon, we consider fitting rectangular.

We checked the calculation time of physics simulation as shown in Fig. 4. We performed simulation of picking an object from the pile for four times where each simulation includes same number of same objects with different resolution of convex decomposition. As shown in the figure, as the number of convex objects generated by the convex decomposition increases, calculation time of physics simulation also increases. In the following, we set each object decomposed into 10 rectangular polygons as shown in Fig. 3.

Here, we note that, although the convex decomposition is introduced just for checking collision of physics simulation, it is not used to obtain a simulated depth image of the pile which is an input to the CNN explained in the next section.

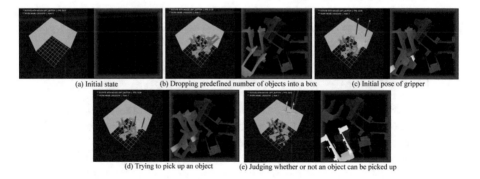

(a) Initial state          (b) Dropping predefined number of objects into a box          (c) Initial pose of gripper

(d) Trying to pick up an object          (e) Judging whether or not an object can be picked up

**Fig. 2.** Physics simulator used in this work where the left side shows the overview of simulation while the right side shows the depth image

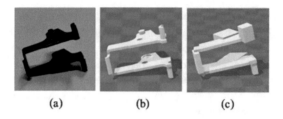

(a)                (b)                (c)

**Fig. 3.** Objects used in a physics simulation where (a) Real object, (b) 3D model used as input depth images of CNN, (c) Approximate model used for interference calculation in simulation.

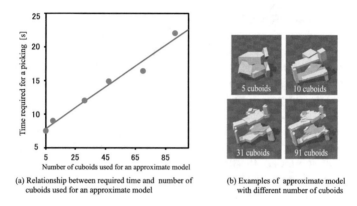

(a) Relationship between required time and number of cuboids used for an approximate model

(b) Examples of approximate model with different number of cuboids

**Fig. 4.** Trade-off between model accuracy and time required for picking

## 4    Learning Based Approach

This section explains our learning based approach for randomized bin-picking introduced in this research.

### 4.1    Convolutional Neural Network

We use CNN (Convolutional Neural Network) [31,32] to predict whether or not a robot can successfully pick an object from the pile. The overview of our CNN is shown in Fig. 5 and Table 1. We use a depth image of the pile ($500 \times 500$ [pixel]) and gripper's pose before picking an object. Since we use a parallel jaw gripper grasping an object with upright posture, a gripper can be expressed by using a segment where two fingers are located at the edge. To reduce the time needed to train the CNN, we consider extracting $250 \times 250$ [pixel] subset of the pile's image. This is an input to the main channel of the CNN. On the other hand, $250 \times 250$ [pixel] image of the segment expressing a gripper's pose is an input to the side channel of the CNN. Our CNN is composed of serially connected convolutional and pooling layers. In the pooling layer, we applied the max pooling of $2 \times 2$ [pixel]. At the end of the convolutional and pooling layers, fully-connected layers is attached. The last layer of the fully-connected ones classifies success/ failure of picking. Success and failure rates are denoted respectively by $y_0$ and $y_1 = 1 - y_0$ by using the following softmax function:

$$y_k = \frac{e^{a_k}}{\sum_{i=1}^{n} e^{a_i}} \tag{1}$$

where $a_k$ denotes weight of the input to the last fully connected layer. Activation function used in convolutional and fully connected layers should avoid the problem of gradient loss. To cope with this problem, we use the following ReLU (Rectified Linear Unit) function [30] as an activation function:

$$f(x) = \max(x, 0) \tag{2}$$

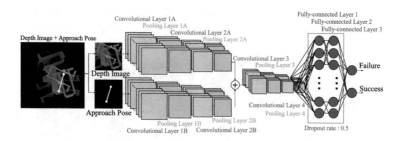

**Fig. 5.** Proposed architecture of CNN

**Table 1.** Details of the proposed CNN

| Layer | Filter | Function | Dropout | Pooling | Output size |
|---|---|---|---|---|---|
| Convolutional Layer 1A·1B | $16 \times 16$ | ReLU | – | $2 \times 2$ | $55 \times 55 \times 32$ |
| Convolutional Layer 2A·2B | $8 \times 8$ | ReLU | – | $2 \times 2$ | $24 \times 24 \times 64$ |
| Convolutional Layer 3 | $5 \times 5$ | ReLU | – | $2 \times 2$ | $10 \times 10 \times 64$ |
| Convolutional Layer 4 | $3 \times 3$ | ReLU | – | $2 \times 2$ | $4 \times 4 \times 64$ |
| Fully-connected Layer 1 | – | ReLU | 0.5 | – | $1 \times 1 \times 1024$ |
| Fully-connected Layer 2 | – | ReLU | 0.5 | – | $1 \times 1 \times 1024$ |
| Fully-connected Layer 3 | – | Softmax | – | – | $1 \times 1 \times 2$ |

### 4.2 Discriminator

Our CNN predicts whether or not a robot can successfully pick an object. Given $250 \times 250$ [pixel] subset of a pile's depth image and a pose of parallel jaw gripper, the CNN outputs the success rate of picking. If the success rate is more than 0.5, we judge that a robot will successfully pick an object. Otherwise, we judge that a robot will fail in picking an object.

### 4.3   Optimum Grasping Pose Detection

To detect the optimum grasping pose, Lenz et al. [19] used a 2 step DNN (Deep Neural Network) where multiple candidates of grasping poses are generated by using a small Neural Network in the first step, and then optimal grasping pose is detected by using a larger Neural Network in the second step. However, this method requires high calculation cost since this method uses Raster scan with changing the size and orientation of a rectangular window and iteratively uses two DNNs. On the other hand, we apply a simple method to detect the grasping pose maximizing the success rate of picking (Fig. 6). Our method uses raster scan with fixed size and orientation of a rectangular window. We consider eight candidates of gripper's orientation corresponding to each rectangular window. By considering $6 \times 6$ candidates of gripper's position, we totally have 288 ($8 \times 6 \times 6$) candidates of gripper's grasping poses. For each gripper's pose, we calculate success rate of picking by using CNN. Among, 288 candidates, we consider calculating a grasping pose with highest success rate of picking.

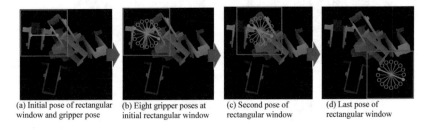

(a) Initial pose of rectangular window and gripper pose      (b) Eight gripper poses at initial rectangular window      (c) Second pose of rectangular window      (d) Last pose of rectangular window

**Fig. 6.** Method for detecting graspable positions

## 5   Collection of Training Data

We performed physics simulation of bin-picking for 6 hours with 15 threads and collected 6000 success data. The failure data is sampled to make the number of failure data be same as the number of success data. 90% of the data is used to train the CNN and remaining 10% is used to verify the trained CNN. By rotating and inverting the depth image included in the training data, we extend the number of training data up to 64800. By using the training data, we trained the CNN shown in Fig. 5 for 17 h.

**Table 2.** Classification results of the verification data

|  |  | Simulation | |
|---|---|---|---|
|  |  | Success | Failure |
| Discriminator | Success | 436(TP) | 134(FP) |
|  | Failure | 164(FN) | 466(TN) |

**Precision=0.765, Recall=0.727, F-value=0.745**

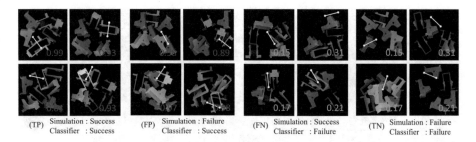

|      | Simulation : Success  Classifier  : Success |      | Simulation : Failure  Classifier  : Success |      | Simulation : Success  Classifier  : Failure |      | Simulation : Failure  Classifier  : Failure |
|------|---------------------------------------------|------|---------------------------------------------|------|---------------------------------------------|------|---------------------------------------------|
| (TP) |                                             | (FP) |                                             | (FN) |                                             | (TN) |                                             |

**Fig. 7.** Classification examples of the verification data. The numerical values indicate the estimated success rate by using the proposed CNN. The values in red indicate a successful picking classification, while the values in blue indicate a failure picking classification.

## 6   Results

### 6.1   Discrimination

As shown in Table 2, we verified the trained CNN by using 1200 verification data including 600 success and 600 failure cases. Figure 7 shows 4 examples included in four classes (TP:True Positive), (TN:True Negative), (FP:False Positive) and (FN:False Negative) shown in Table 2 where red and blue figures respectively show the success and failure cases. We judged the successful cases if the success rate is larger than 0.5. F-value of our discriminator is 0.745 including the cases where a robot successfully picked up an object in spite of the prediction result where a robot fails in picking up an object. We will analyze this prediction error in more detail in the following subsections.

### 6.2   Derivation of Optimum Grasping Pose

By using the trained CNN and 20 verification data, we detected the optimum grasping pose as shown in Fig. 8 where the segments marked in red and yellow shows the optimum grasping pose and grasping poses where the success rate is more than 0.9, respectively. We confirmed that, in all cases, the obtained grasping poses have high graspability index [7]. Figure 9 shows an experimental result where, for given depth image of the pile, we determined the grasping pose by using CNN trained by using physics simulation.

### 6.3   Analysis of Model Approximation

Let us consider the effect of approximation introduced in our physics simulation. We consider fitting a rectangular to each convex decomposed part of a grasped object. While this approximation is used for checking collision among objects and fingers, a simulated depth image is obtained by using object models with the original shape. In our physics simulation, a robot sometimes stably grasps a part of an approximated shaped object where a robot may not be able to

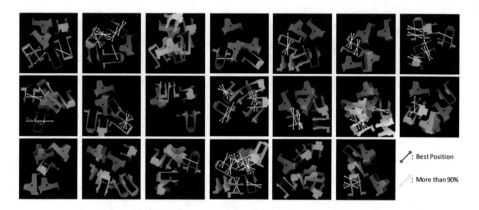

**Fig. 8.** Detection results of the graspable position by using the proposed method

(a) Prediction by using CNN

(b) Picking the target object with highest success rate

(c) Robot successfully picked the target object

**Fig. 9.** Experimental result

stably grasp the same part of an original shaped object. However, even if we use such unrealistic training data caused by the effect of approximation, the effect of approximation may be relaxed if we use the depth image of original shaped object as an input to CNN.

To explain this phenomenon, we collected 200 cases of physics simulation where a robot successfully picked an isolated single object. Among 200 cases, we

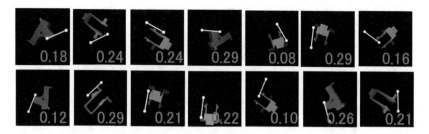

**Fig. 10.** Successful pickings in simulation due to the approximation. Blue values indicate estimated success rate by the proposed CNN.

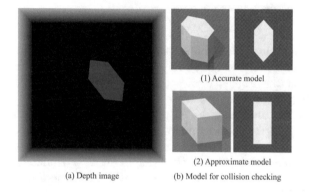

(a) Depth image     (b) Model for collision checking

**Fig. 11.** Successful pickings in simulation due to the approximation. Blue values indicate estimated success rate by the proposed CNN.

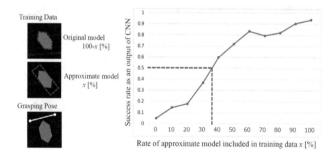

**Fig. 12.** Successful pickings in simulation due to the approximation. Blue values indicate estimated success rate by the proposed CNN.

picked up 15 unrealistic cases as shown in Fig. 10 where a robot stably grasps an object contrary to our expectations. In these cases, a robot stably grasps a part of an object with approximated shape while this part is not included in an object with original shape. The figure also shows the success rate obtained by using the trained CNN. The interesting thing is that the success rate is low in most of the cases in spite of the fact that a robot successfully picks an object in

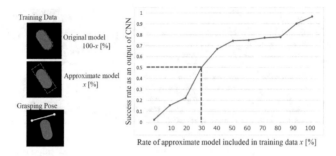

**Fig. 13.** Successful pickings in simulation due to the approximation. Blue values indicate estimated success rate by the proposed CNN.

the physics simulation. This implies that, in the discrimination result shown in Table 2, (FP) and (FN) do not simply show the cases of discrimination errors. We can consider that the effect of approximation is relaxed if we use the depth image including original shaped objects to train the CNN.

Let us consider analyzing the effect of shape approximation in more detail. As shown in Fig. 11, we consider making a robot pick an object placed on a tray. We prepared two kinds of objects to train the CNN where one is approximated by rectangular parallelepiped and the other is not approximated when checking collision. We change the rate of using approximated object when training the CNN. After finished training the CNN, we consider estimating the success rate when a robot trying to pick a part of an object where it is included approximated one and is not included in the original one. The results are shown in Figs. 12 and 13 where a hexagonal prism and an elliptic cylinder are used, respectively. In both cases, if the rate of using approximated object is less than 30%, we can correctly estimate the success of the picking since success is predicted if the success rate is larger than 0.5. This result means that, in randomized bin-picking, we can correctly estimate whether or not a robot can successfully pick an object from the pile if such rough approximation is used in less than 30% of rectangular.

## 7   Conclusions

In this research, we researched the learning based randomized bin-picking. We introduced approximate physics simulation to effectively collect the training data within short period of time. We first formulated the learning based method by using CNN. Then, we obtained the optimum grasping posture of parallel jaw gripper by using CNN. Finally, we showed that the effect of approximation introduced in collision checking is relaxed if we use exact 3D model to generate the depth image of the pile as an input to CNN.

**Acknowledgement.** This research was supported by NEDO (New Energy and Industrial Technology Development Organization).

# References

1. Turkey, M.J.: Automated online measurement of limestone particle size distributions using 3D range data. J. Process Control **21**, 254–262 (2011)
2. Kristensen, S., et al.: Bin-picking with a solid state range camera. Robot. Auton. Syst. **35**, 143–151 (2001)
3. Ghita, O., Whelan, P.F.: A bin picking system based on depth from defocus. J. Mach. Vis. Appl. **13**(4), 234–244 (2003)
4. Kirkegaard, J., Moeslund, T.B.: Bin-picking based on harmonic shape contexts and graph-based matching. In: International Conference on Pattern Recognition, vol. 2, pp. 581–584 (2006)
5. Fuchs, S., et al.: Cooperative bin-picking with time-of-flight camera and impedance controlled DLR lightweight robot III. In: IEEE International Conference on Intelligent Robots and Systems, pp. 4862–4867 (2010)
6. Zuo, A., et al.: A hybrid stereo feature matching algorithm for stereo vision-based bin picking. J. Pattern Recognit. Artif. Intell. **18**(8), 1407–1422 (2004)
7. Domae, Y., et al.: Fast graspability evaluation on single depth maps for bin picking with general grippers. In: IEEE International Conference on Robotics and Automation, pp. 1197–2004 (2014)
8. Dupuis, D.C., et al.: Two-fingered grasp planning for randomized bin-picking. In: Robotics, Science and Systems 2008 Manipulation Workshop (2008)
9. Harada, K., et al.: Probabilistic approach for object bin picking approximated by cylinders. In: IEEE International Conference on Robotics and Automation, pp. 3727–3732 (2013)
10. Harada, K., et al.: Project on development of a robot system for random picking–grasp/manipulation planner for a dual-arm manipulator. In: IEEE/SICE International Symposium on System Integration, pp. 583–589 (2014)
11. Levine, S., et al.: Learning hand-eye coordination for robotic grasping with deep learning and large-scale data collection. In: International Symposium on Experimental Robotics (2016)
12. Harada, K., et al.: Initial experiments on learning-based randomized bin-picking allowing finger contact with neighboring objects. In: IEEE International Conference on Automation Science and Engineering, pp. 1196–1202 (2016)
13. Zeng, A., et al.: Robotic pick-and-place of novel objects in clutter with multi-affordance grasping and cross domain image matching. https://arxiv.org/abs/1710.01330
14. Lin, G., et al.: RefineNet: multi-path refinement networks for high-resolution semantic segmentation. https://arxiv.org/abs/1611.06612
15. Mahler, J., et al.: Dex-Net 2.0: deep learning to plan robust grasps with synthetic point clouds and analytic grasp metrics. https://arxiv.org/abs/1703.09312
16. Bousmalis, K., et al.: Using simulation and domain adaptation to improve efficiency of deep robotic grasping (2017). https://arxiv.org/abs/1709.07857
17. Breiman, L.: Random forests. Mach. Learn. **45**(1), 5–32 (2001)
18. Curtis, N., Xiao, J.: Efficient and effective grasping of novel objects through learning and adapting a knowledge base. In: IEEE International Conference on Intelligent Robots and Systems, pp. 2252–2257 (2008)
19. Lenz, I., Lee, H., Saxena, A.: Deep learning for detecting robotic grasps. Int. J. Robot. Res. **34**(4–5), 705–724 (2015)
20. Pas, A.t., Platt, R.: Using geometry to detect grasps in 3D point clouds. In: International Symposium on Robotics Research (2015)

21. Ekvall, S., Kragic, D.: Learning and evaluation of the approach vector for automatic grasp generation and planning. In: IEEE International Conference on Robotics and Automation (2007)
22. Harada, K., Kaneko, K., Kanehiro, F.: Fast grasp planning for hand/arm systems based on convex model. In: IEEE International Conference on Robotics and Automation, pp. 1162–1168 (2008)
23. Harada, K., et al.: Stability of soft-finger grasp under gravity. In: IEEE International Conference on Robotics and Automation, pp. 883–888 (2014)
24. Thrun, S., et al.: Probabilistic Robotics. MIT Press, Cambridge (2005)
25. Nagata, K., et al.: Picking up and indicated object in a complex environment. In: Proceedings of IEEE/RSJ International Conference on Intelligent Robots and Systems (2010)
26. Aldoma, A., et al.: OUR-CVFH - oriented, unique and repeatable clustered viewpoint feature histogram for object recognition and 6DOF pose estimation. In: Pattern Recognition, pp. 113–122. Springer (2012)
27. Stein, C.M., et al.: Object partitioning using local convexity. In: IEEE International Conference on Computer Vision and Pattern Recognition (2014)
28. PCL-Point Cloud Library. http://pointclouds.org/
29. Mamou, K., Faouzi, G.: A simple and efficient approach for 3D mesh approximate convex decomposition. In: IEEE International Conference on Image Processing, pp. 3501–3504 (2009)
30. LeCun, Y., Bengio, Y., Hinton, G.: Deep learning. Nature **521**, 436–444 (2015)
31. Russakovsky, O., et al.: Imagenet large scale visual recognition challenge. Int. J. Comput. Vis. **115**(3), 211–252 (2015)
32. Grauman, K., Leibe, B.: Visual Object Recognition. Synthesis Lectures on Artificial Intelligence and Machine Learning, vol. 5, no. 2, pp. 1–181 (2011)

# Tool Exchangeable Grasp/Assembly Planner

Kensuke Harada[1,2]([⊠]), Kento Nakayama[1], Weiwei Wan[1,2], Kazuyuki Nagata[2], Natsuki Yamanobe[2], and Ixchel G. Ramirez-Alpizar[1]

[1] Graduate School of Engineering Science, Osaka University, Suita, Japan
[2] Intelligent Systems Research Institute, National Institute of Advanced Industrial Science and Technology (AIST), Toyonaka, Japan
harada@sys.es.osaka-u.ac.jp

**Abstract.** This paper proposes a novel assembly planner for a manipulator which can simultaneously plan assembly sequence, robot motion, grasping configuration, and exchange of grippers. Our assembly planner assumes multiple grippers and can automatically selects a feasible one to assemble a part. For a given AND/OR graph of an assembly task, we consider generating the assembly graph from which assembly motion of a robot can be planned. The edges of the assembly graph are composed of three kinds of paths, i.e., transfer/assembly paths, transit paths and tool exchange paths. In this paper, we first explain the proposed method for planning assembly motion sequence including the function of gripper exchange. Finally, the effectiveness of the proposed method is confirmed through some numerical examples and a physical experiment.

## 1 Introduction

In factory environments, industrial robots are expected to assemble a product. During a robotic assembly process, robotic grippers have to firmly grasp a variety of parts with a variety of physical parameters such as shape, weight and friction coefficient. However, even if we design a gripper to firmly grasp a part, it is not always possible for the same gripper to firmly grasp the other parts with different physical parameters. To cope with this problem, a robotic manipulator used to assemble a product usually equips a tool exchanger at the wrist. By using a tool exchanger, we can selectively use a gripper from multiple candidates. As shown in Fig. 1, we prepared two parallel grippers with different sized fingers. To assemble a toy airplane, a robot first grasps the body and places it on a table. Then, a robot grasps the wing and assembles it to the body. In this example, it is difficult for a robotic gripper to firmly grasp the wing by using the gripper used to grasp the body. Selection of a gripper is often more difficult and complex than this example since a robot has to select a suitable gripper from a set of two-fingered parallel jaw grippers, three-fingered grippers and suction grippers. So far, a gripper use to assemble a part has been selected based on the experience of human workers. On the other hand, this paper aims to construct a grasp/assembly planner which can automatically determine a gripper suitable for a given assembly task.

© Springer Nature Switzerland AG 2019
M. Strand et al. (Eds.): IAS 2018, AISC 867, pp. 799–811, 2019.
https://doi.org/10.1007/978-3-030-01370-7_62

The robotic assembly is a classical topic of robotics extensively researched by many researchers such as [1–6]. However, in most of the previous researches on assembly planners [3–5], grasping posture of a part was assumed to be known. While some researchers such as [9–12] have proposed manipulation planners combined with grasp planners, it is relatively recently where assembly planners combined with grasp planners have proposed [6–8]. Hereafter, we call such assembly planner as the grasp/assembly planner. However, in spite of the fact that the tool exchanging capability is needed for actual assembly tasks, there has been no research on grasp/assembly planner taking the tool exchanging capability into consideration. We believe that this is the first trial on adding a function of automatically selecting a gripper to a grasp/assembly planner. In our previous research, we have proposed a dynamic re grasp graph [8,12] for solving a grasp/manipulation and grasp/assembly planning problems. On the other hand, this research newly assumes multiple grippers for the grasp/assembly planner proposed so far [8]. We show that, by using our proposed grasp/assembly planner, it becomes possible for automatically selecting a gripper from multiple candidates to assemble a part.

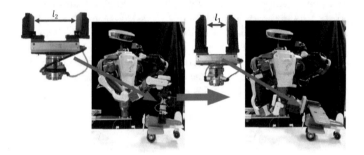

**Fig. 1.** Industrial robots with a tool changer where a gripper was selected depending on each task

The rest of the paper is organized as follows. After discussing previous works in Sect. 2, we show the definitions used in this research in Sect. 3. Section 4 formulates our proposed grasp/assembly planner. In Sect. 5, we confirm the effectiveness of our proposed method through a few numerical examples and a physical experiment where we assume two two-fingered grippers with different size. We show that, according to the shape and the size of a assembled part, our proposed planner can automatically select a feasible one and can complete an assembly task.

## 2   Definitions

Let us consider a product composed of $m$ parts $P = (P_1, \cdots, P_m)$ as shown in Fig. 2. Let $A = (A_1, \cdots, A_n)$ be the assembly of parts as shown in Fig. 3 where

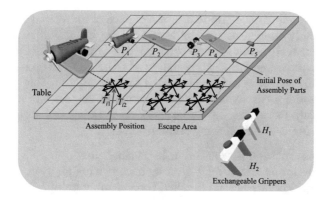

**Fig. 2.** Definition of working area performing product assembly

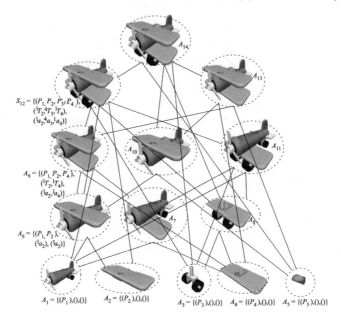

**Fig. 3.** A part of AND/OR graph of a toy airplane

$n \leq \sum_{i=1}^{m} C_i^m$. For example, if the assembly $A_i$ is composed of the parts $P_u$, $P_v$ and $P_w$, it is defined as

$$A_i = \{(P_u, P_v, P_w), ({}^{u}\boldsymbol{T}_v, {}^{u}\boldsymbol{T}_w), ({}^{u}\boldsymbol{a}_v, {}^{u}\boldsymbol{a}_w)\} \tag{1}$$

where ${}^{u}\boldsymbol{T}_v$ denotes a $4 \times 4$ homogenous matrix expressing the pose of the part $P_v$ relative to the part $P_u$, and ${}^{u}\boldsymbol{a}_v$ denotes a 3 dimensional unit vector expressing the approach direction of the part $P_v$ relative to the part $P_u$. On the other hand, if the assembly $A_i$ is composed only of the part $P_i$, it can be defined as follows:

$$A_i = \{(P_i), (), ()\}, \ i = 1, \cdots, m \tag{2}$$

A product assembly is composed of a sequence of individual assembly tasks. Possible assembly sequences can be expressed by using the AND/OR graph $G(A, E)$ [1] where it is composed of the assembly of parts $A$ as the vertices and the edges $E$ connecting them. An example of the AND/OR graph is shown in Fig. 3. Assembly sequence can be generated by searching for this graph.

Let us consider a case where a robot performs a sequence of assembly tasks on a horizontally flat table. Let us consider discretizing the horizontal area of the table. We impose the following assumptions:

**A1:** A robot assembles a product by using a single arm.

**A2:** A robot performs an assembly task by once placing an assembly of parts at one of the grid points hereafter called the assembly point.

Under these assumptions, a robotic gripper picks up an assembly of parts from the table and fit it to another assembly of parts placed at the assembly point.

According to the assumption **A1**, we further impose the following assumption:

**A3:** After finishing an individual assembly task, the assembly of parts is once moved to one of the grid points included in the escape area.

Let us consider preparing $h$ multi-fingered grippers as $H = (H_1, \cdots, H_h)$. In this research, we consider using a grasp planner such as [13, 14] to calculate a grasping posture of a part. We can use any multi-fingered grippers as far as a grasping posture can be calculated. For each pair of a grasped object and a gripper, we consider preparing a database of stable grasping postures. When a robot tries to actually grasp an object, we consider searching for the database to find a stable grasping posture. Let $G_{ij} = (G_{ij1}, \cdots, G_{ijk})$ be a database of grasping postures of the part $P_i$ grasped by the gripper $H_j$ where each element is composed of the wrist's pose with respect to coordinate system fixed to the part $P_i$ and joint angles of each finger.

We additionally impose the following two assumptions:

**A4:** The AND/OR graph is given in advance of planning the assembly motion of a robot.

**A5:** Once a part is assembled, the assembly of parts will not be broken.

As for the assumption **A4**, since there have been a number of researches on automatically generating the AND/OR graph such as [2,3], we can follow their research if we want to automatically generate the AND/OR graph.

## 3   Assembly/Grasp Planner

This section details the grasp/assembly planner proposed in this research.

### 3.1   Placing Pose

To obtain a set of stable placing postures of the assembly $A_i$, we first calculate its convex hull. We consider drawing a line including the assembly $A_i$'s CoG and perpendicular to one of the convex hull's facet. If this line passes through the facet.

Let $T_{ik}$ be the homogenous matrix expressing the $k$-th pose of the assembly $A_i$ stably placed on the table. To determine the homogenous matrix $T_{ik}$, we need the information on (1) the grid point at which the assembly $A_i$ is placed, (2) the facet of the convex hull contacting the table, and (3) rotation of $A_i$ about the table normal. According to these information, we consider multiple candidates of the assembly $A_i$'s placing pose when planning the assembly motion of a robot.

## 3.2   Grasping Posture Set

Let us consider a situation where the assembly $A_i$ is stably placed at one of the grid points. Let us also consider grasping the assembly $A_i$ by using the gripper $H_j$. If the assembly $A_i$ is composed of the parts $P_u, P_v, \cdots, P_w$, a set of grasping postures grasped by the gripper $H_j$ are composed of the elements of the database $G_{uj}, G_{vj}, \cdots, G_{wj}$. For each grasping posture, we consider solving the IK and checking the collision between the robot and the environment. We can obtain a set of IK solvable and collision free database of grasping postures $\hat{G}_{ijk} = (\hat{G}_{ijk1}, \cdots, \hat{G}_{ijkl})$ of the assembly $A_i$ grasped by the gripper $H_j$ where the assembly $A_i$ is stably placed at one of the grid points on the table.

## 3.3   Assembly Graph Search

To plan the motion of a robot to assemble a product, we first search for the AND/OR graph. Then, by using the solution path of the AND/OR graph, we consider constructing the assembly graph where, by searching for the assembly graph, we can generate the motion of a robot to assemble a product. If we failed to find a path of the assembly graph, then we try to find another solution path of the AND/OR graph.

It would be easier for us to understand the structure of the assembly graph if we visualize it by drawing a circle for each $\hat{G}_{ijk}$ and plot the dots on the edge of the circle corresponding to $\hat{G}_{ijk1}, \cdots, \hat{G}_{ijkl}$ (Fig. 4).

By extending the transit/transfer paths which have been introduced in manipulation planners such as [9], we define the following three kinds of edges included in the assembly graph:

*Transit Path*:
    Connect two nodes having the same object placing pose and having the same gripper but having different grasping pose.
*Transfer/assembly Path*:
    Connect two nodes having the same gripper and having the same grasping pose but having different object placing pose.
*Tool Exchange Path*:
    Connect two nodes having different gripper.

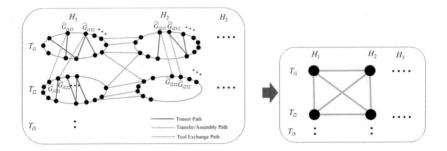

**Fig. 4.** Path definitions of assembly graph and its simplified expression

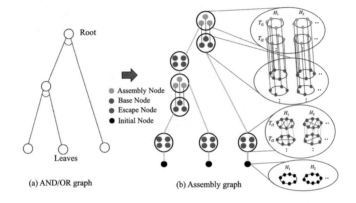

**Fig. 5.** Transformation from AND/OR graph to assembly graph

This visualization method is outlined in the upper side of Fig. 4. The transit paths can be expressed as edges connecting two nodes included in the same circle. The transfer/assembly path can be expressed as edges connecting two nodes included in the different circle but having the same gripper. The tool exchange path can be expressed by edges connecting two nodes included in the different circle, and having different grippers.

Here, we consider the simplified assembly graph as shown in the lower side of Fig. 4. In this expression, each circle of the original manipulation graph is expressed as a single dot. Multiple transfer/assembly paths between two circles are merged into a single bold line. Multiple tool exchange paths between two circles are also merged into a single bold line. This simplified assembly graph does not explicitly include the transit paths.

Next, we consider constructing the assembly graph. Since we imposed the assumptions **A1**, **A2** and **A3**, we consider introducing the following four kinds of nodes included in the assembly graph:

*Base Node*:
An assembly of parts is placed at the assembly point.

*Assembly Node*:

An assembly of parts is fit to another assembly of parts placed at the assembly point.

*Escape Node*:

An assembly of parts is moved to one of the grid points included in the escape area.

*Initial Node*:

A part is placed at the initial position.

Here, for an assembly of parts placed at one of the grid points, we can assume multiple nodes of the assembly graph depending on the rotation of the assembly about the table normal, multiple grasping configurations of the assembly, and multiple grippers grasping the assembly of parts.

To construct the assembly graph from the AND/OR graph, the nodes of the AND/OR graph is replaced by a set of nodes of the assembly graph by the following rules:

- The root node of the AND/OR graph is replaced by a set of base and assembly nodes (Fig. 5) where one of the base nodes and one of the assembly nodes are connected by using the transfer/assembly path.
- The leaf nodes of the AND/OR graph are replaced by the initial and a set of the escape nodes where one of the initial nodes and one of the escape nodes are connected by using a transfer/assembly path and where the escape nodes are connected each other by using a transit and transfer paths.
- The nodes except for the root and the leaves are replaced by a set of base, assembly and escape nodes where one of the assembly nodes and one of the escape nodes are connected by using a transfer/assembly path, and where the escape nodes are connected each other by using a transit and transfer paths.

Here, the assembly nodes are automatically determined by the corresponding base nodes since assembly of parts defined in an assembly node includes the assembly of parts defined in the base node (shown in the dotted line in Fig. 5).

Then, we show a method for searching the assembly graph. In our method, we first search for a solution path of the simplified assembly graph. There are multiple root nodes included in the simplified assembly graph. From each root node, we consider searching for a solution path of the simplified assembly graph by using Dikstra method. Then, we consider selecting a root node where the path cost becomes minimum.

For a solution path of the simplified assembly graph, we consider obtaining a sequence of assembly as will be explained in the next subsection.

For a sequence of assembly tasks, we consider determining the grasping configuration. We first disregard the transit path and try to find grasping configurations with maximum grasping stability index [16]. Our grasp stability index proposed in [16] evaluates the contact area and can be applied for the soft-finger contact model. Thus, the gripper having fingers with large contact area tends to be selected.

After obtaining a sequence of assembly, we check whether or not each edge included in the sequence is a feasible one by using the RRT (Rapidly-exploring Random Tree) algorithm. If RRT algorithm does not find a solution, we consider cutting the corresponding edge and try to find the grasping configuration again.

### 3.4  Assembly Sequence

To perform an product assembly under the assumptions **A1**, **A2** and **A3**, an assembly of parts has to be first placed at the assembly point before it is fit to another assembly of parts. Let us consider the case where a robot assembles the assembly $A_u$ to the assembly $A_v$. After a robot places the assembly $A_u$ to the assembly point, a robot may first exchange the gripper, then picks the assembly $A_v$, and finally assembles it to the assembly $A_u$. However, the solution of the assembly graph obtained in the previous subsection does not include such information. In this subsection, we consider generating an assembly sequence taking the exchange of grippers into consideration. From the assembly graph constructed in the previous subsection, a sequence of assembly is generated by using the following method:

1. Push the root node of the assembly graph to the stack.
2. Iterate the following steps until the stack becomes empty
   (a) Pop the stacked nodes. Connect the path including an assembly node between the stacked node and either an initial or an escape node to the solution path by using the tool exchange path.
   (b) If the last node of the solution path is an escape node, push the escape node to the stack.
   (c) Connect the path between the corresponding base node and either an initial or an escape node to the solution path by using the tool exchange path.
   (d) If the last node of the solution path is an escape node, push the escape node to the stack.

Figure 6 shows how the algorithm shown in the example of Fig. 5 works. As shown in this figure, the base node is scheduled before the assembly node and whole assembly sequence can be performed where an adequate gripper is selected according to the assembly of parts.

## 4  Results

In this section, we show some numerical examples to show the effectiveness of our proposed method. We prepared two two-fingered parallel grippers used to assemble a product. One of the grippers has relatively small contact area where it would be suitable for assembling a small object. On the other hand, the other gripper has relatively large contact area where it would be suitable for assembling a large object.

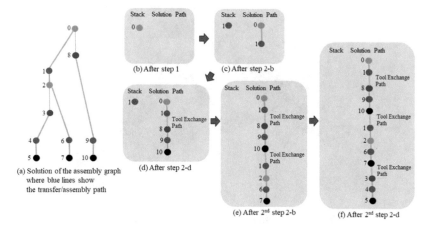

Stack Solution Path

(b) After step 1

(c) After step 2-b

(d) After step 2-d

(e) After 2nd step 2-b

(f) After 2nd step 2-d

(a) Solution of the assembly graph where blue lines show the transfer/assembly path

**Fig. 6.** Searching algorithm of assembly sequence

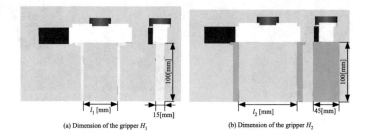

(a) Dimension of the gripper $H_1$

(b) Dimension of the gripper $H_2$

**Fig. 7.** Two fingered parallel grippers used in numerical examples

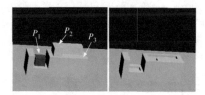

**Fig. 8.** Assembled product used in Example 1

In the first example, the robot tries to assemble a product made of three blocks as shown in Fig. 8. In this example, the width $l_1$ and $l_2$ of the gripper shown in Fig. 7 is set as $0 \leq l_1 \leq 0.06$[m] and $0 \leq l_2 \leq 0.1$[m], respectively. The simplified assembly graph is shown in Fig. 9 where the solution path is expressed by the red transfer paths and blue tool-exchange paths. We used a grasp planner proposed in [13] to calculate the grasping posture. The number of grasping posture included in the database is $\dim(G_{11}) = 347$, $\dim(G_{12}) = 443$, $\dim(G_{21}) = 1221$, $\dim(G_{22}) = 2162$, $\dim(G_{31}) = 12$, and $\dim(G_{32}) = 18$. Some examples of grasping postures are shown in Fig. 10. It took about 2 [min] to

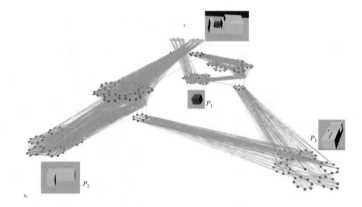

**Fig. 9.** Assembly graph generated in Example 1

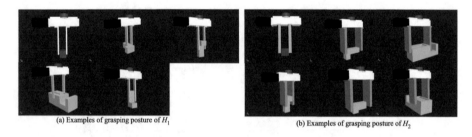

(a) Examples of grasping posture of $H_1$

(b) Examples of grasping posture of $H_2$

**Fig. 10.** Grasping posture used in Example 1

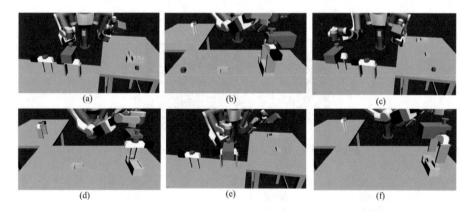

(a)    (b)    (c)

(d)    (e)    (f)

**Fig. 11.** Snapshot of assembly motion of Example 1

calculate the solution path by using the 3.4[GHz] Quad-core PC. Some examples of grasping posture are shown in Fig. 10. The motion of the robot is shown in Fig. 11 where the robot first used the hand $H_2$ to stably pick the part $P_3$. Then, the robot used the hand $H_1$ to pick the part $P_1$ and assembled it to the part $P_3$.

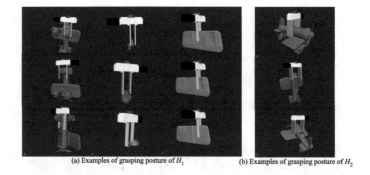

(a) Examples of grasping posture of $H_1$          (b) Examples of grasping posture of $H_2$

**Fig. 12.** Grasping posture used in Example 2

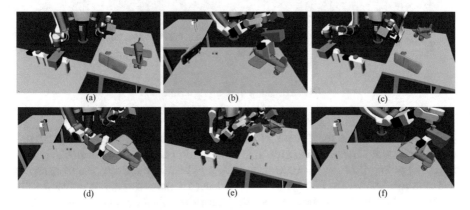

**Fig. 13.** Snapshot of assembly motion of Example 2

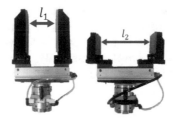

**Fig. 14.** Grippers used for experiment

Here, it is impossible to use the large hand $H_2$ to assemble $P_1$ to the concaved part of $P_3$. Finally, the robot used again the hand $H_2$ to assemble the part $P_2$.

In the second example, we consider the assembly of a toy airplane where its AND/OR graph is shown in Fig. 3. In this example, we consider the assembly problem of three parts: $A_2$, $A_5$ and $A_{11}$. The width $l_1$ and $l_2$ of the gripper shown in Fig. 7 is set as $0 \leq l_1 \leq 0.06$[m] and $0.4 \leq l_2 \leq 0.1$[m], respectively. In this example, the gripper $H_1$ is suitable for grasping a thin object while the

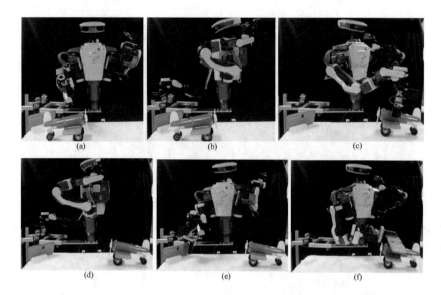

**Fig. 15.** Snapshot of experiment

gripper $H_2$ is suitable for grasping a thick one. The number of grasping posture included in the database is $\dim(G_{11}) = 49$, $\dim(G_{12}) = 58$, $\dim(G_{21}) = 4$, $\dim(G_{22}) = 24$, $\dim(G_{31}) = 0$, and $\dim(G_{32}) = 263$. Some examples of grasping postures are shown in Fig. 12. It took 28[sec] to plan the robot motion. The motion of the robot is shown in Fig. 13 where the robot first used the hand $H_2$ to stably pick the thick $A_{11}$. Then, the robot used the hand $H_1$ to pick the part $A_2$ and $A_5$. Since the same hand is used two individual assembly tasks sequentially connected, the robot does not exchange the gripper.

Finally, we performed experiment on the toy airplane assembly. In this experiment, we used two kinds of parallel jaw gripper as shown in Fig. 14 corresponding to the simulation result of toy airplane assembly. Figure 15 shows experimental result where a robot stably grasps each parts and successfully conducted the assembly task.

## 5   Conclusion

In this paper, we proposed a grasp/assembly planner for a manipulator which can simultaneously plan assembly sequence, robot motion, grasping configuration, and exchange of grippers. For a given AND/OR graph of an assembly task, we generated the assembly graph where its edges are composed of three kinds of paths, i.e., the transfer/assembly path, the transit path and the tool exchange path. We showed numerical examples assuming two kinds of two-fingered parallel grippers where one of the grippers is suitable for grasping a small part and the other is suitable for grasping a large part.

For a future research, we consider conducting a real world experiment. Motion optimization is also considered to be our future research topic.

# References

1. de Mello, L.S.M., Sanderson, A.C.: And/Or graph representation of assembly plans. CMU Research Showcase, CMU-RI-TR-86-8 (1986)
2. de Mello, L.S.M., Sanderson, A.C.: Automatic generation of mechanical assembly sequences. CMU Research Showcase, CMU-RI-TR-88-19 (1988)
3. Kwak, S.J., Hasegawa, T., Chung, S.Y.: A framework on automatic generation of contact state graph for robotic assembly. Adv. Robot. **25**(13–14), 1603–1625 (2011)
4. Wilson, R.H., Latombe, J.-C.: Geometric reasoning about mechanical assembly. Artif. Intell. **71**(2), 371–396 (1994)
5. Heger, F.W., Singh, S.: Robust robotic assembly through contingencies, plan repair and re-planning. In: IEEE International Conference on Robotics and Automation (2010)
6. Thomas, U., Stouraitis, T., Roa, M.A.: Flexible assembly through integrated assembly sequence planning and grasp planning. In: IEEE International Conference on Automation, Science and Engineering (2015)
7. Dogar, M., et al.: Multi-robot grasp planning for sequential assembly operations. In: IEEE International Conference on Robotics and Automation (2015)
8. Wan, W., Harada, K.: Integrated assembly and motion planning using re grasp graph. Robot. Biomimetics **3**, 18 (2016)
9. Siméon, T., Laumond, J.P., Cortés, J., Sahbani, A.: Manipulation planning with probabilistic roadmaps. Int. J. Robot. Res. **23**(7–8), 729–746 (2004)
10. Harada, K., Tsuji, T., Laumond, J.-P.: A manipulation motion planner for dual-arm industrial manipulators. In: IEEE International Conference on Robotics and Automation (2014)
11. Vahrenkamp, N., Asfour, T., Dillmann, R.: Simultaneous grasp and motion planning. Robot. Autom. Mag. **19**, 43 (2012)
12. Wan, W., Harada, K.: Developing and comparing single-arm and dual-arm re grasp. Robot. Autom. Lett. **1**(1), 243–250 (2016)
13. Harada, K., Kaneko, K., Kanehiro, F.: Fast grasp planning for hand/arm systems based on convex model. In: IEEE International Conference on Robotics and Automation (2008)
14. Harada, K., et al.: Grasp planning for parallel grippers with flexibility on its grasping surface. In: IEEE International Conference on Robotics and Biomimetics (2011)
15. Lozano-Pérez, T., Kaelbling, L.P.: A constraint-based method for solving sequential manipulation planning problems. In: IEEE/RSJ International Conference on Intelligent Robots and Systems (2014)
16. Harada, K., et al.: Stability of soft-finger grasp under gravity. In: IEEE International Conference on Robotics and Automation (2014)

# Daily Assistive Robot Uses a Bag for Carrying Objects with Pre-contact Sensing Gripper

Naoya Yamaguchi$^{(\boxtimes)}$, Shun Hasegawa, Kei Okada, and Masayuki Inaba

The University of Tokyo, 7-3-1 Hongo, Bunkyo-ku, Tokyo 113-8656, Japan
`yamaguchi@jsk.imi.i.u-tokyo.ac.jp`

**Abstract.** Since bags are often used at everyday life, it is important for robots to handle bags for daily assistance. Bag manipulation by robots can be divided into the following elements: 1. receiving it from human, 2. transporting it to the designated destination, 3. opening it by manipulating its handles and 4. picking objects from it. However, robots have trouble with handling bags, in that they are usually deformable and their inside are difficult to see. In this research, we propose a motion strategy considering the transformation and the visibility, using robot fingers on which proximity sensors are mounted all around. At the last of this paper, we verify that our method is effective in the 4 movements mentioned above, through an experiment in which a robot carries a bag and takes out the content.

**Keywords:** Motion strategy · Pre-contact · Proximity sensor
Deformable object

## 1 Introduction

In recent years, aging of society with fewer children has become a world-wide problem, and necessity of elderly support is increasing. In response to this social issue, daily assistance by robots has been receiving a lot of attention. Some examples are cleaning up clothes and luggage transportation. Many of these tasks require appropriate handling of storage and transportation equipment, so it is expected that variety of daily assistance will be expanded by robots' handling those equipment. Some examples of the equipment are bins, shelves and bags. At previous researches, picking objects from bins [6] and shelves [7] is achieved by motion strategy based on the internal model constructed by visual recognition beforehand. On the other hand, bags, which are deformable objects, change its shape during manipulation, and the contents of bags are shaken accordingly. Therefore, unlike bins and shelves, recognition of the internal state in advance is not effective for the bag manipulation. In this paper, we propose a method for grasping objects in an obscured, deformable and closed environment, such as bags (Fig. 1).

© Springer Nature Switzerland AG 2019
M. Strand et al. (Eds.): IAS 2018, AISC 867, pp. 812–824, 2019.
https://doi.org/10.1007/978-3-030-01370-7_63

## 1.1  Related Works

**Real-Time Recognition of Environment.** In order to achieve manipulation without advance recognition, robots need to comprehend surroundings in real time. In previous researches, online grasp planning method using a head camera [5] and real-time object tracking method using a hand camera [8] are proposed. However, the method using head cameras is not appropriate in closed environment like bags, in that occlusion between target object and either robot hand or bags happens. Also, the method using hand cameras is not useful, since the camera approaches the object too closely to recognize it during grasping. Therefore, we think that robots can manipulate objects in bags by using not cameras but sensors on its fingers or hands, like humans. In this research, we use a gripper exterior on which proximity sensors are mounted all around and a robot can recognize inside of a bag in real time (Fig. 2).

**Object Search and Grasp.** In closed environments like bags, because it is necessary for robots to avoid collision, non contact sensors are more desirable than contact sensors. In previous research, a robot avoids obstacles and reaches the hand for target place by using proximity sensors mounted on the robot hand [2]. However in this research, the goal joint angles of the robot is well-known, so the proposed method is not effective in a situation where the location of the target object is unknown. In our method, a robot looks for the target object and avoid collision with environment simultaneously. Another research [3] proposes a method for creating physical models of surroundings based on proximity sensors in hands and grasping objects based on the models. However, this method is not effective in deformable environment because the created model will be broken by the deformation. We propose a method which do not need to construct environmental model. Besides this, pre-grasp motion has been researched [4] in order to recognize shape of the target object. In this motion, robot fingers align

**Fig. 1.** A robot is trying to grasp object in a bag.

**Fig. 2.** Gripper with proximity and contact sensor.

with the object surface by using proximity sensors on ball of the fingers. In this research, pre-grasp motion is executed only when the object is in hand, but we expand the method of alignment to previous stage of pre-grasp.

### 1.2   Paper Contribution

We propose a manipulation method for recognizing surroundings in real time, searching for the target object and grasping it, in deformable and closed environment. In addition to this, we check effectiveness of our method through an experiment in which the robot carries an object using a bag. I think that our method can be applicable to more variable situations, such as manipulating deformable objects or groping in occluded locations.

## 2   Daily Assistive Robot System for Carrying Objects with Bags

### 2.1   Overview

Motions required for carrying objects with a bag are below:

1. receiving the bag from human
2. transporting the bag to the designated destination
3. opening the bag by manipulating the bag handle
4. picking objects from the bag

In order for a robot to receive a bag from a human, it is necessary to detect the receipt by some sensors. Next, the robot need to manipulate the bag handle and grasp the suitable position for opening the bag. After opening the bag, the robot put its hand into it and pick objects from the bag. At the stage of picking the contents, the robot cannot see the inside because of the occlusion, so it is desirable for the robot to recognize the interior of the bag from the robot finger. These demands are met with use of sensors on the finger. The whole transportation system is shown in Fig. 3.

**Fig. 3.** Bag Transportation system.

# 3 Pre-contact Manipulation for Daily Assistive Tasks with Bags

In this section, we describe a motion strategy for picking objects from the bag. Among the four motions mentioned in Fig. 3, we give detailed description about '3. open bag' and '4. pick content', which need specially dexterous manipulation. A task system for manipulating the bag is shown in Fig. 4.

'3. open bag' requires two actions as below:

- (3-a) recognizing position of the bag handle and putting the bag handle between the fingers
- (3-b) tracing the bag handle, grasping top of it and opening the bag

'4. pick content' requires two actions as below:

- (4-a) putting the robot hand into the bag and searching for the target object
- (4-b) grasping the object and picking it out of the bag

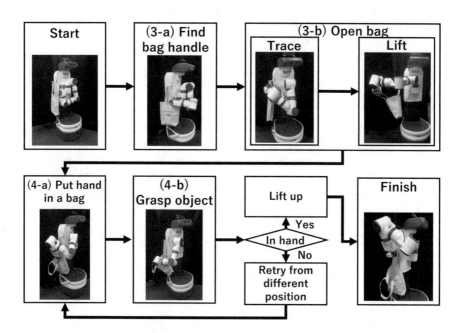

**Fig. 4.** Bag manipulation system.

## 3.1 Finding Bag Handle

Figure 5 shows the whole motion diagram of finding bag handle ((3-a) in Fig. 4), and in the following, we explain each of the elements. First, approximate position of the bag handle is estimated from the positional relationship between the hook on the robot and the bag, so the robot first moves its finger closer to the presumed position (Fig. 6a) and stop it when sensors on back of the finger react (Fig. 6b).

Next, the robot withdraws its hand until the sensors on back of the finger do not react (Fig. 6c). At this state, the robot acquires position of the bag handle.

After this, the robot sticks one of its fingers between two bag handles, taking the gripper size into consideration (Fig. 6d, e, f). The entire circumference arrangement of proximity sensors is effective in judging whether one of the fingers is between two bag handles. Only when sensors on ball and back of one finger and sensors on ball of the other finger react, the robot can estimate that only one bag handle is between the robot fingers (Fig. 6f).

## 3.2   Opening a Bag

Figure 7 is the summary of the whole motion of opening a bag ((3-b) in Fig. 4), and here we describe each of the elements. First, the robot estimates the inclination of the handle by comparing the adjacent sensor value, and aligns fingers with it. Besides, the robot calculates distance from sensors on each finger and moves its hand in order that the handle is at the center of fingers. For example, in the lower right side of Fig. 8a, it can be seen that the left finger is closer to the bag handle than the right.

After the robot moves its hand in such a way that the closest sensors on each finger to the handle show the same distance, it is expected that the bag handle is nearly the center of both fingers (Fig. 8b).

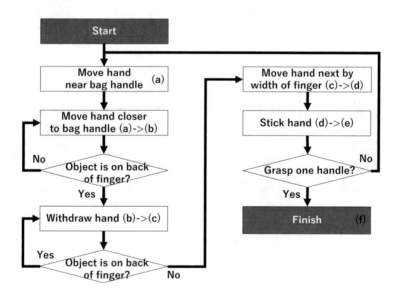

**Fig. 5.** Diagram of finding bag handle, (3-a) in Fig. 4. (a)–(f) in this figure corresponds to Fig. 6

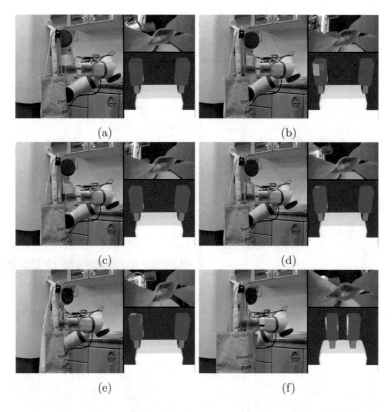

(a)  (b)

(c)  (d)

(e)  (f)

**Fig. 6.** Finding bag handle. The left area is overview. The upper right is a view in the bag. The lower right is visualization of distance estimated from sensor value. The redder the sensor tile is, the closer distance between sensor and other objects is.

In addition to these motions, the robot sticks or withdraws its fingers according to the sensor on front and root of finger. In Fig. 8c, for example, the sensors near the fingertip show stronger reaction than ones near the base of the finger. Judged from this state of sensor reaction, the robot sticks its hand by a certain distance. After this, the sensors react somewhat equally as shown in Fig. 8d. Conversely, the robot withdraws its hand when sensors near the base of the finger show stronger reaction than ones near the fingertip.

By repeating these motions while moving the hand up, the robot can trace the bag handle. The robot judges from well-known length of the bag handle if it reaches the top of the handle. If the robot tries to detach the bag handle from the hook after grasping low side of the handle, the handle may bend and the detachment will result in failure. Tracing the bag handle is important in order to avoid bending of the handle.

At last, the robot lifts it from the hook and open the bag.

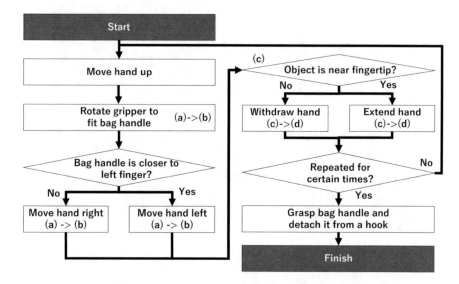

**Fig. 7.** Diagram of opening a bag, (3-b) in Fig. 4. (a)–(d) in this figure corresponds to Fig. 8

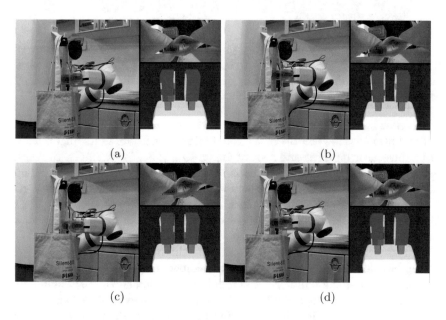

**Fig. 8.** Opening a bag. The left area is overview. The upper right is a view in the bag. The lower right is visualization of the distance calculated from sensor values. The redder the sensor tile is, the shorter the distance between each sensor and other objects is.

### 3.3    Putting the Robot Hand into the Bag

The whole motion of putting the robot hand into the bag ((4-a) in Fig. 4) is shown in Fig. 9, and now we explain each of the elements. In this motion, the robot puts its hand into the bag and searches for the target object.

When the back of robot fingers detects an object, the robot spreads the bag by opening and closing the gripper (Fig. 10a, b, c). Also, when the side of its fingers detects an object, the robot moves its hand away from the bag (Fig. 10d, e).

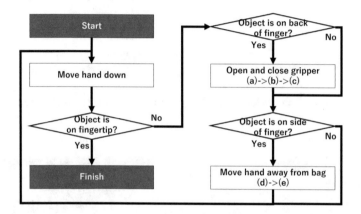

**Fig. 9.** Diagram of putting hand into the bag, (3-c) in Fig. 4. (a)–(e) in this figure corresponds to Fig. 10

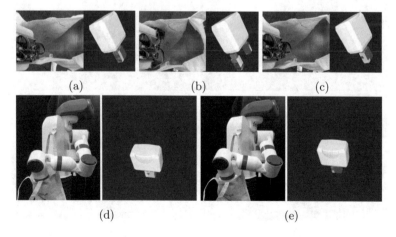

**Fig. 10.** Putting hand in a bag. The left area is an actual picture. The right is visualization of distance estimated from sensor value. The redder the sensor tile is, the shorter distance between sensor and other objects is.

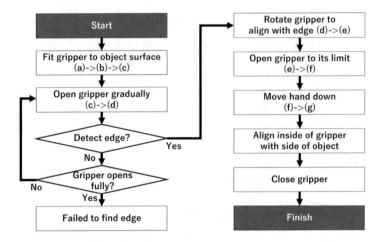

**Fig. 11.** Diagram of object grasping, (3-d) in Fig. 4. (a)–(g) in this figure corresponds to Fig. 12

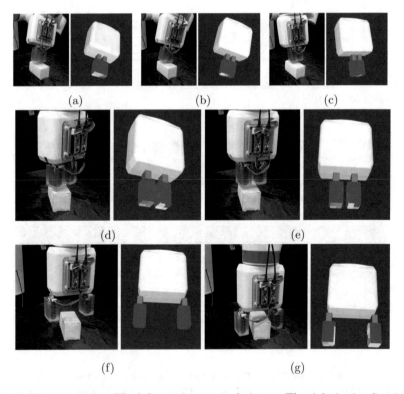

**Fig. 12.** Object grasping. The left area is an actual picture. The right is visualization of distance estimated from sensor value. The redder the sensor tile is, the shorter distance between sensor and other objects is.

By moving the hand into the bag through executing motions mentioned above, the robot can lower its hand to bottom of the bag. When sensors on the fingertip react, the robot judges that it finds the target object to grasp. At this stage, the robot cannot distinguish the bag from other objects.

## 3.4 Object Grasping

The diagram of object grasping ((4-b) in Fig. 4) is shown in Fig. 11, and in the following, we describe each of the elements. As a result of Subsect. 3.3, positional relationship between the hand and the target object is like Fig. 12a. In Fig. 12a, sensors in the far side show a stronger reaction than ones in the near side. At this time, the robot rotates its hand in the PITCH direction, and aligns fingertip with the object surface (Fig. 12b). In Fig. 12b, sensors on the right finger shows shorter distance than the left finger. The robot rotates its hand in the YAW direction (Fig. 12c). By executing these two motions, the robot aligns its fingertip with the surface of the object at two directions. After this state, the robot opens its gripper gradually, and stops opening when it detects object edge. The detection can be judged from the point that reaction of some sensors on either fingertip become weak (Fig. 12d). In this figure, the right finger detects the edge. Next, the robot aligns the edge-detecting finger with the direction of the edge. In Fig. 12d, sensors in the far side shows stronger reaction than ones in the near side at the right finger, which shows that direction of gripper movement and direction of the object edge are not vertical. The robot rotates its hand in the ROLL direction in such a way that they become vertical (Fig. 12e). By doing all these alignments (PITCH, YAW and ROLL), the robot can measure object orientation. Effect of the alignment can be seen at real picture of previous state (Fig. 12a) and following state (Fig. 12e).

Then, the robot opens its gripper fully (Fig. 12f) and moves the hand down (Fig. 12g) until sensors on fingertip react. In Fig. 12g, the robot recognizes that the object is in hand judging from both sensor reactions on ball of two fingers

**Fig. 13.** A hook attached to the side of Fetch.

**Fig. 14.** Bag and its content.

822    N. Yamaguchi et al.

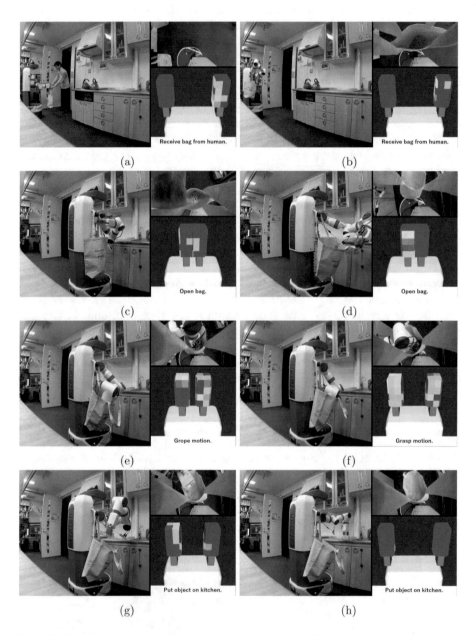

**Fig. 15.** Experiment scene. The left area is overview. The upper right is a view in a bag. The lower right is visualization of distance estimated from sensor value. The redder the sensor tile is, the closer distance between sensor and other objects is.

and distance between the two fingers. At last, the robot aligns inside of the fingers with side of the object by calculating inclination of the plane, and grasp the object. After grasping, the robot can determine whether the overall grasp motion succeeded or not.

## 4 Daily Assistive Robot Uses a Bag for Carrying Objects Experiments

We conducted an experiment in which a robot uses a bag to carry an object, and we selected Fetch [1] as a delivery robot. It has only one arm, so we attached a hook (Fig. 13) on its side and enables it to manipulate bags. The bag and contents used in this experiment are shown in Fig. 14.

First, a human hands over the bag to the robot (Fig. 15a). The robot recognizes receipt of the bag by reaction of sensors on finger, and hangs the bag handle on a hook attached on the side of the robot (Fig. 15b).

Next, the robot moves to kitchen and open the bag by grasping the bag handle (Fig. 15c, d). After this, the robot sticks the hand into the bag (Fig. 15e) and grasps the contents (Fig. 15f). At the manipulation, motion strategy described at Subsects. 3.1 and 3.2 is executed.

At last, the robot lift the hand up (Fig. 15g) and then put the grasped object onto the kitchen (Fig. 15h).

## 5 Conclusion

In this paper, we show the effectiveness of pre-contact and post-contact sensing for bag handling, through the experiment in which a robot manipulate a bag and carries an object. The former is effective in obstacle avoidance, object search and recognition of object orientation in a bag, and the latter is effective in check of grasp result. In the experiment executed in this paper, the robot picks one object from the bag. However, it is desirable for a robot to take out more than one objects from a bag from the viewpoint of efficiency of transportation. For future work, we will construct a system in which a robot distinguishes an object from another object, using more variety of sensors such as tactile sensors.

## References

1. Fetch robotics. http://fetchrobotics.com/
2. Cheung, E., Lumelsky, V.: Proximity sensing in robot manipulator motion planning: system and implementation issues. IEEE Trans. Robot. Autom. 5(6), 740–751 (1989)
3. Hsiao, K., Nangeroni, P, Huber, M., Saxenam, A., Ng, A.Y.: Reactive grasping using optical proximity sensors. In: ICRA (2009)
4. Koyama, K., Suzuki, Y., Ming, A., Shimojo, M: Integrated control of a multi-fingered hand and arm using proximity sensors on the fingertips. In: ICRA (2016)

5. Kragic, D., Miller, A.T., Allen, P.K.: Real-time tracking meets online grasp planning. In: ICRA, vol. 3, pp. 2460–2465. IEEE (2001)
6. Liu, M., Tuzel, O., Veeraraghavan, A., Taguchi, Y., Marks, T., Chellappa, R.: Fast object localization and pose estimation in heavy clutter for robotic bin picking. Int. J. Robot. Res. **31**(8), 951–973 (2012)
7. Wada, K., Sugiura, M., Yanokura, I., Inagaki, Y., Okada, K., Inaba, M.: Pick-and-verify: verification-based highly reliable picking system for various target objects in clutter. Adv. Robot. **31**(6), 311–321 (2017)
8. Wang, H., Liu, Y.-H., Chen, W., Wang, Z.: A new approach to dynamic eye-in-hand visual tracking using nonlinear observers. IEEE/ASME Trans. Mechatron. **16**(2), 387–394 (2011)

# Design of an Adaptive Force Controlled Robotic Polishing System Using Adaptive Fuzzy-PID

Hsien-I Lin$^{(\boxtimes)}$ and Vipul Dubey

Graduate Institute of Automation Technology,
National Taipei University of Technology, Taipei, Taiwan
sofin@ntut.edu.tw, vipul1809@gmail.com

**Abstract.** In this paper, we designed the robotic polishing system for industrial polishing needs. The current robot polishing techniques are available for objects with CAD geometric models, but for an object whose geometric model is unavailable such as ceramic or clay pots, it is still a puzzle to derive robot polishing trajectories and force control. In our paper we designed a polishing system to polish any arbitrary objects by providing a rough robot trajectory and the polishing system will adjust its trajectory to maintain the desired contact force on the object. To perform the desired force tracking, the polishing system adopted an adaptive Fuzzy-PID controller to regulate its parameters and then actuated a stepper motor mounted on the tip center of the robot to the desired contact force. The designed system is capable to track the force with the overshoot less than $0.2\,\mathrm{N}$ and the response time less than $40\,\mathrm{ms}$. The results dictated that the proposed controller is efficient in tracking the desired force with the indication of good polishing quality.

**Keywords:** Robotic polishing system · Adaptive · Fuzzy-PID

## 1 Introduction

In a typical manufacturing process, after the basic machining operations such as milling, drilling, cutting, etc. are performed on an object, its surface has to be smoothened precisely. This finishing process is known as polishing. The polishing process is generally performed by highly skillful workers with several years of experience. However, the crisis of shortage of skillful workers is becoming severe year by year [6]. Manual polishing lacks repeatability and quality control throughout the part surface and is slow [7,9]. Besides, the process is labor intensive and hazardous because of abrasive particles [3]. Hence, replacing manual polishing by automation is an urgent task. Previously, some researches have been conducted in automation of the polishing process. [6] used a high speed camera to acquire the polishing technique of a skillful worker and then the captured motion was replicated by controlling the tool trajectory and posture.

© Springer Nature Switzerland AG 2019
M. Strand et al. (Eds.): IAS 2018, AISC 867, pp. 825–836, 2019.
https://doi.org/10.1007/978-3-030-01370-7_64

The method used for replication was position control but it did not guarantee the polishing forces to be within the desired range. [5] used both position and force control for a robotic arm to perform the polishing process on the PET (Poly Ethylene Terephthalate) bottle mold. The polishing force was controlled by manipulating the serial robotic arm. However, it could not be used in a parallel robotic arm because of the difference in manipulator dynamics. [8] suggested the use of multiple sensors for on-line monitoring of the surface roughness while polishing. [3] proposed to use a pneumatic actuator as a compliant tool for polishing, where the actuator extension and retraction were controlled based on the pressure requirement. [3] used gasbag polishing to enhance the polishing quality. [2] applied a BP neural network with PID to the gasbag polishing to control the force precisely. Since the size of gasbag was quite small, it confined the tolerance of the polishing tool. [7] adopted multiple sensors to decide polishing endpoints merely in rotationally symmetric objects. [1] used a piezoelectric actuator to perform precision polishing suitable for micro force applications. [11] used the force control method for grinding industrial molds.

All the above researches heavily relied on object CAD/CAM models for the position and force control on series/parallel robotic arms. However, it becomes difficult to polish objects whose geometry is not available. To ease the problem, we proposed to use adaptive fuzzy control of PID parameters in a high-quality polishing process. The benefit is that the proposed system does not need the geometric model of an polished object and can be applied to any type of robotic arms because the robotic arm will automatically adjust its rough trajectory to maintain the desired contact force.

## 2    Design and Setup of the Polishing Tool

Several different actuation mechanisms were devised for force control in a task of polishing a workpiece in previous work. [3] used the pneumatic actuation mechanism to control the force output for the polishing tool. The drawback of pneumatic systems is that they occupy much space and are difficult to manage. [2,10] used the gasbag to polish the workpiece. This mechanism has less tolerance for polishing surface waviness because the gasbag can be inflated up to a limited volume. [4] used the voice coil actuation system for force control. The voice coil was used for micro force adjustments. However, the problem with voice coil is that it cannot be used in polishing applications with high force.

Figure 1 shows the proposed robotic polishing system. We used a stepper actuation system to provide force control in polishing because it makes polishing precise by adjusting the force resolution through micro-stepping. The other advantage of a stepper motor is that it moves precisely in an angular position by commanding the number of steps to be moved. It does not need an external rotational position encoder to feedback the exact position of the motor shaft. Apart from this, the stepper actuation mechanism is suitable for high force applications. When the stepper motor rotates, the rotational motion is converted to the translational motion by the ball-screw mechanism, with the result that the exertion is induced on the polishing surface. The stepper motor has 25600 steps for a revolution; thus the rotational resolution is 0.014 degree/step. The higher revolution resolution results in the higher force resolution and the system accuracy is improved. The stepper motor uses PWM (Pulse Width Modulation) signal for actuation. The frequency of the PWM is the controller output to the stepper motor. In our system, the pulse width of PWM is at least greater than 2.5 ms for the system to work. Thus, for the duty cycle of 50%, the PWM operating frequency should be less than $2 \times 10^5$ Hz.

**Fig. 1.** Proposed robotic polishing system.

In the ball-screw mechanism, SGX SFU 1605 Linear guide rail was used. The total moving length for the linear guide rail is 100 mm, the screw diameter is 16 mm and the pitch is 5 mm. The entire motor-linear guide rail setup has the horizontal load capacity of 80 Kg and the vertical load capacity of 30 Kg. The linear guide rail with a longer moving length indicates the more tolerance towards polishing surface waviness and irregularities.

The polishing quality replies on two essential factors. First is how well the polishing force follows the desired one. Second is the desired polishing stop time i.e. if the polishing is stopped before or after the desired stop time, the polishing quality is degraded. For example, [7] used an acoustic emission sensor, a force sensor, and a scattered light sensor to detect the endpoint. In our system we used a Dyn-Pick 200 force sensor. It is a capacitive 6-axis force sensor. The sensor has a nominal rated load capacity of 200 N along all the three axes. To record the force values while polishing, the force sensor was sandwiched between the sensor and polishing tool mountings. The force during the polishing was propagated though the tool, sensed by the force sensor, and then sent to the controller for further processing. In this paper we focused on the polishing of ceramic materials. The polishing tool bit was attached to the tool with the rotational motor actuation whose RPM was adjusted depending on the polishing requirements.

## 3   System Dynamics

The system dynamics will determine all the necessary forces acting on the polishing tool, which affects the polishing quality. Here we discuss the forces acting in horizontal and vertical directions on the polishing tool. In Fig. 2 we can see the forces acting on the polishing tool in horizontal direction i.e. $x - y$ plane. As the robotic arm moves on $x - y$ plane, the friction force is generated opposite to the direction of motion. The friction force acting on the polishing bit can be calculates as:

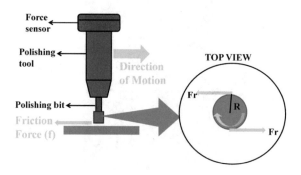

**Fig. 2.** Forces acting in the horizontal plane.

$$f = \mu \times N \tag{1}$$

where $f$ is the friction on the polishing bit, $\mu$ is the coefficient of the kinetic friction, and $N$ is the normal force acting on the publishing surface.

The Fig. 3 shows the free body diagram (FBD) of the polishing tool and polishing surface for the forces in $z-$direction. By analyzing the forces acting on the polishing surface we get:

For polishing surface to be in equilibrium:

$$N = Fp \tag{2}$$

where $Fp$ is the polishing force acting on the surface. Replace $N$ in Eq. (1), we have

$$f = \mu \times Fp \tag{3}$$

Apart from the translational friction force, the frictional torque also acts on the polishing bit as shown in Fig. 2. The two rotational friction forces balance each other and hence do not provide any resultant force in the direction of motion. The rotational friction creates an opposing torque to the rotating polishing bit. This is the resistant torque opposing to the polishing tool rotational motion.

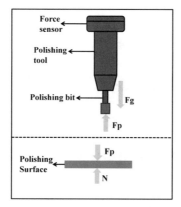

**Fig. 3.** Free body diagram of the polishing tool and surface in $z-$direction where $Fg$ is the force due to gravity.

$$\tau = 2R \times Fr \tag{4}$$

$$Fr = \mu \times Fp \tag{5}$$

Substitute Eq. (5) into Eq. (4), we have:

$$\tau = 2R \times (\mu \times Fp) \tag{6}$$

where $\tau$ is the frictional torque acting on the polishing bit, $R$ is the radius of the polishing bit, and $Fr$ is the friction force due to rotation. The free body diagram of the polishing tool reveals the forces acting on it in the vertical direction. These forces can be used to find the dynamic equation of the polishing tool. Thus, we have

$$Fp - m \times g = m \times a \tag{7}$$

Thus,

$$Fp = m(a + g) \tag{8}$$

where $m$ is the mass of the polishing tool, $g$ is the acceleration due to gravity $(9.81/\text{sec}^2)$, and $a$ is the net acceleration in $z-$direction.

The distance travelled in $z-$direction is derived as:

$$Z = \frac{P}{Sr} \times S \tag{9}$$

where $Z$ is the distance travelled in z-direction, $P$ is the pitch of ball screw (the distance travelled in one full rotation of motor shaft), $Sr$ is the number of steps/revolution, and $S$ is the total number of moved steps.

Take the derivative of Eq. (9),

$$\dot{Z} = \frac{dZ}{dt} = \frac{P}{Sr} \times \frac{dS}{dt} \tag{10}$$

The number of moved steps is equivalent to the number of PWM pules. Thus,

$$Pf = \frac{dS}{dt} \tag{11}$$

where $Pf$ is the PWM pulse frequency. Substitute Eq. (11) into Eq. (10), we have

$$\dot{Z} = \frac{dZ}{dt} = \frac{P}{Sr} \times Pf \tag{12}$$

Take derivative of Eq. (12), it becomes

$$a = \ddot{Z} = \frac{P}{Sr} \times \dot{Pf} \tag{13}$$

Substitute Eq. (13) into Eq. (8), we have

$$Fp = m \left( \frac{P}{Sr} \times \dot{Pf} + g \right) \tag{14}$$

From Eq. (14) we observe that the polishing force depends on the changing rate of the PWM frequency. In this study, we control the polishing force by the change rate of the PWM frequency.

## 4    Fuzzy-PID Controller for Polishing

The input polishing force was digitized by an Analog-Digital converter (ADC) and filtered to remove the high frequency noise. An Equi-ripple low pass filter with pass frequency of 0.2 Hz was used. The filtered force was then calibrated to remove the residual forces and sent to the controller to decide the appropriate PWM frequency to control the velocity of the polishing tool.

For the controller, we first used Proportional Integral Derivative (PID) and then integrated it with fuzzy logistics to design the adaptive fuzzy PID. The governing equation of PID controller is defined as:

$$u(t) = K_p e(t) + K_i \int_0^t e(\tau) d\tau + K_d \frac{d}{dt} e(t) \tag{15}$$

where $u(t)$ is the PID output, $K_p$ is the proportional gain, $K_i$ is the integral gain, $K_d$ is the derivative gain, and $e(t)$ is the error at time $t$.

To achieve the maximum performance the PID coefficients must be tuned accurately. If not, it will lead to system instability, with the result of force overshoot. The parameters that define the effectiveness of the system are the time to reach desired value, overshoot, and the response time. To perform a good polishing, the polishing system should be able to reach the desired force value quickly with less force overshoot. These parameters are affected by the choice of PID parameters. While working at different force values, the system adopts different sets of PID parameters to reach the best performance. Thus, instead of fix the values of PID parameters, we proposed adaptive PID control for the polishing system. To regulate the values of PID parameters, Fuzzy logic was implemented to the proposed system as shown in Fig. 4.

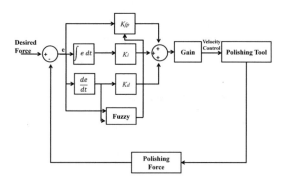

**Fig. 4.** Adaptive Fuzzy-PID block diagram.

Fuzzy control system uses human knowledge to design the system output. At first the input values are defined into different fuzzy sets using linguistic variables and the fuzzy sets are defined using fuzzy membership functions shown in Fig. 5. Next, the inference engine uses the rules defined for each fuzzy membership function using human knowledge to generate the output as in Table 1 where 'e' is absolute value of error and 'rce' is absolute rate of change of error and output corresponds to the value of $K_{fp}$. The output from inference engine is fuzzified and hence not appropriate for practical applications. So, the fuzzified output is converted into a crisp value to be used for real applications. The method used for defuzzification was "Centre of area". The formula for it is shown as below:

$$COA = \frac{\sum_i^N \mu_i x_i}{\sum_i^N \mu_i} \tag{16}$$

where $COA$ is the center of the area value, $\mu_i$ is the membership value corresponding to the input value $x_i$, and $N$ is the fuzzy rule number.

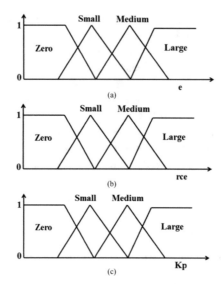

**Fig. 5.** Membership function for (a) absolute error (b) absolute rate of change of error (c) output $K_{fp}$ value.

**Table 1.** Rule base for the fuzzy logic controller.

|     |        | e      |        |        |       |
| --- | ------ | ------ | ------ | ------ | ----- |
|     |        | Zero   | Small  | Medium | Large |
| rce | Zero   | Zero   | Small  | Medium | Large |
|     | Small  | Small  | Medium | Large  | Large |
|     | Medium | Medium | Large  | Large  | Large |
|     | Large  | Large  | Large  | Large  | Large |

## 5    Experimental Setup and Results

### 5.1    Experimental Setup

The robustness and efficiency of our polishing system was tested by performing single pass polishing test on a ceramic plate. Figure 6(a) shows the tested ceramic plate. Figure 6(b) shows that the suitable polishing tool along with appropriate polishing bit was fixed to the tool mounting. We used a WL-800 grinder polishing tool. The power input was DC 18V and the tool could work at variable RPM (revolutions per minute) varying from 5000–18000 RPM. The polishing bit was cylindrical soft material appropriate for the ceramic used in our experiment.

To perform the polishing experiment, we provided a rough trajectory to the Staubli TX-40L robotic arm. Since the ceramic plate used for experimentation was circular with flat surface, we provided a straight line trajectory to our

(a)                                        (b)

**Fig. 6.** (a) Ceramic plate for polishing experiment. (b) WL-800 polishing tool and polishing bit.

robotic manipulator. The robotic arm communicated with the computer through Ethernet to transfer the robotic position. The stepper motor was connected to DM542A driver, which supplied the power to the motor and also the command signal received from the microcontroller. The USB interface was used to transfer and receive motor commands and force readings between the microcontroller and the computer. While the stepper actuation caused the change in the polishing force, the force sensor would detect and send it to the computer through the microcontroller.

## 5.2   Results and Discussion

The results of force response of single pass polishing test using our adaptive force controlled robotic polishing system are discussed in this section. Figure 7 represents the desired force value and the force response using the PID and adaptive fuzzy PID controller. Figure 7 shows the force response of the PID controller at the desired force of 1.5 N. The force rise time from 0.5 N to 1.5 N is 120 milliseconds (ms) without force overshoot observed. Figure 7 also shows the force response of adaptive Fuzzy-PID controller at desired force of 1.5 N. We observed the rise time of 40 ms with negligible force overshoot. The force reaches its desired value quickly without any force fluctuations. After the desired force value is achieved the system tracks the desired force value closely while keeping the system stable.

Figure 8 shows the force response using the PID controller for high force change from 0.5 N to 4 N. The rise time is equal to 80 ms. The force overshoot is 0.3 N. The controller takes some time for the force to stabilize. The force value oscillates around the desired force value. The settling time is 350 ms. Figure 8 also shows the force response of the adaptive Fuzzy-PID controller at desired force of 4 N. The rise time is 40 ms with the overshoot of 0.2 N. The force becomes stable quickly. Once the desired force value is achieved, the system traces the desired force value without any fluctuations.

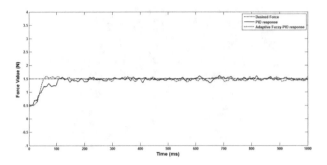

**Fig. 7.** Force response at 1.5 N.

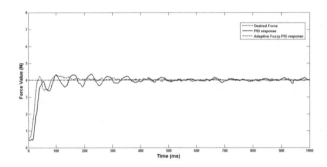

**Fig. 8.** Force response at 4 N.

The second test was performed to analyze the system response to an external force disturbance between the PID and adaptive Fuzzy-PID controllers. The force disturbance was introduced manually after the system has already become stable. Figure 9(a) shows the force disturbance response for the PID controller at a low force value of 1.5 N. It took 200 ms for the PID controller to settle down the disturbance to the desired force value. Figure 9(b) shows that the adaptive Fuzzy-PID controller was also used at 1.5 N. It took 80 ms for the system to cancel the disturbance and make the system stable again. As we can observe, the response was steep and quick to nullify the sudden force change.

From Fig. 10(a), we can see the force disturbance response of the PID controller at high force value of 4 N and the system took 250 ms to settle down to the desired force value. In Fig. 10(b), the adaptive Fuzzy-PID controller was tested at high force value of 4 N and the system took 120 ms to settle down. The response was steep and quick. We can observe that irrespective of operating force value, the adaptive Fuzzy-PID controller provided a quick response to external force disturbance and made it stable.

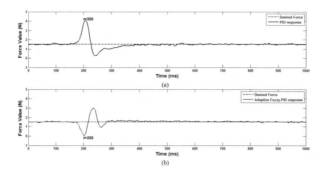

**Fig. 9.** Response to disturbance at 1.5 N.

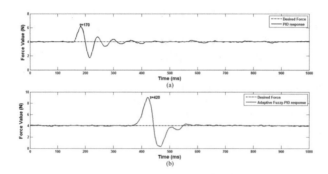

**Fig. 10.** Response to disturbance at 4 N.

## 6   Conclusions

In this paper, the adaptive force controlled robotic polishing system was designed. Previously, [5] used a serial robotic manipulator to perform force control for polishing with the force resolution of 0.5 N. [2] performed polishing with the steady state force error of 2 N using the gasbag mechanism. [5] performed polishing using a macro-mini robot approach with the force overshoot of 3–5 N. [9] performed polishing with the minimum force fluctuation of 2 N. From the results of the proposed system, we can observe that it is possible to control forces with 0.1 N force resolution with the maximum force overshoot of 0.2 N. Force control using for two different controllers, PID and adaptive Fuzzy-PID were compared in the experimental results. The results validates that PID control itself is not self-sufficient for adaptive force control for a wide range of force applications. It is necessary to integrate human experiences to design a better adaptive force controller. Thus, we designed an adaptive Fuzzy-PID controller for the polishing system. We conclude from the results that the designed controller has better performance over the basic PID control. The overall response time is reduced and the system responds quickly to the change in forces. Also the invented controller is more stable and nullifies the external force disturbances quickly. The results demonstrate that our system is efficient to perform precise adaptive force control for polishing. In the future, the adaptive force control algorithm will be implemented to perform precise polishing even without object surface geometries. The polishing tests will be performed on range of objects and their surface roughness will be measured using Coordinate measuring machine (CMM).

# References

1. Guo, J., Suzuki, H., Morita, S.Y., Yamagata, Y., Higuchi, T.: A real-time polishing force control system for ultraprecision finishing of micro-optics. Precis. Eng. **37**(4), 787–792 (2013)
2. Jin, M., Ji, S., Pan, Y., Ao, H., Han, S.: Effect of downward depth and inflation pressure on contact force of gasbag polishing. Precis. Eng. **47**, 81–89 (2017)
3. Liao, L., Xi, F.J., Liu, K.: Modeling and control of automated polishing/deburring process using a dual-purpose compliant toolhead. Int. J. Mach. Tools Manuf. **48**(12), 1454–1463 (2008)
4. Mohammad, A.E.K., Hong, J., Wang, D.: Design of a force-controlled end-effector with low-inertia effect for robotic polishing using macro-mini robot approach. Robot. Comput.-Integr. Manuf. **49**, 54–65 (2018)
5. Nagata, F., Hase, T., Haga, Z., Omoto, M., Watanabe, K.: CAD/CAM-based position/force controller for a mold polishing robot. Mechatronics **17**(4), 207–216 (2007)
6. Oba, Y., Yamada, Y., Igarashi, K., Katsura, S., Kakinuma, Y.: Replication of skilled polishing technique with serial-parallel mechanism polishing machine. Precis. Eng. **45**, 292–300 (2016)
7. Pilnỳ, L., Bissacco, G.: Development of on the machine process monitoring and control strategy in robot assisted polishing. CIRP Ann.-Manuf. Technol. **64**(1), 313–316 (2015)
8. Segreto, T., Karam, S., Teti, R., Ramsing, J.: Cognitive decision making in multiple sensor monitoring of robot assisted polishing. Procedia CIRP **33**, 333–338 (2015)
9. Tian, F., Lv, C., Li, Z., Liu, G.: Modeling and control of robotic automatic polishing for curved surfaces. CIRP J. Manuf. Sci. Technol. **14**, 55–64 (2016)
10. Xian, J.S.J.M.Z., Julong, Z.L.Z.Y.Y.: Novel gasbag polishing technique for freeform mold. Chin. J. Mech. Eng. **8**, 004 (2007)
11. Xie, X., Sun, L.: Force control based robotic grinding system and application. In: 2016 12th World Congress on Intelligent Control and Automation (WCICA), pp. 2552–2555. IEEE (2016)

# Efficient, Collaborative Screw Assembly in a Shared Workspace

Christian Juelg[⊠], Andreas Hermann, Arne Roennau, and Rüdiger Dillmann

FZI Forschungszentrum Informatik,
Haid-und-Neu-Straße 10–14, 76131 Karlsruhe, Germany
{juelg,hermann,roennau,dillmann}@fzi.de

**Abstract.** We provide online dynamic robot task selection in human-robot collaborative contexts through a voxel-based collision avoidance system. This paper describes a partially automated screw assembly work-cell with a worker and a 6-DOF light-weight robot arm. The shared workspace is monitored by 3D point-cloud sensors. The dynamic task selection reduces robot waiting due to obstacles by exploiting situations where multiple tasks are available for the robot. Massively parallel evaluation of voxelized robot trajectories allows online avoidance of blocked paths. The robustness of the screw assembly process is increased through Cartesian compliance based on force-torque sensor feedback.

**Keywords:** Collaborative robots · Human-robot-interaction
Robots for Industry 4.0 · Robot vision

## 1 Introduction

The term human-robot collaboration (HRC) describes a wide range of applications where humans and robots work in close temporal and spatial proximity. Whereas previously industrial robots generally required mechanisms that prevented any physical contact between humans and robots, HRC applications use several measures that allow close collaboration with a robot while maintaining work-place safety for humans [8].

The screw assembly demonstrator described here combines technologies like 3D point-cloud sensors for collision avoidance, force-torque sensor for tactile feedback during manipulation and high-level robotics software that allows flexible and easily extensible robotic solutions for industrial applications.

### 1.1 Motivation

The collaborative screw assembly work-cell described in this work illustrates the potential for partial automation and division of labor enabled by HRC. In this screw assembly scenario, the robot performs the task that is very repetitive and easy to automate, namely the tightening of screws in a set of fixed positions.

© Springer Nature Switzerland AG 2019
M. Strand et al. (Eds.): IAS 2018, AISC 867, pp. 837–848, 2019.
https://doi.org/10.1007/978-3-030-01370-7_65

**Fig. 1.** Collaborative screw assembly demonstrator at Motek 2017 in Stuttgart

The human worker in this scenario performs the tasks that require tactile feedback and that would be more expensive to automate. The cost advantage of partial automation is increased, when the number of units that are produced at a time is low. Partially automated applications require less effort in order to adapt to different work-pieces and processes.

## 1.2   State of the Art

In the past decade, the introduction of light-weight robot (LWR) arms [1] enabled many new human-robot collaborative solutions. One operating principle for LWRs is torque limiting to minimize the risk of injury posed by the robot. The Robot Operating System (ROS) [10] is a communication framework for complex and flexible robotic applications. It allows for scalable communication of loosely coupled components. One example of a flexible service robotics application that combines a light-weight robot and many different robotic skills is the BratWurst Bot [7].

There has been research into adding point-cloud sensors to robot workspaces in order to offer collision avoidance [3]. One approach achieves this by transforming both the environment observations as well as the robot ego-model into a voxel representation in a shared coordinate system. This allows to distinguish between observations of the ego robot and other objects. Additional research evaluated a linear motion prediction capability added onto this generic voxel-based approach [4]. The GPU-Voxels library can test complex planned motions against the current and predicted state of the observed workspace by rendering the planned trajectory as a swept volume and performing massively parallel collision checks. The voxels involved in the swept volume are marked as occupied for each appropriate future time-step, which allows for detailed feedback to the planner. The same data structure can also capture time-step information about predicted motions of observed obstacles. There has also been research into predicting human motions in shared workspaces based on body part tracking [6].

The use of robots in shared workspaces for human-robot collaborations requires several design considerations to maintain design safety [8]. There have been standardization efforts that specify safety requirements for robots in collaborative environments [5].

### 1.3   Problem Description

Screw assembly is a common task in industrial production. We decided on partial automation in order to use the respective skills of humans and robots, namely human dexterity for the screw insertion and repeatability and process monitoring for the robotic screw insertion. In such a scenario, direct collaboration without fences or safety regions that stop the robot whenever the humans enters a fixed area could improve the overall throughput. At the same time, when compared with robotic solutions that do not allow humans in the robot's working area during its operation, LWRs in collaborative mode are limited in their velocity because of safety considerations.

Different from many other approaches to human-robot collaboration, we attempt to constantly choose obstacle-free trajectories for the robot in a dynamic environment.

### 1.4   Goal

The main objectives of the demonstrator are to reliably perform collaborative screw assembly and to show collision avoidance and forward-looking task selection based on sensors monitoring the shared workspace.

With low reaction times, such a system can avoid many collisions and can reduce the periods during which the robot waits for the human operator to clear its planned trajectory.

### 1.5   Concept

Our general concept is to use a LWR arm with a mounted electric screwdriver to tighten the screws. In order to enable force-sensitive manipulation strategies, an add-on force-torque sensor is inserted between the robot end-effector and the screwdriver. The screwdriver controller monitors the torque reached during tightening. Four stereoscopic cameras placed around a table offer redundant depth images about obstacles in the shared workspace. The multi-camera setup reduces occlusion of the human worker by the robot arm. Our dynamic task selection does not involve motion planning. Instead, it performs dynamic, situational trajectory selection. This avoids uncertainty introduced by stochastic planning processes such as [9].

As shown in Fig. 2, our system renders previously recorded joint trajectories from a home position to target positions above each screw head offline to swept volumes at a given voxel grid resolution. We chose to sample each trajectory at eleven discrete time steps in order to achieve a balance between accuracy and

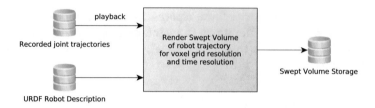

**Fig. 2.** Offline swept volume generation from recorded trajectories

run-time overhead. Figure 10 shows the combined swept volume of one trajectory. During run-time, our system can then perform GPU-accelerated collision checks for all trajectories in parallel. This way the robot can be directed to abort the execution of its current trajectory if the sensor data points to a future collision, as shown in Fig. 3. It is also possible to select a new trajectory from the set of trajectories that are currently collision-free.

In a first iteration, the robot will always attempt to return to the home position by reversing its previous motion. If trajectory execution was canceled because of detected obstacles, the robot executes the already completed part of the trajectory in reverse.

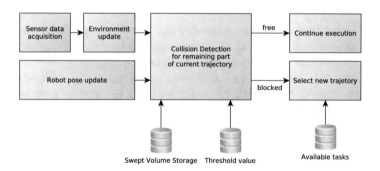

**Fig. 3.** Online environment update and trajectory selection

## 1.6   Structure

In the next section we will present our hardware setup and the specific challenges and solutions we approached. Afterward we show the experiments we performed to evaluate our methods and the screw assembly demonstrator. In the final section we will summarize the work we did and give an outlook on future work that could improve upon the current results.

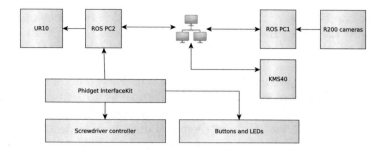

**Fig. 4.** Network topology for the Motek demonstrator

## 2   Approach

### 2.1   Hardware Setup

In our demonstrator we settled on an Universal Robots UR10 6-DOF LWR arm [13]. We used a desktop PC with hexacore Intel Core i7-6850K, 32 GB RAM and a Nvidia Titan Xp GPU with 12 GB VRAM, as well as a Shuttle Barebone PC. Both PCs run ROS Kinetic and Ubuntu 16.04 as well as the GPU-Voxels 1.1 and PCL libraries and are connected as shown in Fig. 4. For the camera setup we used four Intel Realsense R200 stereo-infrared cameras with active infrared projection. As force-torque sensor we used a Weiss KMS40. As screwdriver we employed a Kolver Pluto with a HP Pro control box. To forward the input and output lines of the control box to the PCs, we used a Phidget InterfaceKit 16/16/0, which also connected to the LED buttons that allow input from the human worker during operation.

The robot and cameras were fixed on a table, as can be seen in Fig. 1. The screwdriver we selected is designed for manual use and was not ready to be mounted on the robot. In order to mount the screwdriver on the robot, we used a 3D printed fixture. The same counts for the point-cloud sensors. The sensor fixtures were designed to create overlapping the fields of view in the central working area above the four work-pieces.

### 2.2   Challenges

One of the challenges for our demonstrator was the online capability. In order to be useful, the collision checks for all available trajectories need to run fast enough to avoid collisions and to minimize idle waiting for the robot. A second issue was sensor calibration and the modeling of the collision model for the robot and the known static parts of the environment. If there exist inaccuracies in the extrinsic point-cloud sensor calibration or in the robot collision model, the system could stop the robot even when there are no obstacles blocking the chosen trajectory, or could prevent the robot from selecting trajectories that it could execute without a risk of collision. An adverse factor in this was the presence of flexible cables on the robot arms, whose precise position changes during robot operation. These false positive during collision checks against swept volumes reduce the effectiveness of the overall system and could even permanently block the robot. At the same time any filtering of the observed environment data runs the risk of creating false negative collision checks, where the system disregards real obstacles.

Further challenges are posed by the screw tightening process itself. In order to limit wear on the robot and the work-pieces caused by excessive contact forces, a force sensitive control mode is desirable. At the same time, contact between screwdriver tip and screw head must be maintained in order to reliable tighten the screw without excessive wear of the screw or the screwdriver tip.

## 2.3   Solution

The main goal of the vision component of our robot is to enable dynamic task selection (Fig. 5), so that the robot can work without interfering with the human worker whenever possible. We employed multiple mitigations for false positive collision checks caused by sensor data capturing the robot or parts attached to the robot. Among these are augmented collision models for the robot and the mounted tool. We added virtual collision volumes to the robot collision model that envelop the volume within which the flexible cables could move during operation. Additionally, the tool's collision model was dilated by one voxel in order to increase the minimum acceptable distance between the obstacles and the robot.

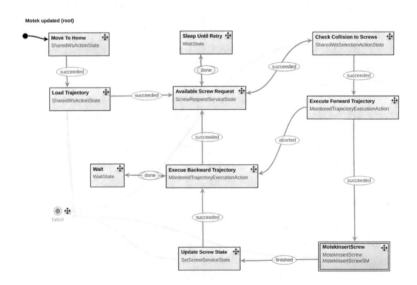

**Fig. 5.** FlexBE top-level state machine controlling the assembly process

We added a density threshold, as well as a collision threshold parameter. The density threshold reduces noise by counting all voxels as empty that are only scarcely populated by the combined point-cloud from all sensors. The collision threshold was added to allow some colliding voxels within a collision check for a swept volume. This addition is a response to sensor noise and discretization artifacts during collision model insertion, as well as to unmodeled parts of the environment. One example for an artifact with a single colliding voxel could be a partly inserted screw at the end-point of a planned trajectory. This solution increases the risk for falsely negative collision checks while potentially increasing the overall robustness of the task execution in a dynamic and partly unmodeled environment.

In order to increase reliability during screwdriver operation, we used the force-torque sensor in combination with a compliant robot controller that allowed us to set a contact force between screwdriver tip and the screw head [12]. We also used the force-torque sensor beforehand to determine whether the screwdriver tip has made contact with the screw head after first moving downwards in compliant mode and then reading the force-torque sensor, as can be seen in Fig. 6.

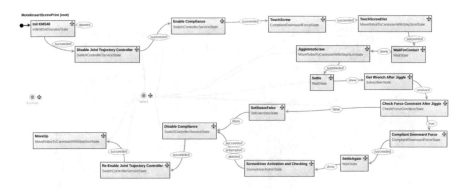

**Fig. 6.** FlexBE nested state machine for screw insertion

In order to provide quality control information, we document the feedback of the screwdriver controller regarding the torque reached during the tightening process. Similarly we report an error to the user whenever the screw head is not detected during the downward movement, as seen in Fig. 7.

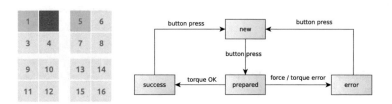

**Fig. 7.** Left: screw status visualization: partially inserted (yellow), successfully tightened (green) and failure (red). Right: screw state diagram.

## 2.4  Highlights

Compared to previous publications based on GPU-Voxels [3,4], we reduced reaction times, increased the number of cameras and the size of the combined point-cloud being processed, as well as adding density filtering to reduce noise. In a first attempt we used a CPU bound sequential method for point-cloud density filtering based on the Point Cloud Library (PCL) [11]. This resulted in increased latency while enforcing a low voxel filter resolution of 5 cm. To improve upon this, we added a density filtering method to GPU-Voxels that uses GPU-accelerated voxellist algorithms to perform the operation faster and with a more detailed filter resolution of 2 cm.

The compliant force-feedback based controller used during the screw tightening phase models a virtual spring connected to the screwdriver tip. This spring can be configured to pull the tool tip downward with a given amount of force, while still responding to lateral forces during screwdriver operation.

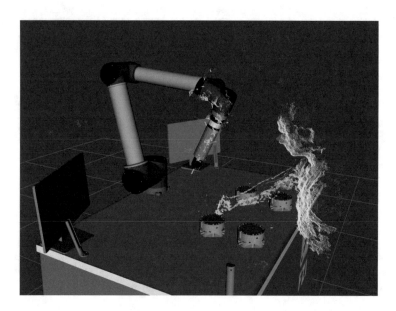

**Fig. 8.** RViz visualization of combined point-clouds. The points are colored according to their source camera.

## 3  Results and Experiments

The approach to collaborative screw assembly outlined in this paper has been demonstrated at the Motek trade fair in Stuttgart in October 2017, as seen in Fig. 1.

## 3.1   Experiments

We encountered several issues with collision checking of the combined point-clouds. Changing intensity of external infrared light-sources, especially daylight, influenced the optimal parameter values for our density filter as well as for the collision threshold.

Another challenge was the extrinsic calibration of the four cameras in relation to each other and to the robot arm. The main source for falsely positive collision checks were depth points of the robot and attached parts that were not filtered and not contained in the voxelized robot collision model. Figure 8 shows the combined point-cloud created from four cameras mounted around the table. The optimization goal for the calibration process was a precise overlap of all depth camera images with the robot ego model and with each other.

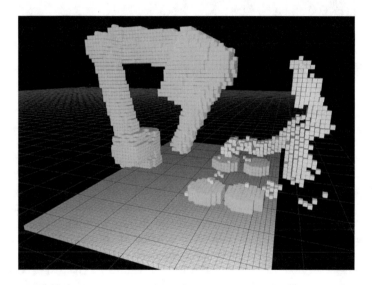

**Fig. 9.** GPU-Voxels visualizer with robot collision model, static and dynamic obstacles

## 3.2   Results

The following timings are for a scene with a single human worker as seen in Figs. 8, 9 and 11 with four Realsense R200 cameras and a voxel grid resolution of 2 cm. The vision system inputs synchronized depth images from the point-cloud sensors at 24 fps, which is lower than the native sensor update rate of 30 fps at a depth resolution of 628 by 468 RGB-D pixels. The transformation to global coordinates of all depth streams combined takes 6.35 ms on average and the following copying to the GPU and density filtering takes 9.70 ms on average. The swept volume rendering process for the pre-recorded trajectories is performed in an online step. Figure 10 shows the visualization of a single trajectory from the home position towards a target position above a screw.

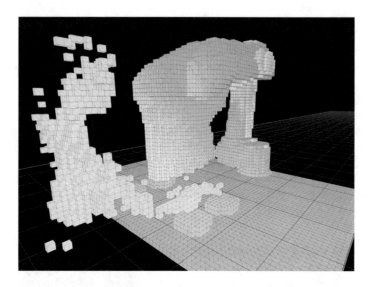

**Fig. 10.** Visualization of rendered swept volume for a single trajectory

A single collision check for the swept volume of the remainder of the current trajectory against the environment takes 13.35 ms on average. Because of this, the reaction time from the moment an obstacle enters the current trajectory until the collision check completes stays below 100 ms on average. Sensor update and collision checking both run at above 10 Hz. This helps avoid frustrating user experiences in HRC scenarios, where the robot reacts to events only after a noticeable delay, creating uncertainty about whether the event was noticed at all [2].

The image series in Fig. 11 shows the behavior of our collision avoidance and dynamic task selection system. As soon as the worker blocks the robot's intended trajectory towards the work-piece on the lower left, the robot reverts back towards the home position and then selects a new available trajectory, this time towards the work-piece in the upper left. The second to last image shows the robot continuing to work in parallel while the worker moves close to it into the middle of the table.

## 3.3    Discussion

Our approach has been shown to add dynamic task selection on the basis of a generic voxel-based online collision avoidance system. The task selection is online capable with sixteen available trajectories. The dynamic task selection can reduce idle waiting times for the robot without excessive false positives by representing robot and environment in a voxel grid with 2 cm resolution. This allows for close spatial and temporal collaboration with the robot.

The reliability of the vision sensors used in our demonstrator is dependent on the environmental lighting conditions. Strong external light-sources can (partially) blind the sensors. Similarly, our methods for noise filtering and robot ego collision model subtraction, as well as sensor occlusion can cause false negatives during collision checks. The collaborative workspace therefore has to remain safe for human workers even without any vision input. A possible mitigation for some of the vision reliability issues is the addition of a watchdog function that periodically checks for the presence of expected elements in the image to detect excessive occlusion and over-exposure of one or more cameras.

The force-based control strategy used in the screw insertion process has proven reliable in the tightening phase. The preceding force-sensitive screw head detection can produce false negatives and false negatives because it checks the contact force at a single moment instead of over a short time window.

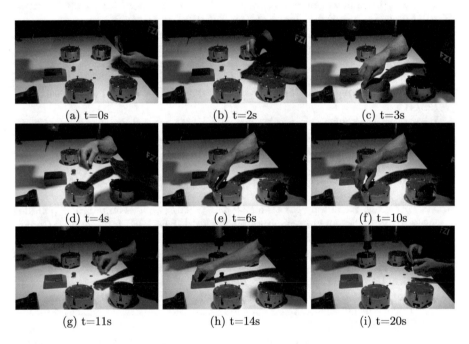

| (a) t=0s | (b) t=2s | (c) t=3s |
| (d) t=4s | (e) t=6s | (f) t=10s |
| (g) t=11s | (h) t=14s | (i) t=20s |

**Fig. 11.** Image sequence: collision avoidance and online task selection allow working in parallel with dynamic task order

## 4 Conclusions and Future Work

In this work we describe a demonstrator that can perform screw assembly tasks with collision avoidance and dynamic task selection in a partially automated scenario with human-robot collaboration. Our approach has been shown to offer online capability for dynamic task selection in a realistic setting. The presented approach is robust and has been demonstrated at a trade fair.

We plan to improve on the current state of the implementation by adding online motion planning, in order to further reduce the probability for idle robot waiting when none of the prerecorded trajectories is collision-free. We also plan to explore online motion recognition and prediction to allow the robot the selection of tasks that can likely be completed without interruption.

**Acknowledgment.** This research was funded in part by the *Baden-Württemberg Stiftung* in the project *KolRob – Kollaborativer, intelligenter Roboterkollege für den Facharbeiter des Mittelstands*.

# References

1. Albu-Schäffer, A., Haddadin, S., Ott, C., Stemmer, A., Wimböck, T., Hirzinger, G.: The DLR lightweight robot: design and control concepts for robots in human environments. Ind. Robot Int. J. **34**(5), 376–385 (2007)
2. Harada, K., Yoshida, E., Yokoi, K.: Motion Planning for Humanoid Robots. Springer Science & Business Media, Heidelberg (2010)
3. Hermann, A., Drews, F., Bauer, J., Klemm, S., Roennau, A., Dillmann, R.: Unified GPU voxel collision detection for mobile manipulation planning. In: 2014 IEEE/RSJ International Conference on Intelligent Robots and Systems (IROS) (2014)
4. Hermann, A., Mauch, F., Fischnaller, K., Klemm, S., Roennau, A., Dillmann, R.: Anticipate your surroundings: predictive collision detection between dynamic obstacles and planned robot trajectories on the GPU. In: 2015 European Conference on Mobile Robots (ECMR), pp. 1–8 (2015)
5. ISO. ISO/TS 15066:2016-02 (e) Robots and robotic devices - Collaborative robots (2016)
6. Mainprice, J., Berenson, D.: Human-robot collaborative manipulation planning using early prediction of human motion. In: 2013 IEEE/RSJ International Conference on Intelligent Robots and Systems (IROS), pp. 299–306 (2013)
7. Mauch, F., Roennau, A., Heppner, G., Buettner, T., Dillmann, R.: Service robots in the field: The BratWurst Bot. In: 2017 18th International Conference on Advanced Robotics, ICAR 2017, pp. 13–19 (2017)
8. Michalos, G., Makris, S., Tsarouchi, P., Guasch, T., Kontovrakis, D., Chryssolouris, G.: Design considerations for safe human-robot collaborative workplaces. Procedia CIRP **37**, 248–253 (2015)
9. Park, C., Pan, J., Manocha, D.: Real-time optimization-based planning in dynamic environments using GPUs. In: 2013 IEEE International Conference on Robotics and Automation (ICRA), pp. 4090–4097. IEEE (2013)
10. Quigley, M., Conley, K., Gerkey, B., Faust, J., Foote, T., Leibs, J., Berger, E., Wheeler, R., Ng, A.Y.: ROS: an open-source Robot Operating System. In: ICRA, vol. 3, p. 5 (2009)
11. Rusu, R.B., Cousins, S.: 3D is here: point cloud library. In: IEEE International Conference on Robotics and Automation, pp. 1 – 4 (2011)
12. Scherzinger, S., Roennau, A.: Forward dynamics compliance control (FDCC): a new approach to cartesian compliance for robotic manipulators. In: 2017 IEEE/RSJ International Conference on Intelligent Robots and Systems (IROS), pp. 4568–4575 (2017)
13. TÜV Nord CERT GmbH. TÜV Safety Certificate UR 3 UR5 UR10 (2015)

# KittingBot: A Mobile Manipulation Robot for Collaborative Kitting in Automotive Logistics

Dmytro Pavlichenko$^{(\boxtimes)}$, Germán Martín García, Seongyong Koo,
and Sven Behnke

Autonomous Intelligent Systems, Computer Science Institute VI, University of Bonn,
Endenicher Allee 19a, 53115 Bonn, Germany
pavlichenko@ais.uni-bonn.de

**Abstract.** Individualized manufacturing of cars requires kitting: the collection of individual sets of part variants for each car. This challenging logistic task is frequently performed manually by warehouseman. We propose a mobile manipulation robotic system for autonomous kitting, building on the Kuka Miiwa platform which consists of an omnidirectional base, a 7 DoF collaborative iiwa manipulator, cameras, and distance sensors. Software modules for detection and pose estimation of transport boxes, part segmentation in these containers, recognition of part variants, grasp generation, and arm trajectory optimization have been developed and integrated. Our system is designed for collaborative kitting, i.e. some parts are collected by warehouseman while other parts are picked by the robot. To address safe human-robot collaboration, fast arm trajectory replanning considering previously unforeseen obstacles is realized. The developed system was evaluated in the European Robotics Challenge 2, where the Miiwa robot demonstrated autonomous kitting, part variant recognition, and avoidance of unforeseen obstacles.

## 1 Introduction

Although robot manipulators and autonomous transport vehicles are widely used in manufacturing, there are still plenty of repetitive tasks, which are performed by human workers. Automation of such tasks would allow for relieving workers from repetitive and dull activities, which may cause harm to their health. Furthermore, automation has the potential to increase productivity and quality.

In this paper, we address the task of kitting in automotive logistics, which is frequently performed manually by warehouseman. Kitting became necessary, because car manufacturing has been individualized. Each customer configures its car, such that for each car sets of part variants must be collected and delivered to the assembly line just in time. Kitting is performed in a large storage area, called automotive supermarket, where all part variants can be collected from transport boxes and pallets. For each manufactured car, an individual order for

© Springer Nature Switzerland AG 2019
M. Strand et al. (Eds.): IAS 2018, AISC 867, pp. 849–864, 2019.
https://doi.org/10.1007/978-3-030-01370-7_66

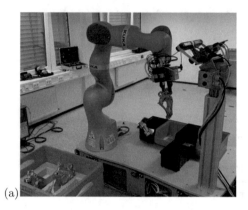
(a)

(b)

**Fig. 1.** (a) Kuka Miiwa (KMR iiwa) robot performing the kitting; (b) Three types of parts used in the kitting task.

the needed part variants is generated. A warehouseman collects the parts in the automotive supermarket and sends them to the assembly line as a kit.

We propose a mobile manipulation robotic system for autonomous kitting, building on the Kuka Miiwa platform [1], which consists of an omnidirectional base, a 7 DoF collaborative iiwa manipulator, cameras, and distance sensors. The robot performing the kitting and three parts representing the kit are shown in Fig. 1.

In order to effectively solve the kitting task, several subtasks must be addressed. First of all, the robot has to navigate precisely within the supermarket in order to reach target locations for part collection. Upon arrival at the location, it is necessary to detect the container with parts and to estimate its pose. The robot has to detect the parts inside the container and must plan suitable grasps. Collision-free arm motions must be planned and executed and the part has to be placed in the kit which is transported by the robot.

Developing an autonomous robotic system for kitting is a challenging task, because of the high degree of variability and uncertainty in each of its subtasks. Our system is designed for collaborative kitting, i.e. some parts are collected by warehouseman while other parts are picked by the robot. To address safe human-robot collaboration, fast arm trajectory replanning considering previously unforeseen obstacles must be realized.

In this paper, we present our approaches for perception and manipulation, as well as system integration and evaluation. We solve the perception task with a robust pipeline, consisting of the steps:

- Container detection and pose estimation,
- Part segmentation and grasp generation, and
- Classification of the grasped part before it is put into the kit—to verify that it is the correct part.

In order to perform manipulation effectively, we utilize an arm trajectory optimization method with a multicomponent cost function, which allows for obtaining feasible arm trajectories within a short time. All developed components were integrated in a KittingBot demonstrator. The developed system was evaluated at the Showcase demonstration within the European Robotics Challenge 2: Shop Floor Logistics and Manipulation[1], where we participated as a challenger team together with the end user Peugeot Citroën Automobiles S.A. (PSA)[2]. We report the success rates for each step of the kitting pipeline as well as the overall runtimes.

## 2   Related Work

In recent years, interest in mobile manipulation robots increased significantly. Many components necessary for building an autonomous kitting system have been developed. Integrating these to a system capable of performing kitting autonomously is challenging, though. In this section, we give overview of autonomous robotic systems for kitting in industrial environments.

An early example of mobile bin picking has been developed by Nieuwenhuisen et al. [2]. They used the cognitive service robot Cosero [3] for grasping unordered parts from a transport box and delivering them to a processing station. Part detection and pose estimation was based on depth measurements of a Kinect camera and the registration of graphs of geometric primitives [4]. Grasp planning utilized local multiresolution representations. The authors report successful mobile bin picking demonstrations in simplified settings, but their robot was far from being strong and robust enough for industrial use.

Krueger et al. [5] proposed a robotic system for automotive kitting within the STAMINA[3] project. The large mobile manipulation robot consists of an industrial manipulator mounted on a heavy automated guided vehicle (AGV) platform. The system utilizes the software control platform SkiROS [6] for high-level control of the mission, which is composed of skills [7]. Each skill solves a specific sub-task: i.e. detecting a part, generating a grasp, etc. [8]. Such an architecture allows for fast definition of the global kitting pipeline for each specific use case, which can be performed by the end user. Crosby et al. [9] developed higher-level task planning within the STAMINA project. SkiROS is used to bridge the gap between low-level robot control and high-level planning. The STAMINA system was tested in a simplified setting within the assembly halls of a car manufacturer, where the robot successfully performed full kitting procedures for kits of one, three, four, and five parts multiple times. Execution speed was slow, though, and safe human-robot collaboration has not been addressed.

Krug et al. [10] introduced APPLE—a system for autonomous picking and palletizing based on a motorized forklift base. The system is equipped with a Kuka iiwa manipulator. The authors propose a grasp representation scheme

---

[1] EuRoC Challenge 2: http://www.euroc-project.eu/index.php?id=challenge_2.
[2] Peugeot Citroën Automobiles S.A.: https://www.groupe-psa.com.
[3] European FP7 project STAMINA: http://stamina-robot.eu.

which allows for redundancy in the target gripper pose [11,12]. This redundancy is exploited by a local, prioritized kinematic controller which generates reactive manipulator motions on-the-fly. The system has a basic safety laser scanner as well as a camera-based system for human detection. Human workers are assumed to be wearing special reflective clothing. Tests showed that the system is capable of performing pick and place tasks in a human-safe manner. An interesting example of such a task was to first pick an empty pallet with a forklift, navigate to the loading zone, load the pallet using the arm, and finally transport the loaded pallet to the destination zone.

A similar task routine is performed in a completely different application domain: medical care. In hospitals, nurses are required to collect necessary supplies and deliver them to the patients. This task creates a constant dull workload for nurses, who could spend the working time in a much more patient-oriented way. Diligent Robotics[4] designed a robot which should perform this routine. The hospital environment in many cases is more challenging than industrial production lines, since the narrow corridors are often crowded with patients.

Srinivasa et al. [13] address mobile manipulation tasks in household environments. HERB—a dual-armed robot with a human-like upper-body—is used for this purpose. The authors compose a manipulation planning module out of several popular planners and trajectory optimizers. This allows to effectively perform complex manipulation tasks. For instance, the approach has been tested with a task when the robot has to load a plate, a bowl, and a glass into a tray. Finally, the tray has to be lifted for further transportation. The last operation required the use of both arms. In order to configure the high-level planner, the user has to specify an action graph.

One unresolved issue with all of the above systems is that due to the large number of different objects, a large variety of grasps is required to safely manipulate them. A robotic system with automatically exchangeable grippers does not seem to be a feasible solution, since there may be dozens of different grippers needed. Another possible solution would be to use two grippers with the flexibility of human hands, but these are not available. Our approach to this issue is collaborative kitting: using a simple and robust robotic gripper for picking parts with simple structure in collaboration with warehouseman who pick more complex or fragile parts.

## 3   System Overview

The developed system is based on the Kuka Miiwa (KMR iiwa) robot. The robot has a compact omnidirectional base with four Mecanum wheels. The omnidirectional drive allows for a very precise and smooth navigation even in the areas with limited free space. The base is equipped with multiple laser scanners on its sides in order to produce 360° distance measurements. On the top surface of the base, the 7 Degrees of Freedom (DoF) collaborative Kuka iiwa arm and a vertical sensor pole with pan-tilt unit (PTU) are mounted. The PTU carries a

---

[4] Diligent Robotics: http://diligentrobots.com.

stereo camera system and a time-of-flight (ToF) depth camera. The components of this system complement each other and thus avoid sensor-specific problems. This sensor system is further referenced as Pan-Tilt Sensor System (PTSS). A stereo camera is attached at the wrist of the iiwa arm. While the PTSS allows to have a global view on the manipulation workspace, the wrist camera allows to measure the manipulated objects more precisely. In order to effectively process the data from all the sensors, the robot has four Core-i7 onboard computers as well as an FPGA for the stereo processing. The top surface of the robot base is flat and has a lot of free space, which is used to place a kit storage system. In our experiments, we use a very simple kit storage system: three plastic boxes.

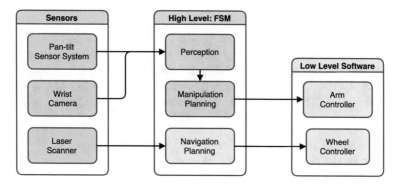

**Fig. 2.** Simplified scheme of the proposed kitting system. Light-blue: Kuka software components. Green: components developed by us.

Figure 2 illustrates the main components of our system and the information flow between them. The Kuka KMR iiwa robot comes with a low-level software stack, as well as a higher-level navigation and mapping stack. We used these components together with ours in order to realize a complete robotic system for autonomous kitting. The highest level of the software stack is represented by a finite state machine (FSM). Its parameters define the whole kitting procedure: how many parts to pick up, where to pick up, where to deliver, etc. The FSM orchesters work of all three main components of our system: perception, manipulation planning, and navigation planning. The perception component uses sensory input from the wrist camera and the PTSS to detect the container with parts, estimate its pose, detect the parts, and define the grasp. The manipulation planning module takes as input raw 3D sensory data for collision avoidance as well as results from the perception module. Finally, the manipulation module produces an arm trajectory to reach the grasp and to deliver the part into the kit. The navigation module performs mapping and path planning, as well as dynamic obstacle avoidance.

## 4    Perception

The location of transport boxes and pallets in the automotive supermarket is known in advance only to a limited degree of precision: boxes are manually placed and their pose can change while picking parts, placing other boxes, etc. Hence, it is necessary to estimate the exact pose of the box in the environment. Similarly, part poses within the containers vary and wrong part variants might be accidentally placed in the containers. In this section, we present the methods used for the perception of containers, segmentation of parts, and part variant recognition.

### 4.1    Container Detection

We use the approach of Holz et al. [14] for the detection and localization of containers in RGB-D data. The method is tailored for finding containers when the upper part of the container is visible. It is based on extracting lines along the edges in the RGB-D image and finding the best fitting models of the container we are looking for. The container detection and localization pipeline is organized in three stages:

- Detect edges in both the color image and the depth image,
- Fit lines to the detected edge points, and
- Sample subsets of lines and fit parametrized models of the containers to the subset.

The best fitting candidate gives both the lines forming the top of the box and the pose of the box.

**Edge Detection.** We follow the approach of Choi et al. [15] for detecting edges in RGB-D data. The method proposes the Canny edge detector for finding edges $E_{RGB}$ in the color image. In the depth image, we inspect the local neighborhood of points, focus on points at depth discontinuities, and identify occluding edges by selecting those points $E_D$ that are closer to the camera. In addition, we efficiently compute local covariance matrices using a method based on integral images [16]. From the local covariance matrices, we compute local surface normals and curvature to obtain convex $E_{conv}$, and concave edges. For the next processing steps, we combine all points at color edges, occluding edges, and convex edges to a set of edge points $E = E_{RGB} \vee E_D \vee E_{conv}, E \subseteq P$, where $P$ is a point cloud.

**Line Detection.** Our line detection approach is based on RANSAC. On each iteration, we select two points, $p$ and $q$, from the set $E$ and compute a line model: point on the line $p$ and direction of the line $q - p$. We then determine all inliers in $E$ which support the line model by having distance to it below threshold $\epsilon_d$. The line model with the largest number of inliers is selected as the detected line $l$. If the number of inliers of line $l$ exceeds the minimum number of inliers, $l$ is added to the set of lines $L$. We then remove the inliers of $l$ from $E$ and continue detecting further lines. If the residual number of points in $E$ falls below a threshold, or the minimum number of inliers for the line segments is not reached, the line detection is stopped.

**Line Validation.** After the line detection, we perform a validation step which is based on two restrictions:

- *Connectivity Restriction.* The inliers of a detected line may lie on different unconnected line segments. While partial occlusions can cause multiple unconnected line segments on the edges of the box, we cluster the inliers and split the detected line into multiple segments in case the box should be fully visible. If the number of points in a cluster falls below the minimum number of inliers for line segments, it is neglected.
- *Length Restriction.* Line segments which are shorter than the shortest edge in the model and longer than the longest edge in the model are neglected. To account for noise, missing edge points, or other errors, a deviation of 20% from these thresholds is allowed.

An example of detected edges and lines is shown in Fig. 3.

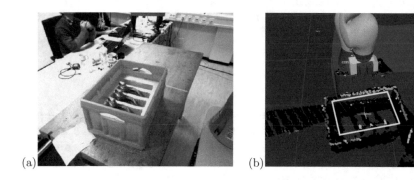

**Fig. 3.** Box detection. (a) Raw image from the pan-tilt camera; (b) Point cloud with detected edge points (cyan) and line segments (random colors). Best matched model is shown as yellow rectangle.

**Model Sampling and Validation.** In order to detect the container, a subset of the detected line segments is selected. We select $N$ line segments where $N$ is the number of line segments in the parametrized model. That is, we sample as many line segments as contained in the model of the container. As a result, we obtain tuples of line segments $(l_0, ..., l_N)$. To avoid repetitively re-checking the same tuples, we use a hash table in which sampled tuples are marked as being processed.

We discard tuples of line segments which are not compatible with the container model. The model contains four edges which are pairwise parallel and perpendicular to each other. If the tuple of sampled line segments is valid, we continue to register the model against the sampled line segments. For the model registration, we sample points from the given parametrized container model in order to obtain a source point cloud $P$ for registration. In addition, we extract the inliers of the sampled segments to form a single target point cloud $Q$ for registration. In contrast to extracting the inliers for the target point cloud, the source point cloud of the model only needs to be sampled once.

Iterative registration algorithms align pairs of 3D point clouds by alternately searching for correspondences between the clouds and minimizing the distances between matches [17]. In order to align a point cloud $P$ with a point cloud $Q$, the ICP algorithm searches for the closest neighbors in $Q$ for points $p_i \in P$ and minimizes the point-to-point distances $d_{ij}^T = q_j - Tp_i$ of the set of found correspondences $C$ in order to find the optimal transformation $T^*$:

$$T^* = \arg\min_T \sum_{(ij)\in X} ||d_{ij}^{(T)}||^2. \tag{1}$$

Finally, we compute a confidence $c$ that is based on the overlap between the model and the sampled line segments: $c = |C|/|P|$, where $|C|$ is the number of corresponding points within a predefined distance tolerance $\epsilon_d$, and $|P|$ is the number of points in the generated model point cloud. In case of a complete overlap, the confidence $c$ is roughly 1. We select the best match to estimate the pose of the container.

## 4.2   Part Segmentation

In order to segment the parts that come in the boxes, we use the estimated pose of the box from the previous step. This gives us an observation pose for the wrist camera above the box, which is used to have a clear view of the inside of the container. We use the detected box borders to extract the points in the obtained point cloud that correspond to the contents of the box.

Engine support variants are segmented using Euclidean clustering on the box content point cloud. The centroid and principal axes of the clusters are used to compute the grasping poses. An example of the segmented parts is shown in Fig. 4.

To determine grasps for the engine pipes in the extracted container content, we cluster the extracted point cloud into cylinders and select the centroid of the highest cylinder as the grasping point. The orientation of the grasp pose is chosen according to the principal axis of the cylinder to be grasped and aligned with the local coordinate frame of the detected container in order to approach the part straight from the top of the container.

(a)                                    (b)

**Fig. 4.** Part segmentation. (a) Raw image from the pan-tilt camera. (b) Three segmented parts (green, blue and red) with their corresponding grasping poses.

## 4.3   Parts Variant Recognition

In real production warehouses, the locations of containers with parts may be mixed up, or a part of a wrong type may accidentally enter a container with other parts. In order to detect such situations, we perform part recognition. The recognition takes place after the part was grasped and lifted up in the air, since in such position it is unoccluded and may be easily observed by the PTSS.

A convolutional neural network, shown in Fig. 5, is used to perform the recognition. The network takes a $64 \times 64 \times 1$ depth image as input. Metal parts are shiny and thus shape features may be not visible on the RGB image. The use of depth information helps to overcome this issue. The first part of the network consists of four convolutional layers, each followed by a pooling layer. The final part consists of four fully connected layers. The last layer outputs two values through a softmax function. The numbers represent the probability of the object belonging to the first and the second class, respectively. We used this network to distinguish between two types of engine supports, as these parts look similar and could be mixed.

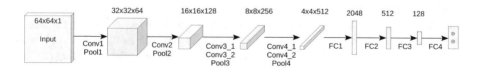

**Fig. 5.** Neural network used to recognize the variant of an engine support part.

In order to train the model, available CAD models of the parts were used to render depth maps. For each variant, 10,000 different poses were used in order to produce synthetic data. To obtain an input to the network in the real world, we project the center of the TCP to the pan-tilt depth image and crop an image window that contains the part.

## 5   Manipulation

Given a grasp pose from the perception module, it is necessary to plan a trajectory for the robotic arm to reach the corresponding pre-grasp pose. The trajectory has to be smooth and must avoid any collisions with the environment or the robot itself. Furthermore, it has to satisfy constraints on orientation of the end-effector. Moreover, the duration of the trajectory has to be as short as possible, since it directly influences the overall time spent for the kit completion. For the same reason, planning time must be short. Finally, it is necessary to constantly track the future part of the trajectory during execution to detect any unforeseen collisions with dynamic objects. In case when future collision is detected, the trajectory has to be replanned as fast as possible and the execution should continue.

To fulfill these requirements, we use our trajectory optimization method [18] which is based on STOMP [19]. The method iteratively samples noisy trajectories around a mean trajectory and evaluates each of them with a cost function. Then the mean is shifted in the direction of reducing costs. Iterations continue until one of the termination criteria is met.

The trajectory $\Theta$ is defined as a sequence of $N$ keyframes in joint space. Start and goal configurations are fixed, as well as the number of keyframes $N$. The cost of the trajectory $\Theta$ is defined as a sum of costs of transitions between adjacent keyframes $\boldsymbol{\theta}_i$:

$$q(\Theta) = \sum_{i=0}^{N-1} q(\boldsymbol{\theta}_i, \boldsymbol{\theta}_{i+1}). \tag{2}$$

We propose a cost function which consists out of five components:

$$
\begin{aligned}
q(\boldsymbol{\theta}_i, \boldsymbol{\theta}_{i+1}) =& q_o(\boldsymbol{\theta}_i, \boldsymbol{\theta}_{i+1}) + q_l(\boldsymbol{\theta}_i, \boldsymbol{\theta}_{i+1}) + q_c(\boldsymbol{\theta}_i, \boldsymbol{\theta}_{i+1}) \\
&+ q_d(\boldsymbol{\theta}_i, \boldsymbol{\theta}_{i+1}) + q_t(\boldsymbol{\theta}_i, \boldsymbol{\theta}_{i+1}),
\end{aligned}
\tag{3}
$$

where $q_o(\boldsymbol{\theta}_i, \boldsymbol{\theta}_{i+1})$ is a component which penalizes being close to obstacles, $q_l(\boldsymbol{\theta}_i, \boldsymbol{\theta}_{i+1})$ penalizes exceeding of joint limits, $q_c(\boldsymbol{\theta}_i, \boldsymbol{\theta}_{i+1})$ penalizes task specific constraints on a gripper position or/and orientation, $q_d(\boldsymbol{\theta}_i, \boldsymbol{\theta}_{i+1})$ penalizes long durations of the transitions between the keyframes and $q_t(\boldsymbol{\theta}_i, \boldsymbol{\theta}_{i+1})$ is a component that penalizes high actuator torques. Each cost component $q_j(.,.)$ is normalized to be within $[0, 1]$ interval and has an importance weight $\lambda_j \in [0, 1]$ attached. This allows to prioritize optimization by manipulating weights $\lambda_j$.

In order to speed up the optimization process, we utilize two phases. During the first phase, a simplified cost function is used. It consists of collision costs $q_o$, joint limit costs $q_l$, and gripper constraints costs $q_c$. As soon as the first valid solution is found, the second phase begins, where the full cost function $q(.,.)$ (as described in Eq. 3) is used. Optimization continues until one of the termination criteria is met.

# 6   Experiments

In order to assess the designed system, we performed several experiments during the EuRoC Showcase evaluation in the lab of the Challenge 2 host DLR Institute of Robotics and Mechatronics in Oberpfaffenhofen, Germany, under severe time constraints and the supervision of judges. The experiments included tests of isolated components as well as full kitting procedures. All experiments have been done on the real robot. In this section, we describe the test procedures and present obtained results.

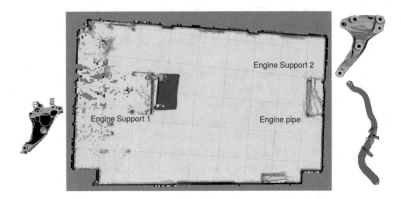

**Fig. 6.** Map of the automotive supermarket. A container with Engine Support 1 parts is located on the first table. Containers with Engine Support 2 parts and Engine Pipes are located on the second table. CAD models of the parts are shown on the sides of the map.

## 6.1   Showcase Setup

The kitting experiment was performed in a simplified supermarket of parts, designed by us. The kit consisted out of three automotive parts, supplied by our end-user partner PSA:

– Engine Support 1: metal part with shiny surface.
– Engine Support 2: metal part with shiny surface, very similar to Engine Support 1.
– Engine Pipe: black flexible pipe made out of rubber.

Each part type is provided in a separate container. Both engine supports were placed in the containers with slots, so that each part is positioned roughly vertically, perpendicular to the bottom of the container. Engine pipes were put into their container without any order, making picking more challenging.

The map of the automotive supermarket as well as CAD models of the parts are show in Fig. 6. Containers with parts were located on tables in opposite sides of the $10 \times 5$ m room. The container with Engine Supports 1 was located on the first table. Containers with Engine supports 2 and Engine pipes were provided on the second table.

## 6.2   Kitting

The procedure of our kitting scenario was defined as follows: the robot starts in the middle of the supermarket. It has to move to the first table and pick up Engine Support 1 and place it in the first kitting compartment on the robot. After that, the robot has to move to the second table and pick up Engine Support 2 and place it in the second kitting compartment. Finally, the robot has to pick up Engine Pipe and place it in the third kitting compartment. To demonstrate that the kit is ready to be delivered to the assembly line, the robot moves away from the table.

In order to demonstrate the capability of our system, we performed two kitting runs, as described above. The robot picking the Engine Support 1 is shown in Fig. 7. Videos of the experiments are available online[5]. We measured the success rate of picking and placing for each part type, as well as the overall runtime. The results are presented in the Table 1. One can observe that placing the parts never failed. Picking parts succeeded on all but one case: picking of Engine Pipe in the first run was not successful. The placing task is much easier than picking, since the positions of the kitting boxes on the robot are known precisely. Picking of the engine pipe failed because the robot attempted to grasp it above the widest part of the pipe. Consequently, the grasp was not firm enough and the part slipped from the gripper.

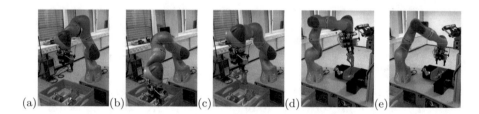

(a)     (b)     (c)     (d)     (e)

**Fig. 7.** Picking of Engine Support 1. (a) Observation pose; (b) Part grasped; (c) Part lifted; (d) Part transported to the drop pose; (e) Part placed into the kitting compartment on the robot.

## 6.3   Additional Experiments

In addition to the complete kitting procedure, we tested several components in isolation. In this subsection we present the obtained results.

---

[5] Experiment video: http://www.ais.uni-bonn.de/videos/IAS_2018_KittingBot.

**Table 1.** Pick and place success rates and runtime of kitting.

|  | Run 1 | Run 2 |
|---|---|---|
| **Parts successfully picked** | 2/3 | 3/3 |
| Engine Support 1 | + | + |
| Engine Support 2 | + | + |
| Engine Pipe | − | + |
| **Successful grasps** | 2/3 | 3/3 |
| **Successful placements** | 2/2 | 3/3 |
| **Runtime [s]** | 759 | 809 |

**Part Variant Recognition.** In this experiment, we demonstrated the capabilities of part variant recognition module. First, we pick up Engine Support 1 and recognize which part is in the gripper. Then we pick up Engine Support 2 and perform the recognition again. The obtained images of the parts are shown in Fig. 8. Both parts were recognized correctly.

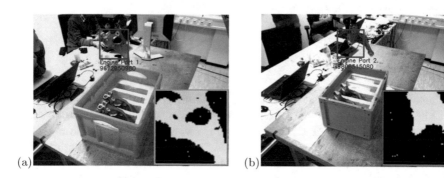

**Fig. 8.** Recognition of the part variant. (a) Raw image of the picked up Engine Support 1; input depth image shown in the bottom-right corner. (b) Raw image of the picked up Engine Support 2; input depth image in the bottom-right corner.

**Unforeseen Collision Avoidance.** To demonstrate the ability of our system to deal with obstacles which appear during trajectory execution, we performed a separate experiment. The robot arm had to move from the observation pose to the pose above the kitting boxes. After the trajectory had been planned and the execution was started, an obstacle was inserted on the way. The system continuously tracks the future part of the trajectory and checks for obstacles during trajectory execution. The future collision was detected, the execution was stopped and the trajectory was replanned, taking the new obstacle into consideration. Both initial and replanned trajectories, as well as the arm avoiding the obstacle are shown in Fig. 9. The replanning took 0.39 s, which in principle allows to perform replanning without stopping the execution, in case the arm does not move too fast and the collision is far enough ahead.

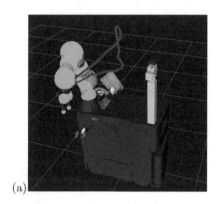
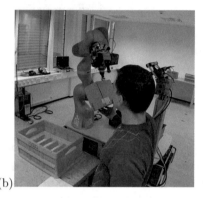

(a)                                    (b)

**Fig. 9.** Replanning of the trajectory to avoid an unforeseen obstacle. (a) The obstacle is inserted during execution. The trajectory is replanned. Red: initial trajectory. Blue: replanned trajectory. (b) Arm, avoiding the new obstacle.

## 7    Conclusion

We have developed a mobile manipulation system for performing autonomous part kitting. We proposed perception software modules which allow to efficiently detect containers, segment the parts therein, and produce grasps. In addition, our system is capable of recognizing part variants. The developed manipulation planner is able to optimize robotic arm trajectories with respect to collisions, joint limits, end-effector constraints, joint torque, and duration. The method allows to perform optimization fast and to replan trajectories in case of possible future collisions due to newly appeared obstacles. We integrated these modules into a Kuka KMR iiwa robot. Together with the Kuka navigation stack, our components formed a system capable of autonomous kitting under guidance of a high-level FSM.

We demonstrated the capabilities of our system in a simplified kitting scenario during the EuRoC Showcase evaluation, in the lab of the challenge host, supervised by judges under severe time constraints. The experiments shown that the perception module can reliably detect containers and segment the parts inside them. Generated gasps were reliable in the most cases, failing only once when grasping an Engine Pipe, which was the hardest part in the kit. The trajectory optimization method shown good performance with short runtimes and allowed to deliver the parts to the kitting compartments in all cases. Real-time supervision of the workspace and online replanning are a suitable basis for collaborative kitting.

**Acknowledgements.** This research received funding from the European Union's Seventh Framework Programme grant agreement no. 608849 (EuRoC). It was performed in collaboration with our end-user partner Peugeot Citroën Automobiles S.A. (PSA). We also gratefully acknowledge the support of the EuRoC Challenge 2 host: DLR Institute of Robotics and Mechatronics in Oberpfaffenhofen, Germany.

# References

1. Dömel, A., Kriegel, S., Kaßecker, M., Brucker, M., Bodenmüller, T., Suppa, M.: Toward fully autonomous mobile manipulation for industrial environments. Int. J. Adv. Robot. Syst. **14** (2017)
2. Nieuwenhuisen, M., Droeschel, D., Holz, D., Stückler, J., Berner, A., Li, J., Klein, R., Behnke, S.: Mobile bin picking with an anthropomorphic service robot. In: IEEE International Conference on Robotics and Automation (ICRA) (2013)
3. Stückler, J., Schwarz, M., Behnke, S.: Mobile manipulation, tool use, and intuitive interaction for cognitive service robot Cosero. Frontiers in Robotics and AI (2016)
4. Berner, A., Li, J., Holz, D., Stückler, J., Behnke, S., Klein, R.: Combining contour and shape primitives for object detection and pose estimation of prefabricated parts. In: IEEE International Conference on Image Processing (ICIP) (2013)
5. Krueger, V., Chazoule, A., Crosby, M., Lasnier, A., Pedersen, M.R., Rovida, F., Nalpantidis, L., Petrick, R., Toscano, C., Veiga, G.: A vertical and cyber-physical integration of cognitive robots in manufacturing. Proc. IEEE **104**, 1114–1127 (2016)
6. Rovida, F., Crosby, M., Holz, D., Polydoros, A.S., Großmann, B., Petrick, R.P.A., Krüger, V.: SkiROS—a skill-based robot control platform on top of ROS. In: Robot Operating System (ROS): The Complete Reference, pp. 121–160 (2017)
7. Pedersen, M.R., Nalpantidis, L., Andersen, R.S., Schou, C., Bøgh , S., Krüger, V., Madsen, O.: Robot skills for manufacturing: from concept to industrial deployment. Robot. Comput. Integr. Manuf., 282–291 (2016)
8. Holz, D., Topalidou-Kyniazopoulou, A., Stückler, J., Behnke, S.: Real-time object detection, localization and verification for fast robotic depalletizing. In: IEEE/RSJ International Conference on Intelligent Robots and Systems (IROS) 2015, pp. 1459–1466 (2015)
9. Crosby, M., Petrick, R., Toscano, C., Dias, R., Rovida, F., Krüger, V.: Integrating mission, logistics, and task planning for skills-based robot control in industrial kitting applications. In: Proceedings of the 34th Workshop of the UK Planning and Scheduling Special Interest Group (PlanSIG), pp. 135–174 (2017)
10. Krug, R., Stoyanov, T., Tincani, V., Andreasson, H., Mosberger, R., Fantoni, G., Lilienthal, A.J.: The next step in robot commissioning: autonomous picking and palletizing. IEEE Robot. Autom. Lett., 546–553 (2016)
11. Berenson, D., Srinivasa, S., Kuffner, J.: Task space regions: a framework for pose-constrained manipulation planning. Int. J. Robot. Res. (IJRR), 1435–1460 (2011)
12. Gienger, M., Toussaint, M., Goerick, C.: Task maps in humanoid robot manipulation. In: IEEE/RSJ International Conference on Intelligent Robots and Systems (IROS) (2008)
13. Srinivasa, S.S., Johnson, A.M., Lee, G., Koval, M.C., Choudhury, S., King, J.E., Dellin, C.M., Harding, M., Butterworth, D.T., Velagapudi, P., Thackston, A.: A system for multi-step mobile manipulation: architecture, algorithms, and experiments. In: 2016 International Symposium on Experimental Robotics, pp. 254–265 (2017)

14. Holz, D., Behnke:, S.: Fast edge-based detection and localization of transport boxes and pallets in RGB-D images for mobile robot bin picking. In: International Symposium on Robotics (ISR) (2016)
15. Choi, C., Trevor, A.J.B., Christensen, H.I.: RGB-D edge detection and edge-based registration. In: IEEE/RSJ International Conference on Intelligent Robots and Systems (IROS) (2013)
16. Holz, D., Holzer, S., Rusu, R.B., Behnke, S.: Real-time plane segmentation using RGB-D cameras. In: RoboCup 2011: Robot Soccer World Cup XV, pp. 306–317 (2012)
17. Holz, D., Ichim, A.E., Tombari, F., Rusu, R.B., Behnke, S.: Registration with the point cloud library: a modular framework for aligning in 3-D. IEEE Robot. Autom. Mag. **22**, 110–124 (2015)
18. Pavlichenko, D., Behnke, S.: Efficient stochastic multicriteria arm trajectory optimization. In: IEEE/RSJ International Conference on Intelligent Robots and Systems (IROS) (2017)
19. Kalakrishnan, M., Chitta, S., Theodorou, E., Pastor, P., Schaal, S.: STOMP: stochastic trajectory optimization for motion planning. In: IEEE International Conference on Robotics and Automation (ICRA) (2011)

# User-Friendly Intuitive Teaching Tool for Easy and Efficient Robot Teaching in Human-Robot Collaboration

Hyunmin Do$^{(\boxtimes)}$, Taeyong Choi, Dong Il Park, Hwi-su Kim, and Chanhun Park

Department of Robotics and Mechatronics, Korea Institute of Machinery
and Materials (KIMM), Daejeon 34103, Korea
`hmdo@kimm.re.kr`

**Abstract.** Production automation by human-robot collaboration has
drawn significant attention due to increasing demands for automation
in the manufacturing process of small electronic products, which were
previously manufactured manually. Accordingly, the research for human-
robot collaboration is being actively conducted and intuitive teaching is
an essential technology to realize easy and efficient teaching of a collaborative robot. This paper proposes an intuitive teaching tool attached to
a robot end effector that can accurately teach motions to a robot manipulator, without being affected by sensor noises. This device consists of
three parts: a motion operation part for teaching six degrees of freedom
motion of the robot, a motion setting part which consists of core functions necessary for teaching and a status display part for displaying the
status of the teaching device and the robot. It is designed to perform
teaching work by a combination of twelve switches which have one-to-
one mapping relation to each degree of freedom of motion. A prototype
has been implemented to verify the performance and has been applied
to an experiment of six degrees of freedom motion teaching with UR5
robot.

**Keywords:** Intuitive teaching tool · Robot teaching
Human-robot collaboration

## 1 Introduction

In mass-production systems such as automobile production lines, robots were
required to perform high-speed, high-precision, simple repetitive tasks. Recently,
however, the use of robots has been widening beyond such simple repetitive
tasks in production sites. Automation using small vertical articulated robots
and dual-arm robots has been introduced into packaging and assembly tasks
for electronic products, which has previously relied upon purely manual work
by human workers [1]. Such kind of complex tasks cannot be performed yet by
robots alone because the cognitive capability of robots is still limited compared
with human workers. Therefore, efforts are being made to increase production

© Springer Nature Switzerland AG 2019
M. Strand et al. (Eds.): IAS 2018, AISC 867, pp. 865–876, 2019.
https://doi.org/10.1007/978-3-030-01370-7_67

efficiency by compensating for mutual deficiencies through collaboration between human and robot. In other words, there is an increasing need for human-robot collaboration in which human workers and robots share the same workspace, rather than the traditional use of robots that separates human and robot by safety fences [2–7]. There are some essential elements required for humans and robots to work together. The most important one is the safety function of robot that prevents injuries caused by collision between the robot and the human [8]. Another important element is an intuitive interface in controlling a robot [9,10]. In the case of human workers, collaborations among several workers, such as giving instructions or requesting a help to co-workers, can be easily performed in the production site. In the case of conventional robots, however, once the job of a robot was programmed by a teaching expert, it is difficult to modify it and almost impossible to respond quickly according to the field situations because an interface of a conventional teach pendant, which is usually used in robot teaching, is not easy to use for a non-specialist. Therefore, an easy, intuitive interface is indispensable in allowing workers to handle robots easily and it would be ideal to use verbal signals. However, it is difficult to implement such kind of interface stably with the current technology. Thus, we focus on the intuitive teaching device and method, which is easy to use for users without professional knowledge about robot and robot teaching. A novel teaching tool with an intuitive user interface was proposed and the proposed teaching tool makes it possible to teach a robot without a conventional teach pendant in similar manner with direct teaching.

The direct teaching is a method in which the user grasps the end-effector of the robot and directly teaches the waypoint to the robot while moving. To implement the direct teaching, an input method for understanding the intention of the user and a method of transferring the teaching instruction to the robot are required. Two methods can be used in large. One is a method of recognizing a six degrees of freedom (DoF) teaching input through a force/torque sensor which is attached to a robot end, and the other one is a method of detecting an external force applied to the robot by the user through a joint torque sensor or the change of current induced in motor in each joint. However, in order to use an external sensor, the sensor must be attached to the robot, which causes an increase in cost of the robot, thus it is difficult to apply such methods to all robots. In addition, there might be a problem that the robot can move differently from teaching command of a user due to drift caused by the noise in the sensor signal. The method of recognizing a teaching command from the current change induced in the motor of each joint is advantageous in that it does not require any additional sensors. Thus it is applied to Universal Robots [11,12] commercially. However, it is difficult to distinguish the current change due to the external human force precisely, and it might be confused with the current change due to external factors such as collision, which makes it difficult to adjust the sensitivity of the teaching. Also, since this method can be used only when the corresponding function is implemented in the robot controller additionally, it is difficult to apply these methods to all robots generally. To solve such problems, we proposed a novel intuitive teaching device capable of six DoF motion teaching without conventional teaching pendant. The proposed device is based on buttons and switches and can be installed on the robot end-effector. Although the direction and the speed of the robot teaching motion in the proposed device are limited as compared with the conventional force/torque sensor-based device, it can be applied to a conventional industrial robot and an intuitive teaching function can be easily implemented at a low cost.

This paper is organized as follows. In Sect. 2, the requirements for intuitive teaching device and the design concept is proposed. In Sect. 3, the detailed mechanism for six DoF motion teaching is proposed. In Sect. 4, intuitive teaching method with the proposed teaching device and the implementation results are suggested. Experimental results with UR5 robot is described in Sect. 5. Finally, in Sect. 6, conclusions and future works are discussed.

## 2   Concept of Intuitive Teaching Device

### 2.1   Requirements

The requirements for the intuitive teaching tool can be outlined as follows:

– Intuitiveness: Users without a professional knowledge about a robot should be able to use a teaching tool easily and program a motion of a robot with a simple explanation. In addition, the functions of a teaching tool should be easily understood. Numerous buttons and program languages of conventional teaching pendant for industrial robot make an operation of a robot difficult for general users and only available for a teaching expert.
– Operability: The buttons and handles must be easily operable and allow users to manipulate them intuitively. Furthermore, teaching device can reflect the user's intention as accurate as possible and thus the motion of a robot should not move in a direction different from the user's intention. In addition, since it is difficult for a user to teach the 6 degree-of-freedom motion of the robot end effector precisely in task space, the function of fixing the motion in specific axis or specific plane during teaching process is required to increase the teaching precision.
– Readability: A status of a robot, which shows the mode of job, should be informed to users. This information allows users to confirm whether the robot is currently teaching mode or playback mode, which type of teaching task is being performed, and whether the waypoints have been saved. This readability can be implemented with color display or simple text.

### 2.2   Design Concept

From this point of view, new type of teaching devices has been proposed. For example, a teaching device in the form of a joystick is developed and attached to the end of the robot. Such device can be used to generate a jog command as an add-on device of the teach pendant. The *ready2_pilot* system developed by KUKA [13] is such a kind of teaching device. The 6D mouse type jog device attached to the teach pendant of the existing KUKA robot is separated as independent device and mounted on the end part of the robot. In ABB, a special type of teaching device based on force-torque sensor is developed and mounted on the end of the articulated robot as an option [14] so that the direct teaching was made possible. In both cases, new devices improve the convenience of teaching work, but they always should be used with teaching pendant together since they only work as an add-on to existing teaching pendant.

Thus we proposed a novel intuitive teaching device capable of six DoF motion teaching without existing teaching pendant. During the teaching operation, the user does not use the teaching pendant but conducts the teaching job by operating only the proposed device attached to the robot end effector. The teaching pendant can be used before and after the teaching job like setting a robot or editing waypoints. The proposed teaching device is largely divided into three parts. The first one is the motion operation part for teaching six degrees of freedom motion of the robot, which can generate intuitive jog motions such as position and orientation motion by detecting jog inputs. Since all inputs are given as digital signals from jog mechanism, the influence of sensor noise can be fundamentally blocked. The second one is the motion setting part which consists of core functions necessary for teaching among functions of existing teaching pendants such as selection of a coordinate system, save of waypoints, and playback of stored waypoints. The third one is the status display part for displaying the status of the teaching device and the robot. In addition, the proposed teaching tool was designed to be attachable to the robot end-effector and thus it is possible to teach a robot in tool coordinate system with grasping the end-effector of the robot. The designed teaching tool is shown in Fig. 1.

## 3   Jog Mechanism for Six DoF Motion Operation

This section describes the jog mechanism for teaching six DoF motion (position and orientation). The design of the jog mechanism is very important and a key

**Fig. 1.** Design of the proposed teaching device

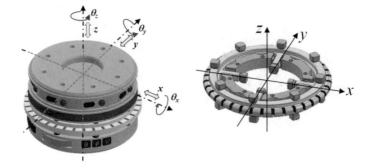

**Fig. 2.** Arrangement of jog dial switches

part for efficient and intuitive teaching. In the previous research [15], the authors proposed a jog mechanism capable of recognizing the user's input by combination of pressed buttons according to the movement of the jog dial. The arrangement of buttons of jog dial mechanism and the motion in the x, y and z-axes and rotation around each axis are illustrated in Fig. 2.

**Table 1.** Operation of teaching handle

| | | | |
|---|---|---|---|
| +x direction | -x direction | $+\theta_x$ direction | $-\theta_x$ direction |
| +y direction | -y direction | $+\theta_y$ direction | $-\theta_y$ direction |
| +z direction | -z direction | $+\theta_z$ direction | $-\theta_z$ direction |

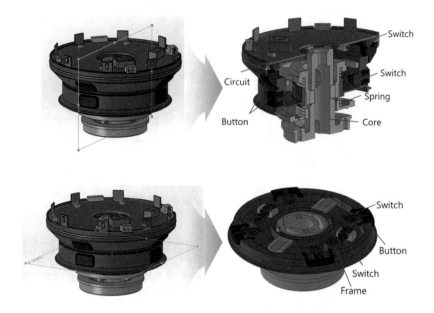

**Fig. 3.** Arrangement of switches of the proposed jog mechanism

Physical buttons are placed inside, above and below a jog dial, which are pressed according to the movement of the jog dial. Thus on/off combinations of the buttons can detect the input for the change of position and orientation of teaching handle. Table 1 shows the combinations of pressed buttons according to the input of teaching handle. Red button means the pressed one.

However, there is a problem in that a partial malfunction of the physical button pressing might occur depending on the assembly quality of the device and the input direction of a teaching handle. In order to improve this, we modified jog mechanism from on/off combination of multiple switches to one switch for one DoF motion teaching. That is, all twelve switches are arranged in consideration of plus and minus direction of six DoF motion and only one switch is pressed at a time. The arrangement of switches of the proposed jog mechanism is shown in Fig. 3. The button pressing in the intended teaching direction can be sensed since a switch is placed below each button.

In order to give a teaching command for position and orientation, buttons for $\pm x\text{-}axis$ and $\pm y\text{-}axis$ directions are arranged at 90° intervals at the lower part, and teaching buttons for $\pm R_x$ and $\pm R_y$ directions are arranged at 90° intervals at the upper part of the teaching device, where $R_x$ means the rotation around $x\text{-}axis$ and $R_y$ means the rotation around $y\text{-}axis$. Buttons for orientation motion are arranged at an inclined position to enhance the intuitiveness and the switch corresponding to each button is fixed to the circuit part at the upper part of the teaching device like Fig. 3. In addition, the teaching device is designed for the frame to move around the core at the center for the motion in the vertical direction and rotation about *z-axis*, In other words, the −*z-axis* direction switch is pressed on the lower flange when the frame of teaching device is lowered and the +*z-axis* direction switch is pressed on the upper flange when it is raised. To get the restoring force to the original position after ±*z-axis* direction motions, springs are installed in the upper and lower directions. In the case of ±$R_z$ direction motion, the frame rotates around the core and the switch, which is located inside the frame, is pressed in accordance with the rotation of the frame. The frame is also restored to the original position after the rotation motion for ±$R_z$ direction. Table 2 shows pressed buttons according to the teaching command.

The proposed mechanism accepts the teaching input through physical buttons. Thus, it has a merit in preventing problems due to sensor noise such as malfunctions of robot by non-intended teaching input and difficulties of teaching due to sensor sensitivity compared to the case of using the force/torque (F/T) sensors attached to the robot end effector. Furthermore, by separating the teaching mechanism from the actuating part of robot completely, misunderstanding of a teaching signal as a collision does not occur during the teaching process. Some users may find it inconvenient that the direction of the teaching command(s) is discrete. However, this is not ultimately a problem because the purpose of the proposed teaching tool is to improve accuracy and intuitiveness in accordance with the predetermined coordinate system such as task coordinate or tool

**Table 2.** Operation of proposed jog mechanism

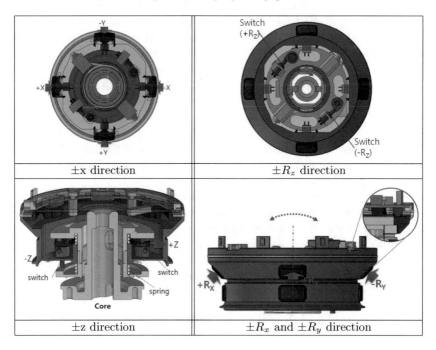

| ±x direction | ±$R_z$ direction |
|:---:|:---:|
| ±z direction | ±$R_x$ and ±$R_y$ direction |

coordinate. In addition, this problem can be solved by improving the sensing precision with an increase of the density of the button.

# 4    Intuitive Teaching Method

This section discusses the teaching method using the proposed intuitive teaching tool. First, the initial setting of the teaching motion is made through the motion setting part. Next, the teaching operation is performed with the teaching handle and then the setting and teaching operations are repeated while checking the status through the status display part.

The motion setting part is composed of buttons for setting core functions required for teaching. Such buttons can be used to select the teaching mode, to select the axis in the case of the joint teaching mode, to save waypoints and load stored waypoints. There are 11 buttons in total in the proposed device and they are classified into 4 groups. The various teaching modes can be decided according to the designated teaching mode and settings. It is possible to select the mode for jog or inching motion in the task coordinate, jog or inching motion in the tool coordinate and jog or inching motion in the joint coordinate. The function of pre-defined motion can be used to teach the robot with fixed orientation of end-effector. In addition, an LED is installed at the top of each button to display the selected mode and current setting status.

The function of each button in the motion setting part is as follows. C button is for selecting the coordinate system among the task coordinate system, tool coordinate system, and joint coordinate system. I button is for changing the jog mode and inching mode including inching step. P button is for activating the pre-defined motion mode. The pre-defined motion is currently set as a motion that changes orientation only without changing the position in the tool coordinate system. T button is for toggling the teaching mode, S button for saving the waypoints and L button for loading the saved waypoints and playing back the paths via waypoints. The buttons R and IN are for resetting the saved waypoints and for entering the interrupt mode to modify the saved waypoints. The buttons + and − are for moving the selected joint in positive and negative direction and J button is for selecting joint in joint coordinated system. Figure 4 shows the motion setting part and the arrangement of buttons.

After the needed setting for teaching was conducted and a user can teach a robot using a motion operation part. The proposed teaching device has buttons corresponding to six DoF as described in the previous section, motion can be generated independently in the directions of $\pm x$, $\pm y$, $\pm z$, $\pm R_x$, $\pm R_y$ and $\pm R_z$ by using twelve switches arranged on the motion operation part in the proposed teaching device. The general flow for implementing task jog mode teaching is as follows and the flow for other modes is very similar too.

(i)  (teaching button(T)) toggle the teaching mode between on and off: Teaching on (LED: On)

(ii) (coordinate button(C)) select the coordinate system among task/tool/joint: Task coordinate system (LED: Red)

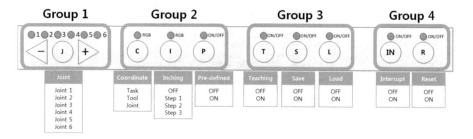

**Fig. 4.** Arrangement of buttons in motion setting part

(iii) (inching mode button(I)) select the jog mode or inching mode including inching step: Inching mode off (LED: Off)
(iv) (intuitive motion teaching) conduct jog motion using 12 switches in motion operation part
(v) (save button(S)) save waypoint: Save (LED: blinking)
(vi) (teaching button(T)) toggle the teaching mode between on and off: Teaching off (LED: Off)
(vii) (load button(L)) load saved waypoints (playback)

Thus a teaching and playback for a pick and place job at tool coordinate system can be implemented like the following flowchart (Fig. 5).

## 5   Experiments

To implement and verify the proposed teaching device and teaching method, a prototype was implemented and shown in Fig. 6. This prototype has motion setting buttons at the bottom; the motion operation part at the middle; the upper LED lamps for the status display; and the top part is laser TOF sensors for distance measurement to prevent collision. For the connection to robot controller, a communication interface supporting RS-485 and Ethernet was developed. UR5 from universal robots was used as testbed and firmware program was implemented according to the communication protocol of universal robots for external device. Pick-and-place motion teaching was conducted to verify the 6 degrees of freedom motion teaching. An experiment setup with UR5 robot and the developed teaching device is shown in Fig. 7.

Teaching mode should be on using the teaching mode button firstly and then the coordinate system should be selected. In this case, tool coordinate system was chosen. In addition, inching mode should be off. Then, the orientation should be set by operating the teaching handle to make a proper pose for picking and placing. In this case, pre-defined motion can be used, i.e., $-z$ direction for picking motion. After setting the desired orientation of robot end effector, a user conducts a position teaching to a robot. The robot motion is implemented according to the button input from the motion operation part. If precise teaching

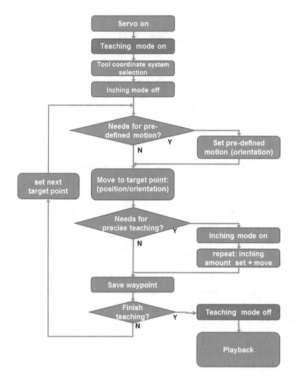

**Fig. 5.** Flowchart for teaching of pick and place motion at tool coordinate

**Fig. 6.** Implementation of the proposed teaching device

is required during teaching work, a user can turn on the inching mode to limit the motion according to the inching amount. When the robot reaches a target point, save button is pressed to save the current position and orientation. Once all waypoints are saved, teaching mode is turned off and the load button is pressed to playback the saved waypoints. Table 3 shows the experimental results for six DoF motion teaching with the proposed teaching tool and UR5 robot. We can see that intuitive teaching work is performed properly.

**Fig. 7.** Experiment setup with UR5 robot and the implemented teaching device

**Table 3.** Experiments for six DoF motion teaching with the proposed teaching tool

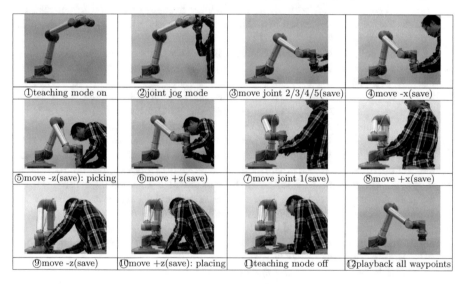

| ①teaching mode on | ②joint jog mode | ③move joint 2/3/4/5(save) | ④move -x(save) |
| ⑤move -z(save): picking | ⑥move +z(save) | ⑦move joint 1(save) | ⑧move +x(save) |
| ⑨move -z(save) | ⑩move +z(save): placing | ⑪teaching mode off | ⑫playback all waypoints |

## 6   Conclusion and Future Research

This paper proposed an intuitive teaching device which can be installed at the robot end effector for the efficient human-robot collaboration. A physical button based jog mechanism for teaching six DoF motion (position and orientation) was designed to generate intuitive jog motions. It makes an accurate teaching possible without being affected by sensor noises. Also intuitive teaching method was suggested using the proposed teaching device. A prototype was implemented and validated with UR5 robot. As a future work, the teaching tool will be improved to enhance the usability and accessibility based on user experience.

**Acknowledgments.** This work was supported by the Ministry of Trade, Industry & Energy and KEIT under program number 10063413.

# References

1. Do, H.M., Choi, T.-Y., Kyung, J.H.: Automation of cell production system for cellular phones using dual-arm robots. Int. J. Adv. Manuf. Technol. **83**, 1349–1360 (2016)
2. Thomas, C., Matthias, B., Kuhlenktter, B.: Human-robot collaboration - new applications in industrial robotics. In: International Conference on Competitive Manufacturing, pp. 293–299 (2016)
3. Coupete, E., Weistroffer, V., Hugues, O.: New challenges for human-robot collaboration in an industrial context. In: Fifth Workshop "Towards a Framework for Joint Action", IEEE International Symposium on Robot and Human Interactive Communication (2016)
4. Bauer, A., Wollherr, D., Buss, M.: Human-robot collaboration: a survey. Int. J. Humanoid Rob. **5**, 47–66 (2008)
5. Johannsmeier, L., Haddadin, S.: A hierarchical human-robot interaction-planning framework for task allocation in collaborative industrial assembly processes. IEEE Rob. Autom. Lett. **2**, 41–48 (2017)
6. Lenz, C., Knoll, A.: Mechanisms and capabilities for human robot collaboration. In: IEEE International Symposium on Robot and Human Interactive Communication, pp. 666–671 (2014)
7. Sheng, W., Thobbi, A., Gu, Y.: An integrated framework for human-robot collaborative manipulation. IEEE Trans. Cybern. **45**, 2030–2041 (2015)
8. ISO/TS 15066:2016. https://www.iso.org/standard/62996.html
9. Schraft, R.D., Christian, M.: The need for an intuitive teaching method for small and medium enterprises. In: Joint Conference on Robotics, 37th International Symposium on Robotics and 4th German Conference on Robotics, pp. 95–105 (2006)
10. Stopp, A., Horstmann, S., Kristensen, S., Kohnert, F.: Towards interactive learning for manufacturing assistants. In: IEEE International Symposium on Robot and Human Interactive Communication, pp. 338–342 (2001)
11. Park, C., Kyung, J., Park, D.I., Gweon, D.-G.: Direct teaching algorithm for a manipulator in a constraint condition using the teaching force shaping method. Adv. Robot. **24**, 1365–1384 (2010)
12. Universal Robots. http://www.universal-robots.com/
13. KUKA ready2_pilot. https://www.kuka.com/en-de/products/robot-systems/ready2_use/kuka-ready2_pilot
14. Who Is Gomtec, the Collaborative Robotics Company Acquired by ABB? https://spectrum.ieee.org/automaton/robotics/industrial-robots/who-is-gomtec-the-collaborative-robotics-company-acquired-by-abb
15. Do, H.M., Kim, H., Park, D.I., Choi, T.Y., Park, C.: Intuitive and safe teaching device for efficient human-robot collaboration. In: IEEE/RSJ International Conference on Intelligent Robots and Systems, pp. 2272 (2017)

# A Critical Reflection on the Expectations About the Impact of Educational Robotics on Problem Solving Capability

Francesca Agatolio[1(✉)], Michele Moro[1], Emanuele Menegatti[1], and Monica Pivetti[2]

[1] University of Padova, Padua, Italy
francesca.agatolio@phd.unipd.it
[2] University "G. d'Annunzio" of Chieti-Pescara, Pescara, Italy
{michele.moro,emanuele.menegatti}@dei.unipd.it, m.pivetti@unich.it

**Abstract.** This paper discuss the outcomes of an experimental course we run during the 2016-17 school year in two secondary junior schools. The aim of the experiment was to validate the use of educational robotics as a *mindtool*, investigating the impact of robotics on problem solving capability. The course lasted about four months and it was evaluated through a self-assessment with pre and post-activity questionnaires. The emerged results encouraged us to reflect about the role of metacognition and the importance to take it into account for the evaluation of problem solving. In the paper, the activity with the students is described and the analysis of the pre and post-activity questionnaires are discussed and conclusions are drawn.

**Keywords:** Educational robotics · Problem solving capability
Learning support tool

## 1 Introduction

Nowadays educational robotics activity has become more and more common in schools of every level. In some countries, the official curriculum includes activities of robotics or of educational robotics. There is a strong feeling that they have a positive effect on students learning and attitude toward science and technology. Many of the works reported in [4] make high claims on the benefits of educational robotics. However, if one gives a closer look at the impact gauged by most experimental studies in educational robotics, those claims are strongly reduced and shrunk. Indeed, the most evident results are restricted to improvements in the student's involvement and to the learning of specific concepts. This comes as no big surprise, because these are the simplest outcomes to be measured. However, to the best of our knowledge, no authors report experimental data showing an improved learning capabilities or improved results in the students' career. So, should we think that educational robotics is not that effective or should we

M. Strand et al. (Eds.): IAS 2018, AISC 867, pp. 877–888, 2019.
https://doi.org/10.1007/978-3-030-01370-7_68

blame other factors? An answer to this question emerged analysing the data collected during an experimental course, which had the aim to validate the use of educational robotics as a *mindtool*, that is a computer-based tool capable of supporting active learning and critical thinking [10]. In particular, we wanted to investigate the impact of using robotics on the problem solving capability, hoping to capture any possible change in the way students approach problems. The course lasted about four months and it was evaluated through a self-assessment pre and post questionnaire. The results were quite unexpected but meaningful: questions about metaskills brought to light some interesting changes but in the opposite direction than we expected. This makes us reflect about the wisdom of our initial expectations, which were close to others mentioned in similar works. In this paper, we try to circumscribe the extent of the impact on the problem solving capability of an educational robotics course lasting a few months. Starting with the work by Mayer about problem solving [12], we give an answer to the following questions: can a short course really affect the problem solving capability? If yes, what kind of effects should we expect? The paper is structured as follow: Sect. 2 contains theoretical premises and a reflection about the results obtained in similar works; Sect. 3 describes participants, setting and activities of the experimental course we carried out; Sect. 4 contains the analysis of the results; Sect. 5 shows an interpretation of the results in view of the metacognitive aspect; Sect. 6 describes the future implications of these experiences in our studies; Sect. 7 contains the conclusion we came to and our final considerations.

## 2    The Uncatchable Problem Solving Capability

One of the reasons that make so difficult to prove the effects of robotics, is that there is no a standard tool to evaluate it. Moreover, the problem solving capability itself is hard to capture: in this paper we define it as the process of detecting and applying the best suited strategies to solve a certain problem, but due to the huge variety of problems it requires diverse types of skills. Therefore when we try to evaluate the impact of educational robotics on problem solving, we should first identify these skills.

### 2.1    The Three Components Model

In [12], Mayer proposed an interesting model about the aspects that contribute to improve the problem solving capability. Starting from the theories of Sternberg [19], he identified the three factors that answer the question "What does a successful problem solver know that an unsuccessful problem solver does not know?". These three components are:

- Cognitive skills: they are the "bricks" necessary to approach a certain problem (e.g. to calculate the perimeter of a polygon you need to know how to measure the length of the sides and how to sum)
- Metacognitive skills (metaskills): they involve knowledge of when and how to use cognitive skills (e.g. the skill to divide a problem in little parts or to use a graphical representation to solve a mathematical problem)

– Will: it concerns the motivational aspect and it is fundamental to affect positively the effort that is required by students to solve a problem.

Fisher in [7] emphasizes a further distinction of metacognitive skills in "cognitive extension" and "metacognitive thinking", in line with the two levels of reflection (low level and high level) proposed by Von Wright [21] and the concepts of "soznanie" (consciousness in a broad sense) and "osoznanie" (conscious awareness) drawn by Vygotskij. The "cognitive extension" covers "how the student is thinking about the content", while the "metacognitive thinking" focus on "the student's thinking about his/her own thinking about the content", with the last being the real metacognition.

## 2.2 A Short Review of Educational Robotics Applied to Problem Solving

The potential of educational robotics in supporting learning is reinforced by the fact that robots are likely the tools that best embody the concept of tangible artefact proposed by Paper [2,13]. Building and programming robots, allow a project-based approach that promote collaborative learning, hands-on experiences and the exploitation of diverse skills, keeping students engaged in school [1,20] and improving their interest in STEAM [14,16]. In particular, Sullivan in [20] reported positive results on how students utilized cognitive skills and science process skills to solve a robotics challenge. Despite these positive aspects, the impact on the problem solving capability is not so clear [4]. Hussain et al. in [9] investigated the impact both on mathematics and problem solving. The results showed a best performance in mathematics for a group of students but no improvements in the problem solving capability. Lindh and Holgersson [11] studied the effect on the ability to solve mathematical and logical problems and they observed significant improvement only for the group of students with medium scores. Barak and Zadok in [3] concluded that students often found inventive solutions to the problems using diverse kinds of heuristic strategies but they were not able to appropriately reflect on the processes they had used. Some positive results on the use of learning strategies are instead reported by Nugent et al. [17]; it is interesting to notice that the instrument used in this case for the evaluation was MSQL [18] that contains a list of items regarding students' use of cognitive and metacognitive strategies.

# 3 Description of the Course

## 3.1 Setting and Participants

The experimental course involved three classes of two different secondary junior schools (11–12 years old) in northern Italy, for an amount of about 55 students. It lasted 4 months, 2 h per week and it was carried on during curricular time in diverse periods of the year. One class was trained from December 2016 to March 2017, the other classes were trained from February 2017 to May 2017. The students used Lego Mindstorms robotics kits, one class used only the EV3 version while the others used both the EV3 and the NXT version.

## 3.2   The Reasons Behind the Choice of Robot LEGO

The choice of the robot kit, that is Lego Mindstorms, was led by didactical considerations, supported also by our previous experience [15]:

- students have often a previous knowledge of the Lego kits;
- it allows to realize activities with a wide range of complexity, without limiting the excellence;
- it allows many interdisciplinary connections;
- it includes a lot of sensors and accessory features such as sounds, lights and a display that makes the robot particularly captivating and funny;
- the icon-based programming environment assures a low initial threshold in the programming aspect.

## 3.3   Method and Activities

The students worked in small groups of 3–4 members each. We started introducing basic concepts such as sensor, actuator, simple movements, rotation, loop and if/else commands. For this purpose, we designed simple activities allowing students to become confident with the basic knowledge required. Then we progressively introduced more complex tasks involving a deeper problem solving capability. Most of the activities were the same designed for the project RoboESL (http://roboesl.eu/), aiming to reduce the early school leaving phenomenon: all of them refer to an easily identifiable context, and they allow extensions in order to not limit the level of excellence. Some examples of activities are (Fig. 1) :

- The patrol: a robot had to tread three times the perimeter of a building whose base is a square. After this first task, which aims to introduce the loop command, we asked students to modify the program for a rectangular base and more complex shapes.
- Go parking: the students had to program the robot in order to place it in the first free slot of N adjacent parking spaces. A variation of this task requires the robot to park in the last free space.
- The three little pigs: in this activity the students had to represents the fable using the robots in whichever way they want. Subsequently we requested them to optimize the code using the minimum number of commands; in order to increase their involvement, we also proposed them to design the scenario and the robots' clothes.

## 3.4   The Questionnaire

The pre-post questionnaire was structured as follow:

- Section 1: a semantic differential with 12 pairs of opposite adjectives to evaluate the students' perception of robotics (e.g. "Robotics is: unlikable/likable, cold/warm, difficult/easy"). The respondent was asked to choose the answer from a 1–5 scale.

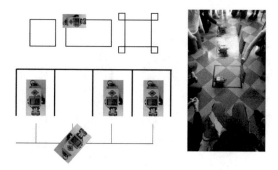

**Fig. 1.** Examples of activities: from the top-left "The patrol", "Go parking" and "Three little pigs"

- Section 2: 21 items to gauge the attitude of students towards learning in general and the usage of robotics in particular. Some of these items examined their approach to problems, which constitutes the metacognitive aspect of problem solving (e.g. "When I solve a problem I am used to go step by step"). The respondent asked using a 5-level Likert scale.
- Section 3: a semantic differential with 4 pairs of opposite adjectives to measure the students' evaluation of the course (e.g. "The lessons were: difficult/easy, useless/useful"). The respondent was asked to choose the answer from a 1–5 scale.

Some of the items are the same used by Hussain et al. in [9].

## 4    Data Analysis

In this section we present the data collected through the questionnaire; we used the two tailed t-test to measure the statistical significance.

### 4.1    The Course's Evaluation

The overall impression about the course is positive: using a 1–5 scale (5 = "completely agree" and 1 = "completely disagree"), 60% of the students answered 4–5 to the question "Did you like the lessons?" and 67% responded 4–5 to the question "Would you like to repeat this experience next year?". This impression was confirmed also by teachers and by the engagement observed during the course. Figure 2 shows the results of the semantic differential related to the question "In your opinion, the lessons were... ?". Even if students felt that the activities were quite difficult, they recognized the utility of them and perceived the lessons as funny.

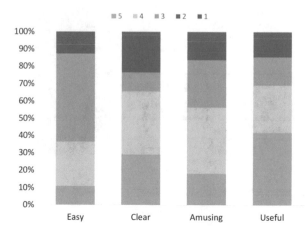

**Fig. 2.** Distribution of the answers "In your opinion, the lessons were...?". 5 is the maximum value

## 4.2 The Perception of Robotics

The semantic differential about the students' perception of robotics reveals a decreasing tendency (Fig. 3) in subsequent values, as confirmed also by our precedent experiences. This is not unexpected: before starting the course most students knew robots only through films, and popular culture; during the course they entered in contact with real robots that have physical limits (item "weak/strong", $t = 1.72$, $p < 0.1$; item "passive/active", $t = 1.8$, $p < 0.1$; item "slow/fast", $t = 1.9$, $p < 0.1$) and are also presented into association to school and effort (item "uncool/cool", $t = 2.08$, $p < 0.05$; item "undesirable/desirable", $t = 1.9$, $p < 0.1$; item "unlikable/likable", $t = 2.55$, $p < 0.05$): as such, this graphic shows they understood what truly means working with robots, and that's positive. It is interesting to observe that the perception of the complexity of the issue hasn't changed (item "difficult/easy").

## 4.3 The Attitude Towards Learning with Robotics

The items about the students' attitude towards the use of robots as a learning tool show interesting results (Fig. 4). Students answered the questions using a 1–5 scale where 5 = "completely agree" and 1 = "completely disagree". The plot (Fig. 4) shows the percentage that said they agree (answers 3–5). A short robotics course apparently doesn't improve the students' self -efficacy (item 6, $t = 2.52$, $p < 0.05$), even if after the course they feel quite secure to be able to realize simple programs for the robot (item 1, $t = 4.8$, $p < 0.01$ ) and to identify eventual mistakes into the program (item 2, $t = 1.69$, $p < 0.1$). Moreover, the 62% answered "completely agree" and "agree very much" to the statement "We learned a lot programming robots". The other items don't show significant changes.

pre
post
Uncool
Not desirable
Unlikable
Difficult
Weak
Far
Passive
Slow
Cold
Unreliable
Cool
Desirable
Likable
Easy
Strong
Close
Active
Fast
Warm
Reliable
1
2
3
4
5

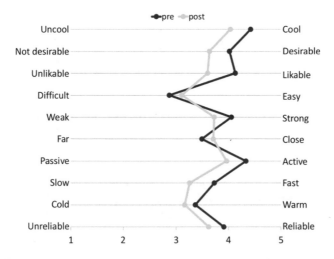

**Fig. 3.** Distribution of the answers about the perception of robotics

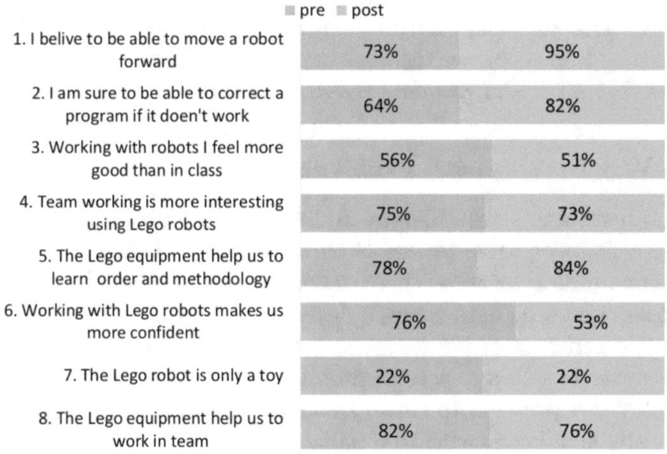

pre
post
1. I belive to be able to move a robot forward
73%
95%
2. I am sure to be able to correct a program if it doen't work
64%
82%
3. Working with robots I feel more good than in class
56%
51%
4. Team working is more interesting using Lego robots
75%
73%
5. The Lego equipment help us to learn order and methodology
78%
84%
6. Working with Lego robots makes us more confident
76%
53%
7. The Lego robot is only a toy
22%
22%
8. The Lego equipment help us to work in team
82%
76%

**Fig. 4.** Distribution of the answers about the students' attitude towards the use of robots

### 4.4 The Team Working Experience

Team working is one of the most positive effects mentioned in relation to educational robotics [5,6]. During our experience, we observed the crucial role of this aspect: students described team working as one of the most stimulating factors, but at the same time as one of the most difficult to manage. Even if there is not statistical significance, the results show that after the course students seem more aware of the difficulties related to work with other students (Fig. 5). It is interesting to note that item 1, that is very similar to item 3, doesn't show substantial variations, probably because it contains the word "group" instead

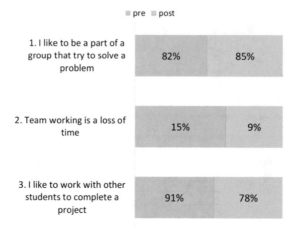

**Fig. 5.** Distribution of the asnwers about the team working experience

of "students". Despite this, students think that working in team is not a loss of time (item 2) and when we discussed about the robotics experience most of them reported team working among the aspects that they appreciated the most.

### 4.5    The Way to Approach Problems

These items investigate the changes in the way students approach a problem regarding the metacognitive aspect of their learning. Before the students filled the post questionnaire, we asked them to answer the questions thinking about the robotics experience. The data shows a decreasing tendency (Fig. 6); in particular items 1,2 point out a negative variation regarding cognitive strategies (item "I am used to go step by step to solve a problem"), persistence and impulse control (item "If there is a problem to solve, I continue to work until I solve it"). Item 1 is statistically significant with $p < 0.05$.

The results point out an unexpected but meaningful scenario that can be summarized with the following points:

- Even if the employment of robots hasn't improved the students' self efficacy, and even if students did perceive the lessons as difficult, they appreciated a lot the course, recognizing they have learnt something new and expressing the wish to repeat the experience.
- The students complained about the team working issues, but they stated they liked very much the opportunity to work with their mates.
- After the course, the percentage of students who think they approach problems in a correct way decreases.

The question that springs to mind is: should we conclude that the several positive effects related to educational robotics are not so real or, perhaps, should we change our perspective about the way that the positive impact occurs? We

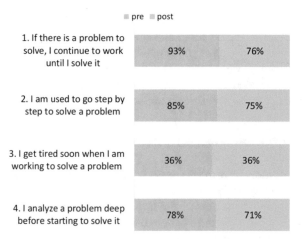

**Fig. 6.** Distribution of the answers about students' behavior when approaching problems

believe that the second option is the correct one and the following paragraph contains the considerations that led us to this conclusion.

## 5   The Essential Step of Awareness

One of the most interesting results that emerges from the data is related to the question "I am used to go step by step to solve a problem". Even if there isn't an evident statistical significance, the gap between pre and post questionnaire is quite wide (pre = 85%, post = 75%) and the items 1 ("If there is a problem to solve, I continue to work until I solve it") and 4 ("I analyze a problem in detail before starting to solve it") seem to support this. We are ready to rule the possibility that during the course students used learning strategies different from the usual. So, it is likely that when they filled the pre questionnaire they had no clear idea about the use of the strategy of splitting of a problem in little parts in order to better face them. When we administered the post questionnaire, we explicitly asked them to refer to the experience they just took part in. Then, the data is the result of a targeted reflection of the way they have approached robotics issues and so, the change between pre and post percentage prove that something has been modified. The same consideration can be done for the data regarding the team working experience. These modifications don't regard directly the problem solving capability, but the self-awareness about a difficulty (team working) or about the lack of a certain cognitive strategy (to split the problem in parts), which is in itself a reflection of their own way to approach a problem: that's a metacognitive process. Indeed, metacognition is described as the individual's own awareness and consideration of his or her cognitive processes and strategies' [7,8]. In [8] Flavell introduces the concept of metacognitive experiences, presenting them as "any conscious cognitive or affective experiences that accompany and pertain to any intellectual enterprise" and reports as example the "feeling that you do not understand something another person just said". Therefore, if we want to prove the impact of educational robotics on the problem solving capability, we have to take into account the fact that any type of improvement is preceded by the development of self-awareness, that manifests itself, among other things, as the capability to recognize our own shortcomings.

## 6   Considerations for the Future

The results that emerged from this experimental course showed us the possibility to carry on our research defining better the goals and redirecting our expectations about the results. We aim to prove the impact of robotics on the problem solving capability taking into account the three components described in Sect. 2, and in particular the metacognitive aspect. Our future steps will be:

- to redesign our course in order to better mediate the metacognitive process (e.g. using workingsheets, helding more time to reflection)
- to identify one or more evaluation tools to investigate how robotics influences the three components of problem solving
- to test the abovementioned tools on a wider sample, comparing the results with a control group.

But, first of all, we ask ourselves these questions: why robotics tool can be particularly suited (1) to support the problem solving capability and (2) to convey metacognitive skills? The answer to the first question is that robotics provides naturally a project based approach in which students are engaged to solve a problem, often related to real situations. Moreover it is widely proved, and our experience confirms it, that using robots increases the students' involvement and "will" is precisely one of the factors required to be a good problem solver. The second question should be further explored but it presents at least two things in its favor. The first is that educational robotics promotes team working and collaborative learning, and therefore the ability to communicate with others. The sharing and argumentation of one's own ideas contribute to the development of language, which Vygotskij considers a fundament of metacognition [22]: in fact, language allows the organization of one's own thoughts and the making of proper considerations about our experiences. The second is that programming a robot provides an immediate feedback on the correctness of the solution. Thanks to this, the individuation and correction of the errors becomes a constitutive part of the task and it isn't relegated to a separate moment.

## 7   Conclusions

When we started the experimental course, we were aware about the difficulty to evaluate the problem solving capability and to obtain interesting values from a training of few months. The results that emerged from similar activities were not encouraging. To evaluate the impact of the course we administered a pre and post questionnaire, composed by a set of items that should have caught eventual changes in different areas: perception of robotics, attitude towards learning and the use of robots, team working experience. Even if only a few values were statistically significant, we noticed a general tendency in opposite direction to what we expected: despite the overall appreciation of the lessons, the values related to the perception of robotics and some other items related to the way students approached problems got worse. This fact suggested us that the robotics activities had an impact on the metacognitive process. Exploring further this hypothesis, we discovered an interesting model about the three components that contribute to improve the problem solving capability: cognitive skills, metacognitive skills and will. The fundamental role of metacognition in problem solving gave us a key for the evaluation of the obtained values: the decreasing of some values can imply a modification in the capability to recognize our own shortcomings and difficulties, which is in itself a development of self-awareness. For this reason, the Mayer model described in Sect. 2 provides great information about how to measure the problem solving capability. The approach of investigating the impact of robotics starting from the elements at the basis of problem solving can help us to better understand the eventual crucial points and reinforce them. Moreover, it is reasonable and realistic to redirect our expectations about the impact of an education robotics course on students, taking into account the necessary steps that lead to the metacognition development.

**Acknowledgments.** This work was partially supported by the University of Padova with the grant ERASM and by EU Erasmus+ project RoboESL

# References

1. Alimisis, D.: Educational robotics: open questions and new challenges. Themes Sci. Technol. Educ. **6**(1), 63–71 (2013)
2. Alimisis, D., Moro, M., Arlegui, J., Pina, A., Frangou, S., Papanikolaou, K.: Robotics & constructivism in education: the terecop project. EuroLogo **40**, 19–24 (2007)
3. Barak, M., Zadok, Y.: Robotics projects and learning concepts in science, technology and problem solving. Int. J. Technol. Des. Educ. **19**(3), 289–307 (2009)
4. Benitti, F.B.V.: Exploring the educational potential of robotics in schools: a systematic review. Comput. Educ. **58**(3), 978–988 (2012)
5. Denis, B., Hubert, S.: Collaborative learning in an educational robotics environment. Comput. Hum. Behav. **17**(5), 465–480 (2001)
6. Eguchi, A.: Educational robotics for promoting 21st century skills. J. Autom. Mob. Robot. Intell. Syst. **8**(1), 5–11 (2014)
7. Fisher, R.: Thinking about thinking: Developing metacognition in children. Early Child Dev. Care **141**(1), 1–15 (1998)
8. Flavell, J.H.: Metacognition and cognitive monitoring: a new area of cognitive-developmental inquiry. Am. Psychol. **34**(10), 906 (1979)
9. Hussain, S., Lindh, J., Shukur, G.: The effect of lego training on pupils' school performance in mathematics, problem solving ability and attitude: Swedish data. J. Educ. Technol. Soc. **9**(3), 182–194 (2006)

10. Jonassen, D.H., Carr, C., Yueh, H.P.: Computers as mindtools for engaging learners in critical thinking. TechTrends **43**(2), 24–32 (1998)
11. Lindh, J., Holgersson, T.: Does lego training stimulate pupils ability to solve logical problems? Comput. Educ. **49**(4), 1097–1111 (2007)
12. Mayer, R.E.: Cognitive, metacognitive, and motivational aspects of problem solving. Instr. Sci. **26**(1), 49–63 (1998)
13. Mikropoulos, T.A., Bellou, I.: Educational robotics as mindtools. Themes Sci. Technol. Educ. **6**(1), 5–14 (2013)
14. Mitnik, R., Nussbaum, M., Soto, A.: An autonomous educational mobile robot mediator. Auton. Robot. **25**(4), 367–382 (2008)
15. Moro, M., Agatolio, F., Menegatti, E.: The development of robotic-enhanced curricula for the roboesl project: premises, objectives, preliminary results. In: Alimisi, R. (ed.) Robotics-based Learning Interventions for Preventing School Failure & Early School Leaving, ROBOESL Conference 2016 Proceedings. EDUMOTIVA (2016)
16. Nugent, G., Barker, B., Grandgenett, N.: The effect of 4-h robotics and geospatial technologies on science, technology, engineering, and mathematics learning and attitudes. In: EdMedia: World Conference on Educational Media and Technology, pp. 447–452. Association for the Advancement of Computing in Education (AACE) (2008)
17. Nugent, G., Barker, B., Grandgenett, N., Adamchuk, V.: The use of digital manipulatives in k-12: robotics, GPS/GIS and programming. In: 39th IEEE Frontiers in Education Conference, FIE 2009, pp. 1–6. IEEE (2009)
18. Pintrich, P.R., et al.: A manual for the use of the motivated strategies for learning questionnaire (MSLQ) (1991)
19. Sternberg, R.J.: Intelligence as thinking and learning skills. Educ. Leadersh. **39**(1), 18–20 (1981)
20. Sullivan, F.R.: Robotics and science literacy: thinking skills, science process skills and systems understanding. J. Res. Sci. Teach. **45**(3), 373–394 (2008)
21. Von Wright, J.: Reflections on reflection. Learn. Instr. **2**(1), 59–68 (1992)
22. Vygotsky, L.S.: The Collected Works of LS Vygotsky: Volume 1: Problems of General Psychology, Including the Volume Thinking and Speech, vol. 1. Springer Science & Business Media, New York (1987)

# Inferring Capabilities by Experimentation

Ashwin Khadke[1]([⊠]) and Manuela Veloso[2]

[1] Robotics Institute, Carnegie Mellon University, Pittsburgh, USA
akhadke@andrew.cmu.edu
[2] Department of Machine Learning, Carnegie Mellon University, Pittsburgh, USA

**Abstract.** We present an approach to enable an autonomous agent (learner) in building a model of a new unknown robot's (subject) performance at a task through experimentation. The subject's appearance can provide cues to its physical as well as cognitive capabilities. Building on these cues, our active experimentation approach learns a model that captures the effect of relevant extrinsic factors on the subject's ability to perform a task. As personal robots become increasingly multifunctional and adaptive, such autonomous agents would find use as tools for humans in determining "What can this robot do?". We applied our algorithm in modelling a NAO and a Pepper robot at two different tasks. We first demonstrate the advantages of our active experimentation approach, then we show the utility of such models in identifying scenarios a robot is well suited for, in performing a task.

## 1 Introduction

Suppose we get a multipurpose domestic robot for our home but are not familiar with all of its functionalities. How do we identify what tasks it can perform? Appearance and specifications of a robot can convey information about its physical as well as cognitive capabilities. Seeing a legged robot equipped with a camera and a microphone could make one wonder if it can climb stairs, recognize faces, detect hand gestures or interpret voice commands. How do we identify which of these appearance-deduced tasks it can actually perform? Moreover, although physical appearances can provide rich cues about a robot's capabilities they are not sufficient to identify the scenarios in which it can function well. The robots Roomba and Braava appear similar and are both used for cleaning floors. But the Roomba can clean carpets and not wet floors, which is exactly the opposite of what Braava can do. For a human working collaboratively with a robot, knowing the robot's strengths and shortcomings is especially useful. A robot's spec sheet provides information about different sensors and actuators. But inevitably how well a robot performs a task depends on the way it is programmed and for a naive user, this is difficult to determine simply based on appearance and specifications. Experimenting with a robot can help identify the scenarios it is well suited for. But experimentation is tedious and intelligent robots are capable of learning new skills and adapting to new scenarios over time. This motivates the

M. Strand et al. (Eds.): IAS 2018, AISC 867, pp. 889–901, 2019.
https://doi.org/10.1007/978-3-030-01370-7_69

need for an autonomous system that can intelligently experiment with robots, identify their skills and quantify their applicability in different scenarios.

In this paper we tackle the problem of an autonomous agent (learner) building a model of a robot (subject) at performing a particular task through experimentation. The outcomes of these experiments can be non-deterministic. We call such models as Capability Models (Sect. 3). We assume the learner can infer tasks a subject can potentially perform from its appearance and, present a method to build a model of the subject at one of these tasks. Apart from the subject's inherent capabilities, certain extrinsic factors may affect its performance at the task. Assuming some of these factors are controllable and the learner can choose values for such factors in the experiments it conducts, we provide an approach to pick values for controllable factors that generate the most informative outcomes (Sect. 4.1). However, knowing the set of extrinsic factors relevant for a particular robot and a task a priori is not always feasible. We present a model refinement method to identify relevant factors from a set of candidates (Sect. 4.2). Further we show that Capability Models can be used in quantifying a robot's ability to perform a task in different scenarios thereby identifying situations a robot is well suited for (Sect. 5).

We applied our algorithm in modelling a NAO robot at the task of kicking a ball. We show that active experimentation leads to faster learning of the subject's model than passively observing it perform and that our model refinement approach correctly identifies the set of relevant factors missing from the model (Sect. 6.1). Furthermore, we learned a Capability Model for a Pepper robot programmed to pick up and clear objects off a table. From the learned model, we identified the types of objects the robot can pick up. We demonstrate how this knowledge is useful in improving performance at a collaborative clear-the-table task (Sect. 6.2).

## 2   Related Work

Earlier works on learning from experimentation either addressed domains that are inherently deterministic [1,2] or assumed that the experiments they conduct are deterministic [3]. Owing to noisy actuation and sensing, outcomes of experiments with robotic systems are non-deterministic and, hence the problem of deducing a robot's capabilities through experimentation is challenging.

Affordance [4] is a relation between a certain effect, a class of objects and certain robot action. Learning affordances [5,6] is similar to learning the physical capabilities of a robot. However, these approaches learn a mapping from motor commands to effects on different objects characterized by raw sensory input. It is difficult for a human to use these models in identifying scenarios a robot is well suited for. Another approach [7] uses visual features (Size, Shape, Color etc.) to characterize objects and learns models that capture the effect (Object speed etc.) of a robot's actions (Tap, Push, etc.) on different objects. However, all of these works [5–7] adopt passive approaches to learn and do not use their existing models to reason about what to explore next. Moreover, they provide no concrete method to quantify a robot's capability.

Modeling a robot's capability can be thought of as learning a forward model [8]. Such models predict the change in a robot's state brought about by its actions. Active approaches to learn forward models exist [9–13]. However, these models only predict the immediate effects of an action i.e. predict the next state given the current state and action and provide no approach to identify from the forward model, the likelihood of successful task execution in different scenarios. An idea to mitigate this problem is to learn a forward model over higher level states [14,15]. But the emphasis in these works is on learning a policy for a task [11,12] or a combination of tasks [10,13] or learning higher level actions (sub-policies) relevant for a task [14].

## 3   Capability Models

Suppose the subject is an anthropomorphic robot (Fig. 3a) and the learner chooses to build a model of the subject at the task of kicking a ball. Extrinsic factors that the learner may consider include, the size of the ball and turf on which the subject is playing. Among these factors, ball size is controllable. An experiment for this task would constitute the learner commanding the subject to kick a ball in certain direction from a particular position and observing the outcome. A robot's perception and actuation is noisy and therefore the outcome is not deterministic. To capture this non-determinism we use a Bayesian Network.

We introduce Capability Model (Fig. 1), a Bayesian Network which consists of three types of nodes namely:

- SIT = CNTX ∪ COMM, is the set of variables that describe the situation in which the subject is performing the task.
  - CNTX is the set of extrinsic factors (context) for the task.
  - COMM is the set of commands given to the subject.
- OUT is the set of variables denoting the outcomes of the task.
- $ATT_o$ is the set of attributes of variable $o \in$ OUT. These variables are not explicitly accounted in the model. Section 4.2 discusses their need and utility.

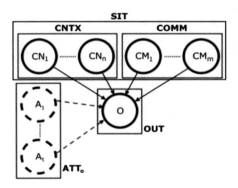

**Fig. 1.** Capability model

We capture the subject's ability to perform a task in the conditional probability tables associated with this Bayesian Network.

Figure 2 presents a Capability Model for the *BallKick* task discussed before using the notation we just introduced. *Position* represents the location of the ball with respect to the subject before it kicks. *KDc* and *KDo* denote the commanded and observed kick direction respectively. *KDo* is *None* if the subject attempts but fails to kick the ball or does not attempt to kick. The set OUT need not always be a singleton. How far a subject kicks a ball could be another outcome for the *BallKick* task.

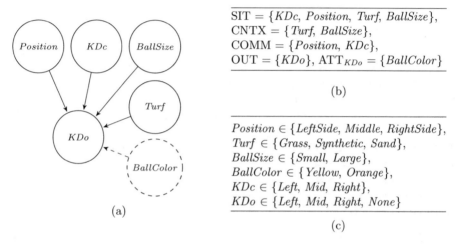

SIT = {*KDc, Position, Turf, BallSize*},
CNTX = {*Turf, BallSize*},
COMM = {*Position, KDc*},
OUT = {*KDo*}, ATT$_{KDo}$ = {*BallColor*}

(b)

*Position* ∈ {*LeftSide, Middle, RightSide*},
*Turf* ∈ {*Grass, Synthetic, Sand*},
*BallSize* ∈ {*Small, Large*},
*BallColor* ∈ {*Yellow, Orange*},
*KDc* ∈ {*Left, Mid, Right*},
*KDo* ∈ {*Left, Mid, Right, None*}

(c)

(a)

**Fig. 2.** Capability Model for the *BallKick* task. (a) depicts the Bayesian Network, (b) shows the type of each variable in the model and, (c) describes the values each variable can take.

## 4    Building Capability Models

Building a model involves identifying the right factors to include in the Bayesian Network and learning the conditional probabilities associated with it. First, we present a method to learn the conditional probabilities assuming the learner knows the right factors and structure of the network is fixed. Later, we describe our method to refine the model if need be. Here we introduce some notation.

- A Bayesian Network is a tuple $(G, \theta)$
    - $G \equiv (V, E)$ is a graph, $V$ is the set of nodes and $E$ is the set of edges.
    - $V = \text{SIT} \cup \text{OUT}$ are random variables with Multinomial distribution
    - $E$ capture the conditional dependencies amongst the nodes $V$
    - $\theta$ parameterize the conditional probability distributions
- $Q \subset \text{SIT}$ are variables a learner can control in an experiment. COMM $\subset$ Q. An instantiation[1] of Q is called a *Query*.

---

[1] An instantiation of a set of random variables is a mapping from variables in the set to values in their domain. *Query* $\leftarrow \cup_{q \in Q}\{q : v_q\}$ where $v_q \in Domain(q)$.

– Instantiations of CNTX, COMM and OUT are called *Context*, *Command* and *Outcome* respectively. *Situation = Context ∪ Command*

## 4.1  Active Learning for Bayesian Networks

We adopt a Bayesian approach to learn parameters and use the algorithm presented in [16] to build a distribution over $\theta$. We describe it here in short.

The algorithm starts with a prior $p(\theta)$ and builds a posterior $p'(\theta)$ by actively experimenting with the subject. In each experiment, it picks a *Query* and requests the subject to perform the task. Variables in SIT $\setminus$ Q either have a fixed value (Eg. *Turf* in the *BallKick* task) or are assigned some value by the environment. A standard Bayesian update on the prior $p$, for the parameters of the conditional distributions identified by the *Situation* i.e. $P(o|Situation)$ $\forall o \in$ OUT, based on the *Outcome* yields the posterior $p'$. The posterior $p'$ becomes the prior for the next experiment.

To generate a *Query* from the current estimate of $p(\theta)$ we need a metric to evaluate how good the current estimate is. We can then quantify the improvement in the estimate brought about by different queries and pick the one which leads to the biggest improvement. Let $\theta^\star$ be the true parameters of the model and $\theta'$ be a point estimate. $\sum_{o \in OUT} D_{KL}(P_{\theta^\star}(o)||P_{\theta'}(o))^2$ denotes the error in point estimate $\theta'$, where $P_{\theta^\star}(o)$ and $P_{\theta'}(o)$ are distributions of variable $o$ parameterized with $\theta^\star$ and $\theta'$ respectively. $\theta^\star$ is not known, but we do have $p(\theta)$, which is our belief of what $\theta^\star$ is given the prior and observations. Error in point estimate $\theta'$ with respect to $p(\theta)$ can be quantified as in Eq. (1)

$$\text{Error}_p(\theta') = \sum_{o \in OUT} \int_\theta D_{KL}(P_\theta(o)||P_{\theta'}(o))p(\theta)d\theta \tag{1}$$

$$\text{ModelError}(p) = \min_{\theta'} \text{Error}_p(\theta') \tag{2}$$

We use ModelError($p$) of a distribution $p$, defined in Eq. (2), as the measure of quality for the estimate $p(\theta)$. Lower the ModelError, better the estimate. We would want to see observations that reduce the ModelError associated with $p(\theta)$ to improve our estimate. But the learner can only control the *Query*. For a particular *Query* we take an expectation over possible observations to evaluate the Expected Posterior Error (EPE) as defined in Eq. (3). In every experiment the learner picks the *Query* with the lowest EPE.

$$\text{EPE}(p, Query) = E_{\Theta \sim p(\theta)}\big(E_{Outcome \sim P_\Theta(OUT|Query)}(\text{ModelError}(p'))\big) \tag{3}$$

In Eq. (3) $p'$ represents the posterior obtained after updating prior $p$ with sample drawn from $P_\Theta(OUT|Query)$. For further details, please refer [16]. $\theta_T$ as defined in Eq. (4) parameterize the conditional probabilities of the Capability Model for task $T$. We compute $\theta_T$ using the learned distribution $p(\theta)$.

---

[2] $D_{KL}(P_1(o)||P_2(o)) = \sum_{v_o \in Domain(o)} P_1(o = v_o)\log\Big(\dfrac{P_1(o = v_o)}{P_2(o = v_o)}\Big)$, $P_1$ and $P_2$ are distributions of $o$.

$$\theta_T = \int_\theta \theta p(\theta) d\theta \tag{4}$$

## 4.2   Model Refinement

Every subject may have a different set of extrinsic factors relevant for a task. If a subject only detects balls of a certain color, a variable *BallColor* should be included in the model for the *BallKick* task. We assume the learner starts with a minimal set CNTX and variables representing some of the relevant factors may be missing. In Sect. 3 we defined $\text{ATT}_o$ to be attributes of variable $o \in \text{OUT}$ not explicitly accounted in the model. We assume that the missing variables, if any, belong to these sets. Including all of them makes the model unnecessarily large and difficult to learn. We need a metric to quantify the dependence of variables in OUT on the attributes to identify the relevant ones.

An attribute $A_j \in \text{ATT}_o$ is relevant if the subject's performance (distribution of $o$) is drastically different for different values of $A_j$ in at least one of the observed *Situation*s. In each experiment, attributes are chosen randomly independent of the *Situation*. After every experiment, the learner computes $\hat{P}(A_j|\text{Situation})$ and $\hat{P}(o|A_j, \text{Situation}) \forall o \in \text{OUT}, \forall A_j \in \text{ATT}_o$. $\hat{P}$ is the estimate of the true distribution from the observations in the past experiments. We use the metric defined in Eq. (5), to quantify the dependence of $o$ on the attribute $A_j$ in a *Situation*.

$$R(o, A_j|\text{Situation}) = \frac{I(o, A_j|\text{Situation})}{\min(H(\hat{P}(o|\text{Situation})), H(\hat{P}(A_j))} \tag{5}$$

$$I(o, A_j|\text{Situation}) = H(\hat{P}(o|\text{Situation})) -$$
$$\sum_{a \in Domain_{valid}(A_j|\text{Situation})} H(\hat{P}(o|A_j = a, \text{Situation})) \hat{P}(A_j = a) \tag{6}$$

In Eq. (6), $H(P)$ is the entropy of distribution $P$ and $Domain_{valid}(A_j|\text{Situation})$ are values of $A_j$ that have been observed sufficiently in a *Situation*. As $A_j$ is sampled independently of the *Situation*, $\hat{P}(A_j) \approx \hat{P}(A_j|\text{Situation})$. Therefore, $I(o, A_j|\text{Situation})$ is approximately the Mutual Information of $o$ and $A_j$ given the *Situation*, and $R(o, A_j|\text{Situation})$ is the Coefficient of Mutual Information. If $R(o, A_j|\text{Situation})$ is greater than threshold $R_{\text{Th}}$, the learner adds $A_j$ to CNTX, updates the parents of variable $o$ in the graph and creates conditional probability tables for the updated graph.

Algorithm 1 outlines the overall method. Function BestQuery (Line 10) computes a *Query* using the approach described in Sect. 4.1. Function IdentifyDependence (Line 16 applies the model refinement method presented above.

## 5   Quantifying Capabilities

To determine how well a robot performs a task in different scenarios, we need a reference that captures the expected performance and, a metric that quantifies

**Algorithm 1. LearnModel**(maxIter, R$_{\text{Th}}$)

---

1  From Domain Knowledge
2      Construct $G \equiv (V, E)$
3      Initialize Q, $p(\theta)$
4  SituationsObserved $\leftarrow \emptyset$
5  Initialize $\hat{P}$
6  $i \leftarrow 0$
7  **while** $i <$ maxIter **do**
8  | $i \leftarrow i + 1$
9  | ATTIncluded $\leftarrow \{\}$
10 | $Query \leftarrow$ **BestQuery**$(p(\theta),$ Q$)$
11 | $Attributes \leftarrow$ **SampleUniform**$\left(\bigcup_{o \in \text{OUT}} \text{ATT}_o\right)$
12 | $Outcome, Situation \leftarrow$ **Experiment**$(Query, Attributes)$
13 | SituationsObserved $\leftarrow$ SituationsObserved $\bigcup Situation$
14 | $p(\theta), \hat{P} \leftarrow$ **Update**$(p(\theta), \hat{P}, Situation, Outcome, Attributes)$
15 | **for** $o \in$ OUT **do**
16 | | ATTIncluded$[o] \leftarrow$
   | | **IdentifyDependence**$(\hat{P}, \text{ATT}_o, \text{SituationsObserved}, \text{R}_{\text{Th}})$
17 | $p(\theta), G, Q \leftarrow$ **Modify**$(p(\theta), G, Q, \text{ATTIncluded})$
18 **return** $(G, p(\theta))$

---

how the robot fares against this standard. We assume the reference for task $T$, is a distribution of the outcome variables conditioned on the commands i.e. $P_{\text{ref}}^T(\text{OUT}|\text{COMM})$. Equation (7) shows a possible reference for *BallKick* task (Fig. 2).

$$P_{\text{ref}}^{BallKick}(KDo|KDc, Position) = \begin{cases} 1 \text{ if } KDo = KDc \\ 0 \text{ otherwise} \end{cases} \tag{7}$$

This reference implies that a robot is expected to always kick in the commanded direction. $Score_T(Context)$, defined in Eq. (9), denotes how well a robot fares at a task $T$ in certain $Context$. Lower score values indicate poor performance. We say a robot functions well in a $Context$, if the score is higher than a threshold.

$$Mismatch(Context) \triangleq \frac{\sum\limits_{Command} D_{KL}\left(P_{\text{ref}}^T(\text{OUT}|Command)||P_{\theta_T}(\text{OUT}|Situation)\right)}{|Domain(\text{COMM})|} \tag{8}$$

$Situation = \text{Context} \cup Command$ (Sect. 4). Thus $Mismatch$ is a function of the $Context$. $|Domain(\text{COMM})|$ is the number of possible $Commands$.

$P_{\theta_T}$ is the distribution parameterized by $\theta_T$. For a task $T$, $\theta_T$ is defined in Eq. (4)

$$Score_T(Context) = \frac{1}{1 + Mismatch(Context)} \tag{9}$$

## 6    Results

We present results of applying our method in modelling a NAO robot (Fig. 3a) performing the task of kicking a ball and a Pepper robot (Fig. 3b) picking up different types of objects and clearing them off a table.

(a) NAO                              (b) Pepper

**Fig. 3.** Subjects for experiments

### 6.1    BallKick Task

Here we demonstrate advantages of the active experimentation approach and our model refinement algorithm. Our approach builds a model, but without knowing the ground truth we cannot determine how good the learned model is. So we programmed the robot to behave according to a predefined model with parameters $\theta_{predefined}$. We use $D_{KL}(P_{\theta_{predefined}}(\text{OUT}) \| P_{\theta_{BallKick}}(\text{OUT}))$ to determine how close the learned and predefined models are. $\theta_{BallKick}$ is defined in Eq. (4), for the *BallKick* task. Figure 4a shows the learner's initial guess of the Bayesian Network.

**Passive Vs Active.** Passively observing a subject, where you do not control the scenarios in which you witness it perform a task, is equivalent to randomly picking *Situations* to build a model. We learned a model using our approach and another one by randomly picking *Situations*. Figure 4b depicts the Bayesian Network for the predefined model. We experimented with a single ball on a synthetic turf and thus variables *Turf* and *BallSize* were dropped in the predefined model. A subtle point to note, the experiments were noisy. The learner chose a particular *Situation*, sampled a direction from $P_{\theta_{predefined}}(KDo|Situation)$ and commanded the robot to kick in this direction. However, owing to noisy perception

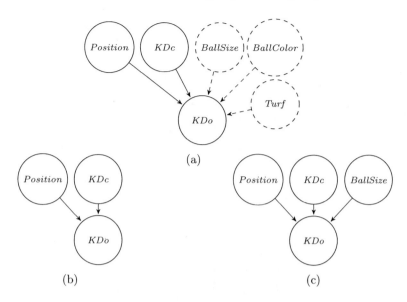

**Fig. 4.** (a) depicts the learner's initial guess of the model and, (b) and (c) depict the Bayesian Network of the predefined model. Nodes are as defined in Fig. 2c

and actuation, sometimes the robot kicked in directions it wasn't commanded to. Figure 5a shows that despite noisy experiments the active approach converged faster.

**Model Refinement.** We programmed the robot to detect balls of any color but only a specific size. If presented with a ball of a different size, the subject would not detect and thus won't kick i.e. *KDo* would be *None*. Figure 4c depicts the Bayesian Network for the predefined model and Fig. 4a shows the learner's initial guess. The learning curve for this model is shown in blue in Fig. 5b. The learner correctly identified the missing variable to be *BallSize*. Once identified, the algorithm resets the conditional probabilities (hence the jump in the trend) and restarts the learning process with the updated model. To show that incorporating relevant attributes yields a model better representative of the subject, we learned another model including *BallSize* from the start. As can be seen in Fig. 5b, the model that included *BallSize* from the start (in green) converged to a lower *KL Divergence* compared to the model that did not (in blue). Moreover, post refinement the trends for both models are almost same.

We tested in simulation, how our approach fares as the number of missing attributes increase. In these experiments the robot could only detect balls of a certain size and the kicked ball would randomly end up in any of the three possible directions i.e. *BallSize* and *Turf* were the missing relevant variables. To simulate noisy experiments we sampled *KDo* uniform randomly 20% of the time. We performed multiple runs and the results in Fig. 5c show that including all the relevant variables gives a better model. More the number of missing vari-

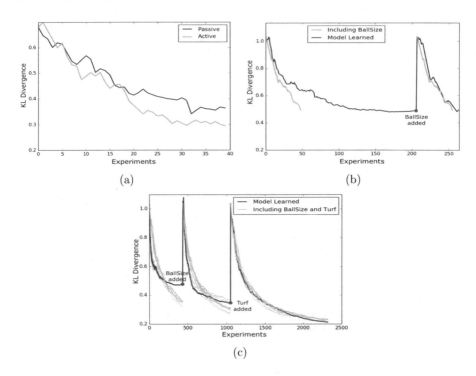

**Fig. 5.** (a), (b) and (c) depict the trend in $D_{KL}(P_{\theta_{predefined}}(KDo)||P_{\theta_{BallKick}}(KDo))$. (a) compares the trends when learned actively vs passively. In (b) and (c), curves in blue depict trends for the learner's model as it identifies relevant factors to include, red points mark the instances when a new variable is added and curves in green depict the trend if the model were initialized with the right variables. (Color figure online)

ables, more the number of experiments needed to identify them all. Moreover, it becomes progressively harder. As a variable gets included in the model, possible *Situations* increase. While evaluating $R(o, A_j|S)$ (Eq. (5)), we only consider the distributions of $o$ conditioned over values of $A_j$ that have been observed more than a certain number of times in the *Situation* S. The active learning algorithm avoids repeating *Situations* and thus it becomes incrementally harder to identify relevant variables.

## 6.2   Pickup Task

We programmed a Pepper robot to detect and pickup objects of three shapes viz. spherical, cubical and cylindrical. We experimented with two sets of weights and two sizes for each shape (in total 12 types of objects). Figure 6 depicts the Capability Model for the *Pickup* task. A *Context* (*Size*, *Shape* and *Weight*) denotes an object type. In every experiment Pepper was asked to pick up a particular type of object with one of its arms.

We performed 2 trials with 70 experiments each to learn the conditional probability tables associated with the Capability Model. We computed $Score_{Pickup}$, as defined in Eq. (9), for each object type using the reference $P_{ref}^{Pickup}$ defined in Eq. (10). We identified object types with score higher than a threshold at the end of both trials, as favourable for Pepper to pickup.

$$P_{ref}^{Pickup}(Pick|Arm) = \begin{cases} 1 \text{ if } Pick = Success \\ 0 \text{ otherwise} \end{cases} \tag{10}$$

Having knowledge of the scenarios a robot is well suited for could help in a collaborative task. To demonstrate this, we employed the robot along with a human to clear a cluttered table. In every experiment, the human cleared all but 4 objects off the table and, the robot had to clear the rest. The robot was allowed 3 tries per object (12 in total). We performed such experiments in two settings. In the first setting, the human randomly selected objects for the robot to pick up and in the second, the human only selected objects of favourable types. We conducted 5 experiments in each setting. Table 1 summarizes the results. Performance at the task is better in terms of number of tries as well as number of objects cleared, when the robot is employed in a favourable scenario.

A video outlining our work can be found at https://youtu.be/_9fm3U80vHE.

**Table 1.** Results of the clear-the-table task ($\mu \pm \sigma$) after 5 experiments per setting.

| Settings | Number of objects cleared | Number of tries |
|---|---|---|
| Objects of any type | $1.8 \pm 0.98$ | $8.8 \pm 1.7$ |
| Objects of favourable types | $3.4 \pm 0.49$ | $6.6 \pm 1.2$ |

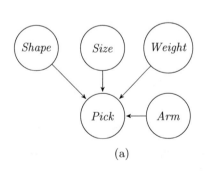

(a)

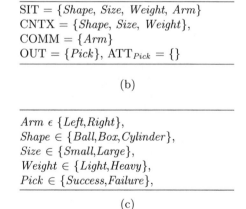

SIT = {*Shape, Size, Weight, Arm*}
CNTX = {*Shape, Size, Weight*},
COMM = {*Arm*}
OUT = {*Pick*}, ATT$_{Pick}$ = {}

(b)

*Arm* $\epsilon$ {*Left, Right*},
*Shape* $\in$ {*Ball, Box, Cylinder*},
*Size* $\in$ {*Small, Large*},
*Weight* $\in$ {*Light, Heavy*},
*Pick* $\in$ {*Success, Failure*},

(c)

**Fig. 6.** Capability Model for the *Pickup* task. (a) depicts the Bayesian Network, (b) shows the type of each variable in the model and, (c) describes the values each variable can take.

# 7   Conclusion and Future Work

Building models of unknown robots becomes all the more relevant as robots become increasingly multifunctional. To the best of our knowledge, the problem of inferring capabilities from appearance and active experimentation is novel and yet unexplored. We presented an algorithm to build capability models from experiments and showed results with a NAO and a Pepper robot at two different tasks. Experimenting with a physical system requires an operator. However, we still need a systematic approach to design experiments. Our algorithm can serve as a tool for humans in determining, "What can this robot do?"

As an extension to this work we intend to develop an approach to draw inferences about the capabilities of a robot from its appearance and specifications. A robot's capabilities may have certain limitations which are intrinsic and others which it may learn to overcome. Enabling the learner to distinguish between the two and experiment accordingly is an interesting future direction. Moreover, we assumed that experimenting with the subject in any *Situation* has the same cost. Depending upon the task this may be quite far from reality. Learning models in such scenarios is another interesting problem.

**Acknowledgements.** This research is partially sponsored by DARPA under agreements FA87501620042 and FA87501720152 and NSF under grant IIS1637927. The views and conclusions contained in this document are those of the authors only.

# References

1. Mitchell, T.M., Utgoff, P.E., Banerji, R.: Readings in knowledge acquisition and learning. In: Buchanan, B.G., Wilkins, D.C. (eds.) Learning by Experimentation: Acquiring and Refining Problem-solving Heuristics (1993)
2. Scott, P.D., Markovitch, S.: Experience selection and problem choice in an exploratory learning system. Mach. Learn. (1993)
3. Gil, Y.: Learning by experimentation: incremental refinement of incomplete planning domains. In: Proceedings of the Eleventh International Conference on International Conference on Machine Learning (1994)
4. Gibson, J.J.: The Theory of Affordances Chapt (1977)
5. Şahin, E., Çakmak, M., R. Doğar, M., Uğur, E., Üçoluk, G.: To afford or not to afford: a new formalization of affordances toward affordance-based robot control. Adapt. Behav. (2007)
6. Dearden, A., Demiris, Y., Kaelbling, L., Saffotti, A.: Learning forward models for robots. In: IJCAI-International Joint Conference in Artificial Intelligence (2005)
7. Montesano, L., Lopes, M., Bernardino, A., Santos-Victor, J.: Learning object affordances: from sensory–motor coordination to imitation. IEEE Trans. Robot. (2008)
8. Jordan, M.I., Rumelhart, D.E.: Forward models: supervised learning with a distal teacher. Cogn. Sci. (1992)
9. Baranès, A., Oudeyer, P.-Y.: R-IAC: Robust intrinsically motivated exploration and active learning. IEEE Trans. Auton. Ment. Dev. (2009)
10. Baranes, A., Oudeyer, P.-Y.: Active learning of inverse models with intrinsically motivated goal exploration in robots. Robot. Auton. Syst. (2013)

11. Wang, C., Hindriks, K.V., Babuska, R.: Active learning of affordances for robot use of household objects. In: IEEE-RAS International Conference on Humanoid Robots (2014)
12. Forestier, S., Oudeyer, P.Y.: Modular active curiosity-driven discovery of tool use. In: IEEE/RSJ International Conference on Intelligent Robots and Systems (2016)
13. Forestier, S., Mollard, Y., Oudeyer, P.: Intrinsically motivated goal exploration processes with automatic curriculum learning. CoRR (2017)
14. Hofer, S., Brock, O.: Coupled learning of action parameters and forward models for manipulation. In: IEEE/RSJ International Conference on Intelligent Robots and Systems (2016)
15. Mugan, J., Kuipers, B.: Autonomous learning of high-level states and actions in continuous environments. IEEE Trans. Auton. Ment. Dev. (2012)
16. Tong, S., Koller, D.: Active learning for parameter estimation in Bayesian networks. In: Proceedings of the 13th International Conference on Neural Information Processing Systems (2000)

# Triggering Robot Hand Reflexes with Human EMG Data Using Spiking Neurons

J. Camilo Vasquez Tieck[1(✉)], Sandro Weber[2], Terrence C. Stewart[3],
Arne Roennau[1], and Rüdiger Dillmann[1]

[1] FZI Research Center for Information Technology, 76131 Karlsruhe, Germany
tieck@fzi.de
[2] TUM Technical University of Munich, 80333 München, Germany
[3] Centre for Theoretical Neuroscience, University of Waterloo,
Waterloo N2L 3G1, Canada

**Abstract.** The interaction of humans and robots (HRI) is of great relevance for the field of neurorobotics as it can provide insights on motor control and sensor processing mechanisms in humans that can be applied to robotics. We propose a spiking neural network (SNN) to trigger motion reflexes on a robotic hand based on human EMG data. The first part of the network takes EMG signals to measure muscle activity, then classify the data to detect which finger is active in the human hand. The second part triggers single finger reflexes using the classification output. The finger reflexes are modeled with motion primitives activated with an oscillator and mapped to the robot kinematic. We evaluated the SNN by having users wear a non-invasive EMG sensor, record a training dataset, and then flex different fingers, one at a time. The muscle activity was recorded using a Myo sensor with eight channels. EMG signals were successfully encoded into spikes as input for the SNN. The classification could detect the active finger to trigger motion generation of finger reflexes. The SNN was able to control a real Schunk SVH robotic hand. Being able to map myo-electric activity to functions of motor control for a task, can provide an interesting interface for robotic applications, and also to study brain functioning. SNN provide a challenging but interesting framework to interact with human data. In future work the approach will be extended to control a robot arm at the same time.

**Keywords:** Human-Robot-Interaction · Humanoid robots
Neurorobotics · Motion representation · EMG classification
Spiking neural networks · Anthropomorphic robot hand

## 1 Introduction

The interaction of humans and robots (HRI) is of great relevance for the field of neurorobotics as it can provide insights on motor control and sensor processing mechanisms in humans that can be applied in robots.

© Springer Nature Switzerland AG 2019
M. Strand et al. (Eds.): IAS 2018, AISC 867, pp. 902–916, 2019.
https://doi.org/10.1007/978-3-030-01370-7_70

Electromyography (EMG) is a common tool in medicine and biomechanics. It is used to monitor and study the electrical activity of the muscles. There are different methods to record EMG signals, and they can be either invasive or non-invasive. Research is being made on processing and classification of EMG signals for clinical diagnoses [6] or prosthetic applications [14].

Nevertheless, there are different hypothesis explaining how does the human motor system work. A wide accepted theory states that the central nervous system uses different base motor components in a hierarchy [4] to generate the full repertoire of motions that we can perform [3].

These base components are formed by specific combination of muscle synergies [7] that are active during a motion, and they are commonly called motor primitives [5]. An approach using these concepts to control a robotic hand was proposed in [22]. Motor primitives can be activated in different ways, for example as a reflex. A reflex is an involuntary response to sensor stimulation and can be either a complete execution or inhibition of a motion. An overview of different spinal reflexes is provided by [15].

In this work, we propose a system to control a robot hand with muscle signals from a human recorded with a non-invasive EMG sensor. We focus on the activation of single finger motions in response to sensor stimuli. A spiking neural network (SNN) [12,16] was implemented to classify EMG data and then trigger the generation of motion as a reflex. An EMG sensor with eight channels was used to record human muscle activity while moving different fingers. First, EMG data was encoded to spikes and the signals were classified to identify the active finger. After that, the activation signal was used to trigger an oscillator to generate motion using a motor primitive. Then, the primitive is mapped to the robot kinematics. Finally, the spikes are decoded to motor commands for the robot.

One highly novel aspect of this paper is the fact that the classification and the generation of the motor primitive is implemented using a spiking neural network. There are two reasons for doing this. First, the real biological system must do something similar to this using spiking neurons. Certainly, in real biology, the classification would not be based on EMG sensor, but would rather be based on neural activity somewhere in the brain, but the classification and generation of movement over time would still need to occur. This means that we can see our system as an initial model of that biological process. Second, there is a pragmatic/engineering reason to implement this system using spiking neurons. Prosthetic applications benefit from low-power hardware implementations, and there is a variety of neuron-inspired low-power computing hardware being developed. For example, the SpiNNaker system [11], Intel's new Loihi chip [8] and IBM's TrueNorth chip [18] all provide extremely energy-efficient spiking neuron hardware. If algorithms can be mapped onto this hardware, then they may be able to be deployed using significantly less power than traditional implementations.

## 2   Methods

The main motivation of this work is to control a robotic hand with human muscle signals using a spiking neural network (SNN). For this purpose we define specific characteristics for the components. The EMG sensor has to be non-invasive, the finger motions are represented with motor primitives triggered at once to resemble reflexes, and the robot hand has to be controllable with a ROS interface. We divide the problem in two parts, that translate to different parts of the same SNN. The first part takes care of the EMG data interface and classification and the second part takes care of the motion generation and robot control. Human EMG data is captured in a non-invasive way. The first part of the SNN classifies the EMG signals to detect which finger was active. An activation signal is then passed on to the second part of the SNN to trigger single finger reflexes on the real robot. A finger reflex is modeled with an oscillator that activates a motor primitive that is mapped to motor commands according to the robot kinematics. In Fig. 1 we present an overview of the main components showing how they interact with each order.

**Fig. 1.** Concept architecture with main components. Human muscle activity is recorded with a non-invasive EMG sensor. The EMG data is encoded to spikes. The first part of the SNN performs a classification to detect which finger was active. The second part of the SNN generates motion and maps it to motor commands considering the robot kinematics.

To generate the spiking neuron models, we used the Neural Engineering Framework [10] and the software package Nengo [1]. This software allows for the creation of large-scale spiking neural networks by breaking the networks down into smaller parts. The connection weights for each sub-part are optimized separately, and then they are combined together into one large neural network. Performing this optimization (i.e. finding connection weights) locally means we can generate large systems without using the traditional neural network approach of optimizing over huge amounts of training data. However, the trade-off is that we must make explicit claims about what each sub-part of the model is doing.

In particular, in order to define a spiking neuron model using Nengo and the NEF, we must break our algorithm down into vectors, functions, and differential equations. The activity of each group of spiking neurons is considered to be a distributed representation of a vector (i.e. we may have 100 spiking neurons representing a 2-dimensional vector). Connections between groups of neurons compute functions on those vectors. That is, the connection weights ensure that if the first group of neurons represents x, then the second group of neurons will represent $y = f(x)$. By changing the connection weights, we change the function being computed. Finally, recurrent connections can be used to implement differential equations. We make use of this here to implement basic movement primitives.

## 2.1  Human EMG Data Interface and Training Data

To record EMG data a single Myo [21] gesture control armband is used. It is made up of eight equally spaced out blocks with non-invasive EMG sensors that provide a sampling rate of 200 Hz. The armband is used around the middle of the forearm as shown in Fig. 2. In order to record consistent data with the sensor, the segment with the LED light has to be placed approximately at the same position. Slight variations after re-wearing the myo on and off didn't have enough influence on the recordings to make the trained network unusable. Retrieving the raw EMG signals is done with the help of a Python API provided by [9]. Each channel encodes the individual measurement as $int8$ values.

**Fig. 2.** The Myo armband sensor placed on the arm and the different EMG channels. When a finger is moved the muscle electric activity is recorded with eight different sensors. The sensor has an indicator so that it can be placed always in a similar way.

For each user a training dataset is required with multiple samples. A sample consist of a continuos sequence of finger activation in one hand. Each finger has to be flexed down for a short period of time and then extended again. This procedure is repeated starting from the thumb to the pinky. The training data has to be recorded as a time continuos EMG stream of all eight channels with appropriate binary labels for the time windows during which a finger was pressed. A sample dataset is provided in Fig. 3.

Notice that individual channels of the EMG sensor have similar activation for different fingers, and thus are not enough to identify the motion of a finger. Therefore, the classification network uses a combination of all eight channels, which provides a unique representation for each finger.

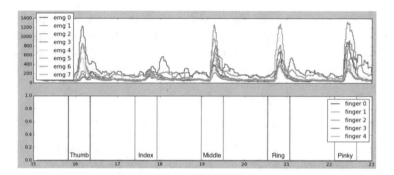

**Fig. 3.** A sample dataset for training with a run of all 5 fingers. From left to right the peaks show EMG activation of the fingers starting with the thumb to the pinky. Each finger is flexed and then extended.

## 2.2   Sub Network for EMG Classification

After recording training data, the first part of the SNN is trained for classification. The detailed architecture for EMG classification is presented in Fig. 4. The raw EMG data from a user using the EMG sensor is feed to the SNN. A population of neurons encodes the signal stream of EMG input to spikes using stochastic population encoding. Then, a second population of neurons is trained offline with a whole training dataset for a user as described above.

The learning rule for offline training the classification population is Prescribed Error Sensitivity (PES) using the labels from the training data serving as error signals $E$. PES is implemented in [2], and was first presented in [17]. The weight updates that PES makes to minimize an error signal during learning can be related to Skipped Back-propagation. For the weights $wij$ from pre-synaptic population $i$ to post-synaptic population $j$, the update rule is defined as

$$\Delta wij = \kappa \alpha_j e_j \cdot E a_i, \tag{1}$$

with $\kappa$ a scalar learning rate, $\alpha$ the gain or scaling factor, $e$ the encoder for the neuron, $E$ the error to minimize, and $a$ the desired activation.

After classification by the second population, the overall signals are low in amplitude and consequently rather close to each other. Therefore, a population is used to refine the classification by amplifying the signals and generating the hand activation signal. An arbitrary defined function scales up activations above a manually set threshold and is used to train this population. The resulting activation signal is passed over to the motion generation part. Examples for classification of all the fingers of the complete procedure are provided in Fig. 11 in the results section. All populations are connected all to all.

## 2.3   Sub Network for Motion Representation for Reflexes

A population takes the hand activation signal from the previous classification to trigger the appropriate finger reflex. We model a reflex as the execution of

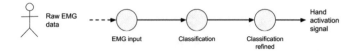

**Fig. 4.** Detailed architecture for EMG classification. Each circle represents a population of spiking neurons. Raw EMG data is recorded from the user and is encoded into spikes. The encoded EMG signals are classified to determine the finger that was activated. The classification signal is refined and amplified to have a clear hand activation signal that is passed to the motion generation part.

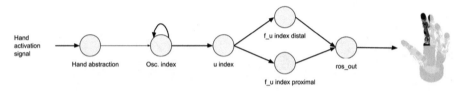

**Fig. 5.** Detailed architecture for motion representation for reflexes. Each circle represents a population of spiking neurons. The hand activation signal is processed by the hand abstraction population to extract individual finger activations. Reflexes are modeled as oscillator that oscillate only one time. The activation is decoded in two components as $u$ and then mapped to a motor primitive with the robot kinematic parameters with $g(f(u))$. Finally, the neural activity is decoded from spikes and send as motor commands to the respective robot finger.

a motor primitive based on a specific stimuli. Accordingly, the motion part of our SNN is divided into reflex activation and motor primitive layers. The whole architecture for the representation of reflexes is presented in Fig. 5.

The reflex activation for each finger is modeled as an oscillator

$$h(\omega) = a \cdot \sin(b\omega \frac{\pi}{2}), \tag{2}$$

with $\omega$ a recurrent connection and $a$ and $b$ the parameters for the amplitude and frequency. The oscillator generates a continuos signal for a finite period that represents the duration of a motion with a start and an end point. By indexing the neurons in the oscillator population, the activity can be mapped to a $2D$ plane. The total neural activity in the population in the plane can be represented with the components $x$ and $y$. We calculate a continuos and normalized signal $u \in [-1, 1]$ as:

$$u = \sin(\arctan(\frac{y}{x})), \tag{3}$$

where $\arctan(\frac{y}{x})$ represents the angle of a vector with components $x$ and $y$. To bound and smooth the signal $sin()$ is applied.

The motor primitive is modeled as a mapping of $u$ to a sequence of joint activations during the period of oscillation, and it can be mapped to one or multiple joints. In Fig. 5 the distal and proximal joints of the index finger are mapped. A mapping

$$f(u) = \frac{sin(u \cdot \pi - \frac{\pi}{2})}{2} + \frac{1}{2}, \tag{4}$$

is defined with a sinusoidal function to have smooth initial an final phases. This characteristic is important when executing the motion in a real robot. The resulting generic primitive for one joint is depicted in Fig. 6a.

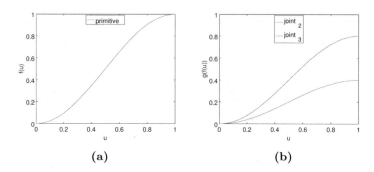

(a)                                    (b)

**Fig. 6.** (a) Mapping of $u$ to the motor primitive with $f(u)$. (b) Complete mapping of $u$ to the robot kinematics with $g(f(u))$.

In general terms, there is no difference between voluntary and reflex motions on the muscular activities (synergies), the difference is in the activation. For voluntary motions it is discrete, and for reflexes it is a complete one time execution.

## 2.4   Mapping to Robot Motor Commands

Finally, in order to actually be able to control the robot, the primitives have to be mapped to the robot kinematics. Which means scaling to the motion interval $(\theta_{max} - \theta_{min})$ and offset $\theta_{min}$ that the joint $\theta$ has. For this purpose we define $g : [0, 1] \rightarrow \mathbb{R}^n$ as a function for each joint as

$$g(f(u)) = f(u) \cdot (\theta_{max} - \theta_{min}) + \theta_{min}, \tag{5}$$

to generate appropriate motor commands.

A schema for the mapping $g$ for a robotic hand is illustrated in Fig. 6b and is defined in the table in Fig. 7. This parametric representation of motions allows us to further combine and change parameters of the motor primitives.

| Joint name | Primitive | $\theta_{min}$ | $\theta_{max}$ |
|------------|-----------|----------------|----------------|
| $joint_1$  | 0         | 0              | 0.7            |
| $joint_2$  | 1         | 0              | 0.4            |
| $joint_3$  | 1         | 0              | 0.8            |

**Fig. 7.** Table for the joint mapping schema. A joint is defined with a name, the associated primitive, and the interval of the joint $\theta_{min}$ and $\theta_{max}$. A primitive can be mapped to one or more joints.

## 2.5    Integration Off All Components

A detailed architecture of the full SNN is presented in Fig. 8. Notice that the primitives for the thumb, ring and pinky are mapped to one actuated joint, whereas the index and middle finger primitives are mapped to two actuated joints.

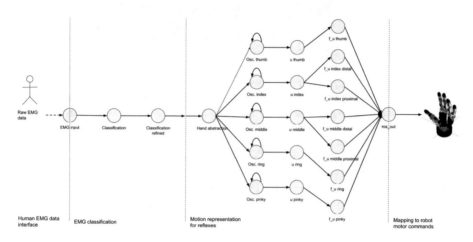

**Fig. 8.** Detailed architecture with EMG and motion sub networks. Each circle represents a population of spiking neurons. The dotted lines divide the conceptual components, which are named on the bottom part according to Fig. 1.

# 3    Results

The experiment setup is depicted in Fig. 9, and consists of a human user, an EMG sensor, a robot hand, and the simulation of the SNN. A user wears the EMG sensor (Myo armband) in the forearm, and the signals are sent via bluetooth to the computer. The computer receives the EMG data and inputs it to the SNN simulation running in Nengo [1]. The computer communicates at the same time with the robot hand (Schunk SVH) via ROS [19]. To control the robot hand the official Schunk ROS driver is used [13].

## 3.1    SNN Implementation

The SNN was implemented in Python with the Nengo simulator using leaky integrate and fire (LIF) neurons. To get an overview of the implementation Fig. 10 presents a view of the whole network running. The structure can be easily mapped to that in Fig. 8. The eight channels human EMG signals are encoded by a population as stochastic spike rates based on their values. After performing offline training with different datasets from the same user, the EMG classification takes place. The classification signal is then passed over to trigger

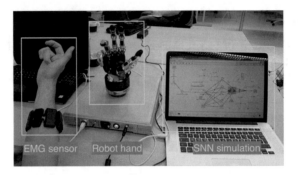

**Fig. 9.** Experiment setup. (*left*) The user wears the EMG sensor (Myo armband) in the forearm. (*right*) The SNN is simulated (Nengo) is a computer connected via bluetooth to the EMG sensor and via ROS to the robot hand. (*middle*) The robot hand (Schunk SVH) is connected to the computer via ROS.

motion generation. The reflexes are implemented as oscillators that activate motor primitives. The motor primitives are mapped to the robot kinematics as defined in the methods section. In the following sections the relevant details of each components are described.

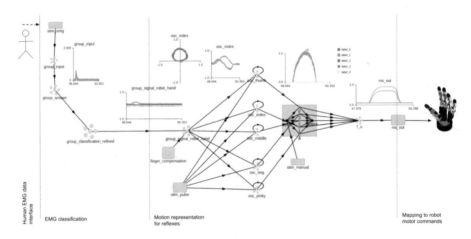

**Fig. 10.** Full SNN pipeline in Nengo. There are four main components: human EMG data capture and manipulation, EMG classification, motion generation, and finally the mapping to the robot. The structure can be easily mapped to that in Fig. 8.

## 3.2   Training Data

For each user a set of training data is required to train offline the SNN as described in Sect. 2.2. Training data for the classification network was recorded in one session lasting 60 s. During that time individual fingers were periodically pressed against the palm of the hand and subsequently returned to a resting pose. The fingers presses occurred in sequence from thumb to index finger with each press lasting between 300 ms and 500 ms. Together with the resting time one cycle took around 7.5 s and a total of 8 cycles were performed. A sample run of the training data with all 5 fingers is presented in Fig. 3. In order to label EMG data, with every finger press simultaneously a keyboard button was pressed indicating the respective finger. The data is labeled in time for each finger, and all eight EMG channels are active.

### 3.3   Processing of EMG Data and Classification

The first group of 800 neurons (EMG input in Fig. 4) was activated with the raw EMG data as *int*8 to convert it to spikes. The second group of 500 neurons (Classification in Fig. 4) was trained with the prerecorded training data at start to give responses of the classified fingers. Then, a third group of 500 neurons (Classification refined in Fig. 4) was used to separate and amplify signals further. A final group of 500 neurons (Hand activation signal, Fig. 4) was trained to give out one single signal for a specific finger and was connected to 5 groups representing the different fingers for the robot hand.

In Fig. 11 we present samples of the SNN classifying the activation of the different fingers. As it can be seen, the eight channels of EMG have different data for each finger. The signals are processed with the SNN, and the activation of the different populations can also be observed. The output of the classification is a dominant activation of one of the populations representing each finger.

### 3.4   Motion Generation

The resulting motion of the robot is presented as a frame sequence in Fig. 13. The data presented corresponds to a reflex motion of the index finger. The corresponding activity of the SNN is presented in Fig. 12 with sufficient information to illustrate the functioning of the SNN. The signal that triggers motion generation comes from the classification part. An oscillator is activated for each finger. Observe the circular activation of the oscillator population when decoded in a plane $XY$ in Fig. 12. From this circular activation $u$ is decoded and mapped to one or more joints in the robot hand. Observe that the mapping in Fig. 12 is performed to two joints of the index finger. Finally the neural activity is decoded and send over ROS to the robot hand.

In order to evaluate the accuracy of the classification we selected one random user. Then he was asked to perform a sample of 50 trials with each finger. The EMG data was feed to the trained network and we counted the classification output for each trial. The results are summarized in Fig. 14. For this user, only the pinky had a 1.0 of accuracy, which means that all the trials where classified correctly. In all other fingers there were false or none classifications.

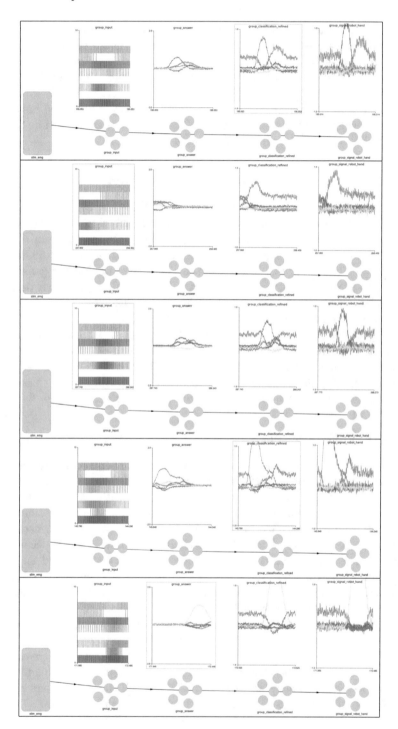

**Fig. 11.** EMG activation and classification for different fingers. The first graph from the left shows the encoded EMG signals to spikes. The second plot shows the classification output. The other plots show the refined classification signal and the hand activation signal respectively with one finger active. From top to bottom thumb, index, middle, ring and pinky are shown. Notice the different activations in the eight channels.

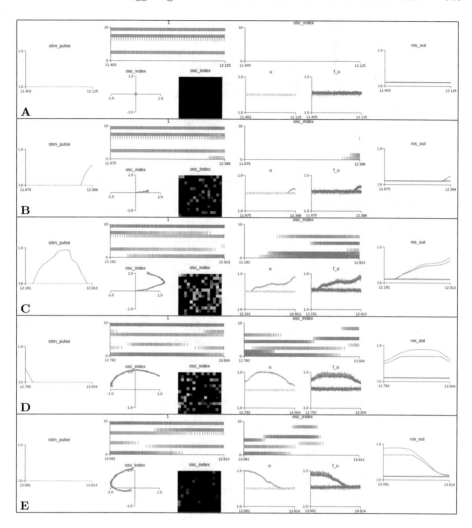

**Fig. 12.** Motion generation from the SNN for a reflex in the index finger. The frames A to E correspond to the fingers from the Thumb to the pinky respectively. From left to right the plots represent: (*left*) Finger activation signal coming form the EMG classification. (*middle left, group of three plots*) Show the activity in the oscillator with a spike train plot, the decoded activity of the population in a plane $XY$, and a raster plot color coded by the neuron's activity. (*middle right, group of three plots*) Show the decoding to $u$, and the mapping $g(f(u))$ to the robot kinematics. Note that the mapping goes to two different joints. (*right*) A continuos plot of the current motor commands being sent to the robot.

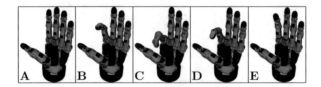

**Fig. 13.** Frame sequence of the index finger motion generated by the SNN activity. The motion presented corresponds to the activation shown in Fig. 12. After receiving the classification signal the reflex is activated and fully executed once.

| Finger moved | Number of trials | Thumb class | Index class | Middle class | Ring class | Pinky class | None class | Accuracy |
|---|---|---|---|---|---|---|---|---|
| Thumb | 50 | 46 | 0 | 0 | 1 | 0 | 3 | 0.92 |
| Index | 50 | 0 | 43 | 1 | 0 | 2 | 4 | 0.86 |
| Middle | 50 | 0 | 3 | 40 | 0 | 0 | 7 | 0.80 |
| Ring | 50 | 0 | 1 | 2 | 44 | 0 | 3 | 0.88 |
| Pinky | 50 | 0 | 0 | 0 | 0 | 50 | 0 | 1.00 |

**Fig. 14.** Table of classification accuracy. An evaluation summary of the classification consisting of 50 trials of each finger for one random user. We counted which finger class was detected and calculated the ratio of accuracy for each finger. Notice that only the pinky was always classified correctly. For the other fingers there are either false detections or none classification at all.

| Joint name | Primitive | $\theta_{min}$ | $\theta_{max}$ |
|---|---|---|---|
| Thumb_Flexion | 0 | 0 | 0.3 |
| Thumb_Opposition | - | - | - |
| Index_Finger_Distal | 1 | 0 | 0.9 |
| Index_Finger_Proximal | 1 | 0 | 0.7 |
| Middle_Finger_Distal | 2 | 0 | 0.9 |
| Middle_Finger_Proximal | 2 | 0 | 0.7 |
| Ring_Finger | 3 | 0 | 0.7 |
| Pinky | 4 | 0 | 0.7 |
| Finger_Spread | - | - | - |

**Fig. 15.** Table. Joint mapping schema. A summary of corresponding topics as joint name, the primitive used for each finger, and the joint angle intervals for each actuated joint. The active joints in the robot hand "Thumb_Opposition" and "Finger_Spread" remained constant all the time.

## 3.5   Interface to the Robot Hand

We describe here the mapping schema that was used with the robot hand (Schunk SVH). The table in Fig. 15 summarizes the data. The "Joint name" column corresponds to the ROS topics described in [13] for the different actuated joints. A different primitive is used for each finger, and the indexing is in column "Primitive". Note that index and middle finger joints are mapped to the same primitives respectively. The "min" and "max" values for each joint complete the table.

## 4 Discussion

We presented a SNN that activates motion reflexes on a robotic hand based on human EMG data. The network classifies the EMG signals to detect finger activation. Based on it, single finger reflexes are triggered. The finger reflexes are modeled with motion primitives and mapped to the robot kinematic.

As can be seen in Figs. 3 and 11, the index finger showed almost no discernable signal in the raw EMG data. For the index finger, the signal is not clear and the output is sometimes ambiguous. As a consequence, the classification step delivers a weak and low activation signal for the index finger that is propagated throughout the following populations and leads to fake classification of other fingers.

The focus of this work was on single finger movements, so data with multiple fingers was not considered, only movements in quick succession of single fingers. The EMG signals were used only to trigger the execution of the reflexes. In future work we want to explore a mechanism to use the EMG signal to perform discrete control of the finger positions.

Using a second EMG sensor could provide additional input from different muscle areas of the arm and improve classification results. Ideally the second EMG sensor would be located close to the wrist [20] closer to the fingers and the hand. It is difficult with the Myo armband because the circumference is too big for that location. A bigger set of training data would also be very likely to improve classification.

**Acknowledgments.** This research has received funding from the European Union's Horizon 2020 Framework Programme for Research and Innovation under the Specific Grant Agreement No. 720270 (Human Brain Project SGA1) and No. 785907 (Human Brain Project SGA2).

## References

1. Bekolay, T., Bergstra, J., Hunsberger, E., DeWolf, T., Stewart, T.C., Rasmussen, D., Choo, X., Voelker, A., Eliasmith, C.: Nengo: a python tool for building large-scale functional brain models. Front. Neuroinform. **7**, 48 (2014)
2. Bekolay, T., Kolbeck, C., Eliasmith, C.: Simultaneous unsupervised and supervised learning of cognitive functions in biologically plausible spiking neural networks. In: Cogsci (2013)
3. Bernstein, N.: The Co-ordination and Regulation of Movements. Pergamon-Press, Oxford (1967)
4. Bizzi, E., Cheung, V., d'Avella, A., Saltiel, P., Tresch, M.: Combining modules for movement. Brain Res. Rev. **57**(1), 125–133 (2008)

5. Chinellato, E., Pobil, A.: The Visual Neuroscience of Robotic Grasping: Achieving Sensorimotor Skills through Dorsal-Ventral Stream Integration. Cognitive Systems Monographs. Springer, Cham (2016)
6. Chowdhury, R.H., Reaz, M.B., Ali, M.A.B.M., Bakar, A.A., Chellappan, K., Chang, T.G.: Surface electromyography signal processing and classification techniques. Sensors 13(9), 12431–12466 (2013)
7. d'Avella, A., Saltiel, P., Bizzi, E.: Combinations of muscle synergies in the construction of a natural motor behavior. Nature Neurosci. 6(3), 300 (2003)
8. Davies, M., Srinivasa, N., Lin, T.H., Chinya, G., Cao, Y., Choday, S.H., Dimou, G., Joshi, P., Imam, N., Jain, S.: Loihi: a neuromorphic manycore processor with on-chip learning. IEEE Micro 8(1), 82–99 (2018)
9. dzhu: Myo python api (2018). https://github.com/dzhu/myo-raw/. Accessed 21 Feb 2018
10. Eliasmith, C., Anderson, C.H.: Neural Engineering: Computation, Representation, and Dynamics in Neurobiological Systems. MIT Press, Cambridge (2003)
11. Furber, S., Temple, S., Brown, A.: High-performance computing for systems of spiking neurons. In: AISB 2006 Workshop. GC5: Archit. Brain Mind (2006)
12. Grüning, A., Bohte, S.M.: Spiking neural networks: principles and challenges. In: ESANN (2014)
13. Heppner, G.: schunk_svh_driver (2018). http://wiki.ros.org/schunk_svh_driver. Accessed 21 Feb 2018
14. Johannes, M.S., Bigelow, J.D., Burck, J.M., Harshbarger, S.D., Kozlowski, M.V., Van Doren, T.: An overview of the developmental process for the modular prosthetic limb. Johns Hopkins APL Tech. Digest 30(3), 207–216 (2011)
15. Knierim, J.: Spinal reflexes and descending motor pathways. Neuroscience Online (2016). Accessed 11 Feb 2018
16. Maass, W.: Networks of spiking neurons: the third generation of neural network models. Neural Netw. 10(9), 1659–1671 (1997)
17. MacNeil, D., Eliasmith, C.: Fine-tuning and the stability of recurrent neural networks. PLoS ONE 6(9), e22885 (2011)
18. Merolla, P.A., Arthur, J.V., Alvarez-Icaza, R., Cassidy, A.S., Sawada, J., Akopyan, F., Jackson, B.L., Imam, N., Guo, C., Nakamura, Y.: A million spiking-neuron integrated circuit with a scalable communication network and interface. Science 345(6197), 668–673 (2014)
19. Quigley, M., Conley, K., Gerkey, B., Faust, J., Foote, T., Leibs, J., Wheeler, R., Ng, A.Y.: ROS: an open-source robot operating system. In: ICRA Workshop on Open Source Software, Kobe, Japan (2009)
20. Tenore, F., Ramos, A., Fahmy, A., Acharya, S., Etienne-Cummings, R., Thakor, N.V.: Towards the control of individual fingers of a prosthetic hand using surface EMG signals. In: EMBC (2007)
21. ThalmicLabs: Myo Diagnostics (2018). http://diagnostics.myo.com/. Accessed 21 Feb 2018
22. Tieck, J.C.V., Donat, H., Kaiser, J., Peric, I., Ulbrich, S., Roennau, A., Zöllner, M., Dillmann, R.: Towards grasping with spiking neural networks for anthropomorphic robot hands. In: ICANN (2017)

# A Localizability Constraint-Based Path Planning Method for Unmanned Aerial Vehicle

Behnam Irani, Weidong Chen$^{(\boxtimes)}$, and Jingchuan Wang

Key Laboratory of System Control and Information Processing,
Department of Automation, Ministry of Education of China,
Shanghai Jiao Tong University, Shanghai 200240, China
wdchen@sjtu.edu.cn

**Abstract.** As unmanned aerial vehicles (UAVs) are used in challenging environments to carry out various complex tasks, a satisfactory level of localization performance is required to ensure safe and reliable operations. 3D laser range finder (LRF)-based localization is a suitable approach in areas where GPS signal is not accesible or unreliable. During navigation, environmental information and map noises at different locations may contribute differently to a UAV's localization process, causing it to have dissimilar ability to localize itself using LRF readings, which is referred to as localizability in this paper. We propose a localizability constraint (LC) based path planning method for UAV, which plans the navigation path according to LRF sensor model to achieve higher localization performance throughout the path. Paths planned with and without LC are compared and discussed through simulations in outdoor urban and wilderness environemnts. We show that the proposed method effectively reduces the localization error along the planned paths.

**Keywords:** Unmanned aerial vehicle · Path planning · Localizability
Localization · Navigation

## 1 Introduction

In recent years, unmanned aerial vehicle (UAV) has increasingly attracted researchers' interest due to its potential to carry out a variety of complex tasks, such as monitoring disaster areas, search and rescue [10], parcel delivery [12] and 3D reconstruction [20]. As UAVs are used in challenging environments, satisfactory level of localization performance is a basic requirement to ensure safe and reliable operations. While GPS receiver is a common on-board device for UAVs to locate their absolute position, the GPS signal is not guaranteed to

This work is supported by the National Key Research and Development Program of China (Grant No. 2017YFB1302200), and supported in part by the Natural Science Foundation of China (Grant No. 61573243 and 61773261).

© Springer Nature Switzerland AG 2019
M. Strand et al. (Eds.): IAS 2018, AISC 867, pp. 917–932, 2019.
https://doi.org/10.1007/978-3-030-01370-7_71

be accessible or reliable in environments such as rugged mountainous terrains, urban canyons and building interiors [17]. 3D laser range finder (LRF)-based localization approach, due to the accuracy of LRF readings, is a suitable alternative for navigation in these environments. In this paper, we focus on global path planning in the presence of localization uncertainty using 3D LRF sensors base on pre-built OctoMaps [13].

The most general formulation of path planning under localization uncertainty is selecting optimal action in a partially observable stochastic domain, which is known as Partially Observable Markov Decision Process (POMDP) [3,14]. While POMDPs are theoretically satisfactory, they may become computationally intractable as the scale of states grows. Although Extended Kalman Filter [24] and Particle Filter [7] based algorithms perform well in many cases, they may still suffer if UAVs navigate in ambiguous (such as long hallway for translation and circle room for rotation), texture-poor or noisy regions in the map. In these cases, localization uncertainty increases drastically due to the lack of sufficient valid LRF readings or mismatching, and therefore resulting in large localization error. During navigation, environmental information and map noises at different locations may contribute differently to UAV's localization process, causing it to have dissimilar ability to localize itself in the environment using LRF readings, which is referred to as localizability in this paper. In the presence of above mentioned regions, it is crucial to take localizability into consideration when planning a path to avoid large localization errors or placing the UAV at risk of failure to perform localization in the pose tracking stage.

To quantify the influence of environmental information on localization process, [5,6] have obtained Fisher's Information Matrix (FIM) using the expected LRF readings and the slope of scanned environmental surfaces in a geometric map, but the influence caused by uncertainty of the map [8] has been neglected. Besides, analytical expression of a geometric map is usually not easy to obtain. Inspired by these works, and having considered map noises, previously we have introduced probabilistic grid map based static localizability estimation matrix (SLM) [16,23]. SLM describes the measure and directional properties of localizability in a quantitative manner without extracting specific observation features, and present the lower-bound of attainable covariance of localization as it was derived using the Cramér-Rao Bound theory [2]. However, the previously proposed SLM only describes the localizability for a robot moving on a 2-dimensional plane, thus further extensions need to be done to describe the localizability for a UAV navigating in a three-dimensional space. Another issue regarding SLM to address is that, as SLM is *orientation-dependent*, the discretization of orientations should also be considered to incorporate it into path planning module. In this paper, we extend the concept using an OctoMap, and incorporate the 3-dimensional SLM into global path planning, with path length as the primary cost and localizability measure as the constraint. By doing so, we propose a localizability constraint (LC) based path planning method for UAV, which determines passable regions on an OctoMap according to the LRF sensor model, and plans paths within this region.

The proposed method assumes that the maps of the environment are partially or fully obtained via techniques such as SLAM, the equipped LRF sensor models are also available so the response of sensors to the environment can be simulated, and that the initial pose of UAV is known. The proposed method promises a satisfactory level of localization performance for the UAV throughout the planned path. Paths planned with and without LC are compared and discussed through simulations in urban canyon and wilderness environments. We show that the proposed method effectively reduces the localization error of the UAV while it navigates along the paths.

The remaining of this paper is organized as follows: Sect. 2 introduces the localizability estimation method. Section 3 describes the LC-based path planning method. Finally, comparisons between paths with and without LC is made via simulation and related discussion is done in Sect. 4.

## 2   Localizability Estimation

### 2.1   Static Localizability Matrix

The PGM-based SLM proposed in [16, 23] is obtained by discretizing FIM, which can reflect a UAV's localizability for any desired pose in a 2-dimensional PGM. We now extend the SLM into 3-dimensional probabilistic OctoMap, with the assumption that the LRF shares the same coordinates with the UAV that it is mounted on. Thus, the pose of UAV in the OctoMap frame is represented as $p = [x, y, z, \varphi, \psi, \theta]$. Given the LRF model, a total of $N_l$ scan rays may be generated by following the scan sequence. The discretized FIM is written as:

$$\hat{L}(p) = \sum_{i=1}^{N_l} \frac{1}{\sigma_i^2} \left(\frac{\Delta r_{iE}}{\Delta p}\right)^{\mathrm{T}} \left(\frac{\Delta r_{iE}}{\Delta p}\right) \tag{1}$$

where

$$\frac{\Delta r_{iE}}{\Delta p} = \left[ \frac{\Delta r_{iE}}{\Delta x} \quad \frac{\Delta r_{iE}}{\Delta y} \quad \frac{\Delta r_{iE}}{\Delta z} \quad \frac{\Delta r_{iE}}{\Delta \varphi} \quad \frac{\Delta r_{iE}}{\Delta \psi} \quad \frac{\Delta r_{iE}}{\Delta \theta} \right] \tag{2}$$

$\sigma_i^2$ is the variance of the $i$-th LRF scan ray, $r_{iE}$ is the expected distance from the UAV to the nearest obstacle along the $i$-th LRF scan ray and it is computed based on the known OctoMap. $r_{iE}$ is calculated using:

$$r_{iE} = \frac{\sum_{j=1}^{s} r_{ij}\mu_{ij}}{\sum_{j=1}^{s} \mu_{ij}} \tag{3}$$

where $r_{ij}$ is the distance between the UAV's position and $j$-th voxel along the direction of $i$-th LRF scan ray. $\mu_{ij}$ is the occupancy probability of the corresponding voxel, and $s$ is the sequence number of the ending voxel. A voxel is considered unoccupied if its occupancy probability is less than a probability threshold $T_s$.

In practise, the roll ($\varphi$) and pitch ($\psi$) angles are often handled by on-board controller and inertial measuring units (IMU), which means their values could be determined without an LRF reading. Thus a simplification can be made based on this fact. We further assume that these two angles are kept at a small range near to the zero degree (parallel to the ground plane) with maximum turning angle of $\Phi/2$ and $\Psi/2$ respectively during navigation. Therefore the components $\varphi$ and $\psi$ may be omitted when computing the SLM for a UAV, but still influencing the LRF scan model. By combining (2) with (1) and substituting $\boldsymbol{p}$ we obtain:

$$\hat{\mathbf{L}}(\boldsymbol{p}) = \sum_{i=1}^{N_1} \frac{1}{\sigma_i^2} \begin{bmatrix} \frac{\Delta r_{iE}^2}{\Delta x^2} & \frac{\Delta r_{iE}^2}{\Delta x \Delta y} & \frac{\Delta r_{iE}^2}{\Delta x \Delta z} & \frac{\Delta r_{iE}^2}{\Delta x \Delta \theta} \\ \frac{\Delta r_{iE}^2}{\Delta x \Delta y} & \frac{\Delta r_{iE}^2}{\Delta y^2} & \frac{\Delta r_{iE}^2}{\Delta y \Delta z} & \frac{\Delta r_{iE}^2}{\Delta y \Delta \theta} \\ \frac{\Delta r_{iE}^2}{\Delta x \Delta z} & \frac{\Delta r_{iE}^2}{\Delta y \Delta z} & \frac{\Delta r_{iE}^2}{\Delta z^2} & \frac{\Delta r_{iE}^2}{\Delta z \Delta \theta} \\ \frac{\Delta r_{iE}^2}{\Delta x \Delta \theta} & \frac{\Delta r_{iE}^2}{\Delta y \Delta \theta} & \frac{\Delta r_{iE}^2}{\Delta z \Delta \theta} & \frac{\Delta r_{iE}^2}{\Delta \theta^2} \end{bmatrix} \tag{4}$$

Given a position, if the information observed by the LRF changes as the orientation of UAV ($\theta$) varies, then the SLM is *orientation-dependent* and needs to be calculated separately for every desired orientation. Otherwise the component $\theta$ in the above equation may also be omitted, which results in:

$$\hat{\mathbf{L}}(\boldsymbol{p}) = \sum_{i=1}^{N_1} \frac{1}{\sigma_i^2} \begin{bmatrix} \frac{\Delta r_{iE}^2}{\Delta x^2} & \frac{\Delta r_{iE}^2}{\Delta x \Delta y} & \frac{\Delta r_{iE}^2}{\Delta x \Delta z} \\ \frac{\Delta r_{iE}^2}{\Delta x \Delta y} & \frac{\Delta r_{iE}^2}{\Delta y^2} & \frac{\Delta r_{iE}^2}{\Delta y \Delta z} \\ \frac{\Delta r_{iE}^2}{\Delta x \Delta z} & \frac{\Delta r_{iE}^2}{\Delta y \Delta z} & \frac{\Delta r_{iE}^2}{\Delta z^2} \end{bmatrix} \tag{5}$$

Figure 1 demonstrates two typical placement approaches of 3D LRF on UAVs. In case of vertical placement, the SLM is depended on $\theta$ and (4) is used for its calculation. In case of horizontal placement, the information observed remains equal as the UAV rotates along the z-axis, therefore (5) is used instead.

As stated earlier, although the angles $\varphi$ and $\psi$ are omitted when computing the SLM, they still influence the LRF scan models by enlarging them from the original ones. Remember that we assumed $\varphi$ and $\psi$ ranges between $[-\Phi/2, +\Phi/2]$ and $[-\Psi/2, +\Psi/2]$ respectively while the UAV navigates, hence the extended LRF scan model may be derived. Figure 2 shows an example of the extended scan models where $\Phi$ and $\Psi$ are both set to $30°$.

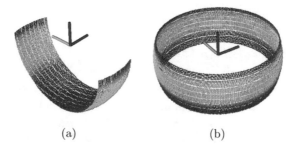

(a)                                    (b)

**Fig. 1.** Original LRF scan models under two typical placement approaches. (a) vertical placement (b) horizontal placement

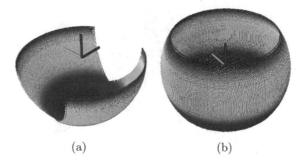

(a)                                    (b)

**Fig. 2.** Extended LRF scan models under two typical placement approaches. (a) vertical placement (b) horizontal placement

## 2.2   Localizability Measure

According to the Cramér-Rao Bound theory, the localization covariance matrix can be estimated using SLM:

$$\mathbf{cov}(\boldsymbol{p}) = \hat{\mathbf{L}}^{-1}(\boldsymbol{p}) \tag{6}$$

Covariance ellipses can be derived from (6) [1], and their area demonstrate the achievable localization accuracy [4] with a smaller area standing for a higher accuracy. For example, the covariance ellipse on x-y plane can be derived from $\hat{\mathbf{L}}_{\mathbf{xy}}^{-1}(\boldsymbol{p})$ which is $\hat{\mathbf{L}}^{-1}(\boldsymbol{p})$ with elements containing only $x$ and $y$. Its semi-major axis and semi-minor axis are $E_{xy}$ and $F_{xy}$. Notice that the area of this covariance ellipse is proportional to the eigenvalues of $\hat{\mathbf{L}}_{\mathbf{xy}}^{-1}(\boldsymbol{p})$ which are $\lambda_x^{-1}$ and $\lambda_y^{-1}$:

$$\pi E_{xy} F_{xy} \propto \lambda_x^{-1} \lambda_y^{-1} = (\lambda_x \lambda_y)^{-1} \equiv [\det(\hat{\mathbf{L}}_{\mathbf{xy}}(\boldsymbol{p}))]^{-1} \tag{7}$$

Similar results can be concluded on other planes. To evaluate and compare the expected localization quality of poses over the pre-built OctoMap, the determinant of the SLM is adopted as the measure [16] and is referred to as localizability measure $l(\boldsymbol{p})$ (LM):

$$l(\boldsymbol{p}) = \det(\hat{\mathbf{L}}(\boldsymbol{p})) \equiv \begin{cases} \lambda_x \lambda_y \lambda_z \lambda_\theta, & \text{if } \theta \text{ considered} \\ \lambda_x \lambda_y \lambda_z, & \text{if } \theta \text{ omitted} \end{cases} \quad (8)$$

$l(\boldsymbol{p})$ is a measure that indicates how discriminative the LRF observation at $\boldsymbol{p}$ in the OctoMap is for determining the UAV's pose. A larger $l(\boldsymbol{p})$ implies smaller area of covariance ellipse and richer observation information for localization. After calculating the LM of voxels over an OctoMap, a corresponding 3-dimensional LM graph is generated.

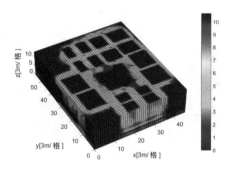

**Fig. 3.** An example LM graph for an urban environment. Blue rectangles are building regions where LM is not calculated.

Figure 3 is an example LM graph. Voxels with lower values indicate lesser environmental information, whereas voxels with higher values represent richer information and better localization performance when the UAV is located at those poses or positions (depending on whether $\theta$ is considered or omitted).

Given the same resolution for a grid and a voxel, the computation of LM in a 3d space is far more expensive than that in a 2d space. For instance, consider a space with the size of $100\,\text{m} \times 100\,\text{m} \times 50\,\text{m}$ and a resolution of $0.1\,\text{m}$. It has a million grids on a 2d plane while the amount of voxels in the same space grows up to 500 millions. To deal with this issue of dimensionality, LM of a 3d space may be computed with a certain voxel interval. For example, if the LM is computed with an interval of 3 voxels, the amount of computation may be reduced by around $3^3 = 27$ times. The computation of LM for a 3d space sized $100\,\text{m} \times 100\,\text{m} \times 50\,\text{m}$ and a resolution of $0.1\,\text{m}$ with an interval of 3 voxels took around 5 h on a PC with Intel i5-3470 3.2 Ghz CPU and 8 GB of RAM.

## 3    Localizability Constraint-Based Path Planning Method

### 3.1    Objective Function

Let $p'$ be a path in an OctoMap which consists of a series of adjacent route points $p$ such that $p' = \{p_1, p_2, \ldots, p_n\}$ where $n$ is the number of points in the path.

$Dis(p_k, p_{k+1})$ is defined as Euclidean distance between points $p_k$ and $p_{k+1}$, and $C(p')$ is defined as cost function of path $p'$:

$$C(p') = \sum_{k=1}^{n-1} Dis(p_k, p_{k+1}) \tag{9}$$

To obtain the shortest path $p'_s$ which links the start point $p_{st}$ and the goal point $p_{gl}$, it is sufficient to calculate the solution $p'$ that meets the following objective function:

$$p'_s = \arg\min_{p'} C(p') \tag{10}$$

## 3.2  Constraints

In the proposed method, $p'_s$ is found within a planning configuration space constrained by the UAV's LM graph(s):

$$s.t. \begin{cases} p_k \in \mathbf{R}_{\text{pass}}, \ \forall p_k \in p'_s \\ \mathbf{R}_{\text{pass}} = \mathbf{R}_{\text{pass\_1}} \bigcup \cdots \bigcup \mathbf{R}_{\text{pass\_M}} \\ \mathbf{R}_{\text{pass\_}m} = \{p | l'(\boldsymbol{p}_m) > T_{\text{bin}}\}, \ m = 1, \ldots, M \end{cases} \tag{11}$$

Definition of related symbols is given in Table 1. By posing this constraint, it can be ensured that all of the route points of $p'_s$ possess desirable $l'(\boldsymbol{p})$ and are all within global passable region $\mathbf{R}_{\text{pass}}$ (refer to Table 1).

## 3.3  Extraction of Local and Global Passable Regions

To derive $\mathbf{R}_{\text{pass}}$ in (11), the LM graph(s) and corresponding $\mathbf{R}_{\text{pass\_}m}$ under orientations 1 to M should be obtained first. The extraction process of $\mathbf{R}_{\text{pass\_}m}$ and $\mathbf{R}_{\text{pass}}$ is illustrated in Fig. 4, and explained in detail as follows:

1. **Point Cloud Generation:** Firstly, point cloud of the environment in which the UAV is going to navigate is produced via method such as LOAM [25].
2. **OctoMap Generation:** As LM is calculated based on the OctoMap of the environment, in this step, OctoMap is generated using point cloud data obtained in the last step.
3. **LM Calculation:** In this step, $l(\boldsymbol{p})$ over the generated OctoMap is calculated to produce an LM graph(s). This is done by applying the method provided in Sect. 2, with a specific sensor model taken into consideration. The sensor model specifies $D_1$, $R_1$ and $N_1$ listed in Table 1. The sensor model and the placement approach of LRF jointly determine whether (4) or (5) should be adopted for SLM calculation and the total number of LM graphs generated(M).

**Table 1.** Definition of symbols

| Symbol | Definition |
|---|---|
| $\theta_i$ | Angular interval. The angular interval used to divide orientations in case SLM is orientation-dependent |
| M | Total number of orientations. In case SLM is orientation-dependent, $M = 360°/\theta_i$. Otherwise, $M = 1$. For example, $M = 8$ if $\theta_i = 45°$ |
| $m$ | a variable that ranges from 1 to M. $m$-th orientation represents angle $\theta = (m-1) \times \theta_i$ |
| $\boldsymbol{p}_m$ | a pose with coordinates of corresponding point $p$ and $m$-th orientation |
| $l'(\boldsymbol{p})$ | Normalized $l(\boldsymbol{p})$ calculated at pose $\boldsymbol{p}$ |
| $T_{bin}$ | Binarization threshold. Used to filter out voxels with low $l'(\boldsymbol{p})$ |
| $\mathbf{R}_{pass\_m}$ | Local passable region under $m$-th orientation. The passable region in the OctoMap for UAV under $m$-th orientation. It is the union set of points with $l'(\boldsymbol{p}_m) > T_{bin}$ |
| $\mathbf{R}_{pass}$ | Global passable region. The final passable region in the OctoMap for UAV. It is the union set of $\mathbf{R}_{pass\_m}$ under orientations 1 to M. Feasible paths are planned within this region in subsequent step |
| $D_l$ | Scan range. Maximum scan range of LRF systems mounted on the UAV |
| $R_l$ | Scan resolution. Scan resolution of LRF systems mounted on the UAV |
| $N_l$ | Number of scan rays. The total number of scan rays of LRF systems mounted on the UAV |

4. **Normalization:** There may appear a noticeable difference in terms of numerical value and order of magnitude between $l(\boldsymbol{p})$s calculated from distinct environments or from a single environment but using dissimilar sensor models. Therefore, a normalization process is required for further processing. One way to do so is using Min-Max normalization method over $l(\boldsymbol{p})$s to produced normalized results $l'(\boldsymbol{p})$s. Values of $l'(\boldsymbol{p})$s are ranged from 0 to 1 with those closer to 1 implying higher and those closer to 0 implying lower localizability.

5. **Binarization:** One way to filter out voxels with low LM in the LM graphs is binarizing $l'(\boldsymbol{p})$s using a threshold $T_{bin}$. $T_{bin}$ is determined by the user: a lower value represents a more conservative filtering strategy and retains more voxels from the original OctoMap, whereas higher value filters the voxels in a more radical manner.

6. $\mathbf{R}_{pass}$**Extraction:** After above steps, in the case of $M = 1$, $\mathbf{R}_{pass}$ is exactly the final extracted $\mathbf{R}_{pass\_1}$. In other cases, $\mathbf{R}_{pass}$ is generated by extracting $\mathbf{R}_{pass\_1}$ to $\mathbf{R}_{pass\_M}$ and finding their union set, as noted in (11).

## 3.4 Optimization Method

In the proposed framework, general concepts of the optimization method are as follows:

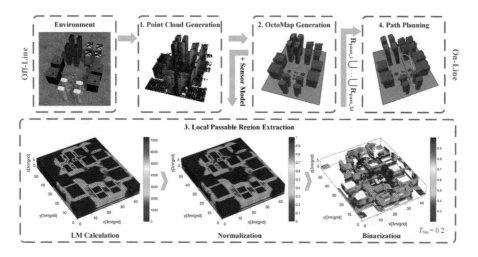

**Fig. 4.** Procedure of the proposed path planning method. The results in step 3 are sliced with z = 20 m for better demonstration.

1. Finding $p'_s$ in (10) is not limited to any specific path planning algorithm. It can be done via implementing algorithms such as A* [11], RRT [15], and their variations.
2. The parameters $\theta_i$ and M are determined based on the sensor model, the placement approach of LRF and the actual need of path planning algorithm. For example, in case the LRF is vertically mounted on the UAV and traditional A* algorithms is used for path planning, M may be set to eight (and therefore $\theta_i = 45°$). This is because traditional A* only explores adjacent points in eight directions (remember that $\varphi$ and $\psi$ are kept at zero degree) of the current *expanding point* in each iteration.
3. During navigation, the UAV should point to the orientation with best localizability (i.e. with highest $l'(\boldsymbol{p})$) whenever possible to achieve better localization performance.
4. If multiple candidate paths exist, the final solution may be determined by the user based on other criteria or preferences.
5. Extracting $\mathbf{R}_{\text{pass}}$ of a new environment which includes step 1, 2 and 3 in Fig. 4 tends to be time consuming, therefore it is mostly computed offline; in real-time path planning, the pre-computed $\mathbf{R}_{\text{pass}}$ is loaded into system before applying the desired path planning algorithm.

## 4    Simulation and Discussion

### 4.1    Simulation Set-Up

Simulations were all performed on a PC (Intel i5-3470 3.2 Ghz CPU, 8 GB RAM) using ROS [21] and Gazebo [9] open source platforms. The quadrotor model used in the experiments is developed based on Hector Quadrotor [18], it is equipped with a simulated on-board IMU and LRF (Velodyne HDL-32 [22]) with a scan rate of 10 Hz. The point clouds of the environments are generated using LOAM and then converted into OctoMap. Traditional A* algorithm is used for path planning. Similar to method in [10], during navigation, relative movement of the UAV between two subsequent LRF scans is roughly estimated by the usage of ICP algorithm from the Point Cloud Library [19] and IMU data, then rectified using a particle filter [7]. Important parameters in simulations are provided in Table 2.

**Table 2.** Important parameters

| Description | Urban | Wilderness |
|---|---|---|
| Placement of LRF | Horizontal | Vertical |
| $D_l$ | 15 m | 50 m |
| $R_l$ | 1° | 0.5° |
| N | 360 | 540 |
| M | 1 | 8 |
| OctoMap size | 150 m × 160 m × 90 m | 220 m × 240 m × 50 m |
| OctoMap voxel size | 1 m × 1 m × 1 m | 1 m × 1 m × 1 m |
| $\Delta x, \Delta y, \Delta z, \Delta\theta_l$ | 2 m, 2 m, 2 m, - | 2 m, 2 m, 2 m, 5° |
| $T_{\text{bin}}$ | 0.2 | 0.2 |

## 4.2   Urban Canyon Environment

The first simulation is carried out in an urban canyon environment (Fig. 5(a)), which includes surrounding regions with big shopping malls and tall office buildings that may cause reflection for GPS signal, and a relatively open park region in the center part. For UAVs deployed in tasks such as parcel delivery, as the overall cost is usually one of the major concern, a low cost LRF is simulated by shortening the scan range of the LRF. The LRF is horizontally mounted in this simulation, thus M = 1 and (5) is used to calculate LM over the OctoMap. After processing the calculated LM graph, the final $\mathbf{R}_{\text{pass}}$ is derived and displayed in Fig. 5(b). Voxels above the park region and the roofs of the buildings are filtered out, as in these voxels no valid information could be observed by the LRF and therefore resulting in low LM.

The paths planned with LC and without LC are shown in Fig. 5(c) and (d). As can be seen from the planning results, instead of taking the shorter path and going through the regions with low LM on the center part of the OctoMap, the path generated by the proposed method stays closer to featured structures and adjusts the height whenever needed. Although the resulting path is obviously longer in this case, compared to the path without LC, it promises better localizability along the path.

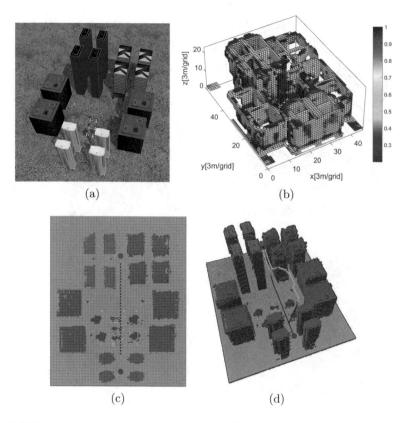

**Fig. 5.** Urban canyon environment, global passable region and path planning results. (a) environment (b) global passable region (c) planning results (vertical view), path without LC (blue), path with LC (green), start (red), goal (purple) (d) planning results (side view), paths colored with height along z-axis.

## 4.3   Wilderness Environment

The second simulation is conducted in a wilderness environment with rugged mountainous terrain (Fig. 6(a)). Mountains with different height spread over the terrain, and a plain region is present in the upper-right area. In this simulation, the LRF is vertically mounted, $M = 8$ and (4) is used to calculate LM over the OctoMap. Similarly, $\mathbf{R}_{pass}$ is derived after going through procedure mentioned in the last section. As shown in Fig. 6(b), the global passable region rises and falls as in the terrain. This is because in the vertical placement of LRF, in order to obtain distinguishable ground features, the UAV must maintain its height within a range that is neither too high or too low. Otherwise the UAV may suffer from lack of valid LRF information within the scan range (too high) or high similarity in observed information (too low).

The paths planned with LC and without LC are shown in Fig. 6(c) and (d). As can be seen from the planning results, the path without LC is nearly a straight

line that traverses the undesired plain region, whereas the path with LC avoids
the mentioned region and possess higher height to maintain its localizability.

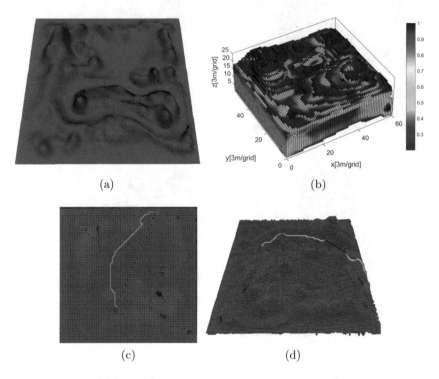

(a)                                                    (b)

(c)                                    (d)

**Fig. 6.** Wilderness environment, global passable region and path planning results. (a)
environment (b) global passable region (c) planning results (vertical view), path without LC (blue), path with LC (green), start (red), goal (purple) (d) planning results
(side view), paths colored with height along z-axis.

## 4.4   Localization Error Analysis

In this subsection, we compare and analyze the localization error between the
paths that are planned with and without LC in the previous subsections. The
UAV tracks the planned path and localization error along the path, which is
defined as the Euclidean distance (always positive) between ground truth (given
by the Gazebo simulator) and estimated pose (output of the particle filter), is
recorded.

Five sets of experiment are conducted for each of the paths, the results are
given in Table 3. The navigation result and the error along the three axes of
experiment No. 1 in both urban and wilderness environment are respectively
presented in Figs. 7 and 8.

The results convey the intuition that the path with LC is much closer to
its ground truth in both environments, while that without LC leads to a far
more noticeable error. Specifically, the localization error of the two paths with

**Table 3.** Localization error ± S.D [m]

| No. | Urban | | Wilderness | |
|-----|-------|---|------------|---|
|     | Without LC | With LC | Without LC | With LC |
| 1 | 9.99 ± 12.97 | 1.29 ± 0.52 | 8.21 ± 10.98 | 1.99 ± 1.33 |
| 2 | 15.32 ± 14.05 | 1.37 ± 0.93 | 9.12 ± 10.53 | 1.81 ± 1.39 |
| 3 | 15.49 ± 17.11 | 1.33 ± 0.64 | 11.20 ± 17.43 | 1.67 ± 1.08 |
| 4 | 20.16 ± 16.34 | 1.65 ± 0.83 | 7.87 ± 11.41 | 2.02 ± 1.53 |
| 5 | 12.29 ± 12.51 | 1.77 ± 1.12 | 9.06 ± 10.16 | 2.11 ± 1.95 |

LC is relatively consistent and kept at a low level through out the path. As for the two paths without LC, the localization error dramatically increases once the UAV enters regions with low LM: in the urban environment, the UAV loses valid perception of its surroundings; in the wilderness environment, the UAV observes highly similar LRF information. In both cases, the UAV cannot correctly estimate the relative movement between two subsequent LRF scans. Consequently, the UAV fails to perform localization as the error accumulates.

By examining Table 3, we see that in all five experiments, the average error of paths with LC is about one-forth to one-tenth of that without LC with a much lower standard deviation. Altogether, the results imply that along the path planned by the proposed method, the UAV always has a better pose tracking performance.

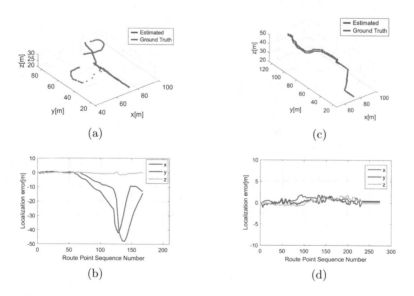

**Fig. 7.** Navigation result and localization error in urban environment. (a) navigation without LC (b) localization error without LC (c) navigation with LC (d) localization error with LC

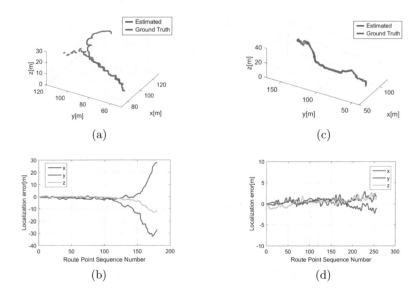

**Fig. 8.** Navigation result and localization error in wilderness environment. (a) navigation without LC (b) localization error without LC (c) navigation with LC (d) localization error with LC

## 5    Conclusion

In this paper, a localizability constraint-based path planning method for UAV is introduced. The proposed method provides a framework of global path planning for UAV, which uses the pre-built OctoMap of the environment, considers the LRF sensor model, and solves for the shortest possible paths with localizability constraint taken into account. One merit of the proposed method is that it is not limited to any specific path planning algorithm in the optimization stage. In the simulation section, the method exhibited its applicability in outdoor GPS-unreliable environments. Through analyzing the simulation results, this conclusion can be made that in situations where there are low LM regions present in the environment, the path planned using the proposed method effectively reduces the localization error and ensures better localization performance as compared to those planned without LC. The resulting paths of the proposed method demonstrated a tendency to stay close to rich-featured regions, which concurs with our intuitive expectation.

## References

1. Bar-Shalom, Y., Li, X.R., Kirubarajan, T.: Estimation with Applications to Tracking and Navigation: Theory Algorithms and Softwar. Wiley, Hoboken (2004)
2. Bobrovsky, B., Zakai, M.: A lower bound on the estimation error for markov processes. IEEE Trans. Autom. Control **20**(6), 785–788 (1975)

3. Candido, S., Hutchinson, S.: Minimum uncertainty robot path planning using a POMDP approach. In: Proceedings of IEEE/RSJ International Conference on Intelligent Robots and Systems, pp. 1408–1413. IEEE (2010)
4. Censi, A.: An accurate closed-form estimate of ICP's covariance. In: 2007 IEEE International Conference on Robotics and Automation, pp. 3167–3172. IEEE (2007)
5. Censi, A.: On achievable accuracy for range-finder localization. In: Proceedings of IEEE International Conference on Robotics and Automation, pp. 4170–4175. IEEE (2007)
6. Censi, A.: On achievable accuracy for pose tracking. In: Proceedings of IEEE International Conference on Robotics and Automation, pp. 1–7. IEEE (2009)
7. Dellaert, F., Fox, D., Burgard, W., Thrun, S.: Monte Carlo localization for mobile robots. In: 1999 Proceedings of IEEE International Conference on Robotics and Automation, vol. 2, pp. 1322–1328 (2002)
8. Diosi, A., Kleeman, L.: Uncertainty of line segments extracted from static SICK PLS laser scans. In: Proceedings. SICK PLS Laser. Australiasian Conference on Robotics and Automation (2003)
9. Gazebo. http://www.gazebosim.org/
10. Grzonka, S., Grisetti, G., Burgard, W.: A fully autonomous indoor quadrotor. IEEE Trans. Robot. **28**(1), 90–100 (2012)
11. Hart, P.E., Nilsson, N.J., Raphael, B.: A formal basis for the heuristic determination of minimum cost paths. IEEE Trans. Syst. Sci. Cybern. **4**(2), 100–107 (1968)
12. Hochstenbach, M., Notteboom, C., Theys, B., De Schutter, J.: Design and control of an unmanned aerial vehicle for autonomous parcel delivery with transition from vertical take-off to forward flight-vertikul, a quadcopter tailsitter. Int. J. Micro Air Veh. **7**(4), 395–405 (2015)
13. Hornung, A., Wurm, K.M., Bennewitz, M., Stachniss, C., Burgard, W.: OctoMap: an efficient probabilistic 3D mapping framework based on octrees. Auton. Robots (2013). http://octomap.github.com
14. Kaelbling, L.P., Littman, M.L., Cassandra, A.R.: Planning and acting in partially observable stochastic domains. Artif. Intell. **101**(1), 99–134 (1998)
15. Lavalle, S.M.: Rapidly-exploring random trees: progress and prospects. Algorithmic and Computational Robotics New Directions, pp. 293–308 (2001)
16. Liu, Z., Chen, W., Wang, Y., Wang, J.: Localizability estimation for mobile robots based on probabilistic grid map and its applications to localization. In: Proceedings of IEEE Conference on Multisensor Fusion and Integration for Intelligent Systems, pp. 46–51. IEEE (2012)
17. Magree, D., Johnson, E.N.: Combined laser and vision-aided inertial navigation for an indoor unmanned aerial vehicle. In: American Control Conference (ACC), pp. 1900–1905. IEEE (2014)
18. Meyer, J., Sendobry, A., Kohlbrecher, S., Klingauf, U., von Stryk, O.: Comprehensive simulation of quadrotor UAVs using ROS and Gazebo. In: 3rd International Conference on Simulation, Modeling and Programming for Autonomous Robots (SIMPAR) (2012, to appear)
19. PCL. http://pointclouds.org/
20. Roca, D., Martínez-Sánchez, J., Lagüela, S., Arias, P.: Novel aerial 3d mapping system based on UAV platforms and 2d laser scanners. J. Sensors **2016**, 8 pages (2016)
21. ROS. http://www.ros.org/
22. Velodyne. http://www.velodynelidar.com/

23. Wang, Y., Chen, W., Wang, J., Wang, H.: Active global localization based on localizability for mobile robots. Robotica **33**(08), 1609–1627 (2015)
24. Zhang, F., Grocholsky, B., Kumar, V.: Formations for localization of robot networks. In: Proceedings of IEEE International Conference on Robotics and Automation, vol. 4, pp. 3369–3374. IEEE (2004)
25. Zhang, J., Singh, S.: LOAM: lidar odometry and mapping in real-time. In: Robotics: Science and Systems Conference, July 2014

# Using IMU Sensor and EKF Algorithm in Attitude Control of a Quad-Rotor Helicopter

Jongwoo An and Jangmyung Lee[✉]

Pusan National University, 2, Busandaehak-ro 63beon-gil, Geumjeong-gu,
Busan, Republic of Korea
{jongwoo7379, jmlee}@pusan.ac.kr

**Abstract.** The level of interest regarding Unmanned Aerial Vehicles (UAVs),
such as a Quad-Rotor, has been increased recently. UAVs can effectively carry
out various monitoring tasks for disasters, life-saving situations, environmental
conditions, traffic congestion and military reconnaissance. This paper presents a
new attitude control method for a quad-rotor based on IMU sensor and EKF
algorithm. The proposed method can enable a quad-rotor to achieve stable
operation in the harsh ocean environment with unexpected disturbance and
dynamic changes.

**Keywords:** EKF · IMU · HDR · Quad-rotor

## 1 Introduction

Recently, researched about Unmanned Aerial Vehicles (UAVs) such as drone, has been
actively carried out according to the development of sensor, material and control
technology.

The UAV is divided into fixed wing and rotary wing according to the wing portion.
The fixed wing used the lift, generated by the difference in the shape of the upper side
and lower side of the wing, so the runway is required for the flight and it is impossible
to stop and vertically ascend and descend. In the other hand, rotary wing used lift,
generated by two or more rotary blade, so it is possible to take off/landing in a narrow
space and stop. For this reason, the drone is used in various fields [1–4].

Thus, as demonstrated by the availability of drones, companies such as Google and
Amazone are launched the service using drone and increasing their interest in diverse
applications.

Drones are being studied not only for flight and reconnaissance, but also for applied
the various missions such as service and life-saving. In order for the drones to perform
their duties smoothly, flight stability must be ensured.

The flight safety of UAVs such as drones is strongly influenced by environmental
factors as Icing, Wake Turbulence, Severe weather, and Wind shear [5]. In order
to activate the industry using drones, researches should be carried out to maintain the
attitude of the aircraft in an abnormal state so that it can fly.

© Springer Nature Switzerland AG 2019
M. Strand et al. (Eds.): IAS 2018, AISC 867, pp. 933–942, 2019.
https://doi.org/10.1007/978-3-030-01370-7_72

Among these studies, NASA's Generic Transport Model (GTM) and SUPRA (Simulation of Upset Recovery in Aviation) project are used to quantitatively analyze abnormal flight models and use them for flight status recovery research [6, 7].

In addition, studies are being conducted to diagnose sensor failures by estimating the state variables of drones, to improve the reliability of attitude sensors, and to apply multi-sensor fusion techniques to unmanned aircraft sensor systems [8] (Fig. 1).

**Fig. 1.** Various applications of drone

In this paper, proposed a method to check the flight status of a drone through the single IMU sensor and to estimate a drone's attitude for flight status recovery using Extended Kalman filter (EKF).

Next, in Sect. 2, described the basic modeling of the quad-rotor. In Sect. 3, described the proposed algorithm. In Sect. 4, experiment environment and results. Finally, conclusion.

## 2   Modeling of Quad-Rotor

To model the Quad-rotor's dynamics, we need to know the antibody coordinate system and the inertial coordinate system as shown in Fig. 2.

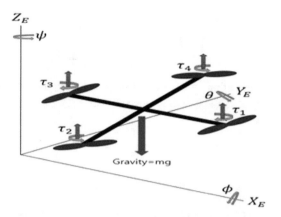

**Fig. 2.** Relation between navigation and body coordinates

The relation between the antibody coordinate system and the inertial coordinate system of the Quad-rotor is represented by the following rotation matrix [9, 10].

$$
R_x(\phi) = \begin{bmatrix} 1 & 0 & 0 \\ 0 & \cos\phi & -\sin\phi \\ 0 & \sin\phi & \cos\phi \end{bmatrix}, \quad R_y(\phi) = \begin{bmatrix} \cos\phi & 0 & -\sin\phi \\ 0 & 1 & 0 \\ \sin\phi & 0 & \cos\phi \end{bmatrix}
$$

$$
R_z(\psi) = \begin{bmatrix} \cos\psi & -\sin\psi & 0 \\ -\sin\psi & \cos\psi & 0 \\ 0 & 0 & 1 \end{bmatrix}
$$

(1)

$$
R_b^E = R_z(\psi)R_y(\phi)R_x(\phi) \tag{2}
$$

The position and Euler angles of the Quad-rotor in the inertial coordinate system are expressed in the following formulas, such as the velocity and angular velocity on the antibody coordinate system.

$$
p_E = \begin{bmatrix} x & y & z \end{bmatrix}^T \tag{3}
$$

$$
\eta_E = \begin{bmatrix} \phi & \theta & \psi \end{bmatrix}^T \tag{4}
$$

$$
v_b = \begin{bmatrix} v_x & v_y & v_z \end{bmatrix}^T \tag{5}
$$

$$
\dot{p}_E = R_b^E v_b \tag{6}
$$

$$
\omega_b = R_\omega \dot{\eta}_E \tag{7}
$$

$p_E$ is the position of the robot, $\eta_E$ is the Euler angle, $v_b$ is the velocity, $\dot{p}_E$ is the angular velocity, $\omega_b$ is a matrix expressing the relationship between the velocity component of the Euler angles and the angular velocity vector of the antibody coordinate system on the inertial coordinate system.

$$R_\omega = \begin{bmatrix} 1 & 0 & -\sin\phi \\ 0 & \cos\phi & \sin\phi\,\cos\phi \\ 0 & -\sin\phi & \cos\phi\,\sin\phi \end{bmatrix} \tag{8}$$

To obtain the acceleration of the Quad-rotor on the inertial coordinate system and the angular acceleration of the antibody, the following Eqs. 12 and 13 are obtained by differentiating Eqs. 6 and 7.

$$\ddot{p}_E = R_b^E \dot{v}_b + \dot{R}_b^E v_b \tag{9}$$

$$\dot{\omega}_b = R_\omega \ddot{\eta}_E + \dot{K}\dot{\eta}_E \tag{10}$$

The relation between force and moment acting on the quad rotor can be expressed as follows.

$$\sum F_{external} = F + F_g = m\dot{V} + \omega(mV) \tag{11}$$

$$\sum T_{external} = Q + Q_G = I_T\dot{\omega} + \omega(I_T\omega) \tag{12}$$

The meanings of the variables used in the above equation are summarized in Table 1 below.

**Table 1.**  Variable list

| Specification | Content |
|---|---|
| $F_{external}$ | External force |
| $T_{external}$ | External moment |
| $Q$ | Moment for control of Quad-rotor |
| $Q_G$ | Gyro effect |
| $I_T$ | Inertial matrix |
| $m$ | Mass of Quad-rotor |

# 3   Attitude Estimation of Quad-Rotor Using EKF

The basic posture control of the Quad-rotor uses the acceleration sensor, the gyro sensor and the geodetic sensor fuse. To do this, it is necessary to pay attention to the bias error of each sensor. Especially, in the case of gyro sensor, Can be. Therefore, each error is applied to the posture control performance through various sensor fusion process using Heuristic drift reduction(HDR) algorithm and EKF (Fig. 3).

## 3.1 System Configuration

The figure below shows the configuration of the quad-rotor used in this paper.

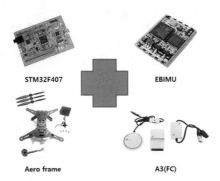

**Fig. 3.** System configuration

The frame of the quad-rotor used Intel's Aero platform frame, and the Flight Controller (FC) used DJI's A3. A3(FC) possible to code-work through API provided by DJI, and can be linked with external system(STM32F407) through UART (Fig. 4).

**Fig. 4.** External system connection

Table 2 below shows the API functions provided by DJI.

**Table 2.** API function provided by DJI

| Function | Command code |
|---|---|
| Get current version | 0xFA 0xFB 0x00 0xFE |
| Activate | 0xFA 0xFB 0x01 0xFE |
| Obtain control | 0xFA 0xFB 0x02 0x01 0xFE |
| Release control | 0xFA 0xFB 0x02 0x00 0xFE |
| Movement control | 0xFA 0xFB 0x04 0x01 Flag x_H x_L y_H y_L z_H z_L yaw_H yaw_L 0xFE |
| Movement control (dry-run) | 0xFA 0xFB 0x04 0x02 Flag x_H x_L y_H y_L z_H z_L yaw_H yaw_L 0xFE |
| Return to home (RTH) | 0xFA 0xFB 0x05 0x01 0xFE |
| Auto take off | 0xFA 0xFB 0x05 0x02 0xFE |
| Auto landing | 0xFA 0xFB 0x05 0x03 0xFE |
| Get broadcast data | 0xFA 0xFB 0x08 0x00 0xFE |

## 3.2　Attitude Control of Quad-Rotor Using IMU

The errors caused by the integration of the gyro measurement values and the drift phenomenon are accumulated due to accumulation of errors when estimating the throttle property of the Quad-rotor using the IMU sensor. The gyro sensor outputs the angular velocity and generally estimates the direction angle of the quad rover by integrating the measured angular velocity. At this time, since a small bias drift is accumulated as an integral, a large error occurs. Therefore, the HDR is applied to minimize the bias drift.

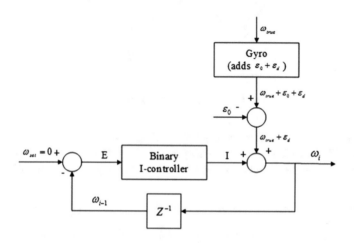

**Fig. 5.**　HDR

Figure 5 shows a block diagram of the HDR algorithm to minimize the drift phenomenon of the gyro sensor.

$\omega_{true}$ is an ideal value that does not include an error, and includes $\varepsilon_0$ and $\varepsilon_d$ when the Gyro sensor measures, and $\varepsilon_0$ is the initial bias value calculated by averaging the measured values of the sensor for a certain period of time, and $\varepsilon_d$ is a bias value that varies slightly with time, which means an object to be removed here.

Since Binary I-controller must respond sensitively to fine errors and insensitive to changes in large values corresponding to changes in actual values generated by rotation of the gyro sensor, only the sign of the error Considering the Binary I-controller, it operates as follows.

$$I_i = I_{i-1} - SIGN(\omega_{i-1})i_c \tag{13}$$

In Eq. 13, $i_c$ is set to a value between 0 and 0.001, and the binary I-controller operates as follows.

$$SIGN(x) = \begin{cases} 1 & for\ x > 0 \\ 0 & for\ x = 0 \\ -1 & for\ x < 0 \end{cases} \tag{14}$$

Even if the bias error is removed by using the HDR as described above, there is a problem that the cumulative error due to integration continues to occur over time.

When estimating the attitude of the Quad-rotor by using the gyro sensor, the cumulative error increases because it depends on the instantaneous rotational angular velocity, and it can't be solved by the HDR algorithm alone.

To solve this problem, an error is corrected using an acceleration sensor and a geomagnetic sensor.

At acceleration, information of roll and pitch can be obtained.

$$\phi_a = \tan^{-1}\left(\frac{a_y}{a_z}\right) \tag{15}$$

$$\theta_a = \sin^{-1}\left(-\frac{a_x}{\sqrt{a_y^2 + a_z^2}}\right) \tag{16}$$

If acceleration sensor is used as above, roll and pitch can be compensated. However, it is impossible to compensate for the yaw, and the posture and direction should still be estimated depending on the yaw value measured by the gyro sensor.

To compensate for this, we use geomagnetic sensors. The geomagnetic sensor is called an electronic compass, and it is a sensor that can measure the earth's magnetic field and can detect the azimuth angle.

It is always parallel to the Earth's surface while facing the Earth's magnetic north pole.

The X and Y components of the earth's magnetic field determine the azimuth angle. The equation below is to measure the azimuth of the magnetic sensor.

$$\psi_c = \tan^{-1}\left(\frac{Y_H}{X_H}\right) \tag{17}$$

As mentioned above, the roll, pitch, and yaw values of the quad-rotor can be obtained through the acceleration sensor and the geomagnetic sensor. To integrate these values with Roll, Pitch, Yaw of geomagnetic sensor, it is necessary to fuse them.

In this research, we used an EKF, which is a kind of Kalman Filter(KF). EKF is widely used for estimating nonlinear conditions such as GPS and navigation, and is also suitable for nonlinear applications such as sensor fusion which estimates the attitude and direction of Quad-rotor.

The following figure is a system diagram of the Quad-rotor attitude control algorithm that is robust to disturbance using EKF (Fig. 6).

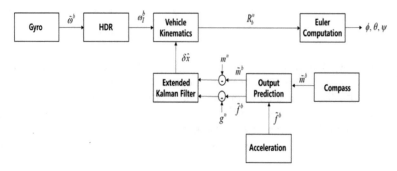

**Fig. 6.** Algorithm

The attitude data of the Quad-rotor measured by the EMIMU is corrected by the EKF in the STM32F407. Based on the corrected attitude data, the commands in the Table 2 are transmitted to the A3 through the UART to control the attitude of the Quad-rotor.

## 4    Experiment and Result

Experiments were carried out under the weather condition of 18.3 mph as shown in the figure below and the hovering state was measured before and after applying the attitude control algorithm (Fig. 7).

**Fig. 7.** Experiment environments

Figures 8 and 9 below show the Attitude and Position data of the Quad-rotor before and after applying the algorithm.

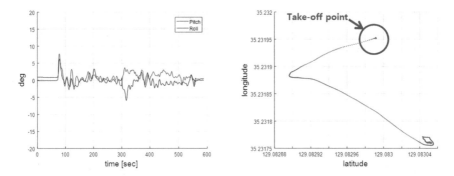

**Fig. 8.**  Before applying the algorithm(Left-Attitude, Right-Position)

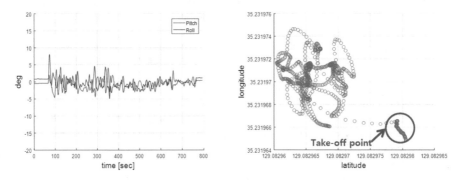

**Fig. 9.**  Before applying the algorithm(Left-Attitude, Right-Position)

Experimental results show that before the application of the algorithm, the quad-rotor does not maintain the hovering state and moves along the wind direction.

After applying the algorithm, it can be seen that it is trying to maintain the hovering state while enduring the disturbance.

# 5   Conclusion

In this paper, we propose an attitude control method of quad-rotor which is robust to disturbance by using IMU sensor and EKF.

Although this study is still in its infancy, we have not been able to carry out experiments in a sea breeze environment like sea environment, but we confirmed the possibility of the proposed algorithm through experiments.

In the future, we will study the performance improvement of the attitude control algorithm to reduce the variation of the position of the quad-rotor and study the stable operation of the quad-rotor in the marine environment.

**Acknowledgements.** This research was supported by the Ministry of Trade, Industry & Energy (MOTIE), Korea, under the Industry Convergence Liaison Robotics Creative Graduates Education Program supervised by KIAT(N0001126).

This research was a part of the project titled 'Developments of Drone Docking System equipped Leisure Boat for Drone surfing and operating', funded by Ministry of Oceans and Fisheries, Korea.

# References

1. Woo, M.K., Park, O., Kim, S.K., Suk, J.Y., Kim, Y.D.: Real-time aircraft upset detection and prevention based on extended Kalman filter. J. Korean Soc. Aeronaut. Space Sci. **45**(9), 724–733 (2017)
2. Cho, S.W., Lee, D.S., Jung, Y.D., Lee, U.H., Shim, H.C.: Development of a cooperative heterogeneous unmanned system for delivery services. J. Inst. Control Robot. Syst. (in Korean) **20**(12), 1181–1188 (2014)
3. Lee, H.B., Moon, S.W., Kim, W.J., Kim, H.J.: Cooperative surveillance and boundary tracking with multiple quadrotor UAVs. J. Inst. Control Robot. Syst. (in Korean) **19**(5), 423–428 (2013)
4. Kim, D.H., Shin, J.H., Kim, J.D.: Design and implementation of Wi-Fi based drone to save people in maritime. J. Korea Inst. Inf. Commun. Eng. (JKIICE) **21**(1), 53–60 (2017)
5. Sung, S.M., Lee, J.O.: Accuracy assessment of parcel boundary surveying with a fixed-wing UAV versus rotary-wing UAV. J. Korean Soc. Surv. Geodesy Photogram. Cartography **35**(6), 535–543 (2017)
6. Engelbrecht, J., et al.: Bifurcation analysis and simulation of stall and spin recovery for large transport aircraft. In: AIAA Atmospheric Flight Mechanics Conference, p. 4801 (2012)
7. Abramov, N., et al.: Pushing ahead-SUPRA airplane model for upset recovery. In: AIAA Modeling and Simulation Technologies Conference, p. 4631 (2012)
8. Jie, C., Patton, R.J.: Robust Model-Based Fault Diagnosis for Dynamic Systems. Springer Science & Business Media, Boston (2012)
9. Elnagar, A.: Motion prediction of moving objects. In: Proceedings of the Second IASTED International Conference of Control and Application, pp. 448–451 (1999)
10. Kim, K.J., Yu, H.Y., Lee, J.M.: Dynamic object tracking of a quad-rotor with image processing and an extended Kalman filter. J. Inst. Control Robot. Syst. **21**(7), 641–647 (2015)

# Unmanned Aerial Vehicles in Wireless Sensor Networks: Automated Sensor Deployment and Mobile Sink Nodes

Juan Marchal Gomez$^{(\boxtimes)}$, Thomas Wiedemann, and Dmitriy Shutin

German Aerospace Center (DLR), Institute of Communication and Navigation,
Oberpfaffenhofen, Germany
Juan.marchalgomez@dlr.de

**Abstract.** This paper describes the design and implementation of a heterogeneous multi-agent system consisting of a wireless sensor network and a mobile agent (UAV) for automatic network deployment and data collection beyond the communication range of individual sensors in the network. A UAV is able to transport and release sensors at specific locations that might be dangerous or inaccessible to humans. The network can be deployed fast when manual placement is a risk. The measured data is likewise automatically collected by a UAV following an optimal route obtained by solving the Vehicle Routing Problem (VRP) that covers all sensor locations. Since the route of the UAV is pre-planned, sensor nodes are able to turn on the radio interface exactly when the UAV is close by and thus save energy. More precisely, the power consumption of the sensor node is improved in two ways: (i) multihop routing is not required, which saves battery life in sensors, and (ii) the transmission power can be reduced because the UAV is able to go near the sensor. The whole system was tested in an outdoor experiment showing promising results.

**Keywords:** Wireless Sensor Network · Unmanned Aerial Vehicle
Power efficiency

## 1 Introduction

Wireless Sensor Networks (WSN) can be used for quite diverse monitoring tasks. Especially for observing spacial distributed phenomena, WSN are the means of choice. Precision agriculture uses WSNs to achieve a better management of the used resources [9]. In addition, WSNs will be used in Smart Cities to offer better services to the citizens like better waste management [3].

A typical structure of a WSN is as follows. A large number of sensor nodes is scattered inside or near the phenomenon of interest to measure the desired variables. These sensor nodes send the collected information to a sink node directly or using multi-hop routing protocols. The sink node sends the information to a task manager node where the data is processed and presented to the final user [2]. Sensor nodes usually do not have an access to the electrical grid so they are

© Springer Nature Switzerland AG 2019
M. Strand et al. (Eds.): IAS 2018, AISC 867, pp. 943–953, 2019.
https://doi.org/10.1007/978-3-030-01370-7_73

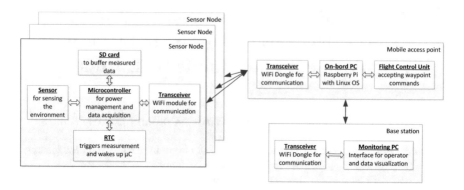

**Fig. 1.** System architecture diagram

battery powered. They thus need to be designed to consume as little energy as possible to extend the battery duration.

The main parts that consume most of the energy in a sensor node are the transceiver, the processing unit and the sensors itself. While it is possible to find low power microprocessors and sensors in the market which are suitable for wireless sensor network design, in general the transceiver is the most energy consuming part. The consumption of the transceiver depends on the used protocol, the data rate and the transmission power. In literature one of the most used protocols in WSNs is ZigBee [4–6]. Other important protocols used in WSN because of their low power consumption are Bluetooth Low Energy (BLE), LoRa, Sigfox or 6LoWPAN. It is also possible to use more power consuming technologies like WiFi or 4G LTE if higher data rates are needed or, in the case of 4G, if direct Internet access is necessary.

Usually, the information is routed using a multi-hop protocol. This approach present two problems: (i) the transmission power needs to be high enough to connect with the nearest node and (ii) the information is send several times until it reaches the destination which also consumes a lot of power in the overall network. Furthermore, the power consumption is not equally distributed between nodes: nodes that are near the sink node will spend more power than the other nodes in the network.

In this paper a mobile sink node is proposed to reduce the power consumption in the sensor node transceiver. This architecture was first introduced in [11] with the name of SEnsor Networks with Mobile Agents (SENMA). In [8] the purposed architecture is used for randomly deployed WSN. The mobile access point is able to wake up all the sensor nodes in a certain radius with RF signals. Therefore, the mobile access point has to cover the whole area to collect the information from all sensors because the locations are not known. In [1] the WSN is divided into several clusters, that are individually maintained by one mobile sink node. Each cluster possesses a central cluster head node collecting all the information from the nodes in the same cluster. In case when nodes are too far away from the cluster head, intermediate nodes act as relays and route the data to the cluster head. Thus, the mobile node has to collect the information only from the cluster head. Also, this mobile node can be used for network maintenance, for example, to replace nodes without battery. In [10] a mobile node is used in a real implementation for an agricultural application. The mobile access node is carried by a fixed-wing plane. The transceiver used for communication in the sensor node is in a low power sleeping mode most of the time and it is woken up by an RF signal of another low power receiver. The communication is limited to a very short time window, because of the dynamic speed constraints of the plane. Thus, the amount of data that can be transmitted is restricted.

The system developed in this paper is designed for low density WSNs, where the sensor node locations are predefined by an operator. We suppose that the distance between sensor nodes is too large for direct communication among the nodes. We consider a UAV that is able to distribute the sensor nodes in the region of interest and it is able to carry an access point. Specifically, we make use of a rotatory UAV which is able to hover steady in a defined position. In this way, the sensors are placed with high accuracy and it is possible for the UAV to stop at each location for data transmission. The UAV is programmed to visit all the sensor node locations following a near-optimal path to collect the measured data. The route optimization algorithm incorporates the UAV's battery duration, the time for data transmission and the velocity of the UAV. Several routes are created where only one battery per route is needed. The individual routes can be executed in serial using only one UAV. Alternatively, in a multi-agent approach several UAVs can follow the routes in parallel, reducing in this way the overall time for data collection. In general, the power consumption of a WSN node is dominated by two factors: (i) the duty cycle of transceiver, (ii) communication range of the transceiver (i.e. higher distances require a higher transmission power). Therefore, we reduce the power consumption in our approach as follows. Once the routes are created, the arrival time of the UAV for each sensor location is estimated. This information is used to turn on the radio only during a short period of time when the UAV is in the communication range. To this end, we make use of a Real Time Clock (RTC) in the sensor nodes. This setup allows us to reduce the duty cycle of the transceiver to a minimum. Further, the UAV is able to approach the sensor node location very closely which also reduces the necessary transmission power. To avoid delays or synchronization problems, the UAV arrives shortly before the radio of the sensor node is turned on and provides the access point some time buffer.

## 2  System Architecture

Figure 1 shows the three main parts of our WSN: (i) the sensor nodes, (ii) the UAV and the (iii) base station. The sensor nodes are distributed over predefined locations by the UAV. They measure desired variables and store them on SD card. A WiFi transceiver is used to transmit the measured values. The UAV also incorporates a WiFi dongle and act as an access point for the sensor nodes. The UAV collects the information from all sensor nodes and send it to the base

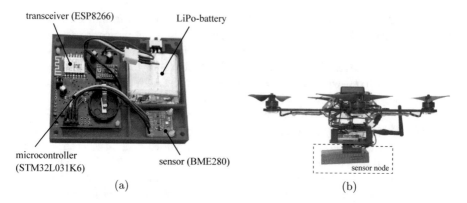

**Fig. 2.** In Fig. 2a the hardware setup of the sensor node is shown, where Fig. 2b depicts the sensor node mounted to the UAV.

station where the final user can visualize the data. The base station is located near the landing point of the UAV to receive the information. The used sensor and the transceivers can be changed to adapt the system to different scenarios.

The workflow of the systems is as follows. An operator defines the locations for the sensor nodes in the base station. Further, he selects the time when the UAVs are going to start gathering the measured data. Based on nodes' locations and constraints of the battery duration, the base station calculates one route or - in case the battery lifetime is not sufficient to reach all nodes - several routes.

The generated routes also contain information about the arrival time of the UAV to each sensor node. The routes are then sent to the UAVs so they can start placing the sensors. Moreover, the radios of the sensor nodes are configured to wake up at the specific time, when the UAV is going to arrive to collect the measured data. In this paper we employ micro UAVs with less than 1 kg take-off weight. So each UAV is able to carry only one sensor node at a time.

When the sensors are deployed, the UAVs wait at the initial location. They will start to go over all the sensor nodes at the specified time to download the measurements. The UAV can stop as near as possible to the sensor to the reduce the required transmission power in the sensor nodes. Each individual sensor node is configured to turn on the radio when the UAV arrives using the RTC. The UAVs have a WiFi access point and the sensor nodes try to connect to the network as soon as they wake up. If a connection is not possible, the transceiver will wake up periodically until a connection is possible. This reconnection time is configurable. In a first step, the sensor node will try to reconnect a selectable number of times in short intervals of time. If a connection is not possible - for example the UAV missed this node - the transceiver will wake up after a longer time. In this way, the UAV has the chance to come back later to collect the measurements without disturbing the behavior of the whole network. If there are more routes than UAVs, they will stop at the initial point after each route to change the batteries and continue with the next route. Further, after each route, the UAVs send the information to the base station, where it is processed to present the results to the final users.

# 3   Individual Components

## 3.1   Sensor Node Hardware

The sensor node needs to be light because it is going to be transported by a UAV. In addition, it needs to have low power consumption, an RTC, SD Card, a transceiver and communication ports to connect sensors. A prototype sensor node has been developed to test the proposed concept. The processing unit is implemented on the microcontroller STM32L031K6 from STMicroelectronics. It has a small form factor, several communication protocols (I2C, UART, SPI), analog input and an RTC. The internal architecture is ARM Cortex-M0+ which provide several low power modes. The RTC is used to drive the measurement process and to turn on the transceiver. The sensor node also incorporates an SD Card to store the data until the UAV arrives to collect the information.

As a transceiver we make use of the ESP8266 device. A ready to use board costs less than 2\$ in small quantities, which makes it suitable for large scale WSNs. Although it uses WiFi protocol, which means a high power consumption compared with other protocols, the communication data rate is high. Therefore, the amount of information to be transferred can be quite high. For our prototype, we attached a BME280 sensor to each node. This sensor is able to measure temperature, humidity and pressure. For the power supply, a small (1000 mAh) one cell LiPo battery is used. The overall weight of one sensor node is approximately 76 g, including the case and the battery. The weigh was an important design parameter, since it is constrained by the maximum possible payload of the employed UAV. The hardware setup can be observed in Fig. 2a.

## 3.2   Releasing Mechanism

We developed a simple mechanism to transport and deploy the sensor with a UAV (see also 2b). It contains a DC motor with a screwed axle driven by an H-bridge motor driver. Figure 3a shows a picture of the mechanism. In Fig. 3b we show a detachable sensor holder. It can be attached to any sensor box and its detachment is then controlled by the motor axle. In this way, it is possible to release the sensor node in an automated fashion controlled by the UAV's onboard computer, when the UAV is hovering over the target location. In addition, the motor can be steered manually by an operator to mount the sensor node to the UAV.

The mechanism was successfully tested in several indoor as well as outdoor scenarios. When it is possible to land the UAV, the sensor can be released on the ground. When landing is not possible, it is necessary to drop the sensor from a certain height. In soft terrains, like high grass, the sensor can be released from around 3 m without special protection. For higher altitudes, we used a parachute

(a)                                                    (b)

**Fig. 3.** In Fig. 3a is shown the motor and the electronics that are used in the releasing mechanism. Figure 3b depicts the piece that can be attached to a sensor node enclosure and screwed in the motor.

attached to the sensor node to reduce the impact velocity. Figure 4 shows the release of a sensor node from approximately 20 m over ground with a parachute. Nevertheless, the use of a parachute is not always possible as there are environmental factor that cannot be controlled. With strong winds, the parachute can bring the sensor off the target location. Also, if there are obstacles like trees or buildings, the sensor node can potentially collide with them.

Those experiments also showed that the flight controller was able to cope with the sudden change in weight of the UAV, whenever a sensor node was released.

### 3.3   Unmanned Aerial Vehicle

The employed UAV is the model Hummingbird manufactured by AscTec. It is a micro UAV that can carry a limited payload of 200 g. It can fly up to 15 m/s and for about 10 min with full payload. To localize itself, the UAV uses a GPS module. In order to run custom algorithms, a Raspberry Pi is attached to the UAV and connected trough a serial port to the flight controller. The Raspberry Pi can then be used to send waypoint commands to the flight controller to reach the target location.

The benefit of using this small UAVs is that the costs of the whole system can be reduced to a minimum. In complex environments it is possible that a UAV is damaged. Cheap (and disposable) UAVs are thus of an advantage in this case. Besides, reducing the costs allows us to use several agents to reduce the data collection time.

### 3.4   Optimal Route Calculation

As we mentioned, the data is collected with the UAV hovering over each sensor node. The sensor node only turns on the radio when the UAV is in close proximity in order to save battery. To know when the UAV is close to the sensor node the arrival time of the UAV is scheduled. The flight time of the UAVs is limited

**Fig. 4.** Sensor with parachute released from the UAV

**Fig. 5.** Routes generated with the Vehicle Routing Problem solver.

because they are battery powered, so the replacement of the battery will influence the UAV path scheduling. Furthermore, replacing the battery is a manual task that introduces more delay in the data acquisition. To reduce the battery replacements and minimize the travel time, an optimal route is calculated. Moreover, some applications requires a certain delay in the data acquisition that has to be considered also in the problem.

Under the above mentioned constraints, the Vehicle Routing Problem (VRP) offers a possible solution. It is an NP-Hard combinatorial problem. In its basic form it is used to design an optimal set of routes to serve a set of customers with a fleet of vehicles. Several variants of the VRP are available in literature. The one that is selected for this approach is the Distance Constrained VRP, which means that the routes are limited in distance. The distance constraint is translated to time using the velocity of the UAV, since the constraint is the battery life of the UAV. The time for data transfer between the sensor node and the UAV also needs to be included in the problem to obtain realistic results. This time is supposed to be the same for all sensor nodes and can be added to the time that the UAV needs to go to the next point.

The VRP can be formulated as an integer linear programming problem as described in [12]. Let $G = (V, A)$ be a complete graph with $V = \{0, ..., n\}$ denoting vertices of the graph and $A$ the arc set. The vertex 0 correspond to the initial point where all vehicles start, also called depot. The vertices with index $i = 1, ..., n$ to the customers or sensor nodes in our case. Figure 5 shows a sample graph, with the blue points the denoting vertices, gray lines representing all possible connection between vertices and the red lines showing the solution to the problem.

The cost associated with each arc $(i, j) \in A$ is represented as $c_{ij}$ and denotes the cost of going from node $i$ to node $j$. In the considered problem, the cost is the time to go from one point to another. The UAV velocity is supposed constant. Then, by assuming a known the Euclidean distance between points, it becomes possible to calculate the required time. The transmission time of the node $i$ is then added to the cost. The resulting optimization problem is then shown in Eqs. (1)–(7).

$$min \sum_{i \in V} \sum_{j \in V} c_{ij} x_{ij} \tag{1}$$

subject to

$$\sum_{i \in V} x_{ij} = 1 \quad \forall j \in V \setminus \{0\}, \tag{2}$$

$$\sum_{i \in V} x_{ij} = 1 \quad \forall i \in V \setminus \{0\}, \tag{3}$$

$$\sum_{i \in V} x_{i0} = K, \tag{4}$$

$$\sum_{j \in V} x_{0j} = K, \tag{5}$$

$$\sum_{i \in S} \sum_{\substack{j > i \\ j \in S}} x_{ij} \leq |S| - r(s) \quad \forall S \subseteq V \setminus \{0\}, S \neq \emptyset \tag{6}$$

$$x_{ij} \in \{0, 1\} \quad \forall i, j \in V \tag{7}$$

The binary variable $x_{ij}$ is equal to 1 if arc $(i, j) \in A$ belongs to the optimal solutions and 0 otherwise. In Fig. 5 can be observed that when $x_{ij}$ is equal to 1, the arc is drawn with red and gray otherwise. Equations (2) and (3) establish that only one arc enters and leaves each vertex. Equations (4) and (5) constraints that $K$ vehicles will leave and enter the depot. The Eq. (6) describes the vehicle distance constraint and the connectivity of the solution. In other words, no route can be created that is not connected to the depot. Function r(S) is the minimum number of vehicles needed to serve a subset of customers S. It is this equation where the battery duration is introduced. To obtain the value of this function, a multi-agent Traveling Salesman Problem (m-TSP) is solved with heuristics. Note that the number of constrains defined in (6) increases exponentially with the number of customers.

Based on (1)–(7) formulation, it is possible to solve the problem optimally with several methods, including Branch and Bound, Cutting Planes and Branch and Cut. For further readings on exact solvers refer to [12]. For the purpose of this paper, a near optimal solution is sufficient. Heuristic methods can provide good solutions in reasonable time [12]. In [7] several approximate algorithms are shown to solve the problem. We implemented an algorithm based on a set of libraries called OR-tools developed by Google[1]. These libraries allow solving several Operation Research problem, including the VRP. We use a heuristic solver, which implements the most widely used algorithms in literature. The configuration is as follows. The solution is found in two steps, first step is to create a first solution, and the second, is to improve the initial solution. These methods are called in literature two-steps heuristics. The selected heuristic for the initial solution is called cheapest arc. The improvements heuristics are selected automatically by the solver.

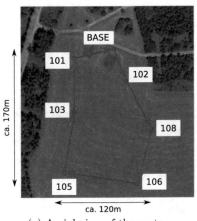

(a) Aerial view of the route.

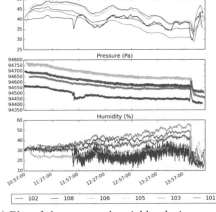

(b) Plot of the measured variables during one experiment.

**Fig. 6.** Grid plots for a specified time.

## 4    System Validation

In order to validate our system we carried out an experiment in an open field scenario. The purpose of this experiment was to prove that our system works as expected. The actual measured data by the BME280 sensor were not of interest and out of scope for this paper. Therefore, we present only uncalibrated raw data measurements.

Six sensor nodes were distributed over an area roughly 170 m × 120 m. The locations were chosen in such a way that they could be covered by one route of the UAV without changing the battery. The sensor nodes' locations together with the node ids are shown in Fig. 6a.

The sensor node measured temperature, pressure, and humidity with a frequency of 1 Hz. The experiment lasted more than 3 h during which the sensors were measuring data. The UAV collected the information every 20 min. During

---

[1] Google Optimization Tools, February 2018, https://developers.google.com/optimization/.

this time, some rounds were skipped in order to test the reconnection capabilities of the system.

Figure 6b visualizes the data that were obtained during this example experiment. As it can be observed, the data was successfully collected from the sensors.

In a second experiment lasting 4 h, the power consumption of the sensor nodes was investigated. The drawn energy was 33 mAh, which makes an average consumption of 8.25 mA at 3.7 V. Accordingly, with the battery capacity of 1000 mAh, a sensor node could be operated for approximately 4 days in this configuration. At this point we like to note that the configuration, i.e. the measurement frequency, period for data collection by the UAV, size of data etc., highly depends on a particular application. The configuration in the experiments was chosen as realistic as possible, while keeping the effort and time to carry out the experiments appropriate.

## 5   Conclusions

In this paper it has been proved that the use of UAVs as mobile sink nodes can effectively reduce the power consumption of the sensor nodes in WSNs. The route calculation algorithm offers good quality solutions in a reasonable time using heuristic methods to solve the VRP. The synchronization between the arrival of the UAV and the wake up of the transceiver was successful in all experiments. Besides, the distribution mechanism can provide a fast deployment of sensor nodes, for example in emergency situations.

## References

1. Abdelhakim, M., Liang, Y., Li, T.: Mobile access coordinated wireless sensor networks - design and analysis. IEEE Trans. Signal Inf. Process. Netw. 3(1), 172–186 (2017)
2. Akyildiz, I.F., Su, W., Sankarasubramaniam, Y., Cayirci, E.: A survey on sensor networks. IEEE Commun. Mag. 40(8), 102–105 (2002)
3. Anagnostopoulos, T., Zaslavsky, A., Kolomvatsos, K., Medvedev, A., Amirian, P., Morley, J., Hadjieftymiades, S.: Challenges and opportunities of waste management in IoT-enabled smart cities: a survey. IEEE Trans. Sustain. Comput. 2(3), 275–289 (2017)
4. Cao-hoang, T., Duy, C.N.: Environment monitoring system for agricultural application based on wireless sensor network. In: 2017 Seventh International Conference on Information Science and Technology (ICIST), pp. 99–102, April 2017
5. Chen, C., Jun-ming, X., Hui-fang, G.: Polluted water monitoring based on wireless sensor. In: 2011 International Conference on Electronics, Communications and Control (ICECC), pp. 961–963, September 2011
6. Gea, T., Paradells, J., Lamarca, M., Roldn, D.: Smart cities as an application of internet of things: experiences and lessons learnt in Barcelona. In: 2013 Seventh International Conference on Innovative Mobile and Internet Services in Ubiquitous Computing, pp. 552–557, July 2013
7. Lin, C., Choy, K., Ho, G., Chung, S., Lam, H.: Survey of green vehicle routing problem: past and future trends. Expert Syst. Appl. 41(4), 1118–1138 (2014)

8. Mergen, G., Zhao, Q., Tong, L.: Sensor networks with mobile access: energy and capacity considerations. IEEE Trans. Commun. **54**(11), 2033–2044 (2006)
9. Ojha, T., Misra, S., Raghuwanshi, N.S.: Wireless sensor networks for agriculture: the state-of-the-art in practice and future challenges. Comput. Electron. Agric. **118**(Suppl C), 66–84 (2015). http://www.sciencedirect.com/science/article/pii/S0168169915002379
10. Polo, J., Hornero, G., Duijneveld, C., Garca, A., Casas, O.: Design of a low-cost wireless sensor network with UAV mobile node for agricultural applications. Comput. Electron. Agric. **119**(Suppl C), 19–32 (2015). http://www.sciencedirect.com/science/article/pii/S0168169915002999
11. Tong, L., Zhao, Q., Adireddy, S.: Sensor networks with mobile agents. In: IEEE Military Communications Conference 2003, MILCOM 2003, vol. 1, pp. 688–693, October 2003
12. Toth, P., Vigo, D.: The Vehicle Routing Problem. SIAM, Philadelphia (2002)

# A Multi-layer Autonomous Vehicle and Simulation Validation Ecosystem Axis: ZalaZONE

Zsolt Szalay[1(✉)], Zoltán Hamar[2], and Peter Simon[2]

[1] Department of Automotive Technologies, Faculty of Transportation Engineering and Vehicle Engineering, Budapest University of Technology and Economics, 6 Stoczek Street, Budapest 1111, Hungary
zsolt.szalay@gjt.bme.hu
[2] Automotive Proving Ground Zala, P.O.B. 91, Zalaegerszeg 8900, Hungary
{zoltan.hamar,peter.simon}@apz.hu

**Abstract.** Developing autonomous driving technologies is complex and needs to be approached from multiple directions. However, these vectors of development need to intersect at one point and our answer is ZalaZONE. This proving ground design is unique when it comes to the development of automated driving technologies and has grass-root origins in higher education and industry. However, the proving ground is not the total solution, but acts as a catalyst and synergistic hub for the different directional vectors addressing autonomous vehicle development. ZalaZONE in its role as an education and research centre from the secondary to the highest trinary educations levels supports this basic progressive vector and creates the first layer. A second layer concerns the proving ground's core function, being used not only for classical dynamic testing, but focusing on the evaluation of the most troublesome "use cases" that simulation indicates. The evaluation of these indicated "use cases" will required autonomous vehicles to be tested at their dynamic limits in a 100% controlled and safe environment. Similarly, the proving ground is designed to calibrate and validate current and future simulation systems, also a significant technological metamorphosis. Public road testing is a third layer or point of axis supported by the PG. Legislative changes in Hungary offer a flexible regulatory environment within Europe for autonomous vehicle testing and this allows the PG to be a one stop hub for launching test in various environments, from live urban to the usage of a tri-national testbed road network in the Hungarian, Slovenian, Austrian enclave. Fourth and fifth layers nurtured by ZalaZONE are "sample vehicles fleet services" and "crowd-sourced traffic cloud" data collection and analysis. This paper will discuss all these 5 complex and integrated axis points and how ZalaZONE provides this synergistic role in this new Multi-layer autonomous vehicle and simulation validation Ecosystem.

**Keywords:** Autonomous vehicle
Connected and automated vehicle technology · Simulation
Testing and validation · Homologation · Proving ground
Public road testing · Education and research · Crowd sourced traffic cloud

© Springer Nature Switzerland AG 2019
M. Strand et al. (Eds.): IAS 2018, AISC 867, pp. 954–963, 2019.
https://doi.org/10.1007/978-3-030-01370-7_74

# 1   Introduction

The connected and automated vehicle (CAV) will no longer be a separate entity like the classical vehicle but be embedded into the traffic and surrounding transportation ecosystem causing a very real challenge for the automotive industry. This phenomenon is especially seen above Level-3 CAVs when the human driver is getting out of the control loop and replaced by complex applied algorithms where traditional vehicle validation will not work. AD technological changes require in themselves a disruptive validation approach regarding testing and validation solutions because classical vehicle dynamics testing will simply be not enough (Li et al. 2016; Szalay 2016). The resulting complex control design's function set must be the focus of new testing and validation processes even at its component level. For example, sensors, actuators, and AI algorithms all need their own test methods and respective test environments where the vehicle's perception capabilities need to be addressed. Such perception capabilities include the interaction to other vehicles, other road users and the surrounding infrastructure and do not exclude communication technologies of both the cellular 5G and the dedicated short range V2X based ITS G5 spectrum (Gear 2030).

The highly realistic and reproducible testing of autonomous driving is one of the biggest development difficulties for CAV validation. There are many approaches (Zlocki 2016; Nyerges and Szalay 2017) for the investigation of testing functions but these have several deficiencies that makes these methods unsuitable to cover all the important levels of perception or actuation. The earliest approaches where based on system development and the related V-model (Passchier et al. 2015). Potentially, the multiple levels of simulation, laboratory, proving ground, limited public road tests and public road tests support solutions for CAV development and technical validation. ZalaZONE, the Hungarian proving ground, supports the multiple levels of CAV development and validation by allowing the most complex test environments to be realised. The proving ground has incorporated into its operation the RECAR program's (REsearch Center for Autonomous Road vehicles) systematic approach to meet the new challenges of CAV testing and validation (Szalay et al. 2017). The RECAR approach is illustrated below in a pyramid format and shows the multiple level CAV development stages with the respective related test and validation types (Fig. 1).

Proving grounds often lack the dedicated facilities to put CAVs truly to the test, but ZalaZONE presents a unique proposition by integrating classic vehicle dynamics testing with complex CAV requirements. This independent testing facility with grassroots origins derived by its co-operation with education and industry, initiated the proving ground project and defined its basic specifications. This cooperation did not only include automotive engineers but also those from the ICT sector who joined the project to both support autonomous and connected strategies (Szalay et al. 2017).

The proving ground design supports the validation testing and homologation services for all passenger cars, urban people movers, special fleet vehicles, buses and other commercial vehicles, including EV models up to 40 tons and 26 m long. ZalaZONE will support the R&D divisions of OEMs and Tier 1 suppliers and also companies from the telecommunication technologies sector, but additional services are planned to coach

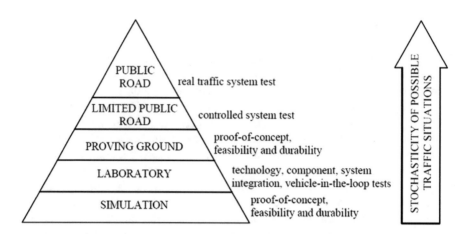

**Fig. 1.** Autonomous vehicle testing & validation pyramid (Szalay 2016)

start-ups and SME's from both sectors in the creation of a Multi-layer Autonomous Vehicle and Simulation Validation Ecosystem.

## 2   Education and Research

ZalaZONE has incorporate higher education into it organization and business models by accommodating R&D project work from partner universities in and outside Hungary. The scope of higher education also includes the integration of RECAR, REsearch Center for Autonomous Road vehicles, (http://recar.bme.hu/eng/) established by BME, ELTE, and MTA SZTAKI Hungarian high education institutions and promotes multidisciplinary cooperation among the academic sphere and industrial partners (Bosch, Knorr-Bremse, Continental). Together with ZalaZONE they support the dual education programs of dedicated BSc/BEng and MSc courses, with special focus on basic and advanced research in artificial intelligence, co-operative control, cyber security and driverless technologies. The specific programs are, an Autonomous Vehicle Control Engineer MSc (in English), a Computer Science for Autonomous Systems MSc (in English), and a Vehicle Test Engineer BEng. This education cooperation has led to the development of a dedicated university research centre adjacent to ZalaZONE allowing it to be a sandbox for university-based R&D&I with the goal to map and exploit the academic and industrial synergies in order to achieve a higher-level and broader span of research and resulting cost-effective technical solutions as well as train highly-qualified engineering professionals. It is expected that this locally developed know-how will create a critical mass of future engineers at ZalaZONE and in the surrounding region who are well-trained experts on the multiple specific research and technical domains concerning CAV testing and validation.

## 3 Proving Ground

ZalaZONE is an autonomous, connected and electric vehicle testing environment and includes a dynamic platform, a special surface braking module, a high-speed handling course, a network of connecting roads of real world road quality, an urban or smart city environment created for the testing and validation of CAVs. Noting that CAV testing is not just the testing of the vehicle dynamics but also of the info-communication technology it encompasses (Szalay et al. 2018), ZalaZONE is not just an automotive but also an ICT technology testing and demonstration centre that will allow for the testing of new disruptive IT based technologies that appear in current and future transportation ecosystems. Figure 2 illustrates and outlines the complete proving ground concept with all its functional modules, together allowing for the greatest variation of testing environments critical for the testing and validation of CAVs.

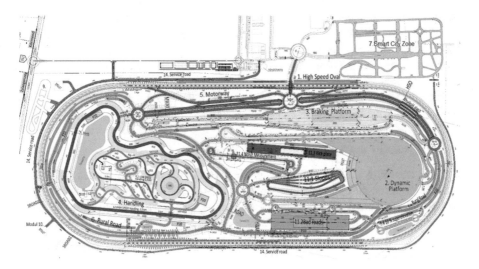

**Fig. 2.** The overview of the ZalaZONE automotive and ICT proving ground

### 3.1 Dynamic Platform

This 300 m diameter ultra-flat asphalt surface has two acceleration lanes with lengths of 760 m and 400 m and is surrounded by a 20 m wide FIA emergency area. The surface can be watered to reach lower $\mu$ with the eastern acceleration lane being a special watered basalt surface. The basalt foundation of over 80 cm will allow for 40-ton vehicles. This module suitable for special autonomous vehicle test cases like platooning at free trajectory, cooperative vehicle control at high and medium $\mu$ with different trajectories at stability limits, as well as fix position obstacles or Euro NCAP scenario testing.

## 3.2   Braking Platform

The braking platform has 8 different surfaces that supported autonomous vehicle test cases include platooning at physical limits; drive through or braking on various surfaces at high speed, cooperative vehicle control at physical limits, and moving or static obstacle avoidance at various speeds during ABS, ATC, ESP activity.

## 3.3   Smart City Zone

CAV testing has been mostly considered as only and urban test[bed] activity up to now (Zhao and Peng 2017). Although the design of the proving ground allows autonomous driving tests on all the classic test modules, ZalaZONE will also include a dedicated urban smart city zone complex currently planned for 16 hectares (Szalay et al. 2017). Consisting of five urban topographies, the CAV development zone is completely compatible for controllable real-world traffic situation simulation, critical for CAV testing and development. The smart city zone includes a parking area and a commercial vehicle logistics centre, a downtown city centre with building facades up to 10 m high, a high-speed four-lane urban zone with a low-μ section, a residential area with elevation changes and building façades, finally a special three-way intersection area. It should be noted "T" intersections are problematic for both CAV sensors as well as AI testing and training. Other infrastructure located within the urban smart city zone includes a rain tunnel, a GNSS shadow tunnel and streetcar tracks.

## 3.4   High and Low Speed Handling Courses

Both courses have various changing terrain topography and V2X coverage for communication tests. Supported autonomous vehicle test cases include platooning at medium speeds in diverse topography terrain and cooperative vehicle control testing in diverse topography with limited visibility.

## 3.5   Motorway Section

The motorway test pilot function is a directions of CAV development based on ADAS. ZalaZONE has planned a motorway module to safely and realistically examining motorway exit and entrance scenarios, high-speed lane change assist, emergency braking, platooning situations, and others. As a motorway section is not common on any proving grounds, the ZalaZONE motorway with its 100 m long real tunnel provides a unique option for that.

## 3.6   Rural Roads

Most contemporary proving grounds only have simple cost effective connecting road between their classical dynamic modules. However, with CAV testing there is a critical need to have the ability of continuously test the vehicles as they move from one track element to another and on such a road surface that represents a real world (EU standard) driving environment.

### 3.7    High-Speed Oval

The high-speed oval supports autonomous vehicle test cases include high speed platooning at high speed motorway situations, high speed cooperative vehicle control,
usage of fix position and moving obstacles, and V2I/V2V communication tests at high
vehicle speeds.

### 3.8    Communication Network

ZalaZONE is covered by both cellular 5G and V2X based ITS G5 communication
networks. The ITS G5 network has three levels, an ITS G5 basic V2X test environment, a V2X developer environment, and a third R&D level. The 5G cellular test
network is available for future ITS applications. Both networks have a redundant layout
for parallel customer networks.

### 3.9    Control Concept

ZalaZONE's special Scenario-in-the Loop concept is an advance control loop where
the complete self-driving functional chain including perception is engaged during the
testing. The elements and the operating principle of SciL is depicted in Fig. 3.

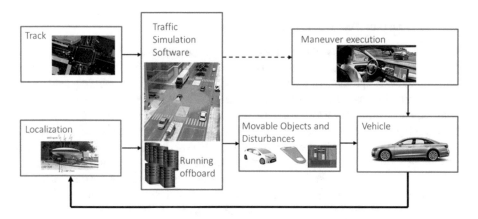

**Fig. 3.**  The scenario-in-the loop concept of ZalaZONE

The entire process is controlled by traffic simulation software which runs separately
from the vehicle, offboard in the control tower. The software has two main inputs, the
precise description of the real test track in a 3D software environment, and the
localization data from the VUT (Vehicle Under Test). With this information the traffic
simulation software can calculate the trajectory and timing of the disturbance targets
such as a VRU or a GVT. The autonomous vehicle can interact with these targets
without bypassing its perception layer. During testing the localization data of the VUT
is acquired with a high sample rate and sent to the traffic simulation software with low
latency, allowing the software to recalculate the scenario in real-time.

# 4 Public Road and Urban Environment

Safe and controlled proving ground testing is a 100% requirement, but CAVs also need to be analysed in a public road environment before and during their homologation process. The ability to first test on a PG and then quickly move the testing to a public road offers many benefits not yet realised in other testing environments.

Within the development of an attractive vehicle testing environment the modification of the Hungarian legislation on April 12, 2017 was an important milestone. Specifically, modifications of the KöHÉM decree 5/1990 (IV.12.) on the technical examination of road vehicles, 6/1990 (IV.12) on technical conditions of road vehicles for entering the traffic, and the NFM decree 11/2017 (IV.12) on testing of vehicles for development purposes allowed for the testing of vehicles with automated functions on Hungary's public road infrastructure without territorial or timing limitations. The new law laid out in simple black and white the steps required for the "Vehicle Developer Company", the "Vehicle Under Test" and the "Personnel (Test Driver)" for maximum safety with the goal of good research. In the process of developing the new Hungarian law, the self-certification method was adopted an in so giving all responsibility for testing on the shoulders of the highest-ranking officer at the vehicle developer's firm. At ZalaZONE this public road testing application process is supported and offered in our basic services.

## 4.1 Real Urban Test Environment

After test vehicles are proven to be safe in the controlled testing environment (Zala-ZONE), they can go out into the city of Zalaegerszeg limited public road testing environment. This pertains to both passenger and commercial vehicles. This network of roads within the city of Zalaegerszeg in many ways replicates the urban smart city environment within ZalaZONE but in a less controlled form. This city section of Zalaegerszeg is called our external smart city and includes the set-up of sensors and other equipment on the live urban roads to the most innovated ITS standards and beyond. Additionally, modified regulations are being considered as to the time of testing during the day to exclude vulnerable traffic participants: pedestrians, cyclists, motorcycles, etc. and increase safety. This smart city area will also incorporate the testing of connectivity to the cellular and V2X networks and make use of features that are not in the PG's urban area such as additional bridges, tunnels, and railway crossings. The planned limited public road testing area in Zalaegerszeg is adjacent to ZalaZONE providing an added service to the customers when requested for key testing and validation levels to be examined.

## 4.2 National and Cross-Border Test Loops

Without a doubt CAV testing must also be extended to public roads since the full extent of traffic scenarios or use cases cannot be obtained without them. However, vehicle evaluation must take place before granting such testing from a safety point of view. Such evaluation should include legal, simulation, controlled proving ground, and limited public road analysis. Below Fig. 4 show the promoted routes we offer. M1

(Budapest-Győr) - M85, M86, M9 (Győr-Zalaegerszeg) - R76, M7 (Zalaegerszeg-Budapest). Currently the Budapest Győr section of the M1 highway is already equipped with ITS-G5 road side units (RSU) to support connected car function testing as well.

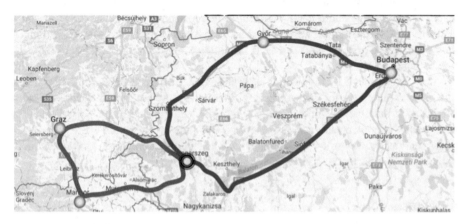

**Fig. 4.** The proposed national and international test loops on public roads (Szalay 2016)

At the international level, to support the automotive industry testing needs, it is not just a legal issue but also a question of infrastructure that must be addressed. As mentioned, since legal public road testing have been allowed as of April 2017, Hungary is offering one of the most flexible regulatory environments in Europe when it comes to CAV testing. This will allow development teams to continue their testing on the public roads around the proving ground and within the city of Zalaegerszeg. Highlighted public roads outside the city of Zalaegerszeg as well as rural roads and modern four-lane highways are available for public road testing. Additionally, just across the local borders in adjacent areas of Slovenia and Austria exist available public roads for AD vehicles, thus incorporating ZalaZONE into one of the only European proving ground focusing on CAV testing that is integrated into a tri-national public road testing environment. Within this approach ZalaZONE is fully complementary to the Graz centred Austrian initiative (Gigler 2017) called ALP.LAB (Austrian Light Vehicle Proving Region for Automated Driving) where each reinforces the others' activities inside the Graz-Maribor-Zalaegerszeg triangle.

The planned M76 expressway (2020–2022) connecting Zalaegerszeg to the M7 motorway will be a smart road in an extent that is technically possible by the newest state-of-the-art technologies and in so be part to the ZalaZONE test environment. Here high-speed testing will be allowed by the systematically closing of half the highway in section up to 10 km long.

## 5    Sampling Vehicles with Crowdsourced Traffic Cloud

ZalaZONE and its partners will be incorporating multiple L1 and L2 vehicles equipped with sensors for data collection and aggregation purposes. The developed data aggregation system will allow the perception of traffic environments with relative high dynamics and act as a basis for HD map creation (Barsi et al. 2017) with static, semi-static, semi-dynamic and dynamic content to be represented. This type R&D stream is critical for CAVs to fulfil their potential technical and societal purposes at the highest levels and also a goal for ZalaZONE in providing a multi-layered ecosystem for CAV deployment.

The data gathered by the fleet of sampling vehicles are transmitted to traffic cloud, that is also prepared for data reception from other crowdsourced vehicle sensors. Providing a suitable IT infrastructure for traffic and transportation Cloud and Fog Computing, extended with an HD Map with dynamic layers would serve as an additional unique framework for the validation of highly automated and autonomous vehicles on public roads in Hungary.

## 6    Conclusion

In Hungary a unique automotive proving ground is being built called ZalaZONE. With its very strong resources from conventional vehicle testing in combination with its in-depth consideration of CAV testing in its design, the PG provides a safe and controlled site to test CAV at their critical dynamic limits, something that will never be an option on public roads. This testing niche is the key and central strength that ZalaZONE provides to CAV development. It is in this role that the PG serves as the heart of the CAV ecosystem where it is complemented by devoted higher education, university-based research, as well as in-city, highway and motorway public road testing on both a national level and extended to an international trilateral cross-border testing environment. With these focal or axis points and their synergies, ZalaZONE formulates an exceptionally complex multi-layered autonomous vehicle and simulation validation ecosystem on a global scale.

## References

Waldrop, M.M.: Autonomous vehicles: no drivers required. Nature **518**(7537), 20–23 (2015). https://doi.org/10.1038/518020a

Li, L., Huang, W-L., Liu, Y., Zheng, N-N., Wang, F-Y.: Intelligence testing for autonomous vehicles: a new approach. IEEE Trans. Intell. Veh., 2379–8858 (2016)

Szalay, Z.: Structure and architecture problems of autonomous road vehicle testing and validation. In: Proceedings of the 15th Mini Conference on Vehicle System Dynamics, Identification and Anomalies: VSDIA 2016, pp. 229–236 (2016). ISBN:978-963-313-266-1

GEAR2030 High Level Group on the Competitiveness and Sustainable Growth of the Automotive Industry in the European Union FINAL REPORT – 2017, DG GROW – European Union (2017). http://www.europarl.europa.eu/cmsdata/141562/GEAR%202030%-20Final%20Report.pdf. Accessed 5 May 2018

Zlocki, A., Fahrenkrog, F., Eckstein, L.: Holistic approach for design and evaluation of automated driving. In: Autonomous Vehicle Test & Development Symposium 2016, 2nd June 2016, Stuttgart, Germany (2016)

Nyerges, A., Szalay, Z.: A new approach for the testing and validation of connected and automated vehicles. In: 34th International Colloquium on Advanced Manufacturing and Repairing Technologies in Vehicle Industry, 17–19 May 2017, Visegrád, Hungary (2017)

Passchier, I., Vugt, G., Tideman, M.: An integral approach to autonomous and cooperative vehicles development and testing. In: 2015 IEEE 18th International Conference on Intelligent Transportation Systems (2015). https://doi.org/10.1109/itsc.2015.66; 978-1-4673-6596-3/15

Szalay, Z., Esztergár-Kiss, D., Tettamanti, T., Gáspár, P., Varga, I.: RECAR: hungarian research centre for autonomous road vehicles is on the way. ERCIM News (109), 27–29 (2017)

Bansal, P., Kockelman, K.M.: Forecasting Americans' long-term adoption of connected and autonomous vehicle technologies. Transp. Res. Part A: Policy Pract. 95, 49–63 (2017). https://doi.org/10.1016/j.tra.2016.10.013

Szalay, Z., Nyerges, A., Hamar, Z., Hesz, M.: Technical specification methodology for an automotive proving ground dedicated to connected and automated vehicles. Periodica Polytech. Transp. Eng. 45(3), 168–174 (2017)

Szalay, Z., Tettamanti, T., Esztergár-Kiss, D., Varga, I., Bartolini, C.: Development of a test track for driverless cars: vehicle design, track configuration, and liability considerations. Periodica Polytech. Transp. Eng. 46(1), 29–35 (2018)

Zhao, D., Peng, H.: From the lab to the street: solving the challenge of accelerating automated vehicle testing. University of Michigan. https://mcity.umich.edu/wp-content/uploads/2017/05/Mcity-White-Paper_Accelerated-AV-Testing.pdf. Accessed 5 May 2018

Department for Transport "The pathway to driverless cars: a code of practice for testing", Great Mister House, 33 Horseferry Road, London SW1P4DR, July 2015. https://assets.publishing.service.gov.uk/government/uploads/system/uploads/attachment_data/file/446316/pathway-driverless-cars.pdf. Accessed 5 May 2018

Hungarian legislation, NFM decree 11/2017. (IV.12) on the modification of the 5/1990 and 6/1990 (IV. 12.) KöHÉM Regulation. https://net.jogtar.hu/jogszabaly?docid=A1700011.NFM&timeshift=fffffff4&txtreferer=00000001.TXT. Accessed 5 May 2018

Hungarian legislation, Regulation 5/1990 (applicable on the 22th August 2017). https://net.jogtar.hu/jr/gen/hjegy_doc.cgi?docid=99000005.koh. Accessed 5 May 2018

Hungarian legislation, Regulation 6/1990 (applicable on the 22th August 2017). https://net.jogtar.hu/jr/gen/hjegy_doc.cgi?docid=99000006.koh. Accessed 5 May 2018

Gigler, B.: Self-driving cars: Europe's most diverse test environment under development in Styria. https://www.tugraz.at/en/tu-graz/services/news-stories/tu-graz-news/singleview/article/selbstfahrende-autos-europas-vielfaeltigste-testumgebung-entsteht-in-der-steiermark/. Accessed 5 May 2018

Barsi, A., Poto, V., Somogyi, A., Lovas, T., Tihanyi, V., Szalay, Z.: Supporting autonomous vehicles by creating HD maps. Production Engineering Archives 16, 43–46 (2017). ISSN 2353-5156

# Towards Large Scale Urban Traffic Reference Data: Smart Infrastructure in the Test Area Autonomous Driving Baden-Württemberg

Tobias Fleck[1], Karam Daaboul[1], Michael Weber[1], Philip Schörner[1], Marek Wehmer[1], Jens Doll[1], Stefan Orf[1], Nico Sußmann[2], Christian Hubschneider[1(✉)], Marc René Zofka[1], Florian Kuhnt[1], Ralf Kohlhaas[1], Ingmar Baumgart[1], Raoul Zöllner[2], and J. Marius Zöllner[1]

[1] FZI Research Center for Information Technology, Karlsruhe, Germany
hubschneider@fzi.de

[2] Heilbronn University of Applied Sciences, Heilbronn, Germany

**Abstract.** This paper presents the concept, realization and evaluation of a flexible and scalable setup for smart infrastructure at the example of the Test Area Autonomous Driving Baden-Württemberg.

In verification and validation of autonomous driving systems, there exists a gap between virtual validation and real road tests: Simulation provides an easy and efficient way to assess a system's performance under a variety of environmental constraints, but is restricted to model assumptions and scenarios, which might ignore important aspects. Whereas expensive real road tests promise an unexpected environment for statistical evaluation of traffic scenarios, but lack of observability. Our setup for smart infrastructure is supposed to close the gap by tackling this issue by observing and providing reference data of traffic scenarios for application in different testing and evaluation settings.

We present the approach of implementing a distributed intelligent infrastructure capable of handling traffic light states, road topology and especially information about locally observed traffic participants. The data is provided online via Vehicle-to-X (V2X) communication for live testing and sensor range extension as well as offline via a backend for high-precision analysis and application of machine learning techniques. To obtain information about traffic participants, a camera based object tracking was realised. To cope with the high amount of information to be transmitted via V2X and to use the available bandwidth optimally, the standard for broadcasting vehicle information is modified by applying a form of data compression through prioritization.

The setup is initially evaluated at a large intersection in Karlsruhe, Germany.

© Springer Nature Switzerland AG 2019
M. Strand et al. (Eds.): IAS 2018, AISC 867, pp. 964–982, 2019.
https://doi.org/10.1007/978-3-030-01370-7_75

# 1   Introduction

In order to pave the way for autonomous vehicles in different application fields such as individual or public transport and logistics, public test beds equipped with intelligent infrastructure are necessary to provide reference data and assist in test cases. The observed traffic situations in terms of realistic pedestrian or vehicle behavior, weather conditions, etc. then supports the development phase under different aspects. These test beds with intelligent infrastructure are supposed to close the gap between virtual and real testing approaches for the release of autonomous driving systems.

In the *Testfeld Autonomes Fahren Baden-Württemberg*[1] (German for *Test Area Autonomous Driving Baden-Württemberg*, short: *Test Area*) selected streets of different type and complexity are equipped with intelligent infrastructure. This infrastructure is used to observe, process and communicate the present traffic situation as Fig. 1 shows. This enables to support the development and testing process of autonomous vehicles in a threefold manner:

1. The generated environment model is applicable for **online and offline evaluation** of the automated vehicle's environment model, typically captured from a restricted system's sensor view.
2. The observed traffic situations can be used as data pool for **algorithm development**, e.g to cluster and identify situation aspects for subsequent evaluation and validation techniques, such as scenario mining, improving models and test runs in simulation or as training data for machine learning.
3. The smart infrastructure is able to **extend the sensor range of automated vehicles** by supporting them with additional online information about the current traffic state, e.g. providing object lists or traffic light states. With the help of this data, sensor occlusion of participating vehicles can be compensated or at least mitigated.

With these applications in mind, we propose a concept for cost-efficient large scale intelligent infrastructure using decentralized local road units with low cost sensors and a central backend. The approach is used in the realization of the *Test Area Autonomous Driving Baden-Württemberg* within the cities of Karlsruhe, Heilbronn and Bruchsal. An early evaluation is given using a complex intersection in Karlsruhe.

The remainder of this paper is structured as follows: In Sect. 2 we give a brief overview over related work, considering existing test areas and smart intersections. Section 3 covers the overall concept, followed by a detailed explanation of the different algorithmic components (Sect. 4). We present a preliminary evaluation of our concept in Sect. 5 and conclude in Sect. 6 with open and ensuing research questions.

---

[1] For more information about the *Test Area Autonomous Driving Baden-Württemberg* see https://taf-bw.de/en.

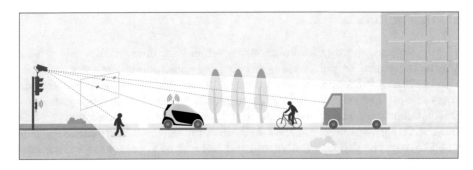

**Fig. 1.** Simplified overview of the smart infrastructure: Traffic participants are detected and tracked over space and time. The information is processed as anonymous, individual objects and provided online via V2X communication and stored offline for further processing. Thereby, the provided information enables the performance evaluation of automated vehicles (blue) or can be sent via V2X as additional sensor input to enhance their functionality.

## 2    Related Work

Several field tests were conducted within Europe focusing on testing *Cooperative Intelligent Transport System* (C-ITS) use cases and deployments on public roads between 2014 and 2017 [11]. The term C-ITS refers to connected and automated mobility, with a strong emphasis on interacting vehicles and road infrastructure. The project *PRE-DRIVE C2X* [29] between 2008 and 2010 aimed at the establishment of a common European architecture for general V2X communication for future field operational tests (FOTs). It was continued between 2011 and 2014 with the project *DRIVE C2X* for the assessment of the technology in concrete cooperative driving functions within the field operational tests. The simTD [16] project in Germany evaluated C-ITS use cases and applications on public roads in different scenarios, along with a general architecture for C-ITS services. The SCOOP@F project evaluates the use of V2X equipment on a large scale on public roads in France [5]. The European Union finances the C-ROADS platform with the goal of coordinating the C-ITS rollout across the EU and developing specifications to ensure interoperability. A wider C-ITS deployment in general is envisioned to begin in 2019 [27]. The communication infrastructure used in this work conforms to standardized C-ITS-Protocols as much as possible with the goal of integrating well with a future large scale C-ITS rollout strategy.

Our approach differentiates from structured and self-contained proving grounds and dedicated test tracks, such as *MCity Test Facility* [3] or the closed-cource *Toyota Research Institute Automated Vehicle Test Facility* [4], in which only controlled experiments are conducted. Test in public traffic are desirable in order to evaluate a vehicle under unforeseen phenomena, such as non-modelable traffic participant behavior. At the same time observations by reference systems are needed.

In former publications mainly standalone data from intersections has been labeled manually in order to develop and evaluate algorithms for recognition tasks, such as the data provided by [18] for the development of tracking algorithms [19]. But mostly common datasets on the base of abstracted environment data are missing, which enable deriving elaborate behavior models and scenarios for further analysis or application in simulation. An initial attempt has been made in the German joint research project Ko-PER, where an intersection in Aschaffenburg has been equipped with multiple camera and lidar sensors. A part of the published dataset [28] was annotated by hand, in order to provide a set of object labels. The extensive use of multiple high precision sensors is promising for gaining high precision data. Though, for the use case of a persistent and widely spread deployment on different intersections and places, the scalability is limited by the high expenses of such a setup.

Besides the mentioned efforts, in Germany there are, as of today 15 ongoing initiatives to provide so called digital test beds on motorways and in cities but also covering cross-border test areas [2]. In Austria, the alliance *ALP.Lab GmbH* [1] by automotive supplier companies, AVL and Magna, and scientific partners, Joanneum Research, TU Graz and Virtual Vehicle, has developed a platform for testing and verifying the components and systems of automated driving in diverse and complex scenarios including facilities for data recording and processing along public roads and privat grounds as well as comprehensive virtual testing environments.

## 3   Concept

Since the Test Area Autonomous Driving Baden-Württemberg focuses on all three applications described in Sect. 1 (sensor extension, online/offline evaluation and algorithm development) several requirements have to be considered to develop an efficient overall system.

Our approach is depicted in Fig. 2: For offline evaluation and longterm algorithm development a backend in a data center is used while local processing is used to deliver online information to the local automated vehicles. The local processing is implemented in local road units that span over areas such as a connecting street or an intersection. The local environment is observed by the local road units using object sensors. To comply with privacy regulations, data with personal information such as image data is anonymized as early as possible and not delivered to the backend nor communicated via V2X. Likewise, the anonymized data is also smaller and easier to deliver to the backend over a long distance. For a large scale deployment we focus on cheap sensors, such as 2D image cameras, but additional sensors such as radar or lidar can be integrated. Besides the observed objects, the local road units additionally have knowledge about the traffic light states and a local map of the road topology. This allows on the one hand to transmit a coherent, complete, processed view of the current local traffic scene to the backend, on the other hand local communication is possible without relying on the backend.

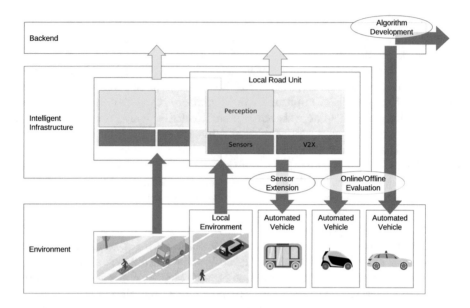

**Fig. 2.** Overview of the concept and applications in the Test Area Autonomous Driving Baden-Württemberg. Local road units perceive the environment and deliver processed data to the backend and local automated vehicles. The three applications sensor extension, online/offline evaluation and algorithm development (using recorded data as training data) are depicted.

Thus, the key components of the approach are the local road units. They can be deployed as permanent installations or mobile solutions for temporary experiments. In this work we focus on permanent installations in urban areas. These are initially deployed in the Test Area Autonomous Driving Baden-Württemberg which will be extended with more infrastructure during the next years.

### 3.1 Local Intelligent Infrastructure

A visualization of the concept of the local road units is shown in Fig. 3. We see the units as small intelligent agents observing and communicating with their environment. Thus, we follow the typical perception-cognition-action approach [25] but with a focus on perception. This allows concepts and interfaces to be reused from automated driving implementations (compare [20]) including the utilization of a simulation framework during development [31].

In order to perceive the traffic situation, multiple optical camera sensors are attached to large poles with overlapping fields of view. The multi-camera setup has to be calibrated to determine the exact intrinsic and extrinsic parameters for retrieving correct geometric information from the images, see Sect. 4.1. The undistorted images are used as input for Convolutional Neural Networks (CNN) to detect and classify vehicles, pedestrians and cyclists in the 2D image space, see Sect. 4.2. For high precision offline processing the output of the single CNN

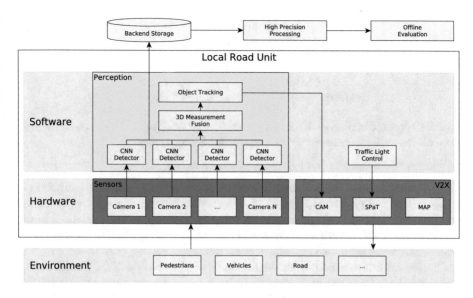

**Fig. 3.** Components inside the local road units including interfaces to the environment and the backend. The local road unit is divided into hardware components as sensors and V2X communication unit and software components for the processing of sensor data and the traffic light control.

detectors can be transmitted to the backend storage. This allows maximum flexibility in processing while not offending against privacy regulations and keeping a low amount of data (compared to saving the actual raw sensor data). In order to obtain 3D objects in world coordinates for online usage, the classification hypotheses are then fused to single 3D measurements (Sect. 4.3) and tracked over time by a Bayesian filter associating a series of measurements with potential vehicles, yielding spatio-temporal tracks for perceived traffic participants (Sect. 4.4). Recognized traffic participants are represented as a list of abstract objects with a predefined set of features. It is transformed to a standards-compliant set of messages and wirelessly broadcasted via Vehicle-to-X communication. This enables users of the Test Area to receive the traffic state in-place and at the same time represents a C-ITS pilot test site for all vehicles. We describe the communication in detail in Sect. 4.5.

## 3.2   Perception Problem Factorization

The overall estimation problem of the perception module can be formulated as estimating the state of all objects $X$ given all sensor measurements with uncertainties $M$ and background knowledge $\theta$ (e.g. intrinsic/extrinsic calibration, learned detection model parameters), see Fig. 4:

$$X \sim P(X|M,\theta)$$

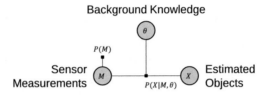

**Fig. 4.** Object estimation problem statement: The set of objects $X$ has to be estimated, given all sensor measurements $M$ and background knowledge $\theta$.

This problem is too complex to be modeled in its entirety. Thus we factorize the problem into subproblems (Fig. 5) that can either be parametrized by expert knowledge or learned using machine learning techniques, as we explained in our previous work [17]. Multi-Sensor Multi-Target Tracking (MTT) is a research topic for decades with various approaches. Recent work [22] tries to formalize all estimation approaches including the big fields of vector-type MTT methods (e.g. Kalman Filter, Interacting Multiple Model Filter) and set-type MTT methods (e.g. PHD Filter). This formalization can be recognized in Fig. 5 with the difference that we include the whole preprocessing chain and use an early sensor fusion.

## 4    Implementation

In the following subsections the implementation details for our processing pipeline are described, ranging from camera calibration to temporal multi-object tracking.

### 4.1    Camera Calibration

In order to register detections of all cameras in a common reference coordinate system, a proper camera calibration is needed. This includes estimating the intrinsics of all cameras respectively and to estimate the extrinsic calibration of all cameras relative to a reference coordinate system all objects and vehicles should be tracked in. The camera calibration is assumed to be static over time and thus is only performed once a-priori to the processing pipeline.

Due to manual focus adjustments all cameras are calibrated intrinsically after mounting. This is commonly done by using calibration patterns with geometric forms that can easily be detected by computer vision algorithms. The intrinsic parameters are calculated by minimizing the reprojection error iteratively with the Levenberg-Marquardt algorithm [30]. Extrinsic calibration can also be achieved by using calibration patterns. This would require a single pattern visible from all cameras or multiple patterns which themselves must be extrinsically calibrated. Both cases are infeasible for a large scale intersection as they either require temporarily disturbing traffic or high-cost equipment. To overcome this problem, we use the a-priori high precision maps of the test area for reference

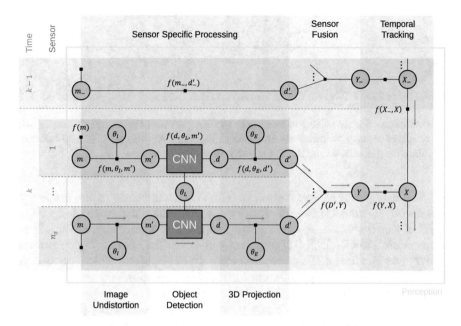

**Fig. 5.** Our approach of factorizing the perception problem $P(X|M,\theta)$ into subproblems: Per timestep the detections $D' = \{d'^1, \ldots, d'^{n_s}\}$ of a sensor specific preprocessing are fused to 3D measurements $Y$. A Bayesian filter tracks object estimates $X$ over time using the Markov assumption. The sensor specific processing consists of image undistortion using the sensor-specific intrinsic parameters, CNN-based 2D object detection and conversion to a common 3D reference coordinate system using the extrinsic parameters and assuming a flat ground plane. The notation is inspired by [15,22].

point localization. Together with the intrinsic parameters the thereby known real-world positions of features in camera images can be used to estimate the extrinsic calibration for each camera. Dominant features like lane markings or poles of traffic signs that are visible in multiple camera images and are recorded in the a-priori map are manually selected to be used in the calibration process (see Fig. 6). The exact position of these features in the reference coordinate system is known through the a-priori map and also in the image coordinate system. The pose of the camera is then iteratively estimated by minimizing the sum of squared distances between image points and projections of the real-world points, again using the Levenberg-Marquardt algorithm. The calibration then allows to calculate the real-world positions of objects in the camera image.

## 4.2   Object Detection

One of the key tasks in the implemented processing pipeline (Fig. 5) is detecting and classifying vehicles in undistorted monocular camera images $m'$. The results $d$ are two-dimensional object surrounding bounding boxes in image coordinates with a probability distribution over predefined object classes. In recent

**Fig. 6.** Manual feature selection in camera images (left and right) and correspondence with a reference map (middle) for extrinsic calibration. Additionally, on the left image an calibration pattern for intrinsic camera calibration is visible.

years, *Convolutional Neural Networks (CNNs)* became state of the art in solving such object detection tasks. There exist two different groups of algorithms for 2D object detection. Proposal based approaches like Region-CNN [13] and its successors Fast R-CNN [12] and Faster R-CNN [24] apply a region proposal algorithm on the image. This proposes candidates of objects which are feed into a classifier in a second step. Different to this pipelined approach, the other group of algorithms directly detects and classifies objects within one single network in a single step. Examples for this group are Overfeat [26], YOLO [23] and SSD. These algorithms usually do classification and object position estimation for each region within an image. The position estimations for objects are typically filtered by confidence values and afterwards clustered or a non-maximum suppression is applied. We approach the detection problem by using an implementation of Mask R-CNN [14], an artificial convolutional neural network which is state-of-the-art in image detection, trained on publicly available datasets ([9,21]). The output $d$ includes a probabilistic estimation of the classification but no uncertainty measures on the position estimates.

### 4.3   Multi Sensor Association and Fusion

Given detected classified bounding boxes $d$ in image coordinates from the object detection module, an association and fusion module generates appropriate position measurements $Y$ in a predefined 3D reference coordinate system (Fig. 5).

For each camera the detected objects in image coordinates are projected to a parametric model of the ground plane. The model can be a simple hyperplane in most cases or a more precise elevation model if necessary. The projection is achieved by intersecting a parametric line from the camera origin through the pixels of the referring image detection, with the parametric ground model, leading to a 3D position measurement $d'$ (Fig. 1). Thus, every camera leads to a set of position measurements relative to the reference frame the cameras are calibrated against. To take into account the measurement uncertainties (that are not delivered by the CNN detector), zero mean Gaussian noise is assumed on the detections $d$. The projection from detections to the ground plane model distorts the distribution. As a first approximation the projected measurements $d'$ are assumed to have again Gaussian distributed noise with the following properties: the eigenvectors of the covariance matrices are aligned with the direction of projection and moreover uncertainty on the position measurement is high in the projection direction, but low on the perpendicular direction to it. To deal with calibration errors, we add additive zero mean Gaussian noise on each measurement that increases with the distance of the measurement to the camera.

In a second step, measurements from each set $d'^1, \ldots, d'^{n_s}$ are mutually associated by computing a minimum cost matching between the computed sets using the euclidean distance of the position measurements as a cost function. Once measurements from different sets have been associated, they are fused into one measurement considering their individual noise model. To prevent the fusion of measurements that are far apart, a common gating step is included before the fusion. Figure 7 shows an example scene with an illustration of the resulting measurement covariances of $Y$.

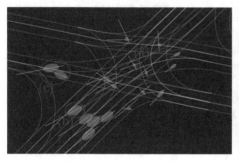 

(i) Generated 3D measurements from a given detection in the image frame.

(ii) The corresponding monocular camera image with 3D measurements projected back into the image.

**Fig. 7.** Example data from the measurement fusion process: Green line segments represent parametric lines that are intersected with a ground plane model. Green ellipses correspond to covariance matrices of the relating measurements. The red circles mark three representative cases the fusion process might lead to: (a) The measurements of two cameras were successfully fused. (b) The object is only recognized from one camera, therefore there is no fusion of measurements. (c) The two cameras have recognized the object, but the resulting intersections with the ground plane are too far apart and therefore no fusion takes place.

### 4.4    Temporal Association and Multi-Target Tracking

Once the vehicles have been located in the reference frame, the next task is to maintain their identity and determine their individual trajectories. To solve these problems, a filter for tracking multiple objects $X = \{x^1, \ldots, x^{n_t}\}$ and an algorithm for assigning measurements $Y = \{y^1, \ldots, y^{n_s}\}$ to tracks are required. The state vector $x^i$ of each tracked vehicle contains the position of the object, its linear and angular velocity and its linear acceleration and the measurement vector $y^i$ describes the position of a possible vehicle in 3D space. The estimation and prediction of the state vector is done using an *interacting multiple model filter (IMM)* [7] with three prediction models: *constant turn-rate and velocity (CTRV)* [6], *constant acceleration (CA)* [7] and *constant location (CL)* [7]. This way, all common movement types for vehicles at urban roads and intersections are incorporated neatly into the estimation process.

Association of measurements to tracks is done using a *global nearest neighbor (GNN)* approach. Herefore, an optimal bipartite matching between current track hypotheses and received measurements is computed using the hungarian algorithm [8] with a distance measure based on the covariance noise model of the estimated tracks $X$ and the measurements $Y$. A gating procedure completes the assignment and supports track management.

A track can have different management states: *valid, invalid* and *potential track*. Each possible track starts as a potential track and when the uncertainty on its estimated state vector deceeds a certain threshold, the track is set to be valid. When a track or track hypothesis is not updated with measurements, for instance when the track leaves the field of view, the uncertainty of the state estimate increases until a parametrized threshold is exceeded then the state is set to invalid.

The result is a list of tracked objects per time step and can be transmitted to the Vehicle-to-X communication module.

## 4.5   Vehicle-To-X Communication

In addition to collecting tracking information in the backend for offline analysis, an online low-latency transmission of the current traffic situation is enabled. Thus making all tracked vehicles available for autonomous vehicles in the local environment of the intelligent infrastructure. Instead of using a proprietary protocol existing V2X standards are leveraged to communicate the intersection state to nearby vehicles. By adhering to the European ITS-G5 specifications we reduce the additional setup required by users of the Test Area. An overview of the used architecture is given in Fig. 8.

Sending V2X messages for other vehicles is not strictly covered by existing V2X standards, but since very few consumer vehicles are equipped to send or receive such V2X messages, the benefits for testing within a more complete V2X setup by emulating V2X capabilities of other traffic participants outweighs the interference with the standard. Also, it is expected that more and more consumer vehicles will possess V2X capabilities in the future, which future version of the Vehicle-to-X pipeline need to consider. One possible solution would be the detection of vehicles actively sending V2X messages and their removal from the internal transmission list.

The European Telecommunications Standards Institute (ETSI) specifies a protocol stack along with a message format for inter-vehicle information broadcasts containing the position, heading, movement state and general information about the sending vehicle [10]. These broadcasts are named *cooperative aware-*

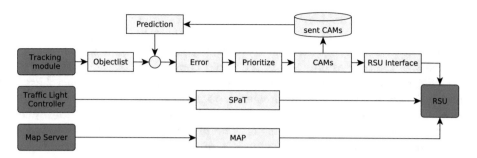

**Fig. 8.** Overview of Vehicle-to-X communication: Sending information about tracked vehicles via CAMs using a RSU-bridge to connect our V2X pipeline to the RSU. Broadcasting traffic lights states in the form of SPaTs and providing the intersection topology through MAP messages is directly done by the traffic light controller. A single road side unit is used as common transmitter.

*ness message* (CAM) and are sent repeatedly by vehicles to inform neighbors about their presence.

Our implementation estimates these features as part of the tracking algorithm and generates CAM messages for every observed vehicle. A deployed V2X roadside unit (RSU) is used to physically send the generated CAMs. In practice our pipeline is limited to a message rate of 400 to 600 CAMs per second. While this is enough for the currently used setup, we need to be prepared for future extensions and requirements. To avoid high latencies and dropped messages in the future, an additional data compression step is applied that uses prioritization and knowledge about the development of the scene to minimize the required amount of messages, which is described at the end of this section.

Beside of the transmission of the traffic participant's behaviour, the RSU also transmits the intersection's topology as well as the current traffic lights' states as depicted in the lower part of Fig. 8. For this the ETSI TS 19 091/ISO TC 204 message standards for signal phase and timing (SPaT) and intersection geometry and topology (MAP) are used. In conjunction, this enables the perception of the intersection's lanes geometry and transitions between them as well as which are currently allowed to be drivable.

Using the allocated bandwidth for CAM messages optimally becomes important if the number of vehicles increases. The main idea of the applied compression is to reduce the transmission of CAM messages based on the estimated error the prediction on the receiving side will observe. For this the sending module keeps track of all CAMs that were already sent and uses the same prediction model a theoretical receiver would use. Assuming no message loss occurs during transmission the error can be estimated by predicting the last sent message and comparing its outcome with the newly available measurement. Comparing the estimated error over all tracked vehicles allows for a message prioritization scheme which minimizes the overall observed error when only a fixed number of entries in the transmission list are actually send over the air. This idea is similar to the decentralized frequency management for vehicle ITS stations mechanism specified in ETSI EN 302 637-2, which controls the CAM frequency based on the vehicle dynamics, but using the knowledge about the complete scenario. An overview of the process is shown in the upper part of Fig. 8.

Information about vehicles with a large derivation from the predicted state are therefore sent with a higher frequency than others. Since the prediction on the receiving side is in general not known, the aim is to estimate the upper bound of the prediction error This is done by using a basic constant velocity predictor, although acceleration values are available within the sent messages. The usage of advanced prediction methods or higher order predictors on the receiving side will invalidate the assumption about the prediction error on the sending side and therefore result in a prioritization scheme that isn't optimal any more, however the prediction error will be in general lower than the sending side estimates and thus the overall accuracy will improve. Using those advanced prediction methods on both sides would further increase the effectiveness of proposed concept on data compression, which is evaluated in Sect. 5.

## 5   Evaluation

In the following we will give a short preliminary evaluation of the proposed concept on the object recognition and V2X communication based on real data from an exemplary local road unit located at a large intersection in Karlsruhe.

### 5.1   Hardware Setup

The exemplary local road unit consists of two cameras, a server for the image processing and tracking software and a *roadside unit (RSU)* for V2X communication. All hardware components are connected via gigabit ethernet switches. In order to transmit the current traffic light status, the RSU is additionally connected to the *traffic light controller (TLC)*. For this we use the second networking interface of the RSU. Except for the (unidirectional) communication between the TLC and the RSU, our processing infrastructure is completely isolated from potentially safety-critical intersection hardware units.

The hardware located in the intersection itself is connected via LAN to a data center for monitoring, administration and data storage purposes, and is not directly accessible from public internet.

### 5.2   Object Recognition

The complete pipeline from object detection to tracked 3D objects is evaluated end-to-end on two datasets from the equipped example intersection. The first dataset was recorded under good weather conditions, while the second dataset contains heavy backlight towards the cameras (Fig. 9). Both datasets are compared to the processed object estimations manually using a projection from the

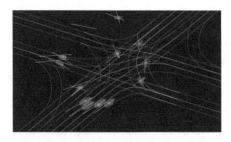

**Fig. 9.** Example tracks from dataset 2 containing strong backlight towards the cameras. Orange boxes visualize tracked objects and green boxes refer to object detections in the input image.

estimated 3D boxes back to the input images. Our analysis takes different cases into account.

The *True Positive Case (TP)* is the case in which the vehicle is detected by at least one of the cameras and the tracker can follow this vehicle from the point at which it enters the intersection to the point where it leaves the field of view. The case in which an object other than a vehicle such as a traffic sign or a pedestrian is detected and tracked as a vehicle is the *False Positive Case (FP)*. If the camera is not able to detect an existing vehicle or the tracker does not track it at all, we have a *False Negative Case (FN)*. The case when the tracker detected a vehicle but was not able to keep its track until the vehicle leaves the field of view, is listed as *Lost Track Id*. Table 1 summarizes the gained results.

**Table 1.** Evaluation results of the implemented object tracking pipeline. Dataset 1 contains good lighting conditions, while heavy backlight against the cameras is present in Dataset 2.

|  | TP | FP | FN |  |
|---|---|---|---|---|
| Setting | Detected vehicle | Detection w/o vehicle | Undetected vehicle | Lost track Id |
| Dataset 1 | 65 | 0 | 1 | 5 |
| Dataset 2 | 88 | 4 | 1 | 11 |

One of the reasons that causes the loss of tracks is that the gained measurements have a low frequency (about 5–6 Hz), so several prediction steps have to be made between two measurements, causing difficulties with the association of measurements to tracks. The high number of False Positives in the second recording is caused by structural false detections of the CNN due to the high backlight towards the camera. This behavior can be mitigated by placing more cameras with different angles into the intersection and by retraining the detection algorithm for such lighting conditions with appropriate data.

## 5.3    Vehicle-to-X

**Correctness.** We confirmed the correctness of the received CAMs visually by deriving oriented bounding boxes and their position from the messages. Afterwards the bounding boxes are reprojected into the camera images and verified that they align with visible vehicles.

**Performance.** The empirically estimated transmission rate of 400 to 600 CAMs per second allows to emulate V2X messages for a maximum of 60 vehicles, assuming a maximum allowed message rate of 10 Hz per vehicle. But to avoid latency during the transmission, a bandwidth of 400 CAMs per second is chosen.

Having 12 ingress lanes, the presented intersection is rather large and carries a lot of traffic at peak times. We estimate from manual observations that a number of 50 vehicles is no exception. However the current perception pipeline is only tracking up to 25 vehicles at the same time, as currently only vehicles on the inner intersection area are recognized. Since this number will increase in the future and to avoid limitations of the service quality, the goal of at least 100 objects to handle at the same time was set.

To utilize the allocated bandwidth optimally, the previously introduced concept for data compression is employed, which takes advantage of the prediction capabilities on the receiver side to reduce the overall error. The maximum improvement depends on the actual degree of unpredictability of other traffic participants. According to the ETSI standard the following frequencies are allowed: For the worst case a 10 Hz rate is needed to cover unpredictable vehicles and a (best case) 1 Hz rate to cover the predictable vehicles as the allowed simplification. Therefore the number of tracked vehicles per second $r_v$, as the sum of number of predictable vehicles $n_{predictable}$ and number of unpredictable vehicles $n_{unpredictable}$ per second, is as follows

$$r_v = \frac{r_{messages}}{q_m} = \frac{400}{q_m} \tag{1}$$

with $r_{messages}$ as the desired number of messages per seconds and $q_m$ being the average number of messages per vehicle

$$q_m = \frac{10 * n_{unpredictable} + n_{predictable}}{n_{unpredictable} + n_{predictable}} \tag{2}$$

$$= 1 + 9 * \frac{n_{unpredictable}}{n_{unpredictable} + n_{predictable}} = 1 + 9 * q_u. \tag{3}$$

We define $q_u$ in Eq. 3 as the relative amount of unpredictable vehicles. To properly handle a rate of $r_v = 100$ tracked vehicles per second crossing the intersection with $r_{messages} = 400$ CAMs per second, a message quotient of $q_m = {}^{400}/_{100} = 4$ is needed, meaning the ratio of unpredictable vehicles $q_u = {}^{(q_m-1)}/_9$ must be lower than 0.33. As $q_u$ depends only on traffic characteristic, it can be measured with empirical data and we assume it is independent of the actual amount of vehicles.

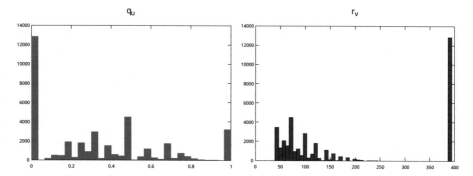

**Fig. 10.** Histogram of values of $q_u$ and $r_v$: The right diagram shows a histogram of $q_u$ with a constant velocity predictor. Values of $q_u = 0$ are mostly situations, where no or one standing vehicle is visible. Values $q_u = 1$ are typically caused by situations with a few vehicles following a turn lane. The left diagram shows the corresponding histogram of $r_v$ values, where $r_v$ is the theoretically limit of vehicles supported by given bandwidth of 400 CAMs per second.

For the empirical measurement of $q_u$ a distance error of 0.5 meters is set as threshold value. Every 250 ms, the difference of the predictor output and the incoming measurement is measured for each tracked vehicle. If the error is larger than the threshold, the vehicle is counted as unpredictable, else it is counted as predictable for this time step. At each time step, we calculate $q_u$ and skip time steps with no tracked vehicles. In recordings of 3 hours an average $q_u$ value of 0.3281 could be seen, which is very close to our target value of 0.33. The value was higher than expected. But since the focus of the sensor setup is designed to be at the center of the intersection, the value still is plausible. In this case, the total number of vehicles was never higher than 25. Still we can not completely rely on the average, as there may be periods where the $q_u$ is higher for longer times, which results in high message latencies during these periods if the number of vehicles is high enough. Figure 10 shows histograms of the $q_u$ and $r_v$ values. While we are well above the 100 vehicles per second mark most of the time, we still have a lot of time steps below this value. Widening the angle of view would bring more standing vehicles into the field of view and therefore yield a lower $q_u$ value, as would using a more complex predictor. Apart from this, the messaging rate using the RSU might need to be improved to scale the system up for future challenges.

# 6   Conclusion

We presented the ideas, concepts and algorithms for a generic, flexible and scalable setup for intelligent infrastructure that can support the development of automated vehicles. We explained it using the application in the Test Area Autonomous Driving Baden-Württemberg and evaluated the object recognition and Vehicle-to-X (V2X) modifications in an exemplary intersection in Karlsruhe.

The intelligent infrastructure consists of cameras used as perception sensors and accompanying V2X communication and is able to provide online as well as offline data from the environment, containing dynamic objects and states of traffic lights. Dynamic objects are detected within multiple camera streams using CNN detectors, associated with the road plane and then temporally tracked using an IMM filter.

Recorded data can be used offline as reference for perception system of automated vehicles currently traversing the intersection on the one hand. This includes evaluating the sensor setup of a vehicle and identifying dead spots within the setup while additionally checking the correctness and consistency of sensors and the ensuing perception and cognition algorithms of vehicles under test. The data could also be used to recreate and modify special, real-world situations in simulation environments. On the other hand, data can be broadcasted via V2X to tackle the occlusion problem for automated vehicles that could lead to potentially dangerous situations by providing additional environment information to and extending the sensor horizon for vehicles, potentially enabling vehicles with smaller sensor setups.

Although a first version of the concept has been implemented and deployed successfully, the concept and individual parts in the processing chain will be further extended and improved. As a next step, dynamic object tracking will be extended to cyclists and pedestrians. Furthermore, increasing the observed area within intersections as well as covering the roads leading into and connecting multiple intersections will provide much more valuable information about detailed traffic movement and is a next step in our rollout.

Another current shortcoming is the manual calibration process to include new cameras. An online recalibration and automated calibration of the cameras will guarantee a persistent quality level over time and an improved elevation model of covered road segments can accommodate non-planar road geometry.

**Acknowledgement.** This work was done within the project "Digitales Testfeld BW für automatisiertes und vernetztes Fahren", referred to as "Testfeld Autonomes Fahren Baden-Württemberg", funded by the Ministry of Transport Baden-Württemberg.

Under the direction of the FZI Research Center for Information Technology, a consortium of the City of Karlsruhe, the Karlsruhe Institute of Technology, Karlsruhe University of Applied Sciences, Heilbronn University of Applied Sciences, the Fraunhofer Institute for Optronics, System Technology and Image Evaluation IOSB and the City of Bruchsal and other associate partners is implementing the development of the Test Area.

# References

1. ALP.Lab GmbH (Austrian Light Vehicle Proving Region for Automated Driving). http://www.alp-lab.at/. Accessed 15 May 2018
2. Federal Ministry of Transport and Digital Infrastructure: Digital Test Beds. http://www.bmvi.de/EN/Topics/Digital-Matters/Digital-Test-Beds/digital-test-beds.html. Accessed 15 May 2018

3. MCity Headquarters: MCity Test Facility. https://mcity.umich.edu/our-work/mcity-test-facility/. Accessed 20 May 2018
4. Toyota Research Institute: Opening in October - Toyota Research Institute Automated Vehicle Test Facility. http://www.tri.global/news/opening-in-october-toyota-research-institute-auto-2018-5-3. Accessed 20 May 2018
5. Aniss, H.: Overview of an ITS Project: SCOOP@F. In: Communication Technologies for Vehicles. Springer International Publishing (2016)
6. Bar-Shalom, Y., Willett, P., Tian, X.: Tracking and Data Fusion: A Handbook of Algorithms. YBS Publishing, Storrs (2011)
7. Barrios, C., Motai, Y.: Predicting Vehicle Trajectory. CRC Press, Boca Raton (2017)
8. Blackman, S., Popoli, R.: Design and Analysis of Modern Tracking Systems. Artech House, Boston (1999)
9. Cordts, M., Omran, M., Ramos, S., Rehfeld, T., Enzweiler, M., Benenson, R., Franke, U., Roth, S., Schiele, B.: The cityscapes dataset for semantic urban scene understanding. In: Computer Vision and Pattern Recognition (CVPR) (2016)
10. EN 302 637-2 V1.3.2; Intelligent Transport Systems (ITS); Vehicular Communications; Basic Set of Applications; Part 2: Specification of Cooperative Awareness Basic Service. Standard, ETSI, Sophia Antipolis Cedex - FRANCE, November 2014
11. European Commission: Cooperative, connected and automated mobility (C-ITS). https://ec.europa.eu/transport/themes/its/c-its_en. Accessed 15 May 2018
12. Girshick, R.: Fast R-CNN. In: International Conference on Computer Vision (ICCV) (2015)
13. Girshick, R., Donahue, J., Darrell, T., Malik, J.: Rich feature hierarchies for accurate object detection and semantic segmentation. In: Computer Vision and Pattern Recognition (CVPR) (2014)
14. He, K., Gkioxari, G., Dollár, P., Girshick, R.: Mask R-CNN. In: International Conference on Computer Vision (ICCV) (2017)
15. Heidenreich, T., Spehr, J., Stiller, C.: LaneSLAM - simultaneous pose and lane estimation using maps with lane-level accuracy. In: International Conference on Intelligent Transportation Systems (ITSC) (2015)
16. Hübner, D., Riegelhuth, G.: A new system architecture for cooperative traffic centres - the SimTD field trial. In: 19th ITS World Congress (2012)
17. Hubschneider, C., Doll, J., Weber, M., Klemm, S., Kuhnt, F., Zöllner, J.M.: Integrating end-to-end learned steering into probabilistic autonomous driving. In: International Conference on Intelligent Transportation Systems (ITSC) (2017)
18. Jodoin, J.P., Bilodeau, G.A., Saunier, N.: Urban Tracker webpage. https://www.jpjodoin.com/urbantracker/dataset.html. Accessed 20 May 2018
19. Jodoin, J.P., Bilodeau, G.A., Saunier, N.: Urban tracker: multiple object tracking in urban mixed traffic. In: Computer Vision (WACV) (2014)
20. Kuhnt, F., Pfeiffer, M., Zimmer, P., Zimmerer, D., Gomer, J.M., Kaiser, V., Kohlhaas, R., Zöllner, J.M.: Robust environment perception for the audi autonomous driving cup. In: International Conference on Intelligent Transportation Systems (ITSC) (2016)
21. Lin, T.Y., Maire, M., Belongie, S., Hays, J., Perona, P., Ramanan, D., Dollár, P., Zitnick, C.L.: Microsoft coco: common objects in context. In: European Conference on Computer Vision (ECCV) (2014) (2014)
22. Meyer, F., Kropfreiter, T., Williams, J.L., Lau, R.A., Hlawatsch, F., Braca, P., Win, M.Z.: Message passing algorithms for scalable multitarget tracking. In: Proceedings of the IEEE, vol. 106 (2018)

23. Redmon, J., Divvala, S., Girshick, R., Farhadi, A.: You only look once: unified, real-time object detection. In: Computer Vision and Pattern Recognition (2016)
24. Ren, S., He, K., Girshick, R., Sun, J.: Faster R-CNN: towards real-time object detection with region proposal networks. In: Neural Information Processing Systems (NIPS) (2017)
25. Russell, S.J., Norvig, P.: Artificial Intelligence: A Modern Approach (2016)
26. Sermanet, P., Eigen, D., Zhang, X., Mathieu, M., Fergus, R., LeCun, Y.: Over-Feat: integrated recognition, localization and detection using convolutional networks (2014)
27. Sjoberg, K., Andres, P., Buburuzan, T., Brakemeier, A.: Cooperative intelligent transport systems in Europe: current deployment status and outlook. IEEE Veh. Technol. Mag. (2017)
28. Strigel, E., Meissner, D., Seeliger, F., Wilking, B., Dietmayer, K.: The Ko-PER Intersection Laserscanner and Video Dataset. In: International Conference on Intelligent Transportation Systems (ITSC) (2014)
29. Tomatis, A., Miche, M., Haeusler, F., Lenardi, M., Bohnert, T.M., Radusch, I.: A test architecture for V-2-X cooperative systems field operational tests. In: International Conference on Intelligent Transport Systems Telecommunications (ITST) (2009)
30. Zhang, Z.: A flexible new technique for camera calibration. IEEE Trans. Pattern Anal. Mach. Intell. (2000)
31. Zofka, M.R., Kuhnt, F., Kohlhaas, R., Zollner, J.M.: Simulation framework for the development of autonomous small scale vehicles for autonomous robots (SIMPAR) (2016)

# Author Index

M. Strand et al. (Eds.): IAS 2018, AISC 867, pp. 983–985, 2019.
https://doi.org/10.1007/978-3-030-01370-7

Printed in the United States
By Bookmasters